IET POWER AND ENERGY SERIES 66

High-Voltage Engineering and Testing

Other volumes in this series:

Volume 1	**Power circuit breaker theory and design** C.H. Flurscheim (Editor)
Volume 4	**Industrial microwave heating** A.C. Metaxas and R.J. Meredith
Volume 7	**Insulators for high voltages** J.S.T. Looms
Volume 8	**Variable frequency AC motor drive systems** D. Finney
Volume 10	**SF$_6$ switchgear** H.M. Ryan and G.R. Jones
Volume 11	**Conduction and induction heating** E.J. Davies
Volume 13	**Statistical techniques for high voltage engineering** W. Hauschild and W. Mosch
Volume 14	**Uninterruptible power supplies** J. Platts and J.D. St Aubyn (Editors)
Volume 15	**Digital protection for power systems** A.T. Johns and S.K. Salman
Volume 16	**Electricity economics and planning** T.W. Berrie
Volume 18	**Vacuum switchgear** A. Greenwood
Volume 19	**Electrical safety: a guide to causes and prevention of hazards** J. Maxwell Adams
Volume 21	**Electricity distribution network design, 2nd edition** E. Lakervi and E.J. Holmes
Volume 22	**Artificial intelligence techniques in power systems** K. Warwick, A.O. Ekwue and R. Aggarwal (Editors)
Volume 24	**Power system commissioning and maintenance practice** K. Harker
Volume 25	**Engineers' handbook of industrial microwave heating** R.J. Meredith
Volume 26	**Small electric motors** H. Moczala et al.
Volume 27	**AC–DC power system analysis** J. Arrillaga and B.C. Smith
Volume 29	**High voltage direct current transmission, 2nd edition** J. Arrillaga
Volume 30	**Flexible AC Transmission Systems (FACTS)** Y-H. Song (Editor)
Volume 31	**Embedded generation** N. Jenkins et al.
Volume 32	**High voltage engineering and testing, 2nd edition** H.M. Ryan (Editor)
Volume 33	**Overvoltage protection of low-voltage systems, revised edition** P. Hasse
Volume 34	**The lightning flash** V. Cooray
Volume 36	**Voltage quality in electrical power systems** J. Schlabbach et al.
Volume 37	**Electrical steels for rotating machines** P. Beckley
Volume 38	**The electric car: development and future of battery, hybrid and fuel-cell cars** M. Westbrook
Volume 39	**Power systems electromagnetic transients simulation** J. Arrillaga and N. Watson
Volume 40	**Advances in high voltage engineering** M. Haddad and D. Warne
Volume 41	**Electrical operation of electrostatic precipitators** K. Parker
Volume 43	**Thermal power plant simulation and control** D. Flynn
Volume 44	**Economic evaluation of projects in the electricity supply industry** H. Khatib
Volume 45	**Propulsion systems for hybrid vehicles** J. Miller
Volume 46	**Distribution switchgear** S. Stewart
Volume 47	**Protection of electricity distribution networks, 2nd edition** J. Gers and E. Holmes
Volume 48	**Wood pole overhead lines** B. Wareing
Volume 49	**Electric fuses, 3rd edition** A. Wright and G. Newbery
Volume 50	**Wind power integration: connection and system operational aspects** B. Fox et al.
Volume 51	**Short circuit currents** J. Schlabbach
Volume 52	**Nuclear power** J. Wood
Volume 53	**Condition assessment of high voltage insulation in power system equipment** R.E. James and Q. Su
Volume 55	**Local energy: distributed generation of heat and power** J. Wood
Volume 56	**Condition monitoring of rotating electrical machines** P. Tavner, L. Ran, J. Penman and H. Sedding
Volume 57	**The control techniques drives and controls handbook, 2nd edition** B. Drury
Volume 58	**Lightning protection** V. Cooray (Editor)
Volume 59	**Ultracapacitor applications** J.M. Miller
Volume 62	**Lightning electromagnetics** V. Cooray
Volume 63	**Energy storage for power systems, 2nd edition** A. Ter-Gazarian
Volume 65	**Protection of electricity distribution networks, 3rd edition** J. Gers
Volume 905	**Power system protection,** 4 volumes

High-Voltage Engineering and Testing

3rd Edition

Edited by Hugh M. Ryan

The Institution of Engineering and Technology

Published by The Institution of Engineering and Technology, London, United Kingdom

The Institution of Engineering and Technology is registered as a Charity in England & Wales (no. 211014) and Scotland (no. SC038698).

First edition © Peter Peregrinus Ltd 1994
Second edition © The Institution of Electrical Engineers 2001
Third edition © The Institution of Engineering and Technology 2013

First published 1994 (0 86341 293 9)
Second edition 2001 (0 85296 775 6)
Reprinted with new cover 2009
Third edition 2013

This publication is copyright under the Berne Convention and the Universal Copyright Convention. All rights reserved. Apart from any fair dealing for the purposes of research or private study, or criticism or review, as permitted under the Copyright, Designs and Patents Act 1988, this publication may be reproduced, stored or transmitted, in any form or by any means, only with the prior permission in writing of the publishers, or in the case of reprographic reproduction in accordance with the terms of licences issued by the Copyright Licensing Agency. Enquiries concerning reproduction outside those terms should be sent to the publisher at the undermentioned address:

The Institution of Engineering and Technology
Michael Faraday House
Six Hills Way, Stevenage
Herts, SG1 2AY, United Kingdom

www.theiet.org

While the authors and publisher believe that the information and guidance given in this work are correct, all parties must rely upon their own skill and judgement when making use of them. Neither the authors nor publisher assumes any liability to anyone for any loss or damage caused by any error or omission in the work, whether such an error or omission is the result of negligence or any other cause. Any and all such liability is disclaimed.

The moral rights of the authors to be identified as authors of this work have been asserted by them in accordance with the Copyright, Designs and Patents Act 1988.

British Library Cataloguing in Publication Data
A catalogue record for this product is available from the British Library

ISBN 978-1-84919-263-7 (hardback)
ISBN 978-1-84919-264-4 (PDF)

Typeset in India by MPS Limited
Printed in the UK by CPI Group (UK) Ltd, Croydon

Dedicated to colleagues, students and clients whom I have worked with over the years at Reyrolle, Sunderland University, IET, IEC, DTI, EPSRC and CIGRE, as well as via research collaborations with utilities and academia (e.g. Universities of Liverpool, Strathclyde, UMIST and Northumbria). A special thanks to my four grandchildren, Alex, Ellie, Lisa and Owen, for the great pleasure they have given me over the past 20 years, and for urging me to finish this book.

[Hugh M. Ryan, Editor, 2013]

Contents

List of contributors		xxv
Preface		xxvii
Introduction		**1**

1 Electric power transmission and distribution systems 17
I.A. Erinmez

- 1.1 Introduction 17
- 1.2 Development of transmission and distribution systems 19
  - 1.2.1 Early developments (1880–1930) 19
  - 1.2.2 The development of the transmission grid concept (1930–90) 21
  - 1.2.3 Global developments 23
  - 1.2.4 Recent developments 1990 to date (or 1990 onwards or post 1990) 26
- 1.3 Structure of transmission and distribution systems 30
  - 1.3.1 Technical factors influencing the structure of transmission and distribution systems 32
  - 1.3.2 Organisational structures 34
- 1.4 Design of transmission and distribution systems 38
  - 1.4.1 Security of supply 39
  - 1.4.2 Quality of supply 39
  - 1.4.3 Stability 40
- 1.5 Operation of transmission and distribution systems 41
  - 1.5.1 Operational planning 41
  - 1.5.2 Extended real-time operation 42
  - 1.5.3 Real-time operation 42
  - 1.5.4 Post real-time operation 43
- 1.6 Future developments and challenges 43
  - 1.6.1 Organisational developments 43
  - 1.6.2 Technical and technological developments 45
  - 1.6.3 Control and communication developments 53
- References 54

2 Insulation co-ordination for AC transmission and distribution systems 57
T. Irwin and H.M. Ryan

- 2.1 Introduction — 57
- 2.2 Classification of dielectric stress — 61
  - 2.2.1 Power frequency voltage — 61
  - 2.2.2 Temporary overvoltages — 61
  - 2.2.3 Switching overvoltages — 63
  - 2.2.4 Lightning overvoltages — 63
- 2.3 Voltage–time characteristics — 63
- 2.4 Rated withstand voltage levels — 64
- 2.5 Factors affecting switching overvoltages — 67
  - 2.5.1 Source configuration — 67
  - 2.5.2 Remanent charge — 68
  - 2.5.3 Transmission line length — 69
  - 2.5.4 Compensation — 70
  - 2.5.5 Circuit-breaker pole scatter — 70
  - 2.5.6 Point-on-wave of circuit-breaker closure — 71
- 2.6 Methods of controlling switching surges — 71
  - 2.6.1 Circuit-breaker pre-insertion resistors — 71
  - 2.6.2 Metal oxide surge arresters — 72
  - 2.6.3 Circuit-breaker point-on-wave control — 77
  - 2.6.4 Comparison of switching overvoltage control methods — 78
  - 2.6.5 Application of insulation co-ordination for ultra high-voltage technology — 79
- 2.7 Factors affecting lightning overvoltages entering substations — 80
  - 2.7.1 Backflashover — 81
  - 2.7.2 Direct strike — 84
  - 2.7.3 Attenuation of lightning overvoltage — 87
- 2.8 Methods of controlling lightning overvoltages — 87
  - 2.8.1 Location of surge arresters — 87
- 2.9 Conclusions — 89
- References — 89

3 Applications of gaseous insulants 93
H.M. Ryan

- 3.1 Introduction — 93
- 3.2 Atmospheric air clearances — 96
  - 3.2.1 Test areas — 96
  - 3.2.2 Sphere gaps — 100
  - 3.2.3 Spark gaps — 100
  - 3.2.4 Overhead lines and conductor bundles — 105
  - 3.2.5 Guidelines for live working — 110
- 3.3 Other gases — 111
- 3.4 Switchgear and GIS — 113
  - 3.4.1 Introduction — 113
  - 3.4.2 Arc extinction media — 116

		3.4.3	General dielectric considerations	119
		3.4.4	Performance under contaminated conditions	130
		3.4.5	GIS service reliability	131
		3.4.6	Gas-insulated transmission lines (GIL)	133
		3.4.7	Vacuum switches	134
	3.5	System modelling for switchgear design applications		136
		3.5.1	Field analysis techniques	136
		3.5.2	Prediction of breakdown voltages	143
	3.6	Summary		148
	Acknowledgements			148
	References			149

4 HVDC and power electronic systems — 153
Gearóid ó hEidhin

	4.1	Introduction		153
	4.2	HVDC transmission – a brief overview		153
	4.3	General principles		155
	4.4	Main components of HVDC links		156
		4.4.1	Thyristor valves	156
		4.4.2	Converter transformer	158
		4.4.3	Control equipment	162
		4.4.4	AC filters and reactive power control	163
		4.4.5	Smoothing reactor and DC filter	165
		4.4.6	Switchgear	166
		4.4.7	Surge arresters	168
		4.4.8	Valve cooling	168
		4.4.9	Auxiliary supplies	169
	4.5	Converter building		169
	4.6	VSC HVDC		172
	4.7	Economics		176
	4.8	Power electronic support for AC systems		177
		4.8.1	Static var compensators (SVCs)	178
		4.8.2	STATCOM	179
		4.8.3	Series compensators	180
		4.8.4	Unified power flow controller (UPFC)	182
	4.9	Power electronics for industrial applications		183
	4.10	Conclusion		183
	References			185

5 The implications of renewable energy on grid networks — 187
Adrian Wilson

	5.1	Introduction		187
	5.2	Drivers for renewable energy		187
		5.2.1	Fossil fuel	187
		5.2.2	Nuclear fuels	188

5.3	UK renewable energy resources and technology		188
	5.3.1	Power transfers	188
	5.3.2	Wind resources	188
	5.3.3	Wave resources	189
	5.3.4	Tidal resources	190
	5.3.5	Biomass resources	193
	5.3.6	Network implications from remote resource locations	193
	5.3.7	Generator technologies – conventional power stations	193
	5.3.8	Generator technologies – full converter wind turbine	195
	5.3.9	Generator technologies – partial converter wind turbine	195
	5.3.10	Generator technologies – wave machines	195
	5.3.11	Generator technologies – tidal machines	195
	5.3.12	Generator technologies – biomass-fed generators	196
	5.3.13	Generator technologies – microgeneration	196
5.4	Renewable generator technologies – network implications		196
	5.4.1	Cost of connections	196
	5.4.2	Voltage rise	197
	5.4.3	Load flow	197
	5.4.4	Fault level	197
	5.4.5	Power quality	198
	5.4.6	Network extensions	198
	5.4.7	Regulation	198
	5.4.8	Grid Code issues	199
5.5	Value of energy		199
5.6	Solutions for renewable energy		200
	5.6.1	Voltage rise	200
	5.6.2	Fault level	200
	5.6.3	Fault current limiters	201
	5.6.4	Principles of superconducting fault current limiters	201
	5.6.5	Fault current limiters	202
	5.6.6	Resistive fault current limiters	202
	5.6.7	Shielded core fault current limiters	203
	5.6.8	Pre-saturated core fault current limiters	203
	5.6.9	Load flow	205
	5.6.10	Energy storage	205
	5.6.11	Distributed intelligence in networks	206
5.7	Conclusions		206
	References		206
	Postscript note – by H.M. Ryan		209
6	**High-voltage cable systems**		**211**
	A.L. Barclay		
6.1	Introduction		211
6.2	Elementary theory		212
	6.2.1	Voltage, electric field and capacitance	212
	6.2.2	Current, magnetic field and inductance	214

6.3	Historical development		214
	6.3.1	Early telegraph cables	215
	6.3.2	Early lighting systems	215
	6.3.3	Flexible cables	215
	6.3.4	Impregnated cables and the renaissance of high voltage	216
	6.3.5	Pressure-assisted (gas) cables	216
	6.3.6	Fluid-filled cables	217
	6.3.7	Polymeric cables	217
	6.3.8	Polypropylene-paper laminate	218
6.4	Features of real cables		218
	6.4.1	The conductor	218
	6.4.2	The insulation system	222
	6.4.3	Multi-core and multi-function cables	228
	6.4.4	Outer layers	229
	6.4.5	Installation	232
6.5	Current ratings		234
	6.5.1	Time-dependent ratings	235
	6.5.2	Factors affecting ratings	236
6.6	Accessories		243
	6.6.1	Terminations	243
	6.6.2	Joints	246
	6.6.3	Other accessories	249
6.7	Direct current and subsea cable systems		249
	6.7.1	Applications of AC and DC transmission	249
	6.7.2	Subsea cable configurations	250
	6.7.3	Insulation systems	252
	6.7.4	Manufacture	254
	6.7.5	Installation	255
	6.7.6	Accessories	255
6.8	Standards		256
6.9	Testing		258
	6.9.1	Development testing	258
	6.9.2	Type testing	259
	6.9.3	Prequalification testing	260
	6.9.4	Factory acceptance testing	261
	6.9.5	After-laying tests	262
	6.9.6	In-service monitoring	266
	6.9.7	Testing for periodic maintenance	267
	6.9.8	Fault location	268
6.10	Future directions		269
References			271

7 Gas-filled interrupters – fundamentals   275
G.R. Jones, M. Seeger and J.W. Spencer
- 7.1 Introduction   275
- 7.2 Principles of current interruption in HV systems   275
  - 7.2.1 System-based effects   277
  - 7.2.2 Circuit-breaker characteristics   280
- 7.3 Arc control and extinction   281
  - 7.3.1 Gas-blast circuit-breakers   283
  - 7.3.2 Electromagnetic circuit-breakers   285
  - 7.3.3 Dielectric recovery   287
- 7.4 Additional performance affecting factors   288
  - 7.4.1 Metallic particles   289
  - 7.4.2 High-frequency electrical transients   290
  - 7.4.3 Trapped charges on PTFE nozzles   290
- 7.5 Other forms of interrupters   292
  - 7.5.1 Domestic circuit-breakers   292
  - 7.5.2 Oil-filled circuit-breakers   293
  - 7.5.3 Vacuum interrupters   293
- 7.6 Future trends   294
  - 7.6.1 Other gases   295
  - 7.6.2 Material ablation and particle clouds   297
  - 7.6.3 New forms of electromagnetic arc control   298
  - 7.6.4 Direct current interruption   299
- References   300

8 Switchgear design, development and service   303
S.M. Ghufran Ali
- 8.1 Introduction   303
  - 8.1.1 SF_6 circuit-breakers   303
  - 8.1.2 Sulphur hexafluoride   305
- 8.2 Interrupter development   305
  - 8.2.1 Two-pressure system   306
  - 8.2.2 Single-pressure puffer-type interrupters   306
- 8.3 Arc interruption   310
  - 8.3.1 Fault current   310
  - 8.3.2 Capacitive and inductive current switching   312
  - 8.3.3 Reactor switching   314
  - 8.3.4 Arc interruption: gas-mixtures   315
- 8.4 Third-generation interrupters   316
- 8.5 Dielectric design and insulators   318
- 8.6 Mechanism   318
- 8.7 SF_6 live- and dead-tank circuit-breakers   319
  - 8.7.1 Basic GIS substation design   319
- 8.8 Opening and closing resistors/metal-oxide surge arresters   324
  - 8.8.1 Opening and closing resistors   324

			xiii
	8.8.2	Closing resistors/metal-oxide surge arresters	325
	8.8.3	Main features of metal-oxide surge arresters (MOSA)	326
8.9	Disconnector switching		327
8.10	Ferroresonance		328
8.11	System monitoring		330
	8.11.1	Monitoring during installation and in service	330
	8.11.2	Continuous monitoring	331
	8.11.3	Periodic monitoring	331
8.12	Insulation co-ordination		332
	8.12.1	Introduction to appendices	332
8.13	Further discussions and conclusions		334
Acknowledgements			335
References			336
Appendices			337
	A	Some historic observations	337
	B	SF$_6$ circuit-breakers in the UK plus a perspective from USA	340
	C	Relevant strategic IEC Standard reports	342
	D	Update of some recent CIGRE activities relating Appendices D and E to Substations (SC B3) and also (SC B5) – compiled from CIGRE publications by H.M. Ryan	343
	E	Residual life concepts, integrated decision processes for substation replacement and an overview of CIGRE work on AM themes – compiled by H.M. Ryan	349

9 Distribution switchgear — 355
B.M. Pryor

9.1	Introduction		355
9.2	Substations		357
	9.2.1	Substation types	357
	9.2.2	Substation layouts	359
9.3	Distribution system configurations		361
	9.3.1	Urban distribution systems	361
	9.3.2	Rural distribution systems	363
9.4	Ratings		364
	9.4.1	Rated current	365
	9.4.2	Rated short-circuit-breaking current	365
	9.4.3	Rated short-circuit-making current	365
	9.4.4	Rated asymmetrical breaking current	365
	9.4.5	Rated short-time current	365
	9.4.6	Rated voltage	366
	9.4.7	Rated insulation withstand levels	366
	9.4.8	Rated transient recovery voltage	366
9.5	Switching equipment		366
	9.5.1	Circuit-breakers	367
	9.5.2	Distribution circuit-breaker types	369
	9.5.3	Disconnectors	376

xiv High-voltage engineering and testing

	9.5.4	Earth switches	378
	9.5.5	Switches	378
	9.5.6	Switch disconnector	379
	9.5.7	Switch fuse	379
	9.5.8	Fuse switch	379
	9.5.9	Fuses	380
	9.5.10	Contactors	382
	9.5.11	Ring main units	382
9.6	Circuit protection devices		383
	9.6.1	Surge arresters	383
	9.6.2	Instrument transformers	385
9.7	Switchgear auxiliary equipment		387
9.8	SF_6 handling and environmental concerns		387
	9.8.1	SF_6 breakdown products	387
	9.8.2	SF_6 environmental concerns	388
9.9	The future (as perceived in 2000 and again in 2011)		389
	9.9.1	The future 1 – by B.M. Pryor in 2000	389
	9.9.2	The future 2 – by H.M. Ryan in 2011	391
Disclaimer			396
Summary			396
References			397
Appendices – by H.M. Ryan			398
	A	Additional CIGRE references	398
	B	Distribution systems and dispersed generation (after CIGRE [8]) – summary of key CIGRE information mainly from SC C6 2011 AGM [8] – compiled by H.M. Ryan	399

10 Differences in performance between SF_6 and vacuum circuit-breakers at distribution voltage levels — 407
S.M. Ghufran Ali

10.1	Introduction	407
10.2	Circuit-breaker	407
10.3	Vacuum circuit-breaker	408
10.4	SF_6 gas circuit-breakers	409
10.5	Puffer circuit-breaker	409
10.6	Rotating-arc circuit-breaker	412
10.7	Auto-expansion circuit-breaker	412
10.8	Operating mechanism	413
10.9	Choice of correct circuit-breaker for special switching duties	413
10.10	Capacitive and inductive current switching	414
10.11	Circuit-breakers for generator circuit switching	415
	10.11.1 DC offset	416
	10.11.2 Current chopping and reignition	416
10.12	Synchronised switching	416
10.13	Conclusions and some future developments	417

		CIGRE perception of future developments: prepared and compiled by H.M. Ryan	418
	Acknowledgements		419
	Bibliography		420
	Appendices		421
	A	Relevant strategic IEC Standard reports	421
	B	The impact of new functionalities on substation design (relating to CIGRE published work) – abridged and compiled by H.M. Ryan	422

11 Life management of electrical plant: a distribution perspective — 425
John Steed

- 11.1 Introduction — 425
- 11.2 Reliability — 426
  - 11.2.1 Sources of data — 426
  - 11.2.2 Typical distribution company requirements for data — 428
  - 11.2.3 Case studies using data — 428
  - 11.2.4 The bath tub curve — 430
  - 11.2.5 Practical example – distribution transformers — 431
  - 11.2.6 Human factors in plant reliability — 432
  - 11.2.7 Conclusions on reliability — 434
- 11.3 Condition monitoring — 435
  - 11.3.1 Definitions — 435
  - 11.3.2 Benefits of condition monitoring — 435
  - 11.3.3 Application to equipment — 436
  - 11.3.4 What condition monitoring information can tell us about asset management — 445
  - 11.3.5 Condition assessment leading to asset replacement — 447
  - 11.3.6 The new working environment – users' requirements — 449
  - 11.3.7 Condition monitoring – the future — 450
- 11.4 Plant maintenance — 450
  - 11.4.1 General techniques — 450
  - 11.4.2 Enhanced maintenance — 452
  - 11.4.3 Reliability-centred maintenance (RCM) — 453
  - 11.4.4 Condition-based maintenance (CBM) — 454
- 11.5 Working plant harder — 455
  - 11.5.1 Towards a risk-based strategy – the reasons why — 455
  - 11.5.2 Risk assessment – FMEA and FMECA — 455
  - 11.5.3 Working switchgear harder — 457
  - 11.5.4 Working transformers harder — 457
- 11.6 Future trends in maintenance — 458
- 11.7 A holistic approach to substation condition assessment — 459
- 11.8 Retrofit, refurbish or replace? — 460
- 11.9 Current challenges — 460
- 11.10 A standard for asset management — 462

xvi High-voltage engineering and testing

		11.11	The impact of smart grids on asset management	463
		11.12	Information management	463
		11.13	Conclusions	464
	References			464

12 High-voltage bushings 467
John S. Graham
- 12.1 Introduction — 467
- 12.2 Types of bushings — 467
  - 12.2.1 Non-condenser bushings — 467
  - 12.2.2 Condenser bushings — 469
- 12.3 Bushing design — 472
  - 12.3.1 Air end clearance — 473
  - 12.3.2 Oil-end clearance — 477
  - 12.3.3 Radial gradients — 478
- 12.4 Bushing applications — 478
  - 12.4.1 Transformer bushings — 478
  - 12.4.2 High-current bushings — 481
  - 12.4.3 Direct connection to switchgear — 481
  - 12.4.4 Switchgear bushings — 483
  - 12.4.5 Direct current bushings — 484
- 12.5 Testing — 486
  - 12.5.1 Capacitance and dielectric dissipation factor measurement — 487
  - 12.5.2 Power frequency withstand and partial discharge measurement — 487
  - 12.5.3 Impulse voltage tests — 489
  - 12.5.4 Thermal stability test — 489
  - 12.5.5 Temperature rise test — 490
  - 12.5.6 Other tests — 490
- 12.6 Maintenance and diagnosis — 490
- References — 492

13 Design of high-voltage power transformers 495
A. White
- 13.1 Introduction — 495
- 13.2 Transformer action — 495
- 13.3 The transformer as a circuit parameter — 497
- 13.4 Core- and shell-form constructions and components — 498
  - 13.4.1 The core — 499
  - 13.4.2 The windings — 503
  - 13.4.3 Cooling — 508
  - 13.4.4 Insulation — 509

				xvii
		13.4.5	Tank	510
		13.4.6	Bushings	510
		13.4.7	On-load tap-changer	511
	13.5	Design features		512
		13.5.1	Dielectric design	512
		13.5.2	Electromagnetic design	515
		13.5.3	Short-circuit forces	517
		13.5.4	Winding thermal design	518
	13.6	Transformer applications		519
		13.6.1	Power station transformers	520
		13.6.2	Transmission system transformers	521
		13.6.3	HVDC convertor transformers	521
		13.6.4	Phase-shifting transformers	523
		13.6.5	Industrial transformers	525
		13.6.6	Railway transformers	526
	13.7	A few predictions of the future		527
	References			528
14	**Transformer user requirements, specifications and testing**			**529**
	S. Ryder and J.A. Lapworth			
	14.1	Introduction		529
	14.2	Specification of user requirements		530
		14.2.1	Need for user specifications	530
		14.2.2	Functional and design specifications	530
		14.2.3	Specifications and standards	531
		14.2.4	Specification content	531
		14.2.5	Guidance on specifications	535
	14.3	Supplier selection		535
		14.3.1	Industry changes	535
		14.3.2	Timing	535
		14.3.3	Format	535
		14.3.4	Aims	536
		14.3.5	Main elements of process	536
	14.4	Testing		538
		14.4.1	Classification of tests	538
		14.4.2	Performance tests	539
		14.4.3	Thermal tests	540
		14.4.4	Dielectric tests	542
		14.4.5	Short-circuit withstand	547
		14.4.6	Condition assessment testing	548
	14.5	Operation and maintenance		548
		14.5.1	Limitations on transformer life	548
		14.5.2	Preventive and corrective maintenance	549
		14.5.3	Time- and condition-based maintenance	549

xviii High-voltage engineering and testing

			14.5.4 Oil tests	550
			14.5.5 Electrical tests	550
	14.6	Concluding remarks		551
	Bibliography			552

15 Basic measuring techniques — 555
Ernst Gockenbach

- 15.1 Introduction — 555
- 15.2 Measuring system — 555
- 15.3 Amplitude measurements — 561
    - 15.3.1 Direct voltage — 561
    - 15.3.2 Alternating voltage — 563
    - 15.3.3 Impulse voltage — 567
    - 15.3.4 Impulse current — 571
- 15.4 Time parameter — 572
- 15.5 Measuring purposes — 573
    - 15.5.1 Dielectric tests — 573
    - 15.5.2 Linearity tests — 573
- 15.6 Summary — 573
- References — 574

16 Basic testing techniques — 575
Ernst Gockenbach

- 16.1 Introduction — 575
- 16.2 Recommendations and definitions — 576
- 16.3 Test voltages — 578
    - 16.3.1 DC voltage — 578
    - 16.3.2 AC voltage — 580
    - 16.3.3 Impulse voltage — 584
- 16.4 Impulse current — 591
- 16.5 Test conditions — 592
- 16.6 Summary — 597
- References — 598

17 Partial discharge measuring technique — 599
Ernst Gockenbach

- 17.1 Introduction — 599
- 17.2 Physical background of partial discharges — 600
- 17.3 Requirements on a partial discharge measuring system — 604
- 17.4 Measuring systems for apparent charge — 606
- 17.5 Calibration of a partial discharge measuring system — 608
- 17.6 Examples of partial discharge measurements — 608
    - 17.6.1 Partial discharge measurement on high-voltage transformers — 609

			xix
	17.6.2	Partial discharge measurement and location on high-voltage cables	609
	17.6.3	Partial discharge measurement on high-voltage gas-insulated substations	611
	17.6.4	Development of recommendation	613
17.7	Summary		614
References			614

18 Digital measuring technique and evaluation procedures — 615
Ernst Gockenbach

18.1	Introduction	615
18.2	Requirements on the recording device	616
18.3	Requirements on the evaluation software	619
18.4	Application of digital recording systems	620
	18.4.1 DC and AC voltage measurements	620
	18.4.2 Impulse voltage or current measurements	621
	18.4.3 Partial discharge measurements	623
18.5	Application examples of evaluation procedures	626
18.6	Conclusions	629
References		629

19 Fundamental aspects of air breakdown — 631
J. Blackett

19.1	Introduction	631
	19.1.1 History	633
	19.1.2 High-voltage laboratory testing	634
19.2	Pre-breakdown discharges	635
	19.2.1 Electron avalanches	635
	19.2.2 Streamer discharges	635
	19.2.3 Leaders	635
19.3	Uniform fields	636
	19.3.1 Electron avalanches in uniform fields	636
19.4	Non-uniform fields	640
	19.4.1 Direct voltage breakdown	641
	19.4.2 Alternating voltage breakdown	642
	19.4.3 Impulse breakdown	644
	19.4.4 Leaders	644
	19.4.5 Sparkover, breakdown, disruptive discharge	645
19.5	The 'U-curve'	647
	19.5.1 The critical time to breakdown	654
19.6	The gap factor	655
	19.6.1 Test procedures	656
	19.6.2 Air gaps of other shapes	658
	19.6.3 Sparkover under alternating voltages	659
	19.6.4 Sparkover under direct voltages	659

xx High-voltage engineering and testing

	19.7	Flashover across insulator surfaces in air	662
	19.8	Atmospheric effects	664
		19.8.1 Introduction	664
		19.8.2 Density effects	664
		19.8.3 Humidity effects	665
		19.8.4 Application of correction factors	665
		19.8.5 Air density correction factor k_1	666
		19.8.6 Humidity factor correction k_2	666
		19.8.7 Other atmospheric effects	668
	19.9	New developments	669
		19.9.1 UHV at high altitudes	669
		19.9.2 Testing transformers	669
		19.9.3 Future work	672
	References		672
20	Condition monitoring of high-voltage transformers		675
	A. White		
	20.1	Introduction	675
	20.2	How do faults develop?	676
	20.3	Which parameters should be monitored?	677
	20.4	Continuous or periodic monitoring	677
	20.5	Online monitoring	678
	20.6	Degrees of sophistication for transformer monitors	678
	20.7	What transformer parameters can be monitored?	681
	20.8	Basic monitors	682
	20.9	On-load tap-changer (OLTC) module	683
	20.10	Insulation module	685
	20.11	Bushing module	685
	20.12	Cooling module	687
	20.13	Advanced features measurements and analyses	688
	20.14	Partial discharge monitoring	689
	20.15	Temperature measurement	690
	20.16	Chemical parameters	690
	20.17	Dielectric parameters	691
	20.18	Conclusions	691
	20.19	Further reading	692
	References		692
21	Integrated substation condition monitoring		693
	T. Irwin, C. Charlson and M. Schuler		
	21.1	Introduction	693
	21.2	The evolution of condition monitoring systems	696
		21.2.1 Periodic monitoring, 1960–1990	697
		21.2.2 Basic discrete online monitoring, 1990–1999	697

	21.2.3	More intelligent discrete monitoring, 2000–2009	698
	21.2.4	Integrated substation condition monitoring (2010 onwards)	698
21.3	Objectives of condition monitoring		700
21.4	Application to key substation equipment		701
21.5	The substation environment		701
21.6	Condition monitoring platform		703
	21.6.1	Data acquisition systems	704
	21.6.2	Sensors and transducers	705
	21.6.3	Conversion module	705
	21.6.4	Interface module	705
	21.6.5	Data acquisition module	705
	21.6.6	Control PC module	705
	21.6.7	Communication module	706
21.7	Data acquisition, analysis and diagnostics		706
	21.7.1	Developing a common monitoring platform	707
	21.7.2	Specification of the node unit data acquisition system	711
21.8	GDM node unit overview		711
	21.8.1	Data collection	712
	21.8.2	Predictive alarms and SF_6 gas inventory	713
	21.8.3	Maintenance	715
21.9	CBM node unit overview		716
	21.9.1	Signals measured and recorded	716
	21.9.2	System alarms	718
	21.9.3	Data storage and display	719
21.10	PDM node unit overview		722
	21.10.1	PDM node unit data collection	724
	21.10.2	Node unit sequencer controls	724
	21.10.3	Node unit noise monitoring and alarm control	725
	21.10.4	Node unit system diagnostics	727
21.11	Power transformer monitoring		727
	21.11.1	Sensors for dissolved gas analysis	731
	21.11.2	Sensors for tap-changer monitoring	732
21.12	Cable monitoring		732
	21.12.1	Sensors for cable monitoring	733
21.13	Surge arrester monitoring		734
21.14	ISCM system overview		736
	21.14.1	Substation gateway	736
	21.14.2	CM module: PD monitoring	740
	21.14.3	CM module: gas density monitoring	744
21.15	ISCM systems going forward		747
21.16	Concluding remarks		748
Acknowledgements			749
References			749

	Appendices		751
	A Gas density monitoring transducer options		751
	B Circuit breaker monitoring data sampling rates		752
	C Partial discharge monitoring data sampling rates		752
	D Disconnector and earth switch monitoring data sampling rates		752

22 Intelligent monitoring of high-voltage equipment with optical fibre sensors and chromatic techniques — 755
G.R. Jones and J.W. Spencer

- 22.1 The nature of intelligent monitoring — 755
- 22.2 The basis of chromatic monitoring — 756
- 22.3 Online monitoring of high-voltage equipment using optical fibre chromatic techniques — 758
  - 22.3.1 Optical fibre chromatic sensors — 758
  - 22.3.2 Examples of chromatic optical fibre sensors for high-voltage systems — 759
  - 22.3.3 Time and frequency domain chromatic processing of optical fibre sensor data — 761
- 22.4 Chromatic assessment of the degradation of high-voltage insulation materials — 768
  - 22.4.1 Chromatic characterisation of partial discharge signals (M. Ragaa) — 768
  - 22.4.2 Offline assessment of high-voltage transformer oils with chromatic techniques (E. Elzazoug and A.G. Deakin) — 773
- 22.5 Conclusions — 780
- Acknowledgements — 781
- References — 781
- Appendix A — 785

23 Some recent ESI developments: environmental, state of art, nuclear, renewables, future trends, smart grids and cyber issues — 787
H.M. Ryan

- Preface — 787
- 23.1 Introduction — 788
- 23.2 International takeovers in UK power sector and possible impacts — 791
  - 23.2.1 Warning: kid gloves treatment — 793
- 23.3 Some aspects of renewable energy development in the UK — 796
  - 23.3.1 Energy-mix and perceived renewable energy costs (2004–10) — 796
  - 23.3.2 Renewable energy vs landscape calculations (*The Sunday Times*, 20/11/11) — 799
  - 23.3.3 UK energy storage: call to build a series of dams to store power from wind turbines (after D. MacKay) (Jonathan Leake, *The Sunday Times*, 18/3/12) — 800

	23.3.4	Press articles: Some very public energy discussions (commentaries on articles by D. Fortson *et al.*, *The Times*, 2010–12)	803
23.4		UK government's recent wind of change	808
23.5		Nuclear power plants: recent events and future prospects	810
	23.5.1	Fukushima nuclear accident: short-term impact on global developments	810
	23.5.2	Future nuclear developments	811
	23.5.3	Future prospects of 'new-build' nuclear plants overseas	815
23.6		Some aspects of carbon trading	819
	23.6.1	Coal-fired to co-fired stations in the UK to avoid paying rising climate taxes (after Danny Fortson, *The Sunday Times*, Energy Environment, 26/2/12)	820
	23.6.2	Frying note: storage energy back-up	823
23.7		A new green technology: carbon capture and storage (CCS) (Tim Webb, *The Times Business Dashboard*, 29/3/12)	825
	23.7.1	Some committed CCS developments worldwide	825
23.8		Recent developments in UK Network/European Grid links	833
	23.8.1	New UK/International DC cable links	833
	23.8.2	Proposal case for a North Sea super grid (NSSG)	833
	23.8.3	Challenges facing AC offshore substations for wind farms and preliminary guidelines for design and construction	839
	23.8.4	Network upgrades and some operational experiences	844
23.9		Some UK operational difficulties with wind farms	845
	23.9.1	Wind farms paid £900,000 to switch off (1)	845
	23.9.2	Storm shut-down is blow to the future of wind turbines (2) (Jon Ungoed-Thomas and Jonathan Leake, *The Sunday Times*, 11/12/11)	846
	23.9.3	Energy speculators now bet on wind farm failures (3) (J. Gillespie, *The Times*, December 2011)	847
	23.9.4	Millions paid to wind farm operators to shut down (4)	848
	23.9.5	Clean energy financial support; impact of Scotland leaving the Union after an independence vote in 2014 (5) (Karl West, *The Sunday Times*, 22/1/2012)	850
	23.9.6	Crown Estate: Scottish assets worth arguing over in independence debate [6] (Deirdre Hipwell, *The Times*, 21/06/2012, pp. 34–5)	851
	23.9.7	Some poor wind farm performance statistics	852
	23.9.8	'Flying wind farms pluck energy out of the blue', states Gillespie in *The Sunday Times* (08/07/12, p. 7)	854
23.10		UK air-defence radar challenged by wind turbines	854
23.11		Noise pollution: wind turbine hum (*The Sunday Times*, 18/12/11)	855
23.12		Balancing fluctuating wind energy with fossil power stations	857

23.13	Future developments including smart grids	858
	23.13.1 US study by Gellings *et al.* from EPRI [30] (1)	860
	23.13.2 Some CIGRE perspectives of energy activities and future development [35] (2)	863
23.14	Discussion and conclusions	877
	23.14.1 Discussion	877
	23.14.2 Conclusions	890
References		891
Appendices		895
	A Cyber-crime and cyber-security	895
	B Cyber-crime	895
	C CIGRE: Treatment of information security for electric power utilities	901

Index   **907**

Contributors

S.M. Ghufran Ali
Former Chief Switchgear
Engineer
PB Power Ltd

A.L. Barclay
Principal Engineer
Kinectrics International Europe

J. Blackett
Independent High-Voltage Consultant

C. Charlson
Team Leader – Diagnostics and
Monitoring
Siemens Transmission and
Distribution Ltd
Infrastructure and Cities Sector
Smart Grid

I.A. Erinmez
Independent Power Systems
Consultant

Ernst Gockenbach
Schering-Institut for High-Voltage
Engineering
Leibniz Universität Hannover

John S. Graham
Chief Engineer - Bushings
Siemens Transmission and
Distribution Ltd

Gearóid ó hEidhin
ALSTOM GRID Power Electronics

T. Irwin
Independent High-Voltage and
Condition Monitoring Consultant

G.R. Jones
Emeritus Professor
Centre for Intelligent Monitoring
Systems
Department of Electrical Engineering
and Electronics
University of Liverpool

John Lapworth
Senior Principal Engineer
Doble Power Test

B.M. Pryor
Power System Services Ltd

H.M. Ryan
McLaren Consulting
Emeritus Professor
Electrical and Electronic Engineering
University of Sunderland

S. Ryder
Principal Engineer
Doble Power Test

M. Schuler
Siemens AG
IC SG EA SYS LM SM

M. Seeger
Senior Principal Scientist
ABB Schweiz AG
Corporate Research
RD - V3

J.W. Spencer
Centre for Intelligent Monitoring Systems
Department of Electrical Engineering and Electronics
University of Liverpool

John Steed
HM Principal Specialist Inspector
Electrical Networks
Health & Safety Executive

A. White
Power Transformer Consultant

Adrian Wilson
Applied Superconductor Ltd

Preface
H.M. Ryan

It is now more than a decade since *High-Voltage Engineering and Testing (HVET)*, 2nd Edition, was published in the IEE Power and Energy Series. The origins of the HVET books, and also the equally successful accompanying International Summer School series by the same title over the period 1993–2008, is briefly recorded and explained at the end of this preface to provide the readers with valuable background information expressed in both historical and technical contexts. In the past decade, significant changes have continued to take place in the electricity supply industry in the UK and worldwide, and many more strategic and very costly network developments anticipated in the very near future will be discussed in this latest edition. There has been much talk in the past few years of 'smart-grids' and 'enhanced intelligent energy networks' of the future, i.e., within the next one to two decades. However major under-investment in the sector for many years, linked to the recent world recession or 'economic downturn', will certainly delay completion and full integration of these diverse/complex/extremely costly proposed technological advances.

This new third edition of HVET will again provide a valuable broad overview of the developments in the sector including renewable energy (windfarms, biomass etc.). Cost, environmental and operational aspects are covered. Modern substation condition monitoring strategies for switchgear, transformers and cables are discussed and new insulation co-ordination (IC) technologies are discussed – adopted using higher performance arresters for new ultra high-voltage AC transmission substations in China, India and Japan (operating at voltages $\geq 1,100$ kV). Fundamental design concepts, special strategic network developments, asset management issues at EHV and other special matters are also discussed.

The book also touches on how network equipment and systems operate and are monitored and managed at this time – and can perhaps best be managed in the future. The important roll of CIGRE in the energy sector via its extensive Study Committee structure (see Table 1, Introduction), and production of Technical Brochures, is also explained. Consider now the first two of several strategic new energy themes discussed in this edition of HVET:

1. Recently, there have been political concerns expressed by MPs and media coverage commenting that Britain's energy markets are 'inherently-flawed' and that 'anti-competitive practices' may be forcing up the costs paid by

consumers. Five of Britain's energy companies are facing mounting pressure to cut fuel prices after recent figures from Ofgem (the industry regulator) showed the average profits they earned, per household, rose 40% one recent winter to the highest figure for five years. This comes at a time of huge profits for the energy companies – several of whom are now (at least partly) owned by overseas companies, for example EDF Energy and GDF Suez (French owned), E.ON and RWE (German owned), Enel (Italian owned). Ofgem indicated recently that profit margins earned by the so-called Big Six Companies – British Gas, Scottish Power, EDF Energy, npower, Scottish and Southern Energy (SSE) and E.ON – increased from £75 per average 'dual-fuel' customer in November 2009 to £105 at the start of February 2010.

Energy bills continue to rise and UK consumers will also be footing the £25–35 billion bill to upgrade the UK's energy network for the next four to five decades. In 2013, Ofgem indicated that householders will be paying off this huge cost, via levies, over a 45 year period (instead of the existing period of 20 years). Consequentially, it is projected, and claimed, that energy bills will reduce soon in the UK.

2. A study from the Energy and Climate Change Committee on the future of Britain's electricity networks has called for the introduction of a more efficient 'smart grid', capable of intelligently managing demand and supply. A member of this committee, P. Tipping, MP, said *'our existing regulatory and policy frameworks, along with grid infra-structure we rely on, were developed to serve the fossil-fuel economy of the twentieth century. The future looks very different'* and called for a review of the British Electricity Trading and Transmission Arrangements, which have formed the foundation for UK power activity since 2005. He also stated that by 2020, the UK network would need to accommodate a more diverse energy mix.[1]

Many strategic aspects will be dealt with in the new third edition of HVET. In addition, speculative new technologies and new techniques perhaps novel today, yet likely to become strategic and standard technology very soon – for next generation systems – are also reported on. One example that could possibly fit this bill is 'fault current limiters', which have been developed/researched for some years but have not as yet achieved the commercial success that many predicted. This situation may well change dramatically in the next decade. Obviously a very strong strategic contender in this category would be 'smart grids', referred to by some as condition monitoring to achieve greatly improved electricity network commercial profitability, management of demand, supply, etc. Electricity grid networks in UK and in many countries are still largely set up as for twentieth-century needs! All areas of the energy sector need to draw on relevant lessons learned in other

---

[1] At that time, the UK energy consumer was already concerned and disgruntled with the realisation that the seemingly ever-escalating domestic energy fuel bills would continue to rise indefinitely – part of these payments contributing towards the highly subsidised future development to the UK's 'diverse-mix' energy network infrastructure.

appropriate sectors, for example telecoms. Technically, this is certainly a very good time for 'smart-grid' changes as everyone is trying to work networks harder and more efficiently. There are 'rich-pickings' available to those who can 'interpret needs' and develop effective intelligent software systems, etc., capable of working energy networks, harder, safer, longer and more efficiently/profitably, and who can anticipate early enough what new techniques/methodologies are likely to play the best strategic and economic roles in managing T&D networks better (technically and economically) in the very near future.

World events and reactions to several other important emerging energy issues warrant and necessitate extended coverage/discussions on renewable and other energy issues in this third edition, partly because of the widespread unrest reported in the UK press/media etc., on several short-term and longer-term energy-related issues. Similarly, because of apparent UK government's indecision concerning nuclear new-build plant vs renewables vs a recent new rush for gas initiative in the 'post'-Fukushima nuclear accident (2011) era – and more significantly, the apparent lack of a coherent and consistent overall UK energy policy – these and other strategic aspects are also covered in this new edition. Very important issues relate to technical, economic and security aspects linked to the current worldwide problems associated with cyber-crime, cyber-hacking, cyber-intrusion, etc. using malicious software. These aspects are covered, mainly in Chapter 23, with an indication given of the scale and frequency of the worldwide cyber issues, and how these issues are currently being dealt with within the energy sector.

Finally, the extensive referencing of CIGRE Technical Brochure publications in this third edition of HVET has been done very deliberately. Perusal of appropriate CIGRE Technical Brochures (TBs) can help empower the reader if he/she uses them as an additional resource when reading refereed IET/IOP/IEEE, etc., publications on similar themes. Sadly, in the UK, many higher degree researchers or engineering degree students doing final year projects have in the past failed to be aware of, or to follow up effectively, extensive and valuable CIGRE TBs and *Electra* paper materials as they often felt little incentive to read, refer to or publish in these 'non-rigorously refereed journals' from a research assessment publishing 'credibility viewpoint'. Fortunately these views are now changing worldwide, and hopefully also in the UK, as there is much valuable technical information to be obtained with this search approach – as this writer has been urging students and academics to do for 25+ years within HVET School and elsewhere.

In summary, the challenges ahead are great and the career opportunities for the next generation of power engineer are very promising – good luck.

Acknowledgements

Significant to the successes of both the HVET course over the period 1993–2008 and the HVET book to date have been the excellent individual expert contributors to both formats, each expert having been particularly active in his/her sector(s), including ESI, IEC Standards, design and manufacture, R&D, consultancy or testing aspects and within the professional bodies IET/IOP/IEC/BSI/CIGRE, etc. The majority of authors in this third edition are also distinguished and active members of CIGRE, WGs, or IEC Committees or the famous Current-Zero-Club, restricted to world experts in Arc-Interruption. All are fully aware of the ongoing dynamic changes in the energy and network sector, worldwide.

This Editor, who was also Chair of the successful IET/HVET Summer School series 1993–2008, is pleased to record his very grateful thanks to all HVET contributors and lecturers for their generous and committed support over the years (1993–2008) in passing on their expertise to the next generation of engineers in this sector.

In particular at this time, he also wishes to express his sincere thanks and grateful appreciation to all contributors to this third edition HVET for giving their time so generously to prepare their valuable and strategic contributions, at a time of heavy work commitments and other duties.

Hugh M. Ryan, Editor HVET 3rd Edition, 2013

Historical background to HVET books, Editions 1, 2 and 3

The first, second and now third editions of this HVET book developed from subject material initially prepared and delivered at the IET HVET International School Courses covering the same subject areas (1993–2008). Interestingly, the International HVET Course series – covering High-Voltage, Engineering and Testing – evolved and developed in the UK following the strongly voiced concerns and wishes of the IEE membership 'at large' during and following two UK IEE meetings on related topics that the writer chaired in 1991, one at CERL, Leatherhead the other at IET Headquarters, Savoy Place, London. At, and subsequent to these events, there were

1. very strong concerns expressed at the diminishing UK expertise in the sector, particularly in HV measurement and traceability/testing
2. the strong wishes expressed by most of the IEE members in the audiences, at both these events, to have suitable new training course(s) established on HV testing, measurement and traceability, etc., developed, organised and made available to facilitate the 'proper and appropriate training' of the next generation of HV experts in the UK and abroad, otherwise it was feared that much of the expertise would be lost.

Note: To put these 1991 IEE membership concerns into perspective, it must be recognised that during the 1980s and 1990s many HV laboratories in both the old and new universities in the UK closed and laboratory space often converted, into new IT suites. Many of the larger machines in undergraduate laboratories were also removed and progressively it was sometimes difficult to distinguish between university electrical departments and computing departments, as both were largely resourced by many computers plus software simulation tools instead of power machines, etc., or HV testing equipment. Similar savage cut-backs in the UK power engineering industrial manufacturing base reduced the popularity of HV power engineering still further at that time.

Disappointingly, in recent years this power engineering downturn situation has continued – fortunately with a few areas of outstanding expertise remaining – culminating in the virtual ending of UK transmission switchgear manufacturing and certain other traditional power engineering manufacturing capabilities. Now, in 2013, the recent closure of a UK Bushing Manufacturing facility and also the Clothier UHV Testing Laboratory complex, both located at Hebburn, UK, has exacerbated the position even further. Consequently, the need for the latest version of this HVET book is again timely, reflecting on traditional, new and anticipated future strategic developments in the power engineering sector.

Historically, this writer (editor of editions 1, 2 and 3 of this HVET book) decided to prepare a report for the IEE Power Divisional Board on a proposed new HVET (High-Voltage Engineering and Testing) course, suggesting a slightly 'broader scope' for this proposed course, after some helpful discussions with colleagues, who were later to become founder members of HVET Course Steering Committee. The IEE Power Division approved this initiative and supported the running this course annually, for the first several years (1993–2002). This International HVET Course Series (1993–2008) proved to be a success (commercially, educationally and technically) with more than 400 delegates coming from more than 30 countries, and the HVET book 2nd Edition appeared in the IEE best seller lists. Therefore, it can truly be stated that this has been a 'bottom-up' initiative, from IEE members in the audiences at the above two technical meetings back in 1991 who were the real 'catalysts' for starting HVET, in both course and book formats. This approach of using delegate feedback continued to prove invaluable when checking, updating and maintaining the HVET course relevance 'year-on-year'.

It had always been the policy of the IET HVET Course Organising Committee (1993–2008) to 'tweak' the course delivery slightly each year and also to introduce new aspects and materials regularly, when felt to be appropriate, reflecting any changes in the ESI and always 'listening and taking on board' from the interactive technical discussions the views of the international delegates, fellow lecturers, etc., regarding any possible improvements to course structure/themes and assessing the interest in possible new area(s) to 'pick up on' for the next year. Indeed, one 'speculative-type' lecture session was usually held each year, to introduce and debate with course delegates one or two different speculative or new technical aspects – or drivers for the future – in anticipation of likely new trends, and to 'touch-on' how changes would affect the operation and economics of electricity networks of the future and equipment. Also considered were the operational implications and how these changes might be 'managed'. This has been done on the HVET course by the course committee over several years, with themes such as advanced condition monitoring, fibre optical monitoring, fibre, new developments in the UHV sector, evolving drivers and strategies for change in Electricity ESIs, global warming, carbon footprints, renewable energy, wind power, tidal, wave, solar, biomass, and so on! Also any relevant strategic updates or anticipated dynamic changes within ESIs or HV testing sectors worldwide were reported to delegates year on year, covering any recent IET/IEEE/IOP/CIGRE publications and including important changes to IEC Standards and relevant CIGRE TBs, Working Group (WG) reports, etc. Brief details of such developments and possible strategic changes were made available to delegates at the school each year and the importance of IET/IEC/CIGRE activities touched on from a perspective of 'empowering delegates' in their chosen area of interest, or even broader issues. This was well received by delegates and who often remained in contact with course lecturers a long time after the school.

The HVET course, and its accompanying book, has been very well received both nationally and internationally for many years. Consider just one tribute: an extract of a recent letter from the IET Chief Executive and Secretary, April 2008, to one retiring founder-member of the HVET Course Steering Committee. It states:

> I am writing to you on behalf of the Institution of Engineering and Technology in recognition of the outstanding contribution you have made over the years to the High Voltage, Engineering and Testing Course. The Course has become our flagship power systems training school which has inspired other sectors in the IET to emulate its formula. It is accepted globally that HVET sets the standard for other organisations providing this training and the sector has the highest regard for its quality and practical delivery. None of this would be possible without the exceptional contribution and energy of the volunteers behind it. It is with sadness that I read that you have had to resign from the steering committee for health reasons on the advice of your Doctor. Your presence among the team and IET staff will be missed as the Course continues to serve the engineering community.

Note: This writer considers the above comments to be an appropriate and accurate tribute to the individual and, in his humble opinion, it could equally well have been written collectively for all the HVET committee and course lecturers, over the entire life of the HVET course up to that time (1993–2008). Many of these individuals have also contributed to all three editions of the HVET book.

This clearly reflects their continuing commitment to passing on their expertise to the next generation of workers in the sector! – To them again, a big thank you.

Introduction
H.M. Ryan

This third edition comprises 23 chapters covering high-voltage engineering and testing themes – with many valuable references describing CIGRE work. Table 1 is set out at the end of this introduction to assist in understanding the range and scope of individual CIGRE Study Committees (SCs) and associated terminology, while Table 2 provides an abridged summary of recent strategic work in the transformer sector by CIGRE (SC A2), which we will return to later.

Chapter 1 provides an authoritative coverage of 'Electric power transmission and distribution systems' by Dr Arslan Erinmez. In this, 'The progressive development from 1880s to date is described mainly with reference to the UK system, as most systems around the world have gone through the same stages of development at and around the same time following technological, political and organisational developments which reflected the trends current at the time'. Global developments are also reviewed together with several key factors, for example technical, organisational structures that heavily influence the development and operation of these networks. Design, security and operational and planning aspects are also considered. Future developments and challenges: organisational, security, technical and technological are discussed, including a large detailed list of conventional and non conventional power electronic thyristor-controlled voltage regulators and other devices (also touched on in Chapters 4, 5 and briefly in Chapter 23 – relating to use with offshore-wind farms). The author also points out that 'although three phase overhead lines is still the usual method of interconnection, in cases where long transmission distances and/or sea crossings are involved, HVDC transmission is economic despite the relatively higher costs of converter and terminal equipment'.

Erinmez comments on HVDC transmission that is also used for interconnecting utilities with different supply frequencies (e.g. 50/60 Hz) and in cases where an asynchronous link between systems is required. 'Back-to-back' HVDC interconnections have also been used with both 'converter stations' situated in the same site. Erinmez (Chapter 1) also reports that the rapid development of computing and communication systems has accelerated development strategies often referred to as the so-called 'smart-grid initiatives'.[1]

The author comments that smart-grid initiatives tend to focus on consumer demand control, remote switching and metering aspects but considers that they currently suffer from inadequately defined objectives as well as protocols and

---

[1] These are discussed further in Chapter 11 and in greater detail in Chapter 23 (section 23.13 and Tables 23.9–23.14).

standards (this editor endorses these remarks). Further, Erinmez considers smart-grid initiatives, at transmission level, are more appropriate for special protection and system control applications, whereas in distribution systems their application will be subject to consumer approval and placement of privacy safeguards. He also warns that the use of Internet-based communications also exposes transmission and distribution utilities to malicious attacks and cyber-hacking activity. These issues, which are likely to require increasing efforts to ensure robustness of systems to such threats, are considered further in Chapter 23, Appendix A.[2]

Erinmez in Chapter 1 points out that NGC is the first and largest fully privatised independent transmission utility in the world that has been subjected to organisational and electricity market driven changes. As a result, it has been 'able to utilise every available technology to address the challenges of facilitating competition and responding to the electricity market place'. He outlines the leading role of NGC in 'FACTS and HVDC development/applications and provides a brief summary of both traditional and the new technology portfolio readily available to utilities' (see section 1.6.2).

The important concept of *insulation co-ordination* (IC) of high-voltage and EHV AC systems is thoroughly covered in Chapter 2 (Irwin and Ryan), while Chapter 3 (Ryan) touches on a few important aspects of IC for UHV AC systems recently covered in CIGRE TBs 546 (2011) and 542 (2013).* The more recent study, reproduced in a summarised form in Table 3.2, takes into account the state-of-the-art technology, with special reference to higher performance surge arresters.

*This review takes into account: the accumulated knowledge of various CIGRE working bodies; recent measured data of very fast temporary overvoltages (VFTO); and air gap dielectric characteristics in collaboration with CIGRE SC A3 and B3.

The emphasis in Chapter 3 relates to the application of gaseous insulants to switchgear, mainly SF6 in GIS/GIL, and outlines criteria for EHV/UHV testing laboratories, dielectric modelling studies enabling optimal dielectric design of circuit-breaker units etc. to be produced and minimum breakdown voltages to be predicted, often without recourse to extensive and costly development testing. Ryan (Chapter 3) also considers that the subtle nuances of the complex fundamental arc-physics measurements in SF6, systematic interrupter high-current performance assessments and development, and effective dielectric design of practical interrupter layouts *fully justified the effective use of the complementary skills of a group of experts in this sector when making decisions regarding the final commercial SF_6 GIS and circuit-breaker designs* that subsequently achieved outstanding interruption and dielectric and in-service performance. These were systematically developed from a four-break interrupter design to a two-break design, and eventually to the design of one-break interrupter in a remarkably short timescale (see Chapter 8).[3]

---

[2] The reader is encouraged to use Chapter 1 as a major 'reference point' source when considering smart grids of the future and cyber-crime, cyber hacking/malicious attacks.

[3] The supporting R&D for these developments, carried out towards the end of the twentieth century, has been extensively reported in the literature, by S.M.G. Ali, G.R. Jones, D. Lightle, H.M. Ryan, *et al.* (see also Chapters 7, 8, 21–23).

Chapters 2, 3 and 8 also discuss strategic aspects relating to UHV substation design; it should be recognised that, because only a few UHV AC transmission networks exist and have only recently entered service worldwide, at 1.1 MV or 1.2 MV (e.g. in China and in India respectively):

1. The background of the technical specifications for substation equipment exceeding 800 kV AC [CIGRE TB 546 (2011)] is less well defined than at lower system voltage levels, where robust standards already exist.
2. Only limited experimental/technical/specification information exists for UHV substation equipment, and two recent CIGRE Technical Brochures TB 546 and TB 542 are currently of strategic importance and will remain so till the time more robust full IEC specifications are developed for UHV systems:
   - TB 546: This study has collated the limited available background UHV information and has presented *interim recommendations* for the international specification and standardisation of UHV equipment [in CIGRE Technical Brochure TB 456-WG A3.22 (2011)];
   - TB 542: This study discusses the insulation co-ordination practices in three UHV AC Systems [TB 542-WG A4.306 (2013)] at:
     [a] the 1100 kV Jindongnan Substation (China),
     [b] the 1100 kV Shin-Haruna Testing Station (Japan) and
     [c] the 1200 kV Bina Testing Station (India).

TB 542, a follow-on study from the work of TB 546 (2011), has described further useful measures and simulation studies. It considers that overvoltage mitigation techniques such as higher performance arresters can drastically reduce lightning overvoltage levels. The CIGRE review (WG A4.306) intimated that it will prepare recommendations, 'such as recent practices of insulation coordination based on the **higher performance surge arresters**, estimation of overvoltage and air clearance, and these will be proposed for future revisions of the application guide IEC 60071-2 (1996) and IEC apparatus standards'.

Insulation co-ordination (IC) is a very complex subject area and one of immense strategic technical/economic importance to the network design and effective operation. Consequently, here again it is always strongly recommended to use the complementary skills of a group of experts in this sector when making final design decisions. Aspects such as IC (including the use made of surge arresters at UHV levels), switching phenomena for circuit-breakers, disconnectors and earthing switches and testing are also considered in Chapters 2 and 8. In situations such as this, refinements and knowledge updates will continue as service experience with UHV systems increases – and until full robust IEC Standards are produced. Again, CIGRE technical activities are considered in many chapters of this book and in Table 1 at the end of this introduction section, the reader is provided with useful backgroud to the wide range of CIGRE technical activities: '**CIGRE key**: SC, Study Committee; WG, Working Group and JWG, Joint Working Group; TF, Task Force; TB, Technical Brochure; TR, Technical Report; SC, Scientific Paper'.

Chapter 3 [CIGRE TB 546] informs the reader that, at present, UHV technology is 'characterised by a need to minimise the sizes, weights, costs and environmental impacts of the overhead lines and substations and hence to develop projects which are feasible from economic, societal and technical points of view'. This *interim utility/CIGRE strategy* is discussed together with a brief explanation of how, by means of 'the application of a number of new technologies and new analysis techniques, utilities are able to reduce the dielectric requirements to values that lead to much smaller structures'.

This results in insulation voltage levels at UHV that are not far from the levels applied at the 800 kV class. For example, in Japan the towers of the UHV OH lines are only 77% of the size that would be necessary if insulation levels would have been extrapolated directly from lower voltage class. Chapter 3 goes on to detail other strategic aspects considered in CIGRE TBs 546 or 542. Three important aspects included in this list, which are also touched on in Chapters 2, 8 and Table 3.2 of Chapter 3, are:

1. the use of *closing resistors* to control slow front overvoltages (SFO)
2. the use of *opening resistors* to reduce opening SFO
3. *damping resistors* to be used in GIS disconnectors to reduce the amplitude of VFTO (very fast transient overvoltage) phenomena which otherwise may exceed the lightning impulse withstand voltage of the switchgear
4. mitigation techniques such as higher performance arresters can dramatically reduce the lightning overvoltage levels [TB 542-WG A4.306 (2013)].

Further, the reader should note that, while some experts are concerned with the present IC situation, everyone should be at least partly reassured that another CIGRE expert Working Group [TB 542-WG A4.306] has had deliberations concerning the vital issue of field testing techniques on UHV substations during construction and operation. An updated CIGRE TB provisionally entitled 'Field Testing Technology on UHV Substation Construction and Operation' was due to be issued by CIGRE by late 2012. However, the work on this proposed document might well have been incorportatcd into TB 452-WG A4.306 (2013).

Chapter 4 by Gearóid ó hEidhin considers HVDC[4] and power electronic systems that are now widely used in modern networks and will certainly find increased application as the industry moves towards smart grids and enhanced transmission and distribution networks of the future, that is within the next one or two decades. This chapter provides comprehensive material and informs the reader how modern HVDC converter stations are designed, general principles (including basic components); details of main components of HVDC links including converter transformer, McNeill HVDC station designs; control systems and AC filters and reactive power control; smoothing reactor and DC filters; switchgear; valve cooling techniques; surge arrester equipment; typical layouts of control building and valve hall; development of voltage-sourced converters (VSCs) and other devices; environmental aspects; adherence to standard

---

[4] The term gas circuit-breakers (GCBs) has recently been used when discussing HVDC systems in the literature.

Introduction   5

specifications and more. Power electronic support for AC systems is also briefly covered in Chapter 4.

Note: A generous list of valuable reference sources is provided for further reading and current/future CIGRE publications are available from IET/IEEE/CIGRE and can be regularly monitored, similarly IET/IEEE/IEC Standards, etc. In particular, CIGRE sources (see Table 1 for details of Study Committees) provide many regular strategic Technical Brochures (TBs) on important worldwide issues in the sector (e.g. see appendices to Chapters 8, 9, 10 and elsewhere in this book).

The theme of 'back-to-back' HVDC interconnections being used, with both converter stations situated in the same site, has already been touched on in Chapter 1. This is considered further in Chapter 4, by the author of 'HVDC and power electronic systems', who points out:

1. If the function of a HVDC transmission scheme is to transfer power over a long distance, then it will invariably use a high direct voltage. Most 'modern' schemes use voltages up to ±800 kV for overhead lines, while cables have been approaching voltages up to ±600 kV progressively over the past 20 years.
2. Converter stations can be roughly characterised into two groups:
   (i) *back-to-back converters* using low direct voltage and high current typically 20–250 kV and 2.5–5 kA
   (ii) *long-distance transmission* schemes using higher direct voltages and more moderate current typically 300–800 kV and 1–4 kA.

Notes:
1. For *back-to-back* schemes, the author points out that the pressure to use HV disappears and the voltage used is the lowest voltage at which the required power can be transferred (within the limitations of the converter valves).
2. *Economic HVDC power transfer*, for example from a remote power source to an urban area: a DC line is significantly cheaper to build than an AC line to carry the same power and additionally, if the distance is great enough, this economy is sufficient to pay for costs of the converter stations at both ends of the line. Chapter 4 quotes the *break-even distance* as approximately 800 km for an overhead line and 50 km for a cable.

Chapter 5 by Adrian Wilson discusses the implications of renewable energy on grid networks and provides a valuable overview of the subject. The reader should also be aware that:

1. Another recent IET book – Power and Energy Series 63, entitled *Energy Storage for Power Systems*, 2nd edn, 2011, by A. G. Ter-Gazarian provides a valuable additional resource and contains 144 references.
2. Chapter 23 of the present book provides several interesting UK press/public perceptions as to the recent effectiveness of renewable energy wind farm supplies in the UK, the high UK energy costs to the British Public, customer fuel poverty and several other strategic aspects relating to the future energy strategy of the UK government including those for new-build nuclear, following on from the aftermath of the Fukushima nuclear accident of 2011.

Basic cable designs and theory are covered in Chapter 6, by A. Barclay, with some brief supplementary offshore wind farm material touching on EU North Sea Energy grid aspirations and other recent applications that plan to use modern low-loss superconducting cable designs being briefly touched on in Chapter 23. Interestingly, CIGRE Table 2, in this introduction, indicates that Japanese R&D has targeted 2020 for the commercial deployment of high-temperature superconducting transformers in power systems.

Chapters 7–10 cover circuit-breaking aspects. Chapter 7, by G.R. Jones, M. Seeger and J. Spencer, provides an excellent comprehensive fundamental treatment of gas-filled interrupters and this is followed, respectively, by thorough treatment and reviews of high-power SF_6 switchgear design development and service (Chapter 8, by S.M. Ghufran Ali, Distribution switchgear; Chapter 9, by B.M. Pryor, Differences in performance between SF_6 and vacuum circuit-breakers; Chapter 10, by S.M. Ghufran Ali).

The subtle nuances of the complex fundamental arc-physics measurements in SF_6 (for example by Professor G.R. Jones and his group at CIMS, Department of Electrical and Electronics, University of Liverpool and by Reyrolle switchgear staff towards the end of the twentieth century), the systematic interrupter high-current performance assessments and development, and effective dielectric design of practical interrupter layouts fully justified the effective use by Reyrolle of the complementary skills of a group of experts in this sector, when making a late decision regarding the final 'worldclass' commercial SF_6 designs that subsequently achieved outstanding circuit-breaker interruption, dielectric and in-service performance. High-power commercial interrupter designs were speedily and systematically developed from four-break interrupter/phase to two-break interrupter/phase and quickly to one-break/phase commercial interrupter design, as reported by S.M. Ghufran Ali, G.R. Jones *et al.* These collaborative developments are considered further in Chapter 3.

It must be stressed that vital partnerships, such as mentioned above, between universities and the power industry, are of even more strategic importance now in the twenty-first century, and in the case of the University of Liverpool it was very pleasing for this writer to note that a major Chinese electrical engineering firm, the Pinggao Group (a direct subsidiary of the Chinese State Grid) is to invest £1.5 million in research at the University of Liverpool over the next five years. Pinggao is one of China's major manufacturers, engaged in the design and production of switchgear and power plant equipment at HV, EHV and UHV voltage levels. It appears that the framework agreement will see the university build on the existing provision of technological support for the development and optimisation of Pinggao's electrical apparatus, in order to supply reliable electrical equipment as China moves towards developing 'smart grids' and enhanced energy networks of the future. Welcoming this agreement while on a visit to University of Liverpool, Quingping Pang, Pinggao, Vice General Manager, ended his statement by stating: 'We hope the technologies developed here will be successfully used for the benefit of all society'.[5]

---

[5] Without doubt this agreement 'provides further robust evidence' of the high quality of expertise available at some UK universities – in this case at the University of Liverpool, UK, Electrical Engineering and Electronics Department/Centre for Intelligent Monitoring Systems.

Much valuable supplementary strategic CIGRE 'switchgear-related' material is also included in the appendices of Chapters 8–10, mostly from appropriate CIGRE (TB) sources, thanks to its effective and wide-ranging worldwide working group activities infrastructure as set in Table 1, at the end of this introduction section. The reader is provided with further strategic CIGRE information in Chapter 23.

Chapter 11, by John Steed, considers 'Life management of electrical plant: a distribution perspective'. Overall Steed presents an interesting distribution perspective of life management of electrical plant. He points out that the effective management of assets to ensure that the user obtains the optimum life for the plant is becoming more vital as electricity distribution systems are worked harder and all equipment need to be reliable. He considers several strategic aspects to substantiate his viewpoints. Steed also discusses the impact of smart grids on asset management; he comments that in the early part of twenty-first century much has been talked about smart grids, i.e. the use of new technologies that will:

1. facilitate the transition to a low-carbon electricity supply system
2. enable an increase in security of supply.

Steed is of the view that as far as (2) is concerned, this relates the challenges of being able to integrate inflexible and/or intermittent generation into the system. He considers the essential elements for this will include:

- a wider use of automation and intelligent systems
- distributed and centralised intelligence, real-time monitoring and diagnostics
- an integrated 'cyber-security' data protection and data privacy safeguards.

Consequently, Steed considers that condition monitoring systems will become increasingly important and are likely to be applied to more equipment especially those identified as *system-critical*, informing the users of impending failures in the system. (*Note*: Three aspects of condition monitoring are discussed exclusively and sequentially in Chapters 20–22.)

Note: At this point, the attention of the reader is again directed towards the fact that frequent reference in this book will be made to CIGRE technical publications. In anticipation of future developments in the power sector CIGRE totally reformed its specialist professional Technical Study Committees in 2002 – determined by worldwide experts in the power sector – as illustrated in Table 1 and considered further in section 23.13.2, linked into discussing 'smart grids' of the future and the CIGRE perception of the main challenges ahead (see also Tables 23.9–23.14).

Chapter 12 by John S. Graham deals comprehensively with 'High-voltage bushings'. This chapter discusses major aspects of bushing design and development for use in equipment in distribution, transmission and including UHV AC and DC transmission systems. AC and DC bushing types are described and design aspects, clearance requirements and bushing applications in substation equipment are considered including transformers, switchgear – including direct connection to switchgear. A wide range of dielectric and other testing strategies are discussed together with required maintenance and diagnosis procedures. Finally, typical faults occurring in practical bushing designs are described and generous reference sources provided for further reading.

Chapter 13, by A. White,[6] provides an authoritative description of the fundamentals of transformer design and Chapter 14, by S. Ryder and J.A. Lapworth, considers 'Transformer user requirements, specifications and testing'.

Chapters 13 and 14 provide much valuable resource material for the reader. They can both be revisited again and again 'in a new light' as new publications from CIGRE, etc., become available – and be perused by the reader prior to revisiting Chapters 13 and 14 and considering the implications of any such new data on the overall picture! Conveniently, CIGRE Study Committee A2 has issued a Study Committee Report, on a transformer efficiency theme, in *Electra*, 2012;**263**(August), pp. 36–40. This has been edited and abridged by this editor and reproduced in Table 2 (at the end of this introduction section) purely to provide the reader with a further insight into transformer developments. Careful consideration of the technical content of this abridged Table 2, if read in conjunction with re-reads of Chapters 13 and 14, will provide the reader with a further strategic insight and empowerment relating to the transformer topic and the direction relevant standards are moving (obviously for advanced study the reader should consult the original CIGRE article).

Turning now to HV measurement and testing themes, these are all covered comprehensively in Chapters 15–18 by Professor Ernst Gockenbach, University of Hannover. Chapters 15 and 16 respectively describe and explain 'Basic measuring techniques and basic testing techniques', while Chapter 17 considers this strategic issues surrounding 'Partial discharge measuring techniques'. In his final theme, Chapter 18, Ernst Gockenbach deals with 'Modern digital measuring techniques and evaluation procedures'.

Finally, linked in to HV testing strategies fully covered in Chapters 14–18, it is important for the reader to be aware of, and appreciate, certain strategic aspects relating to the dielectric breakdown properties and characteristics of atmospheric air. This is very briefly touched on in Chapter 3 but Chapter 19, by J. Blackett, deals exclusively with the 'Fundamental aspects of air breakdown'. This covers pre-breakdown corona discharges and sparkover, the U-curve, gap factor and sparkover test procedures, sparkover voltage characteristics under alternating direct voltages, flashover across insulator surfaces in air, atmospheric effects (density, humidity etc.), and developments in the sector. Generous references are provided to assist further reading.

Chapters 20–22 focus on condition monitoring (CM) approaches:

- *A. White* (Chapter 20), the first on CM, provides an example of the detailed monitoring of a high-voltage substation component *that is a transformer*.

---

[6] Additionally, Chapter 20 by Allen White, one of three chapters dealing specifically with aspects of condition monitoring (i.e. Chapters 20–22), considers *condition monitoring of high-voltage transformers*.

- *Terry Irwin, C. Charlson and M. Schuler* (Chapter 21), the second on CM, explains how monitoring the condition of an entire *substation* incorporating several components transformers switchgear bus bar housings, etc., is being addressed using an integrated substation condition monitoring (ISCM) approach.
- *G.R. Jones and J.W. Spencer* (Chapter 22), the third chapter on CM, provides an insight into future potential developments of *intelligent monitoring* based upon examples of *optical fibre sensors and chromatic techniques* (Centre for Intelligent Monitoring Systems, University of Liverpool approach).

Condition monitoring is now a mature technology, and the reader may find it useful to refer back to the second edition of this book (ISBN 0-85296-775-6, 2001) to appreciate the progress made in the past decade.

The final Chapter 23, by Hugh Ryan, touches on some recent ESI developments and several other strategic issues, including environmental, state of the art, nuclear prospects (post Fukushima 3/2011) in the UK and overseas; national customer concerns in UK regarding energy bills and contributions to new energy investment costs; renewables [wind energy – onshore and offshore aspects, biomass etc.] and certain public perceptions of problems associated with their operation, high costs of fuel bills, current levels of customer fuel poverty in the UK; future trends including offshore wind farms, 'smart-grids' and enhanced networks of the future – presenting in detail two strategic viewpoints as to implications and costs involved; cyber-crime issues; and how CIGRE restructured its specialist Study Committees to meet the challenges ahead. There will be many!

Final comments on smart grid initiatives for the future: The reader will see interesting overview publications in Chapter 23, e.g. by EPRI, US, and CIGRE. However we are still at the early stages. Before considering these documents, the reader should recall again the wise words of Arslan Erinmez, Chapter 1, when he reminded us that:

- at present smart-grid initiatives suffer from inadequately defined objectives as well as protocols and standards,
- there are three other important points concerning transmission and distribution systems:smart grid initiatives:
    (i) smart-grid initiatives at transmission level were more appropriate for special protection and control applications
    (ii) the application of smart-grid initiatives in distribution systems will be subject to consumer approval and placement of privacy safeguards
    (iii) smart-grid initiatives must ensure robustness of transmission and distribution systems to cyber-threats.

Clearly, there will be huge costs involved, much planning required and many technical and other issues to be resolved. There is much still to be done, so many interested parties to satisfy, with differing agendas, and many 'difficult hurdles to clear' before the smart grid becomes a reality.

Table 1 Fields of activities of the CIGRE Study Committees, since 2002 reform*
[after A history of CIGRE, 2011]

SC A1 Rotating electrical machines [www.cigre-a1.org]
Economics, design, construction, test, performance and materials for turbine generators, hydrogenerators, high-power motors and non-conventional machines

SC A2 Transformers [www.cigre-a2.org]
Design, construction, manufacture and operation of all types of power transformers, including industrial power transformers, DC converters and phase-shift transformers, and for all types of reactors and transformer components (bushings, tap-changers, etc.)

SC A3 High-voltage equipment [www.cigre-a3.org]
Theory, design, construction and operation of devices for switching, interrupting and limitation of currents, lightning arrestors, capacitors, insulators of busbars or switchgear, and instrument transformers

SC B1 Insulated cables [www.cigre-b1.org]
Theory, design, applications, manufacture, installation, tests, operation, maintenance and diagnostic techniques for land and submarine AC and DC insulated power cable systems

SC B2 Overhead lines [www.cigre-b2.org]
Design, study of electrical and mechanical characteristics and performance, route selection, construction, operation, management of service life, refurbishment, uprating and upgrading of overhead lines and their component parts, including conductors, earth wires, insulators, pylons, foundations and earthing systems

SC B3 Substations [www.cigre-b3.org]
Design, construction, maintenance and ongoing management of substations and of electrical installations in power stations, excluding generators

SC B4 HVDC and power electronics [www.cigre-b4.org]
Economics, application, planning, design, protection, control, construction and testing of HVDC links and associated equipment. Power electronics for AC systems and power quality improvement and advanced power electronics

SC B5 Protection and automation [www.cigre-b5.org]
Principles, design, application and management of power system protection, substation control, automation, monitoring and recording, including associated internal and external communications, substation metering systems and interfacing for remote control and monitoring

SC C1 System development and economics [www.cigre-c1.org]
Economics and system analysis methods for the development of power systems: methods and tools for static and dynamic analysis, system change issues and study methods in various contexts, and asset management strategies

SC C2 System operation and control [www.cigre-c2.org]
Technical and human resource aspects of operation: methods and tools for frequency, voltage and equipment control, operational planning and real-time security assessment, fault and restoration management, performance evaluation, control centre functionalities and operator training

(Continues)

Table 1 Fields of activities of the CIGRE Study Committees, since 2002 reform*
[after A history of CIGRE, 2011] (Continued)

SC C3	**System environmental performance** [www.cigre-c3.org] Identification and assessment of the environmental impacts of electric power systems and methods used for assessing and managing the environment impact of system equipment
SC C4	**System technical performance** [www.cigre-c4.org] Methods and tools for power system analysis in the following fields: power quality performance, electromagnetic compatibility, lightning characteristics and system interaction, insulation co-ordination, analytical assessment of system security
SC C5	**Electricity markets and regulation** [www.cigre-c5.org] Analysis of different approaches in the organisation of the electrical supply industry: different market structures and products, related techniques and tools, regulation aspects
SC C6	**Distribution systems and dispersed generation** [www.cigre-c6.org] Assessment of technical impact and new requirements, which new distribution features impose on the structure and operation of the system: widespread development of dispersed generation, application of energy storage devices, demand side management, rural electrification
SC D1	**Materials and emerging test techniques** [www.cigre-d1.org] Monitoring and evaluation of new and existing materials for electro-technology, diagnostic techniques and related knowledge rules and emerging technologies with expected impact on the system in medium to long term
SC D2	**Information systems and telecommunication** [www.cigre-d2.org] Principles, economics, design, engineering, performance, operation and maintenance of telecommunication and information networks and services for the electric power industry; monitoring and related technologies

Note: 'Since 1921, CIGRE's priority role and "added value" have always been to facilitate mutual exchanges between all its components: network operators, manufacturers, universities and laboratories. It has continually led to the search for a good balance between the handling of daily problems encountered by its members in doing their jobs and the reflections on the future changes and their equipment. Its role in exchanging information, synthesising state of the art, and serving members and industry has constantly been met by CIGRE throughout events and through publications resulting from the work of its Study Committees.'*

In anticipation of future developments in the power sector, CIGRE totally reformed its specialist professional Technical Study Committees in 2002 – determined by worldwide experts in the power sector – as illustrated in Table 1 and considered further in section 23.13.2, linked into discussing 'smart grids' of the future – and the CIGRE perception of the main challenges ahead (see also Tables 23.9–23.14).*

[*Source: The history of CIGRE (International Council on Large Electric Systems), 2011, p. 169.] www.cigre.org; www.e-cigre.org

CIGRE technical activities are considered in many chapters of this book and in Table 1 of this introduction the reader is now provided with useful background to the wide range of CIGRE technical activities: **CIGRE key**: SC, Study Committee; WG, Working Group and JWG, Joint Working Group; TF, Task Force; TB, Technical Brochure; TR, Technical Report; SC, Scientific Paper.

Table 2   Transformer efficiency: avenues to make a good thing even better*

The comments made in the present HVET book introduction text will now be simply illustrated by briefly touching on a CIGRE Study Committee Report on **Transformer Efficiency** prepared by SC A2 and presented in the name of its Chairman Claude Rajotte in *Electra*, **263**, August 2012, pp. 36–40, prepared to explain, describe and discuss the different aspects of transformer efficiency and **possible avenues** that would allow improvement.

(i) A transformer converts electrical power with exceptional efficiency; yet despite this, Rajotte points out that transformers are with lines and cables among the top contributors to losses in electrical networks and that this topic, i.e. power efficiency, is a strategic topic for CIGRE [1].

(ii) Rajotte goes on to state that transformer efficiency relates to several factors [2], mainly losses but also such factors as noise, type and quantity of materials required, weight, size, overall cost, etc. Improving one of these factors may have adverse impacts on other factors and thus a limited effect on the overall transformer efficiency. This latest CIGRE paper focuses on ways to reduce transformer losses while considering possible side effects influencing transformer efficiency. [The losses include cooling system losses caused by the energy supplying any pumps and fans that the transformer cooling system may require. These are activated differently in some designs.] Examples of losses in a typical modern efficient transformer are presented. It is carefully pointed out that the application and use of a transformer have a direct impact on the losses produced. Transformer applications, e.g.:

Generator transformers. Such transformers are generally designed with generator ratings in mind and are usually operated near full load. For such use, full load losses are the predominant factor.

Transmission transformers. Transmission substations are usually designed with a number of parallel transformers so that an outage of one transformer at any time has no effect on service continuity. Thus, if the number of transformers in a substation is N, the peak load should be scaled for $N - 1$ transformers. If number of parallel transformers is 2 ($N=2$), such a configuration means that each transformer will only be loaded above 50% of its power rating when the other transformer is out of service. In such a configuration, transformers usually operate with small load losses and fairly near their maximum efficiency. However, as the number of parallel transformers (N) increases in a substation, a higher average load is applied to each of the transformer, resulting in higher load losses and therefore lower efficiency.

Converter transformers. Converter transformers generally have specifications similar to generator transformers and often **HVDC and industrial** operate near full load in performing their function. Load losses are thus the predominant factor in most cases. For this application, the effect of harmonics must also be considered since harmonics may impact the distribution of load losses and hence hot spot location and temperature.

Distribution transformers. Distribution transformers are generally designed based on the peak load of the feeder they supply. That load may vary depending on the time of day, day of the week and season. Cold load pick-up may also have to be considered to specify ratings for such transformers.

Avenues to reduce transformer losses. Some simple improvements to operating practice may help minimise transformer losses in an electrical grid. For example, transformer users should avoid leaving unloaded transformers connected to the grid for long periods since this results in quite unnecessary losses. Inversely, prolonged unavailability of one transformer in a parallel group puts more load on the remaining units, again increasing losses. In addition,

(Continues)

Table 2   Transformer efficiency: avenues to make a good thing even better*
(Continued)

---

intelligent cooling control can optimise cooling system use and minimise overall transformer losses. In some ambient temperature and load conditions, it may be advantageous to start the cooling system below the predetermined winding temperature threshold in order to reduce active part temperature and thus overall load losses. Such relatively simple improvements may help reduce losses but have no significant effect on transformer fleet efficiency. However, the five avenues considered below may have a much greater impact.

Avenue 1: Utility power transformer replacement strategy and loss capitalisation. SC A2 informs the reader that many transformers currently in use (i.e. in service) worldwide were installed in the 1950s, 1960s and 1970s and they are gradually approaching the end of their service life due to component ageing, deterioration, technical incompatibility, obsolescence, etc. Modern transformers have significantly lower no-load losses achieved primarily through better core magnetic steel and better core assembly techniques gradually introduced by transformer manufacturers.

Further improvements were achieved through the loss capitalisation approach now used by transformer purchasers. Such an approach broadens the initial cost comparison during transformer procurement to include not only transformer purchase costs but also the cost of losses over the entire life of the transformer. CIGRE/Rajotte/SC A2 considers a typical example – the average transformer core losses for a major North American utility as a function of year of manufacture [3] and the Figure 3 included in the *Electra* article clearly supports the assertion that gradually replacing these transformers, generally starting with the oldest, will 'naturally' improve the efficiency of utility transformer utility fleets. Typically, transformers manufactured in 1960, 1975 and 1990 respectively had approximate initial core losses of 1.5, 0.7 and 0.45 kW/MVA (see original Figure 3 of *Electra* article). Though transformer replacement strategies are clearly driven mainly by reliability concerns, CIGRE SC A2 points out that their positive impact on energy efficiency may be significant.

Avenue 2: New standards to define transformer efficiency classes.

1. SC A2 reports that in several counties there has been a substantial effort to develop new efficiency standards for distribution transformers [4], driven in particular by environmental considerations. IEC is also working on a new standard for power transformer efficiency intended to create different classes of transformers based on loss level and, if possible, to specify a standard calculation method for loss capitalisation.
2. National regulators and utility strategies play an important role since they can influence transformer user purchasing strategies by setting new financial criteria for losses and losses capitalisation. This in turn stimulates manufacturers to make optimum use of today's best technologies in order to optimise design and material selection and therefore significantly improve overall transformer efficiency. CIGRE also reports that a trend towards increasing use of loss capitalisation has begun in several countries.
3. It is essential, however, to understand the effect of reducing losses on other aspects of transformer design. For big transformers, the effects on size and mass may result in transformers that cannot be transported to site or cannot be installed in the existing facility without major structural changes.

Avenue 3: Improvements to reduce no-load losses. The many improvements made to the quality of magnetic steel have had a direct impact on no-load transformer losses. Hysteresis losses can be reduced by selecting higher quality steel and also by simply increasing the

(Continues)

Table 2 Transformer efficiency: avenues to make a good thing even better*
(Continued)

magnetic core cross section. SC A2 points out that the latter approach, however, increases load losses due to the increased length of winding conductors, hence winding resistance (see Avenue 4). A large cross section may nevertheless have a positive effect on transformer noise through adding to transformer weight and cost. For large power transformers, tight transportation constraints (e.g. railway clearances) sometimes set limits to the size and/or weight. Eddy current losses can be reduced by using thinner steel or steel with higher electrical resistance. This type of improvement also increases transformer cost. The magnetic core can also be made of amorphous steel, a vitrified metal alloy manufactured with a very thin lamination that reduces no-load losses by 60–70%. Use of such magnetic steel in distribution transformers is growing in popularity. Unfortunately, such material cannot be applied to power transformers because of its insufficient mechanical strength and labour intensiveness of assembling cores made with it. R&D would thus be required to extend the application of such technology to future power transformers.

Avenue 4: Improvements to reduce load losses. A major improvement implemented several years ago is the use of continuously transposed conductors (CTC). Replacing a massive block of copper as a winding conductor, CTC consists of a set of small insulated conductors each flattened radially to the core in order to reduce eddy losses. SC A2 points out that other good design practices also help reduce load losses, such as the use of magnetic shielding and the control of transformer leakage flux. To further reduce load losses, manufacturers can also increase conductor cross section to reduce conductor resistance. This increases the transformer's weight and, as raw materials become scarcer, its cost. As noted for no-load losses, tight transportation constraints like railway clearances may restrict the size and weight of power transformers, limiting such a practice.

Avenue 5: High-temperature superconducting transformers. High-temperature superconducting (HTS) materials make for a smaller and lighter transformer (possible weight reduction of 50% or even more), which is quieter and more efficient, and has higher overload capability. It is also safer since oil is replaced by gas (e.g. nitrogen) and has a certain degree of current limiting capability. HTS transformers below 10 MVA are now being tested in the USA and Japan. Japan targets 2020 for commercial deployment of HTS transformers in power systems. The overall costs (purchase, maintenance, etc.) and service life of such transformers are as yet poorly understood. Second-generation HTS tapes based on YBCO-coated conductor technology are being developed for transformer applications [5]. Manufacturing technology, quality control, cost and production rate are among the issues being addressed by superconductor manufacturers, according to SC A2. Great progress in these areas has been reported. The long-term objective is to have the cost of HTS conductors approach that of conventional conductors.

Conclusion. The transformer is a very efficient apparatus but remains among the top contributors to losses in an electric network. The topic of transformer efficiency includes not only losses, but also noise, type and quality of material required, weight, size and overall cost, etc. SC A2 informs the reader:

> Improving one of these factors may have adverse effects on other factors and thus only a limited effect on overall transformer efficiency. Moreover, for big transformers, the effects of higher efficiency on size and mass may result in transformers that cannot be transported to site, or cannot be installed in the existing facility without major structural changes. All of these factors must be considered in efforts to enhance transformer efficiency.

(Continues)

Table 2 Transformer efficiency: avenues to make a good thing even better*
(Continued)

- Network efficiency is a key topic for CIGRE. Good operation practices with transformers may help reduce losses, but losses could be further reduced through the avenues discussed in this CIGRE SC A2 paper: utility replacement strategies for power transformers, new standards defining efficiency classes, improvements reducing both the no-load losses and load losses, and the development of high-temperature superconducting transformers. Study Committee SC A2 will continue to explore this topic, and can contribute to improving the overall efficiency of power transformers and hence of electrical networks.

References

1. Southwell P., Negri A. 'Power system component efficiency and energy delivery effectiveness designed for minimal environmental impact'. *CIGRE international symposium*, Bologna, September 2011
2. Boss P. 'Power transformers: Technical developments, trends and challenges'. *My Transfo 2010, Oil & Transformer*, Turin, November 2010
3. CIGRE WG A2.34. 'Guide for Transformer Maintenance'. Technical Brochure, 2011;**445**(February)
4. VITO and Biology Intelligence Service. *Sustainable industry policy-building on the ecodesign directive-energy – using product group analysis. Final Report Lot 2: Distribution and power transformers, study for European Commission DG ENTR unit B1*, January 2011
5. Mehta S., *et al.* 'Power transformers technology review and assessments'. AG A 2.4-Transformer Technology, CIGRE. *Electra*, 2008;**236**(February)

Postscript on fire safety practices by SC A2

A recent CIGRE TB 537 (2013) WG A2.33, 'Guide for Transformer Fire Safety Practices' (See also CIGRE article written by members of SC A2, WG A2.33, Petersen, A., Convenor, *et al. Electra*, 2013;**268**(6):42–49.) It is recognised that, while the risk of a transformer causing a fire is low, but not negligible, the consequences can be very severe. WG A2.33 has studied the key aspects and current practices of transformer fire safety and this TB has been issued as a guide to help transformer designers and users to define and apply best practices of transformer fire safety. The *Electra* article (prepared by members of WG A2.33) summarises the main contents of TB 537. Much useful information is provided in these two documents. The probability of transformer fires vary widely between utilities and between countries. This brochure supplies useful data and provides guidance on how an individual transformer user can calculate the probability of transformer fires for their transformer population based on their failure rate and additional risk-modifying factors.

---

*After Claude Rajotte, CIGRE SC A2 Chairman, on behalf of CIGRE Study Committee A2. (abridged and edited text from original CIGRE article, *Electra*, **263**, August 2012, pp. 36–40).

Chapter 1

Electric power transmission and distribution systems

I.A. Erinmez

1.1 Introduction

Today in all countries in the world that utilise electricity as an efficient source of light and energy, some form of a transmission and distribution system exists. Both systems carry electric current albeit at different voltages and they are connected to each other. They are part of the bulk transport and distribution system essentially delivering electrical energy, converted from primary energy sources, to the end users. The only clear separation between the two systems is based on the perception of their end use and functionality.

Transmission systems provide the bulk transport paths for electrical energy from generation centres located close to the primary energy sources to the major load centres within a large geographical area, thus facilitating economic and efficient bulk power transfer. On the other hand distribution systems are concerned with the delivery of electrical energy to individual customers within a smaller geographical area. In this respect, a distribution system may have a number of delivery points to its major load centres, from one or more transmission systems and/or elements of a transmission system. The final structure of the system is dependent upon the magnitude and the pattern of demand within the geographical area. It is also usual for transmission systems to be interconnected to enable shared economic benefits and operational access to generating capacity.

Generally modern transmission systems transport electrical energy by alternating current (AC) at a frequency of 50 Hz or 60 Hz operating at voltages of 220 kV and above with other standard voltages of 275, 330, 400, 500, 765 and 1,000 kV. As systems develop, each time a new transmission voltage is added the lowest transmission voltage level reverts to play more of a distribution network role. For example the 132 kV transmission network developed in the early days in the UK assumed a distribution role after the addition of 275 and 400 kV transmission networks. It is usual for one or more of these standard voltage networks to be employed with transformers connecting parts of the transmission system at each voltage level.

Generators with output voltages of around 10–25 kV are usually connected onto the transmission system(s) via generator transformers and directly feed electrical energy into one or more transmission lines. The delivery from the transmission to the distribution systems is via transformers located at substations marking the interface boundary between the two energy delivery systems, as shown in Figure 1.1, for UK. Direct current (DC) up to ±800 kV is also used for bulk transmission of electricity either over long distances or across sea crossings where AC transmission is either technically or economically possible. The DC transmission is also used for interconnection of two AC systems or areas each operating at a different nominal frequency or for providing an asynchronous interconnection between two AC systems.

Beyond the transmission interface boundary the distribution system is built in a layered structure with each consecutive layer supplying decreasing amounts of power. The primary voltage level for distribution systems is usually around 132 kV, with other standard distribution voltages at 66, 33, 11, 3.3 and 2.4 kV with 240/110 V at the customer terminals. The distribution system, at its highest voltage level, is directly connected to the transmission system via one or more transformers with sizes dependent upon the magnitude(s) of demand(s) to be supplied and demand supply security considerations.

The distribution system differs from the transmission system also in terms of the complexity of interconnectivity between load centres. The complexity in distribution networks arises from numbers of connections that are required at each voltage level and successive voltage levels, dependent upon the distribution of demand and individual customer demand magnitudes. The interconnectivity needs

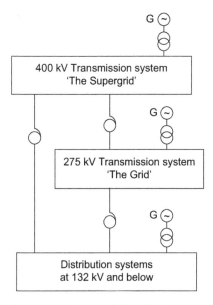

Figure 1.1 Generic transmission and distribution system structure in UK

are less pronounced compared with transmission systems as mainly radial feeders are used at lowest demand levels to pool together sufficient magnitudes of demand that can then be secured by connections to alternate supply points. The distribution system is also required to accommodate local and diverse generation sources such as auto-producers and renewable energy sources.

1.2 Development of transmission and distribution systems

The progressive development of transmission and distribution systems from 1880s to date will be described in the following paragraphs. The description will be mainly with reference to the UK system as most systems around the world have gone through the same stages of development at and around the same time following technological, political and organisational developments, which reflected the trends at that time.

1.2.1 Early developments (1880–1930)

The basic human need for safe, efficient and affordable light was the primary driving force in the development of electric power systems. In 1882 Thomas Edison initiated first large-scale use of electricity by what became known as 'Edison's Illuminating Companies'. The most famous of these was the Pearl Street System, which was a system comprising a load of approximately 400 incandescent lamps invented by Edison each consuming about 83 W, and provided electric lighting for Lower Manhattan.

At about the same time in England, the Holborn Viaduct Generating Station constructed under the provisions of Electricity Lighting Act 1882 provided approximately 60 kW of power for customers who had mixed lighting and motor loads. It is worth noting that both these systems relied on DC generation with the power being directly distributed by underground or surface laid cables. Under the 1882 Act any local authority, company or person licenced by the Board of Trade could install a supply system and an individual consumer had the right to demand a supply without discrimination between consumers.

Throughout the first decade of the electric lighting age, the demand for electricity grew at a phenomenal rate, but towards the end of the nineteenth century came the revolution, which influenced the way in which electricity had since been generated and transmitted. Up to this point in time, many industrial drive systems including the DC generators had been powered by steam engines or complicated mechanical linkages to water wheels. This resulted in the majority of industry dependent on electricity being sited close to riverbanks and bringing its associated pollution problems.

There had been a lot of debate on the merits of DC and AC generation and transmission. However, with the invention of the transformer and the induction motor, the advocates of AC generation and transmission prevailed and gradual development of AC transmission system started in 1896 with the 25 Hz three-phase generation and transmission system from Niagara to Buffalo in the USA.

Gradually, in towns and cities all over the industrial world, generating stations were developed to a point that they were able to supply the demand within their own area. This, however, resulted in a situation where each one of the hundreds of electricity generation and distribution companies defined their own parameters of supply, such as frequency and voltage.

At the turn of the twentieth century, electricity began to be regarded as a product and service that should be available to all. Thus its generation and distribution became the subject of debate between the authorities/companies who produced and sold electricity and their respective Parliaments who laid down the laws for all industry.

There had also been a lot of debate on the fossil fuel reserves available to the electricity industry and on the effects of pollution, which was excessive in urban and new industrial conurbations. In 1918, the Williamson Committee in the UK reported to the President of the Board of Trade that 'the cheap supply of electricity, on town conditions in particular, would be most marked. The reduction of pollution by smoke would result in a lower death rate from bronchial diseases.' The Committee also recommended that the concentration of larger generation units in fewer and bigger power stations was the only solution to reducing the costs of electricity to industry to an absolute minimum. The need for more efficient power stations to help conserve fossil fuel reserves was also recognised. However, lack of compulsory powers in the 1919 Electricity Act handicapped progress at the required scale.

Along with this report and many others produced for the government of the day, it was soon realised that fundamental changes were required to ensure efficient, economical and effective development of the electricity industry. These extensive debates had sown the seeds for the development of electric power systems for 'overall public benefit' in the industrialised nations on both sides of the Atlantic. Lord Weir, a leading UK entrepreneur in his time, was invited to head a committee to investigate 'the national problem of the supply of electrical energy' by the newly elected Conservative government in 1925. Weir completed his report in four months and considered three ways in which the electricity supply industry might operate. These three options were:

1. Power could be bought and sold by a 'Transmission Board', buying only from the cheap suppliers and allowing inefficient plant to close down.
2. A 'Transmission Board' could act as a carrier for electrical power leaving the buying and selling negotiations to the producers and purchasers.
3. The control of all generation should be brought under the Transmission Board.

All of these suggestions were rejected by the government of the day. Nevertheless it is interesting to note that the structure of the present UK Electricity Supply Industry is in fact based on a combination of these recommendations. The additional recommendation of the Committee that was finally adopted for implementation was the establishment of a 'Gridiron' of high-voltage power lines covering the whole country.

1.2.2 The development of the transmission grid concept (1930–90)

A public corporation, the 'Central Electricity Authority (CEA)' was formed under the Electricity Supply Act 1926 to erect the necessary transmission lines that would interconnect a limited number of 'selected' power stations where generation of electricity would be concentrated. The Board would purchase the output of the selected stations and sell it on to the local distribution undertakings. For the first time in the UK, customers benefited from the use of the cheapest power generation stations supplying their electrical energy needs. Between the Electricity Supply Act 1926 and the Second World War, the idea of 'public ownership' or 'nationalisation' of the electricity supply industry increasingly gained favour.

In 1934, the Fabian Society produced a document entitled 'The Socialisation of the Electricity Supply Industry'. The principles advocated by this document, while causing concern among the municipal companies and being regarded as radical, were adopted in the 1947 Electricity Act. The existing 560 generation and distribution undertakings in England and Wales were formed into a single generation board the 'British Electricity Authority (BEA)' responsible for bulk generation, transmission and central co-ordination/policy direction, with 12 'area electricity boards' responsible for retail distribution of electricity to customers.

The first Transmission Grid System, operating at 132 kV is shown in Figure 1.2. It was planned and built in the early 1930s and by 1948 it interconnected

Figure 1.2 The 132 kV transmission system in 1934

22 High-voltage engineering and testing

Figure 1.3   The 132 kV transmission system in 1948

all major generating sites in England and Wales as shown in Figure 1.3. Over the first decade or so of nationalisation, the central planners of the industry had many claims of poor project management laid at their feet for the inability to build new power stations. Demand was growing so fast that proposals were put forward to divide the grid system into three sections.

Arguments for and against this were many, but the then President of the Institution of Electrical Engineers, T.G.N. Haldane, suggested an increase in the grid voltage. Economic studies carried out then, which are still applicable today, showed that it was cheaper to build power stations on the coalfields and transmit the electricity via high-voltage lines. Common sense prevailed and plans for the construction of the '275 kV supergrid' [1] were put forward in the early 1950s and were implemented by the publicly owned 'Central Electricity Generating Board (CEGB)', which under the Electricity Act 1957 took the responsibility for bulk electricity generation and transmission from CEA. At the same time the 'Electricity Council' taking the co-ordination role from the BEA was formed. Generating stations with larger capacities were built, and these tended to be concentrated around the coalfields and close to large rivers, which facilitated easy access to their fuel and cooling water needs.

By the early 1960s it became evident that even the 275 kV supergrid system, shown in Figure 1.4, would not be capable of carrying the predicted power flows,

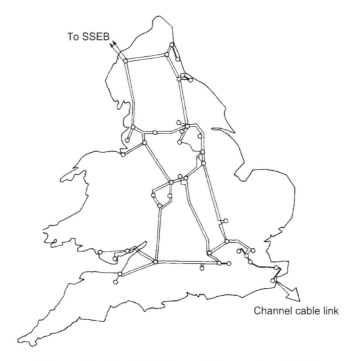

Figure 1.4   The 275 kV transmission system in 1961

and plans for the construction of a further supergrid at 400 kV were accepted [2, 3]. The progress in the construction of this supergrid to adequately address the electrical energy transmission needs in England and Wales is illustrated in Figure 1.5.

While these vast changes to the transmission system within the UK progressed there was very little change to the distribution system structures. The 12 nationalised regional distribution companies formed at the same time as the CEGB took power from the major transmission bulk supply points close to their load centres and distributed it to individual customers. The largest change to the distribution systems was in the number of customers now connected to an electricity supply. In 1948, approximately 9.5 million domestic premises or one quarter of the domestic properties in the UK were connected to a supply of electricity. By the way of comparison this figure had risen to 20 million domestic premises in 1989 [4]. In the same period, domestic electricity consumption alone had risen from 11 TWh to almost 80 TWh.

1.2.3   Global developments

Elsewhere in the world electric power transmission systems developed along similar lines to UK largely influenced by the availability and location of the primary fuel resources in relation to major load centres. Availability of hydro resources far from

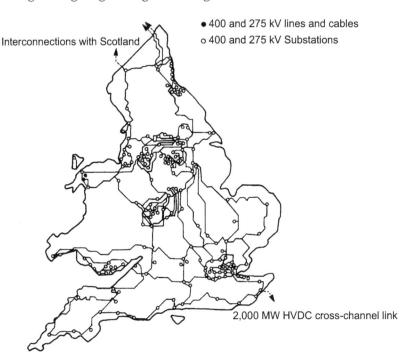

Figure 1.5 The National Grid system in England and Wales today

load centres led to the development of sophisticated high-voltage, long-distance AC and DC transmission systems, while in countries, like the USA, with population centres separated by large distances, development of integrated generation and transmission and distribution systems based on locally available primary energy resources gained favour. These regional systems were interconnected to other similar systems at a later stage to enable sharing of accruable technical and financial benefits. The need for long-distance transmission revived the fortunes of high-voltage DC transmission as the developments in technology increased its reliability and economic competitiveness for a wider spectrum of transmission applications in the late 1960s.

Thus the 'overall public benefit' concepts [5] developed in the first quarter of the twentieth century led to the development of transmission and distribution systems, which were:

1. optimum for a large region or country and not just for a specific region or area
2. based on technology to optimally meet public needs over a long-time spectrum counted in at least quarter to half a century in terms of the longevity of its capital equipment
3. to present least cost penalties in the development of associated generation and distribution systems
4. of sufficient reliability to justify public dependence and confidence
5. at an overall cost to meet the public need and interest.

Across the world the development of transmission systems broadly followed four distinct stages:

1. *isolated plant stage (1885–1910)* – where isolated generating plant fed a specific local area
2. *individual systems stage (1910–35)* – where isolated plant-based systems around a major population centre were connected to each other
3. *regional/national systems stage (1935–60)* – where individual systems were connected
4. *inter-regional/international systems stage (1960–85)* – where AC and DC interconnections between the multiplicities of systems were developed and will continue in the foreseeable future.

The transmission developments went ahead due to one or more of the following 'diversity factors' offering more optimal use of assets for 'overall public benefit':

- demand
- maintenance outage
- fuel source
- generation capacity and transmission outage risk
- generation and demand uncertainty.

The transmission developments produced both public and private company benefits arising from the pooling of generation capacity, generation reserves, scheduling of lowest cost generating plant, making best use of available plant sites and achievement of better supply security at near optimum cost. These developments also enabled management of the security of supply under adverse primary fuel supply conditions resulting from a prolonged coal miners' industrial dispute experienced in mid-1980s [6]. The electricity industry structure, which progressed all the transmission and distribution developments in England and Wales, between 1957 and 1990, is illustrated in Figure 1.6.

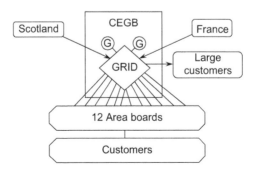

Figure 1.6 The electricity industry structure in England and Wales from 1957 to 1990

1.2.4 Recent developments 1990 to date (or 1990 onwards or post 1990)

The 1983 Electricity Act conferred new rights to privately owned generation companies to:

1. construct new or extend their existing power stations
2. secure purchase of their electricity by CEGB or an electricity board
3. secure their use of the transmission and distribution system.

Under this act the CEGB and the area boards were obliged to set the use of system tariffs and provide access to their systems as well as purchase the electricity produced by the private generators. The governmental and European Community ban on extensive use of natural gas for generation of electricity meant that there was no great rush of private capital to construct large and high capital cost coal, oil or nuclear power stations. In addition, complex tariff structures for the purchase of the generated output and the use of system did not provide adequate incentives. A similar act in France at around the same time also had no success.

In February 1988 the government published proposals to privatise the electricity supply industry by separating the ownership of the generation, transmission and distribution by:

1. ending the CEGB monopoly in generation through splitting its generation activities into separate companies
2. ending the CEGB obligation to provide bulk supplies of electricity
3. permitting existing and potential private generation companies to operate on the transmission and distribution systems
4. maintaining the transmission activities as the National Grid with the retained central role in scheduling and dispatch of generation
5. privatising the 12 area boards as 12 distribution companies each with an obligation to supply in its area with rights to generate.

The privatised electricity industry structure and the successor companies to CEGB shown in Figure 1.7 became effective from 31 March 1990. It is interesting to note that this structure was basically the same as that set out in the Weir Committee Report of 1925 and as amended by the Fabian Society Report of 1934.

Thus although the assets of the transmission and distribution systems and their main functionality remained the same, the concept of electricity trading over these assets has altered from a 'monopoly wholesaler/monopoly retailer' to 'competition in generation and supply via regulated monopolistic transmission and distribution assets' at the customer interface. Competition in generation and supply, together with regulation promoting competition in areas of natural monopoly, i.e. transmission and distribution, is the key factor towards driving down the cost of electricity to the customers.

As the 'owner and operator' of the high-voltage transmission system in England and Wales, the National Grid Company (NGC) would operate within a demanding

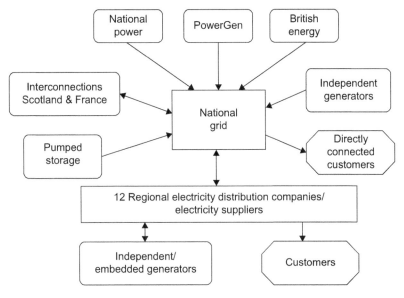

Figure 1.7 The electricity industry structure in England and Wales after privatisation on 31 March 1990

framework of statutory and regulatory requirements. Its two major obligations under the 1989 Electricity Act were established as to:

- develop and maintain an efficient, co-ordinated and economical transmission system
- facilitate competition in the generation and supply of electricity.

These obligations were specified in detail in the transmission licence, which required NGC to:

- schedule and despatch available generation in merit order to meet demand
- establish, implement, comply and ensure compliance with a Grid Code
- develop and operate the transmission system to meet defined security and quality of supply standards, including statutory levels of system frequency and voltage
- contract for ancillary services (reactive power, MW reserve, black start) from most economical sources to meet system security requirements
- set non-discriminatory connection and the use of system charges within regulatory framework
- set charges for the use of interconnectors based on return on capital
- administer the settlement process, particularly the complex computer systems needed to calculate payments due as a result of daily trading in the electricity market or 'Pool'.

It was recognised from the outset of the industry-restructuring process that freedom of access to the transmission and distribution systems would be the key to effective competition in generation and supply. Major national and international debates on the practicalities of ensuring open transmission access have taken place since and still continue particularly on ensuring adequate technical co-ordination between different entities within the structure and maintaining the security of supply.

In England and Wales, the overriding aim has been to provide a 'level playing-field' such that the would-be entrants to the electricity market, irrespective of location or size, are able to connect to and use the transmission system without discrimination or privileges in rights of access. Thus NGC was set up to be operationally independent of generators, suppliers and all users of its system. It was also set up to earn sufficient revenue from the operation of its energy transport system to ensure the capability to invest in new connections and reinforcements, maintain existing assets to required standards and to earn a reasonable return for its shareholders.

Transparency of information is essential to create the right conditions for open access to the transmission system. Therefore, NGC was obliged to make available to existing and prospective users of the transmission system detailed information including:

1. *annual statement of charges* for connection to and use of the transmission system
2. *seven-year statement* providing an annual snapshot of the development of the system in each of the seven years ahead, i.e. the essential background information for investment decisions by generators and suppliers
3. *the Grid Code* providing a comprehensive set of technical and operational requirements for all plants connected to NGC system and for NGC own plant helping to guarantee secure transmission access for all users without discrimination.

NGC is obliged by its licence to provide access to its transmission system to all parties who agree to abide by the 'Grid Code' and to pay appropriate charges for their use of the transmission system. Devising equitable charges, which accurately reflect the costs of providing the NGC transmission system, has been a major challenge, bearing in mind that before the 1989 Electricity Act transmission and generation were operated as an integrated system, the costs of which were recovered through a single, uniform tariff.

NGC use of system charges seeks to reflect the investment costs of providing a system that can accommodate the bulk power transfers which result from regional/local imbalances between generation and demand in England and Wales. Zonally set differentials in the charges aim to create financial incentives for generators to locate new plant near to load centres where there is inadequate generation, and conversely for demand to locate in zones where there is currently a surplus of generation. The transmission charging principles are regularly reviewed and approved by the regulator to ensure appropriate financial messages to the users.

Although NGC provides both technical and financial messages to the users on the optimum areas of the system for the location of new generation and demand, it is, nevertheless, obliged by its licence to provide connection to, and use of, the system to *any applicant*, irrespective of their locational choice. Furthermore, technical and financial terms for connection to the system need to be provided within three months of application by a prospective user, a dramatic change from past practice. The demand of the new commercial environment requires multi-disciplinary teamwork practices, bringing together engineers, lawyers, accountants and economists to ensure a timely and effective response to customer needs.

Since transmission and generation development is no longer centrally co-ordinated, NGC must plan and operate its system, without full knowledge of the intentions of its customers. In contrast with new connections, which may be signalled to NGC some years in advance, closure of older power stations can take place at short notice. This new background of uncertainty, together with regulatory controls on income, means that the need for new investment in the transmission system is subjected to more rigorous scrutiny than ever before and more effort is focused on making existing assets to work harder.

Where the need for new investment is established, procurement practices have changed significantly towards greater flexibility and innovation in transmission system engineering. As an example, NGC is increasingly investing in both traditional and new technologies, e.g. mechanically switched reactive compensation, quadrature boosters, static var compensators (SVCs) and 'flexible AC transmission systems (FACTS)' technologies, for controlling the power flows arising from the continuously changing generation and demand pattern. Application of these technologies also enables deferment or removal of the need for certain costlier system reinforcements. Wherever there is a clear cost benefit, transmission line reconductoring, live line working techniques and online capability monitoring equipment are utilised to improve availability of transmission equipment enabling more efficient and flexible use of the transmission system.

The reorganisation of the industry has also put similar pressures on the 12 privately owned 'regional distribution companies' operating the distribution systems. This leads to a separation of services providing the 'wires' to ensure flow of electricity to the customer on demand and the 'supply' function concerned with the customer being provided with the quality of service level of their choice from a supplier of their choice. The technical design and operation of the wires are subject to rules in the 'Distribution Code' that has been developed along parallel lines to the Grid Code. Changes in one of the technical and operational codes need to be reflected simultaneously in the other code(s). Regulatory pressures to drive the cost of supply to the customer down have affected both functions of the distribution businesses.

When the government initially announced the new industry structure separating ownership of generation, transmission and distribution, there was concern that the security of the supplies might be undermined. Since the restructuring of the industry, however, the new structural arrangements have been dramatically tested on a number of occasions demonstrating that there has been no reduction in supply security levels.

System security has been maintained despite the unbundling of the industry because:

1. all users of the transmission and distribution systems, as a condition of access, agree to abide by the technical and operational rules set out in the Grid Code and the Distribution Code
2. NGC and distribution companies continue to plan and operate the transmission and distribution systems according to the standards of security observed by their predecessors – the CEGB and the area boards
3. as a condition of its licence, NGC purchases the ancillary services (reactive power, black start capability and reserve), which are needed for maintaining system security and stability, from generators.

However, the most important factor is that all companies in the industry have a vested interest in ensuring high security and quality of supply to customers.

1.3 Structure of transmission and distribution systems

Several key factors heavily influence the development and operation of the electricity transmission and distribution systems as described below.

First, since electricity cannot be stored in reasonable quantities, the electricity generated must continually be adjusted to meet the continually varying demand, plus the system losses. Failure to maintain this balance results in a change in system frequency, which could in turn result in widespread demand loss across the system. It is the role of the control engineer to continuously meet the demand by scheduling sufficient generation to maintain the system voltage and frequency. An illustration of the dimensions of this task is given in Figures 1.8 and 1.9, which show impact of major events on demand and typical seasonal demand curves.

Second, the demand for electricity is growing. Although in some parts of the world, electrical energy consumption may have slowed down or declined during certain periods of economic uncertainty, the overall trend still shows annual increase in the 1–10% range. This means that the transmission and distribution system planners have to accurately anticipate this growth several years in advance to ensure the networks are capable of meeting the increased requirements. Furthermore, the long economic life of the transmission and distribution equipment requires particular attention to selection of investments and technologies best suited to meet the system need over such periods.

Finally, since electricity is generated from primary fuel sources, the nature and location of these resources are important factors in influencing the structure of a

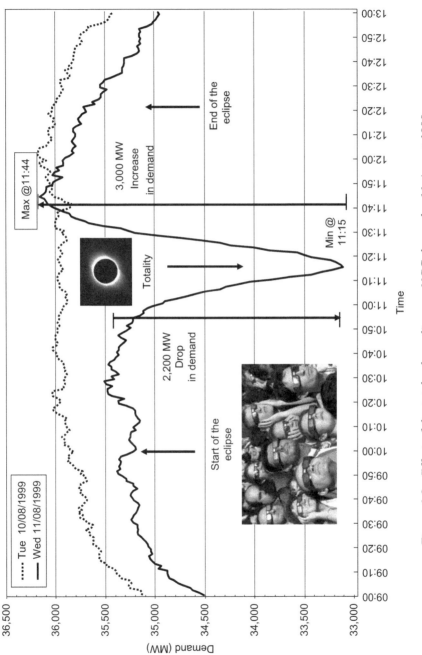

Figure 1.8 Effect of the total solar eclipse on NGC demand – 11 August 1999

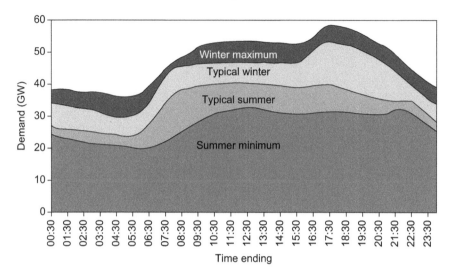

Figure 1.9 Typical seasonal power demand variations
Source: National Grid – National Electricity Transmission System Seven Year Statement May 2010

transmission and distribution system. This is due to the fact that transportation of energy in the form of electricity over transmission lines is more economic than the transportation of other primary energy sources. Large power stations also have a requirement for large amounts of cooling water for use within the steam cycle and thus tend to be sited on the coast or by large rivers. Furthermore, other environmental factors such as emissions of harmful gases and acceptability of certain types of tower structures in the countryside are increasingly playing an important part in the siting, physical characteristics and installation principles of generation, transmission and distribution systems.

1.3.1 Technical factors influencing the structure of transmission and distribution systems

Modern AC transmission and distribution systems transport the electricity generated at a frequency of 50 Hz or 60 Hz from the generators connected via generator transformers. The choice of either frequency was governed by the necessity to ensure that there is no annoying visual flicker in incandescent lighting. This choice was also compatible with the manufacture of transmission and distribution equipment based on technologies at near optimal overall cost to meet public needs and present least cost penalties in the establishment of such systems. The DC transmission systems within the AC transmission utilise converters at their points of connection with the AC system. These operate in the rectifier mode for conversion to DC and inverter mode for re-conversion back to AC. Due to the specific capabilities of the converters the DC connections are capable of connecting two AC systems operating at different nominal frequencies or providing asynchronous connection between the systems.

One of the most important characteristics of a transmission or distribution system is the security of supply which, by implication, means a supply of electricity that is continuous and of the required quantity and quality especially in terms of frequency and voltage. This in turn means that the generation, transmission and distribution systems must have sufficient in-built flexibility by design to maintain supplies under conditions of plant breakdown or weather induced failures for a wide range of demand conditions.

The frequency of the system is governed by the speed at which the generating plant operates. The quality of the frequency generated is conditional on the transmission system operator scheduling sufficient generation to meet the demand plus the losses of the transmission and distribution systems. This quality is further assured by holding generation in reserve for immediate availability in the event of unplanned plant losses. In this way, the level of the frequency can be maintained to predetermined levels. In the UK, the statutory frequency is 50 Hz with $\pm 1\%$ permitted variation. The DC interconnection with France provides an asynchronous connection in that the systems at either side of the connections are isolated from the frequency variations in the other system.

The other important characteristic of transmission and distribution systems is the voltage quality, which is also governed by the generating plant. While the frequency of a system will be substantially constant at all points, the voltage levels will differ at different locations on the system. This difference is governed by the capacitive and inductive characteristics of the 'wires' i.e. the overhead lines and underground cables, forming the systems and the power factor of customer demand. At times of low power flow through the wires, the capacitive effect dominates and causes the voltage at the end of transmission and distribution lines to rise. At times of high power flow, the inductive effect dominates and the voltage tends to fall along the length of a line. Through the installation of capacitive or inductive compensation plant termed 'reactive compensation' it is possible to limit variations in voltage beyond pre-defined limits. This ensures the customer equipment does not suffer any undue interruptions and the customer does not have to invest undue amount in purchasing equipment designed to operate over very wide voltage variations. Statutory voltage variations permitted on transmission and distribution systems are usually around $\pm 5\%$ and $\pm 10\%$ respectively. The manual and automatic control facilities put in place to achieve compliance with these performance criteria are shown in Figure 1.10.

The above technical principles and characteristics are applicable to virtually any transmission and distribution system around the world. The only difference between transmission and distribution utilities is related to the organisational structures of the companies that control the various types of systems.

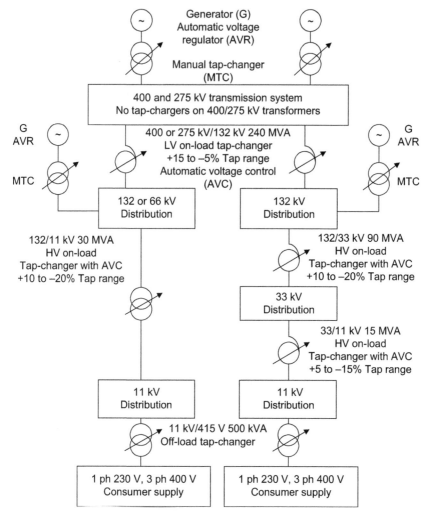

Figure 1.10 Transmission and distribution system voltage control facilities

1.3.2 Organisational structures

In essence, the electricity supply industries or utilities around the world based on their structure and ownership can be classified under four main headings:

1. vertically integrated
2. horizontally separated
3. privately owned
4. publicly owned.

In a vertically integrated industry, the whole industry, i.e. generation, transmission and distribution, is controlled by a single authority. This authority may be wholly or partially owned publicly or privately. Essentially it is able to carry out integrated generation, transmission and distribution planning and operation within a defined geographical area. After the UK privatisation in 1990, the only vertically integrated electricity utilities were two in Scotland (Scottish Power and Scottish Hydro Electric) and one in Northern Ireland (Northern Ireland Electricity), and all these entities were in private ownership.

For the privately owned utility companies that are vertically integrated within a certain geographical area the regulatory pressures on the 'natural monopolies' of transmission and distribution result in a clear definition of the costs of each part of the company activity. The traditional single tariff structures in the publicly owned vertically integrated utilities are not designed to clearly differentiate between the costs of generation transmission and distribution activities. Cross-subsidies between these activities are also more common in such utilities.

The prevailing electricity industry structure in England and Wales from 1990 to 2001, with both energy trading and transmission/distribution infrastructure components shown in Figure 1.11, is horizontally separated by virtue of privately owned generation, transmission and distribution companies being separated from each other. This structure allows for the private ownership of all or parts of each function in 'wires' and 'supply'. However, it is the independence of the transmission utility from all others that tends to be the essential mechanism by which the horizontally separated system structure operates in a secure and economic manner.

The natural pivotal role of a transmission company is to provide a transmission and operational infrastructure by which power can be securely transported across the system and hence facilitate the trading between generators, distributors and suppliers. This also allows access for other interconnected electricity utilities outside

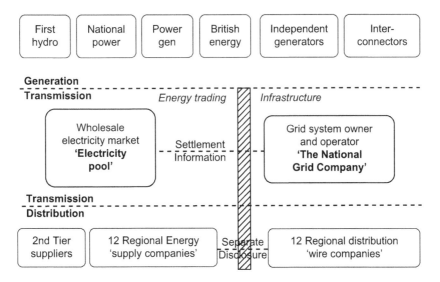

Figure 1.11   Electricity supply industry structure in England and Wales from 1990 to 2001

36 High-voltage engineering and testing

the immediate geographical area. Two further models of horizontally separated industry structures implemented around the world are shown in Figures 1.12 and 1.13 for Argentina and State of Victoria, Australia, respectively.

Transmission system interconnections extend customer access to more sources of electricity supply. Availability of interconnection technologies at economic cost means that systems with different technical characteristics in their choice of voltage and frequency can be interconnected without problems over long distances. The interconnection may be in the form of a high-voltage AC overhead transmission line, as in the case of many interconnections across European countries and

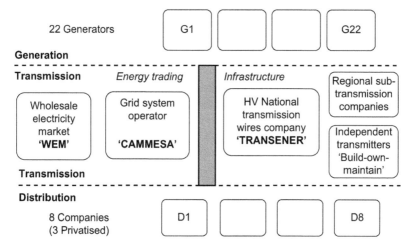

Figure 1.12 The new electricity industry structure in Argentina

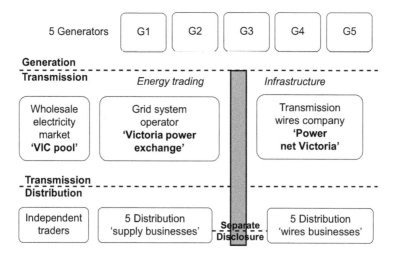

Figure 1.13 The new electricity industry structure in State of Victoria, Australia

utilities, or it may be a DC link as in the case of the cross-channel interconnection of England with France.

Although three-phase AC overhead lines are the usual method of interconnection between utilities, in cases where long transmission distances and/or sea crossings are involved, 'high-voltage direct current (HVDC)' transmission is economic despite the relatively higher costs of converter and terminal equipment. HVDC transmission is also utilised for interconnecting utilities with different supply frequencies, e.g. 50 and 60 Hz and in cases where an asynchronous link between systems is required. The 'back-to-back' HVDC links in Japan, with both their converters situated at the same site, are examples of this type of interconnection.

The horizontally separated industry structures are ideal for introduction of competition in generation and supply. This is more difficult, if not impossible to achieve with vertically integrated structures especially if they are publicly owned. The attempt by the 1983 Electricity Act to oblige CEGB and area boards to set use of system tariffs giving access to other competitive generation sources to its system has failed to attract new connections although such tariffs were produced. A similar attempt in France has also met the same fate due to lack of regulatory/commercial incentives and pressures on the vertically integrated company.

Ownership of the assets forming a transmission and/or distribution is no longer important as the legally binding technical and commercial codes ensure safe, secure and stable operation of a multiplicity of utilities within country and over international boundaries.

Since 1990 the UK electricity industry structure has gone through two further stages of development. First, in 2001, 'New Electricity Trading Arrangements (NETA)' was introduced to eliminate the perceived weaknesses of the 'Pool' trading structure, which was in place from 1990 onwards. In the NETA trading structure shown in Figure 1.14, while the infrastructure arrangements remained unchanged, both generation and demand could make firm bids to a 'balancing market' and would be paid based upon their 'bid price' instead of the 'marginal price' paid to all plants in the previous 'pool' trading structure.

Bilateral contracting	Rolling half-hour trades	Generation and demand balanced by system operator	Real time (1/2 h trading period)
Forwards market	**Power exchange**	**Balancing mechanism**	**Imbalances and settlement**
Year ahead (or earlier)	$T - 24$ h	$T - 3.5$ h	$T - 0$ h

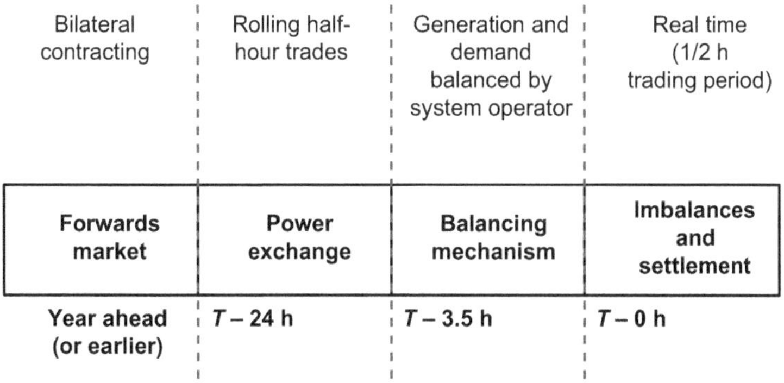

Figure 1.14 New Electricity Trading Arrangements (NETA) in UK in 2001

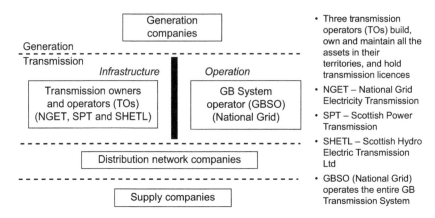

Figure 1.15 Structure of the single electricity market for Great Britain under Energy Act 2004

Second significant development occurred through the 'Energy Act 2004' in July 2004 and created a 'single electricity market for Great Britain (GB – England, Wales and Scotland)' as shown in Figure 1.15. Under this arrangement National Grid as the single 'GB System Operator (GBSO)' would be responsible for:

- contracting for connection/use of system
- generation despatch and balancing
- planning and co-ordination of transmission and generation outages
- transmission charging
- co-ordination of GB transmission investment planning.

Three 'transmission operators (TOs)' of the 'wires' namely:

- National Grid Electricity Transmission (NGET)
- Scottish Power Transmission (SPT)
- Scottish Hydro Electric Transmission Ltd (SHETL).

would be responsible for maintenance, investment planning and carrying out system switching as directed by the system operator GBSO. These infrastructure and trading arrangements are unlikely to be subject to further major development in the foreseeable future.

1.4 Design of transmission and distribution systems

Transmission and distribution systems are designed to meet four basic criteria of electricity supply:

- economy
- security
- quality
- stability.

An economic system is one where adequate and economic sources of electricity exist and the ability to transmit and distribute power to the customer can be efficiently carried out. The security of supply is the ability of the transmission or distribution network to supply customer demand without interruption under operational conditions leading to the loss of defined components within the system including generating plant. The quality of the supply is the extent to which the customer demand is satisfied within the defined quality of supply criteria, e.g. frequency and voltage stability relate to the ability of the system to return to a stable operational state following the occurrence of faults and disturbances defined within the security criteria, which may occasionally affect the system.

In addition, the transmission and distribution lines are designed with sufficient thermal capability to supply the demand and accommodate the power output of generation plant as well as power transfer across system boundaries through the annual, seasonal and daily demand cycles. The amount of power that can be transferred across a system boundary is limited by the thermal rating of the individual circuits and the way in which the power is shared between them. The 'firm' thermal capability, i.e. the capability after the loss of one or two circuits, dependent upon the chosen security criteria, is usually less than the sum of the individual ratings of the remaining circuits. This is because one circuit usually reaches its rating limit before others due to the resulting unevenly distributed power flows.

Seasonal ratings of the circuits also need to be taken into consideration in order to avoid system constraints.

1.4.1 Security of supply

Security of supply criterion is defined at the design stage of transmission and distribution systems. It defines the capacity and flexibility of the systems to maintain supplies under conditions of pre-defined plant breakdown or equipment outages/failures for a wide range of demand conditions. The main parameters defined within the criterion are the:

1. number of individual and/or simultaneous equipment outages or failures
2. magnitude of instantaneous generation/demand losses
3. types of faults and their duration prior to successful clearance by protection.

The systems will then be designed and developed such that they would continue to supply the demand without interruption and within defined quality of supply parameters while operating under such conditions. For outages and failures beyond such parameters some interruption or loss to the quality of supply would be expected. The security standard, thus, determines the overall economics as well as levels of capital and revenue expenditure of the transmission and distribution systems.

1.4.2 Quality of supply

Quality of supply criterion is also defined at the design stage to define the quality of the supply to be maintained in terms of frequency and voltage variations/limits under various normal and disturbed system conditions. In addition other parameters such as voltage dip under plant breakdown or equipment outage/failure conditions, negative sequence current magnitudes, voltage flicker limits, phase unbalance limits and repetitive equipment switching limits are also specified as they have a direct bearing both on the design of the transmission and on distribution systems as well as connected customer equipment.

1.4.3 Stability

Two criteria are usually used to define system stability, which is the ability of the system to return to normal operating conditions following system disturbances defined in the security criteria. 'Transient stability' relates to system conditions following severe disturbances, e.g. a network fault, and 'steady-state and dynamic stability' is concerned with the system response to small disturbances such as the normal random load fluctuations.

Generators will remain transiently stable if following a large disturbance such as a nearby fault, each generator settles down to a new steady operating condition, i.e. continues to operate at the same mean speed. During the fault the electrical output of each generator will be substantially less than the mechanical power input from the prime mover. The excess energy will cause the generator rotor to accelerate and start an electromechanical oscillation or swing against other generators. Provided that the faulted circuits are disconnected quickly typically in 80–100 ms, and adequate transmission remains, generator voltage and speed controls will quickly respond and steady operation will be achieved. If, however, the fault persists or inadequate transmission outlets remain, then large cyclic exchanges of power between the generators on the system will occur. This is likely to cause extensive damage to the generators and initiate system break-up through the forced operation of the system protection.

While large disturbances are relatively infrequent in transmission and distribution systems small disturbances frequently arise as a result of normal load variation and switching operations. These small disturbances cause small oscillations in the power output and terminal voltage to develop between groups of generators within the system. The damping i.e. the reduction in the magnitude of such oscillations is a function of the transmission system design, generator excitation control parameters and technical characteristics of generators and demand. Loss of damping in such oscillations can lead to a continued increase in their magnitude causing loss of the synchronism between generators.

System transient, steady-state and dynamic stability is assured through the use of:

1. fast protection systems, which ensure faults do not persist
2. braking devices, which provide deceleration to generators immediately after the disturbance
3. series compensation devices, which improve the connection adequacy of the transmission system by reducing the system impedance following the disturbance
4. shunt reactive compensation devices, which improve voltage levels at selected points on the system

5. fast generator excitation systems which rapidly restore the ability of the generator to transmit power following a large disturbance
6. fast acting turbine valve controls reducing the mechanical power input to the generator following the disturbance
7. use of 'power system stabilisers' on generator excitation control systems and dynamic shunt compensation devices which improve system damping performance
8. fast adjustments to system configuration.

1.5 Operation of transmission and distribution systems

System operation or system management is concerned with the day-to-day operation of a power system. Based on the nature of activities involved four distinct timescales essential to the safe and economic operation of a power system can be identified as follows:

1. operational planning phase, extending from several years down to a few days ahead
2. extended real-time operation phase, extending from a few days or up to a week to say a few hours ahead
3. real-time operation phase, extending from a few hours ahead to an hour or a few hours after the event
4. post real-time operation phase, covering collection, analysis and archiving of data from actual operation, which may extend from a few hours to a few months after the event.

Different organisational structures are needed to manage the planning and operation of a power system. These organisational structures are dependent upon the overall structure of the electricity supply industry defining the extent of the role and responsibilities of the transmission and distribution entity and its interfaces with all the other entities. The overall industry structure also influences the information flows between and within each entity as well as the working methods and practices adopted. However, the basic functions of operating and managing the system as described below will remain the same regardless of the organisational structure.

1.5.1 Operational planning

The prime objective of operational planning in any organisational structure is ensuring the security of the system in an economical manner by optimal scheduling of constraints such as generation and circuit maintenance outages in a timescale of usually from five years to one year and even down to one week ahead.

The activities in the operational planning timescales can be summarised as:

1. forecasting the real and reactive power demand
2. plant availability forecasting
3. generation and transmission outage planning and optimisation

4. fuel allocation and energy modelling
5. management of protection settings
6. contingency and defence planning
7. specification and control of special protection systems (SPS)
8. preparation and review of operational standards and procedures
9. user specification of operational planning tools and real-time system control, energy management system (EMS) and system control and data acquisition (SCADA) systems.

1.5.2 Extended real-time operation

Extended real time is the period from one or two days prior to real time to about 1 h ahead. Its function is to compliment the operational planning activities by modifying the information received to take account of the latest changes that may have occurred (e.g. in demand level, plant availability and outage conditions) and detailing it to a very high level.

The four main tasks of extended real-time analysis are demand forecasting, generation scheduling, assessment of trading/ancillary service requirements and network security analysis. In each case the level of data required for each task will increase significantly. For example demand forecasting data is required on a half-hourly basis, generator loading data requires synchronising/desynchronising times with outputs over each half hour and network analysis data requires precise details of present network configuration and contingency actions for certain fault conditions.

1.5.3 Real-time operation

The objectives of real-time operation of transmission and distribution systems are:

1. Electricity of adequate security and quality at minimum cost is supplied to all consumers.
2. Necessary access to plant and system for maintenance, repair and new construction is safely provided.
3. Effects of any system and equipment faults and/or disturbances are minimised and system is returned to normal operation as quickly as possible.

These tasks are achieved with the help of the following facilities and/or systems:

1. system control and data acquisition (SCADA) systems
2. energy management system (EMS)
3. state estimation software
4. generation scheduling and despatch software
5. system security and contingency evaluation software
6. system load-flow and overload evaluation software
7. short-circuit level calculation software
8. system stability evaluation software

9. demand management and demand transfer facilities
10. alarm optimisation software
11. automatic reclosure and special protection facilities
12. load restoration facilities
13. contingency and defence plans as well as emergency facilities.

With the advances in computer hardware, communications hardware, graphic display facilities and system analysis and other computing software it is quite usual for the operator to be able to carry out a lot of the necessary load-flow, fault level and stability calculations online on a continuous basis.

1.5.4 Post real-time operation

The post real-time operational phase is essentially a stocktaking phase where the real-time operational facts and actions taken in response to events are analysed with a view to apply the lessons learned to future operation. The statistics of past operation are the most important source of data for estimating future commitments, requirements and actions. They also provide the raw data by which the transmission/distribution utility management can monitor the efficiency of the utility's technical and commercial operation in delivering the level and quality of electricity supply in an economic manner.

A further area of work in this phase is the analysis of system performance especially during adverse operational conditions and abnormal events causing losses of supply or forcing uneconomic operation. Post-event analysis of such events usually through utility staff and external experts is undertaken to determine recommendations to avoid similar recurrences.

Although the operational data collected and in particular the analysis done will be specific to each utility, it is usual for all system data including voltages, active and reactive power flows, generator outputs, demand, frequency and tie-line flows to be routinely recorded in accordance with the SCADA sampling time provisions for every system busbar.

Analysis of abnormal system operational conditions and system disturbances heavily relies on accurate time-tagged system disturbance recorder and fault recorder data collected by such devices. In certain cases where equipment damage has occurred, other data such as climatic data, metallurgical and other forensic data such as oil and gas samples often becomes necessary.

1.6 Future developments and challenges

1.6.1 Organisational developments

The developments of past two decades in the electricity supply industry throughout the world have resulted in the introduction of many different types of privatised structures including, independent power producers, independent transmission and privatised distribution in various combinations based upon the historic development of the industry within the specific region or country. Introduction of regulatory processes to introduce competition in monopolistic transmission and distribution as well as demand for easier access to such systems to take advantage of larger markets and cheaper electricity brought new challenges. The adoption of transparent 'third party access (TPA)' principles resulted in adoption of radical changes in the strategies of transmission and distribution companies in the development of their systems. The uncertainties created by the requirements to connect in short timescales and rapid changes in the generation patterns brought about by new connections and plant retirements could only be countered by the introduction of probabilistic planning and extensive risk analysis in order to meet the security of supply obligations.

Environmental considerations have also resulted in increasingly severe restrictions in the construction of new overhead lines, with mounting pressure for the undergrounding of existing overhead lines. Transmission and distribution utilities are being forced to find ways of working their assets harder and finding methods of utilising the 'wires' to a much higher extent. This has resulted in the development of maintenance, asset life management and risk management techniques as well as utilisation of new technologies utilising power electronic devices categorised as FACTS [7, 8].

Managerial structures have also been subject to change moving away those suitable for centrally planned, vertically integrated systems with a pronounced bias on public service engineering, to those suitable for horizontally structured systems, with commercially oriented public ownership. This changing structure transformed transmission and distribution utilities to the vehicles for facilitating competition in electricity supply. These utilities became increasingly subject to legislation and regulation to ensure access on a transparent basis and adequate addressing of provisions to ensure adequacy and security of supply. Together with their unique presence in the public perception the changes in the accountability of public companies have also played an important role in fashioning the management structures of the utility companies.

The central business district supply failure in Auckland, New Zealand, has focused attention on the consequences of inadequate governance and failure to adequately specify the responsibilities of newly formed utilities [9]. Other large-scale blackouts in the USA, Europe and elsewhere in the world, which affected the population in large areas such as:

1. Western USA on three consecutive occasions within one month in 1996
2. Italy, Scandinavia, Eastern USA and Canada in 2003
3. Southern Brazil in 1999
4. European interconnected system in 2006
5. Brazil and Paraguay in 2009.

have not only demonstrated the vulnerability of interconnected transmission and distribution systems to relatively minor system disturbances but also exposed the lack of robustness in design of such systems, in system security assessments carried out and in their defence plans and mechanisms [10]. Such incidents also brought the high economic and social impact to the forefront. This had led to closer scrutiny of regulatory processes which obliged transmission and distribution entities to limit their "wires" investments without due regard to the economic and social impact of such incidents. The deficiencies in the application technical standards and even basic voltage control mechanisms such as reactive compensation also became very apparent beyond any reasonable doubt.

The organisational concepts developed over the past decade have performed reasonably well especially in cases where the interfaces between various utilities have been adequately defined and mechanisms to address the technical, financial and trading aspects have been adequately addressed. Nevertheless many adjustments to the industry structures have to be made in frequent intervals. There is no doubt further development in refining these processes will need to continue to incorporate the lessons learned from practical implementation. While the pressures will continue to be essentially commercial in their basic nature the successful management of technical requirements will always play a major role in ensuring adequate supply security.

1.6.2 Technical and technological developments

Since AC transmission systems won the debate over DC systems over a century ago, engineers have made efficient use of the AC system to move large amounts of power over great distances with relatively small losses. However, they have been continually challenged with the problem of controlling the power flow, especially during transmission maintenance outage periods as well as in cases of construction/refurbishment outages associated with connecting new generation on the system. The changing generation patterns and the need to ensure better utilisation of transmission and distribution assets increasingly require application of reactive compensation devices, while need to ensure least stranding of assets requires mobility and rapid relocatability of such assets [11]. Increasing use of reactive compensation devices necessitates the development of regional and overall system controls facilitating automatic control of the utilisation of such devices to secure voltage stability and prevent voltage collapse.

The developments in FACTS [7, 8] and HVDC transmission technologies over the past decade offer the transmission and distribution utilities a portfolio of new tools enabling the management of their technical obligations in the security and quality of supply. Application of these technologies alongside the traditional well-developed equipment will increase as the above-mentioned pressures on utilities increase. These technologies have graduated from the first-generation devices utilising mechanical switches and discrete capacitive or inductive components to second-generation devices where the thyristor replaced the mechanical switch to the current third-generation power electronic devices utilising converter technology and new semiconductor devices such as insulated-gate bipolar transistors or IGBTs.

As the first and largest fully privatised independent transmission utility in the world NGC has been subject to the above-mentioned organisational and electricity market-driven changes. This meant that every available technology was put to use to address the challenges of facilitating competition and responding to electricity marketplace. NGC pioneered the use of first-, second- and third-generation reactive compensation devices with the ease of relocatability as an essential characteristic. The following is a brief summary of both traditional and the new technology portfolio readily available to utilities.

1.6.2.1 Shunt reactive compensation (MSC, MSR, TSC and TSR)

The shunt-connected equipment in the form of either mechanically switched shunt capacitors (MSC) or mechanically switched shunt reactors (MSR) to control the voltage at a particular location on the system has been available since the early days of transmission and distribution systems. Switching capacitors into service will raise the voltage, while switching inductors will reduce the voltage. Utilisation of thyristor switches enables rapid switching of shunt capacitors (TSC) and reactors (TSR) within a few milliseconds in comparison with the corresponding closure time for mechanical switches of some 50 ms giving a faster response.

1.6.2.2 Static var compensator (SVC)

SVCs are shunt compensation devices, where the current through the reactor(s) is controlled by thyristors (TCR) and the capacitor(s) is switched in by either mechanical (MSC) or thyristor (TSC) switches as shown in Figure 1.16 [12]. Through an automatic voltage regulator controlling their output, SVCs are capable of providing reactive compensation within their rating limits, at a fast response rate similar to modern generator voltage regulators, to control voltage within specified limits at their point of connection to the system. The dynamic reactive support thus provided can yield significant benefits in system voltage control, dynamic stability and increases in power-transfer capability. It is also possible to construct these devices in a manner making their relocation within a very short period possible to provide the necessary system flexibility without recourse to stranded assets [11].

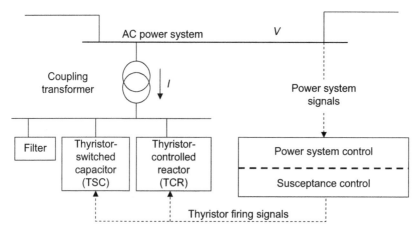

Figure 1.16 Typical static var compensator (SVC)

1.6.2.3 Quadrature booster (QB) or phase-shifting transformer (PST)

These devices are similar to transformers, but have an additional winding on each phase. The connection of the additional windings is arranged such that they are energised from another phase, thus giving a quadrature voltage injection into the three-phase voltage at a particular location. The magnitude of the quadrature voltage is varied via the tap-changing facilities allowing the control of power flow along circuits and power flow sharing between parallel circuits.

1.6.2.4 Series capacitor (SC)

These devices are large capacitors, sited along the length of long transmission lines which help to reduce the series impedance of a line, thus allowing more power flow through that line. Application of series capacitors requires careful study of network resonances at sub-synchronous frequencies, which may coincide with vibrational frequencies of generator shaft systems to avoid damage to generators under certain network operational conditions. Series capacitors can be mechanically switched in and out of service individually or in banks through a bypass switch to reduce or increase the series impedance of the transmission line.

1.6.2.5 Thyristor-controlled series compensators (TCSC and TCSR)

These devices consist of a thyristor-controlled reactor in parallel with a traditional series capacitor (TCSC) or series reactor (TCSR) as shown in Figure 1.17 [7, 8]. By dividing the series capacitor or reactor in to a number of individual banks coarse control of the amount of series compensation is obtained. The thyristor-controlled reactor (TCR) connected in parallel with one or more of such banks introduces fine control by controlling the current flowing through the shunt-connected reactor. The TCSC or TCSR is thus able to dynamically vary the level of series compensation. In the case of TCSC this control can be used to prevent sub-synchronous resonance

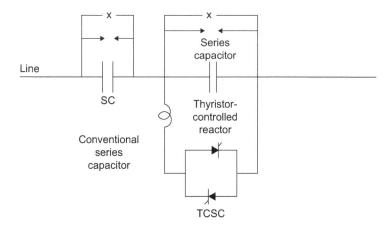

Figure 1.17   Thyristor-controlled series compensator (TCSC)

by de-tuning the network resonance through a controller altering the shunt-connected reactor current.

1.6.2.6 Static synchronous compensator (STATCOM)

This is a shunt-connected reactive compensation device based upon a solid-state synchronous voltage source generating three-phase fundamental frequency voltages whose magnitude and phase angle can be rapidly controlled. Reactive power can be generated or absorbed by changing the amplitude of the voltage source relative to the system voltage. The device is compact as no discrete devices such as reactors and capacitors are used. The control characteristic of the device is similar to that of a rotating synchronous compensator. The response to system voltage changes is, however, very fast as there are no delays other than those involved with measurement of the system voltage. At low system voltages the device continues to supply constant current and thus has a superior voltage-support characteristic compared with conventional SVCs, which are essentially constant impedance devices under such conditions. The typical STATCOM configuration and the comparative performance characteristics of both STATCOM and SVC are illustrated in Figures 1.18 and 1.19 respectively [8, 13].

With the connection of an energy-storage device in place of the DC capacitor the basic STATCOM is also capable of four-quadrant operation that is supplying or absorbing both active and reactive power to the AC system to which it is connected. This capability is the basis for future power system developments incorporating energy-storage technology illustrated in Figure 1.20.

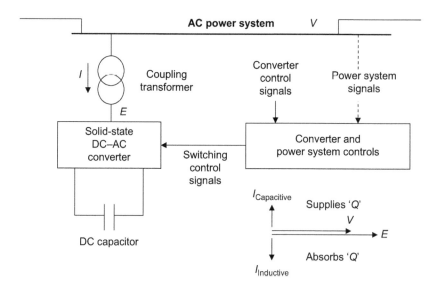

Figure 1.18 Typical STATCOM

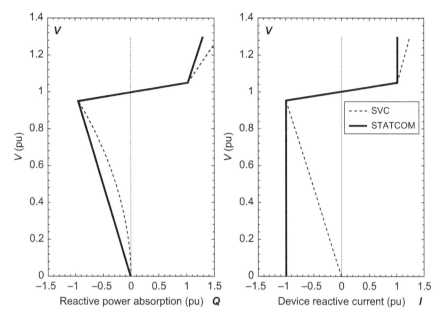

Figure 1.19 Typical V–Q and V–I characteristics of STATCOM and SVC

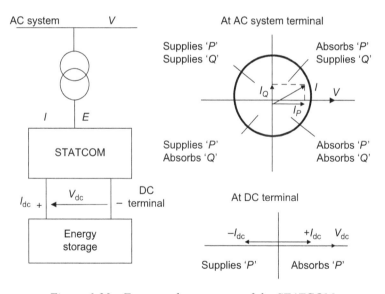

Figure 1.20 Four-quadrant nature of the STATCOM

1.6.2.7 Solid-state series compensator or static synchronous series compensator (SSSC)

The solid-state synchronous voltage source forming the basic building block of a STATCOM can also be used to inject a voltage in series into a transmission line to cancel the voltage drop associated with the line impedance. By controlling the magnitude of the injected voltage to be proportional to the line current an equivalent effect to that provided by a series capacitor is obtained. The injected voltage can also be varied such that the equivalent effect of adding either series capacitance or series inductance can be obtained. Since this device achieves the equivalent effect without incorporating discrete capacitors or inductors no sub-synchronous resonance can occur [7, 8].

1.6.2.8 Unified power flow controller (UPFC)

The unified power flow controller consists of a shunt-connected STATCOM and a solid-state series compensator, which are jointly controlled such that a series voltage, which can be varied both in magnitude and in phase can be injected into a transmission line as shown in Figure 1.21 [11, 8]. The device is therefore able to continuously vary the mode of the series compensation effect thus achieved enabling very effective handling of power flows especially during system contingency conditions. Since the injected voltage can be varied both in magnitude and in phase without constraint it is also possible to independently control both the active and the reactive power flow on the transmission line giving versatility for both steady-state and dynamic stability improvement.

1.6.2.9 Interphase power controller (IPC)

This is a series-connected power flow controller, consisting of inductive and capacitive elements in each phase, each of which is subjected to separately phase-shifted voltages. The internal phase shifts are implemented in a way to give each terminal a voltage-dependent current-source characteristic. By adjusting the phase shifts the level of power through the transmission line can be controlled.

1.6.2.10 Thyristor-controlled braking resistor (TCBR)

Thyristor-controlled braking resistor is a shunt-connected, thyristor-switched resistor usually located at or close to generator output terminals. The device is

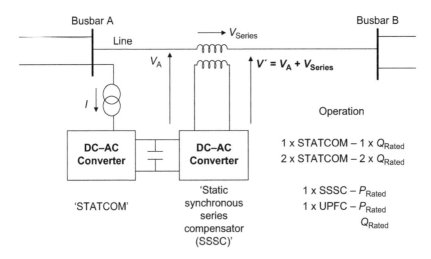

Figure 1.21 Unified power controller (UPFC)

usually switched in following a fault to control the accelerating power of the generator and switched out when the acceleration is reduced to a preset level ensuring return of the generator to stable operation.

1.6.2.11 Thyristor-controlled phase-shifting transformer (TCPST)

This is a phase-shifting transformer or a quadrature booster as described in section 1.6.2.3 whose output is adjusted by thyristor switches giving a much more rapid change of quadrature voltage and hence phase angle than the traditional mechanical tap-changer-controlled quadrature booster.

1.6.2.12 Thyristor-switched series compensation (TSSC or TSSR)

This compensator consists of a number of traditional series capacitor or reactor banks connected to a transmission line with thyristor switches connected in parallel to each bank. By controlling the switches a step-wise control of the series capacitive or inductive reactance is achieved in a rapid manner.

1.6.2.13 Interline power flow controller (IPFC)

By back-to-back coupling of two static synchronous series compensators (SSSCs), each connected to a different transmission line, it is possible to control the active and reactive power flow sharing between the two transmission lines as shown in Figure 1.22. This provides great flexibility in system operation by forcing the power to flow in the less loaded lines especially under system contingency conditions [8, 11].

1.6.2.14 Static synchronous compensator (STATCOM) for arc furnace and flicker control

SVCs have been used for arc furnace and flicker compensation due to non-linear loads for over three decades. However, the inherent and well-understood limits of the SVC only permit reduction of flicker by a factor close to 2. However, ever increasing sizes of arc furnaces and non-linear loads and their connection to weaker system points or points of high equivalent system impedance require reduction of flicker by factors larger than 2 to meet utility power quality requirements. In this respect, application of

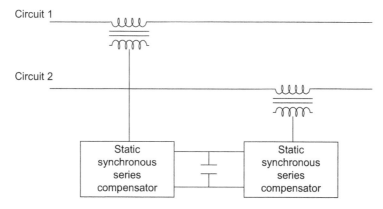

Figure 1.22 Interline power flow controller (IPFC)

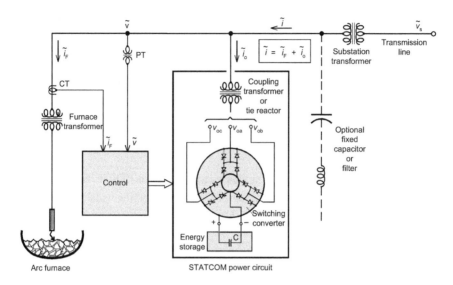

Figure 1.23   Arc furnace and flicker compensation by STATCOM

further SVCs or SVCs of higher ratings does not offer any improvement on their known performance limits of flicker reduction by a factor close to 2 [14–16].

One of the recent examples of an arc furnace application was the connection of a ±82 MVAr STATCOM at 33 kV on the Tornio 110 kV busbar of the FinGrid system with a 160 MVA arc furnace installation. Flicker reduction factors of 4.6–5.0 times have been achieved with the furnace operating at high- and medium-load cycles respectively [16]. Figure 1.23 illustrates the principles of STATCOM application for arc furnace and flicker compensation of non-linear loads connected to the power system.

1.6.2.15   Thyristor-controlled voltage regulators

There are a number of non-conventional methods of voltage regulation using thyristor-based power electronics [17] including:

1. voltage-controlled static var compensator (VCSVC)
2. voltage-controlled thyristor-switched capacitor (VCTSC)
3. voltage-controlled thyristor-switched reactor (VCTSR)
4. shunt capacitor bank series shorting (CAPS)
5. delta-star switching of capacitors and reactors
6. thyristor-based tap-changers.

These methods offer some advantages over the conventional methods of voltage regulation and are described in detail in Reference 17.

1.6.2.16   HVDC technology

HVDC technology is utilised where non-synchronous ties between different utilities or parts of the same transmission system are required. The most obvious cases of this are where the two systems have different frequencies or voltage levels, or are separated by long sea crossings. HVDC can also be used within an AC transmission system to deliver power between two regions directly or in parallel operation with AC transmission. Furthermore, AC overhead lines can be converted to DC operation enabling additional flow of power without the need for reconductoring or installation of new transmission circuits.

Advances in HVDC technologies such as 'voltage-sourced converters (VSCs)' also offer many interesting applications such as current limiters, solid-state circuit-breakers, transfer switches, dynamic voltage restorers for addressing distribution system problems [18].

Further advances in HVDC technologies such as 'capacitor commutated converters (CCC)' [19] and 'VSC transmission' [20] have been successfully applied to HVDC interconnections and point the way to further future developments.

1.6.3 Control and communication developments

Increasing use of reactive compensation devices and solid-state component-based technologies and the need to control voltage and frequency to much higher quality standards are the driving forces behind the control system developments. At a component level increasing reliability of digital processors has led to almost complete dominance of such devices over the conventional analogue-type control systems.

Increasingly the digital control systems are being installed for protection, indications, measurements, monitoring and control purposes. Furthermore, these devices are being integrated to perform these functions in a single control unit utilising the same system quantities. While this has the promise of savings in terms of installed equipment it requires application of special measures to achieve the same level of dependability obtainable from the existing separate functional arrangements. Lack of formalised standards for digital information exchange between different manufacturers equipment is likely to hamper progress.

Increasing dependence on reactive compensation devices to achieve better system utilisation creates its own control challenges. In many transmission systems adequate voltage control is no longer possible if the control is based purely on local power system quantities. In addition to the immediate fast primary voltage control response provided by generator automatic voltage regulators in the 1–5 s timescale, an automatic secondary voltage control response based upon voltage measurements at a number of points within a specific region is required to ensure stable system behaviour in the 5 s to 10 min timescale. This then needs to be augmented by a tertiary voltage control system ensuring readjustment of target voltages in critical parts of the system at around 10 min intervals.

Use of fast-response power electronic devices requires co-ordinated control of a portfolio of shunt and series devices in a manner not only to make most efficient use of full device capabilities but also to ensure non-conflicting control action between these devices, which may lead to unstable system operation. This particular subject area is at present in its infancy with potential for rapid developments.

Developments in the application of fibre optic-based communication technologies have facilitated increasing use of digital control systems as well as increased functional integration of protection, indications, measurements, monitoring and control. There is no doubt that while many control concepts hitherto considered are too complex and difficult to achieve, such as the remote control of Substations, they are now becoming a reality thanks to the application of advanced technologies. The new challenge is in management of these systems to achieve the ever increasing performance expectations of the society from the transmission and distribution systems and in the avoidance of high-impact supply interruptions and blackouts.

With the rapid development of computing and communication systems it was inevitable that so-called smartgrid initiatives have recently developed at a great pace. These initiatives tend to focus on consumer demand control, remote switching and metering aspects but currently suffer from inadequately defined objectives as well as protocols and standards. At transmission level these initiatives are most appropriate for special protection and system control applications. In distribution systems, however, their application will be subject to consumer approval and placement of privacy safeguards.

Increasing use of Internet-based communications also exposes transmission and distribution utilities to malicious attacks and hacking activity. These issues are likely to require increasing efforts to ensure robustness of systems to such threats.

References

1. Sayers D.P., Forrest J.S., Lane F.J. '275 kV Developments on the British Grid System'. *Proceedings of the IEEE*. May 1952;**99**(Pt. II) (Paper No. 1309S)
2. Booth E.S., Clarke D., Eggington J.L., Forrest J.S. 'The 400 kV grid system for England and Wales'. *Proceedings of the IEEE*. December 1962;**109** (Pt. A, no. 48):493
3. *Electrical Transmission and Distribution Reference Book*. Westinghouse Electric Corporation
4. Electricity Association. *Handbook of Electricity Supply Statistics*. London: Electricity Association; 1989
5. Casazza J.A. 'The development of electric power transmission'. *IEEE Case Histories of Achievement in Science and Technology*. IEEE; 1993
6. Ledger F., Sallis H. *Crisis management in the power industry – an inside story*. Routledge; 1995
7. 'FACTS overview', Joint document by IEEE Power Engineering Society and CIGRE, issued as IEEE Publication 95TP108, April 1995
8. Hingorani N.G., Gyugyi L. *Understanding FACTS – concepts and technology of flexible AC transmission systems*. IEEE Press; 2000
9. *Auckland Power Supply Failure 1998*. The report of the Ministerial inquiry into the Auckland power supply failure, Ministry of Commerce, New Zealand, 1998

10. CIGRE Report no. 316 by Working Task Force TFC 2.02.24 'Defence Plans for Extreme Contingencies', April 2007
11. Gyugyi L. 'Converter Based FACTS Controllers', IEE Colloquium on 'Flexible AC Transmission Systems – The FACTS', IEE Publication 98/500, November 1998
12. Erinmez I.A. (ed.). *Static Var Compensators*, CIGRE report no. 025 by Working Group 38.01, Task Force 2, September 1986
13. Erinmez I.A., Foss A.M. (ed.). *Static Synchronous Compensator (STATCOM)*, CIGRE report no. 144 by Working Group 14.19, September 1999
14. Erinmez I.A. (ed.). *Static Synchronous Compensator (STATCOM) for Arc Furnace and Flicker Compensation*, CIGRE report no. 237 by Working Group B4.19, December 2003
15. Lahtinen M. 'New method for flicker performance evaluation of arc furnace compensator', CIGRE Conference Paper 36-205; CIGRE, Paris, August 2002
16. Grünbaum R., Gustafsson T., Hasler J.-P., Larsson T., Lahtinen M. 'STATCOM, a prerequisite for a melt shop expansion – performance experiences', IEEE Power Tech 2003 Conference, Bologna, Italy, 23–6 June 2003
17. CIGRE Report no. 242 by Working Group B4.35 'Thyristor Controlled Voltage Regulators', February 2004
18. CIGRE Report no. 280 by Working Group B4.33 'HVDC and FACTS for Distribution Systems', October 2005
19. CIGRE Report no. 352 by Working Group B4.34 'Capacitor Commutated Converters (CCC) HVDC Interconnections', June 2008
20. CIGRE Report no. 269 by Working Group B4.37 'VSC Transmission' April 2005
21. Knight R.C., Young D.J., Trainer D.R. 'Relocatable GTO-based static var compensator for NGC substations', CIGRE conference paper 14-106, Paris, 1998

Editor's comments: The reader will observe that many of the references listed above [7, 10, 12–15, 17–21] involve international CIGRE related activities. The following footnote key indicates the diversity of strategic technical CIGRE activities, which also extend to international CIGRE Conferences and *Electra* magazine publications. These will be considered further in later chapters.[1]

---

[1] CIGRE key: SC, Study Committee; WG, Working Group and JWG, Joint Working Group; TF, Task Force; TB, Technical Brochure; TR, Technical Report; SC, Scientific Paper.

Chapter 2

Insulation co-ordination for AC transmission and distribution systems

T. Irwin and H.M. Ryan

2.1 Introduction

The international standard covering insulation co-ordination is IEC 60071, which provides the definitions, principles and rules for insulation co-ordination for three-phase systems having a highest voltage above 1 kV. The collective IEC 60071 document [parts 1–4] also provides an application guide along with a computational guide to insulation co-ordination and the modelling of electrical networks.

According to IEC, insulation co-ordination is the process of selecting the dielectric strength of the system insulation in relation to the overvoltage stresses produced during the lifetime of operation of the system. The various high-voltage components of the transmission system must operate without damage by flashover for all envisaged overvoltages.

The operation of the transmission system will present the system insulation with various types of voltage stress from the continuous voltage at which it must operate to various forms of overvoltage.

Overvoltages can be classified into two groups. The first group is externally generated overvoltages from lightning which manifests itself in two ways: by direct strikes to the lines or substations and by strikes to the towers or overhead shield wires from which backflashover voltages from the towers to the conductors can occur. The second group comprises internally generated overvoltages due to switching of capacitive and electromagnetic loads and travelling waves due to the energisation of transmission lines. Also taken into consideration are power frequency overvoltages which are caused by load rejection or by voltage changes on the two healthy phases when a single phase fault occurs.

Up to about 300 kV experience indicates that the highest voltage stress arises from lightning. For transmission systems above 300 kV the switching overvoltages increase in importance so that at about 550 kV the point has been reached where they are equivalent to that of lightning overvoltages.

Power frequency voltages are important since they affect the rating of protective arresters or co-ordinating gaps which provide a means of controlling the overvoltages for the purpose of insulation co-ordination.

58 *High-voltage engineering and testing*

The complexity of insulation co-ordination may be visualised from the flow diagram shown in Figure 2.1, which indicates some of the factors involved and their interrelationship. The procedure takes an iterative approach where several scenarios for insulation co-ordination may be under consideration. Evaluation of the dielectric stresses imposed on the insulation may be carried out using a range of computer models determined from the system characteristics to enable the selection of the insulation requirement. To assist with the selection of insulation levels IEC 60071 has 'Standard Insulation Levels' for system voltages in the range 300–550 kV, an extract from which is shown in Figure 2.2 and Table 2.1.

This chapter provides a generous list of references [1–21] with a supplementary list [22–32] which may be useful for future guidance.

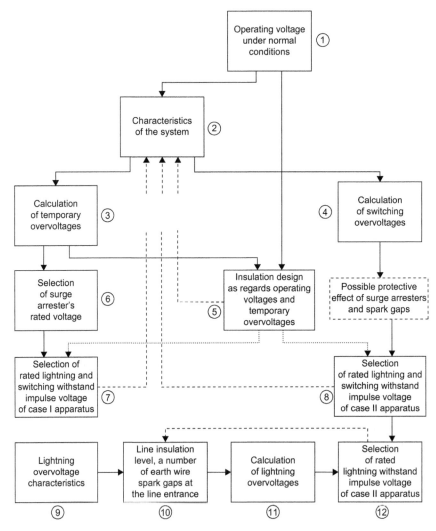

Figure 2.1 Block diagram of insulation co-ordination and design

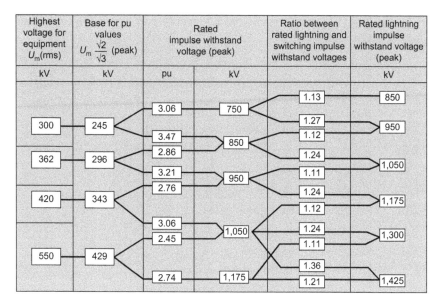

Figure 2.2   Standard insulation levels

Table 2.1   Standard insulation levels from IEC 60071

Highest voltage for equipment U_m	Standard switching impulse withstand voltage			Standard lightning impulse withstand voltage
kV (rms value)	Longitudinal insulation (+) kV (peak value)	Phase-to-earth kV (peak value)	Phase-to-phase (ratio to the phase-to-earth peak value)	kV (peak value)
300	750	750	1.50	850
				950
	750	850	1.50	950
				1,050
362	850	850	1.50	950
				1,050
	850	950	1.50	1,050
				1,175
420	850	850	1.60	1,050
				1,175
	950	950	1.50	1,175
				1,300
	950	1,050	1.50	1,300
				1,425
550	950	950	1.70	1,175
				1,300
	950	1,050	1.60	1,300
				1,425
	950	1,175	1.50	1,425
				1,550

Finally, a list of useful symbols and acronyms used in insulation co-ordination studies is given in Box 2.1.

Box 2.1 Notations: Some useful symbols and acronyms used in this chapter

AIS	Air-insulated substation (open terminal)
Barg	Pressure in bar gauge, i.e. 5 barg = 6 bar absolute
C/B	Circuit-breaker
CFO	Critical flashover voltage U_{50}
GIS	Gas-insulated substation (SF_6)
K_a	Atmospheric correction factor
K_c	Co-ordination factor
K_{cd}	Deterministic co-ordination factor
K_{cs}	Statistical co-ordination factor
K_s	Combined safety factor
K_{tc}	Test conversion factor
LIWL	Lightning impulse withstand level
LP model	Leader progression model
MCOV	Maximum continuous operating voltage for an MOA
MOSA	Metal oxide surge arrester (gapless): formerly referred to as MOA in the literature
Ng	Ground flash density strike to ground per square kilometre
P_{min}	Probability of exceeding a lightning current for the minimum shielding failure current
P_{max}	Probability of exceeding a lightning current for the maximum shielding failure current
PIR	Pre-insertion resistor
POW	Point-on-wave
S_{min}	Strike distance for the minimum shielding failure current
S_{max}	Strike distance for the maximum shielding failure current
SIL	Switching impulse level
SIWL	Switching impulse withstand level
SOV	Switching overvoltage
STD	Standard deviation
Td	Number of thunderstorm days per year (Keraunic level for lightning)
TOV	Temporary overvoltage

$U_{2\%}$	Overvoltage having a 2% probability of being exceeded
U_{50}	50% withstand voltage for self-restoring insulation
U_{90}	90% withstand voltage for self-restoring insulation
U_{100}	100% withstand voltage for non self-restoring insulation
U_{cw}	Co-ordination withstand voltage
U_{e2}	Phase-to-earth overvoltage having a 2% probability of being exceeded
U_m	Highest design voltage [rms phase to phase] for equipment
U_{p2}	Phase-to-phase overvoltage having a 2% probability of being exceeded
Ups	Surge arrester protective level (residual voltage)
U_{rp}	Amplitude of representative overvoltage
U_{rw}	Required withstand voltage
U_w	Standard withstand voltage
VT	Voltage transformer

2.2 Classification of dielectric stress

The following classes of dielectric stresses (see Figure 2.3) may be encountered during the operation of the transmission system:

1. power frequency voltage under normal operating conditions
2. temporary overvoltages
3. switching overvoltages
4. lightning overvoltages.

2.2.1 Power frequency voltage

Under normal operating conditions the power frequency voltage can be expected to vary somewhat but, for the purpose of insulation co-ordination, is considered to be constant and equal to the highest system voltage for the equipment.

2.2.2 Temporary overvoltages

The severity of temporary overvoltages (TOVs) is characterised by amplitude and duration. This overvoltage is a significant factor when considering the application of surge arresters as a means of surge voltage control. Although the surge arrester is not used for the control of TOVs, it results in a considerable increase in the resistive component of leakage current in a metal oxide gapless surge arrester. This results in a temperature rise within the surge arrester and the possibility, if left long enough, of arrester failure.

TOVs arise from (i) earth faults, (ii) load rejection and (iii) resonance and ferroresonance.

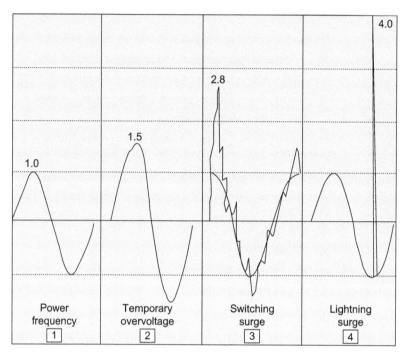

Figure 2.3   Classification of dielectric strength

When considering earth faults, the voltage on the healthy phases will rise during the fault period to values approaching line voltage dependent on neutral earthing arrangements. Typically for 420 kV systems a value of 1.5 times normal system phase voltage can result for periods up to 1 s, by which time the fault is usually cleared.

Load rejection can, on long uncompensated transmission lines, produce voltages of 1.2 times nominal system voltage, due to the Ferranti effect, at the substation end of the line which is disconnected from the source (i.e. remote end). The Ferranti effect is caused by the current flow through the line shunt capacitance which produces a voltage across the source and line inductance which increases with distance from the source end of the line.

TOVs due to resonance and ferroresonance conditions generally arise when circuits with large capacitive elements (transmission lines, cables, etc.) and inductive elements (transformers and shunt reactors) having non-linear magnetising characteristics are energised, or through sudden load changes. Parallel line resonance can also occur during de-energisation of one circuit of a double circuit transmission line which has shunt reactive compensation. The energised line feeds the resonance condition through the intercircuit capacitance and voltages as high as 1.5 times nominal system phase voltage have been recorded on 420 kV systems [1]. This voltage will remain at this level until the line is energised or until the compensating reactor is switched out. Magnetic voltage transformers can also produce ferroresonance conditions but usually these are sub-third-harmonic and the resultant voltage is close to nominal system voltage [2]. This resonant voltage will not present any problem for the insulation but the voltage transformer (VT) primary current will be many times the nominal current and the resultant heat generated in the primary winding would be of prime concern for the VT insulation. However, fundamental ferroresonance can also occur with VTs, and voltages in excess of two times nominal system voltage have been reported. If ferroresonant conditions are indicated during a system study, then design modifications may be considered or steps can be taken to avoid the switching operations that cause them or to minimise the duration by selection of an appropriate protection scheme.

2.2.3 Switching overvoltages

Switching overvoltages according to IEC 60071 can be simulated by a periodic waveform with a front duration of hundreds of microseconds and a tail duration of thousands of microseconds. The waveform shown in Figure 2.3 may be typical of switching surges which have in practice a decaying oscillatory component superimposed on the power frequency waveform. Of major importance are the switching surges produced by line energisation and re-energisation which cause travelling waves on the transmission line and are most severe at the end remote from the switching point. This will be discussed later in more detail. The reader is again reminded that Table 2.1 provides standard insulation levels from IEC 60071.

2.2.4 Lightning overvoltages

Lightning overvoltages according to IEC 60071 can be simulated by an aperiodic wave with a front duration of the order of 1 µs and a tail duration of the order of several tens of microseconds. The lightning surge voltage is produced by a strike to the tower or earth wire which either induces a voltage in the phase conductors or by backflashover of the line co-ordinating gap, injects current into the phase conductor. A direct strike to the phase wire can also occur if the phase wire is not well shielded by the earth wire. The voltage wavefront arriving at the substation can be significantly affected by the line termination, and with cable terminations the wavefront may be drastically reduced to the extent that it more resembles a last switching surge. This will be discussed later in this chapter.

2.3 Voltage–time characteristics

In practice the waveform of lightning overvoltages varies considerably, but for the purpose of testing equipment there is an internationally agreed wave shape used which has a rise time of 1.2 µs and a time of decay of 50 µs to 50% of the maximum amplitude.

Switching overvoltages vary from oscillatory voltages of several tens of thousands of cycles per second to travelling waves with a rise time of up to 1,000 µs. It is now generally accepted that the withstand strength of air insulation is a minimum for a wave with a rise time of approximately 250 µs and with a decay time to 50% of 2,500 µs, although there is a tendency for this minimum to occur at longer

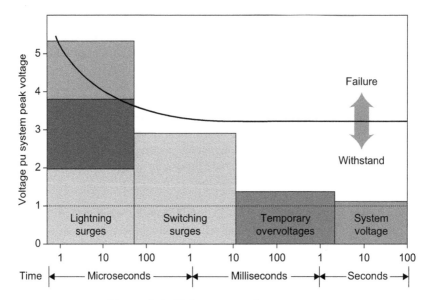

Figure 2.4 Voltage–time characteristics

wavefronts as the air gap increases. Similarly, the minimum under oil withstand strength occurs with a wavefront of approximately 100 μs.

Gas-insulated substations (GIS) have a virtually flat voltage–time (V–t) characteristic from power frequency through to the switching surge range and show an upturn near the 10 μs point.

Figure 2.4 shows a typical V–t characteristic for 420 kV GIS along with a range of dielectric stresses. From the standard insulation levels (IEC 60071) 420 kV equipment can have a 1,425 kV lightning impulse withstand level (LIWL) with a corresponding 1,050 kV switching impulse withstand level (SIWL) based on a rated voltage of 420 kV root mean square (rms), i.e. a phase voltage peak of 343 kV.

The V–t characteristic of insulation used in equipment must be carefully assessed when comparing performance with lightning and switching surges. Also the V–t characteristics of protective devices must be considered; for example co-ordinating gaps in air will not offer practical protection of GIS against switching or lightning surges because the V–t characteristics are totally incompatible.

2.4 Rated withstand voltage levels

The flow chart in Figure 2.5 outlines the procedure for insulation co-ordination taking into account a range of adjustment factors to determine required and rated withstand levels. The process centres on system analysis studies to determine the system overvoltages from which the representative voltages for temporary, slow front and fast front overvoltages can be found. Then a co-ordination and safety factor is applied to the representative overvoltages for internal and external insulation with an additional atmospheric correction factor applied for external

Insulation co-ordination for AC transmission and distribution systems 65

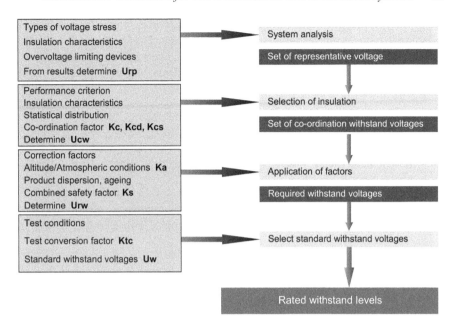

Figure 2.5 Flow chart to determine required and rated withstand levels

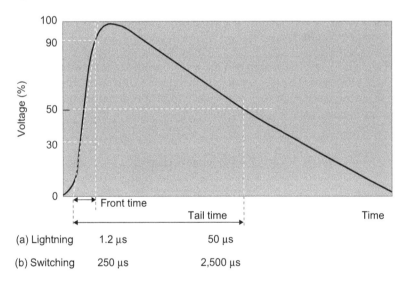

Figure 2.6 Impulse voltages to determine standard insulation levels (a) is for lightning impulse voltage 1.2/50 μs and (b) is for switching impulse voltage 250/2,500 μs

insulation. This determines the 'required withstand' voltages which are used to establish the 'rated withstand' levels for the substation equipment.

The insulation withstand level is determined by using a set of standard waveforms to represent (a) 1.2/50 μs lightning and (b) 250/2,500 μs switching impulse voltages which have defined 'front' and 'tail' (see Figure 2.6).

66 *High-voltage engineering and testing*

Power frequency voltage is also used and is applied by gradually increasing the amplitude to the test level and held, usually for 1 min. The test methods are defined in IEC 60060 where the consequences of test flashovers are discussed.

IEC 60071 provides two insulation ranges depending on highest equipment voltage (U_m):

- Range 1 = 1 kV > U_m ≤ 245 kV.
- Range 2 = U_m > 245 kV.

Within each range there are 'sets of standard insulation levels' for each transmission system voltage:

- Range 1 specifies lightning impulse and power frequency withstand levels.
- Range 2 specifies lightning and switching impulse withstand levels.

A specific set of withstand levels can then be selected for each overvoltage type in order to define the transmission system requirements and provide the 'rated insulation levels'.

The actual levels selected will depend on many factors but include:

- calculated level of each overvoltage
- the type of equipment (number, location, etc.) installed for overvoltage control
- the security of the system – consequences of forced outages
- operating requirements, etc.

Figure 2.7 shows an example of the levels specified for the rated insulation for a 420 kV substation. The levels can only be selected with precedence; for example, as in this case, if the switching overvoltage level was dominant then the 1,050 kV switching impulse level would be selected. This by nature also defines the lightning

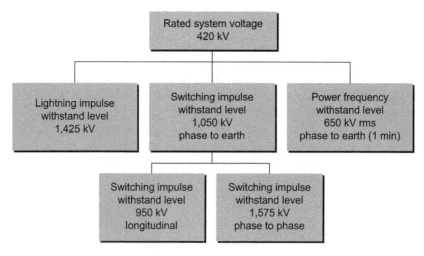

Figure 2.7   Specification of required rated insulation levels

impulse level of 1,425 kV. Conversely, selecting 1,050 kV for lightning would mean a withstand for switching impulse in the range 850–950 kV because the V–t characteristic of the insulation requires LIWL to be greater than the SIWL.

The interdependency of the various standard insulation levels has been shown by the example 'voltage–time' characteristic given in Figure 2.4. In this case, it must be remembered that the curve does not represent a critical pass/fail dividing line but shows the reference curve which is one of the many curves that can be derived depending on the probability of failure. The reference probability for withstand is taken as the 90% level, i.e. 10% failure rate, and is denoted as the U_{90} withstand level, which can only be determined for 'self-restoring insulation'. Self-restoring insulation is defined as insulation which completely recovers its insulating properties after a disruptive discharge during the application of test voltages. SF_6 gas, oil and air insulating systems are classed as self-restoring (*Note*: SF_6 and oil do not self-restore after power arc flashover during in-service fault conditions). The test methods for determining the statistical withstand characteristics are described in IEC 60060.

Certain insulating systems are classed as 'non self-restoring', which basically means that after a 'test voltage' flashover in the laboratory the insulation becomes permanently damaged, usually with a further reduction in the flashover voltage. Achieving a statistical withstand performance for this type of insulation is not possible, so the withstand level for this equipment is determined as a maximum which must not be exceeded, e.g. solid insulation in bushings.

2.5 Factors affecting switching overvoltages

When energising or re-energising transmission lines severe overvoltage can be generated [3]. The overvoltage magnitude is dependent on many factors including the transmission line length, the transmission line impedances, the degree and location of compensation, the circuit-breaker characteristics, the feeding source configuration and the existence of remanent charge from prior energisation of the transmission line.

The magnitude of these switching transients is the main factor determining the insulation levels for extra high voltage (EHV) and ultra high voltage (UHV) transmission systems; consequently reduction in their severity has obvious economic advantages. The two methods used to achieve substantial reduction in transmission line energising overvoltages are resistor insertion [4] and controlled closing of the energising circuit-breaker close to voltage zeros [5]. There have, however, been instances where both methods have been combined for overvoltage reduction in a UHV system [6]. Alternatively, for systems up to 550 kV, gapless metal oxide surge arrestors can also be used to reduce the phase-to-earth switching surge overvoltages to below 2.4 pu of system nominal phase voltage peak.

2.5.1 Source configuration

Source networks can very crudely be split into two types: (a) those with purely inductive sources, i.e. where no lines are connected to the energising busbar, e.g. a remote hydro station, and (b) those with complex sources, i.e. with lines feeding the energising busbar or a mixture of lines and local generation.

Much work has been done over the years on both types of sources [7–9]. Trends are more easily defined with the simple lumped source, and normally slight increases in overvoltages are obtained with increase in source fault level except where resonance conditions are approached with very low source fault levels. With the complex source configuration no general trends exist due to a large number of interacting parameters in the source network.

2.5.2 Remanent charge

The remanent charge on a transmission line prior to its reclosing has a significant effect on the overvoltages produced. The value of trapped charge is very much dependent on the equipment permanently connected to the line, as this determines the decay mechanism (see Figure 2.8).

If no wound VTs, power transformers or reactors are connected, the line holds its trapped charge, the only losses being due to corona and leakage and thus the decay is very much weather dependent. In good weather conditions the time constant of the decay is of the order of 10–100 s so that no appreciable discharge will occur in an automatic reclosure sequence.

With a power transformer connected to the transmission line the trapped electromagnetic energy oscillates within the resistor, inductor and capacitor (RLC)

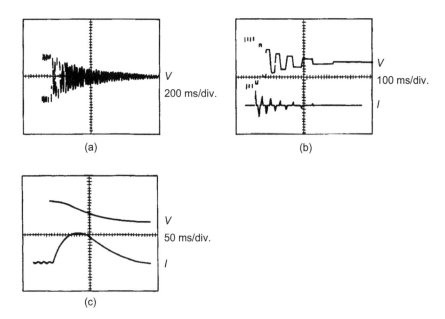

Figure 2.8 Decay of trapped charge by (a) reactor, (b) power transformer, and (c) voltage transformer

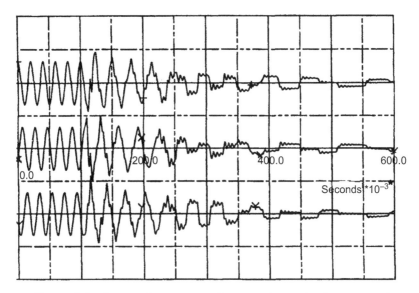

Figure 2.9   Decay of power transformer voltages (with saturation)

circuit formed by the line and transformer. The transformer oscillates between the saturated and unsaturated state resulting in a decay in some 5–50 cycles of the supply frequency. Figure 2.9 shows a computer simulation of the three-phase decay of trapped charge with initial overvoltage when the circuit-breaker opens due to mutual coupling effects.

The decay mechanism with a wound magnetic voltage transformer is similar to the power transformer but the damping is much more effective due to the high winding resistance, of the order of several tens of k-ohms, of the voltage transformer. Most of the stored energy on the line will be dissipated in the first hysteresis loop of the voltage transformer core mainly due to copper loss, although conditions have been encountered with long transmission lines and wound VTs in SF_6-insulated substations where complete dissipation requires five cycles of the supply frequency.

When a shunt reactor is connected to the line, on de-energisation an oscillation exists determined by the line capacitance and the reactor similar to the power transformer. In this case, however, there is no saturation and the oscillations are slowly damped due to high reactor Q-value. In addition, 'beat effects' are introduced due to mutual coupling effects. The damping time constant is of the order of seconds; thus only slight decay in the trapped charge occurs within a practical high-speed auto-reclose time sequence.

2.5.3   Transmission line length

The length of transmission line being energised affects the overvoltage magnitude in that the longer the line the greater is the steady-state open circuit receiving end voltage (Ferranti rise) on which the high-frequency transients, the frequency of which is determined by the line length, are superimposed.

The frequency of the switching surge can be approximated by

$$f = \frac{1}{4T}$$

where T is the transit time of the line, $\approx$ {line length (in km)/0.3} μs, assuming the surge travels along the line at the speed of light.

2.5.4 Compensation

Shunt reactor compensation has a two-fold effect when situated at the transmission line receiving end, both aspects of which contribute to a reduction in the severity of the energising overvoltages. The reactor reduces the magnitude of the Ferranti rise along the line by negating the effect of a portion of the line shunt capacitance and presents a line termination other than an open circuit to any travelling waves from the transmission line sending end. For a 420 kV line the capacitive line charging current is approximately 1 A/km (0.25 MVA per phase). A 200 km line would typically require 40 MVAr of shunt reactive compensation per phase depending on the system operational requirements.

2.5.5 Circuit-breaker pole scatter

Circuit-breakers will seldom produce a simultaneous close onto a three-phase transmission line for two reasons:

1. The circuit-breaker mechanism closes the contacts at high speed and mechanical tolerances will give a spread of closing times between the three phases. Typically this may be of the order of 3–5 ms between the first and last poles to close.
2. Depending on the point-on-wave (POW) at which the circuit-breaker initiates a close, the phase with the highest instantaneous value of power frequency voltage will pre-arc first, just before contact touch.

The pole scatter effect produces voltage through mutual coupling from the first phase to close on the other two phases. This pre-charging effect then produces a voltage greater than phase voltage across the contacts of the other two phases of the circuit-breaker. This in turn forces a greater than 1 pu step voltage to be applied as the other phases close. Certain critical points can be reached depending on the pole scatter time and POW of closure, e.g. when the second pole to close occurs at $2T$, $4T$, etc., for the transmission line, i.e. the point at which the switching surge on the first phase to close returns to the sending end of the line. By studying the various combinations of pole scatter, points of maximum/near minimum overvoltages can be determined. With transient network analyser (TNA) studies, pole scatter diagrams can be created showing the effect of incremental changes in pole scatter and the maximum overvoltage position located. With computer analysis, using programs such as EMTP, a statistical approach is normally adopted using random POW (uniform distribution) with Gaussian distribution for pole scatter. From work previously carried out with TNA using 500 operations with random POW and circuit-breaker pole scatter determined from typical distribution curves, the 2% probability value (i.e. the overvoltage value which will be exceeded for 1 in 50 operations) was approximately 15% below the maximum value derived using the conventional (maximum) method.

2.5.6 Point-on-wave of circuit-breaker closure

The magnitude of the transient voltage is very much dependent on the instantaneous value of power frequency voltage at which the circuit-breaker closes. If all three poles of the circuit-breaker get closed at voltage zeros, then only a very small transient voltage would occur.

2.6 Methods of controlling switching surges

On systems of 400 kV and above, the energisation of long transmission lines (200 km and greater) is commonly required. Voltages above 4 pu of the normal system phase voltage peak have been shown to occur by TNA and computer studies. Methods, therefore, have been employed to reduce these overvoltages to 2.5 pu or less to achieve economic design of the transmission line and substation. At 420 kV the overvoltages can be well controlled using metal oxide surge arresters (MOSA) at the sending and receiving ends of the line. At 550 kV, circuit-breaker pre-insertion resistors (PIRs) have been used with great effect but result in complicated contact arrangements on the circuit-breaker with appropriate increase in maintenance. For 550 kV systems with line lengths below 300 km, MOSA can give an acceptable voltage profile along the line length, with the maximum voltage occurring at the line midpoint. With lengths near 300 km and above, additional surge arresters can be placed at the line midpoints. Recent developments in controlled POW switching have introduced microprocessor-based technology for circuit-breaker operation control. This can be very effective, particularly when used in conjunction with MOSA for transmission line overvoltage control. However, in all cases the substation voltages can be adequately controlled with the surge arresters at the line ends only. Dependent on the line design and the acceptable risk of failure for the 550 kV line, the midpoint surge arrester may not be required even with 300 km lines.

2.6.1 Circuit-breaker pre-insertion resistors

2.6.1.1 Single-stage resistor insertion

Energising a transmission line through single-stage resistors results in the waves transmitted along the line being reduced in magnitude and hence the overvoltages at the receiving end being less severe. Resistors used in this way must also be removed from circuit and the removal of the resistors also initiates travelling waves which create overvoltages. Figure 2.10 illustrates the severity of the overvoltages produced in the initial energising operations through single-stage resistors and the subsequent resistor shorting operations indicating that there is an optimum PIR

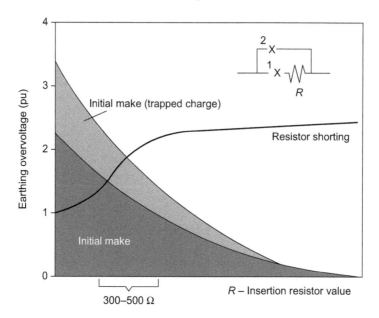

Figure 2.10 Overvoltage reduction with closing resistors

value where the overvoltages produced by the initial energisation and the subsequent resistor removal are equal.

The optimum resistor value varies with different system conditions but is typically in the range 300–500 Ω.

2.6.1.2 Insertion time and pole scatter

The overvoltages produced when energising transmission lines are relatively insensitive to the resistor insertion time. However, if the insertion time is less than the circuit-breaker pole scatter plus twice the transmission line transit time an increase in the overvoltage results, especially in the cases when remanent charge exists. Figure 2.11 illustrates the results from a series of studies to investigate the relationship between pole scatter and insertion time. They show that there is significant increase in the overvoltage if one phase is energised through its resistor and the resistor shorted out before another phase has been energised for the first time as the damping effect of the resistor on the mutually induced voltage is ineffective. The effect is most pronounced in the regions of small insertion resistor values where the mutually coupled transient components are greater.

2.6.2 Metal oxide surge arresters

2.6.2.1 Selection of surge arrester rating

The TOV level and duration must be carefully considered before selecting the rating of the surge arrester (MOSA). The surge arrester must be capable of

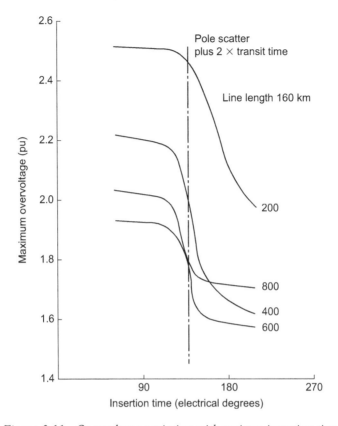

Figure 2.11   Overvoltage variation with resistor insertion time

withstanding, from thermal constraints, the TOV which in most circumstances determines the surge arrester–rated value. From the rated value stems the protective or voltage limiting characteristic of the surge arrester – the higher the rating, the higher the limiting or residual voltage the arrester will have.

Thermal constraints are very important with MOSA since if the rating is too low, TOV may cause excessive heating resulting in thermal instability with a runaway condition being produced and subsequent failure. Figure 2.12 shows typical TOV capability for MOSA. The energy capability of the surge arresters is usually expressed as kJ/kV of arrester rating. The maximum continuous operating voltage is considered as 80% of the rated voltage. Typically for 420 kV systems an arrester rating of 360 kV will be used, which gives a maximum continuous operating voltage of 1.25 pu of nominal system voltage. From the curve given in Figure 2.12, a TOV of 1.5 pu can be withstood for approximately 20 min with a 360 kV MOSA. The surge arrester voltage–current characteristic exhibits an extremely non-linear relationship once the 'knee' point voltage has been exceeded, which causes large increases in current for small voltage increase (see Figure 2.13). Typically for a 444 kV rated arrester (TOV capability of 1.7 pu for

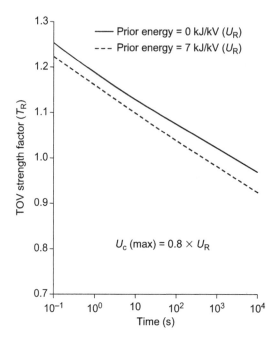

Figure 2.12 TOV capability for typical surge arrester, expressed in multiples of $U_R(T_R)$

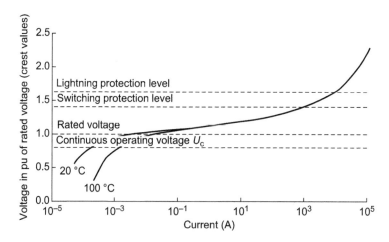

Figure 2.13 Typical voltage–current characteristics for ZnO arresters

10 s on a 550 kV system) a residual voltage change from 860 to 1,220 kV produces a current change from 1 to 40 kA. This arrester will also limit switching surges at the substation to approximately 2.4 pu on 550 kV systems. The rise time of the switching surge is relatively slow in comparison to the transit time of the

2.6.2.2 Application of MOSA for switching surge reduction

For the 420 kV transmission system shown in Figure 2.14 the maximum switching surge is reduced from 2.15 to 2.03 pu (see Figure 2.15) by locating a 396 kV MOSA at the line end. This is, not surprisingly, a very small voltage reduction for a level of voltage which, even without MOSA application, is well within the SIWL (3.0 pu) and would not justify the cost of surge arrester application. For this condition, the surge arrester draws a peak current of 190 A with a pulse width of 0.5 ms. Smaller current pulses are also evident for the other oscillations at the power frequency voltage peaks.

For the same transmission system Figure 2.16 shows the maximum overvoltage with trapped charge which is well in excess of the substation SIWL. For this condition a reduction from 4.28 to 2.35 pu in the switching overvoltage can be achieved with a 396 kV rated MOSA. For this case the surge arrester draws a 2 kA current pulse with 0.5 ms width. By comparison of the two conditions, i.e. with and without surge arresters, not only is the amplitude significantly reduced but the oscillations are much more quickly damped. For the trapped charge (worst case) condition a safety factor (SIWL/protection level of MOSA) of 1.27 has been achieved (IEC* 60071 recommends a minimum of 1.15).

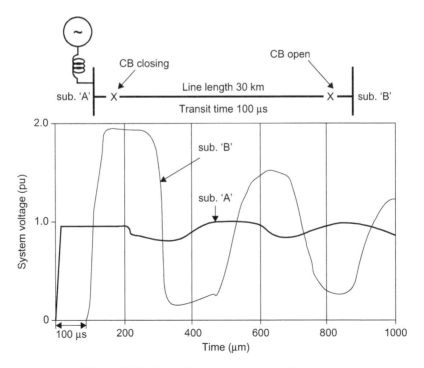

Figure 2.14 Switching surges – travelling waves

76   High-voltage engineering and testing

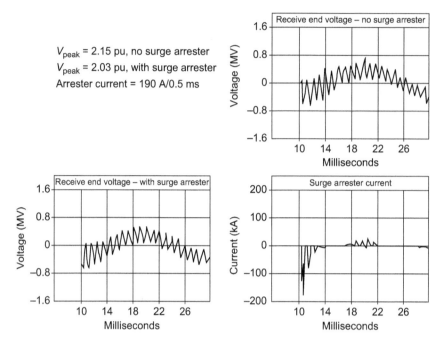

Figure 2.15   *Control of switching surges using surge arresters (low-level surge)*

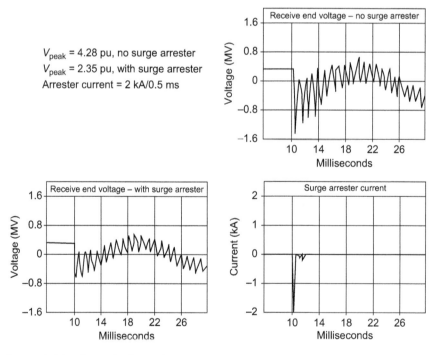

Figure 2.16   *Control of switching surges using surge arresters (high-level surge)*

2.6.3 Circuit-breaker point-on-wave control

One method of switching overvoltage control which has been investigated over many years but recently has seen growing applications [10] is that of POW-controlled switching. With modern circuit-breakers and the use of microprocessor technology, accurate POW control can now be achieved. The controller must be capable of compensative and adaptive control to allow for changes in operating conditions for the circuit-breaker as well as ageing effects.

It can be used for many applications:

- reactor switching
- capacitor bank switching
- transformer switching
- transmission line switching.

Each application has its own 'ideal switching' point and can be used for de-energising as well as energising conditions.

For the purpose of transmission line energisation (see Figure 2.17), the ideal closing point is the system power frequency voltage zeros – for re-energisation during auto-reclose operations the 'ideal' point will vary depending on the line-trapped charge conditions and is basically adjusted to a point at which there is zero voltage or minimum voltage across the contacts of the circuit-breaker.

Switching overvoltages can be reduced to levels below 2 pu provided the required timing accuracy of the circuit-breaker and controller can be achieved.

2.6.3.1 Circuit-breaker controller requirements

The controller must be capable of monitoring the control parameters of the circuit-breaker such as

- DC auxiliary voltage
- stored mechanism energy

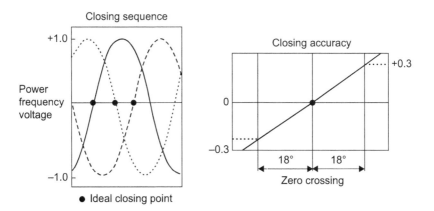

Figure 2.17 Point-on-wave switching control-transmission line energisation

78 High-voltage engineering and testing

- temperature
- system frequency

and compensate for variations in these parameters.

Also, for adaptive purposes, the previous operating times, speeds and achieved accuracy must be recorded and taken into account when calculating the predicted operating time.

When controlling the closing point, account must be taken of pre-arcing between the circuit-breaker contacts. Pre-arcing effectively closes the circuit-breaker contacts before actual contact touch and can shorten the closing time by as much as 3 ms under certain conditions. The required timing accuracy of ±1 ms means that a combined standard deviation for the circuit-breaker and controller of less than 0.4 ms must be achieved unless a higher risk of the switching overvoltages in exceeding 2 pu can be allowed.

If POW closing is used in conjunction with MOSAs (as is the usual case since surge arresters are required for lightning surges), then the substation switching overvoltages will be well controlled and the major concern will then be the switching overvoltage profile on the transmission line.

2.6.4 Comparison of switching overvoltage control methods

When no control methods are used, the 2% probability voltage can exceed 4 pu at the receive substation, and overvoltages in excess of 3 pu can occur for 50% of the line energisations.

All three switching overvoltage control methods (PIR, MOSA and POW) provide acceptable switching overvoltage control for the substation with the 2% probability voltage reduced to less than 2.0 pu.

However, when considering the line switching overvoltage performance, the line overvoltage profile has to be taken into account. In Figure 2.18 the curves have been plotted using the 2% probability voltages for each of the control methods and combinations. Various possibilities are shown here with the highest overvoltages occurring near the line midpoints for the MOSA conditions. The *PIR* control shows the highest overvoltages at the receiving end of the line.

Most line designs will be able to give a satisfactory performance with a 2% overvoltage less than 2.2 pu, e.g. on 550 kV systems. This can be accommodated by all the control methods if high-speed auto-reclose is not required. With high-speed auto-reclose, the 'MOSA only' condition produces an unacceptable profile with virtually 90% of the line exceeding the 2.2 pu overvoltage. Line midpoint surge arresters can be considered to improve the profile or additional methods of control will have to be used. For example magnetic voltage transformers at the line ends can be used in place of capacitive coupled VTs so that the line-trapped charge will decay to a low value within the reclose period; alternatively, POW control or PIRs will have to be used.

When surge arresters are used with *PIR* and POW controls then the switching surge current taken by the arrester is much reduced, which can assist with its TOV

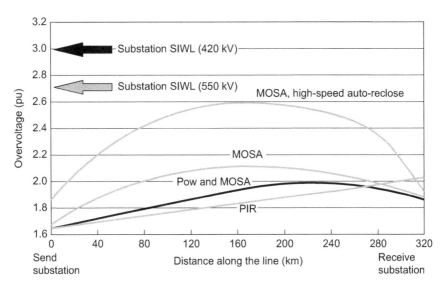

Figure 2.18 Comparison of control methods showing line switching overvoltage profile

handling capability – particularly for systems using multiple auto-reclose strategies. This may allow the application of lower rated surge arresters and therefore further enhance the overvoltage control.

2.6.5 Application of insulation co-ordination for ultra high-voltage technology

We have seen that switching and lightning overvoltage reach similar levels when operating transmission systems at 550 kV. However, when operating at ultra high voltages (UHV) exceeding 800 kV, the switching operation becomes dominant. CIGRE has recently published Technical Brochure TB 456 (2011) [21], which gives background information and recommendations for international standards for substation equipment exceeding 800 kV AC. Service experience at this voltage level is limited but it is expected that the number of UHV projects will increase, and indeed in China, three substations have been operating a single circuit overhead line at 1,100 kV since 2009.

UHV technology centres on the need to minimise size, weight and costs by more widespread use of techniques already discussed in this chapter. For example:

- multi-column surge arresters and/or multiple arresters in parallel in order to reduce the residual overvoltage
- application of circuit-breaker resistors for both closing and opening operations
- extending the switching resistor application to include GIS disconnectors in order to reduce the very fast transient overvoltage (VFTO) phenomena by use

of damping resistors. In this case the internally generated GIS (or VFTO) may well exceed the LIWL if control methods are not used
- extending the surge arrester application to include the reduction of transient recovery voltages (TRV) during circuit-breaker opening operations.

2.7 Factors affecting lightning overvoltages entering substations

The magnitude and rate of rise of overvoltages due to lightning strikes on transmission lines are important considerations for substation insulation and the strategy adopted for limiting these overvoltages.

Having determined the insulation required for the line, it is usual to find that the lightning withstand level is in excess of commercially available LIWL levels of the substation equipment. Thus, unless precautions are taken, overvoltages entering the station can cause undue insulation failure. Surge arresters can be situated at the line entrance but consideration must be given to the voltage profile as the surge travels through the substation. Alternatively, consideration may be given to using rod gaps set to operate marginally below the station LIWL and SIWL levels, and fitted to the first three or four towers. However, consideration must be given to the $V-t$ characteristic of the substation equipment in comparison to that of line gaps. For example 132 and 420 kV GIS cannot be adequately protected by line co-ordinating gaps. Figure 2.19 shows a comparison of line gaps and a metal oxide surge arrester in relation to the standard insulation levels for a 420 kV GIS and shows that the 2 m line gap is totally ineffective as a method of reducing the incoming surge voltages.

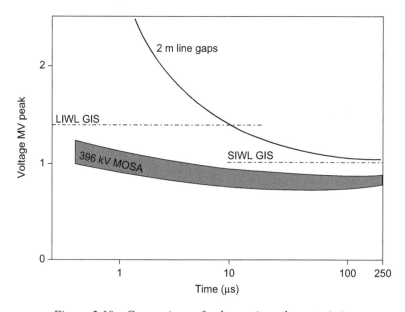

Figure 2.19 Comparison of voltage–time characteristics

Insulation co-ordination for AC transmission and distribution systems 81

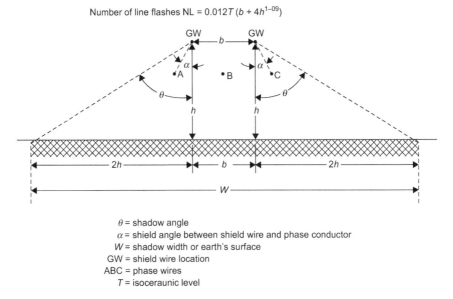

θ = shadow angle
α = shield angle between shield wire and phase conductor
W = shadow width or earth's surface
GW = shield wire location
ABC = phase wires
T = isoceraunic level

Figure 2.20   Model for line flash calculations

The number of strikes to transmission lines is generally accepted to be related to the isoceraunic level which is defined as a number of days in a year in which thunder is heard at a given location. Assumptions are made in relating this isoceraunic level to the number of strikes to towers and earth wires; Reference 11 provides methods of calculating the number of line flashes – see Figure 2.20. The number of strikes is directly proportional to the isoceraunic level, i.e. a level of 20 entails twice the number of strikes as a level of 10.

The calculation method is based on the number of ground flashes that would occur to the area of ground shielded by the transmission line. Two possibilities exist for generating lightning overvoltages on the line conductors – the 'backflashover' and the 'direct' strike. Figure 2.21 shows a typical 420 kV single circuit tower illustrating the two strike conditions and gives the tower and line parameters.

2.7.1   Backflashover

A backflashover occurs as a result of the tower or shield wire being struck by lightning, where the resultant lightning current passes to earth via the tower steelwork, causing a voltage difference between the tower cross-arms and the line conductors. The magnitude of this current can vary from a few kiloamperes to over 200 kA. The statistical data for amplitude and steepness of lightning currents is given in Figures 2.22 and 2.23, derived from Anderson and Eriksson [12].

Due to the height of the tower and rate of rise of current, a travelling wave can be set up on the tower. The combination of shield wire and tower surge impedance (see Figure 2.24) and lightning current impulse will produce a voltage at the tower top which is oscillatory due to successive reflections from the tower base. When the

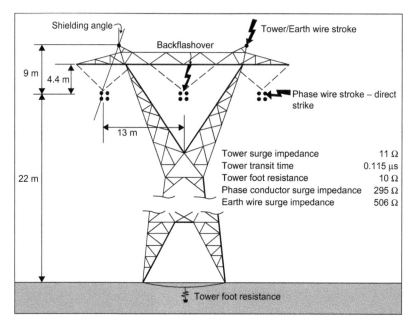

Figure 2.21   Lightning surges

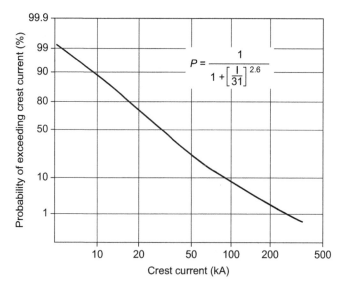

Figure 2.22   Amplitude of lightning stroke current

surge current arrives at the base of the tower it dissipates through the tower grounding to earth. An additional voltage is produced at the tower foot which is dependent on the grounding impedance. Figure 2.25 shows typical voltages with the tower travelling wave voltage superimposed on the tower foot voltage. The first

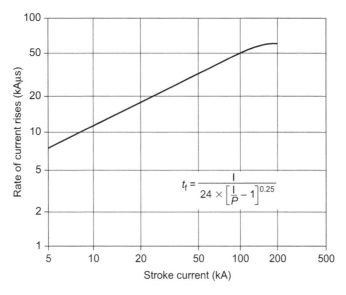

Figure 2.23   Lightning stroke current steepness

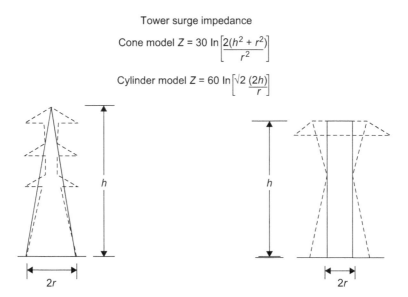

Figure 2.24   Tower surge impedance

voltage pulse width can be estimated by doubling the tower transit time (typically 0.2–0.4 µs). Not all of the tower top voltage will appear across the line insulator because there is some reduction due to the position of the insulator on the cross-arm, and also voltage will be mutually coupled from the shield wire to the phase wire.

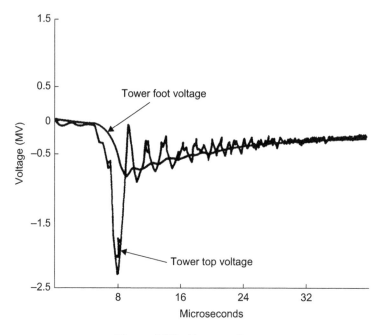

Figure 2.25   Tower voltages

So the voltage that appears across the line insulator co-ordinating gap will be similar but marginally smaller (85%) than the tower top voltage.

Depending on the V–t characteristics of the line co-ordinating gap, the backflashover (i.e. from tower to line) may occur near the peak of the voltage pulse or on the surge tail. Test data for line co-ordinating gaps is limited for 'non-standard' lightning impulse voltages. However, work has been done [13] to establish models of the line gap flashover mechanism. Leader progression models have been proposed which can be used to assess the time to flashover for these wave shapes. Figure 2.26 shows a line gap flashover from a standard 1.2/50 μs lightning impulse voltage and illustrates a 'completed' flashover on one of the gaps with leaders only partially bridging the second gap. It is important to note that as the tower foot resistance is increased the more dominant the tower foot voltage will become to a point where for short towers the voltage wave shape across the line gap will approach that of the 'standard' impulse.

2.7.2   Direct strike

Most transmission line towers will be equipped with shielding wires. In the tower shown in Figure 2.21 there are two shield wires. The purpose of these wires is to divert the lightning stroke away from the phase wire and thus provide shielding. Any lightning strike which can penetrate the shield is termed a 'direct strike' or 'shielding failure'. The electrogeometric model proposed in Reference 11 and shown in Figure 2.27 is a simplified model of the shielding failure mechanism for

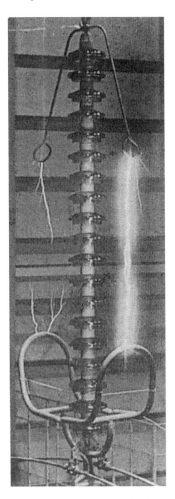

Figure 2.26   Line gap flashover

one shield wire and one phase conductor. As a flash approaches within a certain distance S of the line and earth, it is influenced by what is below it and jumps the distance S to make contact. The distance S is called the strike distance and it is a key concept in the electrogeometric theory. The strike distance is a function of charge and hence current in the channel of the approaching flash. Use of the equation given in Figure 2.27 requires the calculation of S_{max} and S_{min} which then relate to I_{max} and I_{min}, the corresponding stroke currents. The probabilities for I_{max} and I_{min} can then be determined (P_{max}, P_{min}) along with the unshielded width X_s; if $X_s = 0$ then shielding failure will not occur. The objective when designing the line shielding is to minimise X_s. Table 2.2 compares the results of the electrogeometric model calculations for the shielding performance of two towers with identical conductor and shield wire configuration but of different heights. A corresponding

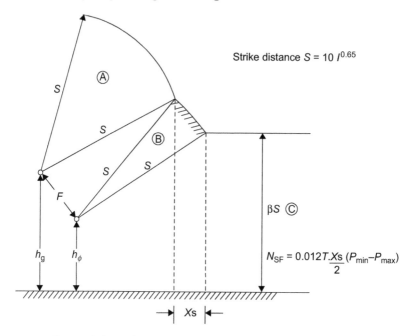

Figure 2.27   Electrogeometric model for shielding failures

Table 2.2   *Lightning performance of transmission lines*

	Tower height DC 132 kV	
	30 m	60 m
I_{min}	5 kA	17 kA
I_{max}	28 kA	84 kA
Number of line/tower flashes	19/100 km/year	39/100 km/year
Number of shielding failures	0.2/100 km/year	0.65/100 km/year
Probable maximun E/W stroke current	93 kA	126 kA
Annual thunder days	13	13

increase in the number of line flashes and shielding failures is indicated with the taller tower. Also the maximum shielding failure current is three times that for the smaller tower.

For the purpose of insulation co-ordination the direct strike may not warrant further investigation if the transmission line is effectively shielded, particularly in the last 5 km of line approaching the substation. Considering the data from Table 2.2, a shielding failure rate of 0.2/100 km/year would mean that a direct strike inside the last 5 km of line would occur once in 100 years or one in three chances during the life of the substation. When assessing the risk of failure for the substation, however, a sum of all the probabilities for each substation line is required (i.e. six lines would give two surges from direct strikes in the life of the substation).

2.7.3 Attenuation of lightning overvoltage

As the lightning surge travels towards the substation from the struck point the wavefront above the corona inception voltage will be retarded by corona loss. Skin effect on the line conductors will cause further attenuation due to the high-frequency nature of the surge. It is usual therefore to consider lightning strikes that are 'close-in' (within 3 km) to the substation when assessing surge arrester requirements and the associated risk of failure of the substation.

2.8 Methods of controlling lightning overvoltages

For well-shielded transmission lines, the backflashover condition, close to the substation, is of prime concern for determining the location and number of surge arresters required to achieve insulation co-ordination of the substation for lightning surges. The risk of a backflashover can be reduced by keeping the tower foot impedances to a minimum, particularly close to the substation (first five to seven towers). The terminal tower is usually bonded to the substation earth mat and will have a very low grounding impedance (1 Ω). However, the procedure for 'gapping' down on the first three or four towers where line co-ordinating gaps are reduced in an attempt to reduce incoming voltage surges will increase the risk of a 'close-in' backflashover.

2.8.1 Location of surge arresters

Considering the system shown in Figure 2.28, where the transmission line is directly connected to a 420 kV GIS, a computer model can be created to take into account the parameters previously discussed. A transient study would reveal the level of

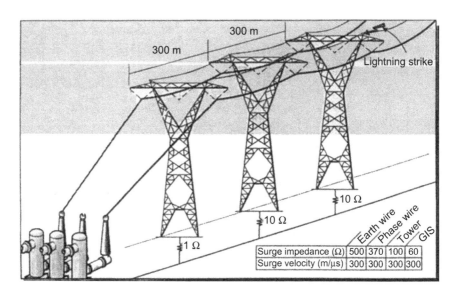

Figure 2.28   System schematic diagram

lightning stroke current required to cause a backflashover. Then according to the number of line flashes/100 km/year calculated for the transmission line and by using the probability curve for lightning current amplitude, a return time for this stroke current can be assessed (i.e. 1 in 400 years, 1 in 10 years, etc.) in, say, the first kilometre of the line. The voltage then arriving at the substation can be evaluated and compared with the LIWL for the substation equipment. The open-circuit-breaker condition must be studied here, since if the line circuit-breaker is open the surge voltage will 'double up' at the open terminal. Various levels of stroke current can be simulated at different tower locations and the resultant substation overvoltages can be assessed. If it is considered that the LIWL of the substation will be exceeded or that there is insufficient margin between the calculated surge levels and the LIWL to produce an acceptable risk, then surge arrester protection must be applied.

The rating of the MOSA will have been assessed from TOV requirements, and from the manufacturer's data a surge arrester model can be included in the system model. Repeating the various studies will reveal the protective level of the arrester and from this the safety factor for this system configuration can be assessed [14–18]. IEC 60071 recommends a safety factor of 1.25 for 420 kV equipment (safety factor = LIWL/protective level). The surge arrester current calculated for this condition should be the 'worst' case and can therefore be used to assess the nominal discharge current requirement of the surge arrester (5, 10 or 20 kA). (IEC 60091-1 is the international standard for surge arresters [16], and an accompanying guide is available which contains detailed information on the application of surge arresters.)

To make full use of the MOSA protective level the arrester should be placed as close as possible to the equipment being protected. In the case of the open line circuit-breaker this may well be 10–20 m distance. Dependent on the rate of rise of the surge voltage, a voltage greater than the residual voltage at the surge arrester location will be experienced at the terminals of the open-circuit-breaker. This must be taken into account when assessing the substation overvoltage. Figure 2.29 illustrates the surge voltage profile of the GIS with the line circuit-breaker closed.

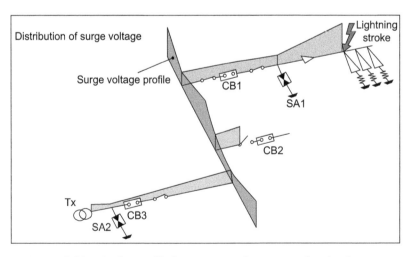

Figure 2.29 *Analysis of lightning surge for gas-insulated substation*

It shows that additional surge arresters may be required because of the distances involved in the layout of the substation. It then follows that surge arresters have a 'protective length' [14] which is sensitive to the rate of rise of the incoming surge voltage, and this must be taken into consideration when assessing the *lightning* overvoltage on equipment remote from the surge arrester.

2.9 Conclusions

This chapter has introduced the important concept of insulation co-ordination of high-voltage systems. It is vital for any engineer working in, or planning to work in, the electrical power industry to be aware of design choices regarding electrical stresses, insulation levels, service performance, etc., together with testing procedures and the importance of IEC Standard for a wide range of equipment [19]. These aspects will be considered further in later chapters. The topics have also been discussed in several major publications in recent years (e.g. *Electra*, CIGRE, IEC, IEEE). (For updates visit http://www.global.ihs.com)

The reader should be aware that increasing environmental and aesthetic concerns regarding the utilisation of overhead electrical power transmission lines have resulted in critical evaluation of underground alternatives and, in particular, at opportunities to further reduce the cost ratio between overhead and underground distribution and transmission systems [20]. A joint CIGRE Working Group has been established to evaluate the influence of transient overvoltages on alternating current (AC) cable insulation design. Briefly, if one considers the inherent costs associated with cable systems (cable and accessories) then, by reducing the required withstand capability with improved protection devices and network design philosophy, one could develop a more economic underground system [20]. However, further insulation co-ordination studies will be required to determine appropriate safety factors and acceptable failure rates as covered by Reference 14.

References

1. Clericci A., Al Rashed S.A., Al Sohaibani S.N. *380 kV system from Riyadh to Quassim: studies of temporary overvoltages and methods to prevent them.* CIGRE; 1990, Paris, Conference paper, SC 22.06
2. Germany N., Mastero S., Vroman J. *Review of ferroresonance phenomena in high voltage power system.* CIGRE; 1974, Paris, Paper 33-18
3. Ritchie W.M., Irwin T. 'Limitation of transmission line energising overvoltages by resistor insertion'. *IEEE Conference Publications.* 1979;**182**
4. Johnson I.B., Titus C.H., Wilson D.D., Hedman D.E. 'Switching of extra-high voltage circuits II – surge reduction with circuitbreaker resistors'. *IEEE Transactions.* 1964;**PAS-I96**:1204
5. Maury E. *Synchronous closing of 525 and 765 kV circuit breakers. A means of reducing switching surges on unloaded lines.* CIGRE; 1966, Paris, Paper 143
6. Stemler G.E. 'BPA's field test evaluation of 500 kV PCSs rated to limit line switching overvoltages to 1.5 per unit'. *IEEE Transactions.* 1966;**PAS-95**(l)

7. Bickford J.P., El-dewieny R.M.K. 'Energisation of transmission lines from inductive sources'. *Proceedings of IEEE*. 1973;**120**(8):883–90
8. Bickford J.P., El-dewieny R.M.K. 'Energisation of transmission lines from mixed sources'. *Proceedings of IEEE*. 1973;**112**(5):355–60
9. Catenacci E., Palva V. 'Switching overvoltages in EHV and UHV systems with special reference to closing and reclosing transmission lines'. *Electra*. 1973;**30**:70–122
10. Dalziel I., Foreman P., Irwin T., Jones C.I., Nurse S., Robson A. *Application of controlled switching in high voltage systems*. CIGRE; 1996, Paris, Paper 13-305
11. Anderson J.G. 'Lightning performance of transmission lines', in *Transmission Lines Reference Book 345 kV and Above* (2nd edn). Palo Alto, CA: EPRI; 1982
12. Anderson R.B., Eriksson A.J. *A summary of lightning parameters for engineering applications*. CIGRE; 1980, Paris, Paper 33-06
13. Watson W., Ryan H.M., Flynn A., Irwin T. *The voltage-time characteristics of long air gaps and the protection of substations against lightning*. CIGRE; 1986, Paris, Paper 15-05
14. IEC 60071-1, 1993. 'Insulation coordination, Part 1: terms, definitions, principles and rules'. Contains the definitions of terms to be found here. It gives the series of standard values for the rated withstand voltages and the recommended combination between these and the highest voltage for equipment
15. IEC 60071-2, 1996. 'Insulation coordination – Application guide'. Gives guidance on the procedures involved in insulation co-ordination as defined in IEC 60077-1
16. IEC 60099-1, 1991. 'Surge arresters, Part 1: non-linear resistor type gapped arresters for AC systems'. Specifies the test requirements and maximum protection levels for this arrester type
17. IEC 60099-4, 1991. 'Surge arresters, Part 4: metal-oxide arresters without gaps for AC systems'. Specifies the test requirements and maximum protection levels for this arrester type
18. IEC 60099-5, 1996. 'Surge arresters, Part 3: application guide to IEC 99-1 and IEC 60099-4'. Gives guidance for the selection of surge arresters depending on the overvoltage stress conditions of the system
19. Ryan H.M., Whiskard J. 'Design and operating perspective of a British UHV laboratory'. *IEEE Proceedings A*. 1986;**113**(8):501–21
20. Rosevear R.D., *et al*. 'Insulation co-ordination for HVAC underground cable systems'. Joint CIGRE Working Group 21/33 report, *Electra*. 2001;**196**:43–5
21. 'Background of technical specifications for substation equipment exceeding 800 kV AC'. CIGRE Technical Brochure TB 456-WG A3.22, 2011 (see also *Electra*, **255**, April, 2011, pp. 60–69; same title, prepared by WG A.3.22). This second TB further develops the background information of TB 362 (2008) and where appropriate presents interim recommendations for the international specification and standardisation of UHV equipment. (Note that beyond this topic,

the field test technology on UHV Substations during construction and operation requires investigation. A Technical Brochure is under preparation and should be issued by CIGRE – late 2012.) A3 high voltage equipment (www.cigre-a3.org); (http://www.cigre.org; http://www.e-cigre.org)[1]

22. Colapret E.E., Reid W.E. 'Effects of faults and shunt reactor parameters on parallel resonance'. *IEEE Transactions on Power Apparatus and Systems*. 1981;**PAS-100**(February 2):572–84

23. CIGRE, Working Group 01 (Lightning) of SC 33 (Overvoltages and Insulation Co-ordination). 'Guide to procedures for estimating the lightning performance of transmission lines', October 1991 [See also Chapter 23 for details of revised CIGRE SCs, post 2002.][1]

24. Hara T., Yasuda Y., Hirakawa Y., Shiraishi K. 'Sensitivity analysis on grounding models for 500 kV transmission lines'. *The Institute of Electrical Engineers of Japan*. 2001;**121-B**(10):1386–93

25. Gustavo Carrasco H., Alessandro Villa R. 'Lightning performance of transmission line Las Clarits – Santa Elena Up 230 kV', *International Conference on Power Systems Transients – IPST*; New Orleans, LA, 2003

26. Martinez-Velasco J.A., Castro-Aranda F. 'Parametric analysis of the lightning performance of overhead transmission lines using an electromagnetic transients', *Program, International Conference on Power Systems Transients – IPST*; New Orleans, LA, 2003

27. Hancock B., Irwin T., Finn J.S., Sabri W.S. 'Verification of system design by tests', *CIGRE China Symposium 2003, 'Large Interconnected Systems'*

28. IEC 60071 – 4 Edition 1.0, 2004–2006. Insulation co-ordination – Part 4: 'Computational guide to insulation co-ordination and modelling of electrical networks'

29. IPST 2007. 'Points to consider regarding the insulation coordination of GIS substations with cable connections to overhead lines'; Osborne M.M., Xemard A., Prikler A., Martinez J.A.

30. Mork B.A., Gonzalez F., Dmitry I., Stuehm D.L., Mitra J. 'Hybrid transformer model for transient simulation – part I: development and parameters'. *IEEE Transactions on Power Apparatus and Systems.* 2007;**22**(January 1): 248–55

31. Mohamad N.N., Rajab R. 'Correlation between steady state and impulse earth resistance values', *American Journal of Applied Sciences*. 2009;**6**(6): 1139–42

32. Ab Kadir M.Z.A., Sardi J., Wan Ahmad W.F., Hizam H., Jasni J. 'Evaluation of a 132 kV transmission line performance via transient modelling approach'. *European Journal of Scientific Research*. 2009;**29**(4):533–9. ISSN 1450-216X

---

[1] CIGRE key: SC, Study Committee; WG, Working Group and JWG, Joint Working Group; TF, Task Force; TB, Technical Brochure; TR, Technical Report; SC, Scientific Paper; Other valuable CIGRE publications: AGMs of SCs, *Electra* (technical magazine bimonthly), international conferences worldwide (including Paris, biannually), and joint conference events with other international organisations including IEC, IEEE, IET, CIRED, etc.

Chapter 3
Applications of gaseous insulants to switchgear
H.M. Ryan

3.1 Introduction

This chapter introduces the subject of gaseous insulation and provides information relating to the application of gaseous insulants to high-voltage systems [1–49]. It examines atmospheric air and compressed gases and illustrates how, by linking available experimental test data from such sources with a knowledge of the 'effectiveness' of various practical gas-gap clearances, the designer can achieve reliable insulation design. The chapter also briefly discusses the need for extra high-voltage (EHV) and ultra high-voltage (UHV) test areas or laboratories. Evidence is presented of how laboratory studies, on representative insulation systems and electrode arrangements, provide the designer with choices relating to electrical stresses, clearance levels, service performance and testing procedures. Gas-insulated substations (GIS) using sulphur hexafluoride (SF_6) gaseous insulation have been used in transmission systems worldwide for more than 45 years. The service reliability of this class of equipment is of paramount importance. In addition, the chapter presents a large amount of experimental breakdown information on SF_6 and briefly reviews the application of field computation strategies in support of GIS and other equipment designs. Several of the major factors influencing the insulation design and in-service behaviour and reliability of SF_6 gaseous and epoxy resin support insulations, as used in GIS equipment, will also be considered.

Atmospheric air is the most abundant dielectric material, which has played a vital role in providing a basic insulating function in almost all electrical components and equipment. However, because it has a comparatively low dielectric strength, large electrical clearances are required in air for high-voltage applications such as overhead line bundle conductor or tower design or open-type EHV switchyards. Air clearances will be discussed in section 3.2. Much research activity has taken place during the past 60 years to investigate compressed air and to develop other gases with even better characteristics than air. Some important characteristics of compressed gaseous insulation, including air, SF_6 and other gases, will be touched on in sections 3.3 and 3.4 and considered further in Chapter 7.

As discussed elsewhere [1–5], the dielectric strength of air and other gaseous insulation is a function of gas density (δ). Advantage has been taken of this fact for many years in transmission switchgear, where high-pressure gases have been used to provide essential insulation and arc interruption characteristics in heavy-duty interrupters. The concept of high-pressure gaseous insulation has now been extended into combined metal-clad switchgear and connecting cables, termed gas-insulated switchgear (GIS). Strategic aspects relating to gaseous insulation systems as used in GIS will be discussed in section 3.4.

Sometimes, gas-insulated cables are referred to as gas-insulated lines (GIL), although the term GIL is now more widely adopted. Until recently, the high cost of GIL as compared to cables restricted the use of GIL to distances of up to about 3 km. This may possibly change, as a current programme within CIGRE (i.e. a JWG) is investigating and analysing factors for investment decisions relating to the case of *GIL vs cables for AC transmission*. (A CIGRE Technical Brochure (TB) covering this aspect should be published in 2014.)

Since gas does not offer mechanical support for live conductors, switches, etc., it is used in combination with solid insulation. The gas/solid interface region must be looked at very carefully from an electrical design viewpoint, otherwise design weaknesses could result in problems in service, with the possibility of consequential breakdown or flashover of equipment. Even when the very best design principles have been adopted, breakdowns can still occur in air/solid interfaces under adverse environmental conditions such as pollution, fog, ice, snow or humidity. These aspects are outside the scope of this chapter and are dealt with elsewhere [5–8].

Whatever dielectric medium is used, be it solid, liquid or gas or combinations thereof, it is essential that careful consideration should be given to the electrostatic field design of insulating components to produce compact, efficient, economic and reliable designs of high-voltage equipment, which will give long and trouble-free life in service. To this end, designers rely on analytical field analysis techniques to model electrostatic and other aspects of specific designs and, when appropriate, to optimise the shape and disposition of components. The general availability of commercial packages to carry out 'advanced' digital field techniques during the past 35 years has greatly assisted equipment designers.

Section 3.5 gives a brief indication of the purpose, scope and application of field analysis methods to provide valuable design information from extensive SF_6 gaseous insulation studies relating to simple electrode arrangements through to EHV equipment designs, including gas bushings, circuit-breaker designs and GIS.

A list of commonly used terms and organisations referred to in this chapter is set out in Box 3.1.

Box 3.1 Terms and organisations

AIS   Air-insulated substation (open terminal)
AC LD AC long duration-induced voltage test [applied in certain UHV projects]
BSI   British Standards Institution (UK)

CIGRE	International Council on Large Electric Systems (CIGRE Central Office – Paris)*
CB	Circuit-breaker
COV	Continuous operating voltage
CT	Current transformer
DS	Disconnector switch
EDF	Abbreviated name of a major French Energy company, which operates in France, UK and elsewhere. (its UK activities are discussed further in Chapter 23)
EHV	Extra high-voltage
GIL	Gas-insulated lines (i.e. gas cables [usually a mixture N_2 - SF_6])
GIS	Gas-insulated switchgear (SF_6) or Gas-insulated substation (SF_6)
HVET	High-voltage engineering and testing (name of this book and its subject grouping)
	Note: Most of the authors of this book have also been closely involved for many years with the very successful IET, International Vacation School on HVET, which has been running in the UK since 1993.
IEC	International Electrotechnical Commission (Standards Worldwide)
IEEE	Institute of Electrical and Electronic Engineering (USA)
IET	Institution of Engineering and Technology (UK)
LIWV	Lightning impulse withstand voltage
MOSA	Metal oxide surge arrester (gapless) (high performance surge arrester used in UHV systems)
NGC	National Grid Energy Company (UK)
SF_6	Sulphur hexafluoride gas (used worldwide as a gaseous insulant in GIS and GIL equipment)
SFO	Slow front overvoltage
SIWV	Switching impulse withstand voltage
SOV	Switching overvoltage
TCV	Trapped charge voltage
TOV	Temporary overvoltage
UHV	Ultra high-voltage [substations/lines/systems/equipment/projects etc]
VT	Voltage transformer
VFTO	Very fast transient overvoltage

*CIGRE key: SC, Study Committee; WG, Working Group and JWG, Joint Working Group; TF, Task Force; JTF: Joint Task Force; TB: Technical Brochure; TR, Technical Report; SC, Scientific Paper.

3.2 Atmospheric air clearances

3.2.1 Test areas

In dry, unpolluted areas, air clearances between live equipment and earth can be defined reliably. Legg [9] has described the minimum clearances necessary for high-voltage test areas. The majority of high-voltage tests on equipment are carried out indoors to avoid large variations in atmospheric conditions. When high-voltage laboratories are constructed, the test voltages to be used are generally known and the physical size of the laboratory is determined by the dimensions of the voltage-generating plant, the test objects and the clearances required to prevent flashover to the walls of the laboratory (see also Chapters 15–19). The flashover voltage of rod–rod and rod–plane gaps as a function of the gap spacing [9] is illustrated in Figure 3.1. From these graphs, the required dimensions of laboratories can be estimated. It will be noted that the lowest flashover voltages are experienced with positive switching surges and 50 Hz voltages, applied to rod–plane gaps. For this reason, the performance of a rod–plane gap under positive switching surges may be usefully taken as a datum (for required laboratory clearances) because all other combinations of gap geometry and voltage waveform produce higher flashover voltages. The sparkover of any gap is subject to statistical variation, and some knowledge of this is required before clearances can be estimated for a laboratory, etc. This essential consideration of statistical aspects also applies in day-to-day testing or measuring within laboratories (see sections 3.3 and 3.4). The breakdown voltages reproduced in Figures 3.1 and 3.2 relate to 50% flashover levels [9]. That is, if a large number of voltage tests at this level were applied to the gap, one-half of these tests would be expected to cause flashover. Breakdown to the walls of a laboratory or test-chamber must be avoided; therefore, greater clearances than these are required.

In Figure 3.2, the 50% flashover voltage is 1,200 kV for this particular gap. The voltage corresponding to the 16% probability line (1,090 kV) is one standard deviation (σ) less than V_{50}, i.e. $V_{50} - \sigma$. The voltage corresponding to the 2% line is $V_{50} - 2\sigma$ (980 kV) and that to the 0.1% line is $V_{50} - 3\sigma$ (870 kV). In practice, one takes the withstand voltage of the gap as 2 or 3 standard deviations less than the

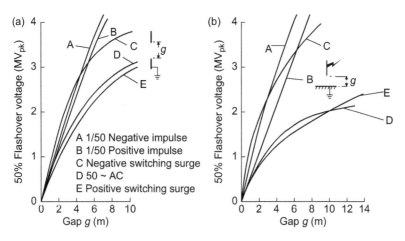

Figure 3.1 Fifty per cent flashover voltage for (a) rod–rod and (b) rod–plane gaps

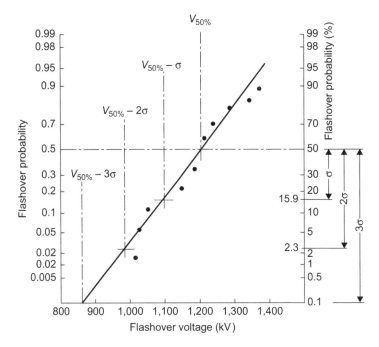

Figure 3.2 Illustrating probability of flashover

50% flashover voltage, i.e. $V_{50} - 2\sigma$ or $V_{50} - 3\sigma$. The withstand voltage of the gap represented in Figure 3.2 is therefore 980 kV or 870 kV. For 1/50 μs impulse waves, the standard deviation is of the order of 1–4%, the lower values occurring at the highest voltages, whereas for switching surges the standard deviation is of the order of 5–12% and increases with voltage. For 50 cycle voltages up to 1,500 kV RMS, the standard deviation is approximately:

%	
1.5–4	For rod–plane gaps
0.5–2.5	For rod–rod gaps
1.0–4	For dry insulators
1.5–6	For wet insulators

Having established the withstand voltage for various gaps – or conversely the gap required to withstand the various test voltages – the dimensions of a laboratory can be estimated. Even clearances based on $V_{50} - 3\sigma$ are often increased for two reasons. First, the walls of the laboratory can influence the flashover voltage and invalidate the test. Second, clearances are required for radio interference or partial discharge measurements. During partial discharge tests, one may be looking for very small internal discharges (often as small as 1 pC) in the test object so that spurious discharges from the circuit (busbars, transformers, etc.) must be an order smaller. If the clearances are too small, even though they are adequate to prevent flashover, then small discharges may flow to the walls of the laboratory, thereby preventing accurate measurements. Test area dimensions are given in Table 3.1. For very high-voltage test equipment (>3 MV), it may be uneconomic to build an enclosure with adequate clearances and the test equipment can be moved and used outdoors.

Ryan and Whiskard [10] describe the design, planning and supervision of the construction of a new UHV testing laboratory (see Figure 3.3). The main purpose

Table 3.1 Approximate dimensions of test hall for 50 Hz three-phase tests [9]

Transformer voltage rating	HT to earth clearance, X				Nominal test room dimensions in feet			Diameter of discharge free busbars	
	Minimum practical		Discharge free at full voltage						
(kV RMS)	(ft)	(m)	(ft)	(m)	L	W	H	(in.)	(cm)
100	1.3	0.4	2	0.6	15	10	10	2	5
250	3	0.9	5	1.5	20	15	15	4	10
500	7	2.1	10	3	40	25	20	8	20
800	13	3.9	20	6	50	40	35	12	30
1,000	19	5.8	25	7.6	75	55	50	15	40

L, length; W, width; H, height.

Figure 3.3 Major items of laboratory equipment [10]
      (a) View of 2 MV transformer, 4 MV impulse generator and UHV divider and test object in test hall
      (b) View of 2 MV sphere-gaps, control room and galleries

of this laboratory was to provide a major facility in the UK for the development of switchgear rated up to 765 kV, and probably higher after changes to 'interim technical specifications' issued for substation equipment exceeding 800 kV AC (CIGRE TB 456-WG A3.22) [11], based on recent technical decisions applied for the 1,100 kV line in China (commissioned in 2009), requiring lower insulation levels.

The authors [10] report on some of the dielectric research undertaken to achieve this position. Following the opening of the Clothier laboratory in 1970, these workers were closely connected with the development of new ranges of open-terminal and metal-clad SF_6 switchgear, rated up to 550 kV and for fault current levels up to 63 kA (see Chapter 8). These switchgear design activities were supported by extensive dielectric research studies, which enabled the major factors influencing the insulation integrity of practical equipment to be determined. Much of this work has already been extensively reported in the literature. Ryan and Whiskard [10] outlines:

1. the criteria used in designing the Clothier UHV laboratory, and present a critical appraisal of the facilities during the first 15 years of its operational life
2. some significant laboratory activities, including examples of studies on various switchgear and non-switchgear components for systems up to 765 kV, all having been subjected to rigorous dielectric proving tests
3. the use of specific EHV and UHV test procedures (e.g. climatic, artificial rainfall and mixed voltage testing, gas handling expertise) together with the assimilation of important technical methodologies/factors, which have influenced the dielectric design and testing of commercial GIS and other equipment.

Chapters 15–18, respectively, by Ernst Gockenbach discuss basic measuring techniques, basic testing techniques, partial discharge measuring techniques and digital measuring techniques.

In Chapter 19 of the second edition of this book (2001), the late R.C. Hughes considered the strategically important theme of traceable measurements in high-voltage tests as it was perceived at that time. Bob Hughes had been extremely active in strategic developments within IEC/CIGRE in this sector for many years. Ernst and Bob also lectured at and were very strong supporters of the IET HVET International School series, always anxious to pass on their expertise to the next generation.

The reader should be aware that modern UHV laboratories exist in several countries, including Brazil, Canada, China, France, Germany, Italy, Japan, UK and USA. Numerous 'quality' technical papers have been presented by experts from these organisations (see IEC/CIGRE/IEEE/IET publications).

3.2.1.1 Historic notes

Turning back historically to UK power engineering manufacturing and UHV testing capabilities:

1. Transmission transformer and switchgear manufacturing has now all but disappeared in the UK. In the past two decades, the UHV laboratory testing capabilities in the UK have been virtually removed following closures of the NGC Leatherhead UHV testing laboratory complex (formerly CEGB/CERL), the British Short Circuit Testing Station facility and, more recently, the

100 *High-voltage engineering and testing*

Clothier UHV Testing Laboratory at Hebburn, Tyne and Wear (formerly owned by Reyrolle and later operated by other companies).
2. Bearing in mind a significant upturn in UHV system developments worldwide in recent years, with transmission systems now operating overseas at voltages up to 1,200 kV, the failure to develop a UK National UHV Testing Facility, as was provisionally mooted and evaluated in 1980s and 1990s, now looks like a water-shed for the demise of much of the traditional UK power engineering design, manufacturing and UHV testing/research expertise in Britain.
3. Sadly, recent worldwide advances in surge arrester technology, linked with changes to insulation coordination strategies and reduced UHV equipment testing levels [11] would have now made the Clothier Laboratory suitable for the majority of dielectric testing, up to latest UHV transmission system requirements.

3.2.2 Sphere gaps

A convenient use of airgaps is as a calibration for high voltages. It is useful in that no voltage dividers or electronic equipment is required (which can malfunction), but suffers from lack of accuracy, ±3% (this covers the whole range of voltage). Spark gaps are recognised in IEC52, 1960. The sphere gap is accurate and reproducible over its range, which extends to gaps equal to 0.5 × sphere diameter, after which it loses accuracy due to field distortion. It is also necessary to take account of humidity, temperature and pressure, all of which affect the breakdown voltage. The sphere gap records peak voltage and can therefore be used for AC or impulse voltages with only slight variations for positive or negative pulses. For very fast impulses (<10 μs), it is necessary to irradiate the gap with a radioactive source in one sphere (or use a UV source) to provide the necessary initiating electrons.

Modern EHV and UHV laboratories are capable of measuring voltages to high accuracy (see Chapter 15), being equipped with precision voltage dividers and used in conjunction with the 'latest' digital measuring techniques incorporating sophisticated software support. However, in industrial laboratory test facilities, it is still common for sphere gaps to be used (e.g. Figure 3.3(b)) to calibrate another measuring system because it can be easily verified by the inspectors. The revision document of IEC 52, 1960, is currently under consideration by the national committees. No change has been made to the extensive tabulated values of 'sparking voltage against gap', since existing values are considered to constitute 'a consensus standard of measurement', which has been unchanged for >75 years.

Sphere gap sparking voltages given in the tables can be considered to have uncertainties of ±3% for a 95% confidence level. A spark gap can be regarded as 'fit-for-purpose' provided its flashover performance characteristic 'conforms', i.e. to be less than a specified value. Otherwise, a high value of σ (obtained in a test series) indicates something is wrong (e.g. dirt, dust, insufficient source of ionising ions), while a sufficiently low value of σ indicates that particular sphere gap is acceptable for calibration use.

3.2.3 Spark gaps

Spark gaps are used for protection of equipment against high-voltage surges, for all voltage ranges, the advantage being that the dielectric is recoverable. Historically,

Applications of gaseous insulants to switchgear 101

spark gaps range from the common GPO arrester, having low-pressure N_2 in a ceramic body, with voltage ratings of 150–1,000 V and current ratings of 10^3 amperes, to low-pressure gas or vacuum 'bottles' with voltage ratings of 1–50 kV and current ratings of 50 kA for railway equipment protection, for example. The simplest and most common spark gap is the rod-gap, used on outdoor equipment such as transformers, substations and bushings. Its characteristics are well documented (e.g. Figure 3.4); its simplicity is offset by a variable breakdown voltage with respect to time and polarity. Special designs with improved characteristics have been developed recently for EHV and UHV systems. For equipment protected by a rod-gap, it is important to know its 'breakdown voltage against time' characteristic relative to that of the equipment being protected to ensure that the rod-gap operates first, under all surge conditions. Such protection is known as insulation co-ordination, as covered by IEC Technical Committee 28 and discussed and references detailed in Chapter 2 (e.g. see also Figures 3.5 and 3.6).

Many workers have derived empirical relationships to correlate breakdown voltage with gap spacing for rod–plane gaps [1]. For example Ryan and Whiskard,

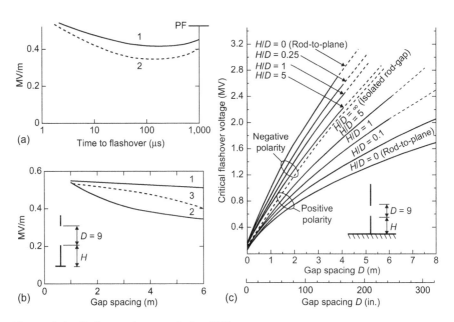

Figure 3.4 Rod-gap characteristics [12]
(a) Typical relationships for critical flashover voltage per metre as a function of time to flashover (3 m gap): 1 rod–rod gap; 2 conductor–plane gap; PF, power frequency CFO
(b) Typical relationships for flashover voltage per metre as a function of gap spacing: 1 1.2/50 μs impulse; 2 200/2,000 μs impulse (rod–rod, H/D = 1.0, positive dry); 3 power frequency
(c) Switching surge flashover strength of rod–rod and rod–plane gaps (courtesy Edison Electric Institute); wavefronts of 100–200 μs; vapour pressure = 12.5 mean Hg

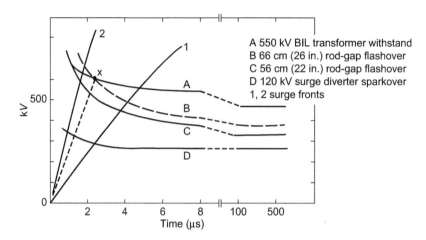

Figure 3.5 Transformer protection by rod-gap and surge arrester [12]

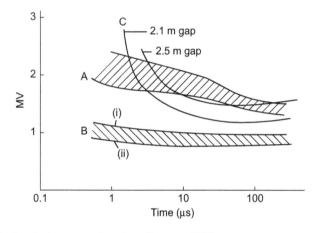

Figure 3.6 Insulation co-ordination diagram [10]
(A) Typical V/t characteristics of 420 kV GIS
(B) Representative characteristics of 396 kV ZnO arrester
(C) Lower limit curve of a line gap for specific gap settings:
(i) upper limit 40 kA amplitude and (ii) lower limit 3 kA amplitude

working in the Clothier laboratory [10] reconfirmed results of earlier Russian workers by studying the 50 Hz breakdown characteristic of three rod–plane electrode systems (i.e. (i) a 1.27 cm square-cut rod, (ii) a 0.2 cm diameter hemispherically ended rod and (iii) a 0.5 cm diameter sphere). It was originally established by Tikhodeyev and Tushnov (1958) – see Reference 1, p. 617 – that the breakdown voltage (V_s), MV_{pk}, could be predicted, to within 1%, for air gaps g in the range ($1 \leq g \leq 9$ m), using an expression of the form

$$V_s = 0.0798 + 0.4779\,g - 0.0334\,g^2 + 0.0007\,g^3 \text{ MV}_{pk}$$

where g is the gap in metres. A simpler abridged equation ($V_s = 1.62\, g^{1/3} - 1.1$) was found to be accurate to $\pm 5\%$. Ryan and Powell had also found that, for rod–plane gaps in the range 1–5 m, the measured 50 Hz breakdown voltages lay within 3.5% of the values given by the full equation above when (i) the end of the rod was square cut and within 3% when it was hemispherically terminated (ii). The corresponding results with electrode system (iii) gave breakdown results some several percent higher than those given by the above full equation. It is worth emphasising that, although the form of the rod termination had little effect on the breakdown voltage, the three electrode systems exhibited widely different values of corona onset voltage, for corresponding gap settings [1].

The role of rod-gaps and surge-arresters, widely used as overvoltage protective devices, is summarised elsewhere [12]. Voltage–time 'withstand' characteristics exist for different types of insulation when subjected to various types of overvoltage. The various time regions refer to particular types or shapes of test voltage. The 'withstand' levels used for self-restoring insulation are at a suitably chosen level below the 'V_{50}/time' characteristic (a selected multiple of σ below), while characteristics for 'non self-restoring' insulation are obtained from withstand characteristics corresponding to the specified test voltages for the various categories of test voltage. The example reported by Diesendorf [12], reproduced in Figure 3.5, illustrates that protection can be achieved in any time region in which the protective characteristic lies below the withstand characteristics of the insulation concerned. (*Note*: Clearly, point X is an intercept point for curve A and curve B.) The 120 kV surge arrester (curve D) protects the transformer over the entire time-range, whereas the 26 in. (660 mm) rod-gap (B) protects the transformer only against surges with front slopes less than OX. Steeper surges would cause the insulation to break down before the rod-gap could operate. To achieve reliable protection, comprehensive test data must exist for both the protective devices and the insulation components of any system [13].

Figure 3.6 provides further results from extensive insulation coordination studies [10, 14], on this occasion for 420 kV GIS assemblies; V/t characteristics for GIS and for corresponding 396 kV (rated) ZnO arresters and standard line-gaps are compared. This investigation looked into the strategic question of insulation co-ordination for GIS. ZnO arresters are now used worldwide to provide robust protection to modern transmission systems [15]. It is also interesting to observe in passing the remarkable expansion of the application of metal oxide arresters to overhead lines at lower system voltages in recent years. The voltage class of these applications ranges from several kilovolts to EHV/UHV voltage.

At present, only a few UHV AC transmission networks are in service worldwide at 1.1 MV or 1.2 MV [11] (e.g. in China and India). Thus, the background of the technical specifications for sub-station equipment exceeding 800 kV AC is less well defined than at lower system voltage levels, where robust standards exist. Consequently, as only limited experimental/technical/specification information exists for UHV substation equipment, a CIGRE study has collated the background information for specifications [11] and has presented recommendations for the international specification and standardisation of UHV equipment in TB 456-WG A3.22 (2011).

Aspects such as insulation coordination (including the use made of surge arresters at UHV levels), switching phenomena for circuit-breakers, disconnectors and earthing switches and testing are addressed. Certain minor and limited modifications are made to data in an earlier CIGRE TB 362-WG A3.22 (2008) [11]. In situations such as this refinements and modifications will continue as service experience with UHV systems increases – and until full robust IEC Standard(s) are available.

Strategically UHV technology is, at present [11], 'characterised by a need to minimise the sizes, weights, costs and environmental impacts of the overhead lines and substations and hence to develop projects which are feasible from an economic, societal and technical point of view'.

It is further stated that 'The UHV voltages presently standardised by IEC are some 50% higher than those for system voltages of the 800 kV class'; however, since insulation strength per metre decreases with the length of the air gap (particularly for switching impulses under wet conditions), 'simple extrapolation of the dielectric requirements would lead to disproportionately large structures'.

By means of the application of a number of new technologies and new analysis techniques, utilities [took the view that they] are able to reduce the dielectric requirements to values that lead to much smaller structures. This results in insulation levels that are not far from the levels applied at the 800 kV class. In Japan, the towers of the UHV OH lines are only 77% of the size that would be necessary if insulation levels had been extrapolated directly from the lower voltage class. The Working Group (WG) authors state [11] that:

- The technologies used to reduce the insulation levels include surge arresters with a lower ratio between LIPL/SIPL and COV (continuous operating voltage); and by applying multi-column arresters and/or a number of arresters in parallel the ratio can be further decreased.
- 'Closing resistors' are used to control slow front overvoltages (SFO) during closing and re-closing overhead lines.
- SOVs generated in healthy lines on the source side of a fault-clearing circuit-breaker are typically most severe for low-probability events such as two- and three-phase ground faults.
- Despite their low probability of occurrence in UHV systems, 'opening resistors' are used to reduce opening SFOs due to the potential consequences of successive breakdown, which may affect the whole system.
- Techniques such as the application of transmission line arresters (TLA) and/or controlled switching may be used to control SFOs in future.
- Shielding of OHLs, improved earth return conditions and other counter measures against back-flashover lead to a better lightning-withstand performance.
- When necessary, *damping resistors* in GIS disconnectors can reduce the amplitude of VFTO – a phenomenon that may otherwise exceed the LIWV of the switchgear.
- The WG authors [11] drew attention to curves which showed reduced margins between LIWL and LIPL and between SIWL and SIPL and help to explain why the insulation levels for UHV are not far from those for 800 kV equipment.

- The authors [11] also report, 'by applying advanced calculation and simulation techniques, utilities are able to assess the critical conditions and events that need to be considered when designing UHV systems and ensure that these are consistent with their policies regarding design risk'.

This is a complex subject area and one of immense strategic technical and economic importance to network design and effective operation. Consequently, it is always vital to use the complementary skills of a group of experts in this sector when making a final decision.

Finally, the reader should note that another CIGRE Working Group document [49] has recently been published. It reviews further studies concerning field testing techniques on UHV substations during construction and operation, in Japan, India and China. A new CIGRE TB 542-WG C4.306, 'Insulation coordination for UHV AC systems,' has been prepared which significantly 'advances' the state of knowledge on this topic for UHV substations and lines. This interim document will have an important impact on insulation coordination strategies, for UHV systems, until the rigorous new revisions to the IEC 28 standard are introduced in a few years time. A self explanatory 'abridged account' of this TB 542 review study as summarised in CIGRE *Electra* [49], is provided in Table 3.2.

Taking into account the accumulated knowledge of various CIGRE working bodies, this review has evaluated recent measured data of VFTO (very fast transient overvoltages) and air-gap dielectric characteristics. Special reference is made to the higher performance surge arresters MOSA) used in these UHV studies.

3.2.4 Overhead lines and conductor bundles

Various classifications of dielectric stress may be encountered during the operation of any equipment ranging from (i) sustained normal power frequency voltages, (ii) temporary overvoltages, (iii) switching over-voltages to (iv) lightning overvoltages. These are represented in the laboratory, for test purposes, by standard test waves: power frequency, switching impulse and lightning voltages. The strength of external insulation is dependent on geometric factors, air density, temperature, humidity (appropriate correction factors are available), precipitation and contamination (due to natural or industrial pollution). The presence of ice and snow can influence the flashover performance of outdoor insulators (see Reports of CIGRE Task Force 33.04.09 [7, 8]). Ice and snow accretions can present serious problems in many cold regions of the world; there is well-documented evidence that the number of single phase-faults increases substantially during cold precipitation and after accretion when followed by a rise in air temperature. Further research is required to provide a better understanding of such flashover events.

Before effective insulation co-ordination can be achieved, a large amount of experimental test data is required on arrangements such as support insulators (various shapes), rod–plane gaps (for protective purposes), rod–plane, ring–plane gaps, etc., which approximate to practical conditions (e.g. air insulator strings for transmission lines (vertical, horizontal, vee-strings, disc, longrod, fog-type, etc.)). A variety of typical characteristics for various arrangements have been summarised and presented by Diesendorf [12, 13]. Examples are reproduced in Figure 3.7.

Table 3.2 Insulation coordination for UHV AC systems

Introduction: Since Reference 11 was published, another CIGRE Working Group, i.e. WG C4.306, has reviewed and discussed insulation coordination practices for voltages in the UHV AC range and for voltages exceeding 800 kV AC, taking into account the state-of-the-art technology, with special reference to higher performance surge arresters (MOSA). **WG C4.306**, under the convenorship of E. Zaima, Japan [49], utilised the accumulated knowledge of various relevant CIGRE working bodies, and this CIGRE study was accomplished in collaboration with related **SCs A3** and **B3**, involving [WGs. A3.22, A3.28, B3.22 and B3.29].

It must be emphasised that economical and highly reliable UHV and EHV transmission lines and substation equipment are of strategic importance and, especially at UHV with larger installations/equipment involved, it is vital that due account is given to environmental considerations. It is essential with UHV AC systems that safe air clearance distances exist. These may be much greater than those applied to EHV systems because, as is well known, the dielectric strength for switching the overvoltage does not increase linearly with the increase in air gap length.

The latest CIGRE study involved rigorous investigation of recent measured data of very fast transient overvoltages (VFTOs) and air gap dielectric breakdown characteristics. This latest report [49] outlined recent practices on insulation coordination for UHV and 800 kV systems. Common countermeasures in suppressing overvoltage, for each UHV project, include:

- use of higher performance metal oxide surge arresters (MOSA)
- suppressing overvoltage using circuit breakers with closing and/or opening with pre-insertion resistors.

The latest EMTP or other analyzing technology is routinely used to simulate and thoroughly study particular overvoltages in various strategic conditions/situations.

Overvoltages in UHV range

Temporary overvoltage (TOV): Amplitude of TOV levels and the energy absorption of the surge arrester in the UHV systems are relatively higher than those in EHV systems.

The representative levels of temporary overvoltages (TOVs) in the UHV systems of different countries vary slightly within the range 1.3 and 1.5 pu.

The energy absorption of the surge arrester for the TOV of the UHV system is higher than that for the 800 kV and EHV systems, being specified as from 40 to 55 MJ.

Switching overvoltage: The waveforms of the switching overvoltages occurring in UHV transmission lines are estimated and their front times are more than several hundred microseconds higher than those in EHV systems. Representative levels of switching overvoltages range from 1.6 to 1.7 pu. Mitigation measures such as a circuit-breaker (CB) with an opening and/or closing resistor, shunt reactor and appropriate MOSA installation are verified to be effective against system overvoltages.

Lightning overvoltage: The ratios of representative overvoltages to the LIPL of a surge arrester for transformers in UHV systems range from 1.0 to 1.7 pu, while the ratios for other equipment range from 1.0 to 1.36 pu. The WG comment that the ratios for UHV systems are likely to be smaller than those for 800 kV and the lower voltage systems, because the overvoltage mitigation techniques such as higher performance arresters can drastically reduce the lightning overvoltage levels.

Very fast transient overvoltage (VFTO): The new on-site test has been systematically measured in China. The cumulative trapped charge voltage (TCV) distribution leads to the conclusion, that the value of -1 pu is a conservative assumption. The assumption of a more

Table 3.2 (*Continued*)

realistic value of TCV can lead to a more realistic VFTO. The application of a slow operating disconnector does not produce a high VFTO, but increases both the sparking time and the number of sparks during the disconnector switch (DS) operation process. The damping resistor installed in the DS can significantly suppress the amplitude and gradient of VFTO caused by the switching of the DS.

Evaluation of overvoltages
Higher performance surge arrester: The WG C4.306 have carefully investigated the main characteristics of MOSA for the UHV projects undertaken worldwide. It was found that the UHV higher performance MOSAs have a higher protective performance and this was considered to be a decisive factor for the UHV power transmission insulation design.

Evaluation of overvoltage waveform: The WG comment that the actual field overvoltages of the non-standard lightning impulse waveforms were analysed and the insulating characteristics of the SF_6 gas and the oil-filled transformer elements for these actual overvoltages 'were clarified' to convert the waveforms into the standard lightning impulse waveform. The WG observed that the evaluation method could be possible [justified] in some cases, and it was considered that lower withstand voltages could be used.

Switching overvoltage mitigation measures for future UHV systems
The WG comment that the study results illustrate the combined contribution of the line end arresters and closing resistors or a controller to the switching overvoltage limitation and those can, it suggests, certainly be viewed as indicative of performance. The accurate representation of the arrester and controller characteristics, resistor insertion and the line configuration is considered essential in order to achieve a valid design basis for lines that meet dependability requirements.

Review on insulation coordination of air gaps in UHV range
WG C4.306 has reviewed the insulation coordination of air gaps in the UHV range to provide hints and recommendations for IEC TC28 (Technical Committee), based on the recent investigations and subsequent analysis on the UHV projects from Japan, India and China. Transferring the U50 measurements from India and China to a withstand voltage U10 by calculating and comparing these results with the recommended air gap clearances of IEC 60071-1 show that the **Conductor-Structure clearances** are rather close to the measurements and it supports the validity of the clearances selected in IEC 60071-1 for the UHV levels.

The much larger clearances necessary for the **Rod-Structure air gaps** in IEC, 'fit more or less' to the measurements obtained for the tower window configurations, but a correlation should be handled with care due to the various parameters which cannot yet be 'figured out' from these measurements.

For the actual configurations in a tower with and without insulators, the critical time to peak is found to be near to the peak of the standard wave shape.

Selection of insulation levels
Safety factor: The WG C4.306 explains that the safety factor compensates for the differences in the equipment assembly and the dispersion in the product quality. The safety factor currently quoted in IEC 60071-2 represents the overall value obtained from field experiences available when the standard was adopted some 20 years ago. The safety factors used in practice range from 1.00 to 1.32 while IEC 60071-2 currently recommends the range 1.05–1.15. In recent years, WG observe that information on the insulation deterioration characteristics due to ageing has continued to be accumulated, while quality control, assembly and installation technology of equipment are three aspects that have been significantly

(*Continues*)

Table 3.2 (Continued)

improved. Based on such developments, the WG suggest, for due consideration, that one option might be to allow some range for the recommended value of the safety factor instead of merely using it at a constant fixed value.

Lightning impulse withstand voltage (LIWV): The WG state that for UHV Projects, LIWV levels adopted are either 1950 kV or 2250 kV [for transformers] and 2250 kV or 2400 kV [for other substation equipment].

Note that the LIWV of other equipment is higher than that for transformers.

As for the ratios between LIWV and the representative overvoltage, a value of 1.02 (for other equipment) is the smallest while 1.42 is the largest.

Switching impulse withstand voltage (SIWV): For UHV projects (generally) each country sets: SIWV levels of 1425 kV or 1800 kV for transformers and 1550 kV or 1800 kV for other substation equipment. From the perspectives of the ratios between SIWV and the representative overvoltage, the smallest value is 1.09 and the largest value is 1.29 for transformers. The differences among the countries are less significant (the WG comment) than those for the LIWV. The representative overvoltages range between 1.46 and 1.69 pu, which, in general, are significantly lower than those in the lower voltage systems.

Very fast transient overvoltage (VFTO): Special measures to mitigate the VFTO are required in cases where the calculated VFTO caused by the GIS-disconnector switching is higher than the withstand level.

(The WG inform that Japanese and Chinese experiences are presented in the TB.542 document.)

A three-step procedure is proposed by WG C4.306, as a general insulation coordination approach:

1. **Calculation of VFTO** (peak and rise time)
2. **Comparison of calculated VFTO values with the LIWV level** by using the coordination factor, safety factor and test conversion factor (exact values of these factors cannot yet be provided)
3. **Decision and definition** concerning the necessity of introducing measures to reduce the VFTO level.

Power frequency voltage test

Japanese, Chinese and Indian UHV projects apply a long-duration induced AC voltage test (ACLD) for the UHV power transformers, whose time sequence is nearly the same for these projects. According to IEC 60071-1, the TOVs should be basically covered by the SIWV.

Following this in **'the way of IEC 60071-1'**, the standard SIWL equivalent to TOV range from 2383 to 2566 kV for these UHV projects and they require a standard switching impulse withstand voltages of a least 2400 kV in the UHV systems. Since SIWVs greater than LIWVs are economically and technically unrealistic and not feasible the WG is of the view that the power frequency voltage test, as an alternative test, is available to verify the withstand capability concerning TOV.

Details are:

	China			Japan			India	
	LIWV	SIWV		LIWV	SIWV		LIWV	SIWV
Transformers	2250 kV	1800 kV	Transformers	1950 kV	1425 kV	Transformers	2250 kV	1800 kV
GIS	2400 kV	1800 kV	GIS	2250 kV	1550 kV	MTS	2400 kV	1800 kV
MOSA	V20 kA = 1620 kA		MOSA	V20 kA = 1620 kV		MOSA	V20 kA = 1700 kV	
(a) 1100 kV Jindongnan substation			(b) 1100 kV Shin-Haruna Testing Station			(c) 1200 kV BinaTesting Station		

Table 3.2 (Continued)

*[The original article also provides photographs of each installation plus [i] a few U50 flashover voltage and time to peak results for 7 and 10 m gaps, for the tower configuration in India and [ii] some laboratory-measured trapped charge voltages during disconnector switch [DS] switching under laboratory conditions (Comparison between measured and calculated values at 1100 kV).]

Conclusion: WG C4.306 has reviewed and discussed insulation coordination practices in the UHV system taking into account the state-of-the-art technology. Recommendations, such as recent practices of insulation coordination based on the higher performance surge arrester, estimation of overvoltage and air clearance will be proposed for the application guide IEC 60071-2 (1996) and apparatus standards.

Note: [Numerous worldwide technical contributions are generally made, from various interested sectors, during the 'transparent IEC standards revision process' and the views of this CIGRE WG C4.306 will undoubtedly carry much weight.]
Source: After Zaima, E. et al. (CIGRE Electra WG C4.306) [49].

The mean breakdown stress, i.e. V/g, in large practical gaps (Figure 3.4) is typically 500 V/mm, somewhat reduced from the 2 kV/mm in sphere gaps. This can decrease even further in wet weather, such that breakdown of air under lightning conditions can occur at stresses as low as 20 V/mm. In overhead line design [12], it is necessary to design conductor bundles such that the stress at the surface of the bundle is less than that to initiate corona, otherwise radio noise would be generated and losses and corrosion would increase. An indication of surface stress levels of

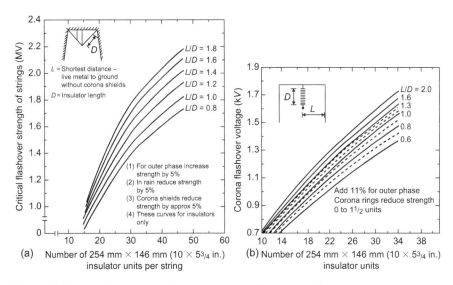

Figure 3.7 Switching surge characteristics of string insulators
(a) Flashover strength; (b) along tangent strings
See EHV transmission line reference book, Edison Electric Institute, New York, 1968

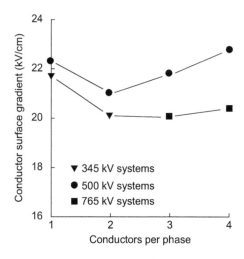

Figure 3.8 Typical surface gradients on overhead transmission line conductors. See EHV transmission line reference book, Edison Electric Institute, New York, 1968

various UHV conductor bundle designs is given in Figure 3.8. Special problems to be overcome before effective overvoltage protection can be achieved in practice are considered elsewhere [12].

Historically, it should be noted that for many years, during the IET School on High Voltage Engineering and Testing, a visit was made to the Clothier UHV Laboratory, Hebburn, Tyne and Wear, UK, to witness examples of pre-breakdown discharge activities in large practical gaps for AC, direct current (DC) and impulse voltages. This workshop complemented lectures by Messrs, Hughes, Hurst, Gockenbach, Allen and Blackett and was relevant to the materials covered in Chapters 15–19.

3.2.5 Guidelines for live working

Live working [16] allows necessary maintenance and repairs to be performed on high-voltage systems without the necessity for interruption of the power supply to customer(s). To achieve effective live working, safety of workers and total reliability of the operation are vital. To this end, 'safe working distances', i.e. the air insulation distances to be maintained in the various 'working configurations' in order that the risk of occurrence of flashovers, which (i) could possibly be harmful to all personnel working nearby and (ii) could cause disturbances to the system, are sufficiently short.

Safe working distances can be determined by applying the principles of insulation co-ordination, which is based on [16] three main elements:

1. knowledge of stresses, which occur at worksite
2. knowledge of the electrical strength of the worksite insulation when subjected to such stresses
3. assessment of the probability of occurrence of insulation failures in the situation under consideration of stresses and strength.

A recent CIGRE Brochure, summarised in Reference 16, deals with the insulation in live working and also addresses special considerations, namely safety aspects related to the exposure of workers to electric and magnetic fields (EMF). It provides:

a. a general subject overview
b. a discussion on voltage stresses at the worksite and the effect of the adoption of overvoltage limiting measures and devices
c. dielectric withstand characteristics of the worksite insulation, starting from the reference conditions and addressing the influence of conducting bodies at floating potentials of damaged insulators and insulating tools and of atmospheric conditions and altitude
d. procedure for evaluation of the risk of failure during live working and for the assessment of working distances (case study examples are given of application)
e. assessment of minimum working distances as from viewpoint of exposure of workers to EMF.

In live working, the main concern is the performance under transient overvoltages. These activities are usually restricted to periods of 'good' weather. Consequently, lightning overvoltages are generally not important, and there is normally little risk of power frequency pollution flashover [16].

Finally, from both health and safety and environmental perspectives, the power industry has always carefully considered 50/60 Hz magnetic field effects on the human central nervous system and results over the years, based on many studies and publications, have been totally reassuring to the power utility sector. The reader's attention is drawn to two recent articles well worth reading:

1. 'Multi-modalites investigation of 60 Hz magnetic field effects on the human central nervous System', *Electra*. 2011;**256**(June):4–18 (with 45 references). Contributors: A. Legros, J. Miller, J. Modolo, M. Corbacio, J. Robertson, D. Goulet, J. Lambrozo, M. Plante, M. Souques, F.S. Prato and A.W. Thomas.
2. 'High and extra high voltage lines, health and the environment', Invited Paper, *Electra*. 2011;**256**(June):20–3, by Daniel Raoul (Senator of MAINE-ET-LOIRE (France District)).

Note: Should the reader wish to pursue this topic further there have been many publications over the years including IEEE, IET CIGRE and UK Health and Safety Executive.

3.3 Other gases

The availability and characteristics of air make it the most used gaseous insulant. In enclosed equipment, e.g. heavy duty gas-blast circuit-breakers, by using air at high gas pressures, the dielectric performance, etc., could be greatly enhanced. However, because of limitations, gases other than air have been considered. The most popular of these has been SF_6, which has dielectric strength about twice as good as air and also offers excellent thermal and arc interruption characteristics (see Chapter 7). Despite this, the search continues for other gases with improved characteristics.

112 High-voltage engineering and testing

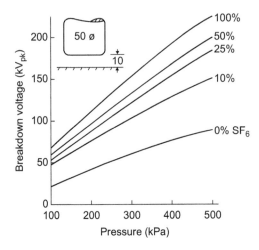

Figure 3.9 Uniform-field breakdown voltages for SF_6–air mixtures

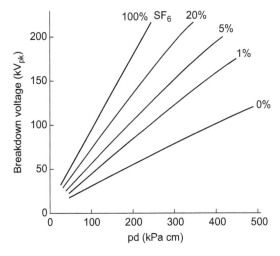

Figure 3.10 Paschen curves for SF_6–N_2 mixtures at 20 °C (50 Hz)

There has been considerable research activity (e.g. see Reference 17) to find alternative gases and gas mixtures, which may have comparable breakdown strengths as SF_6 (see Figures 3.9 and 3.10) but can offer technical, environmental and economic advantages. Binary mixtures with inexpensive gases like N_2, air, CO_2 and N_2O have been under continuous investigation to arrive at efficient and economical mixtures and obtain a better understanding of the dielectric processes involved. The mixture of SF_6 and N_2 is the only mixture to achieve significant commercial application in switchgear, mainly to overcome problems associated with low-ambient temperature conditions (in Canada, for example).

Environmental constraints are now assuming greater significance and these have resulted in great attention being necessary for packaging, transport and reclaiming of used SF_6 and renewed interest in the choice of N_2/SF_6 mixtures for GIS. The attraction of such mixtures is that N_2 gas predominates, enabling the amount of SF_6 used to be kept to a minimum. Much work was carried out in the 1970s and 1980s by switchgear manufacturers, universities and others, and dielectric characteristics were obtained for a wide range of electrode configurations, some similar to those shown in Figures 3.9–3.11. These studies revealed that the dielectric withstand characteristics of N_2 increase rapidly with only a small percentage of SF_6 present. The commercial and environmental implication is that N_2/SF_6 mixtures with <20% SF_6 can achieve a 'significant' dielectric withstand performance for a greatly reduced SF_6 gas requirement (see also 'Guide for SF_6 mixtures', CIGRE Report 163, August 2000, by Working Group 23.02; Task Force 01).

In the present extremely environmentally conscious climate (e.g. concern regarding the depletion of Earth-shielding ozone by chlorofluorocarbons (CFCs) and the resulting large 'ozone hole' having opened up over Antarctica covering up to 11 million square miles), a major reduction of SF_6 gas requirement for large substations appears attractive to many people and this could well provide the 'incentive' for a strategic re-evaluation (see Figure 3.11) of N_2/SF_6 mixtures for GIS. Similarly, present-day difficulties during planning/commissioning of certain new overhead line(s), due partly to delays caused by consideration of a plethora of 'alleged' environmental concerns raised by small protest groups, have resulted in serious consideration being given to the adoption of an alternative line technology, the so-called gas-insulated line (GIL) approach. N_2/SF_6 gas mixtures have again been suggested and evaluation studies are currently being carried out by EDF and others in Europe.

Ternary gaseous dielectrics have been extensively investigated, such as N_2+SF_6 or CHF_3+SF_6, in conjunction with perfluorocarbons. Fluorocarbons, when mixed with SF_6, may exhibit interesting synergistic effects, which at present are not fully explained. A variety of gaseous insulating systems have been investigated (see CIGRE Report 163, referred to above), comprising multi-component gas mixtures carefully selected on the basis of physicochemical knowledge, especially on the interactions of low-energy electrons with atoms and molecules. Much valuable research work in this field has been carried out by a group, led for many years by L.G. Christophorou, at Oak Ridge National Laboratory, Knoxville, Tennessee, USA.

3.4 Switchgear and GIS

3.4.1 Introduction

High-voltage switchgear, which forms an integral part of any substation (i.e. switching station), is essentially a combination of switching and measuring devices, (CBs, VTs, CTs etc) [2, 19]. The switches (circuit-breakers) connect and disconnect the circuits and the measuring devices (instrument transformers) monitor the system and detect faults. Particular care must be taken of the selection of switchgear since security of power supply is dependent on the reliability of all switching and measuring devices. The reliability of any circuit-breaker depends on insulation security, circuit-breaking capability, mechanical design and current-carrying capacity. Considering modern switchgear practice, circuit-breakers can be broadly classified according to the insulating medium used for arc-extinction: bulk-oil, air-break, SF_6 gas, small oil-volume, air-blast and vacuum.

114   High-voltage engineering and testing

Several types exist for indoor or outdoor service conditions. There are established relationships between types of switchgear selected and the design of the system. Briefly, until about 1980: (i) for voltages up to 11 kV, most circuit-breakers were of the oil-break type, or of the air-break type, using air at atmospheric pressure; (ii) from 11 to 66 kV, oil circuit-breakers were mainly used while, at 132 and 275 kV, the market was shared by oil interrupters (both bulk-oil and small oil-volume) and gas-blast circuit-breakers.

Recent developments have seen a move towards 'oil-free' devices and the increased demand for vacuum and 'rotating-arc' SF_6 circuit-breakers for distribution systems and the increased use of SF_6 circuit-breakers for system voltages

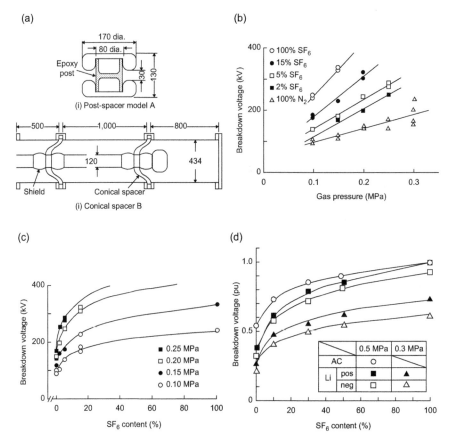

Figure 3.11   Further breakdown characteristics in N_2–SF_6 mixtures [18]
(a) Spacers
(b) Pressure dependency of surface breakdown voltages of post-spacer model A under clean condition; neg. LI, SF_6/N_2 mixed gas
(c) Dependence of negative LI surface breakdown voltages on SF_6 content under clean condition; post-spacer model A, SF_6/N_2 mixed gas
(d) Dependence of breakdown voltages of conical spacer B on SF_6 content under clean condition in SF_6/N_2 mixed gas

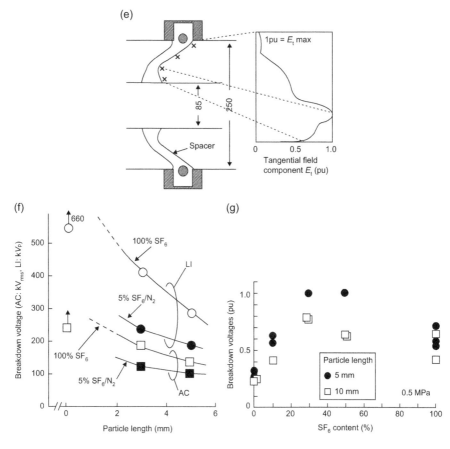

Figure 3.11 Further breakdown characteristics in N_2–SF_6 mixtures [18]
  (e) Conical spacer model C; × = attached particles
  (f) Particle-initiated breakdown voltages of conical spacer C at 0.4 MPa
  (g) Particle-initiated AC breakdown voltages of conical spacer B in SF_6/N_2 mixed gas

66 kV/132 kV, up to and beyond 565 kV. However, it should be appreciated that modular 'small-oil-volume' designs proved to be popular for all but the highest ratings [2] and that earlier generation of 420 kV air-blast circuit-breaker designs, originally developed in the 1960s, is likely to remain in service in the UK grid for at least the first 20 years of this new millennium. Later chapters will discuss transmission and distribution switchgear, substation design, condition monitoring techniques together with the timely and topical themes of 'life management' of network assets and 'working systems harder' [20, 21].

Historically, it is worth noting that, in the 1960s, the second-generation air-blast transmission circuit-breakers for 420 kV had six breaks per phase [2]. As stated in the previous paragraph, many of these designs are still in service in the UK grid. The prediction that such interrupters would develop to higher rating was not fulfilled and the recent trend has been towards compact SF_6 switchgear. By incremental research, it has been possible to develop SF_6 dead-tank circuit-breakers for 550 kV, 63 kA rating, with only two breaks per phase [22, 23]. Recent developments are discussed in Chapters 7–10 and 20–23.

Notes
1. The reader may find it useful to note, and if relevant to read, two CIGRE Technical Brochure documents relating to the appropriate handling and disposal of SF_6 gas, namely (a) TB 430-WG B3.18, 'SF_6 Tightness Guide' (2010) and (b) TB 234-TF B3.02.01, 'SF6 Recycling Guide, Revised Version' (2003). Regarding (b) it is important for the reader to recognise all significant aspects of SF_6 recycling and the essential procedures as to how the gas can be reclaimed to allow its safe reuse.
2. Regarding possible SF_6 impacts on the environment, it should also be noted that CIGRE SC B3.30 considers it necessary to minimise the use of SF_6. In particular, during routine testing of electrical equipment a limited volume of SF_6 may be released. Consequently, a CIGRE WG TB guide is being prepared called 'Guide to Minimising the Use of SF_6 during Routine Testing of Electrical Equipment'. (This Technical Brochure should be issued in the next few months.)

3.4.2 Arc extinction media

The basic construction of any circuit-breaker entails the separation of contacts in an insulating 'fluid'. The insulating 'fluid' that fills the circuit-breaker chamber must fulfill a dual function. First, it must extinguish the arc drawn between the contacts when the circuit-breaker opens, and second, it must provide adequate, and totally reliable, electrical insulation between the contacts and from each contact to earth.

Selection of arc extinction 'fluid' is dependent on the rating and type of circuit-breaker. The insulating media commonly used for circuit-breakers are:

- air (at atmospheric pressure)
- oil (which produces hydrogen for arc-extinction)
- compressed air (at pressure 7 MPa)
- sulphur hexafluoride, SF_6 (at pressure <0.85 MPa)
- vacuum.

The principle of circuit interruption techniques and their application in circuit-breakers, either of single or of multiple break design, is described in Chapters 7–11 and elsewhere [2, 22, 23]. However, it is appropriate to comment briefly here on interruption techniques, relevant to SF_6 circuit-breakers, to illustrate basic principles relevant to the application of gaseous/solid insulation systems. The complex discharge products of SF_6 are discussed in Chapters 7–10. Earlier studies have described extensive research and development [2] on interrupter nozzle performance, relevant to the design of air-blast interrupters. Table 3.3 provides a convenient summary of single- or double-pressure SF_6 interrupter processes [4] together with a brief explanation of live-tank and dead-tank design philosophies.

Table 3.3 SF$_6$ circuit-breaker concepts [4]

Single or double-pressure

The flow of SF$_6$ gas necessary for circuit-breaking may be produced in either of two ways. In both cases, however, the actual process of current interruption is similar, i.e. the arc is established through a nozzle by the separation of the contacts and/or gas flow and is subjected to an 'axial' gas-blast, which abstracts energy from the arc, resulting in extinction at a current zero.

The gas flow may be achieved by the use of a two-pressure system (Figure A) whereby the operation of a blast valve allows gas to flow from a high-pressure reservoir through a nozzle into a low-pressure reservoir. The alternative is a single-pressure system (Figure B) where compression of the gas is caused by movement of a cylinder over a fixed piston (or vice versa). In this system, commonly called the 'puffer' (Figure C), the moving contact system and cylinder are usually joined together so that the movement of only a single component is necessary for arc initiation and interruption.

The single-pressure system is inherently less complicated than the double-pressure system, which requires two separate gas reservoirs with associated seals, a compressor and gas-handling system, and heaters to prevent liquefaction at low temperatures. Furthermore, in the double-pressure system there is the necessity of synchronising the blast-valve and the contact driving systems.

Very high unit interrupting ratings can now be achieved with both systems, though the high-performance single-pressure system requires a more powerful operating mechanism to ensure a sufficiently high velocity for the contact/cylinder arrangement. This,

Figure A   Double-pressure system

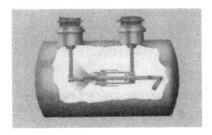

Figure B   Single-pressure system

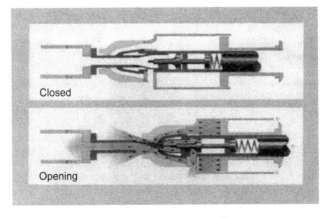

Figure C   Single-pressure puffer system

(*Continues*)

Table 3.3 (Continued)

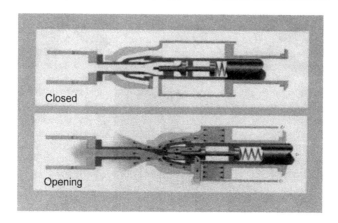

Figure D Partial duo-blast puffer system

however, is easily obtainable with either pneumatic or hydraulic power units, and at the lower performance levels, spring opening mechanisms also offer a practical alternative.

Initially the performance limitations of the single-pressure system meant that earlier types of circuit-breaker incorporated double-pressure interrupters; but the continuing development of the single-pressure puffer interrupter has led to designs that are capable of the highest ratings. It seems likely, therefore, that such interrupter systems will be the basis of most future EHV circuit-breakers. They will also find wider application in the high-voltage distribution field.

In any axial gas-blast interrupter, the unit performance may be significantly improved by using a dual-blast construction in which the arc is drawn through two nozzles. In this construction, the gas now is towards both arc roots, and metal vapour from the contacts is not blown into the arc column. In the partial duo-blast construction (Figure D) one of these nozzles has a smaller diameter, which reduces the amount of gas passed through the nozzles and hence also the mechanical energy input without a significant reduction in performance compared to the full duo-nozzle construction. The Reyrolle puffer interrupters incorporate partial duo-blast constructions.

Live or dead tank
The SF_6 interrupter can be incorporated in either live-tank or dead-tank circuit-breakers (Figure E). The type chosen will depend on economics and/or the type of application; e.g. the dead-tank construction (in which all interrupters are enclosed within an earthed pressure vessel) is essential for use in complete metal-clad installations, although the same breaker may be used with terminal bushings in open-type layouts. In assessing relative economics of live- and dead-tank constructions, the cost of any associated current transformers should be considered. Accommodation for current transformers is, of course, integral in the dead-tank circuit-breaker, but separate post-type units are usually necessary with the live-tank design for outdoor installations. With indoor installations the current transformers may be accommodated in the through-wall bushings or cable-sealing ends.

Table 3.3 (Continued)

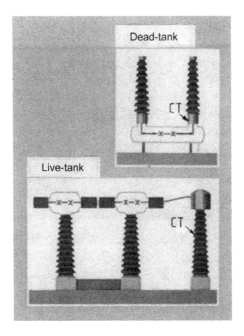

Figure E Live- and dead-tank circuit-breakers (Courtesy NEI Reyrolle Ltd)

Generally it might be expected that at the lower end of the voltage/current scale, where only one or two series breaks per phase are required, i.e. up to 420 kV the live-tank construction is probably more economical; but at higher voltages with three or four (or more) series breaks, the dead-tank construction tends to have an economic advantage because of the reduced amount of external insulation. It is worthwhile noting that dead-tank circuit-breakers are inherently more suitable for areas subject to earthquakes.

3.4.3 General dielectric considerations

Because of the widespread use of SF_6-insulated GIS (see Figures 3.12 and 3.13), it is appropriate to touch briefly on some general points relating to the dielectric performance of gas-gaps and gas/solid insulation interfaces in SF_6 under clean and contaminated conditions. The literature on this subject is vast, and extensive bibliographies exist (References 14, 25–30, for example), listing some important papers, which can be arbitrarily grouped into topic areas: particulate contamination and detection, diagnostic techniques, operating experience, breakdown studies, internal-arcing, spacer experience and dielectric discharge performance.

3.4.3.1 Dielectric withstand capabilities, gas-gap data

Figure 3.14 shows a selection of typical SF_6 breakdown characteristics for the electrode systems illustrated in Table 3.4 [14]. From the vast amount of breakdown data

120 *High-voltage engineering and testing*

Typical duplicate bus circuit

1. Circuit-breaker
2. Interruptor
3. Hydraulic mechanism
4. Disconnecter (circuit)
5. Disconnecter (bus selector)
6. MES
7. FMES
8. Gas console
9. Current transformers
10. Bursting disk
11. Voltage transformer
12. Reserve busbar
13. Main busbar

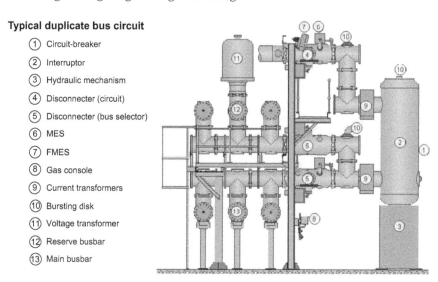

Figure 3.12 Sectional view through a typical 420/550 kV GIS duplicate busbar circuit (Courtesy VA Tech Reyrolle)

Figure 3.13 Example of 420 kV GIS equipment on a recent substation project in the Middle East (Courtesy VA Tech Reyrolle)

Table 3.4 Electrode systems [14]

Electrode system	Description	Shape	Size (mm)	HV electrode material
A, B, C	Concentric/eccentric cylinders		$R/r = 190/70$ $S = 0, g = 120$ $S = 20, g = 100$ $S = 30, g = 90$	Aluminium
D	Perturbed cylinders	$R_2/R_1 = 242.5/112.5$ $g = 130$	—	Stainless steel
E	Perturbed cylinders	$R_2/R_1 = 317.5/250$ $g = 67.5$	—	Brass
F	Perturbed cylinders	$R_2/R_1 = 242.5/125$ $g = 117.5$	—	Copper
G	Support* barrier		—	
H	GISRIG		$g \approx 160$	
I	Concentric[†] cylinders		$R = 224$ $r = 64$ $g = 160$	Aluminium
J	Eccentric cylinders		$R = 285$ $r = 31.5$ $g = 40$	Aluminium

*Various resin formulations have been studied, including, silica, alumina, bauxite and dolomite filled systems.
[†]Using corrugated outer cylinder.

122 High-voltage engineering and testing

Waveshape	Electrode systems	Waveshape	Electrode systems
1 L1 +ve	J	7 SWI +ve	E
2 L1 −ve	J	8 SWI −ve	E
3 L1 +ve	A	9 LI +ve	C
4 L1 −ve	A	10 LI +ve	C
5 L1 +ve	E	11 AC 50 Hz	I,G,H
6 L1 −ve	E	12 LI −ve	I,G,H

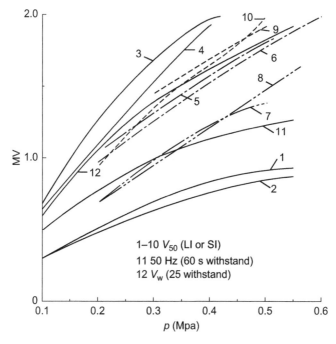

1–10 V_{50} (LI or SI)
11 50 Hz (60 s withstand)
12 V_w (25 withstand)

Figure 3.14 Example of SF_6 breakdown characteristics [14]

accrued, the limits of lightning and switching impulse withstand gradient performance for a family of large practical GIS gas-gap type configurations were produced and are illustrated in Figure 3.15 for SF_6 pressures in the range $0.1 < p < 0.6$ MPa. Experimental determination of critical breakdown (E_{50}) and highest withstand (E_w) gradient values was obtained using similar test techniques to those adopted previously [14] (i.e. $E = V/\eta g$; see section 3.5.2). It should be emphasised that these curves present typical design-type gradient relationships and encompass the results obtained for a large family of coaxial cylinder and perturbed electrode configurations for gas-gaps in the range $50 < g < 180$ mm. Theoretical breakdown gradient levels are also given in Figure 3.15(a–d); it is readily apparent that practical results deviate from theory as SF_6 gas pressure increases [14]. A corresponding withstand curve for 50 Hz conditions is also shown in both Figures 3.14 and 3.15.

Applications of gaseous insulants to switchgear 123

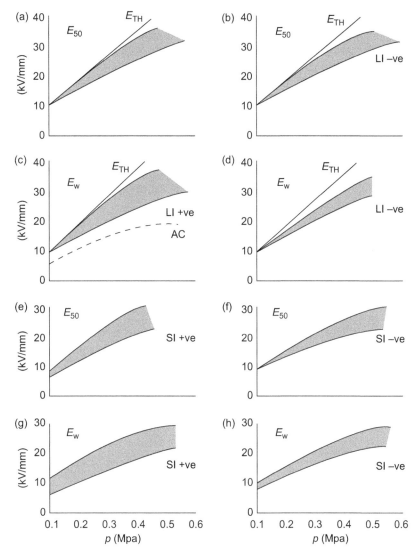

Figure 3.15 Typical limits of 50% breakdown gradient (E_{50}) and critical withstand gradient (E_w) on SF_6 pressure for large coaxial and perturbed cylindrical electrode systems under clean condition [14]
Curves (a–d): lightning impulse (LI) waveshape (1.2/50 μs)
Curves (e–h): switching impulse (SI) waveshape (250/2,500 μs)
(------) shown in curve (c): the lower limiting 50 Hz withstand characteristic (E_w)
E_{50} data: curves a, b, e and f; E_w data: curves c, d, g and h

To provide a better understanding of statistical variations possible with 50 Hz dielectric breakdown characteristics, relating to practical GIS-type assemblies, it is relevant to consider the following results [14]. Figures 3.16–3.18 show the variation in individual short-term 50 Hz breakdown performance, with repeated sparking, for large perturbed and unperturbed cylindrical electrode systems. For comparison, the corresponding 60 s highest withstand levels (V_w) are also given. Some salient findings emerge.

1. Figure 3.16, curves (a–c), relates to a 50 Hz voltage test sequence for three 'successive' test series for a concentric cylindrical electrode system.
    i. Curve (a) illustrates that whereas the highest '60 s withstand level' V_w was 480.8 kV_{pk}, the sequence of '25 instantaneous' breakdown levels was significantly higher, being within the range 520–595 kV_{pk}. For the purposes of comparison, the theoretical critical breakdown voltage level (V_{TH}) is 621.6 kV_{pk}, based on a critical $(E/p)_{lim} = 89$ kV mm^{-1} MPa^{-1}.
    ii. Curves (b) and (c) show corresponding results for two repeat series. Here, it can be seen that the corresponding '60 s withstand' levels (V_w) increased significantly to 565.7 and 594 kV_{pk}, respectively, as compared to the level of 480.8 kV_{pk} obtainable in the first test series, i.e. curve (a).
    iii. Similarly, it can be seen from the sequence of individual 'instantaneous breakdown' levels (V_b) that, despite occasional low-level breakdowns, a significant 'conditioning' effect has taken place and, in curve (c), approximately 10 of the 25 'instantaneous breakdown' levels have achieved breakdown values of ≈ 600 kV_{pk}, i.e. within 3.5% of the theoretical limiting breakdown level.

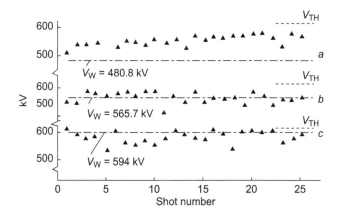

Figure 3.16 Sequence of 50 Hz breakdown levels in SF_6 for concentric cylinder electrode system F (SF_6 pressure: 0.1 MPa)
--- V_{TH} Theoretical breakdown level (621.6 kV)
(Based on $(E/p)_{lim} = 89$ kV mm^{-1} MPa^{-1})
▲ Individual spark breakdown
*p 'PIP' – partial (incomplete) breakdown
V_w Maximum (1 min) withstand level, established immediately prior to test runs a, b or c, respectively

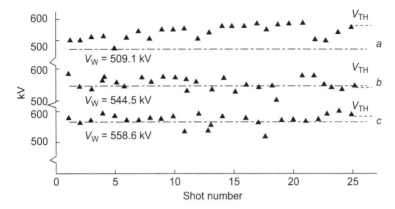

Figure 3.17 Sequence of 50 Hz breakdown levels in SF_6 for perturbed electrode system F (SF_6 pressure: 0.1 MPa)
--- V_{TH} Theoretical breakdown level (574 kV)
(Based on $(E/p)_{lim} = 89$ kV mm^{-1} MPa^{-1})
▲   Individual spark breakdown
*p   'PIP' – partial (incomplete) breakdown
V_w Maximum (1 min) withstand level, established immediately prior to test runs a, b or c, respectively

2. Figure 3.17 shows comparable 50 Hz results for a perturbed concentric cylinder electrode system (F). As before, the results presented in curves (a)–(c), respectively, relate to three complete test series. Once again, the sequence of 'instantaneous breakdown' results exhibits a noticeable voltage 'conditioning' effect, with occasional low-level results, e.g. curve (c), the conditioned level of ≈590 kV_{pk} being within 3% of the theoretical breakdown value. Returning briefly to the sequence of highest '60 s withstand' levels for the test series (a)–(c) (Figure 3.17), it is noted that V_w increases as the test series proceeds, corresponding values being 509.1, 544.5 and 558.6 kV_{pk}, respectively.

3. Figure 3.18 provides comparable results, obtained at higher SF_6 gas pressures. Under these conditions, 'instantaneous breakdown' values were significantly lower than corresponding theoretical levels, e.g. being lower at 0.2 MPa, by between 15% and 31%, for this electrode arrangement.

4. Figure 3.19 shows some further results for a larger concentric cylinder electrode system (I) (see Table 3.4), which illustrate similar trends at 0.1 MPa but show a very significant spread in the SF_6 'instantaneous breakdown' levels as compared to the '60 s withstand' level (V_w) of 795 kV_{RMS}, at the higher gas pressure 0.45 MPa (see curves (a), (d) and (e)). Here, the values of V_w tend to be lower than the theoretical limiting levels V_{TH} by between 18% and 60%, the difference increasing with SF_6 pressure. It should also be noted (curve (a)) that the lowest recorded breakdown level corresponds to a stress figure of ≈ 8 kV_{RMS}/mm, significantly higher than the working stresses for GIS equipment referred to earlier.

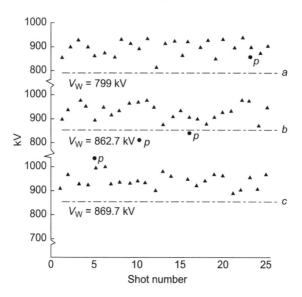

Figure 3.18 Sequence of 50 Hz breakdown levels in SF_6 for perturbed electrode system F (SF_6 pressure: 0.2 MPa)
  -- V_{TH} Theoretical breakdown level (1,148 kV)
  (Based on $(E/p)_{lim} = 89$ kV mm^{-1} MPa^{-1})
  ▲ Individual spark breakdown
  *p 'PIP' – partial (incomplete) breakdown
  V_w Maximum (1 min) withstand level, established immediately prior to test runs a, b or c, respectively

Extensive data exists for SF_6 gas-gaps under clean and contaminated conditions (see References 14, 25–30, for example). It is now established that the presence of particulate contamination of lengths 2–20 mm can reduce the dielectric withstand capabilities of practical gaps by varying amounts up to typically 30%, 40% and 70% for lightning impulse, switching impulse and power frequency conditions, respectively, at working SF_6 pressure. Figure 3.20 illustrates a typical spread in 50 Hz flashover levels for varying degrees of gross contamination and represents the maximum lowering of withstand that can be expected [14].

Barrier performance data: Careful design and assembly of the cast resin support barriers used in SF_6-insulated GIS equipment are vitally important. It should be noted that, for a particular gas pressure, the withstand characteristics of support barriers in SF_6 under clean conditions depend on the particular resin formulation used, the insulation shape and the disposition of stress-relieving fitting, insert, etc. Typical withstand gradient levels of 11.6, 8.7 and 6.6 kV$_{pk}$/mm can be achieved under lightning, switching impulse and 50 Hz short-term voltage conditions, respectively [14].

Applications of gaseous insulants to switchgear 127

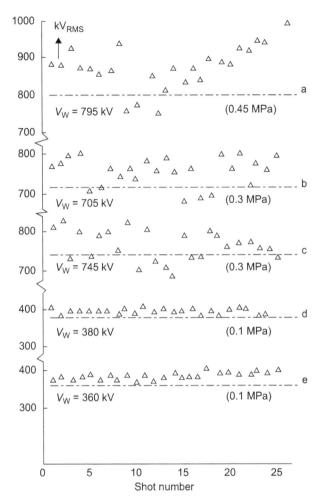

Figure 3.19 Sequence of 50 Hz breakdown levels in SF_6 for large concentric cylinders (electrode system I, SF_6 pressure: 0.1–0.45 MPa)
△ Individual spark breakdown
V_w Maximum (1 min) withstand level, established immediately prior to test run

The presence of particulate contamination can reduce the 50 Hz withstand capability of cast-resin support barriers in SF_6 gas by varying amounts (e.g. up to <30%) depending on particulate size and disposition. For the most onerous dispositions of cast-resin support barriers, the percentage lowering of 'withstand' performance under impulse conditions tends to be much less than that experienced for 50 Hz test conditions, for comparable levels of contamination (see Figures 3.20 and 3.21, for example) at spacer gas-gap interfaces. The reader should compare these

128 High-voltage engineering and testing

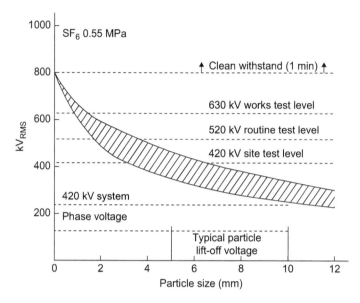

Figure 3.20 Fifty Hertz flashover characteristics of epoxy resin conical spacers under varying degrees of metallic contamination [10]

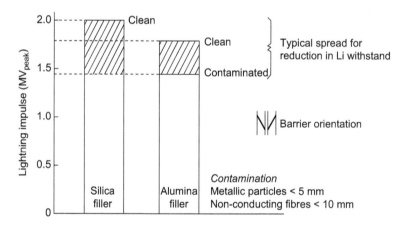

Figure 3.21 Limit of lightning impulse withstand capabilities of epoxy resin conical spacers under clean and contaminated conditions

findings for 100% SF_6 with Figure 3.11, diagrams (e)–(g), relating to particle-initiated breakdown in SF_6/N_2 mixtures. These workers [18] consider that particle-initiated breakdown is more complicated in SF_6/N_2 mixed gas than in 100% SF_6 gas and they recommend that further studies are necessary to achieve a better understanding of the discharge mechanisms.

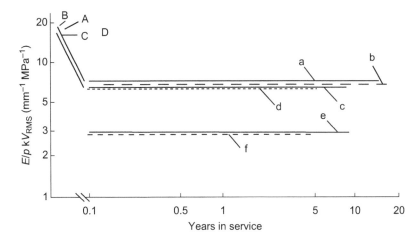

Figure 3.22 Normalised 50 Hz service E/p levels for SF_6-insulated switchgear equipment

    a, A   420 kV post-type current transformers
    b, B   300 kV post-type current transformers
    e, c, C   300 kV GIS equipment
    f, d, D   420 kV GIS equipment
    a, b, c, d, e   service stress levels
    A, B, C, D   factory 50 Hz 1 min test level at working pressure
Working pressure:
    a, A   0.38 MPa
    b, B   0.38 MPa
    c, C   0.40 MPa
    d, D   0.55 MPa

3.4.3.2 GIS equipment: power frequency (*V/t*) characteristics

An understanding of voltage gradients that can be safely sustained in service in GIS emerges from extensive laboratory studies, relating specifically to long-time 50 Hz test conditions. The ratio E/p is a convenient measure of the stress applied to GIS equipment in service. Typical normalised working stress/gas pressure ratios (E/p) are 7 and 3 MV/MPa for SF_6-insulated switchgear including instrument transformers, for 300 and 420 kV systems [4, 14], are shown in Figure 3.22.[1] This (2001) figure was originally prepared in 1983, at which time many of these units had been in service for periods up to 25 years and excellent service reliability had been demonstrated. It should be noted that this equipment continues to provide excellent service reliability, some 30+ years after going into service [14]!

---

[1] Figure 3.22 is unchanged from the 2nd Edition (2001); consequently, some units have now been in service for periods up to 30 years.

Regarding 'GIS-State of the Art 2008' Technical Brochure TB 381-WG B3.17, a summary in *Electra*, 2009;**244** (June):50–59, informed the readership that:

1. More than 80,000 bays of GIS have been installed worldwide
2. There are more than 1 million bay-years of GIS in service operation.

It is readily apparent (see Figure 3.22) that the maximum service stress levels for gas-gaps in SF_6-insulated switchgear are relatively low, i.e. <4 kV_{RMS}/mm, when compared to the significantly higher attainable withstand characteristics achievable (see curves, Figure 3.15). The working stress levels shown in Figure 3.22 can be considered to be generally representative of modern GIS installations. By restricting these gradient levels well below the limiting values, deduced from wide-ranging long-term laboratory studies, manufacturers ensure the long-term dielectric integrity of their equipment – provided, of course, that the normal rigorous quality control and in-service condition-monitoring procedures have been maintained.

The importance of component cleanliness, achieved during factory construction, testing, site assembly and commissioning of GIS, and subsequently throughout the entire service life of SF_6-insulated equipment, is now fully appreciated worldwide. Cleanliness, linked with good design, assembly and the introduction of sophisticated 'in-service condition-monitoring' practices are recognised as being vitally important factors to resolve if safe and reliable operation is to be achieved throughout the service lifetime of GIS, which should exceed >40 years. In recent years there has been much progress in the development of effective UHF condition-monitoring techniques [31–33]. These strategic developments are considered further in several of the later chapters.

Critical areas of metalclad designs have been identified, which merit special manufacturing, testing and assembly controls, and these aspects will be further considered in section 3.5 and also in later chapters. Several papers have considered the power frequency (V/t) characteristics of gaseous and solid insulation using model gas-gaps and gas/insulator arrangement for large electrode systems under conditions with 'gross' contamination present and with varying gas pressures [14].

3.4.4 Performance under contaminated conditions

As has been demonstrated above, the achievable design stress and reliability of SF_6-insulated apparatus under normal power frequency service conditions are crucially affected by particulate contamination. Particles in the gas space and on the insulator surface can significantly lower the dielectric strength of the system. In SF_6, the breakdown voltage for a relatively long particle, fixed in contact with the conductor, is considerably higher than that for a number of free conducting particles, over a limited range of pressures. Under power frequency voltage excitation, free conducting particles tend to bounce along the bottom of the enclosure or across the surface of the insulators. The amplitude of individual bounces depends on the particle size and shape, the potential of the system and other random parameters. Particle-initiated breakdown can occur at voltages considerably lower than those required for breakdown due to the roughness of the electrode surfaces and is, in general, lower for negative impulse polarity voltages [28]. Since then, switchgear and CGIT manufacturers – in support of GIS and GIL equipment developments – have undertaken comprehensive evaluation of the dielectric performance characteristics of practical gas-gap arrangements and solid support insulation configurations in SF_6, under both clean and contaminated conditions. Such studies can involve comprehensive laboratory tests to determine, and quantify, the effects of particulate contamination size on the breakdown voltage in SF_6, under conditions representative of both gas-insulated 'backparts', circuit-breakers, disconnect switches, etc., and for GIL configurations, for widely varying experimental conditions.

Numerous interrelated factors can influence the degree to which the presence of particles can lower the dielectric 'withstand capabilities' of GIS under normal service conditions. These depend on:

- length and diameter of particles present
- whether particles are metallic or non-metallic
- quantity and nature of contaminant material (density, etc.)
- position of particles relative to electric field and also to various GIS components
- actual design and physical disposition of GIS components (e.g. whether circuit-breaker, backparts, etc., are mounted horizontally or vertically)
- type of particle movement, when electrostatic forces exceed those of gravity
- working gas pressure of SF_6 or gas mixtures
- effectiveness of particle 'collection' or 'trapping' techniques.

The above factors are now well known and appropriate procedures are implemented to avoid any problems (see section 3.4.3.2). Manufacturers and utilities take immense care in the design, testing, site-commissioning and also the ongoing 'in-service performance monitoring of GIS' throughout the lifetime of the equipment. From a user viewpoint, it is important at the outset to be able to plan ahead and evaluate/decide on the appropriate GIS performance 'data' needs, which must be measured/monitored 'in-service' on which to judge investment and maintenance decisions throughout the lifetime of the equipment. To cover these strategic aspects, very high frequency, UHF, condition monitoring techniques have been developed and installed in many GIS installations worldwide [31–33]. Currently, the need, justification, design, dependability, management of information and future application of monitoring and diagnostic techniques are under critical review by CIGRE [33].

3.4.5 GIS service reliability

A recently issued report [34], briefly outlining the findings of a second major CIGRE survey, states that GIS technology has contributed very effectively to increasing the reliability of new substations and to improving the asset life cycle of existing ones.

> A later extensive study on high-voltage substations in CIGRE Technical Brochure TB 381-WG B3.17, 'GIS State of the Art 2008' (see also *Electra*, 2009, **244**, June, pp. 50–59), provides a database of GIS service experience on SF_6-insulated equipment, covering collected information referring to more than 80,000 bays installed worldwide and 1,000,000 bay-years of in-service service operation. Current figures at the end of 2012 will now be approximately 83,000 bays and 1.3 million bay-years, respectively.

The earlier survey [34] presents a valuable history, providing an analysis of results on installation and GIS major failure reports data, including:

1. general data about GIS installations
2. data concerning GIS failure frequencies:
    i. failure frequency overview, trend of major failure frequency during the GIS lifetime
    ii. failure frequency-comparison with first CIGRE, GIS survey
3. major failure characteristics:
    i. basic characteristic data overvoltage classes (identification of main component, or GIS part involved in the failure, identification of sub-assembly or component responsible for the failure)
    ii. classification of symptoms, cause of the failure, service circumstances, operational circumstances
    iii. type of repair, immediate consequences of the failure, characteristics of the repair, correlation between basic characteristic failure data.

Also considered in this survey were life expectancy, maintenance and environmental issues:

1. life expectancy:
    i. for already installed GIS
    ii. for newly installed GIS
2. maintenance practice:
    i. routine preventive maintenance
    ii. major preventive maintenance
3. extension and uprating
4. environmental issues:
    i. SF_6 handling
    ii. SF_6 leakage rates (footnote 2)
    iii. analysis of failure relating to SF_6 gas
5. electromagnetic phenomena.

It is hoped that the data presented [34] will provide a valuable resource and benchmark for both users and manufacturers operating within the field of GIS substation planning, design, construction and service. This is undoubtedly a valuable document for anyone interested in the application of gaseous insulation. Nevertheless, it must be recognised that the document has certain shortcomings, due to commercial and statistical constraints, which prevent one obtaining a totally 'holistic understanding' of all failure patterns, etc.

Returning to TB 381-WG B3.17, 'GIS State of the Art 2008', this more recent document provides a comprehensive overview relating to the state of the art of GIS

---

[2] The reader can refer to two CIGRE Technical Brochure information documents relating to the appropriate handling and disposal of SF_6 gas, namely (a) TB 430-WG B3.18, 'SF_6 Tightness Guide' (2010) and (b) TB 234-TF B3.02.01, 'SF_6 Recycling Guide, Revised Version' (2003). Regarding (b) it is important to recognise all significant aspects of SF_6 recycling and the essential procedures as to how the gas can be reclaimed to allow its safe reuse.

at 2008: history, technology, solutions and applications including testing and functional specifications. It is intended to help the reader understand the evolution of GIS and thus enable users and suppliers to optimise the future utilisation of GIS.

It is anticipated that 75% of the world's population will live in 'ultra cities of the future' by 2050 and that both GIL and GIS will feature prominently. GIL technology is inbuilt into GIS substation techniques and approximately 750 bay-years of GIS enters service each year. (Approximately 12,000 bays in service for periods up to 30 years.) (*Note*: see Figure 3.22.) The reader is encouraged to monitor the recent strategic UHV and GIS developments worldwide, including the design and commissioning, in 2009, of GIS for use with 1,100 kV lines in China.

3.4.6 Gas-insulated transmission lines (GIL)

An invited paper by Koch and Schoeffner in *Electra*, 2003 [35], provides the reader with a useful overview of European developments with gas-insulated transmission lines (GIL). A brief personal historical overview is followed by a summary of the technical development, design criteria and existing and future applications. Briefly, the scope of the article is:

1 Introduction
2 Development Phase
2.1 Phase 1-Feasibility Study
2.2 Phase 2-Long Duration Test
    Directly Laid GIL
    Tunnel Laid GIL
2.3 Result of the Development Phase

3 Applications
3.1 Tunnel
3.2 Above Ground
3.3 Present and Future Applications
4 Conclusions
5 References (21 References supplied)

Photographs
1. HV Test Set Up for 420 kV GIL, at EDF Lab
2. Testing of the GIL
    (a) HV Tests, LI, SI and 50 Hz and Type Tests
    (b) For Short circuit and Arc Fault Test
3. Directly Buried GIL
4. [Laying GIL into Trench and Backfilling]
    [420/520 kV GIL; Tunnel for Long Term Testing]
5. First Application of 2nd Generation GIL
6. Sai Noi, Bangkok-Overview of Substation

Technical Data: Much strategic technical data was obtained in these GIL developments:
Successful tests on 420 kV GIL: Rated voltage 420 kV; rated current 3.15 kA, Isc = 63 kA for 0.5 s (Overload capacity $2.2I_N$ for 10 s and 1.9 times I_N for 1 h);
LIWV 1,425 kV; SIWV 1,050 kV; (gas mixture SF_6 maximum 20% N_2 minimum 80%); power frequency withstand voltage 630 kV.
Successful tests on 300 kV GIL: Rated voltage 300 kV; rated current 2,000 kA; LIWV 1,050 kV; SIWV 850 kV; (Palexpo, Geneva)
Power frequency withstand voltage 460 kV; short-circuit current 50 kA for 1 s. (Gas mixture 80% N_2; 20% SF_6; gas pressure 6 bar.)

The technical advantages of this gas-insulated design are explained and reasons for adopting mixtures of nitrogen and SF_6, using SF_6 as an additive to enhance dielectric performance of N_2, are also explained. The authors state that by using nitrogen as a majority insulation gas and therefore reducing the installation cost by 50%, in these tests at EDF laboratories, this was really the start of the second generation of GIL (at about the end of the twentieth century). (However, this is somewhat misleading as some Europeans refer to GIL as first generation merely if it is filled with 100% SF_6 as the insulation gas!)

The reader may find the technical information in CIGRE TB 218 (2003), 'Gas Insulated Transmission Lines (GIL)', to be of assistance: in the design of circuits using GIL and descriptions of GIL; characteristics as compared to O/H lines and cables; continuous and short-term current rating of directly buried GIL; insulation and overvoltage performances from short- and long-term viewpoints; reliability, environmental, economic aspects, etc.; together with an awareness of recent current GIL projects and R&D programmes (often this is also available on the Internet).

Prospects for GIL are promising. One of the most important drivers could well be the emergence of huge 'ultra cities of the future' and the challenges this will bring to bulk energy transportation of energy/power interconnections.

Gas-insulated transmission lines have been used for the transmission of high power ratings for more than 40 years and the technology appears sound and gas leakage minimal. Alternative methods are used for the laying out of GIL in practice; namely, they can be above ground, installed on steel structures or directly buried. As far as this writer is aware no major failures have been reported worldwide. Until recently, the high cost as compared to other transmission systems, e.g. cables, has restricted usage of GIL to short transmission distances of less than about 3 km.

This situation could well change in the next few years. A new CIGRE Joint Working Group comprising Study Committees B1 and B2, i.e. CIGRE JWG/B1// B2, is currently investigating and analysing factors for investment decisions relating to the case for *GIL vs Cable for AC Transmission*. This is considered strategically worthwhile and is becoming increasingly important as a result of forthcoming changes to transmission networks. Participants in this JWG will undertake such an analysis. It is planned to issue the results in a Technical Brochure in 2014 entitled 'Factors for Investment Decisions of GIL vs Cables for AC Transmission'.

3.4.7 Vacuum switches

Vacuum, at better than 10 torr, has an electric strength of typically 30 kV/mm. Under arcing conditions, gas for ionisation is provided by molten metal at the arc root of the electrodes and vacuum breakdown is very dependent on electrode conditions. Arc interruption in vacuum circuit-breakers (VCBs) is therefore achieved (Reece [2, Chapter 8]) by cooling the arc root quickly to suppress the hot spot. This is achieved by rotating the arc root rapidly under its own magnetic field and by using electrode materials of high boiling point and good thermal conductivity. The high electric strength of vacuum ensures that once the arc is suppressed, usually at the first current zero, no re-ignition occurs and dielectric recovery is achieved within a few microseconds.

Applications of gaseous insulants to switchgear 135

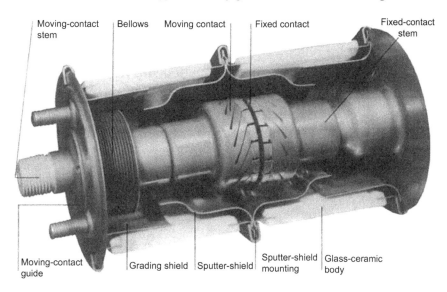

Figure 3.23 Section view of vacuum interrupter (Courtesy Vacuum Interrupters Ltd and [4, Chapter 6])

The contacts are housed in a sealed glass or ceramic bottle (Figure 3.23) with a moving metal bellows and are maintenance-free. Excellent service performance has been demonstrated for fault currents of <40 kA and 33 kV operation. VCBs are ideal for use where many breaker operations are required, e.g. in railways and arc-furnaces. The impulse level (150 kV) on the small gap (10 mm) limits the working voltage of a single bottle, but bottles have been stacked in series to provide 32 kV breakers (Reece [2]). The major use of VCBs is in 11, 33 or 66 kV applications for distribution use where an 'oil-free', 'maintenance-free' breaker is required. Typical applications of VCBs and SF_6 interrupters are discussed further by Pryor and Ali respectively, in Chapters 9 and 10 of this book.

Sometimes vacuum interrupters are used, with the designers opting for SF_6 gas to provide the insulation surrounding the interrupter. The vacuum recloser, type GRV, by Whip and Bourne, after Blower [36], is a good example of this, with the pole-mounted recloser connected to an overhead line. Comparable limited applications, for system voltage levels of 132 kV, were to be found in the USA some years ago: hybrid VCB designs having four vacuum interrupters connected in series.

Note: The reader should be aware of certain strategic current work carried out by CIGRE Study Committee SC A3 High-voltage equipment [*Electra*. 2013;**266**(February):30–4]. In its 2012 Annual Report, Working Group A3.27 stated that it is currently 'investigating the impact of application of vacuum switchgear at transmission voltages'. It is reported that, 'an inventory of state-of-the-art of high-voltage vacuum switchgear applications has confirmed that more than 8,000 units have been in service in Japan and few in China and the United States, all with very good service records'. New products appear on the market frequently up to 145 kV. SC A3, in its 2012 Annual Report, states that, 'there is no reason to assume that the excellent reliability performance demonstrated by distribution vacuum circuit breakers (VCBs) cannot be extrapolated to transmission voltages'. It appears that the primary reason for utilities to be interested in VCBs at this time is, 'the excellent operation capability with frequent operation and

136 High-voltage engineering and testing

associated reduced maintenance work during its lifetime as well as the absence of SF_6'. Not unexpectedly, potential users see issues concerning capacitive switching and dielectric performance (late breakdown and dielectric breakdown scatter as compared with SF6 technology). Current chopping transients and X-ray generation are not considered to be an issue. CIGRE claim that this work is an example of good collaboration with various experts from CIGRE and CIRED organisations. (CIRED covers distribution equipment/systems and organises many technical conferences.)

3.5 System modelling for switchgear design applications

This section provides a brief overview of field modelling techniques developed by Ryan and colleagues many years ago specifically to be able to predict the dielectric performance of commercial GIL, GIS and other switchgear. It was first necessary to be able to analyse various fields analytically by the in-house development and use of numerical methods. After this had been achieved effectively, it was then possible to undertake extensive structured dielectric testing and develop and use alternative voltage breakdown prediction techniques to estimate breakdown of practical 'switchgear-type' components, e.g. backparts for GIS. A brief coverage of the technique is now given, together with strategic examples of how the method proved very effective to predict the dielectric performance of GIS circuit-breaker designs.

This was fully successful thanks to the financial support of the then CEGB via its PERSC-funded programme, which supported a vast amount of EHV dielectric development testing on GIS-type systems (as illustrated in Table 3.4 and Figures 3.14–3.20). The switchgear company's strategic success in reducing the number of CB breaks/phase (e.g. see Figure 3.28) was as a result of Mr S.M.G. Ali and his group being able to fully utilise all the available complementary skills of the dielectric group, including also those participating in a collaborative study [22] with Professor G.R. Jones and his team at the Centre for Intelligent Monitoring Systems (CIS), University of Liverpool.

3.5.1 Field analysis techniques

The electrostatic fields of high-voltage equipment are satisfied by the well-known Laplacian equation which takes the form

$$\nabla^2 \varphi = \frac{\partial^2 \varphi}{\partial x^2} + \frac{\partial^2 \varphi}{\partial y^2} + \frac{\partial^2 \varphi}{\partial z^2} = 0$$

where φ is the potential at any point in the Cartesian co-ordinate system x, y and z. The field distribution in any design is dependent on the shape, size and disposition of the electrodes and insulation and also, in general, on the permittivities of the insulating materials used.

One criterion that can be applied to the evaluation of the electrostatic design of high-voltage equipment is the ratio of the average to maximum field. This parameter, termed the utilisation factor η, is considered below, together with details of a related parameter termed the normalised effective electrode separation. Brief reference is made in section 3.5.1.3 to a simple approximate two-dimensional method, which can often be used to estimate maximum voltage gradients for complicated arrangements, which cannot be solved directly by available numerical methods.

Analytical solutions of the Laplace equation can only be obtained for relatively simple electrode systems, where the conducting surfaces (electrodes) are cylinders, spheres, spheroids or other surfaces conforming to equipotentials surrounding some simple charge distribution. Generally, the multiplicity of boundary conditions for the complicated contours encountered in high-voltage equipment means that analytical solutions of the potential are not possible. Because of this difficulty, several approximation methods have been investigated, the more important of these being (i) analogue methods and (ii) numerical methods [23, 37].

3.5.1.1 Utilisation factor approach

In the evaluation of the electrostatic design of high-voltage equipment, an important consideration is the effectiveness with which the available space has been used. For the region that encompasses the minimum distance between conductors, the ratio of average to maximum electric stress (E_{av}/E_{max}) is a useful criterion. This ratio, termed the utilisation factor η, measures the inferiority of the field system in comparison with that between infinite plane parallel electrodes [38], where $\eta = 1$. In the literature, the reciprocal term $1/\eta$ called the field factor (f) is sometimes preferred. Tables of η values exist [38, 39] for several standard electrode systems for wide ranges of so-called geometric characteristics p and q, where $p = (r+g)/r$ and $q = R/r$. Subscripts 2 and 3, respectively, are used to denote two- and three-dimensional systems. Appropriate formulas are given elsewhere, use being made of existing solutions of Laplace's equations [37–44]. These calculations assume constant permittivity but techniques can be extended to multi-dielectric problems.

Some calculated values of η are also presented in graphical form in Figure 3.24. It is immediately evident that η for each geometry tends to decrease as the geometric characteristic $p = (r+g)/r$ increases. This can be explained by the increased divergence of the field as p increases. Furthermore, it is noted that, for a particular value p, η for a three-dimensional geometry is lower than that for the corresponding two-dimensional geometry (e.g. compare cylinder-plane and sphere-plane, curves 2 and 5, Figure 3.24). Once again, this may be attributed to the greater field divergence in the three-dimensional arrangements. Brief mention will be made later to the simple relationship existing between η values for corresponding two- and three-dimensional systems. To summarise, utilisation factors, or field factors, may be calculated for many practical arrangements as a result of precise analysis either for simple geometries or for more difficult configurations, by approximate numerical or analogue techniques together with appropriate difference equations discussed elsewhere [39].

3.5.1.2 Efficiency factor concept

For certain design problems, conditions of fixed axial distance must be complied with. The utilisation factor has the limitation that it does not take into account the space necessary to accommodate the electrode geometry. To allow for this, η can be multiplied by the electrode separation g. The product of these terms gives the 'effective electrode separation' α of the system. If, now, α is divided by some

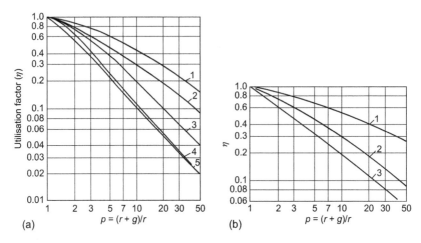

Figure 3.24 Dependence of η upon geometric characteristic p for a few simple geometries
(a) Curves: 1 cylinder–cylinder
2 cylinder–plane
3 sphere–sphere (symmetrical supply)
4 sphere–sphere (unsymmetrical supply)
5 sphere–plane
(b) Curves: 1 hyperbolic cylinder
2 hyperboloid (points)
3 hyperboloid (point)–plane

measure of the total dimension available, a new quantity λ is obtained, which is really the normalised effective electrode separation of the system. The efficiency factor concept (λ) is sometimes useful when designing high-voltage equipment, particularly when conditions of fixed axial distance d exist [38]. For a particular operating voltage it is possible to keep the maximum stress at the electrode surface to a minimum by selecting the best size, shape and disposition of electrodes for conditions of constant d (see Figure 3.25). Under these conditions $\lambda = \hat{\lambda}$ and $\lambda E_m = V/\hat{\lambda} d$.

All the results referred to above assume a single value for the dielectric constant. If two or more dielectrics are being considered, then the concepts of utilisation factor and efficiency factor could be applied to each dielectric separately. These techniques have been widely used to assist in electrostatic aspects of GIS design for many years.

3.5.1.3 Approximate two/three-dimensional concept

The electrode arrangements considered above contained at least one axis of symmetry. In many practical arrangements, however, such symmetry does not exist. In the absence of symmetry, the problem of making accurate pronouncements

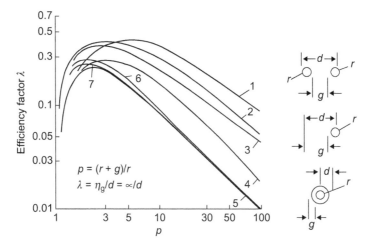

*Figure 3.25 Dependence of λ upon p for several standard geometries
Curve 1 = parallel cylinders; curve 2 = cylinder–plane;
curve 3 = concentric cylinders; curve 4 = sphere–sphere
(symmetrical supply); curve 5 = sphere–sphere (unsymmetrical
supply); curve 6 = sphere–plane; curve 7 = concentric spheres*

on maximum field strengths for practical three-dimensional systems often becomes complicated since it is sometimes either exceedingly difficult or even impossible to make single realistic numerical or analogue model representation without introducing appreciable errors.

It would be useful if, e.g., one could obtain even an approximate solution for a particular practical three-dimensional representations (having conductors similar to axial and radial sections of the original three-dimensional conducting surfaces), which, in general, can be more easily solved. Boag and later Galloway *et al.* [37] investigated the possibilities of this simple approach and have analysed numerous simple configurations. Ryan has produced useful correction curves (see Figure 3.26), which can be used with acceptable accuracy.

3.5.1.4 Numerical methods

Numerical methods of solution, which express the Laplacian equation in finite terms, provide a powerful means of calculating the electric fields of practical arrangements. A great deal of published literature exists relating to this subject. For a useful background overview of this sector, the reader should refer to the technical literature, which contains many useful practical applications of the numerical modelling techniques now readily available to support gaseous/solid/liquid insulation studies, covering electrostatic, electro-magnetic, thermal and arc interruption modelling and equipment design. A few sources are given as follows:

1. In contrast to other workers who have used relaxation techniques Ryan and colleagues [37, 44] and Ryan [40] developed a computer program to solve the

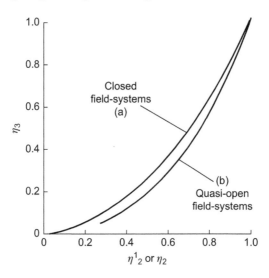

Figure 3.26 Correction curves [40]
(a) Applies for concentric cylinders and concentric spheres; cylinder–plane and sphere–plane; hyperbolic cylinder–plane and hyperboloid plane; parabolic cylinders and paraboloids of revolution
(b) Applies for systems shown in Figure 3.25

Laplace equation in two dimensions and three dimensions with one axis of symmetry by an exact non-iterative method. This method was selected after various techniques for solving the resultant set of simultaneous equations had been studied. The main details of this program, together with numerous illustrations of its extensive application in support of switchgear design, are given in References 37 and 44. Two important examples of such work shown in Figures 3.27 and 3.28 will be considered in the next two sections.

2. Binns and Randall [45] have described an over-relaxation method and give details of various accelerated finite difference formulas used. In this investigation, potential gradients were calculated around a spherical high-voltage electrode separated from an earthed plane of a recessed dielectric slab. The potential gradient transitions were determined and analysed at the point where the surface of the recessed dielectric slab meets the sphere surface.

3. Storey and Billings [46] have described a successive over-relaxation method suitable for determining axially symmetric field distributions. They also discuss a method for the determination of the three-dimensional electric field distribution in a curved bushing.

4. The ISH Symposia series (International symposium on High Voltage Engineering) provides an excellent resource to keep abreast of the latest work in the HV sector, including numerical field techniques. This symposia series has been held at

regular intervals over more than 27 years: Munich (1972), Zurich (1975), Athens (1983), Braunsweig (1987), New Orleans (1989), Dresden (1991), Yokohama (1993), Gratz (1995), Montreal (1997), London (1999) and Bangalore (2001).

Since the work [37, 40, 44] was carried out, much valuable information has appeared in the literature regarding finite element, finite difference, boundary integral and other variants. Numerous commercial field analysis packages are now available. Indeed, further examples are presented in later chapters. A detailed discussion of relaxation methods, difference formulas, accelerating factors, etc., is outside the scope of this chapter. A strategic message from this writer is that, whatever package is to be used, care should be exercised to confirm its effectiveness and the achievable accuracy, by means of simple validation studies. Once this has been demonstrated and carefully verified, the techniques successfully used by Ryan and others can be adopted with confidence.

3.5.1.5 GIS backparts and insulating spacer design

With the widespread interest in gas-insulated cables, busbars, etc., considerable attention has been given to the design of spacer shapes for such equipment. The examples summarised in Figure 3.27 illustrate the extensive use made of analytical field techniques, during early insulation development work on an EHV, SF_6-insulated metal-clad switchgear installation [23].

A general historic point: building up experience working with GIS
1. In the early days of SF_6 switchgear in the UK, young engineers such as the author of this chapter, when carrying out early dielectric HV research studies, design and development, learned 'first-hand' from their own experience/ mistakes regarding research aspects, product development and practical experiences (e.g. proper handling of gases/gas mixtures, making sure pressurised chambers did not leak, chemical and moisture ingress aspects) as they moved towards commercially viable products within their company.
2. Fortunately, much has changed in the past three or four decades as today's entrants to the industry, and indeed all, can now readily be 'empowered' and their knowledge base become fully updated relatively quickly by carefully assimilating strategic information/facts from the wealth of appropriate documents now available in the literature, e.g. in CIGRE Technical Brochures and other publications/sector support documentation.
3. Specialist CIGRE Working Groups (WGs) participate in worldwide collaboration and exchanges of strategic formation between utilities, manufactures, testing laboratories, research laboratories, consultants and standardisation authorities. Experts compile CIGRE TBs or Reports produced by each WG and these are issued for the industry.

For example the effective gas tightness of SF_6-insulated power equipment, from factory assembly, routine and type testing, on-site testing and in-service performance/monitoring has improved incrementally over the years with ongoing R and D. Currently, such equipment has an expected lifetime of up to 40 years. This must be

142 High-voltage engineering and testing

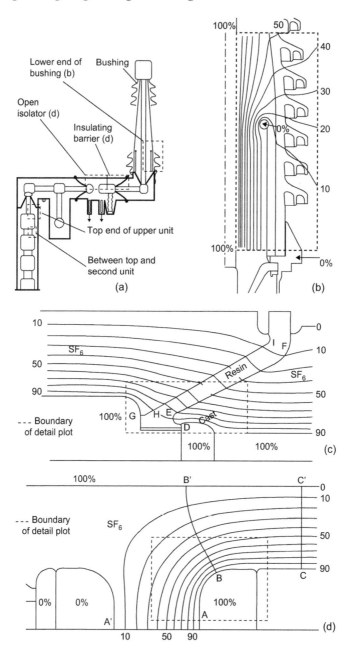

Figure 3.27 Computer field study of an early 300 kV metal-clad switchgear design [23]
(a) Part of single phase layout; (b) field plot at lower end of bushing;
(c) field plot of barrier supporting isolator; (d) field plot in vicinity of open isolator

achieved while maintaining both the functionality of the equipment and the protection of the environment. This now also includes the final decommissioning phase.

A recent CIGRE Technical Brochure provides the reader with valuable a 'SF_6 Tightness Guide' [1], which reviews and discusses significant aspects of measuring and ensuring gas tightness of SF_6 equipment, etc. Access to such documentation is straightforward and can prove very helpful to the young engineer'. (More information supplied elsewhere in book.)

3.5.1.6 GIS interrupter design

In the 1980s, the trend with GIS equipment was towards more compact designs and interrupters with fewer breaks per phase. As a consequence, the dielectric design and performance of SF_6 circuit-breakers became more critical. Careful shaping of stress shields, optimisation of stress in gas-gaps and the careful selection and profiling of insulation materials, including support insulation (e.g. spacers/barriers), were required. All these and other aspects were thoroughly considered at the design stage, using available software tools, e.g. computer field analysis programs, which had been carefully developed, verified and used extensively to study, critically analyse and optimise design stresses in GIS and other switchgear designs, for over 35 years [15, 37, 44]. Earlier papers have described field studies during the sequential development of four-, two- and one-break, heavy duty, high-performance SF_6 puffer circuit-breakers, for voltages up to the highest system ratings [23]. Figure 3.28 shows typical field plots for four- and two-break designs. Briefly, after extensive field studies, closely linked with HV and high-power development testing for a four-break design, it was possible to critically analyse all results, before 'sequentially' repeating the exercise for two-break design and later for a one-break design. However, the vital difference was that the extensive database gathered, prior to and during development of the four-break interrupter design, enabled the dimensions, profiles, etc., for the two-break, and subsequently a one-break, design to be carefully, and indeed appropriately, selected prior to costly testing. Indeed, it could be argued that this successful development represented a good example of 'rapid-prototyping' or even 'virtual-prototyping' [23]. In any event, it is hoped that this brief example provides the reader with evidence of the strategic use, with significant savings regarding product development time and cost, which can be gained from the effective use of available 'design databases' together with analytical tools, such as the field analysis modelling techniques referred to in this chapter.

3.5.2 Prediction of breakdown voltages

3.5.2.1 Empirical approach

In 1961, Ryan first investigated the feasibility of using simple estimation methods of predicting the minimum breakdown voltage levels of non-uniform field configurations in gas-insulated equipment.

Initially, the method suggested by Schwaiger was considered and later extended [38–44] to more complex electrode configurations by incorporating simple

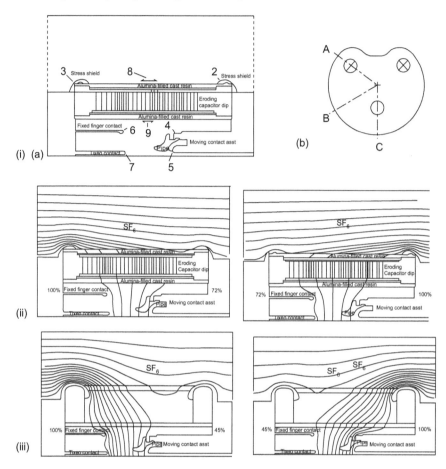

Figure 3.28 E-field design studies relating to the development of heavy-duty SF_6 interrupters in the 1980s [23]
(i) Identification of regions of interest: (a) general location of regions 2–9; (b) plan view, illustrating three radial sections, A, B and C
(ii) Typical E-field plots: first interrupter of four-break design; through capacitor (section A)
(iii) Typical E-field plots: first interrupter of two-break design; through empty support tube (section B)

perturbation principles. The method does not consider breakdown mechanisms but is based on a simple discharge-law concept [43].

By ignoring space charge effects, the breakdown voltage V is given by the relationship

$$V_s = E\eta g \qquad (3.1)$$

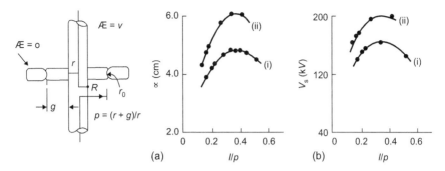

Figure 3.29 Dependence of α and 50 Hz V_s on l/p for conductor and earthed plate electrode system in air at STP [43]
(a) Electrostatic field study ($\alpha = \eta g$)
(b) High-voltage study:
    (i) $R = 4.5$ in., $r_0 = 1.0$ in. (using a range of values of conductor r)
    (ii) $R = 5.5$ in., $r_0 = 1.0$ in. (using a range of values of conductor r)
— Predicted $V_s = E_s \eta g$, where $E_s = 27.2 + 13.35/\sqrt{r}$ (kV/cm) and r and g are in centimetres
• Experimental

where E and g are the appropriate breakdown gradient and gap dimension, respectively, and η, derived from Laplacian field analysis, is the utilisation factor (ratio of average to maximum voltage gradient (E_{av}/E_m)). Initially, standard electrode geometries amenable to precise electrostatic field solution were considered, and extensive tabulated field data has been published [38, 39]. Results of early investigations (e.g. Figure 3.29) firmly established the usefulness of this simple concept for estimating minimum breakdown voltage levels.

In the past, the only major limitation to the application of this empirical approach to switchgear insulation design evaluation has been the lack of reliable breakdown information. Ryan and Watson [27] and Ryan and colleagues [42] have made significant contributions in this area and have shown that critical breakdown gradients (E_{50}) and highest withstand gradients (E_w) in air and SF_6 can be accurately represented by relationships of the form

$$rE = K_1 pr + K_2 \qquad (3.2)$$

where K_1 and K_2 are constants, $p = [(r+g)/r]$ and r is the radius of the inner cylinder or sphere. The real virtue of this simple empirical breakdown estimation technique is the fact that one can often consider practical electrode systems to be perturbations of the same simple geometry [40].

The simple electrode system shown in Figure 3.29 provides a good example to illustrate the effectiveness of this simple empirical breakdown estimation technique. An estimate of the anticipated breakdown voltage characteristics of this arrangement (Figure 3.29) is required, in atmospheric air under standard conditions [$T = 20\,°C$, $P = 760$ mm]. We can consider this electrode system to be a 'perturbed' coaxial cylinder system, which will break down when the electric gradient (E_s) at the inner cylinder (along the plane of minimum air gap) reaches the critical breakdown gradient value – known for the corresponding unperturbed coaxial (and concentric) cylinder system, to occur in air at standard temperature and pressure conditions when $E = E_s = 27.2 + 13.35\sqrt{r}$, where r is the radius of the inner cylinder in centimetres. Predicted and experimental results agreed to within ±5% for all sizes of inner conductor (radius r) considered.

Predicted and experimental sparking voltages have also been compared for numerous gaseous insulants, for a wide range of gas temperatures and pressures and for many different electrode systems, using simple perturbation field techniques. Excellent agreement has been demonstrated (e.g. see Table 3.5). For example designers of GIS switchgear in the 1980s had access to a vast reliable database relating to the characteristics of numerous insulating materials and also SF_6 to assist them to achieve effective, competitive and reliable designs as a result of much dielectric research and development activities (see e.g. References 1, 14 and 23).

3.5.2.2 Semi-empirical approach

As an alternative to the above empirical method, the author and his colleagues have investigated a so-called semi-empirical breakdown estimation method, based on streamer theory of gas breakdown [1]. For air, the criterion for minimum breakdown voltage was given by Pederson [47, 48] as

$$\ln \alpha_x + \int_0^x \alpha\,dx = \ln \alpha + \alpha x \qquad (3.3)$$

Table 3.5 *Breakdown data for hemispherically ended rod/plate arrangement ($r = 12.7$ mm) [42]*

	Gap g (Mm)	Breakdown voltage at relative density (kV)		
		3	5	7
Experimental		100	158	210
Empirical[a]	20	102	158	214
Semi-empirical		103	162	220
Experimental		127	200	259
Empirical[a]	40	126	197	267
Semi-empirical		129	203	276
Experimental		144	227	288
Empirical[a]	80	143	222	301
Semi-empirical		142	227	310

[a] Estimate using (3.1), together with equation of form $rE = K_1 pr + K_2$ derived from concentric sphere hemisphere data.

where a_x is the numerical value of a, Townsend's first ionisation coefficient, at the head of the avalanche, of length x, at which the critical ion number is reached in an electron avalanche; in a non-uniform field, streamers are formed, resulting in corona or breakdown.

The left-hand side of (3.3) is evaluated from the electrostatic field distribution, obtained using a general field analysis program for practical arrangements or from precise equations for standard geometries [41, 42]. The right-hand side of (3.3) is computed from existing empirical breakdown potential gradient data for uniform fields of gap x. The values of a used in both sides of (3.3) for the examples discussed in previous studies [41, 42] have been published by earlier workers [1, 42].

This semi-empirical approach was extended to SF_6 using a modified form of (3.3), namely

$$\int_0^x (\alpha - \eta) dx = k = 18 \tag{3.4}$$

where α and η are Townsend's first ionisation coefficient and attachment coefficient, respectively, in SF_6. An essential prerequisite of Pederson's semi-empirical approach for any particular set of conditions is information concerning the following parameters:

1. uniform field breakdown potential gradients
2. α and η values
3. potential-gradient distribution across the non-uniform gap

General purpose digital computer programs have been developed, which are capable of solving equations of the form of (3.3) to predict, with acceptable accuracy, the minimum breakdown voltage, V, for a wide range of standard and practical electrode arrangements in air, N_2 and SF_6, for pressures in the range 1–5 atmospheres [41, 42, 44] and for air at high temperatures.

Similarity and **Paschen's law** type relationships were also studied. For example with uniform field gaps of 0.5–2.0 cm and temperatures 20–1,100 °C, Powell and Ryan obtained results for the DC (static breakdown) voltage, kV, of air, which fitted a relationship of the form

$$V_s = 24.49(\partial g) + \frac{6.61}{\sqrt{\partial g}} \quad (kV) \text{ to within } +5\% \text{ and } -2\%$$

Here, the same constants were selected as given by Boyd et al. from their earlier study at standard atmospheric conditions, where the relative air density $\partial = 1.0$, pressure $p = 1{,}013$ mbar and $T = 293$ K (20 °C).

An example of these techniques is given in Table 3.5, which compares experimental and estimated breakdown voltages in air, at relative air densities (δ) of 3, 5 and 7, respectively, for a hemispherically ended rod/plate arrangement [radius $r = 12.7$ mm]. It is clearly seen that both these estimation methods are capable of predicting breakdown voltage levels to within ±5% over the range of gaps and air pressures considered.

3.6 Summary

This chapter has provided a general introduction to the application of gaseous insulation systems. Although it has only been possible to touch very briefly on some major aspects, important strategic issues will be further developed in other chapters focusing on special areas. An indication has been given in this chapter of numerical field techniques, which have found widespread application in the insulation design of GIS and other switchgear for many years. The simple breakdown estimation methods (empirical and semi-empirical) by Ryan et al., which are extensions of the work by Schwaiger (1954) and Pederson (1967) [47, 48] together with an available experimental database obtained from extensive Paschen's Law/similarity type studies in gaseous insulants, have been thoroughly developed to such a degree that minimum breakdown voltages of practical GIS design layouts (as well as a host of gas-gap arrangements; see References 1 and 5) can be estimated to within a few per cent, at the design stage, often without recourse to expensive development testing. Such derived voltages are generally the minimal withstand levels attainable under practical conditions. Undoubtedly, there is still considerable scope in the future to predict dielectric performance by utilising advanced planning simulation tools incorporating genetic algorithms and artificial intelligence techniques, linked to various 'extensive databases' relating to breakdown characteristics of gaseous insulation, including equipment service performance data (e.g. from surveys similar to Reference 34).

Again, it should be emphasised that the outstanding strategic developments in switchgear designs reducing from 4 to 1 breaks/phase at this time, as discussed in this chapter, were due to the complementary skills of the switchgear designers and development team, together with the dielectric contribution and importantly the excellent collaboration with Professor G.R. Jones and his team at the Centre for Intelligent Monitoring Systems, Department of Electrical Engineering and Electronics, University of Liverpool, UK.

Finally, it is anticipated that the continuing, and growing, environmental concerns will influence the development of the next generation of gas-insulated switchgear but a commercial replacement for 100% SF_6 gas, for interruption purposes at the higher ratings, still seems very remote!

Acknowledgements

The author wishes to thank the Directors of NEI Reyrolle Ltd and later VA TECH REYROLLE, before Reyrolle became an 'affiliate' to other large international companies for permission to publish, earlier papers, which have been extensively referred to in the studies reported in this chapter. He also gratefully acknowledges the assistance given by many of his former colleagues and for their contributions and generous support over the years.

For about two decades, the IET, High Voltage Engineering and Testing, HVET, International Summer School was held in Newcastle upon Tyne, UK, during a period of major company changes within the UK Power industry. The ongoing generosity of Reyrolle and sequentially, the 'various changing company ownership affiliations' over the years, e.g. Rolls Royce, The Bushing Company, Siemens (-the last Clothier laboratory operator being NaREC-) all continued with strong support of the IET, HVET School series, by making the Clothier laboratory and factory resources available for visits by International HVET School delegates from more than 20 countries, which was greatly appreciated by all.

This author and the HVET Series Organising Committee again acknowledge their sincere thanks to the facility owners for this ongoing generosity and support. Undoubtedly, these industrial visits to the Clothier laboratory/adjacent industrial complex contributed greatly to 'delegate empowerment' and to the great success of this highly rated course series over the years.

References

1. Blair D.T.A. 'Breakdown voltage characteristics', in Meek J.M. and Craggs J.D. (eds.). *Electrical Breakdown of Gases*. Chichester: Wiley; 1978. pp. 533–653 (Chapter 6)
2. Flurscheim C.H. (ed.). *Power circuit-breaker theory and design*. Volume 1 in IET Power and Energy Series, London. Peter Perigrinus Ltd; 1982 (also included Reece M.P., Chapter 8; Vacuum Circuit Breakers, pp. 320–52)
3. Raether H. *Electron avalanches and breakdown in gases*. London: Butterworth; 1964
4. Bradwell A. (ed.). *Electrical insulation*. Series 2 in the IET Electrical and Electronic Materials and Devices Series, London. Peter Perigrinus Ltd; 1983
5. Waters R.T. 'Spark breakdown in non uniform fields' in Meek J.M. and Craggs J.D. (eds.). *Electrical Breakdown of Gases*. Chichester: Wiley; 1978. pp. 385–532 (Chapter 5)
6. Looms J.S.T. *Insulators for high voltages*. Volume 7 in IET Power and Energy Series, London. Peter Perigrinus Ltd; 1990
7. CIGRE Task Force 33.04.09 Report. 'Influence of ice and snow on the flashover performance of outdoor insulators, Part I: Effects of ice'. *Electra*. 1999;**187**(December):91–111
8. CIGRE Task Force 33.04.09 Report. 'Influence of ice and snow on the flashover performance of outdoor insulators, Part II: Effects of snow'. *Electra*. 2000;**188**(February):55–69
9. Legg D. *High-Voltage Testing Techniques*. Internal Reyrolle research report; Newcastle upon Tyne; 1970
10. Ryan H.M., Whiskard J. 'Design and operation perspective of a British UHV laboratory'. *IEE Proceedings A*. 1986;**133**(8):501–21 [Note additional material by these two authors on the 50 Hz breakdown of large air gaps is detailed and is available in 'Recent studies in the Clothier laboratory', *Reyrolle Parsons Review*. 1971;1 (Summer(1)), 11–16.]

11. Ito H., et al. 'Background of technical specifications for substation equipment exceeding 800 kV AC'. *Electra*. 2011;**255**(April):60–9
12. Diesendorf W. *Insulation co-ordination in high voltage electric power systems*. London: Butterworths; 1974
13. IEEE Committee Report. 'Sparkover characteristics of high voltage protective gaps'. *IEEE Transactions*. 1974;**PAS-93**:196–205
14. Ryan H.M., Lightle D., Milne D. 'Factors influencing dielectric performance of SF_6 insulated GIS'. *IEEE Transactions*. 1985;**PAS-104**(6):1527–35
15. CIGRE Working Group 33.11, Task Force 03, SCHEI, A., Convenor WG 33.11. 'Application of metal oxide surge arresters to overhead lines'. *Electra*. 1999;**186**(October):83–112
16. CIGRE Working Group 33.07 Report. 'Insulation coordination in live working and other special conditions in electrical system. Guidelines for insulation coordination in live working'. *Electra*. 2000;**188**(February):139–43
17. Malik N.H., Qureshi A.H. 'A review of electrical breakdown in mixtures of SF_6 and other gases'. *IEEE Transactions on Electrical Insulations*. 1979; **EI-14**:1–14
18. Takuma T., et al. *Interfacial insulation characteristics in gas mixtures as alternative to SF_6 and application to power equipment*, Paris: CIGRE; 2000, Paper 15–207
19. Ryan H.M., Jones G.R. SF_6 *switchgear*. Volume 10, in IET Power and Energy Series, London. Peter Perigrinus Ltd; 1989
20. Jefferies D. 'Transmission today: lessons from a decade of change'. *Electra*. 1999;**187**(December):9–18. Reproduced opening speech: London CIGRE Symposium: working plant and systems harder: Enhancing the management and performance of plant and power systems', 7–9 June 1999
21. Urwin R.J. 'Engineering challenges in a competitive electricity market'. Keynote address, ISH: High voltage symposium; London, 22–27 August 1999, Paper 5.366.SO
22. Ali S.M.G., Ryan H.M., Lightle D., Shimmin D., Taylor S., Jones G.R. 'High power short circuit studies on a commercial 420 kV, 60 kA puffer circuit-breaker'. *IEEE Transactions*. 1985;**PAS-104**(2):459–68
23. Ryan H.M. 'GIS barriers: application of field computation strategies to switchgear'. IEE Colloquim on field modelling: applications to high voltage power apparatus; London, 17 January 1996
24. Ragaller K. (ed.). *Current interruption in high voltage networks*. New York: Plenum Press; 1978
25. Mosch W., Hauschild W. *High voltage insulation with sulphur hexafluoride*. Berlin: VEB Verlag Technik Berlin; 1979
26. Boggs S.A., Chu F.Y., Hick M.A., Rishworth A.B., Trolliet B., Vigreux J. *Prospect of improving the reliability and maintainability of EHV gas insulated substations*. Paris: CIGRE; 1982, Paper 23–10
27. Ryan H.M., Watson W.L. *Impulse breakdown characteristics in SF_6 for non-uniform field gaps*. Paris: CIGRE; 1978, Paper 15–01

28. Laghari J.R. 'Review – a review of particle contaminated gas breakdown'. *IEEE Transactions on Electrical Insulation*. 1981;**EL-16**:388
29. Ishikawa M., Hattiri T. 'Voltage-time characteristics of particle initiated breakdown in SF_6 gas'. *3rd International Symposium on Gaseous dielectrics*; Knoxville. 1982, Paper No. 28
30. Eteiba M.G., Rizk A.M. 'Voltage-time characteristics of particle-initiated impulse breakdown in SF_6 and SF_6–N_2'. *IEEE Transactions on Power Apparatus and Systems*. 1983;**PAS-102**(5):1352–60
31. Hampton B.F. Pearson J.S., Jones C.J., Irwin T., Welch I.M., Pryor B.M. *Experience and progress with UHF diagnostics in GIS\CIGRE*. Paris; 1992, Paper 15/23–03
32. Jones C.J., Hall W.B., Jones C.J., Fang M.T.C., Wiseall S.S. *Recent development in theoretical modelling and monitoring techniques for high voltage circuit-breakers*. Paris: CIGRE; 1994, Paper 13–109
33. CIGRE Working Group 13.09. 'Monitoring and diagnostic techniques for switching equipment'. *Electra*. 1999;**184**(June):27
34. CIGRE Working Group 23.02, Task Force 02 Report. 'Report on the second international survey on high voltage gas insulated substations (GIS) service experience'. *Electra*. 2000;**188**(February):127
35. Koch H., Schoeffner G. 'Gas-insulated transmission line (GIL): an overview'. *Electra*. 2003;**211**(December):8–17
36. Blower R. 'Progress in distribution switchgear'. *Power Engineering Journal*. 2000;**14**(6):260–3
37. Galloway R.H., Ryan H.M., and Scott M.F. 'Calculation of electric fields by digital computer'. *Proceedings of IEE*. 1967;**114**(6):824–9
38. Ryan H.M., Walley C.A. 'Field auxiliary factors for simple electrode geometries'. *Proceedings of IEE*. 1967;**114**(10):1529–34
39. Mattingley J.M., Ryan H.M. 'Potential and potential-gradient for standard and practical electrode systems'. *Proceedings of IEE*. 1971;**118**(5):720–32
40. Ryan H.M. 'Prediction of electric fields and breakdown voltage levels for practical 3-dimensional field problems'. *Eighth Universities Power Engineering Conference*; University of Bath, 1973
41. Blackett J., Mattingley J.M., Ryan H.M. 'Breakdown voltage estimation in gases using semi-empirical concept'. *IEE Conference Publications*. 1970;**70**:293–97
42. Mattingley J.M., Ryan H.M. 'Breakdown voltage estimation in air and nitrogen'. *NRC Conference on Electrical and Dielectric Phenomena*; Williamsburgh, November 1971
43. Ryan H.M. 'Prediction of alternating sparking voltages for a few simple electrode systems by means of a general discharge-law concept'. *Proceedings of IEE*. 1967;**114**(11):1815–21;(6):830–1
44. Scott M.F., Mattingley J.M., Ryan H.M. 'Computation of electric fields: recent developments and practical applications'. *IEEE Transactions on Electrical Insulations*. 1974;**El-9**(1):18–25

45. Binns D.F., Randall T.J. 'Calculation of potential gradients for a dielectric slab placed between a sphere and a plane'. *Proceedings of IEE*. 1967;**114**(10): 1521–8
46. Storey J.T., Billings M.J. 'Determination of the 3-dimensional electrostatic field of a curved bushing'. *Proceedings of IEE*. 1969;**116**(4):639–43
47. Pederson A. 'Calculation of spark breakdown or corona starting voltages in non-uniform fields'. *IEEE Transactions*. 1967;**PAS-86**:200–6
48. Pederson A. 'Analysis of spark breakdown characteristics for sphere gaps'. *IEEE Transactions*. 1967;**PAS-86**:975–78
49. 'Insulation coordination for UHV AC systems', TB 542, CIGRE WG C4.306 (also summarised in WG C4.306. Zaima E., *et al.** *Electra.*, 2013;**268**(6); 72–9) *37 co-authors; [May be proposed for consideration when the application guide IEC 60071-2 (1996) and IEC apparatus standards are revised]

Chapter 4
HVDC and power electronic systems
Gearóid ó hEidhin

4.1 Introduction

Power electronic devices and systems are becoming an increasingly common feature of power systems. Over the years many power electronic products have been developed and can be used to great effect in the alternating current (AC) transmission and distribution system. Existing examples include:

- high-voltage direct current (HVDC) power transmission systems
- *static var compensators (SVCs)* – dynamic shunt reactive power compensation
- various forms of motor drive systems based on power electronics, widely used in industry and traction applications
- *power supplies* – many of these are based on power electronic devices
- 'STATCOM' (gate turn-off thyristor (GTO) and insulated gate bipolar thyristor (IGBT)-based SVC).

Future developments may include:

- *static series compensator* – dynamic series compensation of transmission lines
- *unified power flow controller (UPFC)* – combined dynamic shunt and series compensators
- electronic tap-changer
- *phase-shifter (also known as a quad booster)* – used to share load between parallel circuits. This is generally a mechanical device but could also be implemented as a power electronic system.

The first item HVDC is generally the largest of the various power electronic systems and makes the most dramatic difference to power systems. This is described in some detail. The other power system devices (SVC, STATCOM, series compensation, UPFC) are described more briefly.

4.2 HVDC transmission – a brief overview

Early electricity distribution was by direct current (DC). Then the transformer was developed, so AC became the universal choice for generation, transmission, distribution and utilisation. However, during the last 40 years many HVDC systems have been built and operated (and a few even decommissioned!). Have we turned the clock back?

No! Modern HVDC systems are fully integrated into the adjoining AC systems and in some instances are fully embedded in an AC system. HVDC systems offer the following enhancements and capabilities to power systems:

- *Economic long-distance power transfer for example from a remote power source to an urban area.* The economy in this instance arises from the fact that a DC transmission line is considerably cheaper to build than the AC power line to carry the same power. If the distance is great enough, this economy more than pays for the HVDC converter stations at both ends. The break-even distance is approximately 800 km for an overhead line and approximately 50 km for a cable.
- *Interconnecting independent power systems.* HVDC provides an asynchronous link between different AC systems which may continue to operate at different frequencies and/or relative phase angles. HVDC allows the operators to influence reactive power and voltage at each terminal of the link and to control the real power transmitted, regardless of the system voltages, frequencies or relative phase angles. This is in contrast to an AC link over which the real and reactive power flows are entirely dependent on the voltage and phase angles at each end of the link. Additionally the two power systems retain independence of operation while obtaining the benefits of interconnection such as mutual support and shared spinning reserve. Very often such links are of zero length and are called 'back-to-back' schemes.
- HVDC power transmission is very controllable and can provide both static and dynamic support for a power system. There is one particular system where the presence of the HVDC system means that the power transfer capacity on the parallel AC interties has been increased by 50%.
- HVDC can be used to provide tight frequency control of one system.
- HVDC systems can damp out oscillations in the attached AC power systems and associated generators.
- Unlike an AC interconnection, HVDC transmission causes no increase in the fault level of the attached AC systems.

If the function of an HVDC transmission scheme is to transfer power over a long distance then it will invariably use a high direct voltage. Most present day schemes use voltages up to ±800 kV for overhead lines, while cables have been approaching ±600 kV progressively over the last 20 years. As for AC transmission, high voltage is used to minimise the current in the link to reduce the I^2R losses in the line or cable. The increase in HVDC voltages has been considered in a number of CIGRE reports such as Reference 1. Even higher voltages are also being considered.

For back-to-back schemes the pressure to use high voltage disappears and the voltage used will be the lowest voltage at which the required power can be transferred, bearing in mind the limitations of the converter valves. The converter valves

HVDC and power electronic systems 155

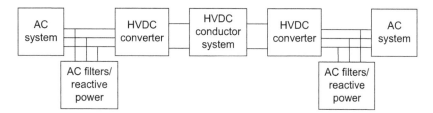

Figure 4.1   HVDC system

are normally the factor which limits HVDC current rating. Thyristors up to 150 mm diameter are now in general use enabling DCs of up to 5,000 A to be used.

Therefore, converter stations can be roughly characterised into two groups:

- back-to-back converters using low direct voltage and high current, typically 20–250 kV and 2.5–5 kA
- long-distance transmission schemes using higher direct voltages and more moderate current, typically 300–800 kV and 1–4 kA.

The basic configuration of an HVDC link is shown in Figure 4.1.

The converter station designer must assemble all of the required components in such a way as to provide them with the necessary environmental protection, whatever access may be necessary, first for construction and later for operation and maintenance, and to provide all the normal supporting services such as auxiliary supplies and cooling. Environmental aspects of an HVDC converter station, as for a conventional HV substation, include:

- audible noise
- visual impact
- radio interference
- electric and magnetic fields.

In designing the converter station, the designer must bear in mind the requirements of particular specifications. Specifications for HVDC transmission schemes vary in length from as little as 10 pages to over 1,000, so it would be idle to suggest that a discussion of practicable length can take into account all possible variations. Therefore, this chapter is only intended to raise the major issues to be considered and not to give a definitive guide to a fixed procedure. More detail is given in References 2–4.

4.3   General principles

Figure 4.2 shows the arrangement of the basic components for a transmission scheme converter terminal. The function and relationship of these components will be dealt with in the following sections.

The HVDC converter consists of a 12-pulse converter connected to the AC system through the converter transformer. Harmonic filters are provided on

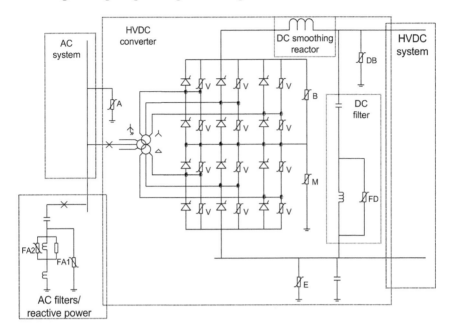

Figure 4.2 Simplified single-line diagram of HVDC station
FA1, FA2 = AC filter arrester; A = AC busbar arrester; V = thyristor valve arrester; B = bridge arrester; M = midpoint arrester; E = neutral arrester or electrode line arrester; DB = DC busbar arrester; FD = DC filter arrester

the AC and DC sides of the converter as required to limit interference to acceptable levels. A smoothing reactor is provided (except on some back-to-back links) to protect the converter from externally imposed impulses on the DC line.

4.4 Main components of HVDC links

4.4.1 Thyristor valves

The heart of a converter station is the thyristor valve which performs the switching function of the converters. Six identical valves are arranged in a three-phase bridge so that full-range rectification or inversion can be achieved by phase angle control. Such an arrangement is commonly referred to as a six-pulse bridge because it requires six firing pulses per fundamental frequency cycle to operate the valves, so it produces a DC side ripple of six times the AC system frequency. Two six-pulse bridges in series on the DC side fed with AC voltages 30° phase shifted from each other form the modern 12-pulse bridge. This design has a lower harmonic output than a six-pulse bridge.

Each valve consists of a number of identical and electrically independent series-connected levels. Typically the number of levels can range from 10 to more than 100, depending on the scheme being considered. Modern thyristor valves are extremely reliable since thyristor redundancy can be included in each valve. This is achieved by increasing the number of series levels by one or two compared with the minimum required to withstand the applied voltage.

The thyristor is fired by sending a light pulse via an optical fibre to the electronics located at each individual thyristor level. On-valve electronic protection protects the thyristor both in the on-state and in the off-state.

All valves are protected against overvoltage by zinc oxide surge arresters connected directly between their terminals (item 'V' in Figure 4.2). Additionally, the thyristors are prevented from reaching excessive voltage in the forward direction by protectively firing them when the voltage exceeds a pre-determined level. This is accomplished by special circuits, using breakover diodes.

The ALSTOM design is built up from assemblies embodying six series-connected thyristors clamped between high-efficiency water-cooled heatsinks, which is called a banded pair. The clamping force is applied by means of a Belleville spring washer arrangement, acting via the special glass fibre resin bands. To ensure that the inrush current is kept within the capability of the thyristors, saturable reactors are used. Other components include the damping capacitor and the water-cooled damping resistors which are also mounted on the thyristor heatsink. The thyristors can be replaced without disturbing water or main electrical connections by means of a simple hydraulic tool. Figure 4.3 shows a six-level assembly elevated for maintenance with the hydraulic tool attached.

The thyristors and their associated components are mounted in a rigid frame to form a tier, several of which can be stacked up to make a mechanically

Figure 4.3 Six-level assembly

158 *High-voltage engineering and testing*

Figure 4.4 View of McNeill valve hall

stable structure. Thyristor valves of modern water-cooled design occupy only a quarter of the volume of the earlier air-cooled design.

The thyristor valves of a 12-pulse converter are usually arranged in three separate stacks with the four valves in each stack physically mounted on top of each other. Each valve stack is called a quadrivalve and serves one phase of the AC supply. Figure 4.4 shows the complete valve arrangement for both converters of the ALSTOM supplied 150 MW McNeill back-to-back scheme in Canada in which each valve comprises only one tier of banded pairs. More detail is given in Reference 5. For higher voltage applications further tiers are stacked to form larger valves.

4.4.2 Converter transformer

The transformer provides the 30° phase displacement between the two six-pulse series-connected sub-converters to provide overall 12-pulse performance. This is achieved by having one star–delta and one star–star connected transformer or star and delta secondary windings on the same transformer. The transformer also provides the required valve side AC voltage to give the optimum DC voltage at normal operating conditions. An online tap-changer is used to maintain this voltage when the AC voltage varies or (in some cases) when the DC loading conditions change. The transformer also provides the isolation between the DC and the AC systems. The impedance of the transformer together with the AC system impedance limits the fault currents experienced by the converter valves. The commutating reactance can normally be assumed to be controlled by the transformer impedance only, since AC harmonic filters maintain near sinusoidal voltage at the AC busbar which supports commutation.

The principal purpose of this section is to illustrate the kind of compromises which arise during the design of a converter. If a design for a given application can be said to be characterised by a single number, that number is probably the commutating reactance, because it influences the design of many major components of a scheme, and also its interface with the AC network. Thus, choosing the most economic reactance is a key decision in converter design.

The steady-state operating conditions of the converter are governed by the standard converter equation, relating direct voltage (V_d), direct current (I_d), converter transformer valve-winding emf (E_{L-L}) and commutating reactance (X_c) with the firing angle (or extinction angle) to which the converter is controlled. From Reference 2 the approximate formula is given below (where I_d and X_c are in pu):

$$V_d = \frac{3\sqrt{2}}{\pi} E_{L-L} \left(\cos \alpha - \frac{I_d X_c}{2} \right) \tag{4.1}$$

It is evident from this equation that achieving a given direct voltage will require smaller E_{L-L} if X_c is decreased. However, to demand very small commutating reactance gives rise to high transformer cost. The precise variation is rather specific to the manufacturing technique used and to the voltage rating of the individual windings, so generalisations are approximate. However, values of X_c smaller than about 15% tend to cause transformer capital cost to rise.

Similarly, the required E_{L-L} will be minimised if α at a rectifier (or γ at an inverter) is minimised. The balance to be achieved here is between minimising the steady-state ratings of the equipment and allowing sufficient margin to permit the minor voltage variations routinely encountered on an operating power system to be absorbed without interfering with power flow. Usually, this is translated into a requirement to accommodate a voltage variation of defined amplitude without encountering control system limits. (In specifications, this is sometimes written as 'without mode change'.) Typically, a compromise will be implemented in which the converter remains within its linear range of operation during AC network voltage variations of around 3%. This commonly leads to the adoption of minimum continuous operating angles of 12° for α and 15° for γ.

Frequently, the minimum values for α and γ occur only when operating at full load, since specifications often require the reactive power exchanged with the AC network at low power transfer to be restricted, and this restriction is often partly obtained by operating the converters at the same E_{L-L} at all currents. Thus, during transfer of low current, the α (or γ) is often larger than the minimum permitted value.

Operating the converters in this way incurs a small amount of additional losses during low power transfer. These losses have to be tolerated by the valve components (not normally a problem) and are often evaluated by customers when bids are assessed.

In the special case of a back-to-back link, both converters are located in the same converter station and they can exert reactive power control much more comprehensively than merely by degrading their power factor at low load.

The direct voltage can be reduced (and the DC increased to maintain the real power transfer) in response to the demands of a closed-loop reactive power controller. From Reference 2 the reactive power absorbed by a converter is approximated by

$$Q = V_d I_d \sqrt{1 - \left(\frac{V_d}{V_{\text{dio}}}\right)^2} \tag{4.2}$$

Therefore, provided the converter valves are designed to withstand the duty, that is if the requirement is stipulated in the specification for a scheme, the converter can operate as an HVDC converter (at rated direct voltage) as a thyristor-controlled reactor (at zero direct voltage) or at any intermediate operating condition which may be convenient. The only absolute limit is that the total mega volt ampere (MVA) at which the converter is asked to operate must not be greater than the rating of the main circuit equipment.

It is necessary to remember that the two converters share the same DC operating conditions; in other words, actions taken to improve operating conditions of one AC network will influence the other in a generally similar way. Thus, while it is possible to do a great deal to minimise the impact of reactive power unbalance (between the converters and the AC harmonic filters) for one of the converters, the behaviour of the other will be similar. Therefore, this feature is best used when one of the AC networks is weak but the other is strong enough to withstand the consequences of favouring the first.

Another aspect of design which is peculiarly relevant to back-to-back HVDC links is the effect of transformer reactance on valve fault current. The current which flows in the event of a converter fault is simply $(V/X_L) \times K$. V is the power frequency emf behind the effective AC network source impedance, X_L is the total inductance between the valves and that emf and K is the constant which quantifies the fault currents peculiar to converters. Typically, X_L is dominated by the converter transformer reactance. Modern power thyristors as used for HVDC have little difficulty in carrying the fault currents which arise from transmission applications, because the DC is rarely much more than 2.5 kA, and economic transformer designs usually produce aggregate values for X_L of 15% on transformer rating or so, yielding typical peak fault currents of around 30 kA. However, valve fault current becomes a much more serious consideration for back-to-back links, because they operate at low voltage. This means that not only is the DC somewhat larger, but also that it is simpler and cheaper to build converter transformers of much lower reactance than a high-reactance transformer. This was overcome at the McNeill converter station by designing converter transformers having an additional winding, to which the AC harmonic filters are connected, shown in Figure 4.5. The fault current to which valves may be subjected is limited by the comparatively large line-winding to valve-winding reactance. The commutation reactance is smaller than the fault limiting impedance, being approximately the reactance between the valve-winding and the AC harmonic filters.

Low commutating reactance causes some increase in the harmonic output from a converter. When the AC harmonic filter resides on an additional winding the effects of this can often be offset by making the reactance between the filter-winding

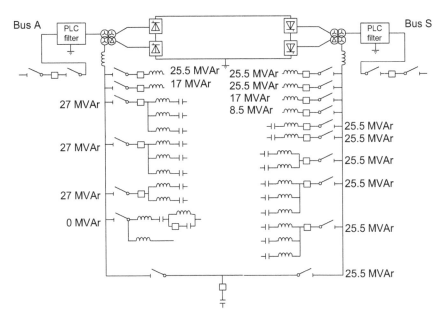

Figure 4.5 Single-line diagram of McNeill HVDC station

and the line-winding act as part of the AC harmonic filter. This was done for the McNeill converter station.

Most of the discussion above concerns the fundamental frequency behaviour of the converter equipment, which determines much of its cost. However, the converter transformer must also tolerate the harmonics arising from converter operation. The difference arises from the difference between the fundamental frequency component and the total rms current. (For an HVDC converter, the ratio between these two quantities is $3:\pi$.) The harmonic content has a greater significance to the transformer designer than is apparent from this ratio because the harmonic currents cause greater heating than do fundamental frequency currents. The transformer design must take into account the anticipated spectrum of harmonic currents, attributing the correct amplitude to each, so that the cumulative heating effect can be estimated.

Converter transformers are invariably provided with on-load tap-changers. Their purpose is to enable the operating conditions imposed on the DC plant to be substantially independent of the AC network conditions. Therefore, the first requirement of such a tap-changer is that it should be capable of compensating for the working voltage range of the AC network. This is typically $\pm 5\%$ (especially on higher voltage networks, say above 200 kV) but is often $\pm 10\%$. The tap-changer must also be able to compensate for any tolerance which the commutating reactance of the finished transformer may exhibit, and it must be provided with a control system which embodies a large enough deadband to avoid hunting. Sometimes there are special circumstances to be taken into account. For example some purchasers require the capability to operate at reduced direct voltage during unfavourable weather, and require some extension of the tap-changer range to make this possible without encountering the large change in reactive power balance which would follow from achieving it by means of firing angle alone. Thus, converter transformers rarely have tap-changer ranges less than 20%, and sometimes they exceed 30%.

The converter transformers can be provided as single-phase or three-phase units. They can be provided as two-winding, three-winding or even four-winding units (as shown in Figure 4.5). The choice of transformer configuration is dictated by transport limits as well as availability considerations. The advantage of using single-phase transformers is that spares can be provided more cheaply, although each three-phase bank becomes more expensive.

4.4.3 Control equipment

The converter control and protection equipment is an area in which the component count is so great that it is necessary to categorise the equipment and to incorporate protective features in the design to obtain correct system performance.

Some of the control equipment influences the behaviour of only a single thyristor level. Since the thyristor valves incorporate redundancy, it is generally unnecessary to incorporate further redundancy in such control equipment.

In cases where it is possible to define what constitutes 'safe' action, it is often possible to incorporate duplication. Thus, many protective features are provided with redundancy in the sense that at least two protective functions react to each main circuit event. This is often adequate to secure the safety of the main circuit hardware, but it may be calculated that this leads to an unacceptably large number of spurious interruptions to transmission. In such cases duplication is insufficient and higher levels of redundancy such as triplication may then be incorporated together with a 'voting' system whose function is to ensure that only the signal presented by a majority of the channels is implemented.

Figure 4.6 is a block diagram showing an example of the different levels of redundancy which may be applied to different functions of the control system. Typically, this will be reinforced by supplying power to each function from at least two independent sources in such a way that failure of one source does not influence the ability of the other to continue to supply the equipment.

With the continued construction of HVDC schemes there are now occasions where multiple HVDC schemes (and possibly even FACTS devices) are used in close proximity if not at the same substation. Since each HVDC (and FACTS device) has its own control system it is important that they act in unison. Considerable work has been done on this subject (see Reference 6), and there are a number of locations around the world where multiple HVDC projects from different manufacturers and owned by different utilities work together successfully.

Operator interfaces take many forms, but the presentation of information on VDUs (and other forms of display screen) controlled by keyboards or other convenient devices such as tracker balls is increasing rapidly for most applications. This is convenient for the converter station designer since it is usually not difficult

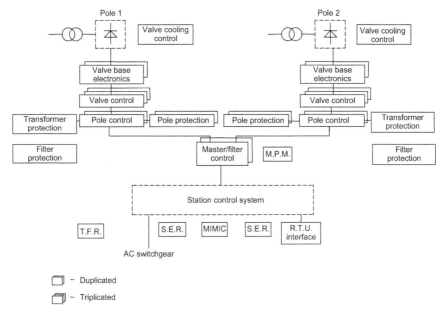

Figure 4.6 Control system block diagram

to provide facilities at a remote location, such as a regional despatching centre, which are identical with those provided in the converter station itself. Inevitably the local facilities will be heavily utilised during commissioning of the equipment, but the present trend is that once the equipment is in service, direct control of its behaviour from the despatch centre becomes the norm.

4.4.4 AC filters and reactive power control

AC filters are provided for two basic reasons:

- to minimise the harmonic interference caused to the AC system by the HVDC converter – this includes waveform distortion and telephone interference effects
- to assist in the reactive power control duties of the HVDC link.

This latter requirement is becoming increasingly significant as HVDC links are being increasingly applied to weaker and weaker AC networks which require them to provide reactive power control.

The harmonic performance is generally achieved using a variety of passive filter types (tuned/damped filters, single or multi-frequency). In the future there is also the prospect of using power electronics to improve the performance of filters either directly to null the current injected from the converters or to retune conventional filters.

The filter characteristics and the number of filter groups provided at a converter station may be influenced by the reliability required of the scheme.

Duplication of specialised filter types can be used to ensure that scheme reliability targets are met. Alternatively, the provision of a single general purpose type of filter may be more cost effective, despite giving slightly inferior performance, since reliability can be achieved with fewer switched groups. The 2,000 MW England/France link uses two different types of filters in each of the four filter groups.

The voltage excursion produced on the AC busbar when sections of AC harmonic filter/shunt capacitance are energised and de-energised is also important. These, being routine events, may occur several times per day. Typical specifications limit the permitted voltage excursion to about 3%, equivalent to a 100 MVA capacitor bank connected to an AC network of short circuit capacity 3,300 MVA. Economy would often result if a larger capacitor bank size could be employed.

The operating conditions of a back-to-back HVDC converter can be adjusted easily to minimise the voltage excursion during switching. Subsequent converter operation can be allowed to relax towards the normal steady state during a time long enough to permit tap-changers to retain control over the converter operating conditions. Thus, energising a capacitor bank would be accompanied by a sharp decrease in direct voltage/increase in DC (to increase the reactive power consumed by the converter). De-energisation of a capacitor would be anticipated by a gradual decrease in direct voltage/increase in DC, which would be released to regain their normal values when capacitor switching has taken place.

If the normal default steady-state operating conditions of the transmission system give rise to reactive power exchange with the AC network which is unacceptable, it may be necessary to increase the reactive power consumption of the converters by altering the steady-state direct voltage and current. This is particularly likely at low power transfer, when the AC harmonic filters required to control harmonic output may provide much more reactive power than is needed to compensate for normal converter action. This technique for limiting the consequences of capacitor switching or steady-state reactive power surplus is most appropriate when the AC network to which the second converter is connected is more capable of accepting reactive power unbalance than is the one which is the subject of control action. There is only one DC circuit, so changes in direct voltage and current which may be imposed to improve conditions on one AC busbar also affect the other.

If the HVDC converter under consideration is one terminal of a long-distance transmission, co-ordination between the two converters to minimise reactive power swings may be more difficult. In particular, such action may depend on the availability of a telecommunication link of adequate security. Thus, for long-distance transmission, the use of the converters to minimise voltage changes arising from switching shunt capacitive elements may be limited to what can be achieved without influencing the direct voltage or current.

However, it should be remembered that use of the converters to control reactive power involves operating the converters at relatively large switching voltages, incurring somewhat increased converter switching losses. Thus the use of converters for controlling reactive power is not free from cost, since these losses are often included in evaluation calculations.

The extent to which the AC harmonic filters should be subdivided is also related to the reliability and availability requirement. Typically, AC harmonic filtering is a service which is shared between two poles. Arrangements should be made to ensure so far as it is possible that no single failure can deprive the transmission of more than one section of AC harmonic filters/shunt capacitor bank.

4.4.5 Smoothing reactor and DC filter

DC filters are necessary only on schemes which have sections of high voltage or neutral overhead line. Neither back-to-back links nor schemes utilising only cable require a DC filter. The main purpose of the DC filters is to reduce the levels of interference induced into open wire telecommunication systems and nearby AC power lines. The filtering of high-voltage DC lines is best achieved by connecting the filter between the high-voltage line and the neutral at the converter station. DC filters can be implemented as either passive or active (power electronics assisted) filters.

For most purposes, it is the total loop inductance of the relevant part of the HVDC circuit which is important. The DC reactor is the supplementary component which is added when the sum of all other inductances is insufficient, or when the inductance from other sources is inconveniently distributed.

The DC reactor contributes to the source impedance through which harmonics are injected into the DC conductor system. If the entire DC conductor system is contained within a valve hall, then harmonic interference arising from coupling to it is minimal, and the DC reactor is not important in this respect. The most extreme illustrations of this logic are the McNeill back-to-back converter station in Alberta (Canada), the Chandrapur, Visakhapatnam and Sasaram back-to-back converter stations in India, and the GCCIA converter station at Al Fadhili in Saudi Arabia which do not use DC reactors at all, see References 5, 7 and 8. In these particular cases, the inductance of the DC circuit is provided almost exclusively by the converter transformers, with a small contribution from the thyristor valves.

These examples do not prove that all back-to-back links can be constructed without DC reactors without further consideration. Before concluding that a DC reactor is unnecessary, the control system must be shown to be stable without it, and the control system must also be capable of limiting harmonic interactions between AC networks to an acceptable level. There are few desirable features which a modern HVDC control system cannot provide by way of restricting and/or damping low-frequency effects on the power system to which it is connected, but there are limitations on the number of features which can be provided simultaneously, since the number of independent variables is limited to two, namely, direct voltage and DC.

The demands made upon HVDC filter circuits are influenced by the inductance of the DC reactor. When passive filter circuits were used exclusively, it was often convenient to use quite large DC reactor inductances, sometimes more than 0.5 H. There are even cases in which it has been convenient to divide the DC reactor into two sections, with an HVDC filter connected between them. With advances in power electronics the application of active HVDC filters has become economic in certain circumstances.

The DC reactor shields the thyristor valves from rapidly changing overvoltages which may be imposed on an HVDC conductor system. It is necessary to determine the distribution of such voltages between the reactor itself and the rest of the converter station in order to calculate the insulation requirements of the reactor. DC reactors are not normally protected by terminal-to-terminal surge arresters in modern converter stations (unless the station design incorporates two such reactors).

Historically, it was believed that the DC reactor inductance had to be selected to prevent the HVDC conductor system from exhibiting fundamental frequency resonance. Modern control systems, based on various implementations of the phase-locked oscillator principle, have made this restriction obsolete.

The earliest designs of smoothing reactors were oil-insulated and oil-cooled reactors. However in recent years the development of dry-type air-cored and air-insulated reactors has been such that this type of design now offers a very competitive alternative to the oil-filled reactors for most applications. Dry-type reactors have as number of disadvantages such as:

- they require a larger area (including magnetic clearance) than the equivalent oil-insulated unit
- an HVDC wall bushing is required
- difficulty with meeting audible noise requirements as the insulation and air flow must not be disrupted
- generally located outdoors, away from the valve hall building.

4.4.6 Switchgear

The AC switchgear is conventional, but its attributes must meet the specific operational requirements of the scheme. For example the circuit-breaker feeding the converter transformers needs to be capable of fast fault clearance in two or three cycles in order to reduce the exposure time of the converter valves to faults. The breakers feeding the AC harmonic filters have to be capable of frequent load switching of capacitors. In most cases, the AC switchgear is conventional SF_6 free-standing outdoor equipment. When space is restricted, such as was the case for the England/France link, indoor compact metalclad gas-insulated switchgear can be installed to minimise the layout area.

Capacitor switching is a particularly onerous duty for circuit-breakers. In capacitive circuits current zero occurs at the instant of peak voltage between the terminals of the capacitor. Since the capacitor takes several minutes to discharge, this voltage is still present one-half period of fundamental frequency later when the voltage applied to the AC network terminals has reversed. Thus, 10 ms after the arc in the switch is extinguished, the voltage between the terminals of the opening switch is 2 pu. If the disconnection takes place during a disturbance, this voltage can be considerably greater. This is the so-called 'double voltage' experience which is applied to switchgear during this duty and introduces the risk of a restrike across the contacts of the circuit-breaker.

For a monopolar DC scheme using sea return, the switchgear on the DC side is extremely simple. Generally disconnectors are provided only to enable the station to be isolated safely from possible incoming surges from the DC overhead line or cable during maintenance outages.

For a bipolar scheme the DC switchgear arrangement may be very extensive and complex to enable maintenance to be carried out on parts of the scheme without the need for a total shutdown. Figure 4.7 shows the switchgear used on one pole of a DC scheme using two series-connected 12-pulse groups in each pole. In this case special converter operating modes or switches capable of commutating currents from and to the converter valves are necessary if operation with only one 12-pulse group is required. Bypass switches with current transfer capability are required for each group to enable maintenance of one group to be carried out while the other remains on-load.

In most cases the switchgear used on the DC side of the converter comprises modified versions of off-load disconnectors designed for AC applications, since the operation of the DC switchgear can be co-ordinated with the DC link controls to ensure that the switchgear is not operated at times when DC current flows through the switch.

However, in special cases such as schemes using parallel connected DC lines, or where permanent use of ground return operation is not permitted, it is possible to provide special DC switchgear capable of transferring DC current to another path.

Figure 4.7 shows the metallic return transfer breaker and the ground return transfer switches which are necessary when changeover between monopolar earth

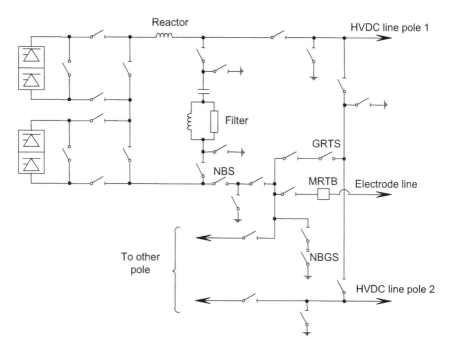

Figure 4.7 Typical DC switchgear arrangement
GRTS = ground return transfer switch; NBGS = neutral bus grounding switch; NBS = neutral bus switch; MRTB = metallic return transfer breaker

return and metallic conductor operation is required without an interruption in DC power flow.

If earth electrodes are used, they will normally be located several kilometres from the converter station. In consequence the neutral connection of the HVDC converter will exhibit appreciable resistance and (more importantly) inductance. This means that during fault conditions overvoltage may occur and precautions have to be taken to ensure that destructive currents cannot occur either in valves or in surge arresters. To avoid such conditions additional inductors may have to be added.

4.4.7 Surge arresters

Surge arresters using metal oxide non-linear resistors are used extensively in HVDC schemes for overvoltage protection (see Figure 4.2). In addition to the surge arresters connected directly across each of the thyristor valves (V), arresters are also used across each six-pulse group (B, M) in order to ensure good voltage distribution between the series-connected groups. Protection of HVDC equipment connected to the DC line is achieved by a surge arrester connected directly to the DC line termination (DB).

The surge arresters use highly non-linear zinc oxide resistor blocks. At normal voltage they conduct less than 1 mA, but at a voltage only 60–70% higher they can conduct appreciable current (such as 1 kA). The arresters are built up from a number of series-connected blocks mounted in a porcelain or a polymer housing. They are very simple in construction.

Surge arresters are also used on the AC side. The busbar phase to earth insulation is protected by surge arresters (A), which are placed close to the transformer terminal. Surge arresters are also applied within the AC harmonic filters (FA1, FA2), connected across the reactors or resistors, where the resulting saving in component insulation levels more than compensates for the cost of providing the surge arresters.

One of the tools used for the study of insulation co-ordination is an AC/DC simulator which provides a real-time representation of the complete system including the converter controls. Studies of a large number of different AC and DC system faults can be performed within a relatively short period using this simulator in order to identify the conditions which produce the worst overvoltages. Digital computer studies complement the simulator studies and are carried out to establish with greater precision the magnitudes of the worst cases, the co-ordination currents and the energy absorbed in the surge arresters.

4.4.8 Valve cooling

Modern HVDC valves are liquid cooled. The liquid offering the best heat transfer properties is water, but pure water has the disadvantage that it freezes. Where exposure to freezing temperatures is anticipated water/glycol mixtures are used. At this level of detail differences emerge between the practices of different suppliers and the requirements of different customers. Water/glycol mixtures may be circulated through the valves themselves or a secondary coolant circuit utilising water/glycol may be used to transfer heat from a primary cooling circuit using fine water. Direct water/glycol cooling is used in the design of the cooling system for the converter valves in the McNeill back-to-back system because the customer specified that no valve damage should occur when all site services were lost in mid-winter at $-50°C$.

Because, first, the coolant forms part of the insulation system and, second, the resistance of the coolant has to be high to minimise leakage currents, it must be maintained in a very pure condition. Deionisers are provided, but the basic design of the cooling system must ensure that the materials used for each individual component is satisfactory in terms of compatibility with all the other materials in the system and to survive for the life of the scheme in the environment of the pure demineralised water or water/glycol coolant.

Figure 4.8 shows a flow diagram for single-circuit valve cooling system. A high degree of redundancy is provided in the cooling circuit to ensure that single component failures do not result in a shutdown of the HVDC converter. For example the coolant circulation pumps are duplicated with one online pump and one standby pump which will be switched on in the event of a failure of the online pump. Similarly, the critical instrumentation is duplicated and redundancy is provided in the air blast coolers in the form of either redundant fans or one or more complete redundant cooler units.

The majority of the cooling system is provided as a frame-mounted unit, fully assembled and tested. This assembly will include the circulating pumps, control valves, the deioniser and much of the instrumentation. This complete assembly can be transported easily to site for connection to the converter valves and to the outdoor heat exchanger. This approach reduces the site installation cost, reduces the commissioning time and gives improved quality control.

4.4.9 Auxiliary supplies

AC auxiliary supplies are normally duplicated (or even triplicated), very often with standby generators as well. The intention is to secure continued power transmission against any single failure, and often to continue to provide some level of power transfer in the event of two unrelated failures. Thus, a power supply arrangement will have two independent sources, each capable of supplying the total load for the whole station via a sectioned distribution busbar. In some schemes where maximum independence between the two poles is required duplicated independent sources are used for each pole, and such an arrangement is illustrated in Figure 4.9.

Even if all auxiliary AC power is lost, sufficient energy is stored in batteries to ensure that safe shutdown of the converter equipment can be accomplished and that the minimum station services including such things as emergency lighting and operation of switchgear continue to be provided with operating power for several hours after AC infeed has ceased.

4.5 Converter building

The valves and the converter transformers are brought together in the valve hall. Wherever possible the converter transformers are arranged along one wall of the valve hall, with their DC-side bushings penetrating into the valve hall where

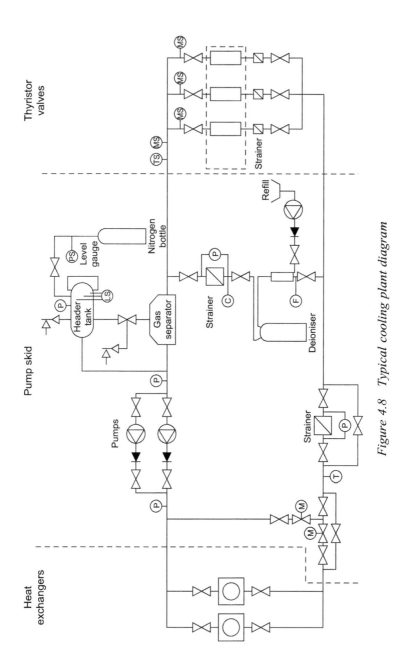

Figure 4.8 Typical cooling plant diagram

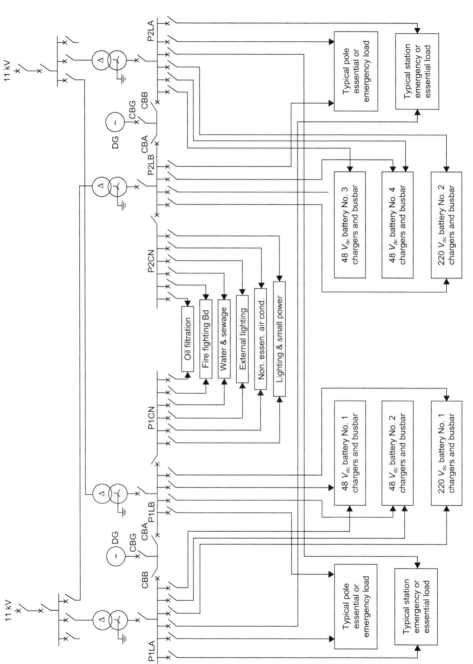

Figure 4.9 AC auxiliary power supply scheme for converter station

the connections to the HVDC valves are made. In parallel with and adjacent to the valves are the surge arresters.

In the case of smaller HVDC schemes (typically only 200 or 300 MW) it is often possible to utilise only a single three-winding converter transformer for each 12-pulse bridge. The transformer would have two valve windings, connected star and delta respectively. Such a converter transformer does not occupy a very long run of valve hall wall and the length would be determined by the valve arrangement. For larger schemes, such as those intended to transmit 500 MW or more, it will usually prove to be necessary to subdivide the converter transformer. The simplest subdivision is into two three-phase transformers, one star connected and the other delta connected. However, if transport restrictions are severe it may not be possible to use three-phase transformers. In this case single-phase units having two or three windings are disposed along the wall of the valve hall. This normally determines the length of the valve hall and is a serious economic consideration.

In addition to providing protection from contamination for the converter transformer valve winding bushings, more extensive contamination protection has been provided on some schemes. For the English terminal of the England/France scheme all the DC equipment that is switchgear, surge arresters, PLC equipment and cable sealing ends were built inside a protected but not air-conditioned enclosure, formed by enclosing the space between the valve halls of a bipole [9]. The DC reactor bushings also protrude into this enclosure. Experience has shown the benefits of this design since no flashovers on DC-associated equipment have occurred on the English terminal during 25 years of operation. Similar practice has been adopted on the Mainland to Jeju Island 300 MW scheme in Korea since severe contamination is possible due to the close proximity of the sea [10]. Environmental protection for the AC filters on the island of Jeju has also been provided by installing the filters inside a building.

The converter building needs to house the following additional items:

- workshops
- control equipment
- all auxiliary supplies and services
- storage space for components
- storage space for access platforms to provide maintenance and repair access to the HVDC valves.

Often purchasers wish to accommodate additional facilities such as offices, meeting rooms and hubs for their telecommunications networks. These features are contract-specific and are too variable to be treated usefully in a general review and Figure 4.10 shows a typical layout of the control building and valve halls for an HVDC scheme.

4.6 VSC HVDC

In a similar manner to the development of STATCOM discussed later there has been a development of voltage-sourced converters (VSC) for HVDC transmission.

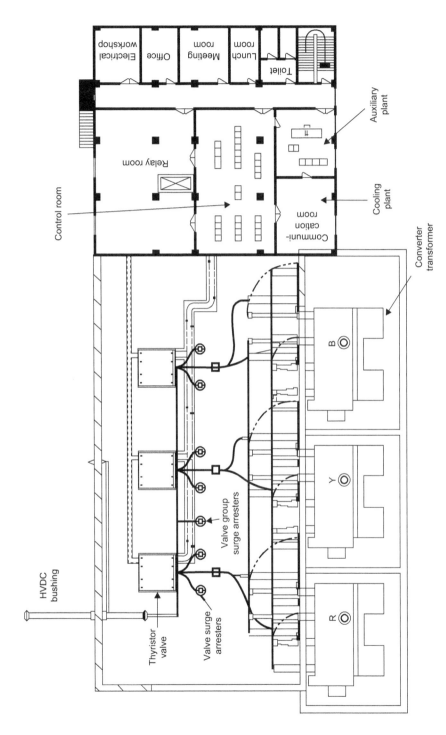

Figure 4.10 Typical layout of the control building and valve hall

174 High-voltage engineering and testing

In the same manner as a line commutated converter (LCC) HVDC system the VSC HVDC system transfers power between the two converters due to the difference in voltage over the DC circuit resistance. However, the VSC interacts with the AC system as a STATCOM but allowing real and reactive power flow as shown in Figures 4.11 and 4.12.

The fundamental concept is that the VSC generates its own voltage with controlled amplitude and phase angle which governs the current through the connecting inductance resulting in real or reactive power flow as though through a transmission line.

There are a number of methods for generating the converter voltage, but two methods are widely used – pulse width modulated (PWM)-based converters and multimodule converters (MMC).

The PWM converter is a conceptually simple converter with two or three levels but switching at a high frequency. This results in high losses and high switching transients and some harmonics.

The alternative MMC is a series of converters which switch at low frequency to build a good approximation to a sinusoidal voltage source as shown in

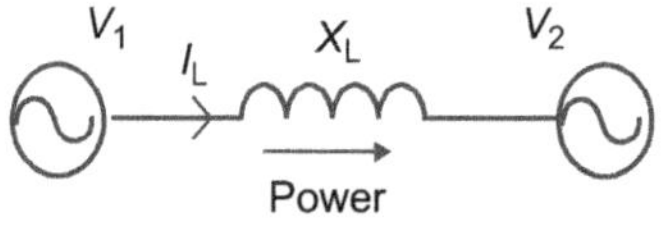

$$\text{Power} = \frac{V_1 V_2}{X_L} \sin \delta$$

V_1 = sending end voltage

V_2 = receiving end voltage

X_L = reactance of transmission network

δ = angle between sending and receiving end voltages

Figure 4.11 VSC interaction with AC system

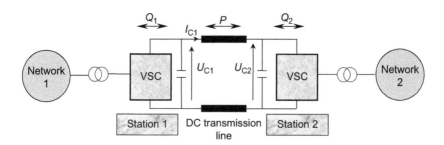

Figure 4.12 VSC transmission

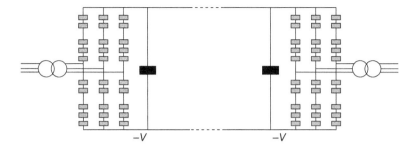

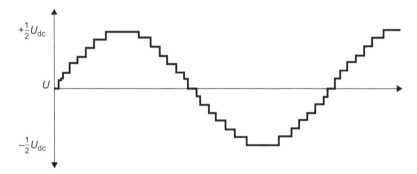

Figure 4.13   VSC converter

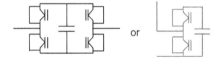

Figure 4.14   VSC modules

Figure 4.13. The principle is that each valve of the bridge can make nearly a sinusoidal voltage waveform by switching an appropriate number of modules in service as required. The individual module detail is given in Figure 4.14.

As a result of the design of the VSC system there is a capability to control both real and reactive power as shown in Figure 4.15. The key point is that without the use of additional shunt reactors or capacitors/filters the VSC converter can import and export real and reactive power.

This greatly increased capability comes at a cost:

- much more complex converter than LCC
- *many more devices* – current thyristors can achieve 5,000 A and 8.5 kV, whereas IGBTs are presently achieving 1,500 A and 4 kV
- increased losses although this gap is narrowing.

176 High-voltage engineering and testing

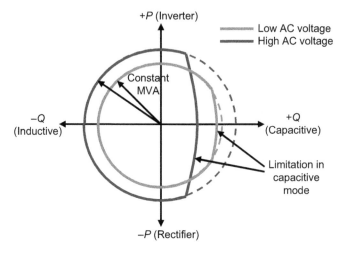

Figure 4.15 VSC capability chart

4.7 Economics

Figure 4.16 illustrates the cost breakdown for a typical converter station. Inevitably, there will be wide variations from this in special circumstances, but it serves to illustrate the basic truth that more than half the cost of a converter station is represented by the HVDC converter itself and the converter transformers. Therefore, it is mainly around these items that the design of the converter station is optimised.

Figure 4.17 shows the approximate distribution of losses throughout a converter station. All power transferred passes via the converter transformers and the converter valves, and they are the components which generate the greatest losses. Since the losses which result from operating the equipment represent lost revenue, purchasers have a legitimate interest in minimising them and evaluate them during

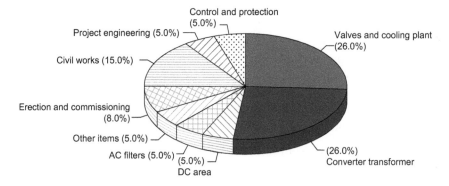

Figure 4.16 Approximate distribution of capital cost of converter station

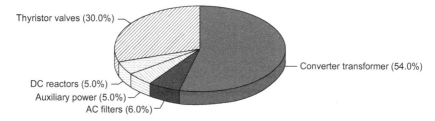

Figure 4.17 Approximate distribution of losses of converter station

bid adjudication at rates declared in the specification. Usually, different rates are applied to losses which remain fixed and those which vary with load.

It is essential to take account of the costs attributed to losses by purchasers, since they vary widely, depending on local circumstances. However, certain general trends can be identified. First, the evaluated cost attributed to losses is comparable with the cost of the major plant incurring them. This makes them important, since, for example, the cost of a converter transformer is commonly several million pounds. Second, the incremental cost of minimising losses may not vary smoothly. Thus it is often found that there is no advantage in using a smaller thyristor for a transmission scheme required to carry DC of only, say, 1,000 A, and it is commonplace to find thyristors similar in size to those used for transmitting 3,500 A DC being used at much lower currents. (Where available, a higher voltage rating would normally be employed, minimising the number of series-connected levels in each valve.) Third, if evaluation rates are high, it may be advantageous to employ energy-saving techniques, such as operating cooling fans at reduced speed during cold weather. In this case, an evaluation has to be made of the impact of such additional complication on the availability of the equipment. To save auxiliary energy in exchange for losing the ability to transmit power reliably would be a bad trade.

4.8 Power electronic support for AC systems

The basic limits of transmission are well known. The thermal capacity in terms of current in a line is usually far above its normal loading. There is a variety of reasons for this:

- At high-voltage conductor size is primarily dictated by corona considerations.
- Unequal sharing between parallel lines (both intentionally and accidentally) results in the effective limit being dictated by the line with lowest rating.
- Circuits must have sufficient capacity to absorb the consequences of failures in parallel circuits.
- The phase angle between the sending end and receiving end voltages is limited by transient stability considerations during fault conditions.
- Loop flows (power which is scheduled to pass through a parallel system of lines) effectively reduce the available capacity of a line as the loop flow takes up the thermal and protective margins in a transmission line.

178 *High-voltage engineering and testing*

It is possible to add a great deal of controllability to an AC system by embedding an HVDC link.

Sometimes the aims are more modest than those which require an HVDC link embedded in an AC system. In such cases a simpler solution than an embedded HVDC link is to use one of the 'FACTS' (flexible AC transmission systems) type of power electronics–based devices mentioned in section 4.1 [see also the many devices considered in Chapter 1, section 1.6.2.1 to 1.6.2.16].

4.8.1 Static var compensators (SVCs)

Traditionally, variable reactive power compensation has been provided by synchronous compensators and saturated reactors. These were used to compensate for the lack of voltage control resulting from transmitting power over long lines. While these performed adequately they are characterised by a lack of flexibility and high capital and maintenance costs.

In recent years SVCs using thyristor switches (Figure 4.18) have been used extensively, particularly in the UK. See References 11 and 12. These SVCs generally consist of some combination of thyristor-switched capacitors (TSCs) and thyristor-controlled reactors (TCRs). The choice of TCR/TSC/filter combination is largely dictated by loss evaluation, harmonic performance requirement and restrictions in the design of individual components. In some installations overseas such as at Watertown, South Dakota, USA, the SVC also controls external switched reactors and capacitor banks [13].

A TCR provides continuously variable reactive power absorption and acts to reduce system voltage as the TCR current increases. However, this results in the generation of harmonic currents from the TCR which can be controlled by configuring the fixed capacitor as a harmonic filter bank.

TSCs are not phase controlled because the stresses caused by such switching of capacitors are excessive. They act as very precisely switched banks and

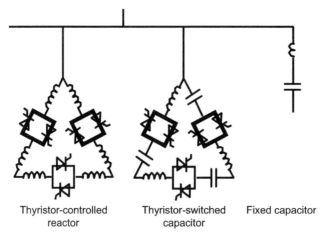

Figure 4.18 *Simplified single-line diagram of TCR/TSC-type SVC*

HVDC and power electronic systems 179

Figure 4.19 View of 225 MVA SVC

consequently do not suffer the usual wear and tear from the repetitive switching which is found in circuit-breaker-switched capacitors. Indeed SVCs consisting of only TSCs which are used to perform swing damping have been supplied to NGC in the UK and in the western USA.

The thyristor valves used for TCRs and TSCs are largely based on the water-cooled, air-insulated, banded pair construction which is also the basic building block of the HVDC thyristor valve. If a suitable connection voltage is not available (typically 2–25 kV) a stepdown transformer is required to connect the SVC to the power system.

A picture of a typical SVC installation is shown in Figure 4.19.

4.8.2 STATCOM

The STATCOM is an SVC based on the GTO or IGBT (see Figure 4.20). This device provides phase-controlled output – both capacitive and inductive – and can respond very quickly to disturbances. The STATCOM also provides a response superior to that of the SVC at low voltage since its output is proportional to voltage which is less variable than the square of the voltage with which the SVC output is proportional.

The basic concept behind the STATCOM is that of a voltage-sourced inverter. The reactive power exchanged with the power system is dictated by the relative magnitude of the inverter voltage (V) and the power system voltage (E). For example when the inverter voltage (V) exceeds the system voltage (E) then current flows 'from' the converter to the system and thus generates reactive power in a manner similar to a capacitor. With this design it is possible to change the STATCOM from fully inductive to fully capacitive in a single half cycle of the power frequency.

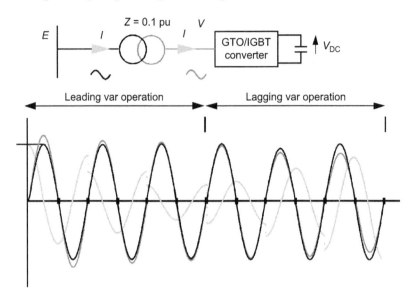

Figure 4.20 Basic theory of STATCOM operation

A STATCOM is very compact, occupying approximately 40% of the area of a comparable thyristor SVC. ALSTOM is currently supplying such a device rated at 225 MVA to NGC in the UK at their 400 kV East Claydon substation [14]. This compensator is also relocatable, which is an important consideration given the changing nature of the electricity network following deregulation.

The STATCOM has a number of other advantages over the thyristor SVC (TCR/TSC):

- Using GTOs or IGBTs (with their ability to 'turn off' at other than zero current) the STATCOM can respond much more quickly than a conventional TCR/TSC SVC.
- The use of GTOs or IGBTs also allows the STATCOM to act as an active filter and reduce significantly the harmonics it produces compared with a TCR of the same rating.
- The STATCOM provides significantly enhanced low-voltage reactive output because it is a controlled device rather than variable impedance. Figure 4.21 shows the extended low-voltage operating range of the STATCOM compared with that of a TCR/TSC-type SVC.

A picture of a typical STATCOM installation is shown in Figure 4.22.

4.8.3 Series compensators

Another solution to compensating long lines has been to insert series capacitors. This introduces a new natural frequency and results in the possibility of resonance with generators and large machines at sub-synchronous frequencies. One solution

HVDC and power electronic systems 181

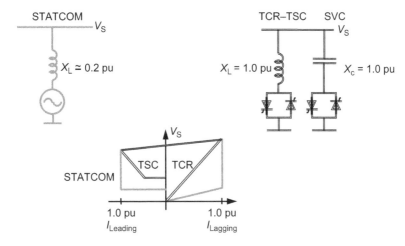

Figure 4.21 Comparison of TCR/TSC-type SVC and STATCOM operation

Figure 4.22 View of 225 MVA STATCOM

has been to effectively add a TCR in parallel with the series capacitor (Figure 4.23). However, this requires an extensive high-voltage insulated platform.

A new solution being investigated presently is to apply a voltage-sourced GTO converter to a series connection (Figure 4.24). This static series compensator presents a voltage source to the transmission line and so removes the sub-synchronous threat. It also has the advantage over the thyristor-controlled series capacitor – that it only requires a transformer and a conventional building (possibly portable) rather than the high-voltage platform.

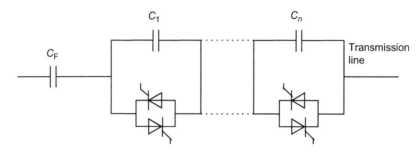

Figure 4.23 Thyristor-controlled series capacitor

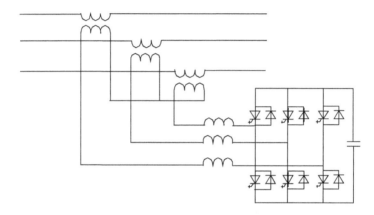

Figure 4.24 Series compensator concept – STATCOM-based variant

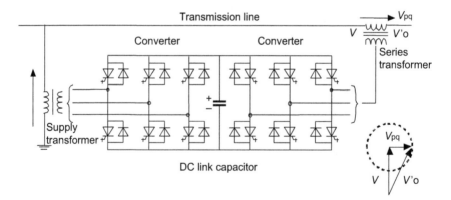

Figure 4.25 Simplified single-line diagram of UPFC

4.8.4 Unified power flow controller (UPFC)

This combines the STATCOM- and GTO-based series compensator in one facility (Figure 4.25). Both transmission angle and voltage come under the influence of this controller, which operates in all four quadrants on the real/reactive power diagram.

The UPFC consists of two voltage-sourced converters, one connected in shunt and one in series to satisfy especially demanding circumstances.

However, the UPFC requires the GTO-controlled capacity of the individual static and series compensators, whereas the user may not always require the series and shunt compensation in the same place.

4.9 Power electronics for industrial applications

In the same way that power electronics can support the power system, they can also support users of power.

Probably the most common use of power electronics in this area is the use of static compensators to compensate for disturbing loads. Typical disturbing loads are mine winders, rolling mills, large drives and arc furnaces. Generally the interface with the utility is governed by a series of power quality requirements such as those in the IEC series [15–17], or more often the national implementation of these standards.

In many cases such SVCs are directly connected to the factory MV busbar (typically in the range 6–35 kV) and controlled to compensate the factory load and not to control the busbar voltage. An example of this is described in Reference 18.

For disturbing loads with a higher frequency response such as an electric arc furnace a fast compensator is required. Very often this is implemented as one or more STATCOM devices such as shown in Figure 4.26. This device is IGBT based and so is capable of high-speed switching to follow the load compensation profile much more closely than a thyristor-based SVC. The performance or a typical installation is shown in Figure 4.27. The improvement is visible and this is shown in the reduction of flicker severity from a Pst of more than 2.0 to 0.5 where 1.0 is generally considered the limit of acceptability.

At MV a great variety of drives are available, many of which include some form of compensation of load, balance or harmonics. In some cases the power electronic devices not only compensate the reactive element of the load but even insert a few cycles of power to enable a load to withstand brief power interruptions. A dynamic voltage restorer (DVR) is an example of such a device.

A major use of power electronics at MV is the connection of wind turbines to the grid. Many modern wind turbines contain VSC DC converters through which all power is converted and processed. Other wind turbines of the DFIG (double fed induction generator) variety use power electronics to supply the field current.

4.10 Conclusion

HVDC is in many respects the ultimate power electronics-based FACTS device which can act within an existing AC network or even replace parts of it. In evaluating the usefulness of HVDC it is important to remember the controllability which it adds to the power system in addition to its bulk power transfer capabilities. HVDC often provides the best of both worlds – support from a neighbouring network without the problems of synchronous interconnection.

184 *High-voltage engineering and testing*

Figure 4.26 Picture of 2 kV 2 MVA cubicle-based STATCOM

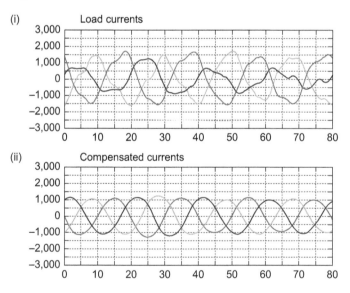

Figure 4.27 STATCOM performance showing:
  (i) load currents and
  (ii) compensated currents

Power electronics-based devices enable better utilisation to be achieved of all the principal attributes of the electricity network – generation, transmission and distribution. Where slow acting devices such as tap-changers and circuit-breakers can be used these devices are usually most economic. However, if fast or controlled response is required for example to prevent system collapse then FACTS power electronic facilities provide a cost-effective solution.

In many cases power electronics is the only way to achieve some functions, particularly with HVDC. Otherwise, they can be justified where the application requires one or more of the following attributes: rapid control action, frequent variation of output, smoothly adjustable output and precise output.

References

1. CIGRE* TB 417-WG B4.45 (2010), 'Technological assessment of 800 kV HVDC applications' (also *Electra*, 2010, **250**(6), pp. 51–59)
2. Kimbark E.W. *Direct current transmission*. New York, NY: Wiley Interscience; 1971
3. Arrillaga J. *High voltage direct current transmission* (Vol. 29). IEE Power and Energy Series. London: The Institution of Electrical Engineers; 1998
4. Uhlmann E. *Power transmission by direct current*. Berlin-Heidelberg: Springer-Verlag; 1975
5. Burgess R.P., Kothari R. 'Design features of the back-to-back HVDC converter connecting the western and eastern Canadian systems'. *IEEE 1989 PES Summer Meeting*
6. CIGRE brochure TB 364-WG B 4.41 (2008), 'Systems with multiple DC infeed'
7. Wheeler J.D., Haddock J.L. 'Chandrapur back to back HVDC scheme in India'. *ICPST Conference*; Beijing, October 1994
8. Barrett B.T., Macleod N.M., Sud S., Al-Mohaisen A.I., Al-Nasser R.S. 'Planning and design of the Al Fadhili 1800 MW HVDC inter-connector in Saudi Arabia'. CIGRE; 2008, Paper B4-113
9. Goddard S.C., Yates J.B., Urwin R.J., Le Du A., Marechal P., Michel R. 'The new 2000 MW interconnection between France and the United Kingdom'. CIGRE; 2008, Paris, Paper 14-09
10. Thanawala H.L., Whitehouse R.S., Kwon G.O., Lee S.J. 'Equipment and control features of Haenam-Cheju HVDC link in South Korea'. *CIGRE 1994 Session*. **August–September 1994**; 28(3)
11. Young D.J., Horwill C., Mukhopadhyay S.B., Haddock J.L., Gardner D. 'City versus country – a comparison of two types of standardised SVC for the National Grid in England'. *AC and DC Transactions IEE*. 1991:248
12. Whitlock I., Young D., Baker M. 'Static var compensators for the city – St John's Wood'. *Modern Power Systems*. 1994(June):45
13. Harrington B.A. 'Improving system stability'. *T & D International*. 1995(4th quarter):20

14. Knight R.C., Young D.J., Trainer D.R. 'Relocatable GTO-based static var compensators for NGC substations'. CIGRE Paper 14-106, Paris; 1998
15. IEC 61000-3-6. 'Electromagnetic compatibility (EMC) – Part 3-6: Limits – Assessment of emission limits for the connection of distorting installations to MV, HV and EHV power systems'
16. IEC 61000-3-7. 'Electromagnetic compatibility (EMC) – Part 3-7: Limits – Assessment of emission limits for the connection of fluctuating installations to MV, HV and EHV power systems'
17. IEC 61000-3-13. 'Electromagnetic compatibility (EMC) – Part 3-7: Limits – Assessment of emission limits for the connection of unbalanced installations to MV, HV and EHV power systems'
18. Horwill C., O'Heidhin G.S. 'AZ to UK, a new life for an SVC', *IEEE Power Engineering Society General Meeting*. 2003;**1**

---

* CIGRE key: SC, Study Committee; WG, Working Group and JWG, Joint Working Group; TF, Task Force; TB, Technical Brochure; TR, Technical Report; SC, Scientific Paper; CIGRE also run conferences tutorials worldwide and publish *Electra* bi-monthly.

Chapter 5

The implications of renewable energy on grid networks

Adrian Wilson

5.1 Introduction

The generation of electricity from renewable sources of energy is a central plank in the government's strategy to reduce the UK carbon emissions into the atmosphere. This chapter considers many of the available generation technologies associated with renewable energy and their differences compared to conventional generation. Finally it considers some of the new network solutions available to allow these technologies to connect to networks in a technically acceptable and cost-effective manner.

5.2 Drivers for renewable energy

5.2.1 Fossil fuel

In the UK, 89.8% of all energy consumed in 2010 came from fossil fuels [1]. For the production of electricity only 7% of the consumed product came from renewable sources in 2010 [2]. The UK government is keen to reduce the UK dependence on fossil fuels. This is in order to reduce the UK dependence on imported fuels and reduce the UK carbon emissions.

An EU directive to reduce energy consumption and carbon emissions by 20% by 2020 has been ratified. A Government Energy White Paper [3] highlighted the following items as evidence of climate change:

- an increase in the levels of atmospheric CO_2 of 30% since the start of the industrial revolution
- an increase in the Earth's temperature of 0.6 °C in the twentieth century
- retreating ice caps on many large mountains like Kilimanjaro
- 1–2 mm rise in sea levels in the twentieth century
- 40% thinning of Arctic sea ice in summer and autumn in recent decades
- global snow down 10% since 1960s
- use of Thames barrier from five times in 1980s to five times per year in the last five years.

Fossil fuel takes many hundreds of millions of years to form, so it is effectively a finite source of energy and inevitably we will have to stop using it at some point.

5.2.2 Nuclear fuels

In the UK, 8.2% of all energy consumed in 2010 came from nuclear fuels [1]. All of this went into the production of electricity, representing 17.4% of all the energy used to produce electricity [2]. The current installed capacity of nuclear power stations is almost 11 GW, although this could drop to 3.7 GW by 2015. Replacement of the UK nuclear power stations is under consideration by government and the decision to replace them will be political. The amount of spent radioactive fuel produced by modern nuclear power stations is quite small; however, it needs to be stored safely over a geological timescale. Nuclear fusion which holds some promise for the future is still many decades away from a commercial reality.

5.3 UK renewable energy resources and technology

5.3.1 Power transfers

Electricity power flows in England and Wales for an average cold spell (ACS) in 2004–5 are shown in Figure 5.1. The figure demonstrates the power flowing on the grid, so for example in Upper North (which is a net importer) there is still a power flow in from Scotland and out to the North area. This figure comes from the Great Britain National Grid Network Operator.

This shows a large import requirement for National Grid Transco's (NGT) Central (covering London, Central Southern England, Norfolk and Southern Wales) and South West (SW England) areas. Excess generation from France, South East England, Northern England and Scotland already supports these power requirements. The NGT network is built to support these power flows, so any additional generation in these locations will need to displace existing generation in these areas (during the winter peak).

5.3.2 Wind resources

Average wind speeds in the UK vary from around 4 m/s to about 10 m/s. The power in the wind varies by the cube of the wind speed, so every doubling in speed increases the potential energy captured by a factor of eight. The costs associated with installation of turbines in all areas are similar (with grid connection being possibly different), so developers tend to develop turbines in windy areas to ensure financially viable schemes.

Figure 5.2 shows the onshore wind resources around the UK demonstrating that the strongest average wind speeds are to be found in the North, around the coasts and on high ground.

Load centres in the UK are not generally located in windy areas, making the use of the electricity grid a necessity to transfer the energy from generation source to the load centre.

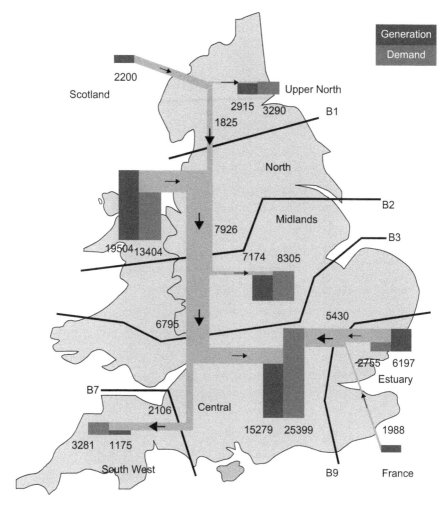

Figure 5.1 Average cold spell power flow (MW) pattern for 2004–5 [4]

5.3.3 Wave resources

Average wavefront energy in the seas around the UK varies from about 1 kW/m to around 70 kW/m of wavefront. The most energetic waves are found in the deep Atlantic Ocean waters with few obstacles in the path of the wave. Figure 5.3 shows the wave resources around the UK demonstrating that the strongest average waves are to be found approaching the North West of Scotland and South West of England.

Load centres in the UK are not generally located close to areas with the best waves thus making the use of the grid a necessity to transfer the energy from generation source to the load centre. The exception here is the South West of England where there is a generation shortfall.

190 High-voltage engineering and testing

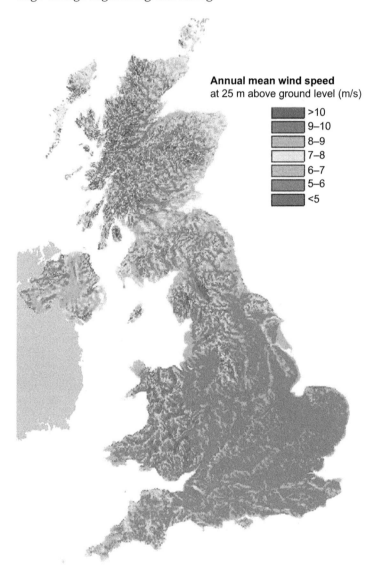

Figure 5.2 Average UK wind speeds from Department of Trade and Industry (DTI) [5]

5.3.4 Tidal resources

Average tidal flow rates around the UK vary from ~1 to ~8 m/s. The power in the tide varies by the cube of the flow rate, so every doubling in speed increases the potential power harnessed by a factor of eight. Figure 5.4 shows the tidal resources around the UK demonstrating that the fastest tides are to be found in shallow or narrow waterways.

The implications of renewable energy on grid networks 191

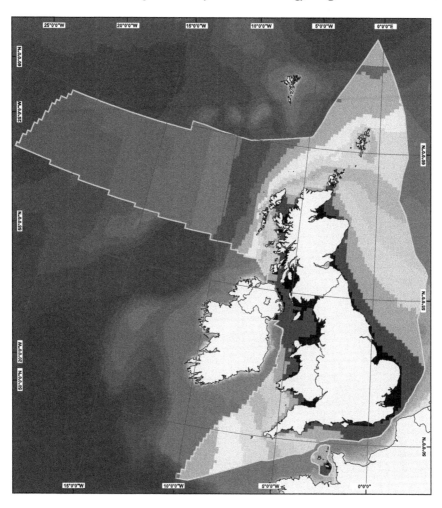

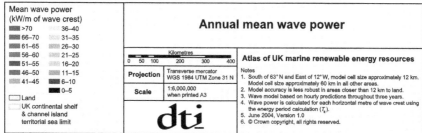

Figure 5.3 Annual mean wave power [6]

192 High-voltage engineering and testing

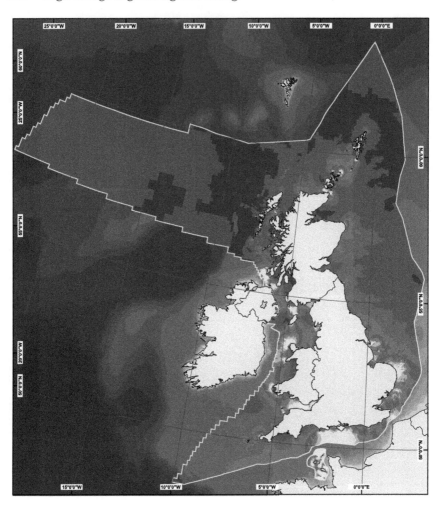

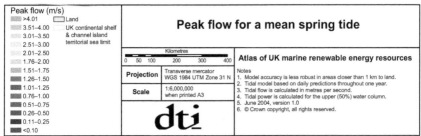

Figure 5.4   Peak flow for a mean spring tide [7]

Load centres in the UK are not too far from many tidal resource areas: South East England, Southampton, and Bristol and South Wales are all close to areas of good tidal flows.

5.3.5 Biomass resources

Biomass resources tend to be from three sources – wood (and grass), waste and oil. Wood-burning stoves and even wood combined heat and power (CHP) units are available for those areas with large wood resources. Quickly growing wood products, for example short rotation coppice willow or Miscanthus grass, can be grown now as biofuels. Natural decomposition can be harnessed for example via landfill gas power stations, and forced decomposition using anaerobic digestion processes can be used to treat waste, for example farm slurry – in the process generating a better fertiliser from the remaining waste solids. Oil seed rape in the UK and nut/palm oils elsewhere in the world can be harnessed to produce fuels which could be used to displace crude oil-based products. Landfill gas and biomass burning power stations are likely to be small (1–3 MW) and widely dispersed in the electricity network. Figure 5.5 shows the location of biomass resources in England.

5.3.6 Network implications from remote resource locations

Wind turbines currently have a maximum size of 7.5 MW (Enercon) and the largest installed farms are approaching conventional power stations. Roscoe Wind Farm, USA, is a 735 MW onshore windfarm, and the Thanet Offshore Wind Project is rated at 300 MW which compare with conventional power stations such as Ferrybridge at 2 GW and Hartlepool Nuclear Power Station at 1,210 MW. Wave, tidal, biomass and most hydropower stations all have similar ratings to wind turbines or wind farms.

Some implications are as follows:

- more generation in the distribution network – voltage rise, fault level, thermal and power quality impacts
- more generation connections
- more network required to carry energy from resource-rich locations, for example linking 'wind belt' (Scotland) to 'coal belt' (Yorkshire/Lincolnshire/Nottinghamshire); planning issues for pylons
- generation tends not to be subject to central dispatch (If it needs to be in the future then communications will be more complex.)
- different generation technology at 1 MW level compared to 1 GW level
- changes to the assumptions regarding availability of generation
- different (worse) utilisation patterns regarding total power flows on the system
- harder to control centrally.

5.3.7 Generator technologies – conventional power stations

The synchronous generator is the absolutely dominating generator type in power systems. It can generate active and reactive power independently and has an important role in voltage control. The synchronizing torques between generators

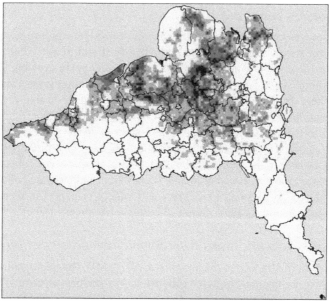
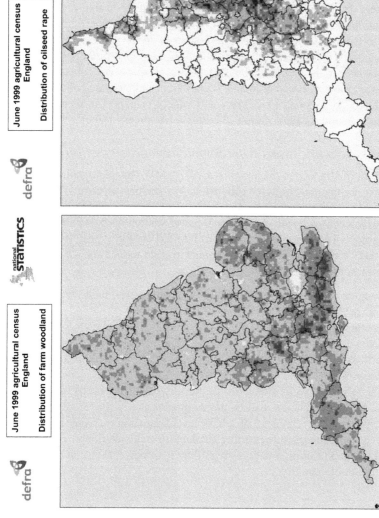

Figure 5.5 English biomass resources [8]

act to keep large power systems together and make all generator rotors rotate synchronously. This rotational speed is what determines the mains frequency, which is kept very close to the nominal value 50 Hz. The reactive generation capability of the synchronous generator can be used to control the voltage at the generator bus. This is done by letting an automatic voltage regulator (AVR) control the rotor field current which in turn determines the internal voltage. The AVR measures the voltage at the generator terminals and adjusts the field current so that terminal voltage is close to a reference value. The entire electricity network is designed on the assumption of synchronous generator-type responses to voltage and frequency excursions.

5.3.8 Generator technologies – full converter wind turbine

Wind turbines such as those from Enercon [9] and Zephyros [10] have synchronous and permanent magnet generators respectively. These generators rotate at a speed related to the wind speed (and the setting of the blade pitch system) and generate variable-voltage, variable-frequency alternating current (AC) electricity. This is rectified and fed to the grid at the correct voltage and frequency via an inverter. The downside of this design is that inverter is quite a costly part of the overall package; however, the design allows for a gearbox-less design.

5.3.9 Generator technologies – partial converter wind turbine

Wind turbines such as those from Vestas [11] and RE Power [12] have asynchronous generators. These generators still rotate at a speed related to the wind speed (and the setting of the blade pitch system); however, the rotor fields of the generator are controlled to ensure the wind turbine output generated is at grid frequency and voltage. The inverter is rated at around 25–30% of the turbine output, reducing the cost of the power electronics in this design.

5.3.10 Generator technologies – wave machines

Wave machine technology has not yet progressed to the point of commercial designs and market leaders. Variable speed technologies seem to be the most suitable and are likely to be taken from wind turbine technologies. There has been some development of linear generators [13] for this market. Energy storage is also necessary, bridging the gap between an energy delivery period of 10–25 s and an energy take-off rate at 50 Hz.

5.3.11 Generator technologies – tidal machines

Tidal machine technology has not yet progressed to the point of commercial designs and market leaders. Variable speed technologies seem to be the most suitable and are likely to be taken from wind turbine technologies. Power point tracking will be necessary adapting the power take off to the 12 h tidal cycle.

5.3.12 Generator technologies – biomass-fed generators

Biomass-fed generators can be governed by the rate of fuel input. This makes the technology suitable for fixed-speed asynchronous (induction) or synchronous generators.

5.3.13 Generator technologies – microgeneration

There are a range of micropower generators of both heat and electricity. Heat-producing technologies include biomass burners (e.g. wood-burning stoves); solar thermal hot water panels which use direct sunlight to heat water; photovoltaic cells that generate electricity direct from sunlight, wind turbines that generate electricity direct from the passing wing, gas or wood powered CHP units that replace a conventional boiler to generate heat and electricity; and heat pumps which use ambient temperature in the air, ground or water to heat a home.

Micropower is rated in the few kilowatt range, so individual installations will make no difference to the national picture; however, they have the potential for mass market adoption, since prices are driven by the economics of mass production. The government is supporting these technologies with feed-in tariffs [14] to try and encourage this market to grow. This has led to bunching of installations where developers have installed units as part of planning permissions or the 'keeping up with the neighbours' mentality has kicked in. A renewable heat obligation is also likely to develop the market further, although with less impact on the electricity market.

Generators at the microgeneration level tend to be inverter connected (although there is at least one induction motor-based gas CHP unit) and have to meet either G83/1 or G59 connection standards [15] which are currently designed to protect the distribution network operator (DNO) networks, not to perform any sort of security of supply function.

5.4 Renewable generator technologies – network implications

5.4.1 Cost of connections

The cost of providing a generator connection varies with the location of the connection and the individual circumstances. In general, it is cheaper for the electricity network companies to provide a large connection than to provide many smaller connections. As the size of the connection drops to the micro level this generality is curtailed and often there is no additional cost of a co-located generator and load; however, as microgeneration adoption increases, even the cost of these small connections will increase.

NGT estimates [16] that a typical supergrid double busbar connection rated at 240 MVA, 400 kV to 132 kV connection point suitable for a traditional power station will cost around £4.34/kW, whereas a 90 MVA, 132 kV to 33 kV connection suitable for a large (onshore) wind farm will cost around £3.82/kW. OFGEM estimates [17] that the average cost of a distributed connection is around £10/kW.

These connections are suitable for the majority of renewable connections. Clearly if the UK is to move away from conventional, large power stations to smaller distributed forms of power generation then the network cost to the UK energy users will increase to cover the additional costs. OFGEM has identified this trend and problem and has launched a number of initiatives to encourage innovation in connections including the Innovation Funding Incentive and Low Carbon Network Fund Schemes [18] where R&D work can be developed and demonstrated so that lower connection costs can be demonstrated. Changes in the way networks are paid for have also been implemented so that the cost of generator connections is now amortised and charged over a longer period of time compared to them being an up front cost to the generator. Shallow and proportional charging regimes [19] have been established so that developers no longer find their development 'the straw that breaks the camel's back' and having to pay disproportionate connection charges.

5.4.2 Voltage rise

As generation is installed in distribution networks, especially those with relatively low loads, when minimum load and maximum generation conditions occur the statutory voltage levels [20] can be exceeded. This requires some remedial work to be carried out to allow the generator to connect. Generally this will be some form of reinforcement to stiffen up the supply so that the proportion of generation swing is no longer significant enough to cause the breach of statutory limits.

Voltage issues can also include the additional wear on tap-change transformers. These transformers are designed to keep the voltage within statutory limits by altering the transformer tapping as the load changes. If distributed generation is also attached then as the generation changes this too will require the transformer to tap to maintain the voltage level. If the load and generation are in sync (e.g. CHP, PV) then tapping will reduce, whereas if they are not linked (e.g. wind, wave, tidal, biomass) then tapping is likely to increase.

5.4.3 Load flow

Generation may overload the thermal ratings of the existing network infrastructure. Transformers and cables are the usual bottlenecks and these would probably require changing to allow unrestricted generation connections. Often it is possible to allow a constrained connection where the generation is turned down or off during network maintenance periods. This often has a minor impact on renewable energy project revenue but is a significant saving on the connection cost.

5.4.4 Fault level

Generation in any network will increase the local fault level. Fault-level contribution will be higher for rotating machines directed directly compared to generation connected through an inverter. This is often worse in urban networks where load levels are already high, so local generation can cause the fault-level rating of switchgear, cables or substation equipment to be exceeded.

5.4.5 Power quality

Power quality issues (flicker, harmonics, electromagnetic radiation, blackouts) can be caused by power electronic-connected devices, high levels of magnetic field, control system problems or uneven power flows caused by a physical phenomenon of the device in question – blade pass frequency effects of wind turbines or uneven power transfers from wave machines for example.

5.4.6 Network extensions

Renewable energy resources are sometimes at some distance from existing networks. Wind farms tend to be on exposed hilly areas or costal locations, and due to the noise of the wind turbines, at least 400 m from the nearest dwelling. Wave and tidal machines will be deployed away from existing networks except in very unusual circumstances. New network connections are often necessary and these require planning approval in the same way as the development itself. Where long extensions are necessary, these can require many different local approvals, or central government approval. Well-organised protesters can cause long delays to the process; the REVOLT group [21] delayed by 12 years the approval of the second Yorkshire line running between Teesside and York. The Highlands before Pylons group [22] have the Beauly to Denny upgrade (a route that takes in some of the Cairngorm national park) in their sights. The Beauly to Denny upgrade will allow the connection of some 2,300 MW [23] of renewable energy in the North of Scotland, an area blessed with significant renewable energy resources.

Offshore network ownership has not yet been determined – should NGC/DNOs own and operate cables out to hubs where multiple wind/wave/tidal farms (of different ownership?) can connect, or should developers own these assets and connect, via the designated land-based 'termination-interface' station, to the selected distribution network? Consultation on these matters is being undertaken by the DTI/OFGEM. HVDC networks, such as the England/France link, have also been considered for offshore wind farms – the break-even point is around 80 km offshore and 800 km onshore.

5.4.7 Regulation

The electricity industry is heavily regulated. OFGEM has put into place measures such as NETA and BETTA that ensure that the supply business is largely competitive. The government has put into place a Renewable Obligation Certificate (ROC) to financially encourage suppliers to meet the government's targets for generation from renewable sources. Distribution and transmission are regulated, including investment levels in new infrastructure. Supply is also regulated. Largely the regulation is to keep the 'business-as-usual' models delivering value for customers and maintaining the reliability and safety of the network. A change in paradigm such as generating significant quantities of power from micro-renewables has found itself being hampered by regulation. The rules for metering, sale of energy and connection to the grid, to name but three issues have been designed for large scale generation. Smaller generators spend disproportionately more on these items and without subsidy the value to be gained from small scale generators, especially micro-generators, is questionable. Further details can be found on the OFGEM website [24].

5.4.8 Grid Code issues

The National Grid Company (NGC) uses a technical document called the Grid Code [25] to govern those connecting to its network. The generators should be capable of the following technical requirements:

- frequency control
- achieving NGC's power/frequency characteristic – this means controlling active power independent of frequency
- reactive power and voltage range (0.95 lead to 0.95 lag)
- frequency range of 47–52 Hz
- fault ride through.

Currently, most renewable generators can achieve the majority of NGC requirements; however, the use of the doubly fed induction machine has resulted in some problems achieving the NGC fault ride through requirements (see Figure 5.6).

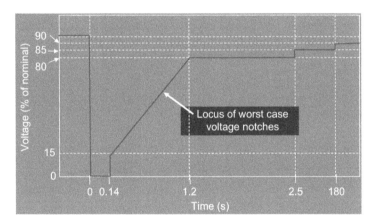

Figure 5.6   Locus of worst case voltage notches [26]

When a voltage depression occurs at the terminals of the doubly fed induction generator, the rotor accelerates. Usually the wind turbine protection detects this and uses the blades to brake the acceleration. When the grid voltage is restored the turbine is no longer capable of returning back to its pre-fault generation level and takes some time to get back to its previous generation level. This is in contravention of the Grid Code where the previous power level is to be delivered on restoration of the grid voltage.

5.5   Value of energy

Electricity has different values at different times of the day. While the majority of domestic customers have a single rate for electricity, dual tariff rates such as

Economy 7 are available. As the customer size increases and their use of electricity forms a significant part of their operating costs, targeted seasonal time of day rates can be deployed by energy suppliers and used to schedule their operations by large customers. For example an 11-part seasonal time of day rate has been used on electric arc steelmaking plants.

For generators, the more flexible and predictable they are, the more value they can achieve for their generation. From lowest to highest value the following types of generation contracts can be struck:

- un-metered spill (micro-generators, worthless)
- non-forecast spill (wind, wave)
- forecast but intermittent generation (tidal)
- base load (landfill gas and nuclear)
- profile following (fossil fuel, large hydro)
- dispatchable (pumped hydro, spinning reserve).

The additional value for renewable energy from Renewable Obligation Certificate (ROC) makes the returns from these poorer contracts viable as an investment.

Non-forecast spill in March 2006 would attract System Sell Price, worth on average £54.27/MWh (range £24.17–414.71/MWh), whereas dispatchable generation would attract System Buy Price which was worth on average £82.90/MWh (range £28.38–513.64/MWh). System Sell and Buy Prices are available from the Elexon website [27].

5.6 Solutions for renewable energy

A report on the solutions to the problems of connection of distributed generation has been commissioned by OFGEM from Mott McDonald. It is available on the Ofgem website [28].

5.6.1 Voltage rise

Potential solutions to voltage rise problems include:

- voltage line regulation
- conductor upgrades
- generation-linked cancellation CTs
- virtual VTs
- FACTS
- active voltage control (GenAVC^{TM}).

5.6.2 Fault level

Potential solutions to fault-level problems include:

- fault current limiters
- network reconfiguration
- increased network impedance

- converter interface technology
- sequential switching
- fault anticipators
- fault-level monitors
- active fault-level management.

5.6.3 Fault current limiters

Adding impedance to the network – inductors or higher impedance transformers – and opening bus couplers or mesh network points all lower prospective fault levels. These solutions have implications for network losses or reliability of supplies. A fault current limiter is a device that operates as it detects rise in current, and needs to operate within the first rising peak of fault cycle. Solutions such as exploding contacts (the IS limiter), power electronic-switched resistors and superconducting fault current limiters are now available. The superconducting design has the advantage of being a fail safe unit – the superconducting property is lost when excessive current flows through the device, adding resistance to the circuit, a physical property of the material, and relying on no other switching devices. A demonstration superconducting device that has been trialled on is a 10 kV, 10 MVA circuit in Germany. More details of the Curl 10 device can be found on the IEA website [29].

5.6.4 Principles of superconducting fault current limiters
5.6.4.1 Superconductivity

In 1911 superconductivity was first observed in mercury by Dutch physicist Heike Kamerlingh Onnes of Leiden University. When he cooled mercury to the temperature of liquid helium, 4 K (-269 °C), its electrical resistance suddenly disappeared. It was necessary for Onnes to come within 4° of the coldest temperature that is theoretically attainable to witness the phenomenon of superconductivity. Later, in 1913, he won a Nobel Prize in physics for his research in this area [30].

5.6.4.2 Expulsion of magnetic fields

The Meissner effect is the expulsion of a magnetic field from a superconductor during its transition to the superconducting state. Walther Meissner and Robert Ochsenfeld discovered the phenomenon in 1933 by measuring the magnetic field distribution outside superconducting tin and lead samples. The samples, in the presence of an applied magnetic field, were cooled below what is called their superconducting transition temperature. Below the transition temperature the samples cancelled all magnetic fields inside, which meant they became perfectly diamagnetic. The experiment demonstrated for the first time that superconductors were more than just perfect conductors and provided a uniquely defining property of the superconducting state [31].

5.6.4.3 Expulsion of magnetic fields

In a weak applied field, a superconductor 'expels' all magnetic flux. It does this by setting up electric currents near its surface. It is the magnetic field of these surface currents that cancels out the applied magnetic field within the bulk of the superconductor. However, near the surface, within a distance called the London penetration depth, the magnetic field is not completely cancelled; this region also contains the electric currents whose field cancels the applied magnetic field within the bulk. Each superconducting material has its own characteristic penetration depth. Because the field expulsion, or cancellation, does not change with time, the currents producing this effect (called persistent currents) do not decay with time. Therefore, the conductivity can be thought of as infinite: a superconductor. Note that a perfect conductor will prevent any change to magnetic flux passing through its surface. This can be explained as ordinary electromagnetic induction and should be distinguished from the Meissner effect [32].

5.6.5 Fault current limiters

In electrical networks occasional short circuits are inevitable and must be allowed for in the design of the electrical equipment. These currents can be 10–20 times larger than the normal circuit current. The addition of too much spinning load (electric motors) or generation to electrical networks can result in substations that can no longer handle the fault current flowing into a short circuit, even though they can accommodate the load current. To allow for the increased fault current, network operators have to increase the impedance of the network. Conventional methods to do this include:

- building a new substation (very expensive)
- fitting higher impedance transformers (costly and extra losses)
- fitting series reactors (quite expensive plus much higher losses)
- using an exploding charge and fuse system limiter (device based upon explosives, not HSE approved, serious downtime post fault)
- running with open bus section (higher losses, risk of customer interruptions)
- running without trip fuses (large-scale outage if a fault occurs).

These solutions have issues in terms of customer interruptions, harmonics and other quality issues and network losses including the associated carbon emissions. The ideal solution is a device that has very low impedance when normal current is flowing and very high resistance when a fault occurs. That device is the superconducting fault current limiter.

There are three basic types of superconducting fault current limiters, resistive, shielded core and pre-saturated core.

5.6.6 Resistive fault current limiters

Resistive fault current limiters utilise the very low impedance of the superconductor to provide a very low impedance link in the substation; however, when fault current flows, the electrical current density of the superconductor is exceeded and the superconductor reverts back to its normal conducting (resistive) state. The resistance added swamps the circuit X/R ratio, killing the direct current (DC) component of the fault, improving the power factor and eliminating the transient recovery voltage. The superconductor, however, in its resistive state quickly becomes hot and the superconductor must be taken out of service to cool back down prior to the circuit being reinstated. Resistive fault current limiters need to pass load current through the superconductor, so a connection between a thermally and electrically conductive copper circuit and the superconductor at cryogenic temperature is necessary. This adds a significant thermal leak load to the cryogenics.

5.6.7 Shielded core fault current limiters

Shielded core fault current limiters utilise the ability of superconductors to repel magnetic fields (see section 5.6.4.2) to hide a magnetic core. One approach is to surround a magnetic core with a superconducting tube which in turn is surrounded by a primary winding. The electromagnetic field from the primary winding is opposed by supercurrents induced in the superconducting winding. When a fault current flows, the induced supercurrent exceeds the current density of the superconducting tube and it quenches. Suddenly the magnetic core appears in the circuit and the fault current is reduced.

One benefit is that the superconductor has no external connections, so only ambient leak must be catered for by the cryogenics; however, the superconductor in its quenched state quickly becomes hot and the superconductor must be taken out of service to cool back down prior to the circuit being reinstated.

The impedance added is primarily inductive, so the DC component is not reduced quickly.

5.6.8 Pre-saturated core fault current limiters

Pre-saturated core fault current limiters use two magnetic circuits (per phase) driven into saturation by a DC winding (see Figure 5.7(a) and (b)). The magnetic circuits have a primary winding from the same phase, with a few turns around one magnetic circuit in one direction and the same number of turns around the second winding in the other direction [33].

When normal current flows, the line current does not drive the cores out of saturation, so the pre-saturated core acts as an air-cored reactor. However, when fault current flows, the magnetic cores are driven out of saturation and impedance is added to the circuit. This comes from one core on the first-half wave and from the other core on the second-half wave.

The flux density from the DC winding required to drive the magnetic core into saturation for a practical fault current limiter is not readily achievable with conventional conductors, nor are the energy and thermal losses associated with the DC winding. A superconducting winding has a current density 150 times that of copper and zero impedance, making this a practical reality.

Pre-saturated core fault current limiters primarily add inductance into the circuit, so the DC component of the fault remains. The superconductor has external connections to the DC power supply which must be catered for by the cryogenics; however, the superconductor does not quench during the fault, so this fault current limiter can instantaneously reset on load.

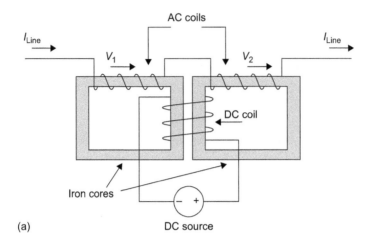
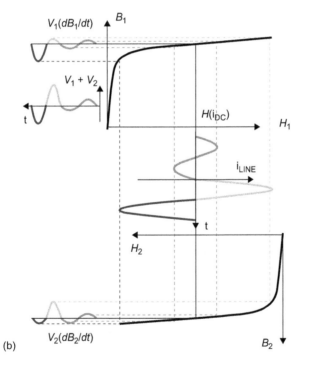

Figure 5.7 *Pre-saturated core fault current limiters*
 (a) illustrates the two magnetic circuits/phase
 (b) current characteristics for unfaulted and faulted conditions

The implications of renewable energy on grid networks 205

All three superconducting technologies can be competitive with conventional solutions in terms of price, losses, size and disruption given the right circumstances.

5.6.9 Load flow

Potential solutions to load flow problems include:

- post flow constraint – inter-tripping
- use of dynamic ratings
- energy storage.

5.6.10 Energy storage

Energy storage has the potential to upgrade the value of renewable energy from System Sell Price to something approaching System Buy Price. It also questions some of the underlying assumptions regarding network design. Storage co-located with generators, especially renewable generators, with a 70% (tidal), 40% (wave) or 30% (wind) load factor compared to peak output, allows for a smaller connection and output from the storage at times of low resources. Storage co-located with loads questions the use of spinning reserve and the redundancy associated with network design. Storage also allows an interlinked island network topology to be practically considered rather than a single system.

Utility scale energy storage technologies have been deployed in several locations around the world. These include the following lead-acid battery installations:

	MW	MWh
CHINO, California	10	40
HELCO, Hawaii	10	15
PREPA, Puerto Rico	20	14
BEWAG, Berlin	8.5	8.5
VERNON, California	3	4.5

Installations using other technologies include:

	MW	MWh
Sodium sulphur	8	57.6
Vanadium redox	0.35	2.0
Hydrogen	0.054	2.6
Compressed air energy storage	110	2,860
Pumped hydro	1,728	8,640

It should be noted that there is over 5 GW of pumped hydro storage capacity in the world.

More details on energy storage can be found in a DTI report titled 'Review of Electrical Energy Storage Technologies and Systems and of their Potential for the UK' [34].

5.6.11 Distributed intelligence in networks

Most network solutions installed by network operators are of the 'fit-and-forget' variety. This tends to lead to very low operating costs and high levels of reliability but with high capital expenditure. Modern network equipment usually comes with a range of operational and conditional information associated with the equipment's status. This information is often not used in the control of the network – if equipment has fit-and-forget functionality then this additional information is not needed to achieve reliability or other objectives.

As network use changes, with power flows reversing and voltage levels changing locally, use of this information in conjunction with new network control algorithms and new network equipment such as storage or demand-side management ready equipment can lead to 'active network management'. Projects in this area include:

- real-time automated reconfiguration
- storage and demand-side management
- large-scale virtual power plant (based on many small generators)
- communications – condition and alarms
- dynamic short-term ratings (e.g. wind farms export on windy days – when there is enhanced cooling for overhead lines increasing their potential current carrying capability).

There is more on active network management and the utilisation of existing intelligence in networks in the DTI report 'Network Management Systems for Active Distribution Networks – a Feasibility Study' [35].

5.7 Conclusions

Renewable energy will be part of the energy mix for the UK since the government and all political parties have committed themselves to the Kyoto agreement.

Renewable energy is from diverse sources and can be captured at a variety of scales – micro to multi-megawatt. Large renewable energy farms could connect to the National Grid; however, most generation schemes will connect at lower voltage levels. The UK best renewable energy resources tend not to be co-located with load, so this will require new distribution and transmission networks.

Renewable energy technology has some technical differences to conventional generation, and the volume and cost of connections necessary to achieve the government's targets for renewable generation are driving innovation into electricity networks. A spin-off effect is to evolve the electricity network into a more cost-effective, actively managed system with lower capital costs and higher utilisation rates, without compromising safety or reliability.

References

1. *Dukes table 1.1.1: Inland consumption of primary fuels and equivalents for energy use, 1970 to 2004.* Available from http://www.decc.gov.uk/assets/decc/11/stats/publications/dukes/2303-dukes-2011-chapter-1-energy.pdf

2. *Dukes table 5.4: Fuel used in generation, found in chart 5.2.* Available from http://www.decc.gov.uk/assets/decc/11/stats/publications/dukes/2307-dukes-2011-chapter-5-electricity.pdf
3. The Energy White Paper. Available from http://www.dti.gov.uk/files/file39387.pdf
4. *Average cold spell power flow pattern for 2004/05 available as Figure 7.1 in the 2004/5 seven year statement.* Available from http://www.nationalgrid.com/uk/
5. *Annual mean wind speeds.* Available from https://restats.decc.gov.uk/cms/annual-mean-wind-speed-map
6. *Annual mean wave power.* Available from http://www.bwea.com/marine/resource.html
7. *Peak flow from a mean spring tide.* Available from http://www.bwea.com/marine/resource.html
8. *English biomass resources.* Available from http://www.defra.gov.uk/
9. *Enercon wind turbine information.* Available from http://www.enercon.de/en-en/66.htm
10. *Zephyros wind turbine information.* Available from http://www.elkraft.ntnu.no/norpie/10956873/Final%20Papers/068%20-%20Norpie%20paper.pdf
11. *Vestas wind turbine information.* Available from http://www.vestas.com/en/wind-power-plants/procurement/turbine-overview.aspx#/vestas-univers
12. *RE power wind turbine information.* Available from http://www.repower.de/wind-power-solutions/wind-turbines/
13. *Linear generator comparison.* Available from http://webarchive.nationalarchives.gov.uk/; /http://www.berr.gov.uk/files/file16064.pdf
14. *Feed in tarriffs.* Available from http://www.fitariffs.co.uk/
15. *G83/1 and G59.* Available from http://www.ena-eng.org/ENA-Docs/EADocs.asp?WCI=SearchResults&DocSubjectID=106
16. *NGC statement of connection charges 2011.* Available from http://www.nationalgrid.com/uk/Gas/Charges/statements/connection/publications/
17. Ofgem connection costs. Ofgem Presentation: 'Rewiring Distribution' given by Martin Crouch on 24 July 2004
18. *OFGEM IFI and LCNF incentives.* Available from http://www.ofgem.gov.uk/Networks/Techn/NetwrkSupp/Innovat/ifi/Pages/ifi.aspx; http://www.ofgem.gov.uk/Networks/ElecDist/lcnf/Pages/lcnf.aspx
19. *Charging changes.* Available from http://www.ofgem.gov.uk/Networks/Connectns/CompinConn/Pages/CompinCnnctns.aspx
20. *Statutory voltage levels (p. 7).* Available from http://www.northernpowergrid.com/som_download.cfm?t=media:documentmedia&i=893&p=file
21. *Website of the anti-pylon group REVOLT.* Available from http://www.revolt.co.uk/
22. *Website of the Highlands Before Pylons group.* Available from http://www.hbp.org.uk/
23. *Scottish & Southern Energy's rationale for the Beauly–Denny upgrade.* Available from http://www.sse.com/BeaulyDenny/

24. *Ofgem consultation on small-scale embedded generation.* Available from http://www.ofgem.gov.uk/Networks/ElecDist/Policy/DistGen/Pages/DistributedGeneration.aspx
25. *NGT Grid Code C.C.6.3.15.* Available from http://www.nationalgrid.com/uk/
26. Causebrook A., Atkinson D.J. 'Assessment of fault ride-through characteristics of grid-coupled synchronous and asynchronous generators with consideration of a decoupled alternative', *Post Graduate Conference PGC2005*. University of Newcastle upon Tyne; 19–20 January 2005
27. *System buy/sell price market data from Elexon.* Available from http://www.bmreports.com/bwx_reporting.htm
28. *Mott McDonald report.* Available from http://www.ofgem.gov.uk/Networks/ElecDist/PriceCntrls/DPCR4/Documents1/6596-RPZ%20IFI%20RIA%20Final.pdf
29. *CURL 10 fault current limiter.* Available from http://ieeexplore.ieee.org/xpl/freeabs_all.jsp?arnumber=1440041
30. *Zero resistance.* Available from http://www.superconductors.org/History.htm
31. *Meissner effect.* Available from http://en.wikipedia.org/wiki/Meissner_effect
32. *Expulsion of magnetic fields – Lenz's law.* Available from http://hyperphysics.phy-astr.gsu.edu/hbase/solids/meis.html
33. *Pre-saturated core fault current limiters.* Available from http://www.zenergypower.com/images/media-coverage/fcl/Modern_Power_Systems.pdf
34. *DTI review of electrical storage technologies.* Available from http://www.bis.gov.uk/assets/bispartners/foresight/docs/energy/energy%20final/baker%20paper-section%202.pdf
35. *Active Network Management.* Available from http://webarchive.nationalarchives.gov.uk/20100919181607/http:/www.ensg.gov.uk/index.php?article=95

Postscript note – by H.M. Ryan

All CIGRE Study Committees (SCs) since the 2002 reform are summarised below, together with contact information for each SC.

Note: A fuller version, 'Fields of activities of the CIGRE Study Committees', is discussed further elsewhere. (See *A History of CIGRE*, 2011. Additional CIGRE and renewable energy information is also provided in Table 23.13.)

A1 Rotating electrical machines (www.cigre-a1.org)
A2 Transformers (www.cigre-a2.org)
A3 High-voltage equipment (www.cigre-a3.org)
B1 Insulated cables (www.cigre-b1.org)
B2 Overhead lines (www.cigre-b2.org)
B3 Substations (www.cigre-b3.org)
B4 HVDC and power electronics (www.cigre-b4.org)
B5 Protection and automation (www.cigre-b5.org)
C1 System development and economics (www.cigre-c1.org)
C2 System operation and control (www.cigre-c2.org)
C3 System environmental performance (www.cigre-c3.org)
C4 System technical performance (www.cigre-c4.org)
C5 Electricity markets and regulation (www.cigre-c5.org)
C6 Distribution systems and dispersed generation (www.cigre-c6.org)
D1 Materials and emerging test techniques (www.cigre-d1.org)
D2 Information systems and telecommunication (www.cigre-d2.org; www.cigre.org; www.e-cigre.org)

Some additional CIGRE reference sources relating to aspects of renewable energy, distribution systems and dispersed generation are included below. Users may find the CIGRE Technical Brochures (TBs) interesting, and they may be consulted for future guidance.[1]

- 'Impact of increasing contribution of dispensed generation on the power system', TB 137, SC 37 WG 37.23, 1999
- 'Distribution systems and dispersed generation', CIGRE, SC C6 Annual Report 2011 (see also *Electra*. 2012;**258**(October):24–28; prepared by Nikos Hatziargyriou)
- 'Connection of generators and other customers–rules and practices', TB 271, SC C6.02, 2005
- 'Operating dispersed generation with ICT (Information & Communication Technology)', TB 311, SC C6 WG C6.06, 2007
- 'Connection criteria at the distribution network for distributed generation', TB 313, SC C6 TF C6.04.01, 2007

---

[1] CIGRE key: SC, Study Committee; WG, Working Group and JWG, Joint Working Group; TF, Task Force; TB, Technical Brochure; TR, Technical Report; SC, Scientific Paper.

- 'Modelling and dynamic behaviour of wind generation as it relates to power system control and dynamic performance', TB 328, SC C4 WG C4.601, 2007 (see also *Electra*. 2007;**233**(August):48–56; same title; prepared by WG C4.601)
- 'Technical and commercial standardisation of DER/microGrid components', TB 423, SC C6 WG C6.10, 2010 (see also *Electra*. 2010;**251**(August):52–59; same title; prepared by WG C6.10)
- 'Grid integration of wind generation', TB 450, SC C6 WG C6.08, 2011 (see also *Electra*. 2011;**254**(February):62–67; same title; prepared by WG C6.08)
- 'Development and operation of active distribution networks', TB 457, SC C6 WG C6.11, 2011 (see also *Electra*. 2011;**255**(April):70–73; same title; prepared by WG C6.11)
- 'Electrical energy storage systems', TB 458, SC C6 WG C6.15, 2011 (see also *Electra*. 2011;**255**(April):74–77; same title; prepared by WG C6.15)
- 'Modelling, new forms of generation and storage', TB 185, TF 38.01.10, 2001
- 'Demand-side integration (DSI)', TB 475, SC C6 WG C6.09, 2011
- 'Status of development and field test experience with high-temperature superconducting power equipment', TB 418, SC D1 WG D1.15, 2010
- 'CIGRE WG 13.10: Functional specification for a fault-current limiter', *Electra*. 2001;**194**:22–29 (contains 10 references and equipment illustrations)
- 'Impact of fault current limiters on existing protection schemes', TB 339, SC A3 WG A3.16, 2008
- 'Fault current limiters–application, principles and testing', TB 239, SC A3 WG A3.10, 2003. Brochure contains four major parts:
  Part A – Fault current limiters–state of the art
  Part B – Functional specification
  Part C – Fault current limiters–system demands
  Part D – Fault current limiters–testing
  Part E – Extensive reference list
  (Original scope of WG covered FCLs in distribution networks and sub-transmission networks up to 145 kV but contributions related to FCLs at transmission systems up to 420 kV have also been considered.)
- 'The impact of renewable energy sources and distributed generation on sub-station protection and automation', TB 421, SC B5 WG B5.34, 2010 (see also *Electra*. 2001;**251**(August))
- 'Guidelines for the design and construction of AC offshore substations for wind power', TB 483, SC B3 WG B3.26, 2011 (see also Chapter 23 of this book for several additional references)
- 'Application and feasibility of fault current limiters in power systems', TB 497, SC A3 WG A3.23, 2012 (see also *Electra*. 2012;**262**(June))

Chapter 6
High-voltage cable systems
A.L. Barclay

6.1 Introduction

Almost all power systems contain a mix of overhead line and cable. The balance between the two can vary dramatically. A number of factors govern the choice between line and cable:

- cost
- route availability
- environmental aspects
- electrical parameters.

In terms of initial cost including installation, high-voltage cable systems are almost always more costly than overhead line systems for the same duty. Naturally, therefore, the designer's first choice is overhead line. The cable alternative is selected only when there is an overriding reason to do so.

The main factor that drives selection of cable is the lack of suitable routes for line. At one end of the scale, air-insulated HV lines would be impractical within buildings such as power stations. Buried cables may bring a lot of flexibility when installation of new equipment requires routes across congested high-voltage air-insulated substations. Bringing power into big-city load centres by overhead line would be unsightly and hazardous. When crossing large bodies of water, the span length of overhead lines soon becomes a limiting factor. In all these cases, it is the technical aspect that drives the choice of cable over overhead line.

Another factor that may drive the choice towards cable is visual amenity. Particularly in areas of natural beauty, public pressure or politics may require an interconnection to be invisible, and therefore to use buried cable rather than overhead line.

One factor that might drive the choice in the opposite direction is the external magnetic field generated by a loaded circuit. A cable may be only 1 m below ground, as opposed to an overhead line 10 or more metres above it, so there is potential for the external field to be higher at ground level. The accepted standard for occupational and public exposure to time-varying (including power frequency) magnetic fields is the ICNIRP Guidelines [1]. At power frequency, the only known effect to be guarded against is heat loading of the central nervous system (the brain and spinal cord) by induced current. To protect the public against such effects with a large factor of safety, the 'reference level' of magnetic field at 50 Hz is 100 μT. In most conventional cable systems at normal loads, fields of this magnitude are only found quite close to the cable surface, so there is no cause for concern above ground.

6.2 Elementary theory

The following discussion covers only the essentials required for an understanding of the rest of the chapter; more detailed coverage is available in the *Electric Cables Handbook* [2].

6.2.1 Voltage, electric field and capacitance

The transmission of electric power requires two things: a voltage between two or more conductors and a current flow along the conductors.

In conventional high-voltage cables, the voltage is applied between a central metallic conductor and an earthed metallic screen that surrounds it. The two components are separated by the insulation system. The electric field is contained within the dielectric part of the insulation system. For now, we will consider only alternating voltages and neglect losses, so that the electric field distribution is governed by dielectric permittivity.

Except at the lowest voltage levels, the conductor has an overall circular shape and the insulation is also circular. This radially symmetric arrangement simplifies the mathematics. We represent the central conductor as a cylinder of radius a and the boundary between the insulation and the outer screen as a cylinder of radius b (Figure 6.1). The electric field is zero for radius r less than a or greater than b.

Now we consider a certain density of electric charge per unit length, Q, to be placed on the inner conductor. The charge distributes itself evenly over the conductor surface at $r = a$ so as to maintain zero electric field within the conductor. An equal and opposite charge per unit length distributes itself on the inner surface of

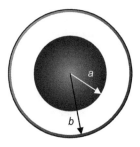

Figure 6.1   Definition of inner and outer radii of dielectric

the outer conductor at $r=b$ so that there is no electric field outside this radius. For this cylindrical symmetry the electric field E (for $a < r < b$) is given by

$$E(r) = \frac{Q}{2\pi\varepsilon_0\varepsilon_r r}$$

where ε_0 is the permittivity of free space, approximately 8.854 pF/m, and ε_r is the relative permittivity of the insulating material. The electric field is a vector: it has a direction that in this case is radially outward. It can be seen that the field is highest at $r=a$, at the conductor surface, and lowest at $r=b$, at the insulation surface. Integrating, the potential difference between the conductor and screen is

$$E(r) = \frac{U}{r \ln(b/a)}$$

The maximum field appears at a and is

$$E_{max} = \frac{U}{a \ln(b/a)}$$

The electric field is associated with stored energy. This is described by the capacitance per unit length

$$C = \frac{Q}{U}$$

where

$$C = \frac{2\pi\varepsilon_0\varepsilon_r}{\ln(b/a)}$$

The only two things that govern the capacitance per unit length, then, are the relative permittivity ε_r of the dielectric material and the ratio b/a between its outer and inner radii. For practical HV cables, the capacitance per unit length is typically 200–500 pF/m. The stored energy E per unit length of cable is given by

$$E = \frac{1}{2}CU^2$$

For alternating current (AC) at frequency f,

$$I = C\frac{dU}{dt} = 2\pi fCU$$

where I is the charging current, the current required to energise a unit length of the cable at the given voltage and frequency. This charging current can be significant for substantial lengths of cable such as manufactured lengths or complete installed circuits, whether U is a test voltage supplied by an alternating-voltage test set, or the system voltage that the circuit sees when in service.

6.2.2 Current, magnetic field and inductance

The voltage applied to the cable corresponds to an electric field within the insulation. In order to transmit power, the cable must also carry current. The current in its turn is associated with a magnetic field. A straight circular conductor of radius a carrying current I is associated with a field

$$B(r) = \frac{\mu_0 I}{2\pi r}$$

for $r \geq a$. This field is also a vector: its direction is circumferential so that the 'lines' of magnetic field form circles around the conductor. The magnetic field is always perpendicular to the electric field (in this first-order theory, neglecting losses).

In real life, conductors are not infinitely long and the current must have a return path to complete the circuit. Two approximations can be considered, both of which are relevant to real cables:

1. The current return path is remote from the cable.
2. The return current flows through an outer conducting layer within the same cable.

In the first case, the magnetic field is as stated above for all $r \geq a$. In other words, the magnetic field extends outside the cable and out to infinity. It does not matter whether the cable has outer metallic layers, as they provide no magnetic screening effect unless return current actually flows in them.

In the second case, consider the return current to flow in a tubular conductor at radius b. This is a two-core cable with concentric conductors, just like a coaxial signal cable. The return current generates its own magnetic field for all $r \geq b$. As the return current is equal and opposite, the magnetic field is also equal and opposite, and the resulting field cancels to zero for $r \geq b$. In other words, the magnetic field is confined within the cable.

The magnetic field is also a form of stored energy. For a coaxial or concentric arrangement where the magnetic field is confined within the cable, the inductance per unit length L is

$$L = \frac{\mu_0}{2\pi} \ln(b/a)$$

where μ_0 is the permeability of free space, exactly $0.4\,\pi\ \mu\text{H/m}$.

6.3 Historical development

The factors governing the design of modern-day cables are well illustrated by a historical survey of the development of cable systems from the earliest days [3]. It will be seen how features of modern cables have emerged as they proved – usually through bitter experience – to be essential.

6.3.1 Early telegraph cables

The first recorded practical application of a recognisable insulated cable is in the eighteenth century for electric telegraphy. As with so many new technologies, the possibilities of the electric telegraph were not recognised at first, and there was little pressure for its development or application. The requirements for early land telegraph cables placed no great demands on the quality of the insulation. However, in the nineteenth century new insulating materials became available with the introduction to Europe of gutta percha, a natural thermoplastic, and then of India rubber. These materials improved the situation so much that useable telegraph cables could be laid between England and France by 1850. Within a decade, considerable routes of land and sea cables were in operation and visionary attempts had been made at crossings of the Atlantic.

6.3.2 Early lighting systems

In parallel with developments in cables, practical dynamos emerged in the 1870s and immediately found application for street lighting by means of Davy's arc lamps. Further development in the next decade included the Siemens alternator, vulcanised India rubber cable insulation, Ferranti's transformer and various other forms of lamp. As lighting installations became more extensive, underground power cables became necessary. It was Ferranti who conceived the bold idea of supplying the whole of central London at high voltage from a generating plant at Deptford, well outside the city. Transformer stations would be built to distribute at low voltage. This concept brought all the advantages of remote bulk generation that we still perceive today: plentiful cooling water, economic seaborne supplies of fuel and removal from the built-up area of sources of pollution, noise and road traffic. What it did require, however, was practical high-voltage cables for the bulk transmission of power from Deptford to the central substations.

Along with the recognisably modern concept of separating transmission from distribution came a recognisably modern cable. It had to be a two-core concentric cable, as it was already realised that allowing return currents to flow through the earth was causing interference to telegraphy. The unprecedentedly high alternating voltage of 10 kV was selected, with a cable insulation thickness of 0.5 in. (13 mm). Based on experience in capacitor manufacture, paper insulation impregnated with the natural resin shellac was selected. The cable was made of rigid lengths, many hundreds of joints having to be made on site.

6.3.3 Flexible cables

The wider application of electric lighting awaited the introduction of a practical flexible cable. Callender's company was importing Trinidad bitumen for road surfacing, and sought to add value to the product by applying it to the electrical industry. A compound described as 'vulcanised bitumen' was extruded onto flexible conductors in a continuous process, resulting in a construction that outperformed anything else. The low-voltage cable found wide application in domestic and commercial installations around the world. An essential contribution to the reliability of the system was the method of installation, in rigid troughs filled with bitumen.

6.3.4 Impregnated cables and the renaissance of high voltage

From 1884, lapped paper insulation impregnated with viscous compound was developed for telegraph cables. Developments in paper processing, lapping and lead sheathing enabled the system to be used for flexible power cables. Paper cables became dominant for high-voltage applications. Tesla's three-phase system prompted the introduction of three-core cables and a change of design: laid-up cores instead of concentric. Conductors were shaped and compacted to keep the complete cable circular. Many other essential features were introduced before 1900: butt gaps between paper tapes to permit bending; vacuum drying and impregnation; and copper wire screens to improve the fault performance of the lead sheaths.

The three laid-up cores of three-phase cables worked well at first. The 'belted' construction had the three insulated cores laid up together, with further insulation between the laid-up assembly and the lead sheath. However, as voltages increased to 33 kV and beyond, the system reached its limit. The limiting factor turned out to be the overlapping electric fields of the three phases, giving rise to a rotating component of field in the space between the three cores. The electric strength of the paper is much less along or across the paper layers than through them. The significant tangential fields in the belted cable, exacerbated by thermally induced void formation, proved fatal.

Hochstadter devised the complete solution to tangential fields: individually screened cores. An earthed outer layer of metallised paper on each core confined the electric field within the respective core, where it was purely radial. The metallised paper was perforated to allow impregnation. This made the paper cable system reliable at 33 kV. Taking this even further, separate lead sheaths could be applied to the individual cores.

The technology soon hit another limit as higher voltage systems became unreliable when cyclically loaded. The problem was again void formation. Thermal expansion under load caused the cable components to pressurise and expand. On cooling, the cable components would contract. The resulting vacuum led to the creation of voids in the compound. Over many such cycles, the lead sheath underwent creepage and was unable to exert enough pressure to close the voids. At the higher voltages then employed, the life of a cable under these conditions could be very short.

A further difficulty was drainage of the viscous oil–rosin compound under gravity in circuits with a height difference, leaving the upper reaches of the cable empty of compound. This was eventually rectified by the development of mass-impregnated non-draining (MIND) compound, described in section 6.4.2.1.

6.3.5 Pressure-assisted (gas) cables

Two main approaches were successful in extending the voltage range of impregnated cables. Both of these involved the use of pressurised gas to exert pressure on the dielectric, directly or indirectly.

One construction is now known as the gas compression cable. Three impregnated paper-insulated screened cores are contained in a common thin, flexible lead sheath to produce an overall triangular shape. Alternatively, as in modern versions of this cable, each core may have its own thin flexible lead sheath. The dielectric is kept under compression, and the formation of voids suppressed, by gas pressure acting on the flexible sheath from the outside. In the most common application of this method, the three-core assembly is pulled into a steel pipe, which retains the gas pressure. In a later development, the rigid external pipe was replaced by a flexible reinforced lead pressure-retaining sheath to produce a self-contained version of the cable.

The other approach was to allow the high-pressure gas itself to permeate pre-impregnated paper-insulated cores. Even in the inevitable voids, the high pressure of the gas would suppress the electrical discharges that would otherwise lead to degradation of the dielectric. A heavily reinforced lead sheath retained the gas pressure to form a self-contained cable. There was sufficient gap between the cores and the sheath to allow the gas to pass freely along the cable. Other designs of gas-filled cable worked on similar principles.

Gas-pressure cables of one sort or another were successful at voltages of 33–132 kV, with occasional application at higher voltages, and many remain in service today despite troublesome gas leaks. However, the future of impregnated paper-insulated pressure systems was not with gas but with oil.

6.3.6 Fluid-filled cables

The concept of the oil-filled cable was first described by Emanuelli in 1917. The key feature is the longitudinal channel, which allows the low-viscosity insulating liquid to fill the cable. Changes in volume are accommodated by an external reservoir that maintains the pressure. Single-core and three-core cables using this technology have been very widely manufactured and installed. As materials and processes improved, the system was applied up to 525 kV and possibly beyond. The reliability of the system has set a benchmark that newer systems may struggle to match. A fuller description of the self-contained fluid-filled system, as it eventually became, is given in section 6.4.2.1.

In a parallel development in the USA, use was made of steel pipeline technology, which was well established there. The oilostatic cable was paper-insulated and impregnated with low-viscosity oil. Three screened but un-sheathed cores were drawn into a common steel pipe, which was then filled with oil and pressurised. This technology became well established in North America as the high-pressure oil-filled (HPOF) system, known outside North America as the pipe-type cable.

6.3.7 Polymeric cables

The earliest low-voltage power cables developed in parallel with telegraph cables, with various waxed and varnished fibres eventually being replaced by rubber insulation. The Second World War disrupted supplies of natural rubber and prompted a transition to polyvinyl chloride (PVC) thermoplastic. However, rubber and PVC were not generally suitable above around 3.3 kV due to their high dielectric losses.

The advantages of polyethylene in having low dielectric loss were recognised. Early experience with polyethylene in North America in particular was unfortunate, with insufficient attention paid to quality of manufacture and installation, water tree degradation and overloading. Its maximum operating temperature was unacceptably low for power cables and the consequences of even momentary overheating were disastrous: above around 65 °C, the thermoplastic polyethylene would soften and melt, the conductor would sag and the cable would fail. This changed with the introduction of crosslinking (vulcanisation) of polyethylene, first by radiation and then by chemical reactions. Just as vulcanisation of rubber opened up all sorts of applications, so it was also with polyethylene. At temperatures where thermoplastic polyethylene would melt, the thermoset cross-linked polyethylene (XLPe) would merely soften, maintaining the integrity of the cable. From around the 1960s onwards, XLPe was applied at increasing rates, higher and higher up the voltage range.

Rubber insulation meanwhile did not entirely fail to penetrate the high-voltage cable sector. A synthetic elastomer, ethylene propylene rubber (EPR), is widely used up to 150 kV in Mediterranean Europe and other markets, as well as for specialised applications such as flexible or subsea cables. The elastomer is heavily filled with mineral fillers, which make essential contributions to its mechanical and electrical performance [4]. The system benefits from greater flexibility, a higher operating temperature and an apparently greater resistance to water tree degradation than XLPe, but suffers from greater dielectric loss, which prevents its application at any higher voltages.

6.3.8 Polypropylene-paper laminate

There was a significant attempt to combine the low dielectric losses of some polymers with the electrical reliability and thermal performance of the fluid-filled system. This was the use of polypropylene-paper laminate (PPL) in place of plain paper in the fluid-filled cable [5]. The dielectric strength was also superior, allowing operation at higher electric stress. However, the introduction of this system came too late to prevent the domination of the market by XLPe.

6.4 Features of real cables

6.4.1 The conductor

In conventional cables, the element that we call the conductor is not in fact a perfect conductor of electric current. The conductor is made of a highly conductive metal, usually copper or aluminium of specified high purity and electrical characteristics. The conductor has a certain geometric cross section and thus a certain resistance R per unit length. As the conductor carries current, this resistance causes a voltage drop and the dissipation of power as heat: the dissipation W per unit length is given by

$$W = I^2 R$$

This power is lost from the electric power transmitted through the cable, and appears as heat in the conductor. The temperature of the conductor rises, and heat is conducted radially outwards through the outer layers of the cable until it reaches the outside world. Under steady conditions, a steady state is eventually reached, where the heat conducted out of the cable balances the dissipation within. At this steady state, the conductor reaches a steady temperature rise of $\Delta\theta$ above ambient. All being well, this temperature rise is such that the materials of the cable can withstand it without undue deterioration throughout their intended life.

The cable is characterised by a thermal resistance T per unit length, such that

$$\Delta\theta = WT = I^2 RT$$

In practice, we need to know the maximum current I that the cable can carry continuously without the conductor exceeding its rated continuous working temperature. If we start with a known external temperature, we have a corresponding temperature rise $\Delta\theta$ across the cable, and can rearrange it so that

$$I = \sqrt{\frac{\Delta\theta}{RT}}$$

In fact, R is not a constant. The two metals commonly used in conductors are copper and aluminium, and both have a significant temperature dependence of electrical resistivity. Characteristics of each are as follows.

	Electrical resistivity S (at 20 °C)	Temperature coefficient (a)
Copper	0.0172 μΩm	0.00393 K^{-1}
Aluminium	0.0283 μΩm	0.00403 K^{-1}

The significant temperature dependence of roughly 0.4% per Kelvin (or per degree Celsius) means that it is critical to specify to what temperature any resistance measurement is referred. It is conventional to specify conductor resistances at 20 °C, while the maximum operating temperature which governs the rating of the cable is much higher, often 90 °C.

A variety of conductor constructions are possible, depending on the overall design of the cable. Some of these are shown in Figure 6.2.

- Most conductors are stranded from layers of wires applied helically, laid in alternating directions to ensure stability.
- If aluminium is selected it can alternatively be extruded as a single solid rod with the required cross section, at the expense of some loss of flexibility.
- Stranded conductors can be compacted after stranding, to reduce their diameter and present a smoother outer surface overall. For the ultimate compaction in a stranded conductor, it can be made from wires pre-formed into individual keystone shapes. Conductors for extruded insulation (at high voltage) are always circular. However, for lapped insulation at sufficiently low operating

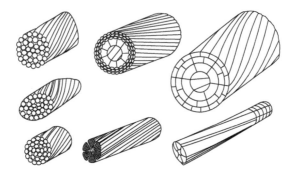

Figure 6.2 Conductor constructions. Clockwise from top left: stranded circular; stranded with keystone duct; fully shaped wires with duct; Milliken transposed conductor; Milliken transposed conductor with helical steel duct; compacted circular; and compacted oval

stress, the conductor is sometimes compacted into oval or even sector shapes so as to reduce the overall size of a multi-core cable, at the expense of an increase in operating stress.

- Conductors for single-core fluid-filled cables incorporate a central fluid duct. In many cables, this is achieved by stranding the conductor onto a pre-formed steel tape helix. Alternatively, there could be an innermost layer of keystone-shaped wires arranged to leave a hollow duct within. In some special applications, a relatively large-diameter central duct is used to allow forced circulation of cable fluid so as to cool the cable in service.

Stranded conductors for extruded insulation unfortunately provide a longitudinal path through the cable; if any water enters the conductor of an XLPe cable, it can run freely along the entire length of the cable and ruin it. Water can enter when cables in storage or transport are inadequately sealed, or when an installed cable is damaged. To minimise the longitudinal spread of water in either case, longitudinal waterblocking is often applied. This is achieved by incorporating water-swellable material during construction of the conductor, either in powder form or as tapes. Of course, solid extruded aluminium conductors need no waterblocking.

In theory, the direct-current (DC) resistance of a conductor is calculated by dividing the resistivity S by the cross-sectional area. In practice, the conductor size quoted for a given cable is a nominal value: what is actually specified is the resistance per unit length (at 20 °C, measured with DC). The values of resistance per unit length are specified in IEC Publication 60228 [6]. For instance, a 1,000 mm^2 copper conductor is required to have a DC resistance per unit length (at 20 °C) of no more than 17.6 μΩ/m (or 17.6 mΩ/km), regardless of the actual geometrical area of the copper. The cable manufacturer will endeavour to give away the least possible amount of expensive metal, while keeping the resistance just the right side of the limit [7].

There is a further complication that this is the DC resistance, while most power cables operate at AC. Whenever current varies with time, the phenomenon of skin effect appears, whereby the current density within the conductor decreases exponentially with depth below the surface. Current tends to flow in a skin near the surface; the skin has an effective thickness, known as the skin depth. Considering an alternating current of frequency f, the skin depth δ is given by

$$\delta = \sqrt{\frac{S}{\pi \mu_0 f}}$$

The higher the frequency of the AC, the shallower is the skin depth, so at very high frequencies the current can be considered to run over the surface. At 50 Hz, the skin depth in copper is around 9 mm, which is small enough to make the skin effect significant for practical conductor sizes. The current density deep within the conductor is less than would be expected from the DC resistivity, and therefore the total current in the conductor is also less. This is expressed by an *AC resistance*, R', which is greater than the DC resistance R, and by a *skin effect factor* y_s such that $\frac{R'}{R} = 1 + y_s$.

The factor y_s can be calculated mathematically for simple cases such as a solid circular conductor, and for practical purposes the same formula applies to a conventionally stranded circular conductor. The mathematics is complex, and in the field of cables a simpler empirical approximation is generally used [8]:

$$y_s = \frac{x_s^4}{192 + 0.8 x_s^4}, \quad \text{where} \quad x_s^2 = \frac{2 f \mu_0}{R}$$

As in other areas of electrical engineering, the problem of skin effect in large conductors has been addressed. In cables, the approach has been to use a conductor comprising a number of stranded subconductors. The current tends to follow the individual wires, so the helical arrangement of the wires within each subconductor tends to force the current into a more uniform distribution than would otherwise apply. In the arrangement known as the Milliken conductor, the subconductors are each compacted into a sectoral shape, so that when laid up together they form an overall circular conductor, as shown in Figure 6.3. Four to six sectors are generally used. They may be laid up on their own, or around a central rod, or (in the case of fluid-filled cables) a central fluid duct, also shown in Figure 6.3. This construction is considerably more complex to manufacture than a plain circular strand, so it becomes economic mainly for larger conductor sizes, typically 1,000 mm^2 and above, where the skin effect is most severe.

The effectiveness of the Milliken construction relies on the current following the individual wires as they zigzag through the body of the conductor. However, if current is able to pass freely between adjacent wires, the effect is lost. The sectors are made by drawing a circular stranded subconductor through a rotating sector-shaped die, so there is inevitably some compaction and the wires are forced into close contact. This difficulty appears to be more severe in XLPe than in fluid-filled cables, possibly due to the pressure exerted on the conductor by the insulation or to the lack of any insulating fluid permeating the conductor. Accordingly, alongside

Figure 6.3 Cross section of a Milliken-conductor paper-insulated fluid-filled corrugated aluminium-sheathed cable

the development of XLPe cables, there have been attempts by manufacturers to reduce the inter-wire conductance, often by either pre-oxidising or pre-enamelling the individual wires before stranding. This is a difficult task, as the insulating layer will have to survive considerable abuse during compaction of the strand into the required sector shape. Furthermore, there has to be an effective mechanical and/or chemical method of removing all the coating where the conductor is to be jointed into an accessory.

If a Milliken conductor were perfectly effective, the AC resistance of the conductor would be expected to equal the DC resistance. On the other hand, if it were completely ineffective, the AC resistance would equal that of a conventional circular strand. Reality lies somewhere in between. The effectiveness is expressed as a *skin effect factor* k_s, and the empirical formula for AC resistance incorporates a modified term

$$x_s^2 = k_s \frac{2f\mu_0}{R}$$

For a circular strand, $k_s = 1$, while for Milliken constructions k_s is (or should be) less than 1 to reflect their enhanced skin effect performance. For fluid-filled cables, standard values for k_s are taken from IEC publication 60287-1-1 [8] and used without too much concern. For XLPe cables, manufacturers quote values of k_s for the purpose of rating calculations and tenders, and it is to be hoped that their values are based on actual measurements. Measurement of the AC resistance of a power cable is not straightforward, and measurement of the inter-wire conductance in finished conductors is highly problematical.

6.4.2 The insulation system

The insulation system is required to fulfil many functions simultaneously:

- to withstand the continuous working voltage and any temporary overvoltages, all without electrical breakdown or undue loss

- to restrain the conductor in its intended position against mechanical forces, including thrusts due to thermal expansion and electromagnetic forces, while being sufficiently flexible to allow the cable to be drummed, undrummed and installed
- to conduct heat radially outwards from the conductor
- to withstand chemical or mechanical degradation, which may be accelerated by temperature, contamination such as moisture, or the electric field, without undue loss of performance over the required lifetime of service.

These requirements inevitably conflict. For instance, increasing the insulation thickness so as to reduce the voltage gradient would increase the thermal resistance and so increase temperature and thermo-mechanical stress.

Present-day designs of HV cable mainly use extruded polymeric dielectrics, of which the dominant material is XLPe. Earlier systems included lapped dielectrics, constructed by applying layers of tapes, their insulation performance relying on impregnation with a fluid (liquid or gas). Extruded insulation has taken over the market, beginning with the lowest voltages, and now including even the highest.

The insulation system for an HV cable, whichever technology it employs, consists of three layers:

- inner semiconducting screen, also called the conductor screen
- dielectric
- outer semiconducting screen, also called the insulation screen, dielectric screen or core screen.

An example of this is shown in Figure 6.4, which shows samples of XLPe insulation made transparent by heat for quality control inspection of the screen interfaces. The electric field is contained within the dielectric by the inner and outer screens, so that no voids, around the conductor strands for instance, are subjected to the field. The interfaces between the screens and the dielectric itself are critical,

Figure 6.4   Samples of XLPE insulation made temporarily transparent for quality control checks by heating in an oil bath

especially the conductor screen interface where the electric stress is highest. The same principles apply to lapped systems.

6.4.2.1 Lapped insulation systems

In lapped insulation systems, the dielectric structure is built up of layers. The layers are relatively thin, so many of them are required to provide the required thickness. In cables, the insulation system must be able to be flexed without damage after manufacture. For this reason, a cable insulation system is made of relatively narrow tapes, generally less than around 20 mm wide. Adjacent tapes do not quite touch: a carefully controlled butt gap is intentionally left between them. This gap is critical to the bending performance of the insulation system, since the tapes must not collide on the inner side of a bend or they will be damaged. The butt gaps are designed so as to allow a specified minimum bending radius.

Butt gaps are essential to the bending performance of the cable, but introduce an electrical weakness. Whatever the impregnating medium, the radial electric strength of the medium on its own in the gap is less than that of the same thickness of impregnated paper. Where the electric field is highest, adjacent to the conductor screen, extra-thin papers are used so that the butt gaps are correspondingly thin. Butt gaps are also staggered between successive layers so that they do not overlap.

Paper cable insulation systems are based on kraft paper, made entirely of new wood pulp from carefully controlled sources. The finished paper consists almost entirely of cellulose fibres. This natural polymer gives plants, including trees, much of their rigidity and strength. In engineering applications, the performance of the paper is closely linked to the average length of the polymer chains. This is expressed as the 'degree of polymerisation' (DP). It is to ensure the highest achievable DP that the paper is made from new wood, to the exclusion of recycled paper or cotton, carefully processed to minimise damage to the cellulose, and to remove all soluble material.

As with other insulation technologies, the inner and outer screens are an integral part of the insulation system. The screens are made of a few layers of paper that is loaded with carbon black, or metallised with aluminium, or both.

In cable manufacture, paper of the required thickness is unwound from the coil on which it is supplied and dried by passing between heated rollers. It is then slit into tapes of the required width and re-wound into coils. Particularly at the highest voltages, the tape coils are kept scrupulously dry, all storage and processing being done in enclosures at controlled humidity. The screening tapes and the various thicknesses of plain paper tapes are wound onto the conductor in the correct sequence to build up the insulation system. The thinnest tapes are used near the conductor where the stress is highest. Each tape is applied at controlled tension, which contributes to the stability of the insulation system under bending. The lapping machine is shown in Figure 6.5.

The dominant lapped technology at the highest voltages is the fluid-filled (or oil-filled) paper insulation system [9]. The paper insulation structure is applied dry, and after sheathing is mass-impregnated under vacuum with a low-viscosity insulating fluid. Originally the fluid used was a low-viscosity mineral oil. By the time that the technology reached its pinnacle, the fluid was not a mineral oil but a

High-voltage cable systems 225

Figure 6.5 A lapping machine for applying paper insulation tapes

synthetic fluid: alkylbenzene, which chemically consists of a benzene ring with a hydrocarbon chain attached to it. The number of hydrocarbon monomers in the chain is variable, but typically the average number is around 12, for which reason the alkylbenzene is often referred to as dodecylbenzene. The alkyl chain may be straight or branched. This fluid is classed as biodegradeable: if spilled in the presence of air it is eventually consumed by natural micro-organisms.

Cable fluid is able to dissolve quantities of gas, including air, so suppressing the formation of voids, which would otherwise sustain partial discharge activity. This makes a significant contribution to the reliability of the system. By contrast, defects causing discharge in the dielectric of an extruded system cannot self-heal but can only grow.

In fluid-filled systems, changes in height along a route are accommodated by the use of stop joints, which divide the route into hydraulically independent sections. The stop joint incorporates a cast resin barrier, which withstands any pressure difference between the two adjacent sections and prevents the flow of fluid between them. Each section is limited both in length and in height difference. The hydraulic pressure at all points within each section is maintained within the allowable operating range by fluid feed/expansion tanks, both under static load and when fluid is admitted to the circuit or expelled from it during changes in load.

The fluid-filled system is highly reliable, but suffers from dielectric losses that limit its use above 400 kV. It was largely to tackle this limitation that the PPL fluid-filled insulation system was introduced in the 1990s. Polypropylene has the required low dielectric loss, sufficient operating temperature and compatibility with cable fluid, while paper has the porosity to allow the fluid to permeate the system. A layer of polypropylene is extruded between two paper tissues to create a sandwich construction, which, in tape form, is used in place of plain paper to make the cable insulation. The system has higher operating stresses and lower dielectric losses than the paper-insulated, fluid-filled system, allowing its application at higher voltages of 500 kV or above and providing a low-loss system at all voltages.

The fluid-filled system has largely been supplanted by XLPe, even at the highest voltages. Despite the inherent reliability of the fluid-filled system, a number of factors drove the transition:

- lower losses
- no requirement for fluid reservoirs, pressure monitoring systems, etc.
- no environmental hazard due to leakage of fluid.

Another variant of the lapped insulation system is the MIND system. The insulation structure is lapped with dry paper. The manufactured length of insulated core, wound on a drum, is then mass-impregnated in a tank. The impregnating compound includes constituents such as microcrystalline wax and has a melting point above the maximum operating temperature of the cable. This means that at room temperature and whenever the cable is in normal operation, the compound is a solid wax. The impregnated core is finally sheathed. The resulting solid cable requires no pressure systems. This construction was limited to around 66 kV in AC applications due to dielectric loss, and has been supplanted by extruded systems. However, it is still in production for DC applications, particularly subsea, especially at the highest operating voltages.

Gas-filled and gas compression cables are briefly described in section 6.3.5. Now all but obsolete, though often still in service, they also operate at lower maximum stress and voltage than fluid-filled.

6.4.2.2 Extruded insulation systems

The most common cable insulation system today is based on polyethylene applied by extrusion. Thermoplastic polyethylene has excellent electrical properties but an unacceptably low softening temperature, which prohibits its use as the primary insulation in power cables. The advantages of polyethylene are retained, and this critical limitation is avoided, by crosslinking, which is the formation of chemical bonds between the polymer chains, also known as vulcanisation. XLPe is not a thermoplastic but a thermoset: it may still soften at a high enough temperature, but the cross links ensure that it maintains its shape without melting. This is critical for cable insulation, which is required to retain the conductor in the centre of the core at all times.

The performance of the extruded insulation system relies critically on the quality of the extrusion [10]. Modern HV XLPe cables are always manufactured on

High-voltage cable systems 227

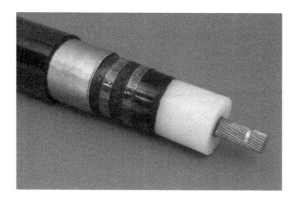

Figure 6.6 An XLPE-insulated lead-sheathed single-core cable

a triple-head extruder, which extrudes the inner and outer semiconducting screens and the dielectric onto the pre-heated conductor in a single pass through the machine. This is the optimum method for creating microscopically smooth interfaces between the dielectric and conducting layers, avoiding concentrations of electric stress, with the minimum opportunity for contamination to enter the system during extrusion. Continuous online monitoring is employed during extrusion, to ensure concentricity, correct thickness and freedom from voids and inclusions. A sample of cable showing the three layers of the insulation system is shown in Figure 6.6.

Crosslinking requires that a crosslinking agent be incorporated in the polyethylene compound, and then activated after extrusion to achieve the crosslinking. In the most common process for high-voltage cables, the crosslinking agent is an oxidising agent such as dicumyl peroxide. On heating, it generates free radicals which promote the formation of the cross links. The crosslinking process generates volatile by-products such as methyl styrene, which have to be removed later by heat treatment. The extrusion process is carried out under pressure to suppress the formation of voids by these volatiles. Early processes used steam heat and pressure to effect the crosslinking, but exposure to moisture is undesirable, and so modern extrusion lines use dry nitrogen as the pressurising medium. The extruded core is subjected to an extended period of heat treatment to drive off volatile crosslinking by-products.

Most HV cables are made on extrusion lines employing catenary continuous vulcanisation (CCV). In this arrangement, the extruded core hangs freely in a catenary curve as it passes through the curing zone. It is kept under controlled tension to maintain the correct curve. The less common alternative is vertical continuous vulcanisation (VCV) whereby the extrusion and crosslinking both take place as the core passes vertically downwards. This is intended to facilitate the manufacture of a core of the highest degree of circularity and concentricity.

Most cables are manufactured using compounds pre-formulated by a material supplier. They include the crosslinking agent and usually other additives to improve the crosslinking process and the dielectric performance of the finished insulation. Various grades are available depending on intended operating voltage. The highest operating electric fields make the highest demands on insulation quality, specifically freedom from inclusions of foreign material and 'ambers,' which are particles of degraded polymer. Inclusions, screen protrusions and voids can initiate electrical trees, which are filamentary damage structures that extend with time and stress. Once they extend nearly across the dielectric, breakdown can be rapid.

Insulation and semiconducting screen compounds are developed together to ensure compatibility with each other. The semiconducting compound incorporates a proportion of carbon black to impart conductivity. At the highest voltages, the semiconducting screens are extruded of crosslinkable XLPe, so that on crosslinking they form a 'bonded screen' with a permanent chemical bond with the dielectric part of the core. During preparation for jointing, the semiconducting core screen is removed mechanically. This is one of the most critical steps in ensuring the quality and reliability of the finished accessory, since the termination of the core screen is likely to be the point at which the electric field (outside the factory-made core insulation) is highest. At the lowest voltages, operating stresses are much lower and the core screen termination becomes less critical. In these cases, 'strippable screens' are employed, which can be simply peeled off without affecting the dielectric part of the core.

XLPe is susceptible to degradation involving moisture. With a combination of moisture and electric stress in the dielectric, electrochemical processes can lead to the creation of branching structures known as water trees. These can lead to the formation of electrical trees and eventual breakdown of the insulation. Control of moisture is essential during extrusion and processing of a cable core, as well as during subsequent storage, transport, installation and jointing.

XLPe insulation now dominates almost all markets and applications for HV cables across the entire voltage range. After some hesitancy in its introduction at the highest voltages, recent developments have taken its application to 500 kV AC [11], where its low-loss performance is a distinct advantage.

In certain markets and applications up to around 150 kV, EPR dielectric is preferred over XLPe. The compound includes a substantial proportion of mineral filler, which contributes significantly to the electrical, mechanical and thermal properties of the insulation [4]. The screens of these cores are generally strippable. EPR cables are widely applied in Mediterranean Europe, where national standards require them. They are also used for subsea cables up to around 36 kV, including 'wet' designs where the core is directly exposed to water.

6.4.3 Multi-core and multi-function cables

The most common construction other than single-core is the three-core cable, for application in three-phase power systems. The characteristics of a three-core cable relative to single cores include:

- simple installation of a single cable as opposed to three
- cable and joints larger and heavier than individual single-core cables and joints

- void spaces between the cores, especially if the overall shape of the cable is to be circular, although for fluid-filled cables the fluid ducts may occupy the voids
- reduced thermal rating, as the phases cannot be spaced apart to improve heat flow
- simple solid bonding of the common metal sheath, due to minimal circulating currents (discussed in section 6.5.2.2)
- reduced losses in the metal sheath
- possibility to require a trifurcating joint and short sections of single core cable, if outdoor terminations are to be fitted
- possibility of phase-to-phase faults.

A recent development is the triplex XLPe cable, used at lower voltages, typically 11 kV. Three complete single-core XLPe cables are laid up during manufacture, so that they can be installed as a single unit. In other respects, they remain single-core cables.

Another possible construction is the two-core concentric. This is applied in single-phase circuits such as rail traction supplies, where the neutral return current flows back to its source through the outer concentric conductor. In this system, the high voltage is applied to the inner conductor; the outer conductor is near earth potential and is insulated accordingly.

High-voltage cables may also carry other services. In three-core cables, there will be plenty of void space which these can fill. Subsea cables often incorporate optical fibres for communication and control, and for monitoring the cable itself. Multi-function cables used in underwater exploration and production of petroleum may incorporate a variety of services such as pipes for liquids or compressed air.

6.4.4 Outer layers

Outside the core of a single-core cable, or the cores of a multi-core, a number of functions are required to create a complete cable:

- longitudinal conductance to allow capacitive charging currents to flow, while maintaining potential close to earth
- longitudinal conductance to allow fault currents to flow under fault conditions
- for pressure-assisted cables, retention of the pressure of insulating liquid or gas
- radial moisture barrier to prevent ingress of moisture to the insulation
- mechanical protection.

These functions, in the varying degrees in which they are required by the cable design, are fulfilled by one or more metal layers.

From the earliest days, impregnated cables have had lead sheaths. This both retains the impregnating medium and excludes moisture. Modern cable sheaths are made of lead alloys with improved performance over pure lead, applied by extrusion. Where the insulation system is pressurised, lead on its own is liable to creep, so bronze reinforcing tapes are applied over it to retain the internal pressure. Often, additional copper wires are required to improve the longitudinal conductance. Where protection against external impact is required, extra armouring may be applied.

230   High-voltage engineering and testing

All the functions of a metal sheath can be performed in a single element if aluminium is used. Compared with lead alloy, modern aluminium alloy has:

- greater conductivity, obviating additional screen wires
- much greater resistance to mechanical fatigue, creep and stress corrosion, obviating the reinforcing tapes on pressurised cables
- lower density, reducing the overall weight of the cable
- poorer corrosion resistance, necessitating more attention to corrosion protection
- greater stiffness, even when corrugated, requiring greater effort to bend the cable during installation
- greater robustness, often obviating any other armour.

Modern aluminium alloy sheaths are applied by continuous extrusion of a tube around the core as it passes through the extruder, so that there is no longitudinal seam. Aluminium sheaths are generally corrugated to allow sufficient flexibility of the cable. The process is shown in Figure 6.7. For fluid-filled cables, the

Figure 6.7 An aluminium extruder and corrugator for applying corrugated seamless aluminium sheathing

corrugations are generally helical, which creates a useful longitudinal path for the flow of cable fluid. However, helical corrugations are a disadvantage for extruded cables, as the longitudinal path allows water to spread along the cable in the event that the sheath is penetrated. For this reason, corrugations on extruded cables are often annular, so no helical path is created.

Disadvantages of corrugated sheaths arise from the corrugations themselves: there is an air space under the corrugations that adds radial thermal resistance, and there is an increase in the overall diameter of the cable. A modern sheath system that addresses these concerns is the foil laminate sheath, of which there are many variants. A metal foil tape, usually aluminium or copper, is applied either helically or longitudinally to provide the radial moisture barrier. The longitudinal or helical seam is either overlapped or suitably sealed, for instance by glueing or welding. The foil is generally laminated to an extruded polymeric layer to impart sufficient strength. The foil has a small cross section and thus little longitudinal conductance; in order to meet short-circuit requirements, additional screen wires, generally copper, are applied. Alternatively, copper tapes may be used alone, with sufficient cross section to meet the longitudinal conductance requirement. The resulting sheath system may be mechanically less robust than an extruded aluminium sheath, for instance. For this reason, these cables are often reserved for installations where the risk of external damage and moisture ingress is low, such as in tunnels that are expected to be kept dry during the life of the cable system.

Before the problems associated with moisture ingress were appreciated, especially in North America, early polymeric cables were installed without a metallic radial moisture barrier at all, just an extruded polymeric jacket. An expectation of a very short service life due to water tree degradation became firmly entrenched in the North American market, and acceptance of the requirement for a radial moisture barrier was very slow.

Where additional mechanical protection is required, armour is applied. The material of choice would be steel in wire or tape form, either galvanised or bitumen coated. However, single-core cables in AC circuits cannot normally be armoured with steel due to prohibitive magnetic hysteresis losses in the steel; in these cases non-magnetic armour such as aluminium wire is used. This does not apply to three-core cables or to DC circuits, neither of which has significant external AC magnetic fields.

Apart from those fulfilled by the metallic layers, there are other functions of a sheath system:

- longitudinal moisture barrier underneath the radial moisture barrier, to prevent spread of water along the cable if the radial moisture barrier is penetrated
- cushioning to prevent indentation by screen or armour wires or by corrugations of the sheath
- insulation of the metallic sheath or armour from earth for control of circulating currents
- insulation of the metallic sheath or armour from earth for control of corrosion
- mechanical protection of the metallic sheath.

These are met by various non-metallic outer layers in a cable:

- extruded bedding layers for screen or armour wires
- woven cushioning tapes, which may be conductive
- water-blocking tapes over the core, or around the screen wires if these are under the metallic sheath
- extruded overall insulation, variously termed the oversheath, the secondary insulation, the serving, jacket or polymeric sheath.

The extruded oversheath may be made of various polymeric materials. PVC, although intrinsically fire retardant, generates highly corrosive and lethal acid smoke if it is involved in fire. Moreover, its manufacture and disposal have considerable environmental impact. More recently, various types of polyethylene have been used, such as medium-density polyethylene (MDPe) thermoplastic or XLPe. Their electrical resistivity, electric breakdown strength and mechanical toughness are far superior to PVC [12]. However, they are highly flammable, so for applications exposed to air such as in buildings or tunnels, a fire-retardant coating is applied either during manufacture or by painting after installation.

In some environments, cable oversheaths are subject to gnawing attack by termites; this can be deterred either by a thin layer of a tough hard material such as polyamide, or by impregnating the oversheath with an insecticide or with a more environmentally acceptable termite deterrent [12].

Since the oversheath has an electrical function, it can and must be electrically tested by applying a test voltage between the metallic sheath and earth. The surface of the cable is often coated with a conductive layer to facilitate such testing, whether on the drum or after installation. In testing a manufactured length on a drum, an earth connection is made to the conductive layer, thereby providing an earth potential along the entire length while test voltage is applied to the metallic sheath. The conductive layer may be a thin layer of carbon-loaded polymer extruded onto the insulating layer, or else a graphite-based paint applied overall. This layer must be removed wherever the cable enters joints or terminations, so as to avoid creating a conductive path to earth from within the accessory. The conductive layer may be hard to distinguish visually from the underlying insulation if both are black. Failure to remove all traces of the conductive layer prior to jointing is a frequent cause of faults from the metallic sheath to earth, which will be detected during commissioning tests after jointing.

Before extruded oversheaths were introduced, cables were generally covered with a serving of woven hessian tape, coated in bitumen, sometimes whitewashed overall. Many lead-sheathed and/or steel-wire-armoured cables with such servings are still in service, although the armour wires may have corroded away completely, but the degree of corrosion protection is completely inadequate for aluminium sheaths.

6.4.5 Installation

Depending on the requirements of the route, a number of methods of installation of varying degrees of sophistication are available [13].

Adjacent to terminations, cables will usually be cleated to structures. For routes within buildings, basements, galleries and tunnels, the entire route may be cleated in this way.

The simplest method of installation cross-country is direct burial. A trench is dug to an appropriate depth, the cable is laid into it and the trench is backfilled. In order to avoid damage to the cable from sharp stones in the soil, screened sand is used around the cable. If the native soil from the trench has unsuitable thermal properties, the backfill may be imported material with known thermal performance. Warning slabs and/or tapes are incorporated into the backfill above the cable so as to give warning to anyone digging in the area subsequently.

It may not be practicable to use this simple method for the entire route, where obstacles such as roads, railway lines, canals and rivers prevent trenching. In these cases, ducts may be installed under the obstacle by means of directional drilling, and the cable later pulled through the duct. Usually, the duct is finally pumped full of a medium such as fluidised bentonite, which goes rigid to provide mechanical restraint of the cable, and also improves the thermal conductance between the cable and the duct when in service.

In urban areas where routes run along carriageways or footways, there may be severe limits imposed by the local government on the length of trench that can be opened at any time. In these and other cases, it is necessary to preinstall substantial lengths of ducts for later pulling of the cable. The ducts are often supplied in discrete lengths, installed by trenching and joined together *in situ*. As installation of the duct is so straightforward, the process is rapid with a minimal length of trench open at any one time. The cables are drawn into the ducts later on, without further excavation. This process also helps to protect against theft of the valuable cable while lying exposed to view in an open trench during installation. These duct installations are frequently left un-filled, which facilitates later replacement of the cable at the expense of increased thermal resistance and consequent reduced rating.

Within substations and similar environments, cables may be cleated in surface troughs. The trough covers may be ventilated to improve cooling, or the covers may be sealed and the cables cooled by forced air circulation. Where the highest ratings are required, the troughs may be backfilled with a medium of suitably high thermal conductivity.

Many special installations use tunnels, for instance crossings of wide estuaries. Another increasing application of tunnels is in city centres, where it may not be feasible to close roads or to lay large cable routes beneath the congestion of existing buried services. It is often impractical for a high-voltage cable circuit to share a tunnel with other services, due to difficulties with induction and also due to the risks both to and from the cable [14]. A dedicated cable tunnel is a substantial investment, but it brings the advantages of ready access for installation, maintenance and later replacement of several successive cable systems during the life of the tunnel. Cables and joints in the tunnel may be cleated to the sides or encased in cement-bound sand, and may be cooled by forced circulation of air or water.

All the above installations are classed as rigid; that is to say, any thermal expansion of the cable is counteracted by rigid fixings, which exert the force necessary to restrain the cable in position. The alternative is flexible installation. For instance, in tunnels, cables may be attached using cleats at spacings of several metres, so that they sag in between cleats. Any thermal expansion (or contraction) results in an increase (or decrease) in the depth of sag, without large forces being exerted.

These methods of installation are not mutually exclusive. In many cases, several or all of them are applied in different parts of a route, depending on the different environments through which the route passes.

Whichever methods of installation are chosen, it is necessary physically to transfer the cable from its drum to its final position. There are a variety of possible methods. Again, more than one method may be used in installing any circuit.

- Where the cable is not too heavy and labour not too expensive, a large crew is spread along the route and the cable is passed from hand to hand.
- More usually, mechanical effort is required, and the simplest method is nose pulling. A pulling fitting is attached to the leading end of the cable, and the cable is pulled off the drum and into position with a winch. The pulling tension is calculated beforehand and then continually monitored, as excessive tension is liable to damage the cable. To avoid damage to the cable by abrasion, it is supported on temporary rollers. Where changes of direction are required, skid plates or vertical rollers are fitted to guide the cable round the bend. They are arranged to minimise the sidewall pressure on the cable, which could also cause damage.

The pulling tension and sidewall pressure are governed by the weight and stiffness of the cable, the length and gradient of the route and the number and radius of bends. Where pulling tension or sidewall pressure would be excessive, more complex methods of pulling are used.

- Bond pulling uses a wire rope (the 'bond'), which is hauled along the route by a winch. It is paid off from a drum, which is provided with a brake so that the tension in the bond can be controlled. At each change of direction, the bond is passed around a pulley known as a snatch block. The cable is tied to the bond at short intervals along its length. The cable cannot pass through the snatch blocks; its ties are un-tied as it approaches each such block, the cable is guided around the required bend and then the ties are re-tied as soon as the cable enters the next straight section. This is shown in Figure 6.8.
- Powered rollers are used; as well as supporting the cable off the ground, they grip it and apply thrust. Many rollers are spaced along the route and all are controlled together to achieve a high precision in movement of the cable.
- For the most demanding installations such as tunnels, mobile installation machines may be used. Such machines pass along the route, lift the cable into position for cleating and apply the required sag for a flexible installation.

6.5 Current ratings

Many mentions of cable cooling in the sections above raise the question of cable thermal rating. The thermal rating (or 'ampacity') of a cable installation is

High-voltage cable systems 235

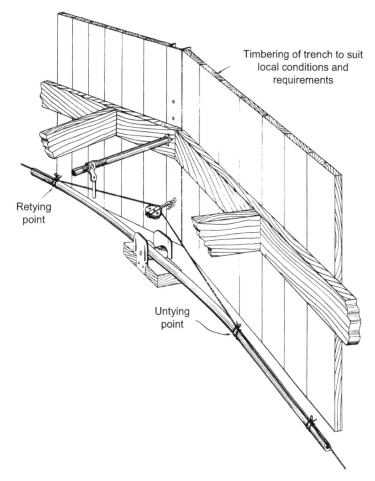

Figure 6.8 Cable installation by bond pulling round a bend

expressed as the maximum current that it can carry without the conductor exceeding a specified temperature, at an assumed value of ambient temperature.

6.5.1 Time-dependent ratings

Various ratings can be considered, including:

- continuous rating, the load that can be carried indefinitely
- cyclic rating, allowing for a daily cycle between two values of load
- short-term transient rating, the load that can be carried for a specified period of minutes or hours, starting from an assumed lower value of temperature or load
- emergency rating, allowing the conductor to reach a temperature somewhat higher than the continuous operating temperature, provided that the cable is designed to allow this for a specified number of hours or days during its operating life

- short-circuit rating, the current that can be carried for a specified very short period such as 1 s, allowing the conductor to rise momentarily to a much higher temperature such as 250 °C from a given starting temperature such as the rated operating temperature.

6.5.1.1 Continuous rating

The continuous rating is relatively straightforward to calculate [8], although the deceptively simple formula.

$$I = \sqrt{\frac{\Delta\theta}{RT}}$$

becomes more complex in practice. The thermal resistance T comprises various terms [15], and allowance is made for losses in addition to I^2R. These items are discussed in section 6.5.2.

6.5.1.2 Short-circuit rating

The other rating that is relatively straightforward to calculate is the short-circuit rating of the conductor or other metallic layer. The duration of the current flow is short, typically 1 s, so that the adiabatic approximation may apply; that is, any heat flow during such a short period is negligible. In this approximation, all the energy dissipated in the conductor per unit length goes to heat the conductor, which has a known heat capacity per unit length. For given starting and finishing temperatures, the allowable dissipation and thus the allowable current can be calculated.

There are cases where heat transfer is not negligible: for longer durations of short circuit or for lightweight components such as small cross-section conductors, screens and sheaths. The calculation then is more complex, and gives higher rated currents than if the adiabatic approximation is used.

Other time-dependent ratings are more complex to calculate, when neither the steady-state nor the adiabatic approximations apply. IEC Publication 60853 [16–18] gives methods for calculating cyclic and emergency ratings.

6.5.2 Factors affecting ratings

6.5.2.1 Temperature rise

The permitted temperature rise $\Delta\theta$ is the difference between the permitted conductor temperature and the ambient temperature. The maximum continuous conductor temperature, both for fluid-filled and for XLPe-insulated systems, is often 90 °C, as enshrined in various national and international standards. For other systems such as EPR, it may be higher.

The ambient temperature is the temperature of the heatsink to which heat must escape. Since heat cannot escape downwards into the ground, it refers to the temperature of the ground surface from which heat can escape by convection through air. In temperate latitudes, the assumed temperature will vary with time of year, resulting in 'winter' and 'summer' ratings.

6.5.2.2 Losses

Power losses are an important factor in cable ratings. However, they are important in themselves as contributions to the operating cost of a cable system in service.

The conductor loss I^2R (where, as discussed previously, R is the AC resistance for AC circuits, at the rated conductor temperature) is not the only source of heat in the cable. Other possible sources are:

- dielectric loss in the insulation
- eddy current losses due to action of magnetic fields on metallic sheath, screen and armour
- losses due to longitudinal currents circulating in the sheath, screen or armour
- hysteresis losses in steel armour
- sunlight or other sources of conducted, convected or radiant heat from outside the cable.

Voltage-driven losses

Dielectric loss is driven by the voltage, not by the current, so will be dissipated whenever the circuit is energised, regardless of load. Thus in cyclically loaded circuits, its economic importance is accordingly greater.

In AC circuits, the dielectric loss per unit length W_d is given by

$$W_d = 2\pi fCU^2 \tan \delta$$

where $\tan \delta$ is the loss tangent or dielectric dissipation factor, that is the proportion of the energy stored in the capacitance that is dissipated as heat per cycle. Dielectric loss was a significant contribution in the fluid-filled paper-insulated system, especially at high stress, which limited the wide application of the system beyond around 400 kV. The fluid-filled PPL-insulated system with its lower losses was a considerable improvement, allowing operation at 500 kV. XLPe insulation has still lower losses, which are almost negligible at typical operating stresses and temperatures.

In any system, if $\tan \delta$ increases with temperature, there is ultimately a potential for a thermal runaway situation to occur. There is a vicious circle, with increased temperatures leading to increased losses and back to further increased temperatures, until the temperature runs away to destruction. This has happened in real situations where the external thermal resistivity increased, through drying-out of the surrounding soil, while the circuit was in operation [19].

Current-driven losses

Eddy-current loss is driven by the magnetic field and can be significant in single-core cables. It occurs in metal components – sheaths, tapes or wires – surrounding the conductor, even when no net longitudinal current can flow in that layer. It is minimised by making each lossy layer as thin as practicable.

Hysteresis loss arises in magnetic materials, such as steel armour, exposed to time-varying magnetic fields. Thus it is potentially significant in single-core cables in AC circuits. In such cases, non-magnetic armour such as aluminium is used instead of steel to avoid this loss.

Circulating current is potentially a significant source of loss when the sheath or armour is bonded to earth at both ends of a cable section. This creates a circuit through earth, and the time-varying magnetic flux linked through this circuit induces an electro-motive force, which drives a circulating current longitudinally along the sheath or armour. It is primarily circuits consisting of single-core cables that suffer from circulating currents, and that induce circulating currents in other services. Current can be induced by 'our' circuit in other nearby conductors, and currents in other circuits can induce circulating currents in 'our' circuit. In such circuits, the solution in essence is not to bond the sheath at both ends. The practicalities of this are explored in section 6.5.2.4.

External gain

Solar gain by cables installed outdoors above ground is rarely a major issue in temperate latitudes, but may be minimised by external sunshades or by applying a reflective coating, either of which will also serve to minimise degradation of the oversheath by the ultraviolet light found in sunlight. Other sources of gain may be nearby cable circuits or other underground services that dissipate heat [20].

6.5.2.3 Thermal resistance

The temperature rise $\Delta\theta$ for a given dissipation W is given by

$$\Delta\theta = WT$$

The thermal resistance per unit length T is made up of several components [15]. Heat from the conductor of a buried cable must pass through a number of layers before it reaches the ground surface where it can be carried away. Each of these layers is associated with an individual thermal resistance. Thermal resistances include:

- the insulation, between the conductor and the outside of the core
- between core and metal sheath, including waterblocking tapes, cushioning tapes, and possibly a radial gap under a corrugated aluminium sheath
- any bedding layers between sheath and armour (if applicable)
- the oversheath
- between the cable surface and the external duct (if applicable)
- the external duct (if applicable)
- the backfill and other external medium between the surface of the cable (or duct) and the ground surface.

The thermal resistance of a radially symmetrical layer is calculated by

$$T = \frac{\rho_T}{2\pi} \ln\left(\frac{d_2}{d_1}\right)$$

where ρ_T is the thermal resistivity of the material (in Kelvin metres per Watt) and d_1 and d_2 are the inner and outer diameters of the layer, respectively. This applies to layers such as the dielectric, bedding layers for screen or armour wires and the oversheath. Metal layers such as sheaths have negligible thermal resistance for our purposes.

The heat dissipated by conductor loss passes through all layers between conductor and ground surface: it experiences the entire thermal resistance of all these layers in series and makes its contribution to $\Delta\theta$ accordingly. The dielectric loss is dissipated within the dielectric; mathematically it can be shown that, on certain assumptions, it experiences the equivalent of only half the dielectric thermal resistance, plus all the thermal resistance contributions further out. Eddy current, circulating current and hysteresis losses are dissipated in the various outer metallic layers and experience only those thermal resistance contributions from further out. Allowing for all this, the rating equation

$$I = \sqrt{\frac{\Delta\theta}{R\,T}}$$

acquires many more terms. Anyone needing to study this in detail should refer to IEC Publication 60287 [8, 15, 19]. However, the heat flows and just some of the possible thermal resistances are represented diagrammatically in Figure 6.9.

6.5.2.4 Optimisation of rating

The thermal rating of a cable system is determined by many factors in the construction and installation of the cable. It is possible in theory to achieve any desired rating by making the conductor cross section large enough, but it is likely to be more practical and economic to use a variety of means to optimise the rating for a smaller cross section.

Reduction of losses
Minimising losses is the most direct way to enhance ratings, and also reduces the cost of the energy lost by the cable in service. Methods to minimise losses for AC circuits include:

- large conductor cross section, to reduce DC resistance
- Milliken construction of conductor to reduce the AC resistance factor
- selection of low-loss insulation system to minimise dielectric loss
- selection of materials and dimensions of metal layers to minimise eddy current losses
- avoidance of magnetic materials in or near single-core cables
- special bonding to reduce circulating currents.

It is special bonding that we will expand on here. It primarily applies to installations of single-core cables in AC circuits. If the sheaths of such cables are bonded to earth at each end, circuits are created via earth. Magnetic flux due to the conductor currents will link through these earth loop circuits and induce currents in them in the opposite direction. The magnitudes of the currents depend on the various inductances and resistances, but could typically be around half of the primary current. This will add significant loss, which will be of both thermal and economic importance.

The most basic form of special bonding is single-end bonding. The cable sheath is bonded to earth at one end, and left free of earth at the other. The electro-motive

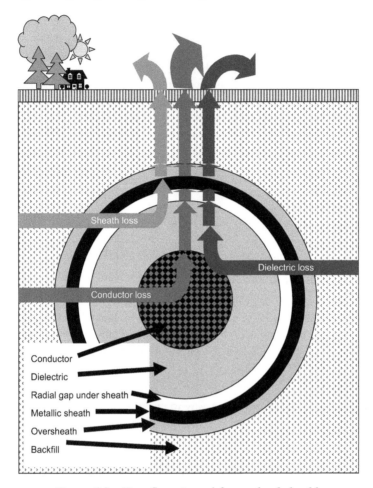

Figure 6.9 Heat flows in and from a loaded cable

force (e.m.f.) induced in each cable appears as an alternating voltage between the free end of the sheath and earth. In most cases, the system is designed so that the standing voltage under normal conditions is sufficiently low not to be hazardous to touch. However, under fault conditions, the currents may be many times higher, and the voltages will be higher in proportion. In order to protect the cable oversheath from excessive voltage, sheath voltage limiters (SVLs) are connected between the sheath and earth at the free end. Modern SVLs are zinc oxide non-linear resistors, with knee voltages of a few kilovolts.

In order that both the cable sheath and the SVLs can be tested, the cable sheath to earth links at one end and the SVLs at the other end are fitted in accessible link boxes, at least at the higher system voltages.

The length of circuit that can be bonded at a single end is limited by the standing voltage appearing at the SVLs. However, the length can be doubled by the simple expedient of bonding the circuit to earth at the mid point, either at a joint or by a direct connection to the cable sheath, and having SVLs at both ends.

Many routes are much too long to be bonded in this way, with many joints, and for these the system of cross-bonding is used. The circuit is grouped into one or several major sections, each consisting of three sections of cable between joints. Ideally the three sections should be of equal length. The cable sheaths are earthed at both ends of each major section, and free of earth at the two intermediate joints, protected from overvoltages by SVLs. At those intermediate joints, the cables are transposed and the cable sheaths cross-connected, so that (for balanced currents in the primary) the e.m.f.'s induced over the length of the major section cancel out. As the e.m.f.'s cancel out, the circulating current is reduced in proportion. The joints must have an insulated flange built in, so that the cable sheaths on either side can be bonded separately. Connections from either side of each joint are brought out to a link box accessible from the ground surface; a two-core concentric bonding lead is often used. There are many variations in detail, but the scheme shown in Figure 6.10 is typical. In practice, the joints are encapsulated in external insulation (not shown here) to insulate their metal components from earth. The link boxes (also not shown) that house the links and SVLs may be in pits under sealed covers, or in kiosks above ground.

For circuits in long tunnels, it is often troublesome to provide connections to earth at every third joint bay. It is increasingly common in these cases to use the system of earth-free continuous cross-bonding. Every joint bay within the tunnel is cross-bonded and protected from overvoltage by SVLs connected across the joint flanges; no connection to earth is required.

Reduction of thermal resistance

Apart from reduction of losses, the other direct means of reducing temperature rise is reduction of the thermal resistance.

The internal thermal resistance is determined by the cable design, its largest component usually being that of the dielectric. The higher the operating electric stress of the dielectric for a given voltage, the thinner the insulation and thus the lower the thermal resistance.

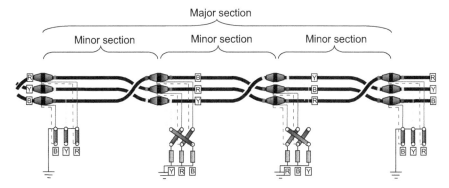

Figure 6.10   One major section of a typical cross-bonded cable circuit

The external thermal resistance is determined by the method of installation. Methods of reducing the thermal resistance include:

Installation method	Method of optimising external thermal resistance
Direct burial	Replacement of backfill with specially selected material or cement-bound sand
Surface troughs	Filling with cement-bound sand or proprietary thermal backfill
Ducts	Filling with pumped grout such as bentonite

In practice, most cable circuits will include several different installation methods. For instance, a direct-buried circuit may include ducted sections under road crossings, the ducts being installed by directional drilling to obviate trenching across the road. Every different combination of installation method, depth of burial or ground condition will give rise to its own rating to be calculated separately, and it is the lowest rating, or the hottest point along the route, that determines the overall rating of the circuit. It is at such points that the effort to reduce thermal resistance will be concentrated. For instance, in the example above with the short ducted sections, the ducts could be grouted with bentonite to eliminate the thermal resistance of the air space in the duct, with the added mechanical benefit of holding the cable rigid within the duct.

Cooling
Where other methods are inadequate to provide the required rating, cooling may be employed. Typically, this involves pipes laid alongside the cables along the route, carrying pumped cooling water. This is usually applied only to the highest rated circuits, especially in urban areas where the width of the installation corridor is limited.

Deep tunnel installations are a particular case. Due to its considerable depth below ground, the thermal resistance between a tunnel and the ground surface may be very large. The thermal time constant of the system will be correspondingly long, perhaps in the order of decades, so the temperature of the tunnel and the surrounding rock increases progressively through the life of the cable. Therefore, it may be necessary to cool the tunnel. The most usual method is to blow air through the tunnel from end to end [11], but in other cases the cables are laid in troughs filled with circulating water or cooled by forced circulation of the cable fluid.

Dynamic rating
Current ratings based on calculations at the design stage are necessarily conservative. There is scope to optimise ratings in service by using knowledge of the actual installation and operational conditions. In particular, temperature measurements along the installed cable route allow hot spots – which will limit the rating – to be detected and measured. Since not all the hot spots will be known in advance, this is best achieved by a scheme of distributed temperature sensing using a continuous sensing element such as an optical fibre. The measured temperatures and

currents are input to a numerical model of the installation, which infers the corresponding conductor temperatures all along the route. The highest conductor temperature will determine the rating. Using knowledge of the cable materials, construction and installation, the current rating can be calculated for the steady state or as a transient rating over any desired term [11].

6.6 Accessories

Cables are useless without their accessories. The main accessories are the joints and terminations that directly connect the cables.

6.6.1 Terminations

As cables are required to carry power from one place to another, the most essential requirement is for terminations. These are the fittings that connect the cable to other equipment. Other names for these include sealing ends, or potheads, a reference to the porcelain insulators.

The main requirements of terminations include:

- low-resistance connection of the cable conductor(s) to other equipment, suitable to pass the load current and any overcurrents without undue temperature rise
- mechanical connection, which may need to be of sufficient rigidity and strength to maintain the cable in position against thermal, magnetic and environmental forces, unless external restraints assist in this
- hydraulic sealing, sometimes to retain fluid or gas within the cable system, but always to exclude water or other foreign matter
- control of the electric field, so as to adapt the very high radial field within the cable to the lower fields in other equipment
- external insulation sufficient to maintain the performance of the insulation system in the environment in which the termination is fitted.

One of the most critical factors in the design of terminations is control of the electric field. Within the cable, the electric field is radial in direction and very high in magnitude. Outside the cable, the electric field is in random directions and generally much lower in magnitude. The termination is required to adapt one field distribution into the other, so that the electric field everywhere can be withstood by the material that experiences it. It is important to avoid electric stress raisers at discontinuities. At the highest voltages, the method of stress control is capacitive: the use of the geometry and permittivity of the components to achieve the desired stress distribution.

The simplest method of geometrical stress control at a cable termination is to flare the outer screen smoothly and progressively, so that the radial electric stress reduces in proportion to the increasing diameter. The core screen of the cable, the outer conductive layer of its manufactured insulation system, is removed and replaced with something that has the required flared shape. It is critical to ensure that there are no discontinuities that would create localised high stresses, and no voids or contamination that would be unduly susceptible to electric stress.

It is easy enough to envisage how this is done for a lapped insulation system such as the oil-filled paper-insulated system. The screen consists of metallised and/or carbon-loaded paper tapes. These screen tapes are removed by hand, then more paper is wound on over the top to create a conical profile and a conductive screen is applied overall.

For extruded systems, a factory-made stress cone is used. These are usually made in EPR or silicone rubber. There are an inner part of insulating rubber and an outer part of conducting rubber; the interface between the two provides the required stress control profile. The core screen of the cable is removed and the cut end chamfered to provide a smooth graduation. The quality of this work is critical to the formation of a defect-free interface with the stress cone. The rubber stress cone is then stretched over the prepared core and slid into the required position.

At the highest voltages, the practice with fluid-filled systems is to use a capacitor cone for capacitive stress control. Just as in bushings, this includes concentric foil electrodes built into a paper structure to achieve highly uniform grading of the electric stress, both within the termination and over its surface. This concept has not been retained in terminations for XLPe.

All accessories are required to be suitable for the cable to which they are fitted, but terminations must also be suitable for the equipment to which the cable is to be connected and for the environment in which they are to operate. Terminations may be classified as in Figure 6.11. The type and size of the external insulator of the termination are determined by this classification.

The commonest termination is the outdoor or air-insulated switchgear (AIS) termination. This is also physically the largest termination, as its external insulation must meet the creepage requirements for outdoor use. The outdoor termination is most often housed in a hollow insulator of the required length, with external sheds to provide the creepage necessary for the required performance. For a long time the common material for the insulator has been porcelain. An extruded tube of 'green' (un-fired) porcelain is turned on a lathe to form the required shed pattern on the outside. The finished form is coated with slip to provide the glazed outer

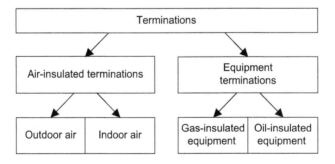

Figure 6.11 Classification of terminations

surface, and with grit to provide a rough surface at each end where the metal fittings will be attached. The part is then dried and fired, which fuses the green material into a finished ceramic insulator, complete with glazed surfaces. The process for making electrical porcelain is normally such that the fired product is practically free of voids. On completion of the fired part, the metal end fittings are attached, traditionally with cement.

More recently, an alternative construction of the external insulator has become common: the composite insulator. This consists of a rigid insulating tube of fibre-reinforced polymer composite, onto which the shed profile of the external insulation is formed in silicone rubber. The composite tube must be manufactured so as to be void-free to avoid degradation by partial discharge. The silicone rubber outer layer must be permanently bonded to the tube so that there is no path for moisture to travel along the interface. The silicone rubber outer surface must also have the hydrophobicity to maintain the performance of its insulation under wet conditions, and the chemical stability to maintain its properties under all the stresses of the weather, including ultraviolet radiation, for the length of its design life. Reasons for choosing composite insulators over porcelain include the following, of which any one may be sufficient to justify the greater cost:

- lighter weight and less fragility, for easier transport, handling and installation
- less vulnerability to mechanical damage in service, whether by accident or vandalism, which in some territories includes gunfire
- reduced hazard to personnel or other equipment if the termination is destroyed by an internal fault, as no razor-sharp chunks of heavy porcelain are thrown around the substation or beyond.

Rigid terminations as described above are self-supporting and rated to withstand forces due to thermal thrusts, short circuits or the weather. Other constructions are possible, such as pre-manufactured one-piece rubber assemblies that include the internal stress control system and the external shedded profile, suitable to be stretched or shrunk onto the prepared cable core in one operation. At the lower voltages, heat-shrink or cold-shrink tubing and external sheds may form the basis of the termination; these are widely used up to 33 kV and less frequently beyond. These non-rigid terminations rely on mechanical support from the equipment to which they are bolted.

Indoor air-insulated terminations are similar to outdoor types, although the minimum creepage version of the external insulator may be selected, as wet-weather performance is not required.

Equipment terminations allow the cable to be terminated directly into equipment filled with insulating oil, such as transformers, or sulphur hexafluoride gas, such as gas-insulated switchgear. The outside of the termination insulator is directly immersed in this oil or gas medium. The insulating performance of the medium is such that a much shorter insulator is used, without the shed profile of outdoor insulation. Often the same insulator is used for oil-immersed and gas-immersed variants. As weather resistance is not required, the insulator is nowadays usually cast in epoxy resin rather than porcelain, which also allows greater precision of its dimensions.

At the lower end of the voltage range, terminations may be of separable designs. The cable termination is housed in a conical insulator, which plugs into a matching socket. A single design of termination is used in multiple applications by choosing a socket in the form of a transformer bushing, to make an equipment termination, or an outdoor insulator to make an air-insulated termination. The same termination plug may be inserted into a transition joint, to connect to an older type of cable, or into a back-to-back insulator to provide a straight joint.

6.6.2 Joints

If cable systems are not to be limited to the lengths of cable that can be manufactured, delivered and installed in one piece, means of joining cable sections together must be provided. Applications for joints include:

- joining manufactured lengths of cable end-to-end to achieve the required route length
- subdividing manufactured lengths to facilitate installation in particularly difficult installation conditions
- diverting an existing installed circuit
- repairing after a fault or external damage, either by inserting a single joint or more commonly by using two joints and a short insertion length of new cable
- transition joints, to join two different cables, commonly for repair or diversion of an existing installed circuit where matching cable is no longer available
- flexible joints in subsea cables, whereby the finished joint is sufficiently similar to the cable in overall diameter, flexibility, tensile strength and other characteristics to allow the cable to be installed in a long continuous length as though the joint were not there
- stop joints, to provide a hydraulic barrier between adjacent hydraulic sections of an oil-filled cable circuit
- trifurcating joints, to split a three-core cable into three single cores, for instance where outdoor terminations are required on a three-core cable system.

Joints are required to withstand all the stresses that affect the cable, not just electrical but also thermal and mechanical. Despite the highest level of attention being paid to quality in manufacture and installation, joints are often the weakest point in any installation.

The most basic requirement for a joint is conductor connectivity. Conductors can be connected by a variety of methods, of which the most straightforward is by mechanical compression in a metal sleeve, usually using a hydraulically actuated compression tool. This is shown 'before' and 'after' in Figure 6.12. Occasionally, connections are made by sweating, a form of soldering. Aluminium conductors are problematical to connect by these methods, and are often instead connected by inert-gas welding. Another alternative is exothermic welding, in which the conductor ends are placed in a mould filled with a pyrotechnic compound; when ignited, it fills the mould with weld metal and so joins the conductors. Conventional

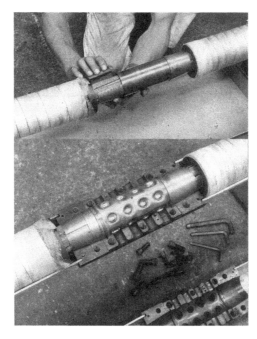

Figure 6.12 Two stages in the joining of conductors by compression

or pyrotechnic welding methods are particularly suitable for making a conductor joint of the same diameter as the original conductor.

The second fundamental requirement for joints is the same as for terminations: control of the electric stress within the dielectric. The stress inside the factory-made cable dielectric, with its cylindrical geometry, is high and purely radial. To make a joint, further insulation must be applied to insulate the conductor connection. The geometry now usually becomes less regular, which can lead to localised stress enhancements. Furthermore, the field-applied insulation of the joint may be less reliable than the factory-manufactured and pre-tested insulation of the cable. The key tasks in the dielectric design of a joint will be:

1. to minimise the maximum electric stress
2. to minimise the electric stress in field-applied insulation, and especially the stress *along* the interfaces between different dielectric components.

The stress distribution can be calculated by finite element methods, with results as shown in Figure 6.13.

As much as possible of the joint insulation is nowadays provided by factory-manufactured assemblies. One widely used approach is to use a single factory-made rubber component with all the stress control profiles moulded in.

Joints are often required to provide an insulating barrier between the cable sheaths on either side. This allows various methods of special bonding to be applied to minimise circulating-current losses. It also allows the secondary insulation to be

248 High-voltage engineering and testing

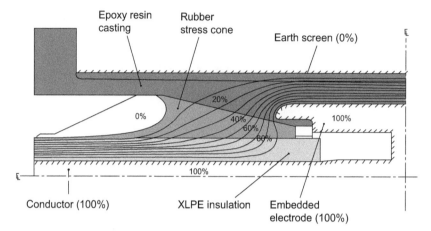

Figure 6.13 Distribution of lines of constant voltage through the cross section of an accessory

Figure 6.14 A joint bay showing joint protection boxes fitted to joints

tested one section at a time so that faults can be located and repaired. A suitable insulating gap has to be left in the screening of the joint, and also in the metal housing of the joint. Joints in metal housings may incorporate a cast epoxy resin insulator ring, separating the two sides.

The outer metal layers of the joint are covered in secondary insulation, which has the same functions as in the cable: electrical insulation, corrosion protection and mechanical protection. The most complete and sophisticated approach is to clamp an insulating container around the joint and fill it with an insulating compound, which may be a thermosetting resin. Joints protected like this and ready to be backfilled are shown in Figure 6.14. A less robust approach is to cover the joint with heat-shrink sleeving. Where special bonding is required, bonding leads are brought out through the secondary insulation to an external link box where the required connections are made.

6.6.3 Other accessories

Accessories that do not contact the cable conductor may be termed secondary accessories. They include the link boxes where bonding connections are made. Within the link box, the bonding leads from the joints are terminated on terminal posts. The links to earth, or the cross-bonding links and connections to the SVLs, are removable for testing. The insulation between the links and the enclosure must be sufficient to withstand all standing voltages, overvoltages and test voltages that it may be exposed to. Additionally, in the event of a flashover within the box, the enclosure must contain the blast without rupture.

Link boxes need to be accessible during the life of the system. In public places, they are often placed in pits, concealed under covers. In other cases, particularly within substations, the links may be housed in small kiosks above ground.

SVLs are small surge arrestors that protect the cable sheath system from overvoltages at points where it is not connected to earth. They are fitted within the link boxes at the cross-bonded joints in cross-bonded systems, or at the non-bonded end of single-end bonded sections. Alternatively, they may be fitted directly across the insulated flange at gas-insulated equipment terminations or at joints open to the air in tunnels. Modern SVLs use zinc oxide non-linear resistors to provide the protection. They have rated operating voltages typically in the range 4.5–9 kV, but can be higher or lower. Modern SVLs are generally housed in an insulating cylinder with two terminals, but an older system used three-phase SVL assemblies housed in a common metal can with flying leads brought out.

Fluid-filled systems require the fluid pressure to be maintained within specified limits under all conditions. External reservoirs are connected, containing stacks of expandable capsules to maintain a nearly constant pressure despite changes in volume. There is monitoring of the pressure, which even today usually involves dial gauges with alarm contacts, although the use of data acquisition systems is increasing. The tanks and associated equipment are generally earthed, so the pipework to the cables has to include an insulating section.

6.7 Direct current and subsea cable systems

6.7.1 Applications of AC and DC transmission

The choice between AC and DC transmission for a given cable link, as with overhead line, is based on a combination of technical and economic factors.

- A DC circuit can interconnect two AC systems, which do not have to be synchronised, and can even operate at different frequencies.
- A DC interconnection allows direct control of the power flow between two points. Modern DC links use the voltage-source converter (VSC) configuration. On the AC side, it allows control of voltage and reactive power, which

can aid system stability at each end of the DC link regardless of the apparent power flow. On the DC side, it allows complete control of the power flow in either direction while maintaining constant polarity. A DC link may thus have the benefit of improving system stability on both sides, and may even be used as a link of zero length to give these benefits where two power networks adjoin.
- The charging current of an AC cable circuit depends on its capacitance and therefore on its length. For very long routes, the magnitude of the charging current required to energise it may be prohibitive. This particularly applies to subsea routes. Subsea circuits longer than around 100 km are almost always DC.
- The absence of charging current, induced currents, skin effect and much of the dielectric loss in the steady state contributes to lower operating losses for a DC circuit. This can justify the choice of DC even in cases where the initial cost of an AC circuit is lower.
- With only two cores (typically) instead of three, a DC cable may be cheaper per kilometre. Set against this is the increased cost of converter stations at each end. This means that, in terms purely of initial cost, a DC circuit may be chosen for a route above a certain length. The critical length depends on the required rating of the cable and on the fluctuating costs of the competing technologies, but may be around 100 km.

6.7.2 Subsea cable configurations

Subsea cables and DC cables are often considered together, as here, but they are not necessarily the same. DC links are used on land, particularly on long routes, and AC circuits are used subsea, particularly on short crossings. In the following text, subsea cables are thought of as being under the sea, but the same applies to crossings of wide river estuaries or large freshwater lakes.

Typical applications for subsea cable systems include the following:

- long-distance DC interconnections between two networks
- DC or AC connections to offshore islands
- short-distance AC river or lake crossings
- AC (or occasionally DC) export connections from offshore generation such as windfarms or wave energy installations
- AC (or occasionally DC) supplies to offshore petrochemical and similar installations.

Such is the financial value of a long subsea installation, that the designer may well have a free hand to choose the optimum operating voltage for the required duty. For AC systems, it is not necessarily optimal to be constrained to use the system voltage found at either or both ends. It may well be preferable to achieve the ideal trade-off between voltage level and conductor size at a different voltage, even at the expense of non-standard transformers at each end. For instance, the cable link from North West England to the Isle of Man, some 104 km long, operates at the unusual voltage of 90 kV AC.

For short crossings of up to around 3 km, there will be a choice between cable and overhead line. Overhead lines can operate to ultra-high voltages, 1,000 kV or more, while very few cable systems operate above 500 kV AC or DC. On the other hand, single-span line crossings of such gaps may require prohibitively high towers, especially if large ships are to pass underneath the centre of the span where the water is also likely to be deepest. For subsea routes requiring a very high current rating, an alternative to subsea cables is a cable tunnel. This allows two or more circuits of conventional land cables to be installed, safe from the hazards of the seabed. At the end of its life, the cable system can be removed and a newer system of the latest technology reinstalled, all without subsea operations. This particularly applies to shorter routes of up to around 2 km such as river crossings. The rivers Severn, Thames and Medway in Southern England are all crossed by dedicated cable tunnels equipped with various forms of cable cooling.

Requirements for low-carbon generation have led to considerable growth in wind energy. Inland sites with suitable wind resources are in finite supply, and many of those are subject to concerns over visual amenity. This has led much wind generation to be installed offshore, in shallow water at first, but now at increasing depths. Wave and tide generation technologies are very much behind wind in terms of implementation, but many different energy conversion systems are under development at the European Marine Energy Centre off the coast of Orkney (UK), as well as other locations. Distributed offshore generation may use very many individual machines. These are typically cabled together in chains, perhaps at 33 kV for modern large wind turbines, and interconnected at substations mounted on offshore platforms. The combined outputs are transformed to a higher voltage for transmission to shore. Generally there are two or more offshore substations and export cables, so that some or all output can still be maintained even if some equipment fails. In the future, larger windfarms further offshore are likely to require DC export connections.

Offshore platforms that are floating, rather than fixed to the seabed, require special flexible designs of power cable to accommodate movement of the platform.

Subsea cables for AC are often three-core cables. The same factors that govern the choice between a three-core unit and three single cores on land also apply subsea:

- simple installation of a single cable as opposed to three
- three-core cable larger and heavier than single-core
- void spaces between the cores; however, these can be occupied by other services
- reduced thermal rating, as the phases cannot be spaced apart to improve heat flow
- ability to use steel armour on three-core, while single-core cables must use non-magnetic armour such as aluminium.

Single-core cables can be spaced apart to improve natural cooling; on the other hand, the increased spacing will lead to increased currents induced in the armour and thus increased losses. If the spacing is sufficient, there may be some protection against all the cables being damaged at once in a single incident such as a ship anchoring on them. Indeed, a spare single-core cable can be laid alongside, so that the circuit can rapidly be restored to service if any one cable is compromised.

Subsea cables for DC can be two-core but are usually single-core. As with AC circuits, single-core cables can be spaced apart for optimum rating and protection from external damage, and a spare single-core cable can be laid for future use. With DC, induced current and armour losses are not a concern, so wide spacings and steel armour can be used. However, one factor that may limit spacing is the external magnetic field due to widely spaced DC-loaded cables: DC magnetic fields will cause interference to the magnetic compasses of ships navigating overhead.

DC-link topologies are dealt with elsewhere in this volume. However, DC links may operate in two-pole mode, with nominally equal and opposite voltages and currents flowing in two cables. Alternatively they may operate in single-pole mode with just one HV cable, either by design or to allow half-power operation if one cable or converter is out of service. In this case, the current must return by another path to complete the circuit. Ideally a metallic return path should be provided, which may be provided within the HV cable, may more usually be an insulated lower voltage cable running alongside the HV cable or may be the HV core of a spare HV cable. Less desirably, the return current may be allowed to flow between the earth electrode systems of the two converter stations through the general mass of earth and seawater, although such currents are liable to accelerate corrosion in the vicinity.

Older generations of AC/DC converter mainly use line-commutated converters (LCC) based on thyristor switches, which are not nearly as versatile as the VSC. In particular, reversal of the direction of power flow is achieved by reversing the polarity of the voltage on the DC link. This reversal imposes a stress on the DC insulation system, which has to be designed and tested to withstand. The difficulty is that space charge stored in the insulation causes a distortion of the stress, which is benign at constant polarity but causes a severe enhancement of the maximum stress when the polarity of the voltage is reversed. Cables for newer VSC systems do not undergo the stress of this reversal and can be designed accordingly.

6.7.3 Insulation systems

Subsea cable installations are difficult to install, maintain and repair, and also of high value economically, operationally and even politically. For all these reasons, reliability is even more critical than with land installations. Those who select, design and operate subsea cable systems will often take a technologically conservative approach, being cautious with newer technologies that are well established on land. This particularly applies to the insulation system. Lapped insulation systems still have a considerable role, especially at the higher voltages, though the use of extruded systems is increasing.

Fluid-filled cables are applicable both to AC and to DC circuits up to around 500 kV, but face certain challenges in subsea applications. The system of stop joints used to control pressure differences in land circuits is not feasible within the long continuous lengths required for typical subsea crossings. When the cable is immersed at depth, its internal pressure is counteracted by the external pressure of water. However, the dynamic condition where load is changing is especially challenging. In a conventional fluid-filled system, large-diameter fluid ducts may be required to keep the hydraulic impedance sufficiently low that the pressure at the centre of the route does not fall below the required minimum value. The system is usable at lengths up to around 50 km. PPL insulation can be used to improve the dielectric performance, especially at the higher voltages.

Alternatively, the Møllehøj self-compensating cable system can be employed. This is a two-core or three-core flat cable, with its cores side-by-side in a flat formation under a common oval-shaped lead alloy sheath. The two flat sides of the sheath are allowed to bulge under the internal pressure, which is retained with the aid of external flexible reinforcing tapes and conventional steel-wire armour. Expansion and contraction of the fluid with load are accommodated by expansion and contraction of the bulge, without the need for bulk longitudinal movement of fluid. Operating at relatively high internal fluid pressure, it can be used in very long lengths.

Dynamic pressure is also a potential problem when handling long lengths of fluid-filled cable during manufacture, loading on board ship and laying. The pressure-retaining sheath must be designed to handle the transient internal pressures that will occur.

A further consideration is the risk of compromise of the pressure-retaining sheath, whether due to external contact, ground movement or electrical cable failure. If seawater is not to enter the cable and ruin a considerable length, the internal pressure of the fluid at the point of rupture must be kept greater than that of the seawater outside. This means that there will be a continuous flow of fluid from the rupture out into the sea. The fluid reservoir must be dimensioned to provide enough fluid from both ends to last until it can be replenished. Moreover, the ejection of a steady stream of fluid into the sea would constitute an environmental incident.

Fluid-filled cables are used in subsea applications, but for the reasons outlined above, the MIND insulation system has certain advantages. The system is described briefly in section 6.4.2.1. This is used in DC subsea applications up to the highest voltages, typically 500 kV. In AC applications it has been supplanted by extruded systems at the lowest voltages, and is in any case unsuitable for the higher AC voltages due to dielectric loss. As a solid construction, it needs no fluid feed/ expansion or monitoring arrangements, and raises no concern over dynamic pressures. This means that it can be applied up to unlimited lengths. Some other installations such as the Cook Strait crossing in New Zealand have avoided the problems of fluid filling by using gas-filled insulation.

XLPe insulation is making considerable inroads into the subsea market. In AC applications, conventional XLPe insulation is used up to around 400 kV, though usually at lower stress than in land systems for greater reliability. XLPe has the lowest dielectric losses of any applicable technology. In DC applications, special XLPe formulations are used. These are used as a system with modern VSC converters, which maintain constant polarity on each cable, even when power flow is reversed. This is an essential requirement for avoiding space charge problems, as severe distortion of the electric field would occur if polarity were reversed. To date, this system has been used up to ±300 kV DC.

EPR insulation is typically used up to 36 kV AC in subsea applications. At this voltage level, a 'wet' design is employed, without any radial water barrier, so that the EPR-insulated cores are in direct contact with the water. The cores may be individually copper tape screened with overall steel-wire armour.

6.7.4 Manufacture

Manufacture of subsea cables involves making very long continuous lengths. Some processes can be slowed or stopped as required, but this does not apply to the extrusion of insulation systems, which must proceed at constant speed without interruption. To this end, the conductor must be continually available to be fed into the extruder at constant speed. This is often facilitated by passing the conductor through an accumulator, which is an arrangement of pulleys that can pull apart to take up slack or pull together again to release it. If the conductor is made in discrete drum lengths, these can be jointed as they are fed into the accumulator input. While the conductor input is stopped for jointing, the accumulator continues to feed conductor to the extruder, allowing it to run uninterruptedly.

If all the successive manufacturing processes are not to run concurrently, part-manufactured long continuous lengths must be stored between processes. This is generally done on turntables. The entire turntable rotates as the cable is fed onto or off of it in layers, which does not impart a twist to the cable, just as if winding onto or off of a drum. Subsea cable factories will generally have at least two large turntables, allowing long lengths to undergo whatever processes are required and then to be stored ready for the next process. Alternatively, cables can be coiled in layers onto a non-moving floor or in a tank. In this process, the cable undergoes one complete twist per turn around the coil. The cable must be designed and constructed to be able to withstand that degree of twist. A cable capable of being twisted, and thus of being handled by coiling, can generally only be twisted in one direction as determined by the direction of lay of its components, so care must be taken to coil it the right way. Cables are transferred between turntables, coiling-down areas and manufacturing processes using roller runways.

A metallic sheath is required to retain the impregnant where applicable, and to exclude moisture; this may be extruded lead or laminate. In three-core cables, there are often three separate sheaths. In three-core cables, it is very common for optical fibres to be incorporated in the interstices for end-to-end communication or for monitoring the cable itself. Moisture also needs to be excluded from the optical fibres, which will have their own metallic sheath.

Subsea cables are usually heavily armoured, both to protect against impact and to impart sufficient tensile strength to withstand the stress of laying. DC and three-core AC cables will generally have steel armour consisting of one or two layers of wires or tapes, protected by bitumen. As the cable will inevitably be installed under tension, careful mechanical design is required to ensure that it does not twist uncontrollably due to lack of torsional balance between its components. Single-core AC cables will have non-magnetic armour such as copper or aluminium wires. Attention must be paid to corrosion protection, as aluminium or even copper will corrode under anaerobic conditions. In order to avoid damaging voltages arising between different outer layers – such as the sheath and the armour – the outer layers are electrically interconnected throughout the cable. The interconnection may be a continuous semiconducting layer all along the cable, or may be physical connections at regular intervals.

Subsea cables may also need protection against *teredo* attack. The *teredo* or shipworm is a genus of marine mollusc that causes damage by boring tunnels into timber and other underwater structures such as cables. Its attacks can be deterred by a layer of copper or brass tape.

6.7.5 Installation

The completed cable is transferred to a ship for laying. There are various options, depending on the magnitude of the task and the sophistication of the methods required. Where the cable is long or large or has to be handled on a turntable, a dedicated cable-laying ship will be used. Such a ship is also required for laying in deep ocean, or for techniques involving divers or underwater trenching machinery. In less demanding installations, cable may be coiled in the hold of a general-purpose cargo ship temporarily fitted with enough cable-handling equipment to load and unload it.

Cables may be laid direct on the sea bed; however, unless the water is deep enough to be beyond the reach of ships' anchors, it is usually buried. Dependent on the nature of the seabed terrain, trenching methods such as rotary mechanical excavators, ploughs or water jets are used, jets being preferred as they are less liable to damage the cable. In shallow water, extra mechanical protection against movement by wave or tide may be required, and the cable route may be covered in rocks to avoid anchor damage.

Cable ships can only operate where there is sufficient depth of water to float. To get a subsea cable onto shore where required at one or both ends requires further techniques. At the start of laying, the ship will typically pass a length of cable from its tanks or turntable into the water, buoyed by floats to keep it off the sea bed. The floating end of the cable is hauled ashore and fixed in position. The floats are removed and the cable is buried across the foreshore as required. The ship can then proceed to lay the rest of the cable through deep water, until it reaches shallow water at the other end, where the same technique may be employed.

6.7.6 Accessories

The principal accessories required for subsea cable systems are:

- in-line factory joints
- transition joints between sea and land cable
- repair joints.

Despite the efforts to manufacture in continuous lengths, there may be occasions when insulated cores, part-manufactured or fully-completed cables must be jointed. This may be planned due to limitations in manufacturing or handling, or unplanned due to mishaps, and may take place in the factory, at the point of loading or on board ship. Special in-line joints are required: the diameter and mechanical properties of the finished joint must be compatible with the handling and laying methods applied to the cable as a whole.

Subsea cables rarely come to shore immediately adjacent to the associated substation or converter station. More usually, unless there is direct connection to an overhead line, the subsea cable will be jointed to a land cable. Land cables do not require the heavy armour of subsea cables and can be laid in more optimal ways, including special bonding and spaced installation, making it more economic to use a conventional land cable to travel any distance on land. The two cable types are connected at transition joints at a suitable position near the shore.

6.8 Standards

The design and construction of cables and cable system components are largely determined by their performance requirements, specifically by the tests that they are required to withstand at various stages of their life cycle.

The most widely accepted standards are promulgated by the International Electrotechnical Commission (IEC). The central standards for HV cables and systems are IEC Publications 60502-2 [21], 60840 [22] and 62067 [23]. IEC 60502 covers extruded cables and accessories up to 30 kV, 60840 up to 150 kV and 62067 up to 500 kV. All the documents include requirements for routine tests, sample tests, type tests and tests after installation. IEC 62067 also introduces the requirement for a prequalification test for 350 days at 1.7 U_0; this idea is also being extended to lower voltage systems.

Fluid-filled cables are specified in the IEC 60141 series [24]. Other IEC standards include IEC Publication 60229 [25] for the outer insulation of cables.

Generic IEC specifications for testing sometimes have to be amplified or modified when testing cables. IEC Publication 60230 [26] provides additional requirements when performing impulse voltage tests to IEC 60060-1 [27] on cables and accessories. IEC Publication 60270 [28] covers partial discharge testing of lumped-circuit test objects and is applicable to relatively short cable assemblies such as type test assemblies. However, in longer cable test objects such as drum lengths during production testing, travelling-wave effects become significant and are dealt with in IEC Publication 60885-3 [29]. Special requirements for on-site tests, including those for VLF waveforms, are given in Publication 60060-3 [30].

Much work leading up to the development of new or revised IEC standards is done under the auspices of CIGRE, the International Conference on Large High-Voltage Electric Systems. Its Study Committees commission international working groups on topics in their respective subject areas. Study Committee B1 (Insulated cables) covers the entire field of insulated cable systems, underground and submarine, AC and DC. CIGRE issues its own guidance, which is widely used in the absence of any more formal standards in areas such as DC and subsea systems. Significant documents in this area, which effectively have the status of standards, include the following:

Electra No. 171, April 1997, 58–66	Recommendations for mechanical tests on submarine cables [31]
Electra No. 189, April 2000, 39–53 and Addendum, *Electra No. 218*, February 2005, 39–45	Recommendations for tests of DC power transmission cables for a rated voltage up to 800 kV [32]
'Technical Brochure 219', 2003	Recommendations for testing of DC extruded cable systems for power transmission up to 250 kV [33]
Electra No. 189, April 2000	Recommendations for testing of long submarine cables with extruded insulation for voltage from 30 (36) to 150 (170) kV [34]

Work is in progress to extend the range of application of the last two mentioned to higher voltages of 500 kV (DC and AC respectively) or more, to reflect progress in extruded cable technology.

European standards are promulgated by CENELEC. Harmonisation documents are generally implementations of the relevant IEC standards, incorporating a range of national stipulations and extensions. European 'EN' standards are usually adopted by BSI (formerly the British Standards Institute) as British Standards.

UK industry standards are promulgated by the Energy Networks Association. The specifications for fluid-filled cables were inherited from former industry bodies. These include Engineering Recommendation C28/4 for 33 kV to 132 kV systems [35] and Engineering Recommendation C47/1 for 275 and 400 kV systems [36]. XLPe systems are specified in a newer generation of standards, principally Technical Specification 09–16 for 132 kV cables [37].

The National Grid Company now operates the backbone of the UK transmission system. It has its own specifications for the full range of equipment, including technical specifications for equipment and schedules of site commissioning tests for on-site testing, of which the most relevant are as follows:

Technical Specification TS2.5, Issue 2, 2009 [38]	Cable systems
Technical Specification TS3.5.2, Issue 2, 2009 [39]	XLPe cable for rated voltages above 33 kV ($U_m = 36$ kV)
Technical Specification TS3.05.07, Issue 6, 2011 [40]	Installation requirements for power and auxiliary cable
Schedule of Site Commissioning Tests SCT36, Issue 6, 2011 [41]	Cable systems

TS3.5.2 for XLPe cables makes reference throughout to IEC publications 60229 for outer coverings, and 60840 and 62067 for cables in the respective

voltage ranges. It has various clarifications and other stipulations. SCT36 specifies a full range of site commissioning tests, generally aligned with the same IEC publications but with numerous extensions to meet practical requirements.

The distribution network operators in the UK do not necessarily adopt ENA standards, and have a variety of test specifications. Network Rail and other infrastructure companies also have their own standards reflecting their special requirements.

American standards are issued by a number of bodies. There is generally no alignment at all with IEC standards. HV XLPe cables among other equipment are specified by the Association of Edison Illuminating Companies (AEIC). Specification CS8 covers the lower end of the voltage range [42] and CS9 the higher end [43]. These two specifications have recently superseded older specifications CS5, CS6 and CS7. Numerous standards for cables, cable components and test methods are also issued by the Insulated Cable Engineers Association (ICEA). Specifications for cables at different voltage ranges include S-93-639 [44] and S-108-720-2004 [45]. There is some relation between AEIC and ICEA standards. IEEE Standard 48 specifies a full range of tests for cable terminations [46]. The IEEE also issues a range of very useful guidance on various topics: the IEEE Standard 400 series [47–49] covers field testing of power cable systems, including Part 3 on partial discharge (PD) measurements [49]. Some test methods for materials are specified by the ASTM. One that is of relevance is ASTM D5334-08, which covers the needle probe method used for *in situ* measurements of the thermal resistivity of soils and backfills [50]. There is an overall standards body, ANSI, which adopts certain standards as US National Standards.

6.9 Testing

Testing is required at almost all stages of the cable life cycle:

- development
- type testing
- prequalification
- routine production testing
- after-laying tests
- in-service monitoring
- testing for periodic maintenance
- fault location.

6.9.1 Development testing

In development testing, the manufacturers of new systems of cable and accessories will satisfy themselves as to their performance before offering them for sale. Samples of cables may be tested for short-term alternating voltage or impulse voltage breakdown with rapid turnaround by terminating in water terminations. This may be followed by various tests on assemblies with accessories fitted, often culminating in a type test to the desired specification.

6.9.2 Type testing

Cables and systems conforming to particular standards are generally required to undergo type tests. Type tests may be conducted in the laboratories of the manufacturer, with witnessing by a customer or impartial expert, or in an independent test laboratory.

Type tests prescribed by cable standards generally consist of sequences of electrical tests and materials tests. The electrical tests are performed on a short length of the cable, generally up to 20 m, fitted with the accessories. The cable must have completed routine production testing and been subjected to a bending test, whereby it is wound onto and off of a drum of specified diameter, generally three times in each direction. The major elements of a typical electrical type test sequence are generally:

a. the loading cycle test, whereby the cable undergoes a specified number of cycles to take the conductor to a specified temperature above its rated working temperature and return it to ambient temperature, while the test assembly is energised at a specified voltage above its rated working voltage
b. the impulse voltage withstand test, in which the assembly must withstand a specified number of lightning impulses at specified levels, with the conductor held at a specified temperature above rated working temperature.

Alternating-voltage test sources for cable type testing are generally series-resonant systems with output currents between 1 and 5 A. These provide a highly pure sine wave voltage waveform with minimum input power requirement and very low follow-through power in the event of a fault.

Lightning impulse tests to IEC publication 60230 [26] allow the timing requirements of IEC 60060-1 [27] to be somewhat relaxed in view of the capacitance of typical cable test assemblies: the front time T_1 may be as long as 5 μs.

Cable heating is usually provided by inducing current in the conductor, with the two ends joined together to complete a circuit. Further heat may be provided by inducing current in a similar way in the sheath or screen, or by other methods such as immersing in a heated bath. A type test assembly of 132 kV cable and accessories, fitted for heating by induction both in the conductor and in the sheath, is shown in Figure 6.15.

Other tests may also be required as part of the sequence, including partial discharge, capacitance and dissipation factor measurements, an overvoltage withstand test, and for the highest voltages a switching impulse voltage withstand test.

Partial discharge (PD) measurements on high-voltage cable test assemblies present various challenges and opportunities. The cable test assembly can make a very effective antenna for the pickup of radio-frequency interference, especially if its ends are linked for conductor heating. Its substantial capacitance can make it impractical to provide the ideal PD coupling capacitor of much higher capacitance. However, there is a significant possibility when the cable sheath is divided at the mid point, perhaps by the joint under test: the balanced PD detection circuit. The PD detection quadrupole is connected differentially between the two halves of

Figure 6.15 Type test in progress on a system of cable, terminations and joints

the test object, forming a detection circuit that is highly sensitive to internal PD but highly insensitive to external interference.

IEC Publication 60229 has test requirements for the cable outer insulation [25]. These include attacks by specified sharp and abrasive objects, followed by a long-term heat cycling test while immersed in salt water, followed finally by dissection and examination for signs of corrosion.

Type tests on DC cables, including subsea, are covered by CIGRE recommendations [32, 33]. For subsea cables there are requirements for extensive mechanical tests: a tensile bending test, a tensile test and an external water pressure test [31]. The impulse voltage test is required to be conducted with the impulse voltage superimposed on a direct voltage. Latest guidance gives different requirements for such tests, depending whether the cable is for use in LCC or VSC systems, as their electrical characteristics are different [33].

6.9.3 Prequalification testing

As XLPe cable systems have been applied at ever higher operating voltages, users have had to face the prospect of purchasing and installing systems with no operational track record. It is to give increased confidence in this situation that prequalification test programmes have been developed. Prequalification tests are now called for in IEC Publication 62067, which covers the top end of the XLPe voltage range [23]. A prequalification test is performed on an installation of cables and accessories, typically 100 m in length, installed as they would be in service.

The installation is subjected to a specified test regime, typically involving continuous overvoltage combined with loading cycles. The duration of this test is typically one year. On completion, the installation may be subjected to a final test, typically an impulse test. Satisfactory completion of this prequalification may give the purchaser increased confidence that many of the design and installation issues that could cause premature failure have been addressed.

6.9.4 Factory acceptance testing

Factory acceptance tests serve to give confidence in the quality of newly manufactured cables, both to the user and to the manufacturer. Central to the test programme is the high-voltage test. For AC extruded cables, this includes two elements:

- an overvoltage withstand test, typically for 1 h, which is intended to convert latent defects into faults
- a partial discharge measurement, intended to detect those latent defects that do not convert into faults during the test period.

Specifications for cables invariably include detailed requirements for the factory acceptance test. As well as the high-voltage test, they include measurements such as the physical dimensions, conductor resistance, capacitance and dielectric dissipation factor. In addition, there are extensive sample tests focused on the quality of the materials.

Cables for AC systems are tested with alternating voltage. Production lengths of cables have substantial capacitance, which requires considerable reactive power to energise at high voltage. However, as they have very low dielectric losses, they require relatively little apparent power. Thus they are almost pure capacitive loads, ideal for energisation with a resonant test system. In such a system, resonance is achieved between the capacitance of the load and the inductance of the test system. Test systems for production cable test generally use a mechanically tuned reactor to adjust the resonant frequency for a given capacitance to equal the supply frequency, 50 Hz or 60 Hz. Once this condition is achieved, a relatively small injection of power at unity power factor will be multiplied to provide the very large reactive power requirement. Resonant systems for this duty are large; Figure 6.16 shows one rated for an output of 31 A at 650 kV.

Cables for DC systems are conventionally tested with direct voltage. However, a CIGRE recommendation is that consideration should be given to testing *extruded* DC cables at alternating voltage where technically feasible, as the test will be more searching [33]. Testing cables for subsea installation is particularly critical, as there may be limited possibilities for testing after installation.

The PD test becomes problematical for production lengths of cables. IEC Publication 60270 is applicable to lumped-circuit test objects [28], and can be applied without difficulty to typical cable type test assemblies up to around 20 m in length. However, it does not take account of the distributed-parameter nature of test objects such as cables, which causes difficulties at the much longer production lengths. These issues are addressed in the related publication, IEC 60885-3 [29].

Figure 6.16   Resonant test system for 650 kV, 31 A AC

It deals with attenuation of the travelling-wave PD pulses along the cable as they travel from their source to the measuring instrument. It also deals with reflections of the travelling-wave signals from the ends of the cable. Additional equipment and expertise are needed to meet these requirements.

There is also usually a high-voltage test on the oversheath of each manufactured cable drum length. The semiconducting layer on the cable surface is earthed, and a test voltage is applied to the metal sheath. It is very difficult to perform this test meaningfully if there is no semiconducting layer, which could mean that defects in the oversheath are only discovered once the cable is installed.

6.9.5   After-laying tests

Many cable specifications have requirements for tests after installation. The purposes of these tests are various:

- to detect damage to the cable during installation
- to prove the quality of the field-installed accessories
- to demonstrate compliance with the design
- to measure the electrical characteristics of the complete circuit.

The cable itself may suffer damage due to the rigours of installation. This damage can be detected by a voltage test on the oversheath: if the outer layer is intact, it is an indication that all inner layers are also undamaged. The cable installer will often perform a voltage test on the oversheath immediately after each section of cable has been installed and covered, but before jointing the accessories or reinstating the road or ground surface. If oversheath damage is discovered at this stage, it can be located and repaired immediately with minimum cost and delay.

On completion of the entire installation, a final voltage test on the outer insulation system is performed. A typical requirement is to withstand 10 kV direct voltage for 1 min. This tests the entire system, including the oversheaths, the secondary insulation of the joints, the bonding leads and the link boxes. Circuits with longitudinally insulated joints are tested one section at a time, which simultaneously tests the longitudinal insulating barrier of each joint.

The test on the primary insulation system is the high-voltage withstand test, with a PD test where applicable. The focus of this test is generally understood to be on the field-installed accessories, the potential weak points in any installation [51]. The test voltage is intended to stress any latent defects so that they develop rapidly and convert into faults during the test period [52]. The sensitivity of this test is improved by PD measurement, which will detect the degradation process in certain defects before the final fault is developed. The choice of voltage waveform, level and duration is intended to detect those defects that could cause early failure, but without using up too much of the life of the insulation system. Taken to extremes, too much overvoltage will destroy perfectly usable installations that would have given long trouble-free service at their operating voltage, which is to the advantage neither of the manufacturer nor of the user.

The cable sections themselves will have been manufactured under a comprehensive quality system and will finally have undergone rigorous testing before delivery, including a voltage withstand test and PD test. It is not the primary purpose of testing after installation to repeat these tests. However, the effectiveness of the factory tests is reliant on the competence of the manufacturer, both in their quality systems and in their high-voltage and PD testing expertise. It is not unknown for defects in newly manufactured cable sections to be detected by tests after installation.

PD measurement on installed long cable routes is particularly challenging. The focus of PD measurement is generally taken to be the field-installed accessories [51]. PD signals travelling along power cables are attenuated progressively with distance, due mainly to losses in the semiconducting screens. In circuits with many joints, attenuation makes it necessary to measure PD at many or all of the joints, not just at the terminations. Unless there are internal PD sensors in the accessories, PD signals are usually detected using radio-frequency current transformers attached to the bonding leads at accessories. The PD signals are typically acquired by digital acquisition units, which transmit their data over fibre optic links to a central monitor. There is a spectrum of possible approaches to measurement and interpretation [53]:

1. measurement on very short circuits entirely according to IEC 60270
2. measurement on longer circuits by techniques based on IEC 60270 as modified by IEC 60885-3
3. measurements based on IEC 60270 but optimised by relaxing certain requirements such as frequency limits

4. measurements without the pseudo-integration that is the essence of IEC 60270, but still using calibration techniques that lead to a form of apparent charge measurement
5. qualitative measurements not attempting to indicate the apparent charge, but supported by a sensitivity assessment.

There are a number of possible test voltage waveforms. Most of these are discussed at greater length in IEEE Standard 400-2001 [47].

- For a long time the only choice was direct voltage. Test durations are frequently 15 min at a voltage substantially in excess of the peak value of the maximum alternating voltage in operation. Direct voltage has the advantage for highly capacitive test objects of requiring relatively low input power and thus lightweight equipment. This waveform was used with success on all generations of lapped cable systems. However, when it came to test XLPe systems, early experience raised concern. It is generally accepted that, at high enough stress, charge is injected into the dielectric where it becomes trapped. The trapped charge distorts the static voltage distribution within the dielectric, which can cause rapid failure when the system is then energised at alternating voltage. This has prompted a move to other test voltage waveforms for XLPe systems. Direct-voltage tests are applied to low-stress systems up to around 33 kV without causing problems, but alternative waveforms are available even in this voltage range.
- An alternating-voltage test on anything as capacitive as a cable installation requires considerable reactive power. The most straightforward way to provide this is by an online test, also called a soak test: simply connect the circuit under test to the live system. Some cable specifications allow for a 24 hour online test as an alternative to an off-line test. An advantages of this method is that it is quick and cheap, especially at short notice or where access for large test equipment is difficult. If combined with effective online PD measurement, it becomes a more sensitive test. Disadvantages of online testing include lack of control of voltage, difficulties discriminating PD from noise and safety, especially in the event of a failure on test.
- If the problems of online testing are to be avoided, an off-line test must employ a test voltage source. Off-line testing facilitates PD measurement, avoiding many of the difficulties and dangers of online testing. The brute-force approach would be to use a test transformer able to supply the required charging current, but a transformer with useful rating would be so large and heavy as to be practically immobile. Alternating voltages for cable testing at 66 kV and above are almost always provided by series-resonant systems. Typical factory test systems use mechanically tuned inductors, which are too large, heavy, delicate and inefficient for mobile use. The dominant technology for field cable testing is now the variable-frequency series-resonant system. A reactor of fixed inductance, typically 16 H, is mounted in a transportable steel tank with an output bushing. When connected in series with the capacitance of a typical cable installation, the *LC* circuit has a resonant frequency usually in the range

from 20 to 300 Hz. An electronic power converter finds this resonant frequency and provides input power at a relatively low voltage. This input voltage is multiplied by the *LC* resonant circuit to provide the required test voltage. The magnetic and mechanical losses of the fixed-inductance reactor are so low that the multiplication factor is very high, typically 100–200. A system with the typical 16 H inductor has a rated voltage of 260 kV and a maximum output current of 83 A. Tests are generally for 1 h, at an overvoltage ratio dependent on the system voltage. For long circuits of high capacitance, the frequency of the test waveform may be substantially less than 50 Hz. In this case, there are concerns that any defects will see substantially fewer than the expected number of voltage cycles, and therefore experience correspondingly less ageing during the test period, with the consequent risk of a severe defect NOT leading to failure on test [52]. With this in mind, there are moves to increase the required duration of tests below 50 Hz in inverse proportion to the frequency, so that the number of cycles does not fall below a minimum.

- Before variable-frequency resonant systems became available, concerns over space charge at the lower end of the voltage range led to the adoption of very low frequency (VLF) test voltage sources. Such sets use a controlled direct-voltage source to charge and discharge the capacitance under test cyclically, so providing an alternating voltage at very low frequency. The capacitive charging current is reduced in proportion to the frequency, so the input power requirement is correspondingly low. Operating frequency is typically 0.1 Hz, but can be even lower if required when testing longer lengths and higher capacitances. The waveform may be an approximation to a sine wave, or may be some form of rectangular wave. PD and the dissipation factor tan δ can be measured for diagnostics. A detailed discussion is given in IEEE Standard 400.2-2004 [48]. The system is widely used up to around 33 kV, and sometimes higher, especially in North America where it enjoys considerable success. However, its use is not universal, with direct voltage also being widely employed up to 33 kV.

- There are concerns whether PD measurements at VLF are comparable with those made at power frequency. It was largely to address this that damped-wave test sources were introduced. Like a variable-frequency resonant system, it employs a fixed reactor that resonates with the capacitance under test. Unlike a variable-frequency resonant system, it does not provide continuous output, so does not need to be highly thermally rated. This makes the reactor much more compact and lightweight. The oscillation is excited by charging the cable with DC to the required peak voltage, and then discharging it through the reactor. An oscillation is set up at the resonant frequency, decaying at a rate depending on the losses. During the oscillation, PD may be measured. It is generally agreed that this technique provides a test voltage source for PD measurements, but not an effective withstand test.

- It is possible to perform PD measurements under impulse voltage. The impulse is generated by charging one core with direct voltage, and then discharging it into the core under test. However, the peak value of the impulse voltage decays away as it travels along the cable, so faraway parts of the circuit may be under-tested.

The DC resistance of the conductors may be measured from end to end. An instrument providing a test current of 10 A is usually used. The AC impedance of the circuit may also be measured by injecting AC, at typically 100 A, and measuring the resulting voltages.

For subsea circuits, a time-domain reflectometer trace of each core is taken and recorded, to facilitate subsequent fault-locating.

The thermal conductivity of backfill may be tested, either by direct measurement or by measuring the density of compaction where the characteristics of the material are already known.

There will be final visual and electrical checks that the bonding system is correctly connected, and measurement of the contact resistances of all bolted-up connections.

6.9.6 In-service monitoring

The temperature of a power cable under load can be monitored, most conveniently by a distributed temperature-sensing system. In these systems, the temperature-sensing element is an optical fibre, which runs along the cable route. It would be desirable to monitor the temperature where it is highest, at the conductor, but this presents practical difficulties. Instead, the sensing fibre is generally incorporated in the outer layers of the cable during manufacture, typically in place of one of the screen wires, or attached to the outside of the cable during or after installation. The fibre is connected to a laser-based measuring system, which can use the characteristics of the back-scattered light to infer the temperature at intervals along the fibre, often with a spatial resolution down to 1 m. The use of these systems to manage the current rating of the cable system in real time is discussed under 'dynamic rating', above.

PD measurements can be made off-line using a test voltage source, or online with the circuit energised from the power system [51]. Online PD monitoring gives the opportunity to measure PD under the actual operating conditions of the cable system, including voltage variations and temperature cycles. However, there are a number of challenges to be overcome, including:

- coupling, since conventional laboratory methods of coupling are not applicable
- hazardous voltages on the cable sheath system
- attenuation of PD signals travelling along the cable, which often necessitates coupling PD signals at each and every accessory to achieve good sensitivity
- data communication from many remote positions
- interpretation of PD signals, automatically or by human expert
- long-term power supply to electronics
- induced voltages in any circuits near to the power cable circuit, especially under fault conditions.

There are a number of approaches to PD monitoring on cable systems, not all of which attempt to comply with IEC 60270. The focus of PD monitoring is usually the accessories; in this case it is possible to use very high measuring frequencies so that PD signals are severely attenuated along the HV cable, improving noise rejection. Sometimes highly sensitive PD sensors may be incorporated within accessories themselves. More commonly, PD signals are coupled by attaching radio-frequency current transformers to the bonding leads at accessories. PD signals are typically acquired by distributed digital units, which transmit their data over fibre optic links, just as for commissioning testing. Providing permanent power to these units can be problematical, but there are solutions that tap a small amount of power from the HV circuit using a ring current transformer on the HV cable.

Pressurised cable systems include low-pressure alarms to detect the pressure loss on leakage of the fluid or gas. Leakage may be due to perforation of the metal sheath through corrosion or external contact.

Extruded systems do not rely on internal pressure, so perforation of the metal sheath may go undetected. Water may enter the cable and begin to degrade the insulation. It is possible to include a water-sensing cable under the metallic sheath; on contact with water its resistance changes, which can be detected remotely.

SVLs are subject to deterioration in service, particularly due to repeated or severe overvoltages, which can cause them to fail catastrophically. One simple way of monitoring these in service is to use an optical fibre running along the cable route and wrapped around each SVL in turn. If any SVL ruptures, it will break the fibre; the loss of signal along the fibre will trigger an alarm, and the break can be located by optical time-domain reflectometry.

6.9.7 Testing for periodic maintenance

High-voltage cable systems generally have insulated sheaths, the insulation being provided by the secondary or external insulation system including the cable oversheath. If the performance and reliability of the cable system are to be maintained, this secondary insulation system must be kept in serviceable condition. The insulation may be damaged by external contact or degraded by cyclic thermal movement. Older circuits with PVC oversheaths are more vulnerable than more modern circuits with harder, tougher compounds such as polyethylene. In order to test for, locate and repair any faults in this insulation, and the circuit must be taken out of service periodically. The interval between tests will depend on the rate at which faults appear and on the importance of the circuit, but may be once every few years at first.

The basic test for secondary faults is to repeat the sheath voltage test. It is not generally appropriate to re-apply the full test voltage to a circuit that has been in service; a lower voltage such as 4 kV for 1 min is often used. If the test voltage is withstood by every section of circuit, the circuit can be returned to service. Otherwise, the faults must be located and if possible repaired. The main method of fault location is the step-potential method. First, it is necessary to inject pulsed DC from a suitable test set into the faulty section of sheath, so that it flows out through the fault to earth. As it flows through the surrounding soil, it creates a voltage gradient or electric field. Then, at the ground surface, the pulsed radial electric field can be detected using a sensitive DC voltmeter and a pair of probes. Once the electric-field pattern has been detected, its centre can quickly be found with great precision.

There are supplementary techniques, which can guide the test engineer more quickly to the correct search area: a DC Wheatstone bridge circuit may be used to find the distance to the fault, or the magnetic field due to the fact that pulsating test current may be traced. Where cables are installed in insulating plastic ducts, this insulation will disrupt the current flows and make the step-potential method difficult or impossible; in this case these supplementary techniques may be the only ones available.

When a cable is taken out of service for serving tests, it is also opportune to test the SVLs for deterioration in their DC characteristics.

In fluid-filled systems, samples of the fluid can be drawn from time to time from sampling ports provided at the accessories. These samples can then be subjected to dissolved gas analysis. The gases dissolved in the fluid give indications of the deterioration process occurring in the insulation, just as with other oil-filled paper-insulated plant such as transformers.

6.9.8 Fault location

Primary faults are most often shunt faults, which compromise the high-voltage insulation between one or more phase conductors and the metal sheath or screen. Dependent on the nature of the defect giving rise to the fault, it may develop undetected for a considerable time before it breaks down, at which time it passes sufficient current to be detected by the circuit protection. It may be possible to re-energise the circuit, in which case the breakdown is likely to recur, or it may be impossible if a permanent fault has developed. In either case, it is necessary to find the location of the fault so that a repair can be effected.

Another possible fault configuration is the series fault, in which one or more conductors become open circuit or high resistance. This may occur in conjunction with a shunt fault, where part of the conductor has been vapourised due to the fault current, or where the cable has been completely severed. Alternatively, it may occur on its own, where ground movement causes a joint to be pulled apart.

A wide range of techniques are applicable to locate primary faults [54], including the following:

- Time-domain reflectometry (TDR) may show certain low-resistance shunt faults or high-resistance series faults, but is difficult in circuits with joints.
- Measurement of capacitance, comparing the faulty phase with a good phase, may indicate the distance to an open-circuit series fault as a percentage of the circuit length.
- Ratiometric techniques include a whole range of low-voltage and high-voltage bridge circuits, in which the two lengths of cable, either side of the fault, become two 'unknown' arms of a Wheatstone bridge.
- Route-tracing techniques using audio- or radio-frequency current injected into the circuit may be adapted so that the injected current flows through the fault;

then when the receiver is passed along the cable route, the fault may be detected as a loss of signal or as a disturbance in the signal.
- The chemical signature of arc-damaged cable insulation may be detected above ground by scanning with a suitable detector.

The most widely applicable techniques rely on a high-voltage surge generator. This instrument contains a bank of capacitors that are charged at high voltage and then switched into the circuit so that the stored energy discharges into the fault. Techniques based on this equipment include the following:

- The impulse current technique relies on capturing the transient current generated when the fault breaks down, including multiple reflections as the transient propagates back and forth along the cable. The time between reflections gives the distance to the fault. Many variations of this technique are possible. It is very widely applicable to almost all faults.
- The technique variously called the arc reflection or secondary impulse method relies on the surge generator to create a momentary arc at the fault. While the arc is running, a time-domain reflectometer captures one or more traces, which show the arc as a temporary low-resistance shunt fault. This technique has the advantage that anyone familiar with TDR can interpret the traces.
- The magnetic pulse due to the breakdown current may be traced above ground with a suitable instrument, which may show a loss or disturbance of the signal near the fault position.
- In most cases, the final pinpointing is done using the discharge noise. Sometimes this can be detected by the unaided ear, but for buried circuits a search instrument equipped with a ground microphone is usually needed. The process is greatly improved if the instrument is also equipped with a search coil for the magnetic pulse, especially if a digital timer is used to show the time delay between the magnetic pulse and the acoustic pulse, so indicating the distance from the microphone to the fault.

Fluid and gas leaks from pressurised systems require a specialised form of fault location. With fluid-filled systems, pre-location is possible by measuring flow rates and pressure drops, analogous to DC electrical bridge techniques. It is also possible to sectionalise a fluid-filled system by applying a freezing collar, filled with liquid nitrogen, to the cable. Final pinpointing may be done by injecting a chemical tag into the cable fluid and detecting its minute traces above ground at the leak site [55]. Similarly, nitrogen leaks from gas-pressurised systems can be pinpointed by injecting a small concentration of tracer gas into the system and scanning above ground with a suitable detector.

6.10 Future directions

If the history of extruded cable insulation is a guide to the future, it will be characterised by increasing reliability at increasing operating stress. Incremental approaches to achieving this involve the development of materials and their processing to improve purity and smoothness. But this can only be taken so far, and will not lead to dramatic increases in operating temperature. Major improvements in the performance of cables are likely to come from novel materials. Intrinsically conducting polymers may be blended into the dielectric to achieve precise control of its conductivity, particularly for direct-voltage application. Certain nanomaterials blended into the dielectric appear to offer vast improvements in voltage endurance (Stevens G.C., private communication, 2011).

Most of the losses in power cables are due to the resistance of the conductor. The phenomenon of superconductivity in metals, whereby the resistivity falls to zero at low enough temperatures, was discovered in 1911. Metallic superconductors have to be cooled to liquid-helium temperatures (-269 °C) to superconduct. There have been many applications of superconductors, but it was not until the 1970s that there were full-scale trials of superconducting high-voltage power cable systems. They were technically successful but never found widespread use. Then came the discovery of what is now called 'high-temperature superconductivity' [56]. In certain ceramic compounds such as yttrium barium copper oxide (YBCO), superconductivity is maintained at liquid-nitrogen temperatures (-196 °C) even in very large magnetic fields. YBCO can be made into a thin film, laminated onto a metal substrate to make a flexible tape [57]. Bundles of the tape can then be used to make equipment such as power cable conductors. These offer the prospect of complete freedom from conductor losses in DC systems and considerable reduction in conductor losses in AC systems [58]. Set against this is the cost of operating the required cryogenic refrigeration plant. As very high conductor currents can be passed, useful power can be transmitted at relatively low voltages up to around 50 kV AC or DC. Practical superconducting power cables and systems are now available, though the applications to justify their use are rare so far. Some standardisation work has been carried out by CIGRE [59].

Another route to very high ratings is the gas-insulated line (GIL). This is similar to a section of gas-insulated substation: a tubular aluminium conductor supported on insulators in the centre of a larger pressure-retaining aluminium tube. The insulating gas is mainly nitrogen with a small proportion of SF_6. The system has a number of advantages over conventional cables:

- direct connection to gas-insulated switchgear
- very high current rating
- absence of flammable materials
- partly self-restoring insulation
- low external magnetic field.

These advantages are at the expense of considerably greater overall diameter. The system has found application mainly for short but highly rated connections within power plants or substations. To date, GIL is mainly installed by field assembly of prefabricated rigid sections. The alternative approach, a flexible system in continuous lengths that can be coiled like a cable, is problematical. Even with field-welded joints in the outer tube, it is claimed that modern systems are practically leak-free.

An alternative approach to improving ratings is to increase the maximum operating temperature. This is at present limited to around 90 °C in XLPe. Polyethylene has to be cross-linked to achieve even this temperature, which brings with it the problems of the volatile crosslinking by-products. There are thermoplastic polymer blends, which can operate at much higher temperatures without melting, and thus without the need for crosslinking [56]. These offer the prospect of highly reliable high-temperature insulation systems for a future generation of power cables.

References

1. International Commission on Non-Ionizing Radiation Protection. 'ICNIRP guidelines for limiting exposure to time-varying electric, magnetic and electromagnetic fields (up to 300 GHZ)'. *Health Physics*. 1998;**74**(4): 494–522
2. Moore G.F. (ed.). 'Basic electrical theory applicable to cable design', in *Electric cables handbook*. 3rd edn. Oxford: Blackwell Science; 1997. pp. 9–32
3. Black R.M. *The history of electric wires and cables*. London: Peter Peregrinus Ltd; 1983
4. Arhardt R.J. 'The chemistry of ethylene propylene insulation, Part I'. *IEEE Electrical Insulation Magazine*. 1993;**9**(5):31–34; 'The chemistry of ethylene propylene insulation – Part II'. *IEEE Electrical Insulation Magazine*. 1993; **9**(6):11–14
5. Endersby T.M., Gregory B., Swingler S.G. 'Polypropylene paper laminate oil filled cable and accessories for EHV application', *3rd IEE international conference on power cables and accessories 10 kV–500 kV*. IEE Conference Publication 382; 1993
6. IEC 60228 ed. 3.0 (2004-11), 'Conductors of insulated cables'
7. Moore G.F. (ed.). 1997, 'Conductors', in *Electric cables handbook*. 3rd edn. Oxford: Blackwell Science; 1997. pp. 69–80
8. IEC 60287-1-1 ed. 2.0 (2006-12), 'Electric cables – calculation of the current rating – Part 1-1: current rating equations (100% load factor) and calculation of losses – General'
9. Moore G.F. (ed.). 'Self-contained fluid-filled cables', in *Electric cables handbook*. 3rd edn. Oxford: Blackwell Science; 1997. pp. 440–455
10. Hampton N., Hartlein R., Lennartson H., Orton H., Ramachandran R. 'Long-life XLPE insulated power cable', Jicable 2007-C5-1-5. 2007
11. Argaut P., Bjorlow-Larsen K., Zaccone E., Gustafsson A., Schell F., Waschk V. 'Large projects of EHV underground cable systems', Jicable 2007-A2-1. 2007
12. Moore G.F. (ed.). 'Armour and protective finishes', in *Electric cables handbook*. 3rd edn. Oxford: Blackwell Science; 1997. pp. 81–91
13. Moore G.F. (ed.). 'Installation of transmission cables', in *Electric cables handbook*. 3rd edn. Oxford: Blackwell Science; 1997. pp. 549–561

14. Boone W., de Wild F. 'HV power cables installed in multi purpose tunnels, a challengeable option!' JICABLE'07 – 7th International Conference on Power Insulated Cables, paper 2007-A2-6. 2007
15. IEC 60287-2-1 ed. 1.2 (2006-05). 'Electric cables – calculation of the current rating – Part 2-1: thermal resistance – calculation of thermal resistance'. IEC publications
16. IEC 60853-1 ed. 1.0 (2008-10). 'Calculation of the cyclic and emergency current rating of cables – Part 1: cyclic rating factor for cables up to and including 18/30 (36) kV'. IEC publications
17. IEC 60853-2 ed. 1.0 (2008-10). 'Calculation of the cyclic and emergency current rating of cables – Part 2: cyclic rating of cables greater than 18/30 (36) kV and emergency ratings for cables of all voltages'. IEC publications
18. IEC 60853-3 ed. 1.0 (2002-02). 'Calculation of the cyclic and emergency current rating of cables – Part 3: cyclic rating factor for cables of all voltages, with partial drying of the soil'. IEC publications
19. Kent H., Bucea G. *Inquiry into the Auckland Power Failure – Technical Report on Cable Failures*. Report to Ministerial Enquiry, New Zealand: New Zealand government publications; 1998
20. IEC 60287-3-3 ed. 1.0 (2007-05). 'Electric cables – calculation of the current rating – Part 3-3: sections on operating conditions – cables crossing external heat sources'. IEC publications
21. IEC 60502-2 Second edition 2005-03. 'Power cables with extruded insulation and their accessories for rated voltages from 1 kV ($U_m = 1,2$ kV) up to 30 kV ($U_m = 36$ kV) – Part 2: cables for rated voltages from 6 kV ($U_m = 7,2$ kV) up to 30 kV ($U_m = 36$ kV)'. IEC publications
22. IEC 60840 Third edition 2004-04. 'Power cables with extruded insulation and their accessories for rated voltages above 30 kV ($U_m = 36$ kV) up to 150 kV ($U_m = 170$ kV) – test methods and requirements'. IEC publications
23. IEC 62067 ed. 1.1 (2006). 'Power cables with extruded insulation and their accessories for rated voltages above 150 kV ($U_m = 170$ kV) up to 500 kV ($U_m = 550$ kV) – test methods and requirements'. IEC publications
24. IEC 60141-1 (1993-08). 'Tests on oil-filled and gas-pressure cables and their accessories Part 1: oil-filled, paper-insulated, metal-sheathed cables and accessories for alternating voltages up to and including 400 kV'. IEC publications
25. IEC 60229 ed. 3.0 (2007-10). 'Electric cables – tests on extruded oversheaths with a special protective function'. IEC publications
26. IEC 60230 ed. 1.0 (1966-01). 'Impulse tests on cables and their accessories'. IEC publications
27. IEC Publication 60060-1, 1990. 'Guide on high-voltage test techniques – Part 1. General'. IEC publications
28. IEC 60270 ed. 3.0 (2000-12). 'High-voltage test techniques – partial discharge measurements'. IEC publications
29. IEC 60885-3 ed. 1.0 (1988-07). 'Electrical test methods for electric cables. Part 3: test methods for partial discharge measurements on lengths of extruded power cables'. IEC publications

30. IEC Publication 60060-3, 2006. 'High-voltage test techniques – Part 3: definitions and requirements for on-site testing'. IEC publications
31. CIGRE. 'Recommendations for mechanical tests on sub-marine cables', *Electra* No. 171. 1997(April): 58–66
32. CIGRE. 'Recommendations for tests of DC power transmission cables for a rated voltage up to 800 kV', *Electra No.* 189. 2005(April): 39–53; 'Addendum', *Electra* No. 218. 2005(February): 39–45
33. CIGRE. 'Recommendations for testing of DC extruded cable systems for power transmission up to 250 kV'. *Technical Brochure 219.* 2003
34. CIGRE. 'Recommendations for testing of long submarine cables with extruded insulation for voltage from 30 (36) to 150 (170) kV', *Electra* No. 189. 2000(April)
35. EAER C28/4, 1975. Type approval tests for impregnated paper insulated gas pressure and oil filled power cable systems from 33 kV to 132 kV inclusive. Electricity Association publications
36. EAER C47/1, 1975. 'Type approval tests for single core impregnated paper insulated gas pressure and oil filled power cable systems for 275 kV and 400 kV', Engineering Recommendation C47/1. Electricity Association publications
37. EATS 09-16 (1983) Issue 1, 'Testing specification for metallic sheathed power cables with extruded cross-linked Polyethylene insulation and accessories for system voltages of 66 kV and 132 kV'. Electricity Association publications
38. National Grid, Technical Specification TS2.5, 'Cable Systems', National Grid Company publications. Issue 2, 2009
39. National Grid, Technical Specification TS3.05.02, 'XLPE cable for rated voltages above 33 kV ($U_m = 36$ kV)', National Grid Company publications. Issue 2, 2009
40. National Grid, Technical Specification TS3.05.07, 'Installation requirements for power and auxiliary cable', National Grid Company Publications. Issue 6, 2011
41. National Grid, Schedule of Site Commissioning Tests SCT36, 'Cable Systems', National Grid Company publications. Issue 6, 2011
42. AEIC CS8-07 (3rd edn) Specification for extruded dielectric shielded power cables rated 5 through 46 kV. The Association of Edison Illuminating Companies Publications
43. AEIC CS9-06 (1st edn) Specification for extruded insulation power cables and their accessories rated above 46 kV through 345 KV ac. The Association of Edison Illuminating Companies Publications
44. Insulated Cable Engineers Association, 2006, Publication No. ICEA S-93-639 (NEMA Standards Publication No. WC 74-2006), '5-46 kV shielded power cable for use in transmission and distribution of electric energy'. The Insulated Cable Engineers Association Publications
45. ICEA S-108-720-2004, 'Standard for extruded insulation power cables rated above 46 through 345 kV'. The Insulated Cable Engineers Association Publications

274 *High-voltage engineering and testing*

46. IEEE Std 48-2009, 'IEEE standard for test procedures and requirements for alternating-current cable terminations used on shielded cables having laminated insulation rated 2.5 kV through 765 kV or extruded insulation rated 2.5 kV through 500 kV'. The Institute of Electrical and Electronics Engineers publications
47. IEEE Std 400-2001, 'IEEE guide for field testing and evaluation of the insulation of shielded power cable systems'. The Institute of Electrical and Electronics Engineers publications
48. IEEE Std 400.2-2004, 'IEEE guide for field testing of shielded power cable systems using very low frequency (VLF)'. The Institute of Electrical and Electronics Engineers publications
49. IEEE Std 400.3-2006, 'IEEE guide for partial discharge testing of shielded power cable systems in a field environment'. The Institute of Electrical and Electronics Engineers publications
50. ASTM D5334-08, 'Standard test method for determination of thermal conductivity of soil and soft rock by thermal needle probe procedure'. ASTM International publication
51. CIGRE Working Group 21-16. 'Partial discharge detection in installed HV extruded cable systems', *CIGRE Technical Brochure 182*; 2001
52. Fenger M. 'Experiences with commissioning testing of HV & EHV cable systems: the influence of voltage level and duration for identifying life limiting defects', *2012 Proceedings of the IEEE DEIS international symposium on electrical insulation*. Conference Record of the 2012 IEEE International Symposium on Electrical Insulation (ISEI); 2012
53. CIGRE Working Group B1-28. 'On-site partial discharge assessment of HV and EHV cable systems', *CIGRE Technical Brochure*; 2013
54. Clegg B. *Underground cable fault location*. 3rd edn. Ware: CFL division of BCC Electrical Engineering and Training Consultancy
55. Fairhurst M., Keelan P. 'Oil filled cable leak location using PFT (Per-fluorocarbon tracing) – non-intrusive oil filled cable leak location', JICABLE'07 – 7th International Conference on Power Insulated Cables, paper 2007-A1.3. Jicable A.1.3, 2011
56. Grant P.M. 'Down the path of least resistance', *Physics World*. 2011(April): 18–25
57. Moore G.F. (ed.). 'High temperature superconductors', in *Electric cables handbook*. 3rd edn. Oxford: Blackwell Science; 1997. pp. 666–675
58. Moore G.F. (ed.). 'High temperature superconducting cables', in *Electric Cables Handbook*. 3rd edn. Oxford: Blackwell Science; 1997. pp. 676–682
59. CIGRE Working Group 21-20. 'High temperature superconducting (HTS) cable systems', *Electra* No. 208. 2003(June): 51–57

Chapter 7
Gas-filled interrupters – fundamentals
G.R. Jones, M. Seeger and J.W. Spencer

7.1 Introduction

The interruption of current in an electric power network needs to be considered with respect to the system operating voltage and the type of system components and structure. The operating voltage affects the type of interrupter chosen for duty, while the system components and structure (e.g. the extent to which the network is inductive) will influence the detailed design of the interrupter unit because of the voltage transients produced during the current interruption process.

The current interruption process is considered mainly with respect to SF_6 interrupters since these interrupters remain the main candidates for interrupting currents on extra high-voltage (EHV) networks while also having been used at medium and high voltages. Other types of interrupters – air, oil and vacuum – have a lower voltage-withstand capability as shown in Figure 7.1. It is noteworthy that both air and SF_6 at elevated pressures have more than twice the breakdown voltage of vacuum and that with SF_6 only about a quarter of the gas pressure compared with air is needed. However, other considerations than breakdown voltage are important for the current interruption process so that at lower voltages there are ranges in which vacuum, in particular, can offer advantages.

This chapter first considers the basic principles of current interruption and the influence of the type of fault produced by the connected network. The electrical characteristics of circuit-breakers are described in References 1 and 2. The formation of an electric arc [3] during the current interruption process is indicated and the significance of the manner in which the arc is controlled and then extinguished is explained. Various means for achieving such control and with different arcing media are considered. Some basic performance limiting factors are discussed.

The chapter concludes by discussing some evolving trends in response to various modern requirements such as environmental issues.

7.2 Principles of current interruption in HV systems

Methods for interrupting current in high-voltage systems rely upon introducing a non-conducting gap into a metallic conductor. To date, this has been achieved by

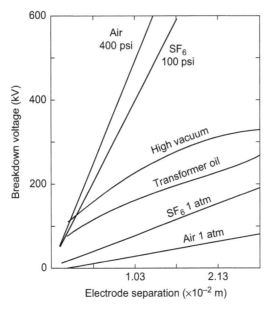

Figure 7.1 Breakdown voltages of various interrupter media

mechanically separating two metallic contacts so that the gap so formed is automatically filled by a liquid, a gas or even vacuum. In practice, such inherently insulating media may sustain a variety of different electrical discharges which then prevent electrical isolation of the two separated conductors being achieved.

There are three major phases to the current interruption process. First, as the contacts separate, an arc discharge is inevitably formed across the contact gap [1–3]. The problem of current interruption is then one of controlling the discharge during the high current phase and then quenching it in the presence of the high voltage provided by the power network which can sustain a current flow through the discharge. Since this physical situation is governed by a competition between the electric power input due to the high-voltage current and the thermal plus radiation losses from the electric arc, this phase of the current interruption process is known as the 'thermal recovery phase' and is typically of a few microseconds duration.

The final phase of current interruption relates to the removal of the effects of arcing which may remain for many milliseconds after the current ceases to flow, even under the most favourable conditions. The problem then is one of ensuring that the contact geometry and materials are capable of withstanding the highest voltage which can be generated by the system without electrical breakdown occurring in the interrupter.

An intermediate phase of the current interruption process bridges the gap between the thermal recovery phase and the breakdown withstand phase. During this phase the remnant effects of the arcing have cleared sufficiently to ensure thermal recovery but insufficiently to avoid a reduction in dielectric strength. This is known as the 'dielectric recovery phase'.

Based upon this understanding, current interruption technology is concerned on the one hand with the control and extinction of the various discharges which may occur, while on the other it relates to the connected system and the manner in which it produces post-current interruption voltage waveforms and magnitudes.

7.2.1 System-based effects

It can be advantageous for interrupting a current to do so when the current naturally passes through zero in a controlled manner. With an alternating current of 50/60 Hz frequency, the current reduces naturally to zero once every half-cycle (Figure 7.2), at which point current interruption can be sought. This represents the minimum naturally occurring rate of current decay (di/dt) for the current wave, so that for conventional power systems, which are inherently inductive, the induced voltage following current interruption is minimised. Consequently, the contact gap is electrically less severely stressed transiently during both the thermal and dielectric recovery phases.

The voltage transients of most interest in interrupting current in high-voltage transmission systems are those produced by short-circuit faults and short line faults [1]. Short-circuit faults occur close to the circuit-breaker (Figure 7.3). These produce the most onerous fault currents. In this case, the restrike voltage consists of a high-frequency (ωn) oscillation governed by the system inductance (L) and capacitance (C) superimposed on an exponentially decaying component governed by the system resistance (R):

$$V_t = V_0 \left[1 - e^{(-R/Lt)\cos(\omega nt)}\right] \tag{7.1}$$

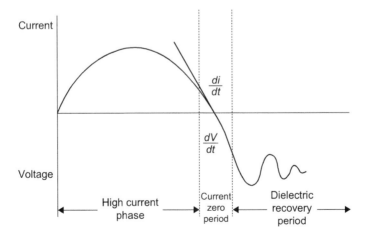

Figure 7.2 Voltage and current waveforms during the current interruption process

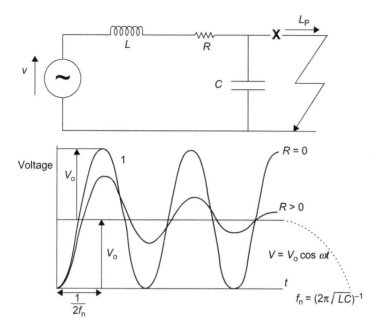

Figure 7.3   Short-circuit fault – onerous fault current

The maximum fault current is [1]

$$I_F = \left(\frac{V_0}{\omega L}\right) \sin(\omega t) \tag{7.2}$$

and the maximum possible voltage across the circuit-breaker is twice the supply voltage. The system resistance R reduces this maximum voltage to lower values.

Short line faults occur on transmission lines a few kilometres from the circuit-breaker (Figure 7.4) and constitute the most onerous transient recovery voltages. The voltage across the circuit-breaker is the sum of the line-side (V_L) voltage and the source side (V_s) voltage which occur at two different frequencies f_L and f_s respectively (Figure 7.4). At current zero dV/dt is typically about 10–20 kV/μs.

The situation is made more complicated in three-phase systems because current zero occurs at different times in each phase, implying that the fault is interrupted at different times leading to different voltage stresses across the interrupter units in each phase.

Apart from the symmetrical sinusoidal current waveform with its natural current zero, other current interruption situations exist (Figure 7.5). For instance, the sinusoidal waveform may be superimposed on a steady current to form an asymmetric wave with major and minor loops which cause different circuit-breaker stresses. A related condition which occurs with generator faults corresponds to

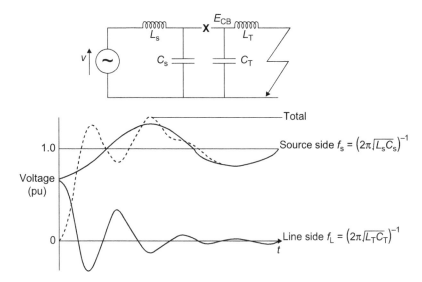

Figure 7.4 Short line fault – onerous transient recovery voltages

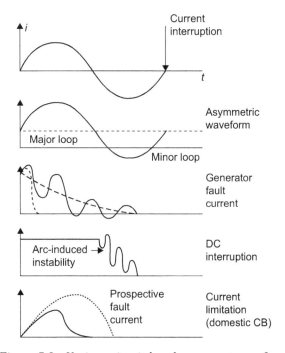

Figure 7.5 Various circuit-breaker current waveforms

the power frequency wave being superimposed on an exponentially decaying component (Figure 7.5) so that zero current crossing may be delayed for several half-cycles. A further situation is the interruption of direct current (DC) faults which is achieved by inducing an oscillatory current via arc instability in the circuit-breaker and some external oscillatory circuit to force the current to pass through zero eventually (Figure 7.5). At the lower domestic voltages, current limitation can be conveniently induced, leading to an earlier and slower approach to zero current than that which occurs naturally and with the additional benefit of reducing the energy absorption demands made of the interrupter module (Figure 7.5).

Finally, there are situations whereby high-frequency (kHz–MHz) currents can be induced and these need to be tolerated by the circuit-breaker. These occur, for instance, when switching on-load inductors. In this case the line-side configurations shown in Figure 7.4 are replaced by a lumped inductor load.

7.2.2 Circuit-breaker characteristics

The basic characteristics of a circuit-breaker relate respectively to the thermal and dielectric recovery phases.

The thermal recovery characteristic is in the form of a critical boundary separating fail and clear conditions on a rate of rise of recovery voltage (dV/dt) and rate of decay of current (di/dt) diagram (Figure 7.6) [1, 2]. Typically, the boundary is given by the relationship

$$\frac{dV}{dt} = \text{const.} \left(\frac{di}{dt}\right)^{(-n)} \tag{7.3}$$

with $n = 1 \rightarrow 4.6$ depending upon the type of circuit-breaker and other factors. The thermal recovery performance may be improved by increasing the pressure of the

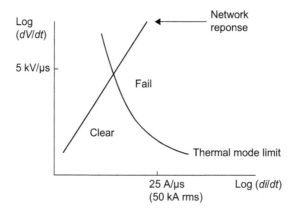

Figure 7.6 Typical thermal recovery characteristic of a gas-filled interrupter – rate of rise of recovery voltage post-current zero versus rate of current decay pre-current zero

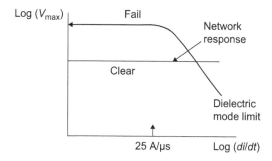

Figure 7.7 Critical recovery voltage versus rate of current decay pre-current zero

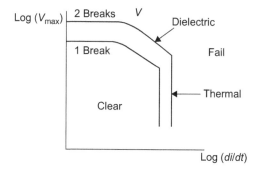

Figure 7.8 Overall interrupter characteristic – superposition of thermal and dielectric characteristics

circuit-breaker gas, the nature of the gas and the geometry of the interrupter or by the number of contact gaps (interrupter units) connected in series.

For the dielectric recovery regime the characteristic is represented by the critical boundary separating successful clear and fail conditions on a maximum restrike voltage (V_{max}) and rate of decay of current (di/dt) (Figure 7.7). The dielectric recovery performance may be improved by similar measures as the thermal interruption capability. However, the quantitative effect of such measures will usually be different for both interruption regimes.

By combining the thermal and dielectric recovery characteristics, the overall limiting curves for circuit-breaker performance are obtained (Figure 7.8). These show that typically the performance is limited at low values of di/dt by dielectric failure and at higher values of di/dt by thermal failure. Increasing the number of arc gaps improves the performance.

7.3 Arc control and extinction

An important aspect of good circuit-breaker design and operation is to provide proper arc control and efficient arc quenching such that the specified voltage-withstand capability is achieved rapidly after the current interruption. The arc quenching needs to be optimised to avoid too rapid a current reduction rate since too high a value of di/dt can produce excessive overvoltages due to the inductive nature of the connected network, which may result in restrikes. Such current reduction produced by over-rapid arc quenching is known as 'current chopping' and needs to be avoided.

Appropriate arc control and quenching rely upon the removal of thermal energy sustaining the arc plasma and produced by joule heating through the flow of the fault current and the system voltage (iV). Such heating produces temperatures of the order of 20,000 K in the arc plasma [3] whereby the arc is intensely luminous (Figure 7.9(a)) producing typically megawatts of radiated power. Consequently the thermal energy stored in the arc plasma is dissipated via not only the physical processes of convection and conduction but also radiation (Figure 7.9(b)).

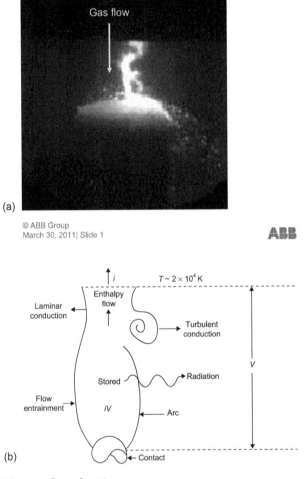

Figure 7.9   Nature of an electric arc
  (a) Appearance of an electric arc during an overhead line fault
  (b) Physical processes governing the control and quenching of an electric arc plasma column in a gaseous medium

These processes may be enhanced through the choice of arcing medium (SF_6, air, oil, vacuum, etc.) and the method of arc confinement and control (e.g. use of gas flow, electromagnetic fields and arc splitting plates).

At temperatures up to 20,000 K the composition of a gas within which the arc is produced can vary substantially with molecules dissociating into atomic and ionic fragments as well as electrons. An example of the variation of such dissociated components with temperature [1] is shown in Figure 7.10 for SF_6. This illustrates the complexity of the components which exist in an SF_6 arc discharge and which contribute to the storage and removal of energy from the arc.

Arc control involves taking advantage of these properties via the use of an imposed flow of gas or via a superimposed electromagnetic field.

7.3.1 Gas-blast circuit-breakers

Circuit-breakers utilising an imposed flow of gas are known as gas-blast circuit-breakers. In these circuit-breakers the current-carrying interrupter contacts are separated along the axis of a nozzle through which a supersonic flow of gas is produced by having sufficient difference in the gas pressures on either side of the nozzle. Such a flow of gas produces strong convection to constrain the arc and remove the stored thermal energy.

Rather than storing the flow-producing gas indefinitely at high pressure within the circuit-breaker until a fault occurs, the gas may be compressed transiently by a piston action ('bicycle pump') simultaneous to separating the interrupter contacts when fault current interruption is required. Circuit-breakers which operate in this

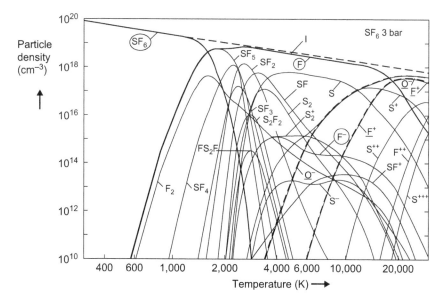

Figure 7.10 Equilibrium concentration of SF_6 dissociation products as a function of temperature

manner are known as 'puffer circuit-breakers' [2]. They may be configured with two identical nozzles in tandem, back to back to provide a duo-flow (Figure 7.11(a)) or with two differently sized nozzles to provide a partial duo-flow (Figure 7.11(b)).

A typical thermal recovery characteristic of such a circuit-breaker (dashed curve) is compared with that of a two-pressure circuit-breaker (continuous curve) in Figure 7.12. These characteristics indicate that the performance of the puffer

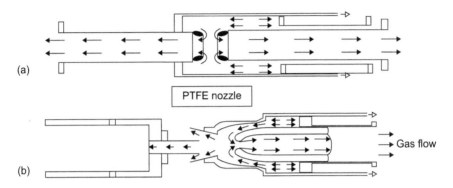

Figure 7.11 Structures of gas-blast puffer interrupters
(a) Duo-blast configuration
(b) Partial duo-blast configuration

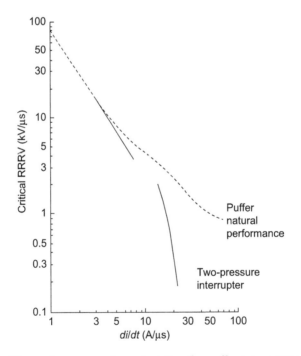

Figure 7.12 Thermal recovery characteristic of a puffer interrupter compared with two-pressure interrupters

circuit-breaker is better maintained as the current decay rate (and hence peak fault current) increases compared with the two-pressure circuit-breaker. This results from two effects. First, the arc initially throttles the gas flow by blocking the nozzle so that there is an additional upstream pressure built up. This subsequently sustains a higher gas flow transiently once the arc decreases in diameter as the fault current approaches zero at the end of a half-cycle. Second, the presence of the arc itself can produce gas compression due to arc-induced gas heating within the confined volume of the interrupter. This is especially used in 'self-blast' circuit-breakers, which require only a minimum of drive energy and which are, therefore, more efficient than puffer circuit-breakers.

The empirical expression $(dV/dt):(di/dt)$ of (7.3) for the thermal recovery characteristic, therefore, needs to be amended to apply to puffer or self-blast interrupters by making the gradient parameter 'n' a function of the static gas pressure (p), incremental pressure rise due to piston action (dp) and peak fault current (I_{pk}) (to account for arc throttling plus heating) [2].

7.3.2 Electromagnetic circuit-breakers

The operation of high-voltage electromagnetic circuit-breakers has hitherto involved spinning the arc through the interaction of a Lorentz force produced by a B field from a coil carrying a fault current and the fault current itself flowing through the arc [2]. The arc may be spun azimuthally (Figure 7.13(a)) or helically (Figure 7.13(b)). Thus by contrast with the gas-blast circuit-breakers, the arc control and arc quenching convection is generated by driving the arc itself through the surrounding gas rather than vice versa.

The empirical expression $(dV/dt):(di/dt)$ of (7.3) for the thermal recovery characteristic is amended to apply to such electromagnetic interrupters by making the gradient parameter 'n' a function not only of the background gas pressure (p) but also of the B field and the phase angle (ψ) between the time-varying B field and the fault current [2].

Typical $(dV/dt):(di/dt)$ characteristic curves for rotary arc-type interrupters are shown in Figure 7.14 for a series of constant B field and phase angle (ψ) values [2]. In practice the B field magnitude increases with the fault current so that the characteristic of a real circuit-breaker forms a locus across the constant B field curves shown in Figure 7.14 (much as the characteristic for the puffer breaker forms a locus across the constant pressure curves of Figure 7.12). Also shown for comparison in Figure 7.14 is a characteristic (dashed line) for a two-pressure (non-puffer) gas-blast interrupter.

The electromagnetically driven arcs produce additional effects which aid arc plasma quenching and current interruption:

- The thermal energy from the electromagnetically driven arc is dissipated in the surrounding gas so producing an increase in gas pressure and can be used in conjunction with a puffer arrangement to further enhance the latter's pressurisation.

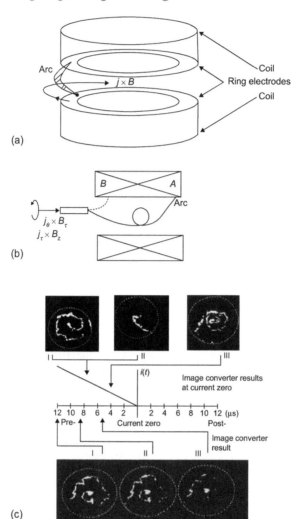

Figure 7.13 Electromagnetic interrupters
(a) Rotating arc interrupter
(b) Helical arc interrupter
(c) High-speed photographs of a helical arc

- The electromagnetically driven helical arcs have the capability to pump gas along the axis of the tube formed by the annular contact (Figure 7.13(b)) via two processes [4]:
  - the 'fan action' of the rotating arc within the tubular contact inside the B field producing coil
  - the reciprocating 'piston action' of the arc helix penetrating axially into the annular contact and repeatedly short-circuiting itself.

Each mechanism can produce flow rates of about 0.2 kg/s. The 'fan action' dominates at lower arc rotation speeds while the 'piston action' dominates at higher

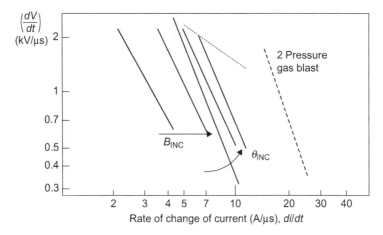

Figure 7.14 Effect of various parameters on the thermal recovery characteristics of rotary arc interrupters ((·······) constant B field and (----) particular unit)

speeds [4]. The significance of such gas pumping action is that the SF_6 dielectrically weakened by arc overheating locally is removed by this induced unidirectional flow from the dielectric sensitive localities within the interrupter.

7.3.3 Dielectric recovery

The dielectric strength of a gas is determined by the gas type and the (particle number) density. The dielectric recovery in gas circuit-breakers is therefore determined by complex spatial and temporal variations of gas pressure and composition, gas temperature and electric field strength between the contacts. The withstand voltage during the dielectric recovery increases with the gas pressure and decreases with temperature and electric field strength. The efficient control of all these parameters is an important design aspect of gas circuit-breakers. With the aid of modern computational fluid dynamic (CFD) simulations the development of gas pressure and temperature within the interrupter can be predicted with reasonable accuracy [5]. However, an important aspect remains, which is the complex dissociation chemistry of SF_6 shortly after current zero (Figure 7.10) [1] and which is only partially understood. Theoretical estimates of the time variation of concentration of dissociation by-products following arcing (and based on unconfirmed equilibrium assumptions) show the complex nature of the dissociated components and their variation with time when subjected to voltage stresses (Figure 7.15(a)). The temperature of the dissociated SF_6 in the post-arc column is deduced to recover via a number of steps which are governed by the thermal capacity and conduction properties of the dissociated by-products. Estimates of the recovery of dielectric strength then yield characteristics having a step-like nature as shown in Figure 7.15(b) [6]. The first fast rise of the recovery is caused by the rapid cooling of the post-arc channel to temperatures close to 2,000 K. At such temperatures recombination to SF_6 occurs (Figure 7.19) and the electric conductivity is sufficiently low such that the residual effects of the arc channel vanish and a capacitive electric field distribution develops [7].

288 High-voltage engineering and testing

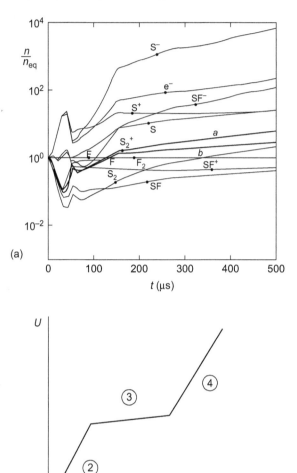

Figure 7.15 Time variation of dielectric recovery conditions
(a) SF_6 dissociation by-products
(b) Voltage withstand as a function of time

7.4 Additional performance affecting factors

Fundamental factors which govern circuit-breaker performance under relatively ideal conditions have been hitherto described. For instance, the circuit-breaker thermal recovery characteristics of Figures 7.7 and 7.14 are, in principle, predictable from the physical phenomena (thermal convection, conduction, radiation, etc.) [1, 2] shown in Figure 7.9(b) using the laws of conservation of mass, momentum and energy provided the boundary conditions and material properties are known. However, in practice, complicating effects arise from a lack of knowledge of the material properties (e.g. enthalpy, radiation transport, etc.) of the arcing medium which follows from the complex dissociation chemistry (Figures 7.10 and 7.15(a)) and the entrainment of foreign species from, for example, the interrupter contacts and nozzle and affects the fundamental plasma properties in a complicated manner. In addition to these factors additional complex phenomena have emerged which without proper designs may limit the performance of SF_6 circuit-breakers in other ways.

7.4.1 Metallic particles

At high current levels (several kiloamperes) tungsten from the sintered copper–tungsten contacts of the interrupter may be ejected into the arc gas as particles (Figure 7.9(a)). This occurs because of tungsten's higher evaporation temperature than copper, which leads to copper evaporating preferentially and leaving tungsten as a fragile matrix. The presence of such particulate tungsten has been detected spectroscopically in the gap between contacts during the current zero period of current interruption. Figure 7.16 shows results indicating whether tungsten was present in the contact gap close to current zero and whether the interrupter failed to clear. Such effects can be minimised via the introduction of the duo-flow interrupter geometry (Figure 7.11) which produces a flushing action of the contact gap because of the resulting contra-flow which occurs (Figure 7.9(a)). There is experimental evidence that luminous particles may persist for long periods after current zero in SF_6, possibly due to some chemical reactivity so that, although such particles may be removed from the contact gap itself, they may nonetheless exist for longer times within the interrupter volume. As a consequence these particles

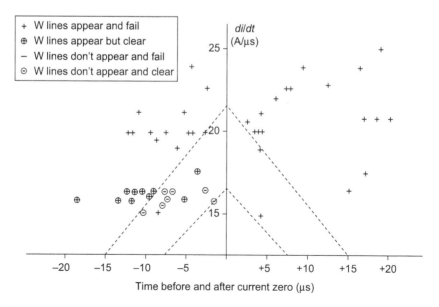

Figure 7.16 Deleterious effect of tungsten metal particles from the interrupter contacts on the thermal recovery of a contact gap

may degrade the dielectric recovery of the interrupter by reducing its voltage-withstand capability.

7.4.2 High-frequency electrical transients

Two high-frequency electrical effects can occur in interrupters and isolators which may affect the current interruption process. The first is the formation of high-frequency (kHz) electromagnetic waves which can be produced and propagated within SF_6-insulated busbar chambers of a gas-insulated substation (GIS). These waves are only slowly attenuated because of the low-loss nature (low R in Figure 7.3) of the co-axial GIS, which acts as waveguide. One source of such transients is pre-arcing during the closing operation of an isolator or interrupter. Optical investigations of such switch operation show that events with frequencies in excess of 10 MHz may occur [8, 9] and that reflected waves within the GIS system can interact to feed energy back into the arc which has insufficient time to be self-quenching.

A second high-frequency effect occurs during on-load inductor or capacitor switching [10]. In this case, arcing at frequencies in excess of several kilohertz may occur during the closing or opening operations of an interrupter but with peak currents of only a few hundred amperes. Such effects are associated with network resonances close to the interrupter. Thus although during such events, the peak current is low the di/dt at current zero is excessive (due to the high frequency of the waveform) leading to di/dt values higher than those shown on the performance characteristics of Figure 7.12. Since the voltage-withstand capability (dV/dt) decreases with increasing di/dt an interrupter may fail to interrupt such low-amplitude, high-frequency currents. Furthermore, this high di/dt at current zero coupled with the inductive nature of the load is inclined to produce over voltages leading to re striking of the circuit-breaker. This may lead to voltage escalation and damage to network components.

7.4.3 Trapped charges on PTFE nozzles

Experimental investigations [11, 12] have indicated that electric charges may be trapped on and migrate along the surface of the PTFE forming the nozzles of a gas-blast interrupter and that such charges can have a deleterious effect upon the electric field distribution within the interrupter. Figure 7.17(a) shows a typical electric field distribution between the contacts of a gas-blast interrupter with no surface charges on the (PTFE) nozzle, while Figure 7.17(c) shows the extent to which this electric field distribution becomes concentrated close to the upper, positive contact due to − 20 μC of charge on the nozzle. The effect is dependent upon the relative polarity of the contacts and charge as shown in Figure 7.17(b and c). Thus with the appropriate voltage polarity on the contacts, electrical breakdown is enhanced between the downstream (top) contact and along the external wall of the nozzle. The probability of the electrical breakdown occurring outside the nozzle depends on the moving contact

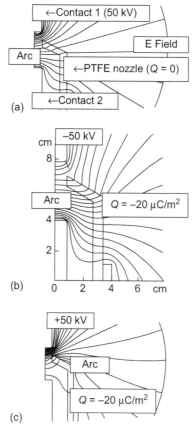

Figure 7.17 Electric field distributions around the contacts and PTFE nozzle of an SF_6 interrupter with and without surface charge on the nozzle
(a) No surface charge
(b) +20 µC surface charge
(c) −20 µC surface charge

position with respect to the exit of the PTFE nozzle [12]. As a result electrical discharge puncturing may occur through the wall of the PTFE nozzle.

The decay with time of electric charge on the surface of PTFE is extremely slow [12]. Under ideal conditions the decay can take well over 400 days (Figure 7.18), but in practice may be shorter due to the presence of ions in the gas and surface conductivity. Such prolonged charge preservation would appear to be a contributory reason for late voltage restrikes which may occur across open contact gaps with some interrupters under particular conditions. To overcome such effects and improve interrupter performance, different types of PTFE materials loaded with different impurities may be utilised in an interrupter.

292 High-voltage engineering and testing

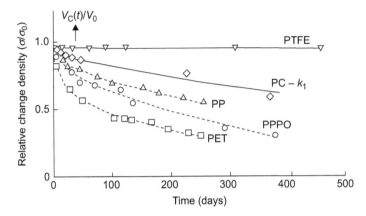

Figure 7.18 *Decay of electric charge on various electrets in a dry atmosphere at room temperature [5]*

7.5 Other forms of interrupters

7.5.1 Domestic circuit-breakers

Lower voltage circuit-breakers such as those used at domestic (250 V) and industrial voltages utilise air as the arc quenching medium in combination with a linear magnetic field drive. The arc is formed between two parallel contacts and driven along these by the electromagnetic force produced by the fault current itself (Figure 7.19). Ultimately it is forced into narrow gaps between a stack of de-ionising plates. The arc is extinguished by the plates absorbing energy from the arc plasma and through the formation of a series of separate arcs between the plates. The formation of a number of anode and cathode spots on the plates increases the voltage drop across the series of electric arcs and because of the limited supply voltage produces current limitation and ultimate extinction (Figure 7.5).

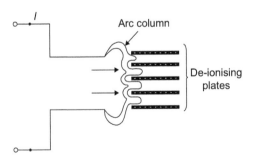

Figure 7.19 *Domestic circuit-breaker with de-ionising plates*

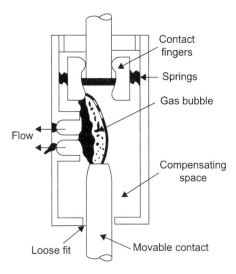

Figure 7.20 Schematic diagram of an oil-filled circuit-breaker

7.5.2 Oil-filled circuit-breakers

Oil-filled circuit-breakers have been widely used at distribution voltage levels. In these interrupters arc quenching is achieved by the oil absorbing energy from the arc plasma, causing it to locally evaporate and so form a gas bubble with good arc extinguishing properties. In some designs there may be a lateral flow oil and the arc driven onto a series of de-ionising vanes similar to the domestic circuit-breaker arrangement (Figure 7.20). Also combinations of axial and lateral flow have been used successfully.

7.5.3 Vacuum interrupters

Vacuum interrupters [13] differ fundamentally from the other types of interrupters in that the arc plasma is diffused and does not rely on a surrounding medium for absorbing its stored energy.

Instead the arc is sustained by the emission of electrons and ions from the circuit-breaker contacts. Current interruption occurs at current zero if the supply of such ions and electrons can be sufficiently rapidly ceased and those already in the contact gap removed. Metal vapour condensing shields may be used to assist the process (Figure 7.21(a)). Thus the successful operation of such an interrupter relies on the prevention of a constricted arc column and electrode spots being formed. This may be assisted through the use of contact designs which rapidly move the arc roots electromagnetically as shown in Figure 7.21(b). Such interrupters have mainly been used at distribution voltage levels. Extension to higher voltages would require the interrupter contacts to be moved further apart to provide the required electrical insulation when the contact gap is fully open (Figure 7.1). Opening such long gaps under high vacuum is difficult, so recourse needs to be made to the use of several interrupters in series to provide the necessary insulation. However, the

294 High-voltage engineering and testing

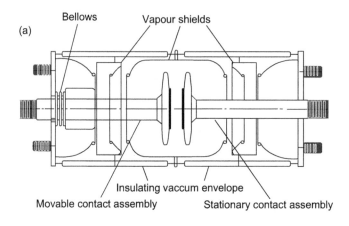

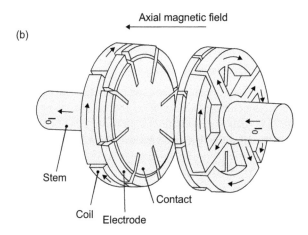

Figure 7.21 Schematic diagrams of vacuum interrupters
   (a) Cross-sectional view
   (b) B field producing contacts

voltage across each interrupter unit after current zero may not be distributed evenly. This may need a large number of interrupters in series and such an arrangement can prove uneconomic.

7.6 Future trends

The advent of SF_6 circuit-breakers led to significant commercial benefits:

- At EHV, the interruption capability per interrupter unit was increased significantly and incorporation into gas-insulated systems was facilitated. The net effect was to reduce substation size with accompanying reduction in land and other costs.
- Storing gases at high pressures for prolonged periods, with all the related problems such as gas tightness and corrosion of seals, has been avoided

through the use of transient pressurisation using rapid piston action and self-pressurisation due to arc-induced gas heating.
- The use of gears to allow the contacts in HV gas circuit-breakers to be moved independently has led to more flexible and optimised designs. Not only linear but also non-linear movements are used.
- At distribution voltages, rotary arc circuit-breakers have had better immunity to current chopping than other types of circuit-breakers and their operation is energy efficient in that no power for gas pressurisation is needed. However, in the recent years, the distribution level has been dominated by vacuum circuit-breakers and this trend is expected to increase.

However, SF_6 is a potent 'greenhouse effect' gas with potential environmental implications. It is 3,900 times more potent than carbon dioxide and has a natural lifetime in the atmosphere of approximately 3,600 years [14]. Since SF_6 is man-made and its major use (~80% every year) is in electrical equipment, consideration has been given to:

- improving control and reclamation
- dilution with another gas
- alternative arc quenching and dielectric media.

Control and reclamation have been stringently improved in recent years, while there have been extensive investigations into dilution with other gases and the use of alternative gases. In 2002 the contribution of SF_6 from electrical equipment amounted to only 0.05% of the total greenhouse gas emissions in the European Union (EU-15) (Figure 7.22).

7.6.1 Other gases

Earlier investigations into other arc extinguishing gases were based upon thermal recovery characteristics of various candidates which are presented in Figure 7.23 [1]. These results show that the $(dV/dt):(di/dt)$ characteristic of SF_6 is substantially superior to those of a range of gases extending from the inert gases (Ar, He) via gases such as N_2, CO_2 and air to other fluorine-based gases (CF_4, C_2F_6). It has also a

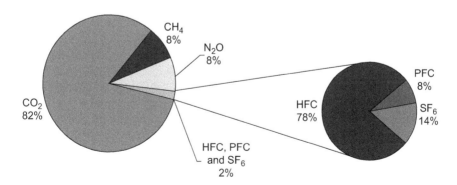

Figure 7.22 Distribution of greenhouse gas emissions in the EU-15 in 2002 [9]

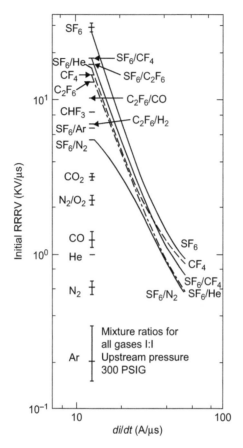

Figure 7.23 Thermal recovery characteristics (dV/dt:di/dt) for various gases in a gas flow interrupter chamber [15]

superior thermal interruption performance compared to its dilution with other gases (N_2, He, CO). Also with respect to dielectric properties SF_6 is unique when taking into account the stability to arcing, the non-toxicity and the zero ozone depletion potential (ODP). Therefore, technologically SF_6 is clearly the option with the highest performance. However, there remains interest in further examining the possible use of some of these gases as alternative arcing media. A systematic study on SF_6 alternatives (Table 7.1) [16] combining all different requirements of a switching gas has shown that promising candidates for replacing SF_6 in circuit-breakers are gases like CO_2 or air. They combine optimum environmental properties with non-toxicity and stability under arcing at, however, reduced interruption capability compared to SF_6. Recent investigations focus on gases such as CO_2 [17] and CF_3I [18]. Although CF_3I has a lower global warming potential than SF_6 (<5 compared to 23,900) and a shorter lifetime in the atmosphere (0.005 compared with 3,600 years), it has a higher boiling point (−22.5 °C compared with −63.9 °C) [18],

Table 7.1 General properties of CF_3I gas and SF_6 gases

Material	CF_3I	SF_6
Molecular mass	195.91	146.05
Characteristic	Colourless	Colourless
	Non-flammable	Non-flammable
Global warming potential (GWP)	Less than 5	23,900
Ozone depleting potential (ODP)	0.0001	0
Life time in atmosphere (year)	0.005	3200
Boiling point (0.1 MPa)	−22.5 °C	−63.9 °C

which makes it necessary to mix it with other gases, for example CO_2. A further disadvantage of CF_3I is its low thermal stability (e.g. [19]), which may lead to hazardous decomposition products.

The possibility of replacing SF_6 by vacuum has received much attention. Single-vacuum chambers of 145 kV have been reported [20]. For EHV applications the dielectric strength of vacuum circuit-breakers falls far short of pressurised SF_6 and air as already shown in Figure 7.1. Special optimisation is necessary and it is still unclear if vacuum circuit-breakers can be economic for such applications.

In the evaluation of possible alternatives not only is it necessary to take account of the direct environmental impact of a switching medium (gas or vacuum), but also the complete environmental impact over the whole life cycle (including production, service and replacement) has to be considered. Under such considerations SF_6 might still be the environmentally preferable solution, since less material and space are needed in comparison to the known alternatives.

7.6.2 Material ablation and particle clouds

The ablation by an electric arc of suitable solid materials (Figure 7.24, [21]), for example PTFE nozzles and shrouds, can increase the gas pressure transiently in an

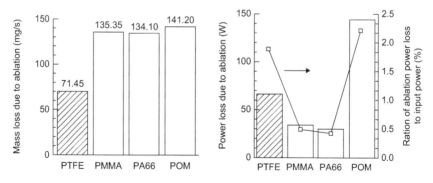

Figure 7.24 Material ablation effects in gas-filled interrupters [21]
  (a) Mass loss due to ablation
  (b) Power loss due to ablation

interrupter which is beneficial for arc control and extinction. But the ablation of PTFE also produces dissociation species which are similar to those of SF_6 [13] (Figure 7.10) and which so provides additional potential benefits for arc quenching [22]. This leads to the possibility that such arc quenching species may be produced transiently for enhancing arc extinction without the need for a background of SF_6 gas. Such current interruption possibilities have been demonstrated in principle by Reference 23, whereby SF_6 equivalent thermal recovery performance has already been demonstrated on a laboratory scale.

Clouds of various polymeric particles suspended in air or nitrogen are also being investigated for arc control and quenching. Unlike the metallic particles produced from evaporating arc contacts (which can have a deleterious effect upon an interrupter's performance (section 7.4.1, Figure 7.16)), such particle clouds have the potential to act as efficient thermal energy absorbers if properly controlled and utilised. Laboratory tests have shown that with nitrogen seeded by such particles enhanced gas pressurisation and convective effects can be produced with good post-arc dielectric strength [24, 25]. When air seeded by polymeric particles is exposed to a fault current arc, the particles may combust to form high pressures of CO_2 from the carbon in the particle material and oxygen in the air [26]. Provided that exothermic chemical reactions can be avoided, the resulting high-pressure CO_2 can be advantageous for arc quenching.

7.6.3 New forms of electromagnetic arc control

A form of electromagnetic arc control and quenching which differs fundamentally from the rotary arc forms shown in Figure 7.13(a and b) is a convoluted arc [27] whose principle of operation is explained in Figure 7.25. The convoluted arc is produced by separating two ring contacts in a spatially distributed magnetic field produced by the fault current flowing through a coil embedded in a PTFE shroud lying along the axis of the two ring contacts. As the contacts are initially separated in the mainly inward radial B field at the end of the coil, a short arc is formed and spins around the ring contacts. Further contact separation leads to the arc lengthening axially and becoming exposed to the mainly azimuthal B field at the central axial location of the coil which compresses the arc against the outer surface of the PTFE shroud. Yet, further contact separation leads to the arc being exposed to the mainly outward radial B field at the other end of the coil which spins the arc section in an opposite direction to the arc section nearest to the other contact. The net effect is for the central part of the arc column to become wrapped around the PTFE shroud circumferentially and electromagnetically trapped in that position until the current reduces to zero. At that point the electromagnetic forces are relaxed and the arc plasma expands radially, so disrupting the arc column and rapidly quenching the plasma. Consequently there is a rapid recovery of the arc gap so that current interruption at zero can be obtained. The electromagnetically convoluted arc also offers an advantage of enhancing PTFE ablation for arc quenching (section 7.6.2) since the electromagnetic forces squeeze the arc plasma column onto the coil containing PTFE shroud.

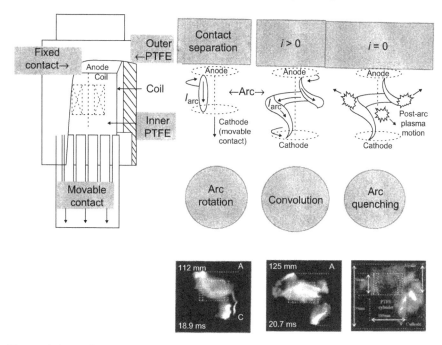

Figure 7.25 Electromagnetic arc convolution for current interruption (schematic of interrupter structure and arc convolution; high-speed photographic images of convoluted arc)

The potential of such an arrangement for current interruption has been demonstrated in laboratory tests [27].

7.6.4 Direct current interruption

With the increasing occurrence of DCs at distribution and domestic levels there would appear to be a need to revisit the problem of efficiently and economically interrupting such currents. The conventional approach of producing oscillatory current variations to force the current artificially to zero (section 7.2.1, Figure 7.5) can involve external inductive and capacitive elements with the accompanying extra costs. However, the use of electromagnetic helical arc units to enhance such current interruption has been considered in Reference 28. Further exploration of the possible use of such electromagnetic arc methods now becomes possible through the emergence of the convoluted arc current interruption (section 7.6.3) and its capability for inducing arc column instability without additional peripheral circuitry. Such considerations and investigations are currently only at an early stage as is the possibility of providing B field excitation independent of the fault current level itself to provide higher B fields for enhancing performance at low fault current levels (Figure 7.26).

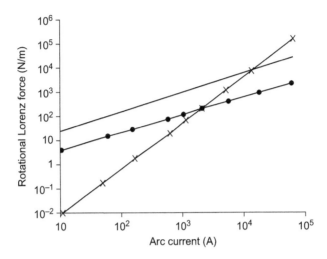

Figure 7.26 Lorentz force for arc rotation in a rotary arc circuit-breaker as a function of fault current [7]
(—×—) Fault current through arc and coil, (—•—) 100 A coil current, (———) 500 A coil current

References

1. Ragaller K. (ed.). *Current interruption in HV networks* (p. 67ff). New York, NY: Plenum Press; 1978 [ISBN 0-306-40007-3]
2. Ryan H.M., Jones G.R. SF_6 *switchgear* (vol. 10). IEE Power Engineering Series. England: Peter Peregrinus Ltd; 1989 [ISBN 0-86341-123-1]
3. Jones G.R. *High pressure arcs in industrial devices*. Cambridge, UK: Cambridge University Press; 1988 [ISBN 0-521-33128-5]
4. Ennis M.G., Jones G.R., Kong M.G., Spencer J.W., Turner D.R. 'A rotating arc gas pump for circuit-breaking and other applications'. *IEEE Transactions on Plasma Science*. 1997;**25**:961–6
5. Franck C.M., Seeger M. 'Application of high current and current zero simulations of high-voltage circuit breakers'. *Contributions to Plasma Physics*. 2006;**46**(10):787–97
6. Schade E., Ragaller K. 'Dielectric recovery period of axially blown SF_6-switchgear arcs', *Proceedings of VII International Conference on Gas Discharges and their Applications*. London; 1982, pp. 100–3 [ISBN 0-906-04886-9]
7. Moll R., Schade E. 'Investigation of the dielectric recovery of SF_6-blown high-voltage switchgear arcs'. *Proceedings of the 3rd International Symposium on High Voltage Engineering*, Milano, 1979, pp. 1–4
8. Simka P. 'An approach to model very fast transients in high voltage circuit-breakers', *IEEE International Symposium on Electrical Insulation*. 2008: 449–52

9. Simka P. 'Complete circuit breaker model for calculating very fast transient voltages', *IEEE International Symposium on Electrical Insulation*, 2010:1–5
10. Looe H.M., Brazier K.J., Huang Y., Coventry P.F., Jones G.R. 'High frequency effects in SF_6 circuit breakers'. *IEEE Transactions on Power Delivery*. 2004;**19**(3):1095–1104
11. Liu W.D., Spencer J.W., Wood J.K., Chaaraoui J.J., Jones G.R. 'Effect of PTFE dielectric properties on high voltage reactor load switching'. *IEEE Proceedings – Science, Measurement and Technology*. 1996;**143**(3):195–200
12. Liu W.D., Chaaraoui J., Wood J.K., Spencer J.W., Jones G.R. 'Parasitic arcing in EHV circuit-breakers'. *IEEE Proceedings A*. 1993:**140**(6):522–8
13. Greenwood A. *Vacuum switchgear* (vol. 18). IEE Power Series. London, UK: Institution of Electrical Engineers; 1994 [ISBN 0-852-96855-8]
14. Wartmann S., Harnisc J. 'Reductions of SF_6 emissions from high and medium voltage electrical equipment in Europe' Ecofys Study No. dm70047.2. Germany; 2005
15. Frind G. *Experimental investigation of limiting curves for current interruption of gas blast breakers.* 1978
16. Niemeyer L. 'A systematic search for insulation gases and their environmental evaluation', *Gaseous Dielectrics VIII*. 1998:459–64
17. Uchii T., Hoshina Y., Mori T., Kawano H., Nakamoto T., Mizoguchi H. 'Investigations on SF_6-free gas circuit breaker adopting CO_2 gas as an alternative arc-quenching and insulating medium', *Gaseous Dielectrics X*. 2004:205–10
18. Katagiri H., Kasuya H., Mizoguchi H., Yanabu S. 'Investigation of the performance of CF_3I gas as a possible substitute for SF_6'. *IEEE Transactions on Dielectrics and Electrical Insulation*. 2008;**15**(5):1424–9
19. Donnelly M.K., Harris R.H., Yang J.C. 'CF_3I stability under storage', NIST Technical Note 1452. Gaithersburg, MD: Building and Fire Research Laboratories, US Department of Commerce; 2004
20. Saitoh H., Ichikawa H., Nishijima A., Matsui Y., Sakaki M., Honma M., *et al.* 'Research and development on 145 kV/40 kA one break vacuum circuit breaker'. *IEEE/PES T&D 2002 Asia Pacific Conference EEE Asia*. 2002;**2**:1465–8
21. Tanaka Y., Kawasaki K., Onchi T., Uesugi Y. 'Numerical investigation on behaviour of ablation arcs confined with different polymer materials'. *Proceedings of XVII International Conference on Gas Discharges and their Applications*: 2008: 161–4
22. Mori T., Spencer J.W., Humphries J.E., Jones G.R. 'Diagnostic measurements on rotary arcs in hollow polymeric cylinders'. *IEEE Transactions on Power Delivery*. 2005;**20**(2):765–71
23. Telfer D.J., Humphries J.E., Spencer J.W., Jones G.R. 'Influence of PTFE on arc quality using an experimental self-pressurised circuit breaker'.

Proceedings of XIV International Conference on Gas Discharges and their Applications. 2002(**1**):91–4 [ISBN 0-9539105-1-2]
24. Looe H.M., Humphries J.E., Spencer J.W. 'Investigation of low-current arc interruption performance of a non-SF$_6$ interrupter unit operating with polymeric materials'. *Proceedings of XVIII International Conference on Gas Discharges and their Applications*; Greifswald, Germany, 2010, p. 1104
25. Brookes R.T., Spencer J.W. 'Pressurisation characteristics of polymeric interruption media'. *Proceedings of XVIII International Conference on Gas Discharges and their Applications*; Greifswald, Germany, 2010, pp. 70–4
26. Brookes R.T., Spencer J.W. 'Influence of combustion on polymer arc interruption', *Proceedings of XVIII International Conference on Gas Discharges and their Applications*; Greifswald, Germany, 2010, pp. 102–6
27. Shpanin L.M., Jones G.R., Humphries J.E., Spencer J.W. 'Current Interruption using electromagnetically convolved electric arcs in gases'. *IEEE Transactions on Power Delivery.* 2009;**24**(4):1924–30
28. Zhang J.R., Jones G.R., Ma Z., Fang M.T.C. 'Interaction of helical arc devices with interconnected networks'. *IEEE Proceedings of Generation, Transmission and Distribution.* 1996;**143**(1):89–95

Chapter 8
Switchgear design, development and service
S.M. Ghufran Ali

8.1 Introduction

This chapter describes the design, development and operation of SF_6 switchgear [1–16]. It also describes how the development of generation and transmission has influenced switchgear evolution (see Appendices A and B). Factors that have contributed to the simplicity of design and increased the reliability of SF_6 switchgear are addressed and the important features of various manufacturers, designs in first-, second- and third-generation interrupters and improvements in circuit-breaker performance are highlighted. It also addresses issues associated with installation and on-site operations and monitoring.

A list of important recent IEC Standards and some CIGRE publications[1] is also included in Appendices C–E together with key supplementary information including certain recent activities of CIGRE Substation, Study Committee SC B3 and also a few of SC B5.

Brief mention is made in Appendix E of residual life concepts and integrated decision processes for substation replacement. Finally, an overview study report of CIGRE work on asset management (AM) themes is also touched on briefly. Table 8.1, provides the reader with an initial indication of the scale of international work carried out worldwide by CIGRE, and the number of CIGRE TBs produced on this AM theme alone [14]. This study is introduced further in section 8.12.1. The originals of these CIGRE and indeed IEC, IET and IEEE documents[1] should always be consulted by the reader for 'empowered' guidance.

8.1.1 SF_6 circuit-breakers

A circuit-breaker is a device that breaks or interrupts the flow of current in a circuit. It is used for controlling and protecting the distribution and transmission of electrical power. It is connected in series with the circuit it is expected to protect. A circuit-breaker has to be capable of successfully:

---

[1] CIGRE key: SC, Study Committee; WG, Working Group and JWG, Joint Working Group; TF, Task Force; TB, Technical Brochure; TR, Technical Report; SC, Scientific Paper. [See also Table 23.13 in Chapter 23.]

Table 8.1 Technical committee activity listing of CIGRE Technical Brochures [14]

Study Committee (SC):	A2	A3 HV	B1	B2	B3	C1	D1	Other
AM topic	Transformer	Equipment	Cables	Lines	Substation	AM	Material	
(i) Condition assessment and monitoring	393 436	083 259 WG A3.06			300 380 381 400		226	SC A1: 437, 386
(ii) End-of-life issues	227	165 368	358	353	252	422	296 409 414	
(iii) Risk management and AM decision making	248				300 472	309 327 422	420	SC C3: 340, 383
(iv) Grid development		335 336		385		176		
(v) Maintenance processes and decision making	445	259 319	279	230	380[a]			
(vi) Collection of asset data and information	298							SC B5: 329 SC D2: 341

After Rijks et al. [14]. 'Overview of CIGRE publications on Asset Management Topics: A Technical Report'. *Electra*. 2012;**262**(June):44–9.
(CIGRE Reviewers: SC C1: Rijks E., Sanchis G., Ford G., Hill E., Tsimberg Y., Krontiris T., Schuett P., Zunec M. and Sand K.)

1. This table provides an extensive listing of strategic CIGRE Technical Brochures (TBs). It is set out as follows: Rows cover one of the six individual subject themes, identified in the first column of the table under the heading AM topic. The first column heading is of Study Committee (SC), while each of next eight columns provides the reader with listings of TBs of particular interest to at least one of the eight SC groups.
2. Each column details individual TBs associated with a particular SC, e.g. A2 Transformer, B1 cables, D1 Material and so on; each row relates to a particular AM topic, i.e. (i)–(vi).
3. It must be emphasised that only a small proportion of the extensive international work carried out by CIGRE has been briefly touched on in Appendices D and E of this chapter.

[*Note*: For detailed CIGRE activity listings, see either Table 8.2 of Introduction to this book or Table 23.13 in Chapter 23.] Here, the CIGRE work has only been reported in abridged format, and edited summaries have been prepared by the editor, who extends the usual caveat that readers working in this industry, or researchers, should always use the full official CIGRE TBs and other official reports, papers etc. The strategic material, only touched on briefly in Appendices D and E, is again presented to the readers on the basis adopted for delegates at the successful IET HVET summer schools for many years, that one has to provide students/readers with 'taster' information of relevant important 'real-world' situations.

[a]This TB 380 is discussed briefly in Reference 15 and also in Appendix B of Chapter 10.

- interrupting (i) any level of current passing through its contacts from a few amperes to its full short-circuit currents, both symmetrical and asymmetrical, at voltages specified in IEC 62271–100 and (ii) up to 25% of full short-circuit currents at twice the phase voltage
- closing up to full short-circuit making current (i.e. $2.5 \times I_{sym}$) at phase voltage and 25% of full making currents at twice the phase voltage
- switching (making or breaking) inductive, capacitive (line, cable or capacitor bank) and reactor currents without producing excessive overvoltages to avoid overstressing the dielectric withstand capabilities of a system
- performing opening and closing operations whenever required
- carrying the normal current assigned to it without overheating any joints or contacts.

This interrupting device becomes more complex as the short-circuit currents and voltages are increased and, at the same time, the fault clearance times are reduced to maintain maximum stability of the system.

A circuit-breaker has four main components: (i) interrupting medium (sulphur hexafluoride gas), (ii) interrupter, (iii) insulators and (iv) mechanism.

8.1.2 Sulphur hexafluoride

The pure SF_6 is odourless and non-toxic but will not support life. Being extremely heavy (4.7 times denser than air) it tends to accumulate in low areas and may cause drowning. It is a gas with unique features, which are particularly suited to switchgear applications. Its high dielectric withstand characteristic is due to its high electron attachment coefficient. The alternating-voltage withstand performance of SF_6 gas at 0.9 bar (g) is comparable with that of insulating oil. SF_6 has the added advantage that its arc voltage characteristic is low, and hence the arc-energy removal requirements are low.

At temperatures above 1,000 °C, SF_6 gas starts to fragment and at an arc-core temperature of about 20,000 °C, the process of dissociation accelerates producing a number of constituent gases including S_2F_{10}, which is highly toxic (see Figures 7.10 and 8.1). However, these recombine very quickly as the temperature starts to fall and the dielectric strength of the gap recovers to its original level in microseconds. This allows several interruptions in quick succession.

The solid arc products consist mainly of metal-fluorides and sulphides with elemental sulphur, carbon and metal oxides. These are acidic and must not be inhaled. Metallic fluorides formed during the arcing do not harm the switchgear components provided the moisture in the interrupting chamber is absorbed by the molecular sieve. The dielectric integrity of the equipment is not impaired by the presence of these fluorides and sulphides

8.2 Interrupter development

The key component in any circuit-breaker is the interrupter. Early extra high-voltage (EHV) SF_6 interrupters were two-pressure type, and these were superseded by the single-pressure puffer type in the early 1970s.

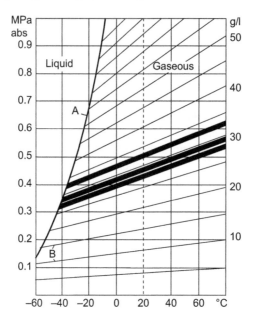

Figure 8.1 Pressure–temperature diagram for SF_6 gas. A is the liquefaction line and B are density lines

8.2.1 Two-pressure system

In the late 1960s and early 1970s, SF_6 EHV interrupters were based on the well-established two-pressure air-blast technology modified to give a closed loop for the exhaust gases. SF_6 gas at high pressure ($\approx$15 bar) was released by the blast valve through a nozzle to a low-pressure reservoir instead of being exhausted to atmosphere. The gas was recycled through filters, and then compressed and stored in the high-pressure reservoir for subsequent operations. Heaters were used to avoid liquefaction of the high-pressure gas at low temperatures.

The relative cost and complexity of design led to the development of inherently simpler and more reliable single-pressure puffer-type interrupters.

8.2.2 Single-pressure puffer-type interrupters

8.2.2.1 First-generation interrupters

The principle of a single-pressure puffer-type interrupter is explained by the operation of universally known device – a 'cycle pump' – where air is compressed by the relative movement of a piston against a cylinder. In a puffer-type interrupter, SF_6 gas in the chamber is compressed by the movement of the cylinder against the stationary piston. This high-pressure gas is then directed across the arc in the downstream region through the converging/diverging nozzle to complete the arc-extinguishing process. The basic arrangement of the puffer-type interrupters can be classified according to the flow of compressed SF_6 gas. These are generally known as mono, partial-duo and duo blast interrupters (Figures 8.2 and 8.3).

Switchgear design, development and service 307

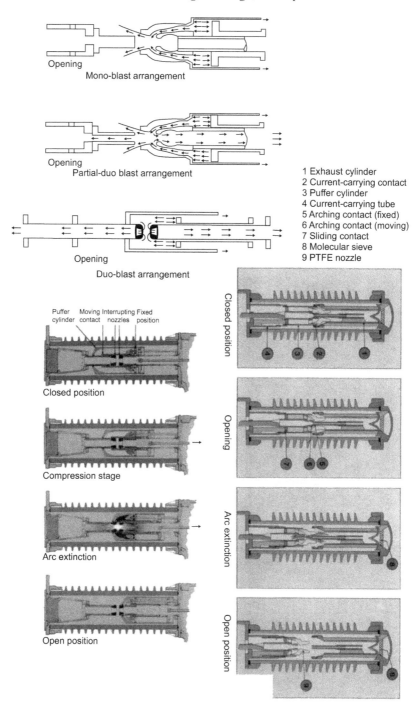

Figure 8.2 Principles of SF_6 puffer-type interrupters

308 High-voltage engineering and testing

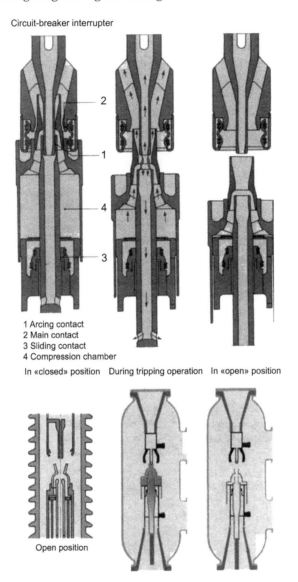

1 Arcing contact
2 Main contact
3 Sliding contact
4 Compression chamber

In «closed» position   During tripping operation   In «open» position

Open position

Figure 8.3 First- and second-generation interrupter circuit-breakers

All types of interrupters are capable of high short-circuit current rating, but superior performance is generally achieved by either partial-duo or duo blast construction because most of the hot gases from the arc are directed away, giving the improved voltage recovery of the contact gap. Most of the puffer-type interrupters have been developed for 50–63 kA ratings, and some even for 80 and 100 kA rating.

For a puffer-type interrupter, retarding forces act on the piston surface as the contacts part. These forces are due to the total pressure-rise generated by compression and heating of SF_6 gas inside the interrupting chamber and are highest with maximum interrupting current and arc duration. Therefore, to provide consistent opening characteristics for all short-circuit currents up to 100% rating, high-energy mechanisms are required.

8.2.2.2 Second-generation interrupters
Worldwide development in the second-generation SF_6 interrupter concentrated on:

- rationalisation of designs
- improving the short-circuit rating of interrupters
- better understanding of the interruption techniques
- improving the life of arcing contacts
- reducing the ablation rate of nozzles by using different nozzle-filling materials.

Most of the present-day SF_6 circuit-breaker designs are virtually maintenance-free. This means that the arcing contacts and nozzles on the interrupters have been designed for long service life.

Most arcing contacts are fitted with copper–tungsten alloy tips. The erosion rate of these tips depends on the grain size of tungsten, the copper-to-tungsten ratio, cintering process and production techniques. The choice of the copper–tungsten alloy is therefore essential for both the erosion rate of the tips and the emission of copper vapour, which influences the recovery rate of the contact gap.

The nozzle is the most important component of a puffer-type interrupter. The interruption characteristic of an interrupter is governed by nozzle geometry, shape, size and nozzle material.

In the western world, at present there have been only nine EHV circuit-breaker manufacturers: ABB, AEG, GEC-Alsthom (now Alstom), Hitachi, Merlin Gerin, Mitsubishi, Reyrolle, Siemens (Reyrolle is now merged with Siemens) and Toshiba.

The nozzles on these circuit-breakers can be classified into two categories – long and short. There is no evidence available to show that at 550 kV the dielectric performance of the long nozzle is superior to that of short nozzle, since most of the designs have achieved 50/63 kA ratings at 420/550/1,200 kV. It is, however, very clear that the rate of nozzle ablation very much depends on the choice of material, which could be either pure (virgin) polytetrafluoroethylene (PTFE) or filled PTFE.

Pure PTFE is white in colour and is most commonly used because of its reasonable price. The rate of ablation of pure PTFE is relatively high and very much depends on grain size, moulding or compacting pressure, cintering procedure and quality of machining and surface finish. It has also been observed that the radiated arc energy penetrates deep in the body, producing carbon molecules. To overcome this, some manufacturers use coloured PTFE, which absorbs the radiated arc energy on the surface and prevents deep penetration.

To ensure consistent performance with reduced rate of ablation and long life, most manufacturers use filled PTFE for high short-circuit current interruption (in the region of 63 kA and above).

There are three types of filling: boron nitride (cream colour), molybdenum (blue colour) and aluminium oxide (white).

310 *High-voltage engineering and testing*

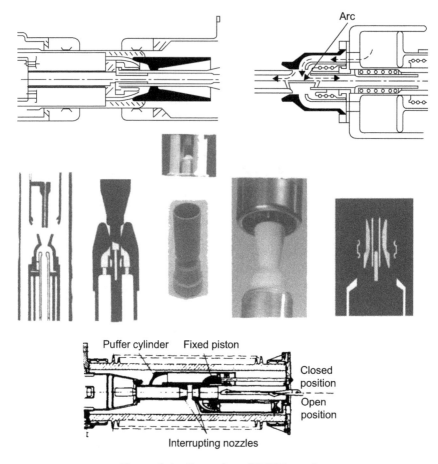

Figure 8.4   Examples of PTFE nozzles

Since the ablation rate of the filled nozzles is low, the change in nozzle throat diameter after about 20 full short-circuit interruptions is normally very small. The pressure-rise characteristic of the interruption hardly changes and therefore the performance of the interrupter remains consistent, giving a long, satisfactory service life.

Filled PTFE material is slightly more expensive than pure PTFE, but its consistent performance and extra-long life justify its use on high current interrupters. Some examples of PTFE nozzles are shown in Figure 8.4.

8.3   Arc interruption

8.3.1   Fault current

Present-day transmission circuit-breaker designs are based on single-pressure puffer-type SF_6 interrupters. The gas flow is produced by the self-generated pressure difference across the nozzle, either by the movement of the cylinder against the stationary piston, or by the simultaneous movement of both the piston and the cylinder. The magnitude and the rate of rise of the no-load pressure rise depend on the interrupter design parameters. These are diameter of the cylinder, nozzle geometry, throat area, swept volume and opening speed.

During fault current interruption, an arc is drawn between the moving and fixed arcing contacts or between the two fixed arcing contacts. The throat area of the nozzle is partially or completely choked by the diameter of the arc, while the arc energy heats up the SF_6 gas in the interrupting chamber, thus causing a substantial pressure rise (Figure 8.5).

The total pressure rise inside the interrupting chamber consists of the no-load pressure-rise and the arcing pressure rises, which produce a considerable pressure difference between the upstream and downstream regions of the nozzle. This pressure difference causes a sonic flow of relatively cold SF_6 gas across the arc. This fast movement of SF_6 gas makes the arc unstable and removes heat energy and in the process cools the arc.

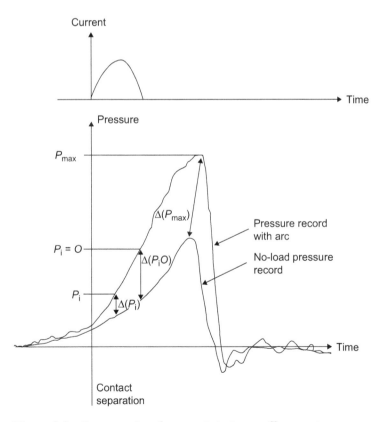

Figure 8.5 Pressure-rise characteristic in a puffer-type interrupter

If the rate of recovery of the contact gap at the instant of current zero is faster than the rate of rise of the recovery voltage (RRRV), the interruption is successful in the thermal region (i.e. first 4–8 μs of the recovery phase), followed by successful recovery voltage withstand in the dielectric region (above 50 μs) and then full dielectric withstand of the AC recovery voltage. This whole process is known as successful fault current interruption.

If, however, the RRRV is faster than the recovery of the gap, then failure occurs either in the thermal region or in the dielectric region after clearing the thermal region.

Over the past 40 years, researchers worldwide have carried out very useful work, which has brought better understanding of the physical processes associated with the arc interruption at and near current zero. Computer models and programmes have been developed, which can accurately predict the performance of an interrupter at, and near, current zero (see Chapter 2). However, in a real interrupter the recovery process in the dielectric region (i.e. above 50 μs) is very complex, since it is influenced by many factors. These include contact and nozzle shapes, arc energy, gas flow, rate and nature of ablation of nozzle material, rate and amount of metal vapour present in the contact gap, dielectric stress of the gap, dielectric stress on the contact tips and the proximity effect of the interrupter housing.

The dielectric aspect of interruption is not yet fully resolved and to the author's knowledge accurate prediction is not possible at present. Siemens successfully tested its GIS CB at 1,100 kV in the KEMA Laboratories in February 2008.

8.3.2 Capacitive and inductive current switching

When a puffer-type interrupter switches small inductive (transformer and reactor) or capacitive (line, cable, capacitor bank) currents, it relies entirely on its no-load pressure-rise characteristic, since there is practically no contribution of pressure from these small-load current arcs.

The magnitude of the no-load pressure rise and the gas flow across the nozzles determine whether the current is interrupted at, before (on falling-current) or after (on rising-current) current zero.

If the current is interrupted at current zero, the interruption is normal and the transient recovery voltages are within the specified values.

However, when premature interruption occurs due to current chopping the interruption is abnormal – it causes high-frequency reignitions and overvoltages. If the interrupter chops the peak current, the voltage doubles instantaneously. If this process is repeated several times due to high-frequency reignitions, the voltage doubling continues with rapid escalation of voltages (Figure 8.6). When these overvoltages exceed the specified dielectric strength for the switchgear, the interrupter and/or other parts of the switchgear may be damaged.

The phenomena of current 'chopping' and 'reignition' are attributed to the design of an interrupter. Most of the EHV interrupters are designed to cope with high fault currents, up to 63 kA in the UK and up to 80 kA in some other countries.

Switchgear design, development and service 313

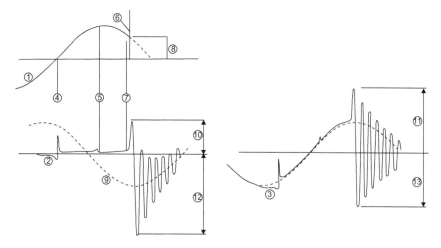

Chopping phenomena in a single-phase circuit

1 Current to interrupt
2 Voltage across circuit-breaker
3 Voltage across inductive load
4 Failed interruption due to reignition (short contact distance)
5 Influence of arc voltage
6 Current instability oscillation leading to current chopping
7 Arc voltage oscillation
8 Effective chopping level
9 Main power frequency voltage
10 Suppression peak, first voltage maximum across circuit-breaker
11 First voltage maximum across inductive load
12 Recovery voltage peak, second voltage maximum across circuit-breaker
13 Second voltage maximum across inductive load

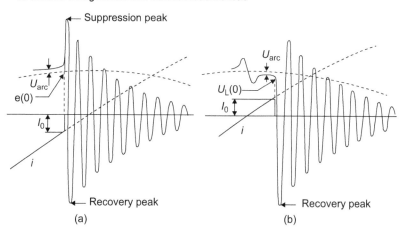

Figure 8.6 *Overvoltages from inductive current chopping and reignitions. Lower diagrams show current chopping (a) before and (b) after natural current zero*

If a design is concentrated only on performance of high currents with high, no-load pressure rise in the interrupting chamber, it will be too efficient for small current and will try to interrupt before its natural current zero. This efficiency sometimes works against it and produces the phenomenon of current chopping and reignitions with adverse consequences.

The interrupter design should therefore incorporate features that cope equally well with small as well as high currents (i.e. softer interruption). It is sometimes desirable to have a softer interrupter to give satisfactory performance for all conditions.

8.3.3 Reactor switching

In a high-voltage system, reactors are used for system VAr compensation. These are connected either directly onto the high-voltage system with EHV circuit-breakers, or to the low-voltage delta, tertiary windings of the auto-transformer (at 12 kV or 33 kV) by MV circuit-breakers. The shunt reactors are frequently switched at least two to three times per day. The SF_6 circuit-breakers for this duty are expected to carry out about 5,000 satisfactory switching operations on one set of contacts and nozzles.

Service experience worldwide has shown that this switching duty can cause difficulty for some circuit-breaker designs, which were earlier tested satisfactorily, on available standard circuits at the major testing stations. It is generally acknowledged that those circuits did not truly represent the site conditions.

The mechanism of reactor current interruption, the phenomena of current chopping and multiple reignitions and the generation of high-frequency oscillatory overvoltages are well understood. It has been established that the high-frequency oscillations are governed by the electrical circuit in a given system, configuration and the interrupter design (i.e. load-side capacitance, load-side inductance, inductance of the busbars, value of parallel capacitance across the interrupter (including the grading capacitance), inductance of the loop formed by the grading capacitors and no-load pressure-rise characteristic of the interrupter).

Since one or all the above parameters could be variable, the issue becomes extremely complex. IEC Subcommittee, CIGRE Working Groups and ANSI have examined various aspects to produce universally acceptable switching test guidelines and a test circuit. The most recent IEC Document IEC 61233 gives a circuit for testing and application guide for assessing the general performance of a circuit-breaker.

Until IEC provides a final solution, it is recommended that site measurements and system studies should be carried out to ensure that as realistic a circuit as possible is used for testing a circuit-breaker on that site. In addition where possible 'R' and 'C' damping circuits and metal-oxide surge arresters should be used to ensure safe operation. Failing that, there could be very costly and serious consequences for the users. Some examples are described below.

The overvoltages produced by chopping and high-frequency reignition can result in the following:

1. For extra high-voltage circuit-breakers:
   i. tracking on the nozzle surface and nozzle puncturing
   ii. dielectric failure on switchgear
   iii. flashover inside the circuit-breaker causing explosion in open-terminal circuit-breakers and severe damage

2. High-voltage circuit-breakers:
   i. catastrophic damage to circuit-breakers
   ii. resonance between the high-frequency overvoltages and transformer windings, possibly causing damage to the tertiary windings, loss of transformer and loss of supply.

8.3.4 Arc interruption: gas-mixtures

Pure SF_6 gas is very efficiently used in the present-day commercial circuit-breakers to interrupt high currents up to 63 and 80 kA. It is, however, an expensive gas and, at common operating pressures of 6 bar (g), it starts to condense at $-20\,°C$ and in its pure form it is not suitable for operation at lower ambient temperatures. When the ambient temperature becomes too low (i.e. -50 to $-60\,°C$) the gas starts to liquefy and the interrupter contains a mixture of SF_6 gas, liquid droplets and fine mist. This mixture affects the gas-flow conditions in the nozzle and transfer of thermal energy from the arc, thus impeding the performance of the breaker.

The need for more efficient and economical transmission switchgear, which can operate satisfactorily at temperatures down to $-50\,°C$ in Canada, Scandanavian countries and Russia, has stimulated the search for new gases and gas-mixtures, as a potential replacement for pure SF_6 gas in puffer-type circuit-breakers. Extensive research in the USA, Japan and Europe has confirmed that at present no single gas tested is found to be superior to pure SF_6 gas in all aspects of dielectric withstand capability and arc interruption.

Several gas-mixtures have been studied with the objective of exploiting the properties of the component gases so that they could be used effectively in switchgear. A survey of some of the published work on various gas-mixtures has shown that the performance of several gas-mixtures appears very promising.

One such mixture is SF_6/nitrogen (N_2). Its dielectric strength at about 15% increased pressure equals that of pure SF_6 gas at normal pressure. The mixture is less sensitive to strong localised electric fields and can be operated at higher pressures up to 6 bar or at lower temperatures down to $-50\,°C$, with potential cost savings of up to 35%. While the dielectric strength of the clean SF_6/N_2 gas mixture is extremely encouraging, its interrupting performance does not compare favourably with pure SF_6 gas. The short-circuit rating of an SF_6/N_2 circuit-breaker at $-50\,°C$ is normally degraded by one level of the IEC standard rating (i.e. a 50 kA circuit-breaker is generally used for 40 kA rating).

Work is in progress worldwide to find a suitable mixture that can be used efficiently down to $-50\,°C$ ambient temperature without 'penalising' its short-circuit or dielectric performance. I understand that some manufacturers have achieved this and hopefully soon the breakers with such gas-mixtures will be commercially available.

8.4 Third-generation interrupters

Puffer-type interrupters require drive mechanisms to provide energy for moving the cylinder of the interrupter at relatively high speeds of 6–9 m/s. The fast movement of the cylinder compresses the SF_6 gas. The high pressure rise upstream of the nozzle due to compression and arc heating of the gas is required for quenching the longest possible arcs associated with both the single-phase to earth fault and the last phase to clear condition of the three-phase fault. This results in very complex and powerful drives, which exert high reaction forces on dashpots, seals, joints, structures and foundation, and affects the reliability and cost of a circuit-breaker. The experience over the last 25 years has shown that the majority of failures on site are mechanical. Therefore, switchgear manufacturers have concentrated their efforts on producing simple interrupting devices and reliable and economical mechanisms. To achieve this, they have addressed the fundamental issue of reducing the retarding forces on the drive mechanisms during an opening stroke. This work has led to the development of third-generation interrupters, which have the following improved design features and economies compared with first- and second-generation interrupters:

a. Reduction of 10–20% in energy has been achieved by optimising the puffer-type interrupter designs, which ensures that the maximum arc duration for the highest current does not exceed 21 ms.
b. Reduction of 50–60% in drive energy has been achieved by utilising skillfully the arc energy to heat the SF_6 gas, thus generating sufficient high pressure to quench the arc and assist the mechanism during the opening stroke.

At least two switchgear manufacturers have successfully used this principle to produce low-energy circuit-breakers (Figure 8.7). The design criteria depend on optimising the volumes of the two chambers of the interrupter:

1. the expansion chamber, to provide the necessary quenching pressure by heating the gas with arc energy
2. the puffer chamber, to provide sufficient gas pressure for clearing the small inductive, capacitive and normal load currents.

The optimum sizes of these chambers are determined by detailed computer studies of the arc input and output energies, temperature profiles, gas flow, the quenching and total pressures.

The main advantages of this design concept are as follows:

a. softer interruption producing low overvoltages for switching small inductive and capacitive currents

Arcing chamber with optimised quenching principle

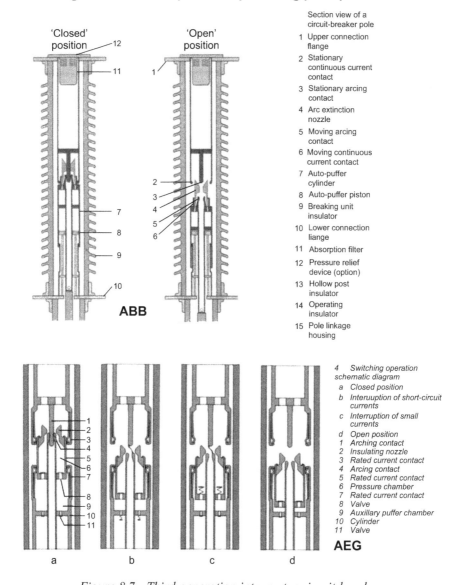

Figure 8.7 Third-generation interrupter circuit-breakers

b. low-energy mechanisms, lighter moving parts, simpler damping devices and reduced loads on foundation and other switchgear components
c. long service life, at least 10,000 trouble-free operations
d. increased reliability and low-cost circuit-breakers.

8.5 Dielectric design and insulators

On EHV circuit-breakers for voltages up to 550 kV, the number of interrupters per phase has decreased from six to one and more recently on UHV, four breaks at 1,200 kV. Therefore, the dielectric performance has become extremely important. To optimise the designs, it is essential to carry out detailed stress analysis studies for all critical components, using sophisticated computer programs and CAD systems.

Most of the major manufacturers have developed their own computer programs, and their designs are based on stress levels from past experience. These CAD techniques are used to optimise the shapes of stress shields, contacts and insulators. They also optimise stress levels on stress shields, gaps, grading capacitors (if any), support insulators and drive rods (see Chapter 3).

In addition to the proper stressing of components, it is extremely important to choose the correct insulating material, since the presence of the degradation products inside the circuit-breaker can damage the silica base insulators. Long-term resistance of insulating material to SF_6 degradation products is essential and most manufacturers have chosen alumina-filled cast resin insulators or the designs that prevent direct contact of degradation products with the surface of silica base insulators. Field computation relating to switchgear design has been covered elsewhere in detail [1].

8.6 Mechanism

The operating mechanism is a very important component of a circuit-breaker. When it operates, it changes the circuit-breaker from a perfect conductor to a perfect insulator within a few milliseconds. A failure of the mechanism could have very serious consequences. It is therefore essential that the mechanism should be extremely reliable and consistent in performance for all operating conditions.

The circuit-breaker may have one or three mechanisms, depending on the operational requirements – either single-phase or three-phase re-closing. The mechanisms fitted to the circuit-breakers are hydraulic, pneumatic or spring, or their combination. The circuit-breaker mechanisms used by manufacturers are grouped as follows:

Pneumatic	Close and pneumatic	Open
Hydraulic	Close and hydraulic	Open
Spring (motor charged)	Close and spring	Open
Hydraulic	Close and spring	Open
Pneumatic	Close and spring	Open

The number of operating sequences and the consistency of closing and opening characteristics generally determine the performance of the mechanism. Although IEC 62271-100 type tests require only 2,000 satisfactory operations to prove its performance, the present tendency is to carry out 5,000 extended trouble-free operation tests to demonstrate the compatibility of these mechanisms with the SF_6 circuit-breakers, which are virtually maintenance-free.

The task of the third-generation SF_6 circuit-breakers, which are fitted with low-energy mechanism and lightweight moving parts, becomes much easier. They satisfactorily perform 10,000 trouble-free operations without any stresses and excessive wear and tear on the moving and fixed parts of the breaker.

8.7 SF_6 live- and dead-tank circuit-breakers

In live-tank (AIS) SF_6 circuit-breakers the interrupters are housed in porcelain insulators. The interrupter heads are live and mounted on support insulators on top of a steel structure to conform with the safety clearances (Figures 8.8a and 8.8b). In dead-tank circuit-breakers the interrupters are housed in an earthed metal tank, usually of aluminium, mild steel or stainless steel, depending on the current rating. The GIS (SF_6) circuit-breakers could be of either vertical or horizontal configuration. Four switchgear manufacturers in the world produce horizontal EHV circuit-breakers, while the rest have vertical designs (Figures 8.9a and 8.9b).

The live-tank circuit-breakers are generally used in open-terminal outdoor substations (Figures 8.8a and 8.8b), while the dead-tank circuit-breakers are used in GIS indoor and outdoor substations (Figures 8.9a and 8.9b). Both types of circuit-breakers have been developed for ratings up to 63 kA and up to 1,200 kV. They have given satisfactory service all over the world during the past three decades. The choice of the type of circuit-breaker, however, depends on many factors. Some of these are:

- cost of the switchgear
- atmospheric pollution
- potential environmental restrictions
- price and availability of the land
- individual preference
- security against third-party damage.

In locations where the price of land is high (i.e. in the centre of cities) the GIS option becomes very attractive because it drastically reduces the overall dimensions of a substation. For example, at 420 kV, the ratio of land required for a GIS installation to that necessary for an open-terminal substation is about 1:8.

8.7.1 Basic GIS substation design

GIS substations are of two types – outdoor and indoor. The basic designs of both indoor and outdoor switchgears are the same but the switchgear for outdoor substations requires additional weather-proofing to suit climatic conditions. Both substation types have been in service throughout the world over the last two decades and have given satisfactory performance.

In Japan, most SF_6 GIS installations are located outdoors without a protective building and have been in service since 1969. In the UK, GEC and Reyrolle SF_6 GIS installations have been in use outdoors since 1976, notably at Neepsend (1976) and Littlebrook (1979). To the author's knowledge there have been no major failures in either country.

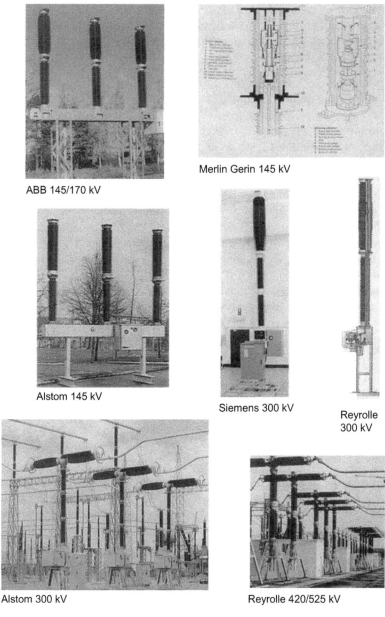

Figure 8.8a   Examples of SF_6 live-tank (AIS) circuit-breaker installations

Recently in the UK, there has been a trend in favour of indoor GIS substations for technical and environmental reasons. SF_6 GIS equipment requires clean, dry and particle-free assembly for safe operation. The assembly and dismantling of outdoor GIS equipment in the UK have been carried out in the past under a portable tent,

Switchgear design, development and service   321

Alstom 420/525 kV

Merlin Gerin 420/525 kV

Reyrolle 420/525 kV

Siemens 800 kV

Siemens 420/525 kV

Figure 8.8b   Examples of SF_6 live-tank (AIS) circuit-breaker installations

322 *High-voltage engineering and testing*

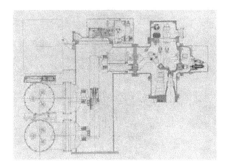

ABB 123 kV

ABB 145/170 kV

ABB 123 - 145/170 kV

ABB 123 kV

Alstom 145 kV

Siemens 145/170 kV

Siemens 145/170 kV

Figure 8.9a   Examples of SF_6 dead-tank (GIS) circuit-breakers

and internal drying of the GIS chambers has been achieved by circulating dry nitrogen. This process can take some days. Therefore, it has become usual practice in the UK to install GIS substations indoors in purpose-built buildings so that assembly, installation and maintenance can be undertaken in controlled conditions.

Switchgear design, development and service 323

ABB 420/525 kV

Alstom 420/525kV

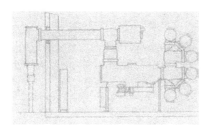
Siemens 420/525 kV

Alstom 420 kV one break

Reyrolle 420/525 kV

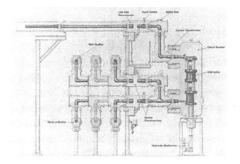
Reyrolle 420/525 kV

Figure 8.9b   Examples of SF_6 dead tank (GIS) circuit-breaker installations

Since a GIS installation consists of an assembly of pipework and steel structures, it is usually regarded as unattractive and unsuitable for environmentally sensitive locations. The visual impact is reduced by housing the GIS in a building.

The switchgear equipment used at open-terminal and GIS substations includes:

- circuit-breaker (already discussed in detail)
- disconnector[2]
- switch disconnector[2]
- earth switch[2]
- current transformer[2]
- voltage transformer[2]
- closing resistors[2]
- surge arresters[2]
- busbars.[2]

The circuit breaker has already been discussed in detail in this chapter, while the basic designs and operations of all the other devices have been discussed elsewhere[2] [2, 5]. However, the following important issues will be discussed here with case examples:

a. closing resistors/metal-oxide surge arresters – to control overvoltages
b. on long-line switching (see also section 2.5)
c. disconnector switching
d. ferroresonance
e. condition monitoring (CM) on site, see also Chapters 20–22.

8.8 Opening and closing resistors/metal-oxide surge arresters

8.8.1 Opening and closing resistors

For many years, opening and/or closing resistors may have been applied to a circuit-breaker in order to reduce switching overvoltages. A CIGRE Technical Brochure entitled 'Background of Technical Specifications for Substation Equipment Exceeding 800 kV AC' discusses recent special type test methods for UHV circuit-breakers equipped with opening resistors (TB 456-WG A3.22 (2009), or see WG A3.22 article in *Electra*, 2011;**255**(April):60–69).

Novel and established techniques are reported, such as:

1. the application of UHV transmission line arresters
2. multi-column arresters and/or a number of arresters in parallel
3. closing resistors used to control slow-front overvoltages (SFOs), during closing and re-closing, of a recently commissioned 1,100 kV overhead line network in China. For additional recent details [16], see section 8.8.3, and Chapter 3, Table 3.2.

The developments and techniques described in this important interim document will eventually be incorporated into future IEC Standards when further

---

[2] See Chapter 16 for strategic definitions relating to insulation co-ordination and other relevant terms, and correction factors, used in IEC Standards on this theme.

service experience is available at the UHV level, as is usual with many CIGRE strategically planned international technical programmes undertaken by experts in the sector.

A recent follow-up study by CIGRE has further reviewed, measured and analytically tested three UHV AC substation projects in China, India and Japan and a new Technical Brochure, TB 542-WG C4-306 has been published in 2013. WG C4.306 has also reviewed this work in an *Electra* article [16]. An abridged version of this review is presented in Chapter 3, Table 3.2, and will be considered further in section 8.8.3.

8.8.2 Closing resistors/metal-oxide surge arresters

The switching and re-energising of long EHV/UHV transmission lines at voltages up to 1,200 kV can generate very high overvoltages, which can overstress equipment insulation and sometimes cause dielectric breakdown. The magnitude of the overvoltages depends on circuit-breaker characteristics, circuit-breaker grading capacitors, circuit parameters and line lengths. The system is normally protected against excessive overvoltages by damping the transient recovery voltage. This is achieved by inserting a specified value of resistance in the line circuit just before the electrical contact is made. For safety, the resistor contacts are then fully opened before the travel of the circuit-breaker main contacts is complete. These precise movements are achieved by mounting the closing resistors on the interrupter assembly in parallel with the main contacts. The moving contacts of the closing resistors are directly connected to the main drive of the interrupters, so that the relative movements can be accurately set. Since the pre-insertion time of the resistor before the main contact touch is very critical, the mechanical drive has to be very precise and positive. The closing resistor drive thus becomes complex, requiring high-energy operating mechanisms for the circuit-breaker.

The pre-insertion time for the closing resistors is determined by detailed network analysis, which takes into consideration all circuit parameters and point-on-wave switching techniques. In most cases the ideal pre-insertion time for optimum overvoltage control is in the region of 10–12 ms.

If we examine a circuit-breaker with closing resistors and compare it with the performance of a standard SF_6 circuit-breaker without closing resistors, we can see that resistors add complexity to the drive mechanism, increase the drive energy requirement of the circuit-breaker mechanism, reduce the reliability of the SF_6 circuit-breaker and increase the cost by 30–40%.

All the above complexities are introduced just for the duration of 12 ms while the circuit-breaker is closing, otherwise the closing resistors remain in the open position.

The author understands that several major utilities have experienced difficulties with failures of closing resistors on air-blast circuit-breakers and have had problems with the long-term reliability of closing resistor drives. Some have already implemented alternative solutions for controlling the switching overvoltages on long lines, by applying metal-oxide surge arresters (at both line ends and in the middle of the line) [4] and by point-on-wave switching.

Surge arrester technology has undergone a radical change in the last 20 years. The undoubted simplicity of the metal-oxide arrester, in which the overvoltage is controlled basically by the arrester's internal non-linear resistance, was initially attractive but there were reservations about the ageing of the resistor 'blocks'. The technology has now advanced to such a stage and sufficient experience has been gained that the metal-oxide arrester is now fully accepted in the industry and is now applied in cases where overvoltage protection is required up to 1,200 kV. This is evidenced by the fact that all manufacturers of surge arresters in the world have changed over completely to metal-oxide arresters. Metal-oxide surge arresters (MOSAs) are tested to IEC 60099-1 and IEC 60099-4.

8.8.3 Main features of metal-oxide surge arresters (MOSA)

MOSAs act continuously, their response time is short, they reduce switching overvoltages as voltages start to build up and, because there are no gaps in the assembly and no arc products, and they have prolonged life.

Because of increased reliability and trouble-free service, experience of MOSAs over the last 20 years including >5 years in UHV AC substation systems, the availability of low protective levels and high discharge energy capabilities of the resistor blocks, several utilities have started to replace the closing resistors with the simple and more economical MOSA devices for controlling overvoltages during energising and de-energising of long lines. ABB installed the first 550 kV GIS circuit-breakers without closing resistors in China. They have been in service for more than 12 years without any trouble. The switching overvoltages at this installation are controlled by MOSAs [4, 5].

Recently, a new TB 542 has been published by CIGRE and has also been reviewed in *Electra* [16] by WG C4.306 that did the evaluation study for this TB. It reviews and discusses insulation coordination practice in the UHV AC range, i.e. at system voltages exceeding 800 kV. It claims to have taken into account the state-of-the-art technology with special reference MOSA. This review has taken into account the accumulated knowledge of various CIGRE working bodies, recent measures and simulated data of VFTO and air gap dielectric characteristics in collaboration with related SC A3 and B3.

A useful abridged overview account of this review [16] is specially reproduced in Chapter 3, Table 3.2, purely for the reader's enlightenment. Much strategic information is provided relating to three substation designs, corresponding insulation levels (LIWV and SIWV) and protection level of the surge arresters used for these three recent UHV AC substation projects in China, India and Japan [16]. Good layout photographs are also available in the *Electra* article regarding substations, bushings and layout of arrester chambers [16]. Special reference has been made in the review to successfully utilise the higher performance of MOSA. The energy absorption of the surge arrester for the temporary overvoltages of these UHV substation projects is specified as being from 40 to 55 MJ, which is higher than those for the 800 kV and EHV network systems.

8.9 Disconnector switching

Disconnectors in GIS installations are used mainly to isolate different sections of busbars either for operational reasons or for safety, during maintenance and refurbishment. They are also used for certain duties, such as load transfer from one busbar to another, off-load connection and disconnection of busbars and circuit-breakers. The switching duties imposed on disconnectors have caused difficulties on some GIS designs in service. These difficulties have resulted in dielectric failure to earth on GIS equipment or dielectric failures on power transformer windings.

At the 1982 Gas Discharge Conference, Yanabu [6] highlighted the switching problems associated with slow moving contacts of a disconnector. Some utilities have observed bright glow on GIS flanges during the disconnector closing and opening operations and reported a few dielectric failures on GIS equipment. Since then, switchgear manufacturers and utilities have continued their investigations. IEC and CIGRE have also taken a very active interest to achieve a better understanding of the disconnector switching phenomena.

During the past two decades, the techniques for high-frequency measurements have improved considerably. With the present sophisticated measuring and recording techniques, the very high-frequency (VHF) transient voltages produced during the disconnector switching operations can now be very accurately recorded and analysed. They also help to explain the switching phenomena.

When the slow moving contacts of a disconnector close or open hundreds of pre-strikes or re-strikes occur between the contacts. These re-strikes generate steep-fronted (4–15 ns) voltage transients, which last for several hundred milliseconds. The magnitude and frequency of the VHF transient voltages depend on:

- contact speed of the disconnector
- SF_6 gas pressure in the disconnector
- dielectric stresses on contact tips, stress shields and contact gap
- circuit parameters, voltage offset, polarity and trapped charge.

This subject has now been extensively discussed worldwide during the past 20 years and the results of investigations have been reported in numerous publications (e.g. from IET, IEEE, ISH, CIGRE).

The recent CIGRE TB 456 (2011), as considered briefly in section 8.8.1, also discusses disconnector and earthing switches/secondary arc extinction, and high-speed grounding switches (HSGS) used in the 1,100 kV transmission network in China and contains much useful technical data including the generation of very fast-transient overvoltages at frequencies up to tens of megahertz. This information provides valuable strategic technical information for the reader; it appears that this phenomenon is more severe at UHV.

The pre-strikes and re-strikes during the closing and opening operations of the disconnector generate very fast, increased voltage transients locally across the contact gap and to the earth, giving rise to transient ground potential rise (TGPR). The VHF transient voltages propagate on both sides of the disconnector as very fast travelling waves into the GIS installations, sometimes causing failure of GIS switchgear and the transformer.

Most switchgear manufacturers now have sufficient experience to incorporate design features, which avoid failures on GIS equipment. The latest IEC Standard provides further guidelines for testing and evaluating the switching performance of disconnectors.

In an installation where the high-voltage side of the transformer is directly connected to the SF_6 metal-clad circuit-breaker by the GIS busbars and disconnectors, a surge arrester is normally connected near the transformer for protection against most of these overvoltages (Figure 8.10).

However, even the present-day fast acting MOSA cannot cope with the VHF transient voltages generated by disconnector switching. They let these fast transients through to the high-voltage windings of the transformer. The continuous overstressing of the transformer windings causes deterioration of the winding insulation. This ultimately can cause dielectric failure to earth inside the transformer tank with very severe consequences such as loss of the transformer, loss of supply and expensive repair.

Discussions with the manufacturers of surge arresters have confirmed that the present-day MOSA technology may not be able to provide adequate protection against these VHF transient voltages in the foreseeable future – so the jury is still out!

The switchgear manufacturers, on the other hand, have had sufficient experience with these switching processes to accurately measure the amplitude, the rate of rise and the durations of these VHF transient voltages. The author firmly believes that by careful design they will be able to dampen or eliminate altogether these overvoltages so that other switchgear equipment such as the transformer on the substation will not be damaged.

Finally, the reader should note the strategic statement appearing in CIGRE TB 456 (2011), relating to metal-oxide surge arresters (MOSA) at UHV: 'Surge arrester designed and applied to limit over voltages to the lowest possible levels is seen to be one of the most important tools for the insulation co-ordination in UHV systems.' (Time will tell when more service experience will be gained at UHV!) Certainly, quick perusal of the recent 2013 TB 542 review of insulation co-ordination (IC) for UHV AC substation projects in China, India and Japan, prepared by more than 37 CIGRE experts worldwide [16], makes one feel that the WG has made some encouraging steps towards UHV IC enlightenment.

8.10 Ferroresonance

Ferroresonance is a well-known phenomenon and it is defined in Chambers' dictionary as follows:

> A special case of paramagnetic resonance, exhibited by ferromagnetic materials. It is explained by simultaneous existence of two different pseudo-stable states for the magnetic material B-H curve each associated

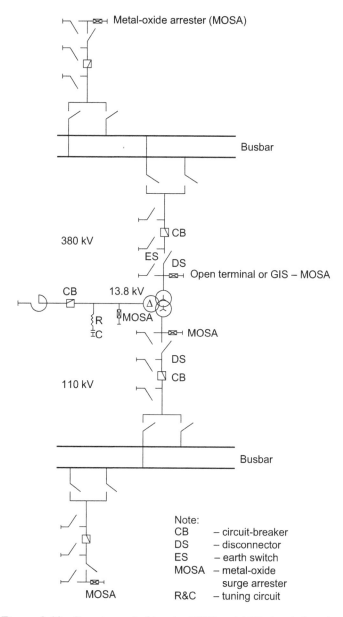

Figure 8.10 Reactor switching by EHV and MV circuit-breakers

with a different magnetisation current for the material. Oscillation between these two states leads to large currents in associated circuitry.

In a switchgear installation where electromagnetic voltage transformers are fitted, the ferroresonance can occur if the conditions are conducive.

Ferroresonance depends on the capacitive coupling between the unearthed, disconnected parts of the switchgear and the remainder of the plant. This happens when the sections of busbar fitted with voltage transformers are left energised through the grading capacitors of the open circuit-breaker; the value of the resultant voltages depends on the values of coupling capacitance and voltage transformer characteristics. The occurrence of ferroresonance is a statistical phenomenon which depends on the switching instants and the remanence effects of the voltage transformer.

The duration for which a voltage transformer may be left energised through the grading capacitors is critical. It is determined by the duty on the voltage transformer and the thermal capacity of its primary windings. In an ideal situation when a bus zone is de-energised, it should be immediately isolated and earthed. This may not always be possible in practice. Therefore, as soon as the switching sequence permits, the disconnectors nearest to the circuit-breaker should be opened and the earthing switches should be closed.

It is essential to analyse the network switching sequence to determine the optimised switching sequences, which minimise the coupling effects and the maximum energising time for the voltage transformers (VTs). It is recommended that ferroresonance damping devices must be fitted to the secondary windings of the VTs on a 'fit and forget basis' for safe operation of the system. The consequences of not carrying out the above recommendations could be quite serious. Several installations have experienced burnt-out voltage transformers and flashovers on busbar-insulating barriers.

8.11 System monitoring

8.11.1 Monitoring during installation and in service

The designs of EHV/UHV SF_6 switchgear are becoming simpler and the numbers of interrupters per phase for the highest system voltages and fault currents (up to 1,200 kV and 63 kA, respectively) are getting fewer and fewer. There were six interrupters per phase for 420 kV in 1976 and only one in 1985/92. In 2009, four-break designs at 1,200 kV entered service in China. The recent CIGRE review [16] provided further valuable operational experiences relating to UHV substation projects in China, India and Japan, which provides still further evidence that the reliability of present-day SF_6 CBs has improved and they are now virtually maintenance-free.

Most SF_6 circuit-breakers are capable of interrupting 20–25 full short-circuit currents and of performing over 10,000 trouble-free mechanical operations. This is the result of:

a. improved arc interruption techniques employed in SF_6 gas and SF_6 gas-mixtures
b. availability of low-erosion rate nozzle materials
c. reduced operating energies with low mechanical stresses on switchgear components
d. computer-aided dielectric stress analysis techniques to optimise the shapes of theoretical components and to obtain low dielectric stresses on contact tips,

SF_6 gaps (across the contacts and to earth), stress shields, insulators and drive rods
e. reduced number of moving parts and dynamic and static seals.

Therefore, the practice of conventional regular maintenance will have to be re-examined. Because the present-day SF_6 circuit-breakers can perform a large number of mechanical operations and have longer contact and nozzle service life, it is not necessary to open a GIS circuit-breaker every six months. Instead of regular maintenance, it is recommended that essential parameters listed below should be monitored, some continuously and others periodically, so that the assessment of switchgear condition can be made.

8.11.2 Continuous monitoring

The necessary parameters to be monitored by fibre-optic diagnostic techniques are:

- current
- voltage
- arcing time
- SF_6 gas pressure
- circuit-breaker contact travel characteristics.

Fibre-optic monitoring equipment has now been fully proven in service. It is stable over a long service period, robust, maintenance-free, easily installed and replaced and easily accessible. These measuring devices are now commercially available. They are accurate, reliable and reasonably priced. Fibre-optic instruments that can be used to see inside the circuit-breaker tank when the switchgear is live are also available.

8.11.3 Periodic monitoring

During the assembly and installation of GIS switchgear on site, care must be taken to ensure that all joints are correctly tightened, all loose particles are removed and all gas chambers are properly cleaned. After assembly, a UHF partial discharge technique may be used to ensure that the whole GIS installation is free from loose particles. Historically, this technique has been very successfully used to locate loose particles within a few hundred millimetres. (Reyrolle and Strathclyde University who developed the technique have published several papers on this subject.) The other techniques of partial discharge measurement have been evaluated by CIGRE WG 15–03 (CIGRE paper 15/23–01, 1992 Paris). These are not discussed here.

The advantage of this technique is that measurements can be made at a relatively low voltage level without overstressing the switchgear insulation. Once the installation is found to be free of any loose particles, it can be safely energised. After the energisation of GIS substation, there is, in the author's view, no need for continuous partial discharge monitoring. Several GIS installations in this country and abroad have been continuously energised for 20–25 years without ultra high-frequency partial discharge (UHF-PD) monitoring devices and have given trouble-free service. The long-term reliability of these sophisticated monitoring devices is still to be proven. However, a periodic check, say every two years, can be made so that the signature prints of the spectrum of discharges can be compared with those obtained just before energisation, and any deterioration in the dielectric integrity of the GIS installation can be detected.

The reader is directed to Chapters 20–22, which provide more recent and detailed treatment and discussion in these and other strategic aspects concerning modern CM strategies, including integrated condition monitoring systems (ICMS).

8.12 Insulation co-ordination

To ensure the safety and reliability of a GIS open-terminal or a hybrid switchgear installation, it is necessary to carry out proper insulation co-ordination of switchgear equipment and complete installation so that the switchgear assembly shall be able to withstand all overvoltages imposed on the system during its service (e.g. see Chapters 2, 3 and 16).[2] The possible sources of overvoltages are:

- *atmospheric overvoltages* – caused by direct lightning strikes and back-flash
- *transient overvoltages* – caused by inductive (reactor), capacitive (line, cable) loads and out-of-phase switching
- *temporary overvoltages* – caused by the resonance of the network and power transformer windings and by ferroresonance in electromagnetic VTs.

The choice of insulation level of equipment is very critical. The design of switchgear should ensure that flashover cannot occur across the open contact gaps with impulse or other overvoltages on one terminal and the out-of-phase alternating current (AC) peak voltage on the other terminal (i.e. the gap sees the sum of the two overvoltage peaks). The design should be complemented by detailed system studies of the switchgear installation to optimise the switching sequence and the number and locations of suitable metal-oxide surge arresters.

8.12.1 Introduction to appendices

Readers will recall (see Chapters 2, 3 and Table 3.2) that higher performance metal-oxide line arresters have been used at 1,100 kV enabling significant reduction to height of tower structures and a lowering of the lightning impulse withstand voltage (LIWV) requirements, etc. There are still a few worries concerning field testing techniques on UHV substations during construction and operation. In the past, such problems were readily overcome for lower voltage systems by carrying out thorough test/evaluation programmes on representative substation layouts – and the same must be done for UHV system voltage levels. However the recent review by CIGRE TB 542-WSG C4-306 dealing with the insulation coordination of UHV AC substation projects in China, India and Japan provides other testing/simulation options [16]. (Table 3.2 of Chapter 3 provides an abridged account of the *Electra* article [16]).

Finally, Appendices A–E* of this chapter provide the following:

- Appendices A and B: some brief historic background material;
- Appendix C: strategic information relating to a current list of IEC Standards;

- Appendices D and E: provide information on some recent studies by CIGRE Study Committees (SCs B3 and B5) on substation-related themes ending with brief comments on three strategic issues:
  a. residual life concepts applied to GIS
  b. integrated decision process for substation equipment replacement (Technical Brochure TB 486-WG B3.06)
  c. CIGRE overview of publications on asset management (AM) themes.

[*Note: Appendices B.1, D and E are prepared by the Editor.]

It should be noted that in Appendix E, aspect E.1.3 includes some of the broad conclusions of this CIGRE Technical Report (TR) in *Electra* (WG B3.06), written by E. Rijks, *et al.* and entitled 'Overview of CIGRE Publications on Asset Management Topics' [14], while in E.1.4 brief comments are given on this Technical Report, relating to WG C1.25. A TB is expected to be issued soon and consequently was not available at the time this appendix was compiled by the Editor.

Below are the six CIGRE AM topics covered and Table 8.1 (see section 8.1) provides a detailed listing of the related Technical Brochures (TBs) issued by CIGRE [14] in each of the technical sectors (e.g. see TB 380 for substations (B3)):

1. condition assessment and monitoring
2. asset end-of-life issues
3. risk management and asset management decision making
4. grid development
5. maintenance processes and decision making
6. collection of asset data and information.

Appendix E shows that the report covers management topics, rotating electrical machines, transformers, high-voltage equipment, substations, substation automation systems, insulated cables, overhead lines, system; system development and electronics. This study provides a valuable list of more than 40 CIGRE Technical Brochures, reproduced in section 8.1, Table 8.1 [14].

It also demonstrates a huge worldwide commitment, as can be seen by the countries involved alone [14] and there many more involved overall in the large range of diverse strategic studies in comprehensive programmes (e.g. see Table 22.13 in Chapter 23 of this book). Factor into this commitment the large participation by workers in industry/academia/standards/consultancy and by IEC/IET/IEEE, etc., and one soon recognises the immense overall participation on inter-related activities.

Table 8.1 is re-created by the Editor from information presented in a CIGRE Technical Report (TR), published in *Electra* (2012) by nine SC C1 reviewers (see Reference 14 and Appendix E).

Note: The listing of important IEC sector standards and some of the strategic CIGRE publications included in Appendices C–E (with more details in Chapters 9,10 and 23) should provide useful initial scoping materials, together with some key supplementary information relating to certain recent activities of CIGRE Substation Study Committee SC B3 and also a few of SC B5.

Health warning: All the above information may be consulted initially for guidance/ reference as necessary; however, it is stressed that the reader should always refer directly to the original IET/IEEE, CIGRE publications and IEC Standard documents for 'empowering' guidance or enlightenment!

8.13 Further discussions and conclusions

Modern SF_6 circuit-breakers (CBs) are simple, reliable and virtually maintenance-free, and some designs have now achieved ratings up to 63 kA at 522 kV with one break per phase. Recently, four breaks per phase designs have been developed for a new 1,100 kV UHV transmission system in China, which entered service early in 2009. Modern SF_6 CB design and performance are now fully proven even up to UHV levels.

However, it would seem from certain recent literature that there is cause for some concern: 'perhaps' insufficiently rigorous UHV laboratory and 'in-service' field test performance evaluation studies have been carried out for substation equipment in this 1,100 kV network in China, and possibly for other UHV transmission networks.

Appropriate 'best technology strategies' may still need to be thoroughly debated, agreed and validated, and UHV substation equipment be rigorously tested at appropriate levels for the particular rated voltage level. It is prudent that all substation components – CBs, bushings, GIS, all cable sections and GIL, disconnector switches, etc. – must be rigorously tested separately and collectively – wherever possible – just as was done to resolve this issue at lower voltages 400/500 kV some 20+ years ago in the UK and elsewhere.

Work using appropriate techniques needs to be completed as soon as possible, particularly bearing in mind following the recent CIGRE concerns relating to:

1. 'Field testing technology on UHV substation construction and operation viewpoints' – this is the title of a CIGRE follow-up Technical Brochure (TB) due to be issued soon after further consideration and deliberation by CIGRE WG A3.22 since the 2009 earlier publication of TB 456 by the same WG which covered only 'interim technical requirements'.
2. Possibly also, the reduced insulation design test levels (LIWL, etc.), and insulation co-ordination (IC) strategies employed for substation regions in this 1,100 kV transmission system in China and for UHV networks elsewhere, which may in general need to be carefully re-appraised as 'certain innovative IC techniques' were supposedly employed in the 1,100 kV network design in China and detailed information has possibly not yet become generally available.

Obviously, a sufficient amount of robust in-service UHV field testing data is required – from the still somewhat sparse number of operational UHV networks worldwide – before 'robust IEC Standards can be "worked-up" and produced for UHV networks'. Meantime, work continues towards achieving robust 'full' IEC Standards and CIGRE TBs. Both of which require to be thoroughly discussed, edited, agreed, officially approved and finally issued.

Returning to EHV networks, circuit-breaker designs with third-generation SF_6 interrupters have reduced the driving energy by 50–60% and have brought increased reliability and further reduction in costs.

Reactor switching causes difficulty for some circuit-breaker designs. Until a realistic test circuit is available to verify the performance of the circuit-breaker, for reactor switching duties it is recommended that metal-oxide surge arresters (MOSA) and R-C tuning circuits (where possible) should be used for added safety.

Owing to increased reliability, high energy discharge capability and trouble-free service over two decades, MOSA are extensively used for insulation co-ordination on the substation and gradually replacing the closing resistors for switching EHV transmission lines [4, 5]. On GIS installations, where electromagnetic voltage transformers are used, ferroresonance can occur and some protection is afforded by fitting ferroresonance damping devices to the secondary windings of the voltage transformers.

Appendices A–E, respectively, provide the reader with:

1. a few aspects of useful historical information
2. SF_6 circuit-breakers in the UK with a perspective of the USA (Appendix B)
3. a valuable listing of relevant IEC Standards for the sector
4. a brief introductory awareness of some of the many valuable strategic 'substation-related' themes recently researched and compiled by the relevant CIGRE Working Groups (WGs)
5. three strategic issues:
    i. residual life concepts applied to GIS
    ii. integrated decision process for substation equipment replacement
    iii. overview of CIGRE publications on asset management (AM) themes.[3]

These may prove particularly helpful to the reader as an initial resource, before moving on and referring directly to the full official (1) IEC Standard(s), (2) CIGRE documentation, (3) refereed papers and (4) international conference documentation. Hopefully, this approach will assist the reader to gain the strategic 'empowerment and discipline' that this book hopes to 'inculcate'.

Finally, strategic sources of photographic and other energy-related material can be obtained by the reader simply by visiting websites of equipment manufacturers, utilities, etc., regarding product range major projects, for most of the subject themes considered in this book; including, for example, the 1,100 kV substation and transmission network development in China in 2009 and much more. Start looking soon![4]

Acknowledgements

The author wishes to thank colleagues at Merz and McLellan Ltd, in particular Mrs E. Adamson, in preparing this chapter and the directors of Merz and McLellan Ltd and Parsons Brinkerhoff for permission to publish it. The views expressed do not necessarily represent the views of Merz and McLellan Ltd and Parsons Brinkerhoff.

Acknowledgement is also made to the following organisations for generously providing technical information and illustrations: ABB, AEG, GEC-Alsthom (now Alstom), MG, NGC, Reyrolle, Siemens and Scottish Power.

---

[3] TB 380 (2009) is also discussed in Appendix to Chapter 10.
[4] The editor gratefully acknowledges the extensive reliance on CIGRE referenced materials in several chapters of this book.

References

1. Ali S.M.G., Ryan H.M. 'Further application of field computation strategies to switchgear design'. ISH-89, *Sixth International Symposium on High voltage engineering*, New Orleans, LA. Paper 27.37
2. Ali S.M.G., Goodwin W.D. 'The design and testing of gas insulated metal-clad switchgear and its application to EHV substation'. *Power Engineering Journal*. London, 1988;**2**(1):17–25
3. Goodwin W.D., Wills A.S. 'The design of outdoor open-type EHV substation'. *Power Engineering Journal*. London, 1987;**1**(2):75–83
4. Eriksson A., Grandl J., Knudsen. *Optimised Line Switching Surge Control Using Circuit-Breakers Without Closing Resistors*. Paris: CIGRE Conference Paper; 1990
5. Schmidt W., Richter B., Schett G. *Metal Oxide Surge Arresters for GIS-Insulated Substations*. Paris: CIGRE Conference Paper; 1992
6. Yanabu S. Paper presented at the 7th International Conference on Gas Discharges and their Applications, London; 1982
7. Barnevik P. 'Electrifying experience: ASEA Group of Sweden' 1983–93
8. Clothier W.H. 'Switchgear stages', G.F. Leybourne, Newcastle; 1933
9. Kahnt R. 'The development of high-voltage engineering – 100 years of AC power transmission'. Siemens
10. Rowlan J. 'Progress in power'. The contribution of Charles Merz and his associates to sixty years of electrical development, 1899–1959
11. Ryan H.M., Jones G.R. 'SF$_6$ switchgear'. Peter Peregrinus Ltd; 1989 (Volume 10 in the IET Power Energy Series, London)
12. Ali S.M.G. 'Switchgear design, development and service'. in Ryan H.M. (ed.). *High Voltage Engineering and Testing*. 2nd Edn. Peter Peregrinus Ltd; 2001 (Volume 32 in the IET Power and Energy Series, London), Chapter 8, pp. 301–334
13. Ryan H.M. 'Circuit-breakers and interruption' in Haddad A. and Warne D. (eds.). *Advances in High Voltage Engineering*. Peter Peregrinus Ltd; 2004 (Volume 40 in the IET Power and Energy Series, London), Chapter 9, pp. 415–476
14. 'Overview of CIGRE Publications on Asset Management Topics: A Technical Report'. *Electra*. 2012;**262**(June):44–9. (Cigre Reviewers: SC C1: Rijks E., Sanchis G., Ford G., Hill E., Tsimberg Y., Krontiris T., Schuett P., Zunec M. and Sand K.)
15. 'The Impact of New Functionalities on Substation Design', CIGRE TB 380-WG B3.01, 2009 (See also WG B3.01. *Electra*. 2009;**244**(June):44–9). Members: Osborne M., [Conv]., Bosshart P., Di Mario C., Finn J., Olovsson H-E., Twomey C., Mackrell A., Mueller L., Yonezawa H. and Sato T.
16. 'Insulation coordination for UHV AC systems', TB 542, CIGRE WG C4.306 (also WG C4.306. *Electra*. 2013;**268**(6);72–9). Some of the technical materials in this TB may be proposed for the application guide IEC 60071-2 (1996) and IEC apparatus standards

Technical information on switchgear, from ABB, AEG, GEC-Alsthom (later Alstom), Merlin Gerin (MG), NGC, Reyrolle, Siemens and Scottish Power.

Notes: Readers may also find the following listing of IEC Standards, CIGRE Technical Brochure and *Electra* information on 'substation-related' themes helpful. (See Appendices C–E, respectively.)

Appendices

A Some historic observations

- Since the 1830s, electricity has been used commercially in the telegraph industry, and many firms in the UK, Europe and America supplied various types of low-voltage equipment to suit.
- In Germany, Siemens & Halke was the largest company, which became interested in higher voltage technology. The others that followed were Allgemeine Elektrische Gesellschaft (AEG), established in 1883 as Deutsche Edison Gesellschaft, Schuckert & Co. and the Union Company. In Switzerland, the leading companies were Oerlikon (1882) and Brown Boveri (1891), and in Sweden, ASEA (1883).
- The use of AC started in lighting plant at the end of the 1880s. Edison, among others, was opposed to the use of AC, which at the time was considered hazardous. Crompton was one of the leading advocates of direct current, while Ferranti supported the cause of AC. However, the research continued and engineers in different countries had a breakthrough almost at the same time in the use of polyphase AC. Later, great progress was made in this technology, in particular with the invention of the transformer and AC motor.
- This subsequently led to the concept of the present-day transmission system, in which electric power at high voltage and low current was transmitted to another place at a distance and transformed back to a reasonable voltage and distributed for local consumption (i.e. generation at one point and consumption some distance away).
- The first successful long-distance transmission of electrical power employing three-phase AC was from Lauffen hydro-electric station. The Lauffen Frankfurt transmission line in Germany, 175 km long, transmitting at 15 kV, 40 Hz with an overall efficiency of 75%, was inaugurated on 24 August 1891.
- Other important dates on the transmission calendar worldwide [7–15] are:

  1911: 110 kV transmission line – Lauchhammer, Riesa, Germany
  1929: 220 kV transmission line – Brauweiler, Hoheneck, Germany
  1932: 287 kV transmission line – Boulder Dam, Los Angeles, USA
  1952: 380 kV transmission line – Harspranget, Halsberg, Sweden
  1965: 735 kV transmission line – Manicouagan, Montreal, Canada
  1985: 1,200 kV transmission line – Ekibastuz, Kokshetau, Kazakhstan (formerly Kokchetav, USSR)
  2009: 1,100 kV transmission line – China[5]
  1,200 kV transmission line – India.

---

[5] A landmark installation – by applying a number of new technologies, new analysis techniques and insulation co-ordination strategies, utilities were able to reduce the dielectric requirements, which led to the use of significantly smaller structures (i.e. by 23%) for TEPCO's 1,100 kV towers.

The main reason for using ever higher voltages was economy of transmission.

Note: Robust IEC Standards are still to be discussed, agreed and issued for UHV systems, so interim criteria are used meantime. A useful reference to read is 'Background of Technical Specifications for Substation Equipment Exceeding 800 kV AC', CIGRE Technical Brochure TB 456-WG A3.22 2009 – covering interim technical requirements. (See also an article of same title, prepared by the Working Group A3.22 members, in *Electra*, 2011;**255**(April):60–9.)

- In the UK electrical industry, Merz and McLellan's contribution stretches back to 1889 when they were the consulting engineers to the North Eastern Electric Company, the pioneer in its field of generation and distribution. This company was regarded as a model of an efficient private utility not only in the UK but throughout the world.
- The foresight of Charles Merz brought about fundamental developments in electrical power supply in the North East, elsewhere in Britain, its overseas dominions and in the USA. He was instrumental in establishing the first 20 kV integrated transmission system of the north east in 1907 and the standardising of the British Supply frequency at 50 Hz instead of 24 and 40 Hz used in different parts of the country.
- In 1907 Merz predicted a saving of 55 million tons of coal each year if UK power supply was operated and managed as an integrated whole. In 1924, he proposed the establishment of a super-tension transmission network for linking up the existing supply areas and developing new ones and to allow interchange of power at one frequency, similar to those already existing at that time in the USA, Canada, Sweden, Japan, France, Germany and the north east coast of England.
- Merz's suggested form of super-tension network was endorsed by the similar IEE proposal 28 years later for the British 275 kV 'supergrid'. His dream became a reality when on 1 April 1948 the British Electricity Authority (BEA), the largest utility in the western world, was created. In 1954 it became the Central Electricity Authority (CEA) in Britain and at the same time came the formation of South of Scotland Electricity Board (SSEB). In 1957 Merz's vision was completed by the formation of the Central Electricity Generating Board (CEGB), with the 400 kV interconnected supergrid in the UK.
- From 1890 to 1960, the transmission voltages increased from 2 to 400 kV. Important dates in UK transmission [7–15] are:

	kV
1890	2.0
1905	5.5
1907	20
1924	60
1926	132
1953	275
1963	400

Switchgear design, development and service 339

The switchgear industry worldwide has kept pace with the increasing demands of both currents and voltages during this period, by developing reliable switchgear for controlling and protecting the electrical networks.

- Merz once again played an important role in bringing Parson, Clothier and Reyrolle together on Tyneside who jointly brought about the electrical revolution in AC generation, transmission and control. Merz, Clothier and Reyrolle pioneered the concept of bulk oil, compound-filled switchgear in UK and jointly developed the first iron-clad switchgear. This was a bulk-oil double plain-break, metal-clad, 5.5 kV circuit-breaker with compound-filled busbars in 1905.

- This brought new standards of safety to the high-voltage distribution system. With continuous improvement in switchgear technology and the efficient use of SF_6 gas, the ultimate goal in circuit-breaker design worldwide has now been achieved in the development of one-break 420 kV (Ali and Goodwin [2] and Goodwin and Wills [3]; Figure A.1) and 550 kV (Suzuki *et al.*, Japan, [4]) circuit-breakers.

Figure A.1   One-break 420/550 kV SF_6 circuit-breakers (1985–92)

340 High-voltage engineering and testing

B SF$_6$ circuit-breakers in the UK plus a perspective from USA

Early transmission in the UK at 132 kV employed bulk-oil circuit-breakers in open-terminal substations but from the mid-1940s air-blast breakers were made in increasing numbers, particularly driven by the establishment of the 275 kV supergrid system. The earliest 420 kV open-terminal circuit-breakers in the UK were first commissioned in the early 1960s and had 12 series air-blast interrupters per phase. SF$_6$-insulated current transformers were produced over the range 145–420 kV in 1950s.

- Following the development and validation of synthetic testing techniques, the recovery voltages available for tests were no longer governed by the maximum direct output of the short-circuit testing stations. This allowed the development of interrupter units with higher breaking capacity. In 1971 two-cycle 420 kV 35 GVA air-blast breakers with six series breaks per phase were installed by GEC and Reyrolle.
- SF$_6$ open-terminal circuit-breakers appeared in Europe from the mid-1960s onwards and the first SF$_6$ GIS installations were commissioned in Europe in the late 1960s. In the early-to-mid-1970s in the UK, 300 kV GIS installations were supplied to CEGB by both GEC and Reyrolle followed by 420 kV GIS substations towards the end of that decade.
- The first high-power SF$_6$ interrupters utilised air-blast technology, modified to give a closed two-pressure system. The relatively high-cost and complex mechanisms of the system led to the development of single-pressure puffer-type interrupters, which were first applied in EHV circuit-breakers in the early 1970s for both open-terminal and GIS installations.

B.1 Overseas SF$_6$ circuit-breaker development (1950s, 1970s and 1990s) – compiled by H.M. Ryan

A CIGRE article in *Electra* (2003) by J.H. Brunke, a past Chair of the old CIGRE Switchgear Technical Committee 13 (SC 13), entitled 'Circuit-Breakers: Past, Present and Future', is worthy of perusal by the reader as it contains a few historic visual perspectives of devices that provide some interesting comparisons on overseas circuit-breaker designs in the 1950s, 1970s and 1990s, and also provide *size reduction* and *reductions in complexity* comparisons for circuit-breaker designs over this same period. These include:

1. Early SF$_6$ circuit-breakers using puffer, two-pressure and thermally assisted puffer technology (1950s) (see Figure 3*).

2. The physical size reductions that occurred in 242 kV circuit-breakers in the 1950s, 1970s and 1990s are clearly evident from perusal of a series of photographs (see Figure 6*).
3. The reduction in complexity between a multi-chamber 500 kV air-blast and single-break 500 kV SF_6 puffer is apparent from photographs (see Figure 5*).
4. HVDC (prototype 500 kV, DC) circuit-breaker using air-blast breaker in parallel with a resonant circuit to create a voltage zero (see Figure 8, after J.H. Brunke, *Electra*. 2003;**208**(June):14–20).
[*See original article for figures/photographs.]

Postscript

- Returning to the switchgear article by J.H. Brunke (*Electra*, 2003), he also describes very effectively the role of professional engineers working within CIGRE/IEC frameworks and how they collectively work together to produce important CIGRE Technical Brochures and also collaborate on IEC Standards and related technical developments.
- Some examples have been given above (and elsewhere in this book) of the valuable IEC and CIGRE documents produced in this manner, and more are listed by J.H. Brunke in the earlier 2003 *Electra* article worked-on during the period of his 'stewardship' of the old CIGRE Switchgear Study Committee 13. These provide the reader with a valuable insight into some of the excellent contributions made by IEC/CIGRE/IEEE/IET/ISH GDC/Current Zero Club, etc., in the switchgear sector.

Few could disagree with the following remarks made by Brunke when he wrote:

> 'The CIGRE/IEC experts past and present who participate in the working groups, and the engineers who designed this equipment should be recognised for the technical contributions they make to this incredible device, the circuit breaker.'

- Many will fully endorse the above complimentary and accurate remarks made by Brunke, expressing his gratitude to the switchgear researchers/designers/ developers/testing staff, contributors to switchgear CIGRE SCs, etc.

Similar gratitude is earned equally by the very large number of contributors in all other technical areas in the energy sector (IET/IEEE/IEC/BSI, etc.) for their ongoing professionalism and dedication.

C Relevant strategic IEC Standard reports

Equipment	IEC Standard
Switchgear	IEC 60050
	IEC 62271-200, -201
	IEC 62271-202, -203
Circuit-breakers	IEC 62271-200, -201
	IEC 62271-202, -203
	IEC 62271-100, -102
	IEC 62271-103, -105
	IEC 62271-107, -110
	IEC 60694
	IEC 61233
Disconnectors	IEC 62271-102, -203
Conductors	IEC 61089
Current transformers	IEC 60185
	IEC 60044-4
Voltage transformers	IEC 60186
	IEC 60358
Line traps	IEC 60353
Substation earthing	IEC 60080
Surge arresters	IEC 60099-1-4
Bushings and insulator	IEC 60137
	IEC 60233
	IEC 60305
	IEC 60383
SF_6	IEC 60376
	IEC 60480
	IEC 61634
Insulation co-ordination	IEC 60071
	IEC 60076-3
Testing	IEC 60168
	IEC 60270
	IEC 60427
	IEC 60060
	IEC 61633
Standard current ratings	IEC 60059

Note: The IEC Standards listed above are provided for guidance only. The reader should be aware that IEC Standard Specifications are reviewed at regular intervals. Consequently, the latest and relevant version of a standard on specific topic/theme must be used. This can be checked directly via IEC (International Electrotechnical Commission) Standards and National Standard organisations, which in the UK is the BSI.

D Update of some recent CIGRE activities relating Appendices D and E to Substations (SC B3) and also (SC B5) – compiled from CIGRE publications by H.M. Ryan

Some recent CIGRE activities for Substation SC B3 and a few from a related SC B5 will now be considered briefly in this section.

D.1 CIGRE Study Committee (SC B3), Substations

Since the second edition of this HVET book was published in 2002 there have been significant incremental changes within the energy industry and also in the substation sectors. By now most of the readers will know that the structure of the electricity grid network is likely to change significantly in the next decade or two with the widespread implementation of distributed generation and new technologies. For example, RES/techniques that will improve efficiency and capacity with the follow-up that these changes will pose many challenges and present opportunities for all involved in their planning and operation – in both transmission and distribution sectors.

Considerable investments costs will be necessary towards further improvements culminating in intelligent or 'Smartgrid' networks in the near future – within the next two decades (see Chapter 23).

As stated in Chapter 23, future transmission networks are being evaluated that involve HVDC technology and long-distance bulk transmission infrastructures or meshed HVDC networks to transmit renewable energy, e.g. wind-turbines from (onshore and offshore) wind farms to load centres. Many varied aspects must be considered.

- CIGRE SC B5 covers protection and automation of substations and the protection scheme requires new strategies, developments and approaches. In fact SC B5 embraces principles, design, application and management of power system protection, substation control, automation, monitoring and recording including associated internal and external communications, substation metering systems and interfacing for remote control and monitoring.[6]
- Interested readers can follow the activities of CIGRE SC B5 and all Working Groups (WGs) to learn of specific challenges, strategies and opportunities relating to various aspects. A selection of recent activities relevant to SC B5 are mentioned below:
    - *Technical Brochure, TB 484* (2011), which was produced by JWG B5/B4.25, has analysed the impact of HVDC stations on protection of AC systems and has provided guidance on designing an appropriate HVAC protection system in the presence of adjacent HVDC systems.
    - *WG B5.14* (2013) studies specific challenges, strategies and opportunities for the application of synchro-phasors to wide area protection and control of modern interconnected power systems. It apparently covers available technologies, their current application in schemes worldwide and also

identifies their complete system requirements. (A CIGRE TB is to be issued.)
- *TB 421-WG B5.34* (2010) analyses 'The Impact on Distributed Generation on Substation Protection and Automation'. 'It is widely recognised that the key aspect "to shape" distribution networks in the future will be the large penetration of distributed generation and the "widespread" introduction of intelligence in the network. This will significantly change procedures and structures of electrical utilities (see also *Electra*. 2010;**251**(August):34–43 (footnote 6))'.

Some restated observations from the later body of work:

Increasing integration of the distributed generation (DG) and with wind farm renewable energy in the power network adds further complexity and challenges to network management. The increase of two-way power flows at sub-transmission and distribution level, the impact of intermittency and the connection to non-optimal parts of the networks could all contribute to major instability. Until recently, the regulatory framework and the connection requirements for DG have been restrictive to protect the integrity of passive radial distribution systems. However, in the past few years, a gradual change of attitudes has been observed, states WG B5.14, more supportive regulations have already been implemented or currently are under revision. Power system protection requirements are adapting to cope with large renewable generation so they can contribute to power system stability.

The CIGRE TB aims and objectives:

1. identify the impact of renewable and distributed energy sources on the substation protection and automation, addressing possible solutions
2. collect the practices of protection schemes and settings in the connection of DG at the sub-transmission and distribution networks from several countries and identify best practices
3. assess the performance of current methods of DG connection requirements
4. contribute to the standardisation of DG connection requirements
5. identify challenges and opportunities of the new protection and generation technologies that permit an effective integration of DG in high-penetration scenarios.

Extensive conclusions are provided 'covering the following headings: impacts on network protection performance; requirements for interface protection and settings; anti-islanding protection; current protection practices in different countries; new

---

[6] For more details (full list of WGs, terms of reference, Strategic Plan of SC B3, etc.) refer to www.cigre-b3.org

capabilities of DG for network integration; trends in protection schemes; trends in automation, monitoring and communications; and islanded systems':

- *WG B5.48* (2015), 'Protection for Developing Network and Different Characteristics of Generation', will evaluate (1) the capability of existing protection to operate correctly and (2) the needs of new protection solutions. SC B5 recognises that perhaps one of the main aspects of the introduction of RES and HVDC links will be the impact on protection of reduced short-circuit contribution from new generation technologies (will be issued as a TB).
- *IEC 61850*, briefly touched on in Appendix to Chapter 23, has been considered in various *Electra* articles since 2000. CIGRE SC B5 considers IEC 61850 to be the key communication solution for the future Grid. It represents the standard which provides a data model on high semantic level and services with high performance and high reliability. It was successfully designed and accepted for substation automation. In addition, the standard has already been extended beyond the limits of substations e.g. for distributed energy resources (DER), hydro and wind power, where first applications are going into operation.

Note: It appears that IEC 61850 is now the most common application worldwide within substations.

The CIGRE challenge is to launch the introduction and utilisation of the process bus concept.

(*Further information*: Check up via SCs, CIGRE sites or *Electra* articles.)

Education and training: SC B5 deals with requirements for education, qualification and continuing professional development (CPD) of engineers in protection and control. Various reports have been published or are due soon:

2011, TB 479 *Synchronous generators*
2011, TB 465 *Transformers*
2013, WG B5.37 *Shunt reactors* (footnote 6) (*No number yet allocated)
2015, WG B5.49 *Shunt capacitors* (footnote 6)
2015, WG B5.44 *Special transformers* (footnote 6).

Summary: 'CIGRE SC B5 is working to facilitate development and application of new technology in order to improve the efficiency of the engineering operations and maintenance of the electric power systems'.

The CIGRE SC B5 has designated priorities for future developments to:

- the impact of integration of renewable energy systems and distributed energy resources (DER) on protection systems
- development and implementation of advanced protection solutions, including solutions based on the new communications technologies
- supporting the development and implementation of new international standards in the domains of substation automation and protection
- tools, concepts and systems for protection and automation life-time management.

Reference: 'How to complete a substation automation system with an IEC 61850 "Process Bus"', *Electra*. 2011;**255**(April):12–24 (Brand, K.P., Brunner, Ch. de Mesmaeker, I).

D.2 CIGRE SC B3 substation

With huge energy mix changes, DER, etc., to reflect sector changes and to maintain current outstanding sector relevancy, in 2002, CIGRE re-formed its SC structure to the format now summarised in Table 23.13.

Today, transmission and distribution substations continue to play a strategic and central role in supplying safe and reliable energy with high availability. In Chapter 23, we have briefly touched on smartgrids Gellings EPRI and networks of the future proposed by CIGRE, in which substations continue to play vital roles.

- As CIGRE SC B3 Substations mentions in its 2011 Annual Report, high-voltage power transmission research and development continues to make advances in new technologies and applications that provide transmission owners with confidence and optimism and additionally give them the reassurances, flexibility, security and stability they need to continue to expand their systems reliably and efficiently.
- As the electric power infrastructure expands and global demand for power grows, suppliers and customers alike will have to communicate and co-operate – *in the UK case, for customers to have an appropriate understanding of/voice on strategic new energy mix alternatives and other developments and contribute significantly to funding the huge costs of essential new technological upgrades and system enhancements* – to ensure that the appropriate technologies are 'eventually' developed and deployed in a sustainable manner.
- At the same time, the sector will have to maintain the availability of the energy delivery system, provide dependable and affordable sources of electricity – *including such strategic issues as adequate storage energy support back-up for wind farms* – and ensure the highest standards of public welfare and safety.
- Information systems and telecommunications have enabled the latest evolutions in the electrical sector. None of the latest evolutions of the electrical sector, the deregulation of the electricity network sector being the major one, would have been possible without the information systems and telecommunications capacities: power exchanges, metering and billing, and security of supply.

The four key strategic directions considered in the CIGRE SC B3 Strategic Plan include one dealing with the impact of new communication standards and smartgrids on existing and new substations:

- T1 new substation concepts
- T2 substation management issues
- T3 lifecycle management and maintenance
- T4 impact of new communications standards and smart-grids on existing and new substations.

D.3 Ongoing CIGRE work on substations (SC B3)

Concepts and developments: CIGRE SC B3 in its 2011 Annual report said:

> 'The activity cluster "*Substation concepts and developments*" is focusing on new concepts with regard to actual changes and developments in the power grid. The impact of new functionalities on the substation design was another aspect considered within this activity cluster.
>
> Mixed technology solutions hybrid (AIS/GIS), compact AIS switchgear, non-conventional instrument transformers, reactive compensation, wind farms, active power flow control, fault current limiters, HVDC, protection, GIL/super-conductors, monitoring, diagnostics, etc., are influencing and changing the design of substations. In particular, the impact of new solutions for substation equipment such as MTS or compact AIS and the impact of the development of distributed power generation such as wind farms on the substation concepts will have increasing importance.'

- 'Obtaining value from substation condition monitoring' (CIGRE TB 462-WG B 3.12 (2011)) is of increasing importance to asset owners in particular, with regard to the implementation of smart grids. Online condition-monitoring provides continuous information of the substation condition, enabling increased performance of the overall network. (See also *Electra*. 2011;**256** (June):50–6, same title.)
- This theme concerning offshore wind farms has been reported briefly in Chapter 23. There are now an increasing number of offshore wind farms. The substations used in offshore wind farms can face harsh environments and unique operating requirements and conditions. SC B3 has set up WGs to look at two separate aspects, as follows:
    - CIGRE has already published TB 483-WG B3.26 (2011) entitled, 'Guidelines for the Design and Construction of AC Offshore Substations for Wind Power Plants'.
    - CIGRE has also recently set up, another WG B3.36 entitled 'Special Considerations for AC Collector Systems and Substations Associated with HVDC Connected Wind Farm Plants'. (This is currently being studied – to be published as a TB.)

Note: The UK and the USA have recently signed an agreement to study offshore wind farms in deep water conditions.

- *CIGRE, SCs B3,* C1 and C2 recognise that substation circuit configurations are a strategic consideration in overall substation design, balancing design constraints associated with operations and maintenance requirements/constraints and the overall system needs with regard to security and functionality.

    A TB JWG B3/C1/C2.14 is due to be issued, late 2012, entitled, 'Circuit Configuration Optimisation'.
- *SC B3 Gas insulated substations:* TB 499-WG B3.17 (2012), 'Residual Life Concepts Applied to HV GIS'

It was decided that because of the high level of GIS installations worldwide and their increasing life-cycle time, residual life concept investigations, applied to HV GIS, were appropriate and could result in an even higher availability and reliability of the grid while at the same time increasing the system efficiency. (See also Pohlink, K., et al. *Electra*. 2012;**262**(June):64–71.)

- SF_6 has been used extensively for many years and is still the subject of much debate. Discussion regarding its usage continues. WG B3.25 is preparing a TB, due to be issued in 2012, entitled 'SF_6 Analysis for AIS, GIS, and MTS Condition Assessment'.
- Regarding the issue – impact on the environment – SC B3 considers it necessary to minimise the use of SF_6. In particular, during routine testing of electrical equipment a limited volume of SF_6 may be released. WG B3.30 is preparing a guide to minimise the use of SF_6 and a TB, entitled, 'Guide to Minimise the Use of SF_6 during Routine Testing of Electrical Equipment' is to be issued late 2012.
- The first 1,100 kV substation in China and 1,200 kV in India (footnote 6) made it necessary to discuss the technical requirements for substations exceeding 800 kV. A Technical Brochure TB 456-WG A3.22 was issued in 2009 covering the interim technical requirements. Further deliberations have been made concerning field testing technique on UHV substations during construction and operation and an updated TB entitled, 'Field Test Technology on UHV Substation Construction and Operation' will be issued late 2012. In fact, a new TB 542 by WG C4.306 [16] has just been published, mid 2013, dealing with insulation coordination of UHV AC Systems in China, India and Japan.

Brief mention can also be made to further Technical Brochures, at various stages of completion, according to CIGRE SC B3:

- A new CIGRE JWG/B1/B3 is in preparation to analyse factors for investment decisions of GIL vs cables for AC transmission. Due to changes in transmission system, the participants of CIGRE SCs B1/B3 consider this analysis is becoming increasingly more important. It is planned to issue a TB entitled, 'Factors for Investment Decision of GIL vs Cables for AC Transmission' in 2014.
- *AIS air-insulated substations*: Existing AIS are recognised as an important asset. Upgrading or up-rating them can increase asset lifecycle time and increase the efficiency of substations. A TB 532-WG B3.23 is scheduled to be issued in 2013, entitled 'Substation Up-rating and Up-grading'. (See also *Electra*, 2013;**267**(April): 40–9.)
- It is recognised that climatic conditions have an impact on the operation of air-insulated substations (AIS). A new TB WG B3.31 is being prepared as various aspects need to be carefully considered during planning, procurement and operation, under severe climatic conditions. The TB WG B3.31 will be entitled 'Air-Insulated Substation Design for Severe Climatic Conditions'.
- Another key issue is to assure a long life span for substations without maintenance problems. A new TB WG B3.32, entitled, 'Saving through Optimised Maintenance of Air-insulated Substations', is being prepared. It is known that

optimised procedures can save money. The findings of the investigations regarding optimised maintenance for AIS will be collected by WG B3.32 and published as a TB in 2014.

- *Substation management*:
  - Practical application of AM information strategies involves the analysis of data and AM tools. The decision process for substation equipment replacement has been explained by the WG and issued as TB 486-WG B3.06 (2011), entitled, 'Integral Decision Process for Substation Equipment Replacement'. (See also *Electra*. 2012;**260**(February):42–9.)
  - TB 472-WTG B3.10 (2011), entitled 'Primary/Secondary System Interface Modelling for Total Asset Performance Optimisation'. (See also *Electra*. 2011;**257**(August):80–7.)
  - SC B3 report that (a) new activities have been initiated to analyse the impact of future grid concepts on the management of substations; (b) first results are expected during 2014; and (c) SC B3 plans to publish a TB on this topic in 2014.

- *Future CIGRE SC B3 activities*: As per SC B3 2011 Annual Report, the networks of the future, super grids and environmental aspects and their impact on substations will have a strong influence on the future activities of SC B3. New working groups are under consideration to work on these challenges. The terms of reference of the new working groups will be discussed and approved in 2012.

Recent Update TB 435-WG C4.2.8 'EMC within power plants and substations', is a review and update of the 1997, CIGRE 124 guide on EMC Power Plants and Substations (document updated based on technological advances and more recent findings).

Note: The CIGRE material reproduced in Appendices D and E stems from this writer's personal experiences (i.e. IEC/CIGRE/IET/HVET/EPSRC/DTI participation over many years) and recent material details are from numerous CIGRE sources (including *Electra* articles or CIGRE-listed website sources). Information is reproduced edited, abridged and summarised in an informal style. In addition, a few useful supplementary personal observations by this writer have been inserted (where considered appropriate, solely for the benefit of the readership of this book as each advances along their chosen career pathway), recalling the caveats referred to in the text and the proven strong beneficial association of being given or directed to relevant strategic information and, naturally, achieving empowerment through knowledge.

E Residual life concepts, integrated decision processes for substation replacement and an overview of CIGRE work on AM themes – compiled by H.M. Ryan

Three strategic CIGRE issues will now be discussed briefly:

1. residual life concepts applied to GIS
2. integral decision process for substation equipment replacement (TB 486-WG B3.06)
3. overview of CIGRE publications on AM topics.

E.1 Residual life concepts applied to HV substations

Some GIS has been in service for more than 40 years and the residual life is becoming an increasingly important issue. Strategically, with more than 12,000 CB Bays installed, there is a need to be able to identify the factors which determine the expected residual life of HV GIS. It is reported that currently there are relatively few bays with an age of more than 30 years but the number will rise significantly in the next decade. Because of this, CIGRE SC B3 is of the view that these factors had to be considered together with associated operational risks so that guidance can be given on the evaluation of the residual life of any particular HV GIS installation.

WG B3.17 was set up specifically with the task of applying residual life concepts to GIS – with options for maintenance procedures, repair, refurbishment, retrofit, to be presented to users of GIS in the TB to be prepared by this WG.

Historic note: During 1960s–1980s UK switchgear companies undertook extensive collaborative R&D programmes with utility companies to gather much strategic information undertaking extensive mechanical/electrical and accelerated life and other tests on equipment. They also had significant databases and large R&D staff who could rapidly respond to and evaluate and assist to resolve certain lifetime-related equipment issues – should they arise. The situation has changed and research on this scale is now not generally undertaken (e.g. very minor changes to cast resin (material formulation) can drastically change insulation performances in cast resin/SF_6 arrangements in GIS equipment).

See a recent *Electra* article, 'Residual Life Concepts Applied to HV GIS Substations', prepared by CIGRE WG B3.17 (see *Electra*. 2012;**262**(June):66–71), who also prepared CIGRE TB 499-WG B3.17 with the same title. This article discusses factors considered (in this TB) together with associated operational risks. Guidance on the evaluation of the residual life of any particular HV GIS installation is offered. Alternative strategies to replacement (i.e. retrofit) can sometimes be a solution.

CIGRE WG B3.17 decided that the TB would also include:

- Impact of GIS materials, design aspects, type testing manufacturing, routine testing, installation, commissioning, maintenance, monitoring and diagnostics together with the operating duties and environmental conditions to which the GIS is subjected. It was also recognised that the determination of end of life can often be influenced by external factors such as changes in future system requirements as well as certain service aspects.
- The scope of the TB includes:
  - identification and discussion of the factors which determine the expected residual life of HV GIS
  - guidance on evaluating the expected residual life of individual installed equipment

- options for extending the residual lifetime (enhanced maintenance, refurbishment vs replacement of individual pieces of equipment and/or the whole HV GIS)
- end-of-life procedure for HV GIS – re-use, disposal including SF_6 gas handling.

The *Electra* article shows two examples*:

1. A photograph of partial retrofit of new circuit-breakers to existing GIS.
2. A photograph of GIS with circuit-breaker drive retrofitted.

It is reported that numerous practical examples (case studies) are included in the TB which demonstrate different aspects of residual life and the choices made by users.

[*See original articles for photographs or figures.]

The *Electra* article comments on recommendations in the brochure:

- To support the assessment process and make available additional options for life extensions recommendations are sent out to original equipment manufacturers (OEMs) and also to users.
- Today, the limits of lifetime cannot be determined for the GIS from available statistics as it is not yet known when and how rapidly the failure rate will increase.
- The CIGRE WG presume that this may differ for each generation of GIS. Consequently, the WG identified a need to gather more information in the future about each and all generations of GIS and collate and observe behaviour over time.
- Predictably the main issues of current concern identified are SF_6 leakage of first-/second-generation GIS and corrosion for outdoor GIS. WG comments that it will be important to observe how these issues develop and evaluate their impact on GIS Residual life. (Further work to better manage corrosion and moisture ingress is recommended by the WG.)
- The extension of GIS – especially older GIS installations can be challenging. Consequently, one recommendation by the WG is to develop a standard interface to allow connection of GIS from different manufacturers. (This could possibly form a topic of future work.)

Note: Should he/she wish to follow-up on any of these aspects, the reader should use the full CIGRE TB 499, 'Residual Life Concepts Applied to HV GIS Substations'.

E.2 Integral decision process for substation equipment replacement *(TB 486-WG B3.06)*

The above title was used by CIGRE WG B3.06 when it presented an article in *Electra*. 2012;**260**(February):42–9, highlighting TB 486-WG B3.06 the same title). This describes an integral decision process for asset replacement as applied to substations. From the outset, one must recognise that the term *substation asset replacement decisions* encapsulates high-voltage and secondary system support structures, etc. Consequently, one has to evaluate the status of the equipment and also to assess the overall condition of substation infrastructure.

The WG B3.06 comment in the *Electra* article that, instead of focusing solely on asset condition, a top-down approach is adopted by incorporating business drivers, means both equipment ageing and changing system operating conditions are considered. Examples are provided of relevant experience and lessons learned from the field, and serve as an implementation guideline for the asset manager when defining the process dealing with equipment replacements. This appears to be a very thorough document.

E.3 Electra: WG B3.06 conclusions

CIGRE WG presented the following broad conclusions:

1. For substation asset replacement decisions it is necessary to adopt a general viewpoint, as non high-voltage infrastructures (e.g. secondary systems, support structures, buildings etc.) are also included in substations. It is therefore necessary to not only evaluate the status of the equipment but also assess the overall condition of substation infrastructure.

    At this point, the article refers to Figure 3*, which comprises two pictures: *Early control & protection system replacement, performed together with the GIS extension as one large project.* RHS picture shows replacement control and protection system. LHS picture shows GIS extension plus original GIS portion.
2. Replacement of an individual asset is often postponed using possibly temporary solutions, although replacement could possibly imply cost savings; however, as soon as the project is initiated due to, for example the capacity plan, the context changes (Figure 4). In this respect, the capacity plan has a strong influence on asset replacement decisions.

    At this point the article refers to Figure 4* which also comprises two pictures: 'Before' picture shows a 150 kV Y style AIS substation. 'After' picture shows total renewal of an AIS vertical modern style integrated with a capacity plan initiated project (very clean modern open style).
3. The integral nature of the decision process described in the TB results from the consideration of both the maintenance needs of existing assets (due to ageing), as well as of the requirements by the capacity plan (due to grid enlargement of structural changes). This integral approach to substation asset replacement is necessary so as to ensure that high level, general business drivers are systematically applied to all levels of individual replacement projects.

    [*See original article for figures.]

E.4 'Overview of CIGRE publications on AM topics' (CIGRE Technical Report) (Electra. 2012;262(June):44-9; written by E. Rijks, et al. SC C1 Reviewers)

Only very brief comments will be offered now on this CIGRE Technical Report relating to WG C1.25. A TB is expected at end 2012.

The six AM topics covered and the scope are outlined below:

Scope	Topics
Rotating electrical machines	1. Condition assessment and monitoring
Transformers	2. Asset end-of-life issues
High-voltage equipment	3. Asset management decision making and risk management
Substations	4. Grid development
Substation automation systems	5. Maintenance processes and decision making
Insulated cables	6. Collection of assets data and information
Overhead lines	7. Conclusion
System	8. References
System development and electronics	

Table 8.1 (see section 8.1 in Chapter 8) provides the reader a valuable list of useful Technical Brochures (TBs). Each column heading relates to an individual SC, while each row covers subject themes (i)–(vi). For example the first row covers (i) condition assessment and monitoring, and so on. All rows relate to AM topics.

The reader should note that Technical Brochure TB 380 [15] (2009) will be discussed briefly in Appendix B of Chapter 10.

- The large number of Technical Brochures prepared in the AM sector gives clear evidence of CIGRE commitment, as it attempts to prepare the energy sector to optimally manage assets, and their associated performance risks and expenditures over their life cycle, for the purpose of achieving the required quality of service in the most cost-effective manner.
- 'In this process, asset managers develop and assess operating and capital budgets, for mid and long terms for asset sustainment and grid development. The timing of when to make these investments within the planning horizon also needs to be determined due to obsolescence, unprofitable use, or unacceptable business risk', state the *Electra* report authors.
- The work undertaken so far by CIGRE is very impressive. This writer is:
  1. surprised that the industry (utilities, manufacturer, consultants) has still to agree to develop a standard interface to facilitate and allow ease of connection of GIS from different manufacturers;
  2. somewhat concerned that perhaps not enough collaborative effort by manufacturers/utilities in the sector has gone into developing ongoing monitoring R&D evaluation strategies in the past 30+ years and producing a robust scientific measurement technique to accurately calibrate and

determine deterioration rates due to rusting, corrosion, moisture ingress, etc. – bearing in mind the vast number of bay-years of GIS/GIL service history – and the exhaustive CM programmes that have been carried out over the years.

In other words, consideration as to any further CIGRE work 'to better manage corrosion and moisture ingress' should perhaps also include the development of a purpose-built and 'calibrated' chemical-type analytical instrument.

[Supplementary recent material in Chapter 8 and in Appendices B.1, D and E compiled by the Editor, H.M. Ryan]

Note: The CIGRE material reproduced in Appendices D and E stems from this writer's personal experiences (i.e. IEC/CIGRE/IET/HVET/EPSRC/DTI participation over many years) and recent material details are from numerous CIGRE sources (including *Electra* articles or CIGRE-listed website sources). Information is reproduced edited, abridged and summarised in an informal style. In addition, a few useful supplementary personal observations by this writer have been inserted (where considered appropriate, solely for the benefit of the readership of this book, as each student advances along their chosen career pathway), recalling the caveats referred to in the text, and the proven strong beneficial association of being given or directed to relevant strategic information and, naturally, achieving empowerment through knowledge!

Chapter 9
Distribution switchgear
B.M. Pryor

9.1 Introduction

Switchgear is a term used to refer to combinations of switching devices and their interconnection with associated control, measurement and protection equipment. It allows the interconnection of various parts of the electrical network by means of transformers, overhead lines or cables to allow control of the flow of electricity within that network from power station to customer. Switchgear is also designed to be able to safely interrupt any faults that might occur in any part of the network to protect the network itself, associated equipment and operational personnel. It also provides, by means of disconnectors, facilities for isolating sections of the network and, with the provision of earthing switches, to allow the safe application of devices to ensure that the isolated sections of the network are earthed and made safe for maintenance activities or possible fault repair.

A combination of busbar circuits, circuit-breakers, switches, fuse switches, disconnectors, earthing devices, terminations and associated control and protection equipment is referred to as a 'substation'. A substation may or may not include a means of voltage conversion, i.e. one or more transformers.

The typical electrical network may comprise generators feeding directly into part of the transmission system, and the transmission system itself may have interconnection to other transmission networks. Such systems are used to allow the flow of large amounts of power from the generating stations to the distribution load centres, which may be many tens or hundreds of kilometres away.

Distribution of the power from the transmission system will be via bulk supply points, which allow the means of tapping off power to feed local communities via a 'distribution' network.

Substations are used at all interconnecting node points, i.e. generator to transmission system, within the transmission system, transmission system to distribution networks and within the distribution networks themselves. A typical electricity supply network is shown in Figure 9.1.

Within the UK generators typically operate at 25 kV and their output is immediately transformed up to the transmission system voltage of either 400 kV or 275 kV. Tee-off points to feed distribution networks are typically rated at 132 kV

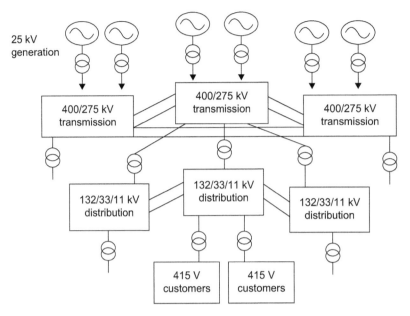

Figure 9.1   Electricity supply network. Interconnection of circuits is by means of substations

or 33 kV. The local distribution network itself, almost universally, operates at 11 kV. Tee-off feeds to customers are transformed from 11 kV to 415 V, with local low-voltage (LV) networks operating at 415 V three-phase.

The foregoing relates to typical power supply networks; however, with deregulation of the electrical power utilities and introduction of greater competition, there is a growing trend to operate high-efficiency small generating plants within the electrical distribution networks, and such networks have to adapt to allow interconnection of these embedded generators. Interconnection, however, is still via local substations.

The subject of this chapter is to discuss the types and combinations of switchgear, which may be encountered within the distribution networks. While in this text the term 'distribution switchgear' will be used, a more recent trend is for it to be referred to as 'medium voltage (MV) switchgear'; the terms are synonymous.

Appendices A and B provide supplementary material relating to CIGRE activities, which often culminate in the preparation and issue of a Technical Brochure (TB). This is widely used within the energy supply sector worldwide. Appendix A lists summary information for a few recent TBs issued by CIGRE Study Committee SC B3, in full designated by CIGRE as B3 Substations (www.cigre-b3.org). Appendix B provides the reader with the scope of activities carried out by one CIGRE Study Committee, namely, SC C6; the designated field of activity set by CIGRE for this SC is specifically 'Distribution systems and dispersed generation' (www.cigre-c6.org). This SC C6 presented a Technical Report at its 2011 AGM and this work has since been published in *Electra*, etc. The annual report is now reproduced in Appendix B in augmented/edited format, additionally including supplementary information as will be touched on later. The reader learns some of the main concerns of SC C6 activities, together with a summary of key information from this AGM. Numerous aspects are covered in this document including CIGRE membership; active distribution networks; strategic direction of CIGRE in general and SC C6 in particular; reports on some completed strategic work are identified; comprehensive information is provided on a TB for a wide range of issues including demand side management methodologies; application of energy storage systems; distributed generation (DG); renewable energy sources (RES); demand response (DR); microgrids; grid integration of wind generation; technologies for rural electrification; planning and optimisation methods for active distribution systems; coping with limits for very high penetrations of renewable energy and much much more.

9.2 Substations

There are a number of methods of construction of substations (supplementary information is provided in Chapter 9 of 2nd Edition by P Fletcher[1]). Common types are described below.

9.2.1 Substation types

9.2.1.1 Open air

Open-air substations comprise separately mounted interconnected switching devices and components, e.g. busbar support insulators, current transformers (CTs), voltage transformers (VTs), cable sealing ends, etc., where atmospheric air provides the main insulation path to earth. The solid insulators clearly have to be designed to be capable of withstanding, for many years, all encountered environmental conditions, e.g. rain, ice, snow, wind loading, temperature variations, pollution, lightning activity and associated switching activity to ensure that dielectric breakdown to earth or between phases is improbable. This generally requires relatively large clearances and, in consequence, open-air substations tend to cover large ground areas and may also be unsightly if located near local communities. Some degree of landscaping is normally required to minimise the visual impact.

9.2.1.2 Metal enclosed

With this arrangement all switching devices and associated components are enclosed in a metallic earthed structure on a per-bay basis and are cable connected. This arrangement has the disadvantage that there is no segregation within the panel; hence, a fault in an instrument transformer, e.g., could readily spread to encompass the switching device and panel interconnecting busbars. Such arrangements are not generally used where high-reliability and high-availability systems are required, i.e.

---

[1] Fletcher, P. 'Transmission substations' in Ryan H.M. (ed.), *High Voltage Engineering and Testing*. 2nd Edn.

at bulk supply points or at primary distribution substations. They are, however, widely used for distribution secondary substations, e.g. ring main units (see section 9.5.11). Such arrangements have the advantage that they may be physically small in size and only require a fraction of the land area required by an equivalent open-air substation. Metal-enclosed substations may be either of the outdoor type, in which case the enclosure must protect the internally connected equipment against all prevailing environmental conditions, or alternatively the substation may be enclosed within a building or low-cost weatherproof housing.

9.2.1.3 Metal-clad

This is a derivative of the metal-enclosed switchgear where all major components per bay are physically segregated from one another by means of earthed metalwork such that a fault in any one compartment cannot readily spread to adjacent compartments. It is vitally important that busbars are unaffected by a fault in any particular circuit panel so that adjacent panels can safely remain in service until appropriate repairs can be executed. Typical segregation is between cable box and instrument transformers, instrument transformer chambers and circuit-breaker and circuit-breaker and busbars. Such arrangements allow the achievement of higher reliability and higher availability systems than may be achieved with metal-enclosed concepts. Most primary switchboards are of the metal-clad type. For economic reasons many such switchboards were originally designed for outdoor use; however, in harsh environmental climates environmental deterioration may readily occur and maintenance costs may be high. It is now common policy for such metal-clad switchgear to be housed in appropriate weatherproofed buildings. A further modern trend is for them to be housed in transportable containers such that, as systems and local load patterns change, they can readily be transferred to more appropriate locations.

9.2.1.4 Insulation enclosed

With this type of construction, all high-voltage (HV) components are enclosed within insulation without external metallic screens. The insulation is generally of the solid moulded form with components interfaced by atmospheric air. Correct dielectric stressing is vital with such combinations of components. While the equipment may generally be physically smaller than metal-clad types of switchgear, the high dielectric stress, under varying humidity conditions, can, if not properly controlled, lead to early failure mechanisms.

Such equipment is not generally utilised within the UK but is, however, employed quite widely within some European countries.

9.2.1.5 Gas-insulated substations (GIS)

A further form of metal-clad substation that is used mainly at transmission voltages is the gas-insulated substation. With this concept SF_6 provides the main dielectric medium between the primary conductor and earth. In view of the very high dielectric strength of SF_6 the dimensions can be made very small. There are two basic concepts: the first is where each phase is physically enclosed within earthed metalwork and the other is where the three phases are enclosed within one chamber. The former has a higher degree of integrity as phase-to-phase faults cannot occur. The latter, however, provides a more economic and more compact concept. Both arrangements are widely used, with the phase-isolated concept being used mainly at higher voltages and the three phases in one tank concept being used at the lower transmission, or high distribution, voltages, i.e. 132 kV. Some manufacturers have, however, utilised the phase-isolated concept down to voltage levels of 33 kV, and there are a number of such installations within the UK.

9.2.1.6 Insulation considerations

Correct insulation design is vital for all substations; this applies not only for open-air types and insulation-enclosed types but also for metal-enclosed and metal-clad types, which employ many insulating components. For example with metal-clad equipment connections must be provided to connect one compartment to adjacent compartments; such connections are usually by means of bushings (see Chapter 8). Insulation has also to be provided between phases and between phase and earth. Typical insulating materials that may be employed are atmospheric air, insulating oil, bituminous compound, oil-impregnated paper (OIP), synthetic resin bonded paper (SRBP) epoxy resin or SF_6. The correct design of insulating interfaces between components and different insulation materials is vital for the long-term service integrity of the equipment.

9.2.2 Substation layouts

9.2.2.1 Primary distribution substations

Primary distribution substations within the UK typically operate at 11 kV and are utilised to supply a relatively large number of consumers within a local area. They may be interconnected with other primary substations or fed directly from a 33/11 kV bulk supply point. Outgoing circuits may feed either rural or urban networks. A primary substation feeding a rural network is commonly of the open-terminal air-insulated design, whereas a primary substation feeding urban networks is more commonly of the indoor meta-clad design.

A common electrical arrangement for either substation is of the single busbar design but with the busbar being split into two sections and interconnected via a bus – section circuit-breaker. There are usually two incoming circuits – one feeding each section of busbar. There may be typically five outgoing circuits feeding either multi-radial networks for overhead rural systems or ring circuits for urban cable-connected networks.

To enable isolation from the system for maintenance purposes it is common for disconnection facilities to be provided on either side of the circuit-breaker (CB). Facilities will also be provided for earthing outgoing or incoming circuits and for earthing each section of busbar. CTs will invariably be fitted within the outgoing circuit for either protection or, less commonly, for tariff metering purposes, the latter facilities usually being provided for large single customers.

A typical single busbar primary layout is shown in Figure 9.2.

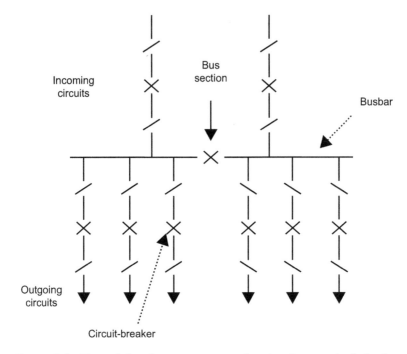

Figure 9.2   Typical distribution primary substation layout: single busbar

9.2.2.2 Bulk supply points

For large important supplies, that is as at bulk supply points, which in the UK typically operate at 132/33 kV, or perhaps for large industrial customers, higher security may be built into the substation design. This is usually achieved by means of the provision of a duplicate busbar system. Such an arrangement is shown in Figure 9.3. This allows the option of four separate sections of busbar to be utilised such that if any one busbar fails, supplies could generally be maintained via the other three busbars. All circuits can be selected to either of the two main busbar sections. Busbar protection is usually provided to ensure very rapid clearance of any faulted section of busbar. However, such arrangements are very costly and are suitable for commercial customers with 'complex load' situations (e.g. chemical processing plant) where continuity of electricity supply is paramount (e.g. see section 1.3.1). Security of supply criteria is established at the network design stage with a knowledge of customer requirements. In more 'usual load situations', as modern distribution network equipment is very reliable, the same degree of availability can often be achieved without having to resort to a second busbar system. It is also necessary to ensure that each section of busbar is normally kept energised. If a section remains de-energised for any period of time then deterioration may occur such that possible busbar failure may result on re-energisation.

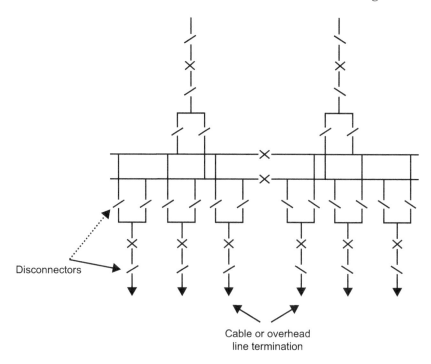

Figure 9.3 Typical distribution primary substation layout: double busbar

9.2.2.3 Secondary substations

Secondary substations are almost universally used in urban networks and are usually connected into ring circuits. The purpose of the secondary substation is to provide a feeding point to local customers at low voltage – typically 415 V three-phase or single phase. Hence they always employ an 11 kV/415 V three-phase transformer, which may have ratings of typically up to 1,000 kVA. The transformer HV is fed from a tee-off from the 11 kV ring main circuit. This ring main equipment is usually of the metal-enclosed form and, while built to be weatherproof, they are now usually enclosed within a small prefabricated weather-resistant housing. The complete substation is usually enclosed within a fenced-off area. It should be noted that the main purpose of the weatherproof housing is to reduce the maintenance requirement otherwise needed for environmentally exposed equipment. The weatherproof housing is not intended to provide safety to the operator or to third parties. Safety is built into the ring main unit itself. The ring main unit is described in more detail in section 9.5.11.

9.3 Distribution system configurations

9.3.1 Urban distribution systems

A typical urban distribution system is shown in Figure 9.4.

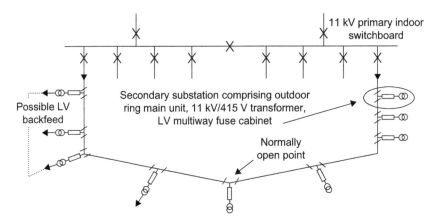

Figure 9.4 Urban distribution system

The system is fed from a primary distribution switchboard as described earlier. Each outgoing feeder on one section of the busbar feeds via an 11 kV cable network to typically 10–12 secondary distribution substations, which are electrically connected within the ring circuit. This ring circuit is connected back to a feeder on the adjacent section of busbar at the same substation. The ring circuit usually has a normally open point. The purpose of this is to minimise the number of customers affected by the faulted section of the ring circuit. Once the faulted circuit has been located it can be isolated and earthed, via the ring main units, to allow safe repair work. In the meantime the normally open point can be closed to re-energise the maximum possible number of customers. The ring system normally operates with the primary switchboard bus-section circuit-breaker in the open position.

Ring main switches need to be manually operated; hence, fault location and customer re-energisation can be time-consuming. A modern tendency is for the ring main switches to be fitted with remotely operable mechanisms such that switching times can be reduced by allowing remote substation facilities or, with modern intelligence systems, to allow automatic de-energisation and isolation of a faulted section and re-energisation up to the point of the fault.

The tee-off point of the ring main unit feeds via an 11 kV/415 V three-phase transformer to an LV fuse board, typically having up to five outgoing circuits which feed directly to large customers or groups of customers. In most of these LV circuits it is possible to achieve an LV backfeed from an adjacent ring main unit, thus maximising the number of customers on supply while fault repairs are in progress.

The general network principles described here operate in most industrialised countries, although some may employ radial as well as ring circuits. Typical distribution voltages used elsewhere may range from 10 to 20 kV. Six point six kilovolt systems were common at one time within the UK but these have now largely been phased out.

9.3.2 Rural distribution systems

A typical rural distribution system is shown in Figure 9.5. The primary switchboard will usually comprise two sections of busbar connected via a bus-section circuit-breaker as described earlier. The number of outgoing feeders tends to be less than that with an urban network with typically six panels. The primary switchboard in this case may be of the open terminal form or in the outdoor weatherproofed form. However, for reasons given earlier it is now more common for these primary switchboards to be enclosed within a brick-type enclosure, in which case conventional indoor metal-clad type switchgear is more commonly used. Most 11 kV rural primary switchboards are of the metal-clad type, whereas at 33 kV open-type busbar-connected switchgear is more common.

Most rural networks in the UK feed overhead line distribution circuits; hence, there is usually a short connection of cable connecting from the switchboard to the terminal pole of the overhead line. It is also common for surge arresters to be fitted at the cable sealing end and mounted physically at the top of the pole at the junction between the cable and the overhead line. Surge arresters reduce the probability of transient induced overvoltages resulting from lightning strikes to the overhead line affecting either the cable or the associated switchgear.

The overhead line feeders are generally of a radial form, although in some more densely populated rural areas interconnection from radial circuits may be achieved to effectively produce a combination of both ring and radial circuits. The radial circuits will usually have numerous spur lines feeding to local small communities. While the radial circuits will comprise a three-phase system, spur circuits often constitute a two-phase (single-phase) supply. It is necessary to ensure that all customers do not lose supply for a fault on any of the radial or spur circuits.

Traditionally most spur circuits have been protected by a device known as an expulsion fuse (see section 9.5.9.1). This comprises an 11 kV rewirable fuse in which, when it operates, its fuse-carrier hinges about a bottom bracket to fall to its

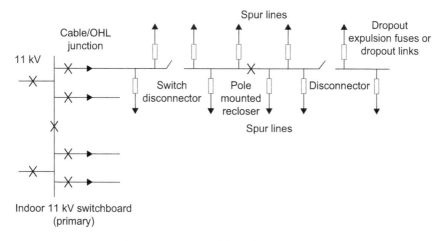

Figure 9.5 Rural distribution system

operated position while at the same time providing a visual indication of isolation. However, many faults on 11 kV overhead line networks are transient in nature, e.g. a tree blowing on to a line, a foreign object on the line, a bird between lines, etc., and once the transient fault has been cleared it is often safe to re-energise the line. With expulsion fuses this can take some time, typically up to 1 h, for engineers to be notified arrive on the site to locate the operated fuse and repair and replace it to reinstate supplies. A more recent development to overcome this problem has been what is referred to as an 'intelligent link'. This is mounted in the expulsion fuse base and counts the passage of fault current to allow the substation primary circuit-breaker to trip and reclose; after three pulses of fault current it will assume that the line has a permanent fault, and an electrically charged circuit will fire a release pin to allow the carrier to drop to an isolated position within the de-energisation time of the primary circuit-breaker.

For particularly long lines it is of concern if all customers experience the temporary loss of supply while the feeding primary circuit-breaker auto-recloses. For this reason it is then more common for a pole-mounted auto-reclosing circuit-breaker to be located some half-way along the overhead line feeder. Such devices used to derive their closing energy from a falling weight type of operating mechanism; such a system clearly needed rewinding after it had completed its charge of typically 10 closing operations. Modern versions of pole-mounted auto-reclose circuit-breakers are solenoid operated and receive their closing energy from the feeding primary circuit at 11 kV. Pole-mounted auto-reclosing circuit-breakers were traditionally of the oil-filled type and were protected against transient over-voltages by means of 'duplex' spark gaps, i.e. two gaps in series to minimise breakdown by the presence of birds or foreign objects. Modern types of pole-mounted auto-reclosers are either of the vacuum-type circuit-breaker or of SF_6-type circuit-breaker and require to be protected against internal breakdown by metal-oxide surge arresters (MOSAs) (see Chapters 2, 3 and 8).

Another device that is commonly fitted in series with the main rural feeder is a disconnector. This is generally manually operated from the bottom of the pole by a long metallic operating drive rod and series-connected insulated rod. The disconnector has no breaking capacity and should therefore be operated only at a time when the line is de-energised to provide a point of isolation such as to allow work to be safely undertaken downstream of the disconnector. A further device that may be fitted is a switch disconnector. This is a device for interrupting load current but not fault current. The switch disconnector usually has some form of rudimentary arc control device to enable it to interrupt the load current flowing in the line, i.e. a switch disconnector can be operated with the line energised. Its interrupting medium is usually atmospheric air and gases produced within its arc control device (see section 9.5.6).

9.4 Ratings

All switching devices have 'ratings' and it is necessary to understand how equipment is rated. A brief description of some of the typical ratings encountered is given in the following subsections.

9.4.1 Rated current

This is the continuous current that equipment is capable of passing without exceeding its temperature rise limits. Such limits are defined in the appropriate international specifications.

Typical rated currents encountered for older equipment would be 400, 800, 1,200 and 2,000 A.

Newer equipment is rated on the basis of preferred numbers, which are related to the 'Reynold' series, with the R10 series being the most commonly used for switching equipment.

Typical current ratings encountered for newer equipment would thus be 630, 800, 1,250, 2,500 A, etc.

9.4.2 Rated short-circuit-breaking current

The rated short-circuit-breaking current is the RMS value of the short-circuit current that a circuit-breaker is capable of breaking at its rated voltage to the prescribed conditions of the appropriate international specification, which for circuit-breakers is IEC 62271-100.

Typical values of rated short-circuit-breaking current to be found on older switchgear equipment are 7.9, 13.1, 18.4 and 26.2 kA. For newer equipment the appropriate R10 ratings are 8, 12.5, 16, 25 and 31.5 kA.

It should be noted that older equipment were commonly rated in terms of their MVA capability. The MVA rating is equivalent to the rated voltage × the rated short-circuit-breaking current × $\sqrt{3}$.

9.4.3 Rated short-circuit-making current

The rated short-circuit-making current is the peak value of the rated short-circuit current that the circuit-breaker is capable of closing against, at 0.1 power factor, i.e. it has to close against the maximum current asymmetry.

The asymmetrical factor commonly used is 1.8. This would thus give a making current of $1.8 \times \sqrt{2} = 2.55 \times$ RMS value of current. For more modern equipment a 'rounded' value of 2.5 is used.

9.4.4 Rated asymmetrical breaking current

This is the value of asymmetrical breaking current that a circuit-breaker is capable of interrupting under typical system asymmetric conditions. While traditionally a value of a direct current (DC) component offset equivalent to 50% of the RMS value of the rated short-circuit-breaking current has been used, the modern tendency is to specify the asymmetric current in terms of the system X/R ratio. Typically, a ratio of 14.5 has been used, but IEC 62271-100 recognises that more onerous conditions may exist and in such cases it is necessary for the user to specify the required value.

9.4.5 Rated short-time current

This is the rated short circuit through current that a switching device and its assembly are capable of safely withstanding for a rated time. The peak value of the rated short-time current must be equivalent to the value of the rated short-circuit-making current.

In the UK for distribution switchgear the rated time is normally 3 s. For transmission switchgear and for distribution switchgear used elsewhere a rating of 1 s might only be required.

9.4.6 Rated voltage

The rated voltage of an item of equipment is the RMS voltage that the equipment is capable of withstanding continuously, and it is the value used to verify all performance criteria. The rated voltage relates to the maximum permissible system voltage, e.g. 10 kV or 11 kV system equipment would have a rated voltage of 12 kV, similarly for 33 kV systems the equipment-rated voltage would be 36 kV.

9.4.7 Rated insulation withstand levels

These are the values of overvoltage that the equipment is capable of withstanding for a short duration. They are usually defined in terms of two criteria for distribution switchgear, i.e. the 1 min power frequency withstand voltage and the rated lightning impulse withstand voltage. For 11 kV equipment, e.g., the rated power frequency overvoltage might be 27 kV RMS for 1 min and the rated lightning impulse voltage may be 75 kV peak, or alternatively 95 kV peak.

The rated lightning impulse voltage comprises a double exponential wave shape having a rise time to its peak value of 1.2 μs and a time to half-life of 50 μs. Electrical power equipment is required to withstand 15 such lightning impulses of both positive and negative polarities. Such tests are clearly intended to simulate lightning-induced overvoltages entering into the switchgear. For transmission switchgear switching overvoltage tests are also required. These are similar to lightning impulse tests but have a wave shape of 250/2,500 μs.

9.4.8 Rated transient recovery voltage

When a circuit rapidly changes from one steady-state condition to another it cannot do so without inducing transient overvoltages and associated current surges. Since circuit-breakers invariably clear at zero current the overvoltage produced at that point will be dependent on the rate of change of current, the circuit inductance and stray capacitance. For 11 kV equipment, e.g., the rated transient recovery voltage would have a peak value of 20.6 kV with a time to its peak value of 60 μs. The rate of rise of voltage is particularly critical and for the values quoted this would equate to a rate of rise of 0.34 kV/μs.

There are many other rated values for switchgear but the ones described above are typical of the information that might be encountered on an equipment rating plate. For further information the reader is referred to IEC 62271-100 High Voltage Alternating Current Circuit-Breakers.

9.5 Switching equipment

The foregoing sections have described distribution networks and some of the electrical switching equipment that may be encountered together with typical equipment-rated values, which may be encountered on the equipment rating plates. These describe the performance criteria. It should be noted that, by themselves, these ratings do not necessarily ensure safety. Safety provisions must be built in to equipment, first in accordance with local regulations and then in accordance with the utilities' own requirements. Similarly it might be necessary for the utility to specify the operational features that are required to be incorporated.

A variety of terms are used to describe such equipment, and it is important that the power engineer fully understands the meaning of these terms and the performance and operational differences between the different types of equipment, which may be encountered.

A brief description of such equipment follows.

9.5.1 Circuit-breakers

A circuit-breaker is a device that, in addition to carrying, and making and breaking, its rated load current, is also capable of interrupting, when energised via suitable protection relays, its rated short-circuit-breaking current. It is also capable of closing against its rated short-circuit-making current.

Extremely large magnetic forces are set up when a circuit-breaker attempts to close against the system short-circuit current, and it is essential that the operating mechanism has sufficient stored energy to impart into the moving contact system to ensure that it fully closes and is 'latched' closed. There are a number of different types of operating mechanisms that may be encountered, these being typically as follows.

9.5.1.1 Manually operated mechanisms

Many older circuit-breakers required direct manual operation where the closing energy and operating force were entirely dependent on the strength and skill of the operator. Failure to close with sufficient force to overcome the short-circuit throw-off forces results in the circuit-breaker contacts opening, or being held in a partially open position, and circuit-breaker failure is inevitable. This is an extremely dangerous situation for the operator. In consequence, manually operated circuit-breakers should be removed from the system or operated only when the system is de-energised. Alternatively, it may be possible for the direct manual operating mechanism to be replaced with a stored energy type of operating mechanism, but in this case further type test verification will be required.

Direct manually operated circuit-breakers should not be closed with the primary system energised.

Incidentally, virtually all circuit-breaker operating mechanisms employ an arrangement of mechanical linkages, toggles and latches. With the circuit-breaker in a closed position, a latch engages within the mechanism of toggle arrangement to mechanically hold the circuit-breaker closed. Tripping is achieved by means of a DC-operated coil supplied from the substation battery. The tripping energy is most commonly derived from springs, which are charged during the closing operation of the circuit-breaker.

9.5.1.2 Independent manual spring-operated mechanisms

In view of the inherent problems with manually operated mechanisms manufacturers found ways of producing a low-cost operating mechanism that was still manually operated but where, during the early part of the closing operation, a closing spring is charged; halfway through the operation the charged springs travel over-centre and closing is achieved by the energy imparted into the springs such that the closing force and speed are independent of the operator. Such mechanisms are generally only used at 11 kV and where the equipment short-circuit rating does not exceed 12 kA.

9.5.1.3 Dependent spring-operated mechanisms

Circuit-breaker closure by manual means with an operator standing in front of the circuit-breaker is not always desirable, particularly as it requires an operator to travel to site for closing the circuit-breaker. Hence, an operator-independent means of closing was required.

This is achieved in a number of ways, the cheapest and often the most preferred way of closing being to prestore energy into closing springs. This can be achieved either manually or by means of a motor gear drive arrangement. The operating mechanism is then held in a 'charged' position by means of a latch, which can be released to allow the circuit-breaker to close on operation of a DC-operated closing coil. When the springs have been released to allow circuit-breaker closure the operating mechanism is mechanically latched in its closed position such that the closing springs can again be recharged, either manually or by remote operation. It is thus possible to 'store' a charge within the closing springs such that if the circuit-breaker trips it is immediately ready for reclosure to provide an auto-reclosing feature.

Spring-charged operating mechanisms are widely used in all circuit-breakers and there is now a modern tendency for them to be used for higher short-circuit ratings and even on transmission circuit-breakers.

9.5.1.4 Solenoid-operated mechanisms

For many years an alternative to spring-operated mechanisms was a solenoid-operated mechanism. This comprises a very large DC-operated solenoid coil whereby the moving solenoid moving plunger directly drives the operating mechanism linkage to close the circuit-breaker. Closing times are generally slow and large battery drains ensue particularly if the circuit-breaker is used for an auto-reclosing duty. Such mechanisms are rarely used these days but may well be encountered on older switchgear up to 132 kV.

With most solenoid-operating mechanisms the circuit-breaker can be 'slow' closed by means of a manual lever operating on the solenoid plunger. There are usually no facilities to prevent such a manual operation on a live circuit.

It is vitally important that a solenoid-operating mechanism should not be used to allow manual closure of a circuit-breaker onto a live circuit.

9.5.1.5 Magnetically operated mechanism

Over recent years there have been very rapid developments in the technologies and manufacturing processes for magnetic materials, so magnetically operated circuit-breakers are returning into fashion. These are being applied particularly for vacuum circuit-breakers where small contact travels are encountered. In this case the closing and hold-on force is derived from a permanent magnetic pole. To achieve closing or opening a small coil is used to bias the pole in one direction or the other. Such mechanisms alleviate the need for toggle linkages and can be produced very economically.

9.5.1.6 Hydraulically operated mechanisms

With this type of operating mechanism the closing energy is stored in an accumulator, which comprises a high-pressure cylinder with a centrally located piston. On one side of the piston is a gas, which is usually nitrogen and on the other side is hydraulic oil. The oil can be fed via a pump to move the piston to compress the nitrogen. Closing energy is thus stored in the compressed nitrogen. A series of valves is provided, the first of which is operated by a closing coil. The first stage valve then allows high-pressure oil into a second-stage valve, which in turn typically operates a third-stage valve to allow large quantities of high-pressure oil to flow to an actuator piston. The piston then moves under the pressure of the oil to drive the circuit-breaker contacts to their closed position. The circuit-breaker is usually held closed by a differential pressure piston. To trip, a valve is operated via the circuit-breaker trip coil to allow the high-pressure oil to be dumped to the low-pressure reservoir. Typically two close–open operations can be stored within the accumulator.

Such operating mechanisms tend to be expensive and while widely used at transmission voltages are seldom encountered at distribution voltages.

A simpler variant is sometimes found at distribution voltages whereby a hydraulic ram is used to charge the springs of a spring-operated mechanism.

9.5.1.7 Pneumatically operating mechanisms

This type of mechanism is similar to the spring-operated mechanism but uses a pneumatically operated piston to drive a conventional toggle/linkage mechanism to close the circuit-breaker. A pressurised air receiver is required to store the energy, which typically is released to the closing piston via a closing coil-operated three-stage valve assembly. While such mechanisms may occasionally be encountered in distribution applications, they are nevertheless widely utilised within transmission switchgear applications.

9.5.2 Distribution circuit-breaker types

The types of distribution circuit-breaker encountered will briefly be described, but detailed description of arc extinction mechanisms is given in Chapter 6.

9.5.2.1 Bulk oil plain break circuit-breaker

This was an early design of circuit-breaker in which all three phases were enclosed within a tank of oil. The circuit-breaker contained no specific means of arc control. Arc extinction was determined solely by the oil characteristics and the pressure rise within the circuit-breaker tank. For three-phase rated short-circuit currents of up to 12 kA such circuit-breakers operated satisfactorily since high pressures were built up within the circuit-breaker tank to assist in the arc extinction process.

Under single-phase or low-current fault conditions, however, insufficient pressure may be built up to extinguish the arc. This results in long spindly arcs being produced, which may continue until the circuit-breaker contacts reach the end of their travel, after which interruption is solely fortuitous and failure may result. In consequence, plain break circuit-breakers should be removed from the system or should only be operated in a de-energised state, that is to be used solely as a disconnector.

Plain break circuit-breakers should be removed from power systems.

9.5.2.2 Bulk oil arc control circuit-breaker

The shortcomings of plain break bulk oil circuit-breakers were soon realised and solid insulated assemblies were developed to enclose the arc produced at each set of interrupter contacts. These are referred to as arc control devices whereby for high short-circuit currents very high pressures are built up in the top section of the arc control device as a result of vapourisation of the oil. The high pressure results in rapid arc extinction. Pressure is released usually via two or more apertures in the region where the high pressure is developed, towards the top of the device. For lower fault currents insufficient pressure is produced in the upper section of the device to extinguish the arc, and much longer arcs are drawn out into the lower, more constrained region of the device. This allows a higher pressure to result in this lower region. There are usually interconnecting chambers within the device, which allow a flow of oil upwards and across the upper region of the arc to assist in cooling and arc extinction. This cross jet flow of oil lends its name to the arc control device, which is commonly referred to as a 'cross jet pot'. Such devices allow satisfactory high short-circuit current ratings to be achieved and at the same time allow safe interruption of both single-phase and low-current fault conditions (see Figure 9.6).

Notwithstanding the satisfactory development of such arc control devices, even in the mid-1930s, plain break circuit-breakers continued to be manufactured up until the early 1960s. This was due to their significantly lower costs.

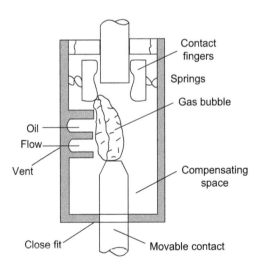

Figure 9.6 Oil circuit-breaker arc control device

9.5.2.3 Minimum oil circuit-breaker

The development of the 'cross jet pot' allowed development of an alternative lower cost form of oil circuit-breaker. This is referred to as a minimum oil circuit-breaker where each phase has its own independent oil chamber, which is of a relatively small volume. As the chamber is at line potential it must be supported from earthed metalwork by some form of insulator and, similarly, its outgoing HV terminal must be insulated. Such devices are referred to as live tank minimum oil circuit-breakers. While for distribution purposes they can be made small and enclosed in metal-clad switchgear constructions, they have one significant disadvantage. In view of the small oil volume, oil carbonisation rapidly occurs on fault interruption and, in consequence, more frequent maintenance is required than would be necessary for a bulk oil-type circuit-breaker. If maintenance is not undertaken at the appropriate time, internal electrical tracking and deterioration can occur on insulating surfaces within the oil chamber which, if not corrected, will lead to catastrophic failure. For this reason minimum oil circuit-breakers have not been widely applied within the UK. They are, however, widely utilised within many European countries.

9.5.2.4 Air-blast circuit-breaker – sequentially operated

Air-blast sequentially operated circuit-breakers are devices that require a blast of high-pressure air to cause extinction of the arc across the opening circuit-breaker contacts. The high-pressure air is usually stored in an air receiver on which are mounted the insulator support and interrupter assemblies. On being required to trip, the operating mechanism will mechanically open a valve, which allows the high-pressure air to enter the contact chamber. This high-pressure air operates a piston connected to the moving contact assembly, which causes the contacts to open and at the same time allows a blast of air to flow across the opening contacts to clear the fault. Following the passage of the flow of air the contacts reclose under the influence of a spring compressed during the opening operation. Such circuit-breakers cannot provide permanently open contacts, and during the arc extinction process it is necessary for a mechanically driven sequential disconnector to operate to provide an open gap before the contacts reclose. Closure of the circuit-breaker is achieved solely by closing the sequential disconnector.

The advantages of such circuit-breakers are that they provide rapid fault clearance under all fault conditions and clearance is usually achieved within one cycle of arcing. They are also ideal for, more economically, providing a means of clearing both high fault levels and low inductive fault currents. Such circuit-breakers are rarely found on distribution networks but may occasionally have been utilised at 11 kV, or more commonly 33 kV, for shunt reactor switching duties.

9.5.2.5 Pressurised head air-blast circuit-breaker

These are devices whereby the contacts are held open by pressurised air and the circuit-breaker does not, in consequence, require a series sequential disconnector. These circuit-breakers are used only at transmission voltages where very rapid fault-clearance times are required. They will not be found in distribution circuits, but may be encountered at 132 kV bulk supply points.

9.5.2.6 Free-air circuit-breaker

This is a device that is used solely at 11 kV and below. It can generally achieve very high current ratings, i.e. 4,000 A and very high short-circuit ratings of typically 50 kA. In consequence it is seldom used in distribution applications and also tends to be very costly. Its main advantage is that it uses nonflammable components and, with its high ratings, is ideally suited for use in auxiliary circuits of power stations. In the oldest UK power stations this is the only form of circuit-breaker that will be found for auxiliary circuits.

Its contacts operate in free air; on opening, an arc is pulled out for the full length of the open contacts. The arc is then transferred to runners that cause it to rise; it then enters an arc chute, which is a device comprising a large number of vertically rising steel plates each being physically separated by a distance of 6 mm or so. As the arc enters the arc chute, it is split up into numerous small arcs in series, each of which is cooled by the plates. The arc rises within these plates and is extinguished by the time it reaches the top of the arc chute. Such circuit-breakers tend to have relatively long fault-clearance times but have been proven to be very reliable (see Figure 9.7).

A further version of this device is commonly found in LV networks, and such devices can be designed to achieve very rapid fault-clearance times by chopping the current to zero even before it reaches its first peak value. These free-air circuit-breakers are additionally referred to as 'current-limiting' circuit-breakers. Current chopping features, however, cannot be achieved for voltages above 1,000 V.

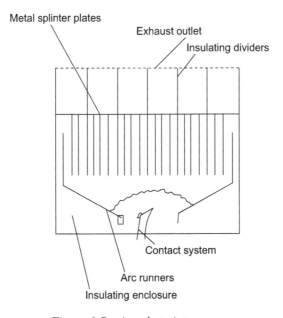

Figure 9.7 Arc chute interrupter

9.5.2.7 Vacuum circuit-breaker

Up until the 1960s there was no real economic alternative to the use of bulk oil circuit-breakers for distribution applications. It had long been known, even from the 1920s, that contacts opening in a vacuum would not sustain an arc. This feature was never employed for circuit-breakers due to the difficulties of providing a sealed chamber that could hold a vacuum, without leakage, for the circuit-breaker lifetime of at least 30 years.

Post–Second World War saw very rapid developments in television technology, which required the tubes to be operated with a sustained vacuum. Television tube technology was thus applied to produce a sealed insulation enclosure for opening the contacts of a vacuum circuit-breaker. Movement of the circuit-breaker moving contacts is achieved by means of stainless steel bellows. The technology was proven to work, but much refinement of contact design and materials was required before the technology became commercially available (see Figure 9.8).

This refinement has continued ever since, to the stage where economic, reliable designs of vacuum interrupters and associated operating mechanisms are widely available and will be found in many 11 and 33 kV distribution switchgear applications.

The designs have been perfected to such an extent that their application can be found even in 415 V circuits.

Unfortunately, it is not physically and economically possible to produce a vacuum interrupter 'bottle' for voltages higher than 33 kV. While vacuum interrupters have been used for high-voltage applications, in particular at 132 kV, these concepts comprise a number of 33 kV interrupters in series, and in consequence vacuum technology at higher voltages is not economic.

9.5.2.8 Dual-pressure SF_6 circuit-breakers

Where SF_6 was first used for circuit-breaker interruption it was used in a concept similar to that of an air-blast circuit-breaker. With this arrangement SF_6 was used at a

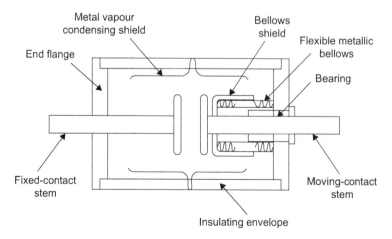

Figure 9.8 Cross-section of a vacuum interrupter

relatively low pressure as an insulating medium across the open contacts and from the contacts to earth. For circuit interruption high-pressure SF_6, stored in a separate pressure vessel, was allowed to blast through the interrupters and across the opening contacts to extinguish the arc in a similar manner to that of an air-blast circuit-breaker. The concept worked very well but it needed a dead tank construction of circuit-breaker with external bushings connecting to the HV circuits. The design was, in consequence, costly and even more so as it required a separate high-pressure SF_6 storage vessel. At the high storage pressure, of typically some 10 bar, the SF_6 would liquefy at low ambient temperatures. To prevent this happening heater tapes were wound around the high-pressure cylinder to ensure that the SF_6 was always in its gaseous state. Failure of the heater, however, could result in circuit-breaker failure, and hence heater-monitoring circuits were required.

In view of these shortcomings, and in view of the rapid developments of other SF_6 interruption technologies, the dual-pressure SF_6 circuit-breaker had only a short time of availability. The concept, however, has proven to be successful and, even now, some 132 kV dual-pressure SF_6 circuit-breakers still remain on the system within the UK.

9.5.2.9 Self-generation gas-blast (puffer) SF_6 circuit-breaker

The next generation of SF_6 circuit-breakers still employed the same principle of arc interruption as the dual-pressure circuit-breaker, but in this case there was no requirement for storage of the high-pressure SF_6. SF_6 within the interrupter was used at a slightly higher pressure than for the dual-pressure circuit-breaker, typically 6–6.5 bar gauge. The high-pressure blast was generated by means of gas compression from a piston within the interrupter during the opening stroke – hence the name self-generating blast or puffer.

There then followed rapid developments of designs of interrupter, nozzle and gas flow principles to the extent where the partial duo-blast concept was successfully developed, allowing higher interruption capabilities to be achieved with a very short arcing time. Such a typical interrupter is shown in Figure 9.9.

The puffer interrupter principle has been widely applied in both distribution and transmission circuit-breaker applications. Its main disadvantage is that it requires a high operating energy and long opening stroke to precompress the SF_6. This necessitates the use of large, costly operating mechanisms, and it is for this reason that for transmission circuit-breakers hydraulic or pneumatically operated mechanisms have been developed. At distribution voltages spring operating mechanisms can still be used, but these are nevertheless relatively costly.

9.5.2.10 Self-pressurising SF_6 circuit-breaker

In view of the need for costly operating mechanisms for the puffer type of circuit-breaker, developments ensued to attempt to use the energy of the arc to create pressure to cause it to extinguish itself. This works very well for interruption of high fault current but does not work so well for interruption of lower levels of fault current where long, thin arcs can be drawn out. To overcome this problem a low-energy piston is still required to provide a flow of SF_6 across this low-energy arc.

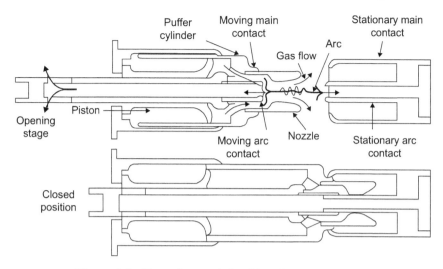

Figure 9.9   Typical partial duo-blast SF_6 in interrupter

Nevertheless, considerably reduced operating forces are required when compared with the puffer type of circuit-breaker. This has led to the development of significantly lower cost operating mechanisms. Spring operating mechanisms are now widely being applied to this type of circuit-breaker at transmission voltages (see section 8.6).

9.5.2.11   Rotating arc circuit-breaker

The puffer type of circuit-breaker proved to be costly compared with the rapidly developing vacuum circuit-breaker technology at distribution voltages and particularly where only low short-circuit rated currents were required. In consequence, an alternative form of SF_6 interruption was developed. The technology allowed the arc to move through the gas rather than the gas moving across the arc. The arc was caused to move by automatic insertion of a coil during the contact opening process. The current passing through this coil caused a magnetic field to be generated, which led to rapid rotation of the arc (see Figure 9.10).

Rotating arc circuit-breakers tend to arc to the full length of their contact stroke and, in consequence, have relatively long arc durations, particularly at low levels of fault current. This is because the lower magnetic field generated by the lower fault current will not cause the arc to rotate so rapidly. At even lower currents arc rotation may cease altogether and arc interruption may be purely fortuitous. This may be particularly critical if the circuit-breaker is required to switch low-energy reactive currents where very high values of transient recovery voltage may be encountered. Such reactor current switching breakers require to be proven by laboratory tests for this duty.

Notwithstanding this potential shortcoming, rotating arc circuit-breakers are very widely applied in distribution switchgear applications and offer very economic solutions. Low-energy cost-effective operating mechanisms can be employed when compared with the puffer principle.

376 High-voltage engineering and testing

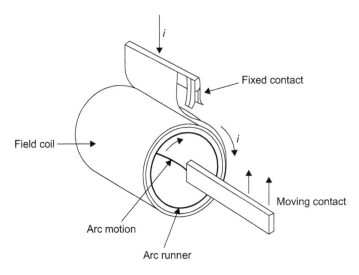

Figure 9.10 SF_6 rotating arc principle

9.5.3 Disconnectors

Disconnectors (isolators) are used to connect or disconnect sections of the circuit to allow the facility for safe earthing of the disconnected circuits to enable maintenance or repair work to be undertaken. They are usually manually operated at distribution voltages but may, at high voltages, be motor operated to allow remote, or automatic, operation.

Disconnectors are not intended to interrupt circuit currents but will be required to interrupt small capacitive currents associated with either open circuit-breakers or adjacent live circuits.

Where they are used on multiple busbar circuits they may also be required to facilitate the live transfer of a circuit from one busbar to another. In such a case they will be required to break the parallel currents circulating in the different sections of the busbar.

When closed, the disconnector must clearly be capable of carrying its rated load current and also its rated short-circuit current. In its open position it must provide isolation against both the rated power frequency voltage and associated switching or lightning impulse voltages that may be superimposed thereon.

Disconnectors used in outdoor substations may be of various types – for distribution applications the most common type encountered is either the centre rotating post type or the rocking post type. The general construction of such disconnectors is shown in Figure 9.11.

For metal-enclosed switchgear, disconnection facilities are generally built into the equipment. For very many years such facilities have been provided but by allowing means of physically unplugging the circuit-breaker from its fixed position. The unplugging process provides a physical gap between circuit-breaker isolating contacts and the fixed contacts of both the circuit and busbar sides of the

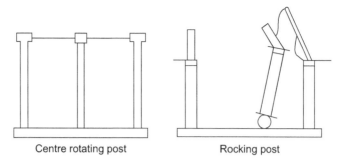

Figure 9.11   Disconnector types: outdoor substations

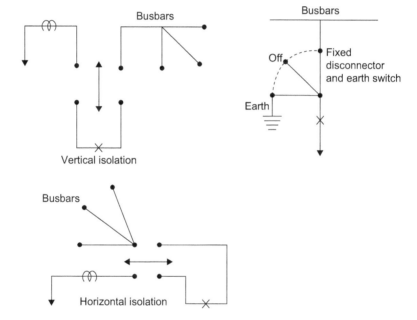

Figure 9.12   Metal-enclosed switchgear: isolation facilities

fixed panel. When isolation is completed metallic shutters automatically close the aperture to the fixed contacts (i.e. spouts). The shutters can then be padlocked or closed for safety purposes. Interlocks are provided to prevent physical damage on attempted reinstallation of an isolated section with padlocked shutters. It should be noted that while it has long been the practice to provide metallic shutters on all such equipment installed in the UK this has not necessarily been the case on switchgear utilised elsewhere.

Circuit-breaker isolation will be by means of either vertical or horizontal movement (see Figure 9.12).

In view of the potential high reliability of both SF_6 and vacuum circuit-breakers, and the considerably reduced need for maintenance, there has been a tendency in recent years to make the circuit-breaker fixed and non withdrawable. In such cases it is necessary to provide separately enclosed disconnectors. These may be provided on both the busbar and circuit sides but the modern tendency is for them to be provided solely on the busbar side. Such disconnectors usually incorporate an earthing switch so that the circuit side can be earthed via the circuit-breaker. Interlocks are required to ensure that the disconnection can only be achieved with the circuit-breaker in its 'off' position. A separate interlock prevents the disconnector from individually being moved to its 'earth on' position. Such an arrangement is also shown in Figure 9.12.

9.5.4 Earth switches

When work is required on outgoing circuits or on substation busbars or circuit-breakers it is necessary, subsequent to isolation, to apply an earth to the circuit to make it safe for work to commence. This may be facilitated by the provision of an earthing device or earth switch. For older metal-clad switchgear it was common practice to provide a separate device that could be plugged into a de-energised spout. Such devices could on occasions be inadvertently plugged into a live circuit, with disastrous results. In consequence, this type of earth device was banned and a device that could be fitted to the circuit-breaker isolating contacts was utilised. This allowed earthing of the circuit via the circuit-breaker, that is a fault-making device. Such devices are widely employed on distribution metal-clad switchgear assemblies. They may also provide the facility via interlocks for checking the correct phase relationship and cable testing.

Earthing should only be applied to metal-dad distribution switchgear by means of a fault-making device.

In view of these inherent problems and the cumbersome nature of the earthing device, manufacturers build earthing features into the metal-clad equipment itself. This is often achieved via transfer earthing, i.e. the movement of the circuit-breaker from the circuit-engaged position to be plugged into an additional two positions to select either circuit to earth or busbar earth. Incidentally, circuit-breaker tripping facilities must be locked off with the circuit-breaker in its earthing location to prevent inadvertent removal of the earth from the system.

For overhead line circuits earthing switches are generally not of the fault-making type. Earthing switches may be provided at either side of a disconnector. With such an arrangement isolation gaps are readily visible and the overhead circuit can be readily tested and checked to ensure that it is de-energised.

For higher voltages, 132 kV and above, the earthing switch may also be required to interrupt the capacitive and inductive currents induced into the earthed line from an adjacent live circuit on the overhead line towers.

9.5.5 Switches

A circuit-breaker is commonly incorrectly referred to as a switch. A switch is a device that must be capable of making and breaking load currents but not overload currents. It is not capable of breaking short-circuit currents but may, in certain circumstances, be required to have a rated short-circuit-making capacity. It must also carry in its closed position its rated short-time current. It is generally power operated to ensure that consistent opening and closing speeds are achieved.

Switches may be air, oil or SF_6 insulated, and they will usually have some form of rudimentary arc control system. They are widely used on distribution networks in combination with either disconnectors or fuses. They have minimal use at transmission voltages.

9.5.6 Switch disconnector

Switch disconnector is a switch that provides in its open position an isolating distance. A typical outdoor switch disconnector arrangement is shown in Figure 9.13. Switch disconnectors, using SF_6 interruption technology, were first used at transmission voltages c. 1970, and still remain on the system. With the rapid development of SF_6 arc interruption technology for circuit-breakers, switch disconnectors no longer offered economic advantages and their use diminished. In the USA, however, more economic versions are again starting to appear and their use is again becoming viable.

9.5.7 Switch fuse

A switch fuse is a combination of one assembly of switch and fuse in series. It is widely used in LV applications but not so much so at distribution voltage applications. The reason for this is that the switch is not protected for a fault between the switch itself and the fuse.

9.5.8 Fuse switch

This is a device whereby the fuse is physically mounted on the moving contact assembly of the switch. This ensures that the switch will always safely close against

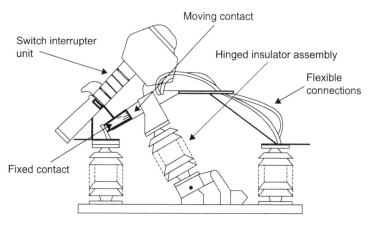

Figure 9.13 Outdoor pole-mounted switch disconnector arrangement

its rated short-circuit-making current, with the fuse operating to clear the fault. As the fuse operation is very rapid, the fault-making current duty of the switch is considerably reduced and more economic designs can ensue.

Fuse switches are very widely used in distribution networks at 11 kV. Their application is limited at 33 kV and non existent at higher voltages.

9.5.9 Fuses

A fuse is a one-shot device capable of carrying its rated current, carrying defined circuit overloads for predetermined times, of clearing overloads in excess of predetermined times and of clearing short-circuit currents caused by system faults. Once operated its fuse-link must be replaced.

9.5.9.1 Expulsion fuse

An expulsion fuse is a semi-enclosed (rewirable fuse) fuse, which is widely used on 11 kV distribution overhead line networks. It has a reduced application at 33 kV. A typical arrangement is shown in Figure 9.14.

The fuse element assembly comprises flexible 'tails' with a small central region comprising the element itself. It is this portion that melts, or volatalises, on fault operation. The element is enclosed in a solid insulated tube, the end fittings of which are held by two spring loaded contacts. On operation, high-pressure gases are developed within the bore of the tube. These expel the ends of the element flexible leads and at the same time assist in arc extinction. The top contact then releases and allows the fuse carrier to swing around its bottom, hinged assembly to give a visual sign of operation and to provide an isolation gap.

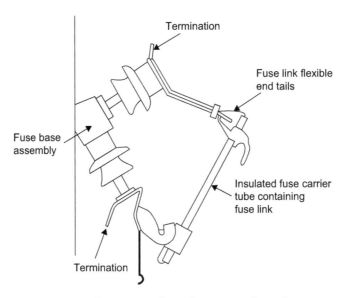

Figure 9.14 Circuit switching device: expulsion fuse type

Expulsion fuses are not current limiting in that they always require a current zero to clear. Arc durations are in consequence long, and arc energies are high. The rated short-circuit-breaking currents are relatively low – typically no greater than 8 kA at 12 kV. When the expulsion fuse operates it expels hot gases from either end of its fuse carrier (hence its name). They usually give a very loud retort on fault operation. They provide a relatively economic means of protecting overhead line spur circuits.

9.5.9.2 Current-limiting fuses

A current-limiting fuse is a sealed device that does not expel any gases on operation. As it is sealed it can also be used, at 11 kV, under oil. Its mode of operation differs considerably from an expulsion fuse in that it will *limit current*, i.e. it will *cut off* a rising short-circuit current before the current reaches its full peak value.

The device comprises a fuse element, usually made from a silver strip. The element may be typically 2 m long at 11 kV and, to accommodate this length within the length of the 250 mm fuse body, it is necessary for it to be wrapped spirally around an insulated star core, usually of ceramic material with a central hollow bore. There may be two or more elements in parallel to increase the rated current capability of the fuse. Each element has numerous regular notches along its length. The purpose of these notches is to ensure that on operation under high short-circuit currents, each notch will volatalise and in so doing will produce a number of arcs in series to build up a back electro magnetic field (EMF) to withstand the system applied and transient voltages. Very rapid arc extinction is achieved by energy being extracted from the arc by the surrounding medium. This medium is in the form of a very fine grained silica sand. Energy from the arc is used to fuse the sand into a glass-like substance known as fulgurite. This fulgurite also surrounds the arc to limit its spread.

For interruption of lower short-circuit currents a different interruption mechanism is used. This comprises a small deposit of lead–tin alloy placed towards the middle of the element length. When the element overheats from an overload current the alloy melts and passes through the molecular structure of the silver to cause the silver to have a significantly lower melting point. This element will thus part at this point, which then allows a long low-energy arc to slowly burn back along the element length until arc extinction occurs. This phenomenon is known as the 'Metcalf effect' or 'M' effect.

The HV current limiting fuse usually has a very thin parallel fuse wire passing through the bore of the star core, which operates subsequent to main element operation. On such operation it causes a small gunpowder charge, located at one end of the fuse cap, to explode. This explosion forces a small pin through the end cap. This is referred to as a striker pin and is used to allow three-phase switch operation on single-phase fuse operation.

A typical construction of an 11 kV current-limiting fuse is shown in Figure 9.15, with diagrammatic representation of its mode of operation being shown in Figure 9.16.

382 High-voltage engineering and testing

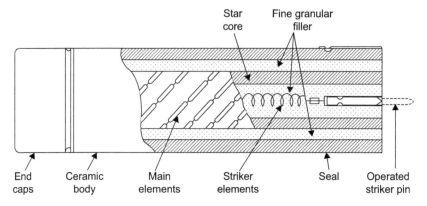

Figure 9.15 Current-limiting fuse: HV construction

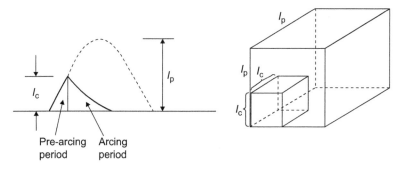

Figure 9.16 Current-limiting fuse: fuse cutoff current

These fuses are widely used under oil at 11 kV in ring main units, and are also widely used in LV applications. They are not generally used in applications above 33 kV.

9.5.10 Contactors

A contactor performs much the same duty as a switch. The main difference is that it is designed to be capable of performing very frequent switching operations and is also capable of interruption of overload conditions and motor stall current conditions. It is widely used in industrial processes for motor operation. It is not generally used in HV distribution circuits but will be encountered in use, having rated voltages of up to 11 kV, in power station auxiliary switchgear applications.

9.5.11 Ring main units

A ring main unit is a metal-enclosed construction of distribution switchgear, which is widely used in urban ring circuits. By itself, and in combination with a transformer and LV switchboard, it forms a secondary distribution substation (see section 9.2.2.3).

The metal-enclosed ring main unit comprises a 'ring' busbar circuit with a switch disconnector in each side of the ring. Teed off from the central busbar is a fuse switch that invariably feeds to an 11 kV/415 V three-phase transformer.

On the ring circuit side of each ring switch disconnector is an earth switch. This provides facilities for earthing the cable at either end to allow fault repair. Subsequent to repair it is common practice to subject the cable to HV tests. Testing facilities are achieved by means of a separate, loose test device, commonly referred to as test prods. Once the cable circuit has been isolated and earthed at both ends it is then possible to release a padlocked, and interlocked, cover to allow the insertion of the test device. With the test device in position it is then possible to open the earth switch to its 'earth off' position. The remote end switch is then placed to its earth off position and the cable test voltage can be applied to the terminals of the device. On completion of the tests the device cannot be removed until the switch has been moved to its earth on position to reapply the circuit earth.

The earth switch cannot be operated until the test access cover has been correctly relocated and is secured in position. Both built-in mechanical interlocks and padlocking facilities allow safe operation to be achieved.

In practice, most ring main switch disconnectors are in fact three position devices which, in addition to the on and off positions, can be moved, after appropriate interlock selection to the earth on position.

The tee-off circuit comprises, in general, a fuse switch whereby the HV fuse-links are mounted on the moving portion of the switch. The HV fuse-links typically have rated currents of up to 80 A and are used to provide both overload and short-circuit protection to the HV side of the transformer.

A low-rated earth switch is provided between the fuse switch and the transformer terminations to allow earthing of the transformer HV in the event of an LV backfeed.

Traditionally, ring main units have been of the oil-filled type using HV fuses on the tee-off circuit, but more laterally these have been superseded by SF_6-insulated devices. An SF_6-insulated ring main unit may use an SF_6 interrupter (circuit-breaker) in its tee-off circuit or possibly a vacuum interrupter. Some still employ HV fuses, but these must be mounted externally in air, which leads to complication and added expense.

A typical ring main layout is shown in Figure 9.17.

9.6 Circuit protection devices

9.6.1 Surge arresters

Surge arresters are devices that dissipate, to earth, a transient overvoltage to protect down line connected equipment from possible dielectric breakdown. There have been major developments in the design of surge arresters in recent years.

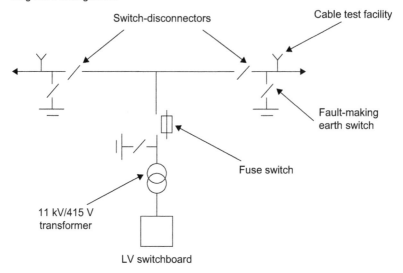

Figure 9.17 Typical distribution secondary substation layout

Traditionally silicone carbide arresters have been used. These are satisfactory for dissipating high overvoltages and high current surges, but are not suitable by themselves for withstanding continuous alternating current (AC) power frequency voltage since they would have to pass continuous current in the order of 100 A. To overcome this problem a number of series capacitive gaps are incorporated. These will reduce the value of the series current that flows to that of the capacitive current of the associated gaps. On operation from high transient overvoltages the series gaps will flashover to insert into the circuit the silicone carbide resistor blocks.

Such arresters require the fault current to pass through a current zero point before they will clear; hence operating times can be in the order of 10 ms, and also rapid deterioration of the gaps ensues.

A more recent development has been the zinc oxide surge arrester. This material passes a very low current at its normal operating voltage, typically less than 1 mA; hence, series gaps are no longer required. This considerably reduces the costs of the arrester and improves its performance and reliability.

Zinc oxide surge arresters, as they do not require a series gap to operate, are very fast in operation, typically less than 1 ms, and do not require the passage of a zero current.

Sealing problems have been widely encountered on silicone carbide type arresters with resultant gap deterioration and eventual failure. Zinc oxide arresters are not so prone to this deterioration mechanism, and improvements in sealing techniques have been achieved such that high-reliability devices can now be supplied.

Figure 9.18 shows the comparison between the characteristics of a silicone carbide arrester and a zinc oxide arrester.

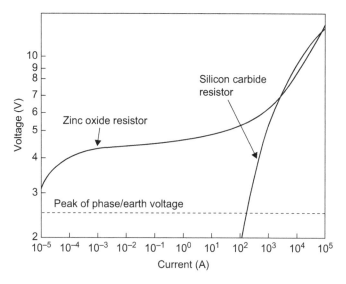

Figure 9.18 Surge arresters: comparison of silicon carbide and zinc oxide resistor characteristics

9.6.2 Instrument transformers

Instrument transformers are devices that produce, at earth potential, a replica in magnitude and phase of the current flowing in the primary circuit and of the voltage across the primary circuit. The former are called current transformers and the latter are referred to as voltage transformers.

9.6.2.1 Current transformers

Current transformers (CTs) are usually in the form of a one-turn bar primary, which passes through the centre of a toroidal core, and around the core is a multiturn secondary winding. As the bar primary is at line potential, insulation must be provided between it and the secondary winding/assembly. For 11 kV or 33 kV metal-clad switchgear this insulation is usually provided by means of an HV bushing, which, at 33 kV, may be of the SRBP construction and more commonly at 11 kV it is of cast epoxy resin construction. It should be remembered that the outside of the HV bushings must be at earth potential to avoid the possibility of electrical discharges between this surface and the earthed CT assembly.

For distribution applications the CT secondary winding usually has a current rating of 5 A, but for applications involving long lengths of secondary leads the current rating may be 1 A to reduce the lead burden.

CTs must have a defined current ratio, i.e. 400/5, 200/1, etc. As the primary circuit current rating decreases it may be that insufficient ampere turns are generated to provide the required secondary output to drive the associated protection equipment, in which case the primary winding may comprise two or more turns.

This adds further complication to the insulation design between the primary and secondary windings, and care is needed to ensure long-term integrity.

Secondary outputs are usually given in terms of volt-amperes, i.e. VA. A typical output is 15 VA. Various classes are defined for both protection and high-accuracy tariff metering. For the latter conventional core material losses are too high and it is necessary to use a nickel alloy material commonly referred to as Mumetal.

For higher voltage applications, e.g. 132 kV and above, the construction is somewhat different and may be either of the hairpin dead tank design or, more latterly, of the bar primary live tank design. In either case insulation between the HV conductor and secondary assemblies is provided by multiple layers of oil-impregnated paper, and the whole assembly is filled with insulating oil. However, detailed description of the construction of these CTs is beyond the scope of this chapter.

9.6.2.2 Voltage transformers

Voltage transformers (VTs) for 11 and 33 kV distribution applications are usually of the three-phase type and have traditionally been of the oil-insulated construction with the whole assembly being enclosed within its own metal-enclosed oil-filled tank. HV connections are achieved via three protruding bushings. These bushings plug into associated spout and fixed contact assemblies on the metal-clad switchgear panel. VTs may be connected to either the circuit side connections or busbar side connections. The VT tanks are mounted on wheels that run on rails on the switchgear to allow them to be either rolled back to an isolated position or rolled inwards to engage the primary circuit contacts.

All such VTs are protected on each phase by an HV fuse often enclosed within the VT bushings. The fuses traditionally have a rating of 0.6 A and have a resistive fuse element to limit fault infeed in the event of VT failure. Without such a fuse dramatic VT failure and consequential substation disruption may result.

A recent trend, particularly at 33 kV, is to utilise single-phase VTs with one end of each winding connected directly to earth. This arrangement is used mainly on resistive earth systems where the single-phase earth fault current is limited to something in the order of 1,000 A.

With such single-phase arrangements it is imperative to ensure that appropriate protection is provided both to the equipment and to personnel in the event of VT failure.

Most modern-day voltage transformers are insulated with cast epoxy resin. The manufacturing technique is, however, critical as any voids incorporated into the moulding process can lead to partial discharge activity and eventual VT failure.

VT secondary voltage outputs are typically rated at 110 V phase to phase, or 63.5 V phase to earth. Output ratings are similarly given in volt-amperes, and a typical rating would be 100 VA per phase. Again high-accuracy devices can be provided for tariff metering purposes.

For outdoor substations weatherproof enclosures can be provided with associated HV bushings.

For higher voltage systems a different principle is used in that a series of capacitors are provided between the HV conductors and earth to form a capacitor divider. Voltage is tapped off from the lower capacitive assembly to feed an electromagnetic type of voltage transformer to provide the secondary voltage. CVT design is outside the scope of this chapter.

9.6.2.3 Alternative voltage and current transducers

Conventional current and voltage transformers occupy a relatively large amount of space on a metal-clad switchboard and by themselves are also relatively expensive. The provision of such devices, therefore, forms a significant cost contribution to the total panel cost. Conventional electromagnetic devices have been required for many years because large outputs are necessary to drive conventional protection relays. With the advent of modern electronics, or digital relays, significantly lower energy infeeds are required. In consequence, this has led to the development of new current and voltage transducer devices. For the low-energy voltage requirement a conventional resistive voltage divider can now be utilised. With current transducers optical techniques are rapidly being developed, as are similar associated techniques for voltage transducers. The benefit of such techniques is that they can be made very small and are more readily accommodated than conventional devices, allowing significant potential cost reduction.

9.7 Switchgear auxiliary equipment

Auxiliary equipment, i.e. the low-voltage part of the switchgear provided for control and monitoring of main components, has traditionally been provided by hard-wired components, conventional switches, terminations and electromechanical relays, with such secondary systems being supplied by the switchgear manufacturer.

There is a rapidly growing trend to the introduction of electronic technologies whereby the equipment and processes may be supplied by alternative suppliers.

The reliability of electronic components has improved significantly in recent years, and their performance is more and more deterministic. Such electronic facilities allow the provision of more automated operation and also provide diagnostic data on the performance of the primary plant.

Once concepts have been perfected significant advantages will be gained by the end user.

International standardisation of such approaches is rapidly occurring and indeed some manufacturers are already incorporating such concepts.

9.8 SF_6 handling and environmental concerns

Over recent years attention has been focused both on potentially hazardous breakdown products of SF_6 and on possible environmental concerns.

9.8.1 *SF_6 breakdown products*

SF_6 decomposes at high temperatures encountered during the arc interruption process into very many constituent products but, on gas cooling, these decomposition products tend to recombine to form SF_6. Any gaseous breakdown products remaining tend to be absorbed by desiccants placed within the gas enclosure. Hence for normal and short-circuit switching operation SF_6 gas decomposition is of little concern and SF_6 gas-filled interrupters can remain in service under normal operating conditions for very many years without the need to be opened up for examination.

In the event of very heavy fault arcing or long-term partial discharge activity some of these breakdown products may remain and if chambers are to be opened up some care is necessary.

Typical breakdown products from sparking or from partial discharges are SF_4, WF_6, SOF_4, SOF_2, SO_2, HF, SO_2 and F_2. Contact erosion and internal arcing might also produce CuF_2, WO_3, WO_2, F_2, WOF_4 and AlF_3.

Most of the reactive decomposition products and their follow-up reactions have toxicities comparable with SO_2 and may constitute a health risk if present in too high a concentration.

Some reactive decomposition products are corrosive, particularly in the presence of moisture, and appropriate handling precautions should be followed on the opening up of such chambers.

The aluminium fluoride breakdown products will be evident by the presence of a white/grey powder lying on insulating and chamber surfaces. These powders are nonconducting and are not critical for insulation performance.

The gaseous breakdown products will be evident by a 'rotten egg' smell produced mainly by SO_2. However, SF_6 arced gas should never be deliberately smelt.

When opening up chambers it is necessary to wear disposable overalls, gloves, face mask or respirator if entering chambers, head protection and appropriate safety footwear. Arc products should be removed by means of a fine filter vacuum cleaner, and contaminated surfaces should be neutralised to prevent acidic corrosion. Before entering chambers the SF_6 must be safely removed, filtered and stored.

SF_6 *must not be deliberately released to the atmosphere.*

There has been concern over recent years that electrical breakdown of SF_6 may produce S_2F_{10}, a very toxic gas. While this has been identified in laboratory experiments it has never, to date, been positively identified in faulted equipment in service. In any event it very rapidly recombines as the temperature drops to the normally found breakdown products. Three years of study by international experts in Canadian Laboratories concluded that S_2F_{10}, even if initially present, was of no practical consequence provided appropriate, normally accepted, handling procedures were followed.

9.8.2 *SF_6 environmental concerns*

High-level balloon measurements of gases in the upper atmosphere over the last 30 years or so have identified the presence of SF_6. Estimates suggest that total quantities in the outer atmosphere could be of the order of 80,000 tonnes

Work thus commenced to ascertain the possible effect on atmospheric contamination of this quantity of gas. Initial work showed that SF_6 was not an ozone-depleting gas as it cannot initiate the ozone catalytic chain, i.e. SF_6 has no effect on ozone depletion.

Work also commenced to ascertain the ability of its molecule to reflect radiation in the infra-red spectrum. This showed it had a very high reflectability in this spectrum and, as such, constituted a potentially major greenhouse gas. Work to ascertain its natural lifetime in the atmosphere estimates this to be between 800 and 3,200 years, i.e. it takes a very long time to naturally decompose.

The above two factors combined make SF_6 the worst known greenhouse gas. Efforts have since continued to assess the effect of the 80,000 tonnes of SF_6 on greenhouse warming. Figures produced in 1995 show that the anthropogenic contribution, i.e. man-made greenhouse effect from SF_6, was less than 0.1%.

In earlier years this effect was not known and there were no restrictions, apart from cost, on the deliberate release of SF_6 to the atmosphere.

Clearly, deliberate release can now no longer be allowed. In addition, handling procedures must be improved to avoid losses during these processes. These issues are addressed in Reference 1, and Reference 2 discusses in more detail the environmental issues. Reference 3 is a further updated paper on environmental concerns.

1. Appendix A provides some further support information relating to:
    a. obtaining value from online substation condition monitoring
    b. SF_6 gas – tightness guide
    c. SF_6 recycling guide (Revised)
    d. evaluation of different switchgear technologies (AIS, MTS and GIS) for rated voltages of 52 KV and above
    e. the impact of new functionalities on substation design
2. Appendix B provides additional information on distribution systems and dispersed generation.

9.9 The future (as perceived in 2000 and again in 2011)

9.9.1 The future 1 – by B.M. Pryor[2] in 2000

Deregulation of the electrical supply industry in many parts of the world is having a significant effect on the way utilities operate. They are now facing new pressures to deliver ever increasing quality and continuity of service, to reduce supply restoration times and to reduce electricity cost to the customer. In parallel with this the maintenance workload is increasing due to utilisation of a wide variety of lower cost plant and due to the ever increasing age of existing equipment. Further factors affecting utilities are increased cost constraints, difficulties in obtaining and justifying the stocking of spare parts and reductions in the numbers of experienced personnel, particularly those with expert knowledge on the equipment.

All utilities need to become more efficient. One way of achieving this is to reduce maintenance costs, which represent a significant part of operational costs, by effective techniques such as condition-based maintenance where condition

---

[2] Author of this chapter in 2nd and 3rd editions.

assessment is achieved by the use of diagnostic monitoring techniques. Developments are such that automated monitoring will be the norm, with deviations from the norm being detected before failure occurs. We could thus end up with a fully automated self-monitoring distribution system.

All these factors are being driven by the customers' need for a high-quality, highly reliable and low-cost supply of electricity. Competition between utilities will force these concepts even further.

There are yet two further additional pressures, which will affect future concepts, the first being the environmental pressures and the second the pressures resulting from new technology developments.

The environmental pressures have arisen over the last 150 years or so, as a result of industrialisation and there are signs that, even within this period of time, considerable environmental deterioration has resulted. Clearly, we cannot continue contaminating the planet to this extent. The public is becoming increasingly more aware of these issues and environmental pressures can no longer be ignored.

Environmental aspects can be significantly improved in many areas of the electrical power industry. There is already pressure to minimise lifetime energy usage. This applies from mining raw materials to manufacture of components, manufacture of assemblies, lifetime energy usage, and energy usage on final disposal. The greater energy usage is usually during the lifetime use of the equipment. At the end of the life of the equipment it must be possible to dispose of all components in an environmentally acceptable way. Energy usage losses can also be minimised by the use of highly efficient local generators, and this is also a further trend that is already occurring.

Very many new technology developments are looming on the horizon. As mentioned earlier there are major developments in the case of magnetic materials; further major developments are also taking place in the use of polymers. These are having applications in energy-storage devices, e.g. batteries, capacitor storage, etc. They are also allowing the fundamental development of new technologies, such as fuel cells, photovoltaic cells, heat pumps, etc. In parallel with this there are also major developments in superconductivity.

All these developments will have a major influence on the way in which we manufacture, deliver and use electrical energy.

In parallel with this, load patterns are tending to change much more rapidly than has been experienced in the past. This is due to inward investment of new technology industries, some of which are proving to be short lived, particularly due to the varying economic climate around the world.

With this background, what sort of systems might we have in 10–15 years' time? A possible scenario is shown in Figure 9.19. If such a scenario is probable, how do we move from where we are at present to the new concepts in the most cost-effective way?

Such requirements for future substations:

- must be relocatable
- must not require manual intervention

Distribution switchgear   391

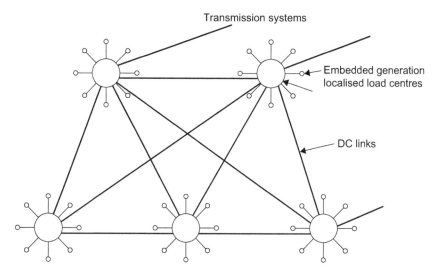

Figure 9.19   The future system (as seen by Pryor in 2000)

- must have a very high degree of reliability
- must have automated operation, only signalling when the system is in an incorrect state
- must have low losses
- must be recyclable
- must only use environmentally compatible components
- must not require maintenance
- must be compatible with communicating to other system nodes
- must maintain quality of supply.

A challenging time is ahead for young engineers!

9.9.2   *The future 2 – by H.M. Ryan[3] in 2011*

This chapter, written by B.M. Pryor, who was very active in IET HVET International Courses plus IET and CIGRE Committees – and particularly active on environmental aspects – for many years, reflects the distribution switchgear sector as seen and written by him in 2000 – the start of the twenty-first century. Although no major design changes have occurred in SF_6 circuit-breakers for transmission networks since 2000, there have been strategic developments in the UHV and asset management fronts as the industry moves towards the 'smartgrid' networks of the future (see Chapter 23). For example, in the years since 2000:

---

[3] Based partly on a wide range of materials reviewed by the writer and updates via CIGRE Study Committees, TBs, and in particular, the activities of SC C6, Distribution systems and dispersed generation as reported from its 2011 AGM [8].

1. Four breaks per phase SF_6 substation designs have been developed for a 1,100 kV UHV transmission system in China, which entered service in 2009 (see Chapter 8). Consequently, SF_6 CB designs have now been proven up to UHV level.
2. The extensive number of TBs prepared by CIGRE on asset management (AM) is evidence of the 'CIGRE commitment to prepare the energy sector to *optimally manage assets, and their associated performance risks and expenditures over their life cycle*, for the purpose of achieving the required quality of service in the most cost effective manner' (see Appendix E, Table 8.1 and Chapter 10).
3. Distribution networks have developed appreciably (see Chapters 5, 8, 10 and 23). We will touch briefly on certain of these strategic developments here in Appendix B and Chapters 10 and 23.

In recent years, there were growing concerns about global warming and climate change issues, particularly any contribution to greenhouse gas emissions that energy production or equipment could possibly have. Attention later moved to utilising renewable energy/wind farm developments and the extensive integration into energy networks of renewable energy sources (RES) and dispersed generation (DG). Initially, many felt that the public – from a global warming perspective – viewed this development of RES positively, that is it had a more positive and 'environmentally friendly public perception of "impacts" power systems might have on our environment', as CIGRE reported in 2011, 'Network of the Future' [4] (*Electra*. 2011;**256**(June):42–9).

Today, however, one certainly cannot take for granted that the UK public perception is the same, after a turbulent period of sustained negative articles and correspondence in the press from energy customers and by numerous energy press reporters, over many concerns, for example high energy bills and rising uncertainty regarding future new-built nuclear plant, worries regarding some wind farm issues and the absence of any coherent UK energy policy. These and other aspects are discussed in Chapter 23.

The energy industry [5] and CIGRE [4] recognised that it was necessary to go 'back to the drawing board' and re-think or revisit previous philosophies. That is 'it had to develop a new mindset' to ensure that it was going to be possible to successfully obtain efficient and robust integration of large shares of RES and DG. The planned introduction of *higher intelligence* to help further improve efficiency, as the energy industry moves towards 'smartgrids' and 'networks of the future', that is within the next one or two decades [4, 5] (see section 23.13), will be achieved at great cost to the consumer in the UK, and will only be achieved effectively with early strategic advanced planning, management and control.

CIGRE has identified 'integration of RES and DG into the grid' as one of several key driving factors, in a very broad scheme – comprising many specialist CIGRE WGs (see Tables 23.13 and 23.14) – involving: driving factors, network development scenarios and relevant activities to achieve its over-reaching aim of contributing to the vision and the development of the future energy supply systems in a global co-ordinated way [4, 5].

9.9.2.1 Some activities 2000–11 – compiled by H.M. Ryan

In the following brief overview, emphasis is given to likely network system developments and this contribution is not merely restricted to switchgear. In 2002 Laycock [6] discussed the topic of intelligent networks, identifying on-coming rapid changes, post liberalisation, as more players are attracted to this sector creating the need for significantly more intelligence in 'active' networks than was necessary with the earlier 'passive' nature of the distribution network design. In 2001, Jenkins [7] wrote a valuable early article on the Impact of Dispersed Generation on Power Systems. In recent years, there have been many conferences with numerous dedicated sessions to discuss, develop and advance knowledge on various aspects of relevant themes such as RES, DS and DG. CIGRE Study Committees have carried out a substantial body of work focusing on various strategic activities, each via specialist WGs, as evidenced [8] by CIGRE, SC C6 WGs. We can return to consider the progress made by these global CIGRE activities later, by reviewing activities listed in Appendix B.

Meantime, very briefly, Laycock [6] argues the need for *greater intelligence* and discusses how engineers and managers in diverse locations are being challenged to find solutions that will drive down costs and encourage better utilisation of distribution sector assets by the appropriate development of intelligent networks. Laycock outlines the application of intelligence, centres of advanced intelligence, protection intelligence and network services (i.e. *on-network*: frequency control, voltage control, waveform control, reactive power supply, dynamic stability, load following, fault level control; and *off-network*: asset management, performance measurement, plant condition monitoring (CM), refurbishment, meter-reading, billing, connection management).

He provides useful diagrams to show changes from *passive* networks through to *active* schemes, for example schemes including open ring with IPP/wind generators and CHP. These aspects are covered further in Chapters 20–22, when considering various mature CM approaches, and in Chapter 23, when discussing 'smartgrids' and 'networks of the future' [4, 5].

(The reader is encouraged to read the original Laycock article [6] for fuller treatment; or can look at certain of Laycock's figures and tabular information – reproduced in Ryan, sections 9.4.2–9.4.4, of *Advances in High Voltage Engineering*, A. Haddad and D. Warne (eds.), IEE Power & Energy Series 40, 2004.)

We will move rapidly on to consider certain recent strategic international developments in the distribution sector. In 2002, CIGRE made significant changes to its specialist Study Committee (SC) structures, reflecting the 'winds of change' regarding the forthcoming huge and costly global developments and changes to the transmission and distribution networks of the future [4, 5]. (Ironically, just as this chapter closes, rumours suggest another 'dash for gas' phase may be upon us in the UK!)

This CIGRE 'intellectual contribution' to network restructuring [4] had to be done strategically and thoroughly, and had to ensure the remit of each Study Committee (SC) was carefully established such that overall the new SCs covered, with only limited appropriate overlaps, all the areas relevant to the needs of the energy sector in forthcoming years. This included new technologies, widespread adoption of high-level intelligence, new built nuclear reactors, increased penetration of RES/DG/storage back-up, etc., an increased energy mix and a widely held anticipation of 'smartgrids' in the future one or two decades ahead. Noticeable strategic changes in the energy sector began to emerge with significant short-, medium- and long-term developments and strategies formulated and discussed (as touched on further in Chapter 23).

EPRI in the USA [5], CIGRE [4] and many other organisations planned thoroughly for the quickly changing energy environment and undertook detailed major technical, economic and logistical evaluation studies [5] for the anticipated 'smartgrids' and CIGRE for 'networks of the future' [4]. Such planning had also to evaluate the ongoing economic implications, environmental consideration/concerns associated with the burgeoning renewable energy wind farm developments *onland* (and later *offshore*), the anticipated widespread up-take of electric and hybrid electric cars in the near future and the possible significant changes and enhancements that will certainly become necessary to transmission and distribution networks.

To compound matters we are currently still in a world recession or economic downturn situation. In 2002, CIGRE reformed all its Study Committee (SC) structures (after, *A History of CIGRE*, 2011).[4,5]

Note: Full version, 'Fields of activities of the CIGRE Study Committees' is also reproduced later, see Chapter 23 (in Table 23.13). Briefly, the present CIGRE list of technical Study Committees (SCs) and website information (post 2002, CIGRE reorganisation) is as follows:

- A1 Rotating electrical machines (www.cigre-a1.org)
- A2 Transformers (www.cigre-a2.org)
- A3 High-voltage equipment (www.cigre-a3.org)
- B1 Insulated cables (www.cigre-b1.org)
- B2 overhead lines (www.cigre-b2.org)
- B3 Substations (www.cigre-b3.org)
- B4 HVDC and power electronics (www.cigre-b4.org)
- B5 Protection and automation (www.cigre-b5.org)
- C1 System development and economics (www.cigre-c1.org)
- C2 System operation and control (www.cigre-c2.org)
- C3 System environmental performance (www.cigre-c3.org)
- C4 System technical performance (www.cigre-c4.org)
- C5 Electricity markets and regulation (www.cigre-c5.org)

---

[4] Since 1921, CIGRE's priority role and 'added value' have always been to facilitate mutual exchanges between all its components: network operators, manufacturers, universities and laboratories (as recently re-stated in *A History of CIGRE*, 2011). It has continually led to the search for a good balance between the handling of daily problems encountered by its members in doing their jobs and reflection on the future changes and their equipment. Its role in exchanging information, synthesising state of the art and serving members and industry has constantly been met by CIGRE through events and through publications resulting from the work of its Study Committees. www.cigre.org; www.e-cigre.org.

[5] CIGRE key: SC, Study Committee; WG, Working Group and JWG, Joint Working Group; TF, Task Force; TB, Technical Brochure; TR, Technical Report; SC, Scientific Paper.

- C6 Distribution systems and dispersed generation (www.cigre-c6.org)
- D1 Materials and emerging test techniques (www.cigre-d1.org)
- D2 Information systems and telecommunication (www.cigre-d2.org)

In Appendix B we will only briefly consider certain recent activities of SC C6 Distribution Systems and Dispensed Generation [8], although the reader should be aware that SC B3 is also relevant (see above). The TBs listed in Appendix A, relating to SC B3, Substations, supplemented by additional materials in Appendix B together with Appendices to Chapters 5, 8 and 10, provide excellent wide-ranging sources of resource material should the reader wish to follow up later.

Finally, moving on now up to the year 2011, Appendix A, which relates to SC B3 Substations, provides additional reference and technical support information up to the present time while Appendix B provides the reader with useful indications/illustrations as to what the technical and operational activities have been within SC C6 since 2002. It is possible to 'deduce' changing trends by carefully examining individual topics/themes of strategic concerns in the industry that have been or are still being actively studied by CIGRE WGs, as at 2011, or looking at new CIGRE Working Groups (WGs) currently being set up – all of which give strong and clear indications (both temporal and technical) as to

1. what themes are currently being, or are about to be, strategically studied by specially selected experts in WGs, within the sector
2. the technical problem areas which appear to have been cleared by the issue of CIGRE Technical Brochures (TBs) or *Electra* articles, prepared by experts in SC C6 WGs since the year 2002, the year when CIGRE re-aligned its specialist Study Committees to meet the strategic needs of the energy sector in the twenty-first century.

Note: Sometimes a TB may be of interim status, in which case it could be followed later by a further TB if, for example, more substantive strategic information/evidence becomes available.

- This should greatly assist the present reader by allowing him/her to 'dip-in' and examine some of the lines of enquiry which have been, are being, or are soon to be studied by specially established WGs – each set up to address specific strategic aspects of work still to be developed or better understood within the sectors covered by CIGRE SC C6. (Having reviewed this material, if appropriate, the reader should be able to follow up selected themes in research publications of IEEE, IET, etc., or even company websites.)
- At this point, he/she should be able to 'position' and 'judge the status of sector knowledge' in each technical theme area considered – and this can subsequently be checked again and again (and updated) over succeeding years. SC C6 Annual Report statements, and references to any TBs issued, together with the summary linked-presentation made by the WGs in *Electra* (or by the reader) in conjunction with the issue of each Technical Brochure is now usually well documented.

Note: Before one gets too carried away with this access to strategic CIGRE TB information we have touched on, such endeavours can often be merely the 'tip of the iceberg'. R&D engineers in the UK switchgear sector, some 40/50 years ago, would routinely be employing comparable tactics, following up on recent relevant worldwide developments in the sector, e.g. refereed technical papers in the IEEE/IET/IOP and overseas foreign language papers on switchgear from Japan, Russia, Germany, etc. (many of which were translated by in-company technical translators). In this manner it was possible to link such information to supplementary CIGRE/IEC information and, with this broad sweep approach, the R&D engineer was able to keep abreast of any developments of interest in their sectors.

Consequently, the emphasis in Appendix B is centred on merely providing the reader with 'a taster', a brief general overview by examining CIGRE, SC C6 activities: 'Distribution systems and dispersed generation', which looks at the current broad CIGRE programme of activities worldwide – within remit of C6 – as they were presented in the 2011 Annual Report of CIGRE, Study Committee C6 [8].

Note: The designated field of activity set by CIGRE for this SC is specifically 'Distribution systems and dispersed generation' (www.cigre-c6.org) covering assessment of technical impact and new requirements which new distribution features impose on the structure and operation of the system, widespread development of dispersed generation and application of energy.

A summary of SC C6 activities can now be reviewed in Appendix B.

Disclaimer

It is again emphasised that while these documents have been compiled and abridged, mainly from CIGRE *Electra* materials, they are arranged in the present format only to provide the reader with a valuable and balanced introductory overview to strategic topics currently being actively studied within CIGRE's unique global infrastructure. Considerable compiling/editing of relevant *Electra* articles was necessary. (Naturally, all enquiries to any publication should be sent via appropriate company addresses.) The writer gratefully acknowledges the kind and generous support he has had from CIGRE colleagues worldwide, over the years.

Summary

Sadly, the *present worldwide economic downturn* and project investment difficulties, together with other constraints, will probably slow down competition of major transmission and distribution network enhancement initiatives. However, this may not all be bad news because, as illustrated by the excellent wide-ranging activities currently being carried out, as briefly sampled in this chapter (i.e. by CIGRE/IEEE/IOP/IET/IEC/manufactures/academics/researchers/experts/ government-funded research studies, etc.), there are still 'some challenges outstanding' concerning major technical and economic issues. Some alternative options for the best way ahead have yet to be evaluated and resolved; robust and timely economic evaluations must surely be vital – and 'realistic cost estimates' secured for the various options before final decisions are made.

Nevertheless, technically the future for the energy sector looks very promising. We must all be aware that developments in the energy sector are always ongoing. All academics involved in engineering, ICT and business sectors must make an effort to keep abreast with strategic developments within the energy sector, and perhaps even more important, they have a responsibility to ensure that all students are made aware of, and monitor evolving developments. Meantime, patience is important to ensure 'we get it right' for this next huge phase of network expansion and enhancement. CIGRE have done an excellent job thus far.

To conclude the chapter, one can do no better than repeat the final rallying call from B.F. Pryor in section 9.9.1: 'A challenging time is ahead for younger engineers!' – and they are again encouraged to consider participating in some of the technical activities of CIGRE SCs and related activities (IET/IOP/IEEE/IEC).

References

1. CIGRE WG 23–10. 'SF$_6$ and the global atmosphere'. *Electra*. 1996;**164**: 121–31
2. CIGRE TASK FORCE 23.10.01. 'SF$_6$ Recycling Guide – Re-use of SF$_6$ Gas in Electrical Power Equipment and Final Disposal'. CIGRE Technical Brochure. TB 234-SC B3 TF B3.02.01 'SF$_6$ Recycling guide', Revised version, 2003
3. Maiss M., Brenninkmeijer C.A.M. 'Atmospheric SF$_6$: Trends, sources and prospects'. *Environmental Science Technology*. 1998;**32**:3077–86
4. CIGRE WG. 'Network of the Future'. Electricity Supply Systems of the Future; Report on behalf of the Technical Committee. *Electra*. 2011;**256**:42–9
5. Gelling C.W. 'Estimating the costs and benefits of the "smartgrid" in the United States'. Invited Paper. *Electra*. 2011;**259**(December):6–14 (This is a recent article by C.W. Gellings (from EPRI, USA). The full study with the same title is available online (at www.epri.com, Estimating the Costs and Benefits of the Smart Grid, TR-1022519)
   - Supporting references provided by Gellings:
     – 'Future Inspection of Overhead Transmission Lines', EPRI, Palo Alto, CA, 2008. 1016921
     – 'Automated Demand Response Tests: An Open ADR Demonstration Project', EPRI, Palo Alto, CA, 2008. 1016062
     – 'Estimating the costs and Benefits of the Smart Grid'. EPRI, Palo Alto, CA, 2011. 1022519
   - A further useful reference:
     – 'Power System Infrastructure for a digital Society: Creating the New Frontiers', Keynote Speech (Montreal Symposium CIGRE/IEEE), Marek Samotyj, Clark Gellings, and Masoud Amin, *Electra*. 2003;**210** (October):23–30

6. Laycock W.J. 'Intelligent networks', *Power Engineering Journal*. 2002: 25–9
7. Jenkins N. 'Impact of dispersed generation on power systems'. Invited Paper. *Electra*. 2001;**199**(December):6–13
8. CIGRE SC C6 - 2011. Annual Report: 'Distribution Systems and Dispersed Generation' (See also *Electra*. 2011;**258**:24–8 (prepared by Nikos Hatziargyriou, same title.))

Appendices – by H.M. Ryan

A   Additional CIGRE references

A few additional CIGRE Technical Brochure (TB) reference sources are given below to assist the reader and provide a possible extra 'introductory' guidance source at a later date.

Note: TBs referred to throughout this book may well be available within the reader's organisation or, in the first instance, a scan of contents of *Electra* editions could be helpful. *Electra* publications are free to CIGRE individual members and any TB can be purchased from CIGRE.

1. TB 462-SC B3 WG B3.12, 'Obtaining Value from On-Line Substation Condition Monitoring', 2011
   'A broad overview of substation condition monitoring is given and the technology currently available is described, together with an analysis of the factors which require to be identified/considered to determine the degree of application necessary to provide optimum "value" solution having taken due account of the improved reliability of modern HV substation equipment.'
2. TB 430-SC B3 WG B3.18, 'SF_6 Tightness Guide', 2010
   'Maintaining gas tightness of electrical power equipment is essential for equipment functionality on the one hand and for protection of the environment on the other. This useful guide reviews significant aspects of measuring and ensuring gas of electrical power equipment containing SF_6. A functional description concerning state-of-the-art test procedures, test methods and instrument devices is performed.'
3. TB 234-SC B3 TF B3.02.01, 'SF_6 Recycling Guide, Revised Version', 2003
   'All significant aspects of the recycling of SF_6 gas used in electrical power equipment are reviewed and essential procedures addressed, ending with a clear proposal as to the required purity standards to which the SF_6 should be reclaimed to allow its safe re-use.'
4. TB 390 B3 WG B3.20, 'Evaluation of Different Switchgear Technologies (AIS, MTS, GIS) for Rated Voltages of 52 kV and Above', 2009
   'New switchgear components have been developed based either on (i) AIS or on (ii) GIS technologies or possibly on a combination of both (iii), i.e. MTS. This CIGRE document provides comparative information of the several

possible arrangements to aid the selection of the most effective equipment solution for a new or existing substation.'

5. TB 380-WG B3.01, 'The impact of New Functionalities on Substation Design', 2009

'Guidelines are given on the technical issues and challenges of introducing new technology into the substation environment, with general overview in main body of document and in fine detail in the appendices. It is aimed as a guide to some of the practicalities and problems to be faced when installing the solution, rather than as a problem solver for system issues.'

B Distribution systems and dispersed generation (after CIGRE [8]) – summary of key CIGRE information mainly from SC C6 2011 AGM [8] – compiled by H.M. Ryan

In its 2011 Annual Report, published [8] in *Electra*. 2011;**258**(October) (prepared by N. Hatziargyriou), CIGRE, Study Committee SC C6, informs the reader of (i) the present main concerns of SC C6 activities and (ii) its membership. (Additional useful information has also been inserted by this writer, for example Technical Brochure (TB) numbers and related *Electra* references plus other useful details together with further TBs and other source material that might be helpful to the reader for future guidance.) A summary of key information from this CIGRE SC C6 AGM is reproduced in the following sections.

B.1 Scope and membership

- Assessment of the technical impacts and requirements that a widespread adoption of distributed/dispersed generation could impose on the structure and operation of the network system
- Assessed together with the practical importance and timing of the related technical impacts and requirements
- Rural electrification, demand-side management methodologies and application of storage also fall within the remit or scope of SC C6
- At this time SC C6 had 24 members, 12 observer members and 70 experts following C6 activities and attending SC C6 meetings. The worldwide membership represents manufacturers, utilities, consultants, universities and research institutes. Young experts are especially welcome to participate in Working Group activities.

B.2 Active distribution networks

- 'Modern systems require increased system security (including security from cyber hacking/infiltration into networks), safety, environmental performance, power quality and energy efficiency impose new requirements for the development and operation at distribution level.

- Distribution grids are progressively being transformed from *passive* to *active networks*, in the sense that decision making and control is distributed and power now flows bi-directionally.
- This type of network eases the participation of distributed generation (DG), renewable energy sources (RES), demand response (DR) and energy storage, creating opportunities, protocols and standards.
- The function of the active distribution network is to efficiently link power sources with consumer demands, allowing both to decide how best to operate in real time. Power flow assessment, voltage control and protection require cost-competitive technologies and new communications systems with Information and communication Technologies (ICT) playing a key role.' Hatziargyriou reminds the reader (*Electra*. 2011;**258**(October):24).
- SC C6 recognises that the co-ordination of 'a multitude' of small resources poses a huge technical challenge and calls for the application of more decentralised, intelligent control techniques. Massive implementation of *smart-metering* will be required to make possible the effective co-ordination of the above-mentioned generation resources.
- Finally, in order to take full advantage of the potential benefits provided by DER and customer participation, novel distribution network architectures are required. 'Microgrids' (also called active cells) and 'virtual power plants' are two concepts that are being actively investigated by SC C6. (Its findings will be published in *Electra*.)

B.3 Strategic direction

The following requirements (1–4) define and set out the agreed strategic direction of CIGRE, SC C6 at the time of the 2011 Annual Report. Clearly, all interested readers are encouraged to follow up progress in SC C6 Working Groups (WGs) year on year.

1. connection and integration of DER, including small-size generators, storage and relevant power electronic devices
2. application of the DER concept as a part of the medium- and long-term evolution of distribution systems (microgrids and active distribution networks)
3. demand management and active customer integration
4. rural electrification.

B.4 Some completed SC C6 work

SC C6 reports that several WGs have recently completed their activities and published their results in e-cigre and in CIGRE *Electra* magazine – or are very close to finalisation. Topics include:

1. Connection and integration of DER, including small-size generators, storage and relevant power electronic devices.

    SC C6 informs the reader that these are the basic components that characterise modern distribution networks and the way they are connected to the

system ('fit and forget' vs 'active integration in management and control') is key to their formation formulation:
- SC C6-TF C6.04.01, 'Connection Criteria at the Distribution Network for Distributed Generation', TB 313-SC C6 WG C6.04.01, 2007

  'The Task Force brochure contains a review of the current connection criteria and protection practices applied in various countries for DGs; a review of existing international standards; simplified methods applied to DG connection in various countries; identifies inadequacies in existing connection practices and describes appropriate connection analysis techniques that could be applied for connection evaluation and formulates relevant recommendations.'
- SC C6-TF C6.04.02, 'Computational Tools and Techniques for Analysis, Design and Validation of Distributed Generation Systems'
- SC C6-WG C6.08, 'Grid Integration of Wind Generation', TB 450-WG C6.08, 2011 (See also *Electra*. 2011;**254**(February):62–7; prepared by WG C6.08; same title.)

  'The increasing amount of fluctuating wind energy will change the planning and operation methods for transmission and distribution systems significantly. This TB describes this challenge and presents solutions for different countries. It is focused on the following topics: power flow control and contingency management, influence on the conventional generation, frequency control, grid stability, reactive power and voltage control.'
- SC C6-WG C6.15, 'Electric Energy Storage Systems', TB 458-SC C6 WG C6.15, 2011

  'Active distribution networks, a recent developed concept considers the move towards the management of active distribution systems, where the DSO integrates the distributed generation, demand response and energy storage into the operation of the system. The technology, its functionalities and state-of-the-art are defined, and the TB concludes with recommendations for actions to promote further development of this concept and its applications.'
- JWG C1-C2-C6.18, 'Coping with Limits for Very High Penetrations of Renewable Energy' (still in progress at this time)

2. Application of the DER concept as a part of the medium- and long-term evolution of distribution systems (microgrids and active distribution networks).

   SC C6 informs the reader that novel distribution network architectures need to be developed, in order to be operated as active distribution networks. Two innovative structures proposed are 'microgrids' and 'virtual power plants'. The way these approaches are integrated in the operation and also medium- and long-term planning procedures is another key issue.
   - SC C6-WG C6.10, 'Technical and Commercial Standardisation of DER/ Microgrids Components' (TB 423-WG C6.10, 2010); (see also *Electra*. 2010;**251**(August):52–9, same title)

'TB (i) reviews commercial and technical communications standards applicable to low-voltage distributed generation technologies and (ii) provides a critical review. It is hoped that the two-fold target will *pave the way* to allow the easy installation of micro sources with plug-and-pay capabilities into existing power networks, with particular focus on microgrids'.

- SC C6-WG C6.11, 'Development and Operation of Active Distribution Networks' (TB 457-SC C6 WG C6.11); (see also, *Electra*. 2011;**255**(April): 70–3, 2011; prepared by WG C6.11, same title)

  'Active distribution networks are a recently developed concept which considers the move towards the management of active distribution systems, where the DSO integrates distributed generation, demand response and energy storage into the operation of the system. The TB defines this technology, its functionalities and the state-of-the-art and concludes with recommendations for actions to promote further development of the concept and its applications.'

3. Demand management and active customer integration.

    SC C6 informs the reader that the new role of the 'prosumer' (*pro*ducer/con*sumer*) is another key issue in the modern distribution network development. Relevant activities have been performed within WG C6.09, 'Demand Side Response'.

4. Rural electrification.

    Although not directly relevant to distribution networks, rural electrification provides interesting insights on how new systems should evolve in areas where networks are not yet fully developed and poses the basic question: should the same route of centralised systems similar to the developed countries be followed or would a more decentralised approach prove more efficient and cost-effective?

    - SC C6-WG C6.16, 'Technologies for Rural Electrification' (WG established in 2009 and work still in progress).
    - *Note*: Meantime WG C6.13 has presented a scoping report, 'Rural Electrification: a Scoping Report', in *Electra*. 2009;**244**(June):8–10. A comprehensive version of the scoping report together with a list of reference documentation can be found on the website of CIGRE's SC C6 (www.cigre-c6.org) under 'publications' and 'other documents'.

B.5 New C6 working bodies

In 2010, SC C6 established five new WGs and one advisory group:

WG C6.19 'Planning and optimisation methods for active distribution systems'.
'WG will identify the requirements of planning methodologies, models for short-, medium-, and long-term planning (i.e. technical, economic and market models) and algorithms for active distribution system expansion/upgrade planning suitable to different scenarios and regulatory framework.' (Expected completion is 2012–13.)

WG C6.20 'Integration of electric vehicles (EV) in electric power systems'.
This WG will mainly identify the impacts of a massive integration of EVs in the future transmission and distribution electricity grids, identify potential smart control approaches based on *smartgrid* concepts, to allow the deployment of EV without major changes in the existing network and power system infrastructures, identify the most appropriate ways to include EV into electricity markets and collect practical experience from utilities, pilot projects and studies and information on standardisation efforts for EV charging systems. (Expected completion is 2012–13.)

WG C6.21 'Smart-metering-state-of-the-art, regulation, standards and future requirements'.
This WG will provide a Technical Brochure on regulatory approaches in different countries, the state-of-the-art on-going or in progress (applicable standards, AMI deployment agenda, principal requirements and objectives, etc.), the most widely used communication standards and protocols and the testing procedures already used for AMI. (Expected completion is 2012–13.)

WG C6.22 'Microgrids evolution roadmap'.
This WG will formally define and clarify the *microgrid concept*, will examine market and regulatory settings for microgrids, control elements and control methods of a microgrid, provide a justification of microgrid deployment and finally a microgrid evolution roadmap including electricity infrastructure replacement scenarios. (Expected completion is 2012–13.)

WG C6.23 'Capacity of distribution feeders for hosting DER'.
This WG will study the limits of distribution feeders for hosting DER and will derive practical guidelines for the connection of DERs without the need to resort to detailed analytical studies. To achieve this it will examine the problems caused by the connection of DER and the distribution level, review national experience, case studies and DER connection standards. It will also investigate the effect of DER, DSM, EVs and network control in increasing hosting capability. It will identify limitations and gaps to adopt DER control at the MV and LV levels. (Expected completion is early 2013.)

CIGRE Advisory Group (AG) on 'Terminology'.
This AG aims to establish a worldwide accepted glossary of terms and definitions related to the SC C6 activities including whole range of distributed resources including storage. The active participation of the customer and the resulting changes brought in the operation and control of the distribution networks that is their evolution from *passive to active* components in the power system.

B.6 Two other events

The 2011 CIGRE SC C6 meeting was held in Bologna in conjunction with CIGRE symposia, 'Electric Power Systems of the Future – Integrating Supergrids and

Microgrid Technologies', September 2011, pp. 13–15. CIGRE SC C6 also supported the symposium 'Novel Solutions of the Information and Communication Technology as the Backbone of Smart Distribution', 12–14 April, Darmstadt, Germany. (An overview of 'projects funded for realisation of 'smart distribution' in Europe and in Germany', was also presented.)

B.7 Other Technical Brochures of interest

TB 475-SC C6 WG C6.09, 'Demand Side Integration (DSI)', 2011. (See also *Electra*. 2011;**257**(August):100–7, prepared by WG C6.09; same title.)

'The WG comment that this is a wide-ranging report, dealing with Demand Side Integration (DSI) which is focused on advancing the efficient and effective use of electricity in support of both power system and utilisation. Demand Side Resources (DSR) are resources on the customer side of the meter that can be relied on to respond in a co-ordinated fashion to electric power system or market conditions.'

TB 458-SC C6 WG C6.15, 'Electric Energy Storage Systems', 2011. (See also *Electra*; **255**(April):74–77, prepared by WG C.15; same title.)

'Renewable energy sources are becoming an important component of the power system structure. Nevertheless, their integration is a challenging task. WG-C6.15 has investigated electric energy-storage systems aiming to evaluate different storage technologies, their commercial backgrounds and the integration and support of power networks with a high penetration of renewable-based generation (RES).'

TB 311-SC C6 WG C6.03, 'Operation Dispersed Generation with ICT (Information & Communication Technology)', 2007.

Finally, *M.D. Galus and G. Andersson* (ETH Zurich – Power Systems Laboratory, Geneva, Switzerland) have prepared a two-part Scientific Paper in *Electra*, June/August 2012.

Part 1, 'Balancing renewable energy sources using vehicle to grid services controlled' (*Electra*. 2012; **262**(June):4–11 lists 22 references). It is well known that wind energy sources are fluctuating in nature. Consequently, 'they introduce challenges to power system operation, which was historically designed for generators with constant infeed'. The authors report that 'these fluctuations are balanced by ancillary services contracted from flexible generators or storages'.

To date, only pumped hydro can provide large-scale solutions for storage. However, these authors are exploring a potential solution for large storage, namely, 'to aggregate plug-in "hybrid electric vehicles" and "electric vehicles" – once they have been adopted in a wide enough scale into virtual storage. The paper investigates how a large fleet of electric cars can be aggregated over a large urban electricity distribution network and how this storage can be used to balance infeed error. The balancing is 'claimed to be achieved through so-called vehicle-to-grid services'.

The model developed by the authors is able to simulate the complete distribution network of the city of Zurich, Switzerland, with a wide-scale adoption of plug-in electric vehicles (PEVs), and their use in electricity networks including controlled charging, that is smart-charging, as well as vehicle-to-grid (V2G) services.

'The PEV behaviour throughout the day is determined by an agent-based transportation simulation model called MATSim,' the authors report.

So far only Part 1 has been published. The reader is encouraged to read Parts 1 and 2 when available. Clearly, this is a very interesting technical concept but in the UK, the penetration of electric and hybrid electric cars is very low at present in many locations and at best, this looks like having a very limited application for a long time – unless the *public are squeezed from using petrol- or diesel- fuelled vehicles, by government edict(s)*. Nevertheless, this writer does hope that this concept study has a successful outcome and may find an application in 'ultracities of the future'.

Section 9.9.2 and appendices on CIGRE activities and publications prepared by Hugh Ryan.

Chapter 10

Differences in performance between SF$_6$ and vacuum circuit-breakers at distribution voltage levels

S.M. Ghufran Ali

10.1 Introduction

This chapter reviews the circuit-breaker designs which are type tested to IEC 62271-200 and IEC 62271-100 for use on distribution voltages up to 52 kV. The reader is provided with a valuable bibliography for future background study [1–16].

The design and service experience of different types of commercially available circuit-breakers are considered. The chapter also discusses some special switching duties and focuses on aspects which are necessary for the selection of circuit-breakers for various duties: for example for switching capacitor banks, capacitive and inductive currents, generators, reactors and synchronised switching of transformers with reactors on the secondary side.

A list of recent relevant IEC Standard reports is provided in Appendix A of this chapter, while Appendix B touches briefly on some recent CIGRE WG studies concerning 'the impact of new functionalities on substation design'.

10.2 Circuit-breaker

A circuit-breaker is an electro-mechanical device which initiates (makes) or interrupts (breaks) the flow of current in a circuit and is used for controlling and protecting the distribution system. It has to be reliable as it may remain dormant in a closed position for a long period and yet, when it receives a trip command signal, it must operate without any hesitation. Depending on its rating, the circuit-breaker has to interrupt currents from as little as 10 A up to its full short-circuit rating which may be 40/50 kA. (A brief bibliography [1–16] is provided at the end of this chapter.)

From 1900 to the early 1970s most of the AC circuit-breakers used oil (bulk or small oil volume (SOV)), air-break and air-blast interrupting techniques. These are now obsolete. Interrupting devices utilising semiconductor technology are still being developed and are not yet available commercially at a competitive price.

Since the mid-1960s vacuum and sulphur hexafluoride (SF$_6$) circuit-breakers have been available to the power supply industry. The reader should be aware that

CIGRE Study Committee (SC A3), High-voltage equipment recently reported (see also *Electra*. 2013;**266**(February):30–4) that it was currently studying the impact of application of vacuum switchgear at transmission voltages. This recent resurgence of interest in vacuum circuit-breakers relates to the excellent operational capability, with frequent operation, and the associated reduced maintenance work required during the lifetime of VCBs together with the absence of SF_6. Some users still need to be convinced and have issues relating to capacitive switching and dielectric performance (i.e. late-breakdown and scatter in dielectric breakdown levels) as compared with SF_6 technology. However, experts involved in this CIGRE study are of the opinion that current chopping transients and X-ray generation will not be issues. (Readers are encouraged to follow progress in this important worldwide collaborative CIGRE programme which will be of strategic interest to both modern distribution and transmission sector users.)

10.3 Vacuum circuit-breaker

A vacuum interrupter is a high-technology, sealed-for-life, ceramic bottle. It has either a vapour condensation shield or a magnetic coil. It has butt contacts of disc or cup-shape designs, a contact gap of about 8–10 mm and a vacuum of 10^{-6} to 10^{-8} torr.

Several contact materials have been tried since 1960. The semi-refractory material, chromium (Cr), together with good conductor material, copper (Cu), has emerged as the best for circuit-breaker application.

The contact shape and material have been developed to reduce:

- contact bounce and contact welding
- contact wear
- chopping current.

The development of commercial vacuum interrupters is a highly involved and extremely costly process. However, once the vacuum bottle is manufactured, it can be mounted in a circuit-breaker, in any position.

In general, vacuum circuit-breakers:

- are safe and reliable
- are compact
- have low contact wear
- require low maintenance.

Vacuum bottles rated at 50 kA, 4,000 A, are now available from several manufacturers. This relatively high rating has been made possible because of the inherent efficiency of vacuum circuit-breaker technology. The dielectric withstand level of a 1 cm gap in a vacuum of 10^{-6} mm of mercury is about 200 kV but increases only very slightly with increase in contact gap. Thus it limits the withstand voltage which can be applied to each break in vacuum.

The drawback facing switchgear designers using vacuum bottles is that they cannot change many design parameters to improve the circuit-breaker ratings. They have to depend on a limited number of vacuum interrupter manufacturers for a bottle to suit the required duty.

Vacuum circuit-breaker designs are now available for ratings of 50 kA to 36 kV and 4,000 A (Figure 10.1). In spite of considerable research and development, coupled with design improvements over the last three decades, it is acknowledged that vacuum circuit-breakers are prone to reignitions and current chopping because of their excellent insulating and interrupting characteristics. The efficiency of the vacuum interrupting device therefore sometimes works against it and causes premature interruption of small inductive currents. Under these circumstances excessive overvoltages are produced, the magnitude of which depends on the high surge impedance of the switched circuit. Therefore, care should be taken when selecting a vacuum circuit-breaker for small inductive and reactor switching duties. Metal-oxide surge arresters (MOSAs) have to be used in this application to provide additional protection for connected equipment.

Some manufacturers produce vacuum bottles up to 145 kV, but these are very expensive and are only used on circuit-breakers for special duties. Historically, for reasons of cost, almost all the switchgear manufacturers in the world limit the use of the vacuum bottles on their general purpose distribution circuit-breakers up to the maximum voltage of 36 kV.

10.4 SF$_6$ gas circuit-breakers

SF$_6$ is a gas with unique features which are particularly suited to switchgear applications. It has been successfully employed in the HV, EHV and UHV (up to 1,200 kV) circuit-breakers for the last 35 years. There is a consensus of opinion that SF$_6$ gas has still much to offer, and therefore research and development continue. There is no single superior gas available at present to replace SF$_6$.

It is worth noting that some current-interrupting designs have combined the excellent dielectric properties of SF$_6$ gas with vacuum technology. One has produced a compact gas-insulated vacuum circuit-breaker design, based on their 170 kV gas-insulated switchgear (GIS)/SF$_6$ circuit-breaker. The 36 kV panel is a single-pole, fixed-mounted metal-enclosed construction in which vacuum bottles are housed in SF$_6$ gas to minimise the clearances to earth. The current is interrupted by the vacuum bottles and the resulting panel has a width of only 600 mm.

The knowledge and experience of SF$_6$ gained over the last four decades have been skillfully and successfully employed in the development of three efficient designs for distribution circuit-breakers. These are based on single-pressure puffer, rotating-arc and auto-expansion techniques.

10.5 Puffer circuit-breaker

The principle of a single-pressure puffer-type interrupter is explained by the operation of a universally known device, a bicycle pump. In this type of interrupter, SF$_6$ gas in the chamber is compressed by the movement of the cylinder against the stationary piston mimicking the operation of the bicycle pump. The high-pressure gas is then directed across the arc in the downstream region through the nozzle to complete the arc-extinguishing process (Figure 10.2).

Figure 10.1   Vacuum circuit-breaker air/SF$_6$ gas insulated

SF$_6$ and vacuum circuit-breakers at distribution voltage levels 411

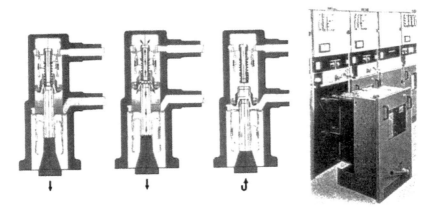

Figure 10.2 SF$_6$ puffer-type circuit-breaker

During an opening operation, the gas pressure-rise generated by compression and heating of SF$_6$ gas by the arc produces retarding forces, acting on the piston surface. High-energy mechanisms are required to overcome these forces and to provide consistent opening characteristics for all short-circuit duties. Puffer circuit-breakers are now well established on distribution systems for ratings up to 50 kA to 36 kV and 4,000 A. The performance of the present-day puffer interrupters has now been considerably improved and some designs operate at SF$_6$ gas pressure only slightly above atmospheric pressure. This relatively low pressure assists to reduce not only the earlier problem of gas-leak through the seals but also the energy requirement for the operating mechanism and therefore has a minimising effect on the cost.

There are a few 12 kV puffer-type circuit-breaker designs operating at 0.6 bar (g) which are capable of interrupting the rated short-circuit-breaking current at 0 bar (g).

Some puffer interrupter designs produce excessive switching overvoltages. For improved interrupter designs, the overvoltages for all switching duties are ≤2.5 pu. Therefore, care should be taken when selecting a circuit-breaker for small inductive and reactor switching duties in service.

Switchgear manufacturers are constantly seeking to improve their circuit-breaker designs to achieve:

- cost and space reductions
- improved arc interruption technique
- increased fault current ratings
- longer contact life
- improved inductive and reactor switching performance producing very low overvoltages
- low-energy and reliable circuit-breaker operating mechanisms
- ultimately a maintenance-free and sealed-for-life circuit-breaker.

As a result of this effort, the rotating-arc and auto-expansion designs have evolved.

10.6 Rotating-arc circuit-breaker

The principle of arc rotation for current interruption is not new and is not confined to SF_6 circuit-breakers. Arc rotation is also used to reduce contact erosion on some vacuum and SF_6 puffer circuit-breaker designs. Switchgear designers use different rotating-arc techniques, one of which utilises an electromagnetic coil. Unlike SF_6 puffer-type circuit-breakers, where the SF_6 gas is blown across in the arc, the arc in the rotating-arc device is rotated at a very high speed by an electromagnetic field. The field is produced by the coil through which the current to be interrupted flows during the period between contact separation and arc extinction (Figure 10.3).

This technique has been used for a SF_6 circuit-breaker design (Figure 10.4), which has ratings up to 40 kA at 12 kV. The device requires a low-energy mechanism and produces overvoltages less than 2 pu for all known switching circuits.

The sealed-for-life, maintenance-free, low operating energy SF_6 and rotating-arc circuit-breakers are now commercially available.

10.7 Auto-expansion circuit-breaker

Auto-expansion is a relatively new interruption technique to be used on distribution circuit-breakers and owes much to the development of third-generation puffer-type interrupters for 170 kV, 40 kA circuit-breakers.

This technique essentially uses the power dissipated by the arc to increase pressure within the expansion chamber. The overpressure generated has the double effect of causing strong turbulence in the gas and of forcing the gas out of the chamber across the arc as soon as the nozzle starts to open, thus extinguishing the arc (Figure 10.5).

The auto-expansion type of circuit-breaker uses a low-energy operating mechanism and produces switching overvoltages up to 2 pu. This technique is used to obtain up to 50 kA rating at 12 kV.

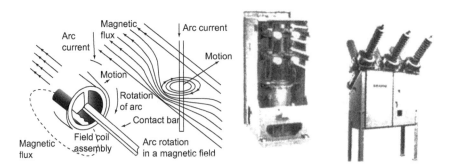

Figure 10.3 SF_6 rotating-arc principle

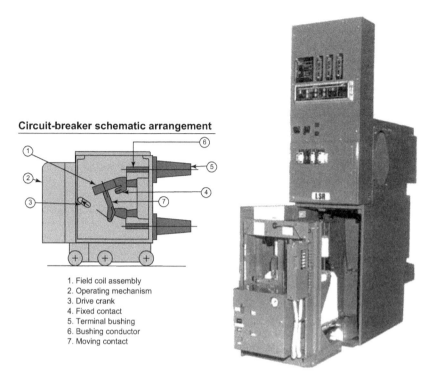

Figure 10.4 SF_6 rotating-arc circuit-breaker

Another design combines the rotating-arc and auto-expansion techniques to produce circuit-breaker design for ratings up to 72.5 kV (Figure 10.6).

10.8 Operating mechanism

The trend towards increased remote control of distribution networks has led to the requirement of either solenoid or motor-charged spring-operated mechanisms, suitable for multishot auto-reclose duty. Low-energy, simple, motor-charged spring mechanisms avoid the extra cost of providing and maintaining battery/rectifier supplies for solenoid operation-type mechanism.

Although IEC 62271-100 requires only 2,000 satisfactory operations to prove its performance, the present tendency is to carry out extended 10,000 trouble-free operations tests to demonstrate compatibility of these mechanisms with the vacuum and SF_6 interrupters which are virtually maintenance free.

10.9 Choice of correct circuit-breaker for special switching duties

System designers and users will need to give careful consideration to the choice and type of circuit-breaker for some of the applications identified in the following sections.

414 High-voltage engineering and testing

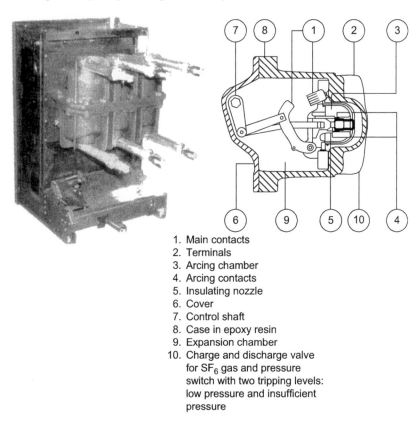

1. Main contacts
2. Terminals
3. Arcing chamber
4. Arcing contacts
5. Insulating nozzle
6. Cover
7. Control shaft
8. Case in epoxy resin
9. Expansion chamber
10. Charge and discharge valve for SF_6 gas and pressure switch with two tripping levels: low pressure and insufficient pressure

Figure 10.5 SF_6 auto-expansion circuit-breaker

10.10 Capacitive and inductive current switching

The phenomena of current chopping and reignition and associated high-frequency oscillatory overvoltages are well understood. It has also been established that the high-frequency oscillations are governed by the electrical parameters of the circuit concerned, circuit configuration and interrupter design.

Sometimes higher frequency voltage oscillations are caused by premature interruption of low capacitive and inductive currents. This premature interruption produces high-frequency reignitions and overvoltages which can give rise to dielectric failures and extensive damage to system equipment.

Historical note: As a result of some circuit-breakers, after having been satisfactorily tested to the existing IEC 62271-100, subsequently failing in service, an IEC working group was established to investigate the problem of higher frequency overvoltages caused by circuit-breaker operation and submit their recommendations. The result of this work was the issue (in 1994) of IEC 61233 to cover

SF$_6$ and vacuum circuit-breakers at distribution voltage levels 415

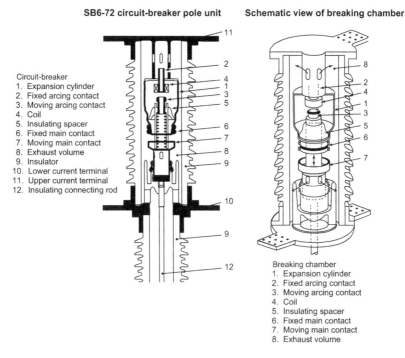

Figure 10.6 SF$_6$ rotating-arc/auto-expansion combined circuit-breaker

inductive current switching. IEC 62271-100, which covers the capacitive current switching requirements, was also circulated for comments. This latter draft document is still under consideration. The final recommendations have been incorporated in IEC 62271-100.

In addition to testing the circuit-breakers to the latest specifications, where possible, damping circuits and metal-oxide surge arresters should be used to ensure safe operation.

10.11 Circuit-breakers for generator circuit switching

Since there is no IEC Standard for generator switching duties, the use of distribution voltage circuit-breakers for generator switching needs careful examination.

As may be expected, a circuit-breaker used for switching distribution circuits may be subject to prospective fault duties in service which are different from the prospective fault duties of a generator circuit-breaker. The difference in fault duty mainly results from the difference in X/R ratio between the distribution circuits and generator circuits as explained in the following sections.

Before choosing a circuit-breaker suitable for generator circuits the following two main requirements will have to be considered.

10.11.1 DC offset

Generator circuits (i.e. the generator and generator transformer combination) usually have a higher X/R ratio (40–70) than the distribution circuit ($X/R = 14$). This difference means that the DC offset of the fault current waveform in the generator circuit will decay much more slowly than the DC offset in a distribution circuit. A current zero in the generator circuit may not be reached until after full contact travel is achieved. Circuit-breaker designs tested to IEC 62271-100 may have difficulty in clearing fault current under these conditions. A key factor in switching generator circuits is the arc resistance which lowers the X/R ratio and so reduces the DC offset. Some types of circuit-breakers have inherently higher arc resistance than others, for example the now-obsolete air-break and the present SF_6 rotating-arc and self-blast circuit-breakers have higher arc resistance than vacuum circuit-breakers. The high resistance of the arc column modifies the time constant of the circuit and helps to quench the arc in a relatively short duration.

In the absence of any IEC Standard for generator circuit-breakers, all circuit-breakers offered for generator switching duty should be tested to IEEE C37-013:1989, which provides guidelines for asymmetrical interrupting capability found when switching generator circuits.

10.11.2 Current chopping and reignition

The amplitudes of overvoltages resulting from current chopping and reignition and their rates of change submit the windings of the inductive equipment including generators to different types of risks. The amplitude of the overvoltage stresses mainly the insulation to ground, whereas the rate of change of overvoltage stresses in particular the inter-turn insulation. It is therefore essential to choose the correct circuit-breaker for switching the inductive currents of the generator circuits.

Circuit-breakers which are prone to current chopping and reignition and produce excessive overvoltages may not be suitable for generator switching applications.

10.12 Synchronised switching

Circuit-breakers with different pole opening and closing points are offered for synchronised switching of capacitor banks or transformers loaded with reactors. The circuit-breakers offered for these duties have intentional pole scatter on the three poles and are operated by one mechanism. For suppressing any switching surges, controlled point-on-wave switching is carried out by an electronic synchronising device.

The application of these modified circuit-breakers is either for some special switching duties or for both the switching and the fault interruption duties as a general purpose circuit-breaker.

The novel nature of pole scattering on a circuit-breaker and the use of synchronising devices appear to be attractive, but before these can be fully accepted, they should be subjected to type testing using the *direct testing method* and their performance proven for all duties.

Test evidence of circuit-breakers of similar design but without any intentional pole scatter (i.e. with simultaneous contact touch and part) should not be applied to the staggered pole circuit-breaker for the following reasons:

(a) Increased pre-arcing on the pole with maximum scatter during closing operations could affect its close–open performance.
(b) Increased arcing during every open operation on the pole with maximum scatter can cause heavy erosion of the contacts, resulting in increased pole scatter, and consequent effect on long-term close–open and open performance.
(c) Increased, unbalanced short-circuit and mechanical stresses on the cranks and the operating rods could cause them to fail.
(d) Testing authorities agree that a test certificate is applicable only and exclusively to the particular circuit-breaker which is subjected to the test series.

If a manufacturer intends to use the maximum pole scatter allowed in Clause 5.101 of IEC 62271-100, then the type test should be performed on the most onerous condition required by IEC.

When choosing a suitable circuit-breaker for these special duties, consideration should also be given to the circuit-breaker designs, which are capable of performing all special switching duties without excessive overvoltages on the system.

10.13 Conclusions and future developments

Modern distribution voltage circuit-breakers are simple and virtually maintenance-free, and some designs have now achieved ratings up to 50 kA at 12 kV.

The vacuum and SF_6 rotating-arc and self-expansion circuit designs up to 36 kV require low-energy operating mechanisms and have a long service life. They provide reliable and low-cost circuit-breakers.

Reactor switching causes difficulty for some circuit-breaker designs. These should be tested to IEC 61233, and for added safety it is recommended that metal-oxide surge arresters and R-C tuning (damping) circuits should be used with vacuum circuit-breakers. It is also essential to choose the correct circuit-breakers with overvoltage factor ≤ 2.2 pu for switching low-inductive current circuits.

Distribution circuit-breakers type-tested to IEC 62271-100 may not be suitable for generator switching circuits. Any breaker offered for this duty should be tested to IEEE C37-013, 1989. Circuit-breakers that produce excessive overvoltages may not be suitable for generator circuit switching operations. Circuit-breakers with intentional pole scatter for synchronised switching should be type-tested on their own, using direct testing techniques.

CIGRE perception of future developments – prepared and compiled by H.M. Ryan

- Strategically, at this time, there is a world economic downturn, on the one hand, and an urgent need for major technical network enhancements, huge investments and much strategic planning as the industry moves towards developing (i) 'smart-grids' and (ii) enhanced 'networks of the future' (see *Electra*. 2011;**259**, **256**, respectively) within the next one or two decades, on the other.
- It is vitally important for the reader to get an overview of the scale of this impact and the huge costs involved, together with the technical implications as to what is required. These aspects are considered in section 23.13, and especially in Tables 23.11–23.14.
- Against this background, it is not surprising that consideration continues to be given, at this time, to asset management (AM) themes such as the one considered in Appendix B, 'the impact of new functionalities on substation design'. Finally, Table 10.1 considers some of the key issues and drivers for the future including SF_6 and vacuum circuit-breakers at distribution voltage levels under the grouping 'increasing capacity and power control' (GIL, phase-shifting transformers, fault current limiters and superconductivity).
- These are all important elements, with fault current limiters and superconductivity having already been touched on in Chapter 5. The present writer considers that 'fault current limiters', which have been evaluated for many years and have not yet had commercial success, may achieve this success in the near future. *So watch this space!*

Table 10.1 Some key issues and main findings

The drivers behind the changes forcing substation development are briefly described along with some of the resolutions to these challenges but the TB concentrates on the direct impact on the substation itself. Based on an industry-wide survey, different themes are presented highlighting the technologies under consideration and the key impacts that need to be considered. Essentially, these can be grouped into:

- *power electronic applications* (FACTS devices, custom power, HVDC)
- *switchgear* (mixed technology switchgear, compact, non-conventional instrument transformers)
- *reactive compensation* (capacitor banks, shunt reactors, SVC, STATCOMs)
- *distributed generation* (integrating windfarms, energy storage)
- *substation automation* (protection, diagnostics and condition monitoring)
- *increasing capacity and power control* (GIL, phase-shifting transformers, fault current limiters and superconductivity)

Figure 1 from Appendix 4, the WG provided as an illustration of the summary of key issues presented in the complete brochure. Apparently each appendix contains specific examples and describes design aspects that are likely to be affected and offers some guidance on how the problem has been dealt with previously.

Figure 1 Example of the summary for reactive compensation (in Appendix 4)	
Appendix 4	Power electronics require reliable control systems and cooling circuits
	Specific maintenance strategy to ensure availability: power electronic auxiliary systems, strategic spares, capacitor cans
Shunt connected equipment to provide steady voltage support and dynamic system compensation post fault	Relocatable design may be necessary if generation patterns change
	Requirement for coordinated control if more than one reactive control device is installed on site
Steady state: fixed/switched	Step voltage, transients and resonances for switched units
Capacitor banks, shunt reactors	Monitoring of dynamic performance
Dynamic: SVC, STATCOM	Magnetic fields and acoustic noise near air-cored reactors
	Circuit-breaker switching duty for mechanically switched units

Main findings

In the *Electra* article the WG Osborne *et al.* stated some of the key observations to emerge during the review as follows

- Do not underestimate the additional cost and complexity of interfacing between new and legacy technology, particularly for secondary systems.
- There is an economic trade-off between capital and operational expenditure to minimise outage duration during construction, commissioning and maintenance.
- To accrue the real benefits of new technology, adoption of different substation practices and maintenance regimes may be required.
- An information system strategy must be established to manage substation data, security, access and software upgrading. The pace of change in the telecommunications and microprocessor industry is such that secondary systems can become technology obsolescent sooner than expected.
- Suitability of the existing substation design: Some technologies require paradigm shift in a utility's approach to substation development.
- New technology should be introduced in a controlled manner.

After *Electra*. 2009;**244**(June):44–9; prepared by Osborne *et al.*, CIGRE, WG B3.01.
Note: For any professional study the reader should always use the official CIGRE document(s).
[Some future developments are written and Appendix B prepared by H.M. Ryan.]

Acknowledgements

The author wishes to thank the following: the Directors of Merz and McLellan Ltd and Parsons Brinckerhoff for permission to publish this paper; and ABB, AEG, GEC-Alsthom, Hawker Siddeley Switchgear, Merlin Gerin, Ottermill, Reyrolle and Siemens for providing technical information and illustrations.

The views expressed in this chapter do not necessarily represent the views of Merz and McLellan Ltd and Parsons Brinckerhoff.

Bibliography

1. Ali S.M.G. 'Switchgear design, development and service'. In: H.M. Ryan (ed.), *High voltage engineering and testing*. IET Power and Energy Series (vol. 17, pp. 199–232). London: Peter Peregrinus Ltd; 1994 (Chapter 7)
2. Blower R.W. 'Factors in the design of sulphur hexafluoride switchgear for distribution system'. *IEEE Transactions on Power Apparatus and Systems*. 1984;**PAS-103**(9):2753–61
3. Blower R.W. 'SF$_6$ switchgear for power station auxiliaries'. *Modstream Power System Supply*. 1987(June)
4. Cornago K.M. 'The four MV breaking techniques'. *Asian Electronics*. 1985 (January)
5. Gibbs I.D., Koch D., Malkin P. 'Investigations of prestriking and current chopping in medium voltage SF$_6$ rotating-arc and vacuum switch-gear'. *IEEE/PES Winter Meeting*; New York, 1 January 1989
6. Hannebert J., Gibbs D. 'Behaviour of the SF$_6$-MV circuit-breakers Fluarc for switching motor starting currents'. *MG Cah. Tech*. 1988;**143**(December)
7. IEC Standards: IEC 298; IEC 56; IEC 694; IEC 1233; Draft IEC 17A; IEC 62271-200; IEC 62271-100; IEC 60694; IEC 61233; IEEE C37-013 (see also Appendix A and the extensive list of IEC Standards, detailed throughout this book)
8. Slade P.G., Long R.W. 'Vacuum technology for medium voltage switching and protection'. *Power Technology International*. 1993
9. Gellings C.W. 'Estimating the costs and benefits of the "smart grid" in the United States'. *Electra*. 2011;**259**(December):6–14 (This is a recent article by C.W. Gellings (from EPRI, US); the full study with the same title is available online at http://www.epri.com, Estimating the Costs and Benefits of the Smart Grid, TR-1022519)
10. CIGRE WG. 'Network of the future. Electricity supply systems of the future'. *Electra*. 2011;**256**(June):42–9
11. Browne E. Jr. *Circuit interruption: Theory and techniques* (vol. 21). A Series of Reference Books and Textbooks: Electrical Engineering and Electronics. New York, NY: Marcel Dekker; 1984
12. Lakervi E., Holmes E.J. *Electricity distribution network design* (vol. 21, 2nd edn). IET Power and Energy Series. London: Peter Perigrinus Ltd; 1995
13. Greenwood A. *Vacuum switchgear* (vol. 18). IET Power and Energy Series. London: Peter Perigrinus Ltd; 1994
14. Blower R.W. *Distribution switchgear*. London: Collins; 1986
15. Wright A., Newbery P.G. *Electric fuses* (vol. 20, 2nd edn). IET Power and Energy Series. London: Peter Perigrinus Ltd; 1994
16. Stewart S. *Distribution switchgear* (vol. 46). IET Power and Energy Series. London: Peter Perigrinus Ltd; 2004 and new cover in 2008

Appendices

A Relevant strategic IEC Standard reports

Equipment	IEC Standard
Switchgear	IEC 60050
	IEC 62271-200, -201
	IEC 62271-202, -203
Circuit-breakers	IEC 62271-200, -201
	IEC 62271-202, -203
	IEC 62271-100, -102
	IEC 62271-103, -105
	IEC 62271-107, -110
	IEC 60694
	IEC 61233
Disconnectors	IEC 62271-102, -203
Conductors	IEC 61089
Current transformers	IEC 60185
	IEC 60044-4
Voltage transformers	IEC 60186
	IEC 60358
Line traps	IEC 60353
Substation earthing	IEC 60080
Surge arresters	IEC 60099-1, -4
Bushings and insulator	IEC 60137
	IEC 60233
	IEC 60305
	IEC 60383
SF_6	IEC 60376
	IEC 60480
	IEC 61634
Insulation Coordination	IEC 60071
	IEC 60076-3
Testing	IEC 60168
	IEC 60270
	IEC 60427
	IEC 60060
	IEC 61633
Standard current ratings	IEC 60059

Note: The IEC Standards listed above are provided for guidance only. The reader should be aware that IEC Standard Specifications are reviewed at regular intervals. Consequently, the latest and relevant version of a standard on specific topic/theme must be used. This can be checked directly via standard specifications from IEC Standards (International Electrotechnical Commission) and National Standard organisations.

B The impact of new functionalities on substation design (relating to CIGRE published work) – abridged and compiled by H.M. Ryan

It should be remembered that the purpose of condition assessment and performance monitoring of equipment is to optimise asset management decision-making and risk reduction. Several strategic asset management (AM) themes have been actively studied and widely discussed recently by CIGRE (within its huge Study Committee structure). In 2012, an overview Technical Committee Report (see *Electra*. 2012;**262**(June): 44–9), 'Overview of CIGRE Publications on Asset Management Topics', provided a valuable, broad and strategic review of AM themes relating to several Technical Study Committee sectors (SCs A–D); complete details are listed in Chapter 23 (Tables 23.13 and 23.14). It is generally recognised that the availability of good, reliable lifetime asset data and information and relevant condition data is the basis for optimal AM decision-making. A comprehensive Technical Brochure should be published by CIGRE by late 2013 (SC C1.25) which will focus on six asset management topics:

1. Condition Assessment and Monitoring
2. End-of-Life Issues
3. Risk Management and AM Decision-Making
4. Grid Development
5. Maintenance Processes and Decision-Making
6. Collection of Asset Data and Information.

It is indeed timely that this comprehensive new Technical Brochure is currently being prepared, as many individual CIGRE studies have already been carried out on these themes; a coordinated update will be welcomed by the energy sector.

Theme 5, Maintenance Processes and Decision-Making, has been studied by CIGRE and several Technical Brochures have already been issued for transformers, high-voltage equipment, cables, transmission lines and substation groups [see also Table 8.1 (Chapter 8)].

In this appendix we will consider only the substation sector and in particular CIGRE TB 380. (See 'The impact of new functionalities on substation design', *Electra*. 2009;**244**(June):44–49; prepared by CIGRE WG B3.01, by M. Osborne (Convener, GB) *et al.*)

This 2009 *Electra* article was written by the Working Group (WG) who prepared TB 380-WG B3.01 (2009) and informs the *Electra* readership of some background 'nuances' to the development of this CIGRE document, namely:

- The TB provides guidelines on the technical issues and challenges of introducing a range of new technologies into the substation environment.
- The TB is set out to provide a general overview of the issues on the main body of the document and fine detail and examples for the specialist in the appendices.
- The *Electra* article goes on to observe that Task Force 4 (*Substation Concepts*) has examined a broad spectrum of technologies and this document provides guidance for developers in implementing whichever solution meets the various challenges facing the transmission or distribution network.

- It is emphasised that this document is not intended as a problem solver for system issues but rather as a guide to some of the practical problems to be faced when installing the final solution selected.
- In this *Electra* article the WG go on to state that, while system aspects of new functionalities and their technical details have been broadly investigated in other texts, this TB aims to fill in gaps in knowledge relating to installing the application into the substation environment.
- The WG comment that one of the factors holding back the implementation of new technology is the lack of experience or confidence in new or different equipment. In addition there is the risk of the actual installation and its effect on the existing substation operation.
- In an attempt to address this issue the document includes a source of advice, experience and recommendations from users who have already implemented the technology.
- The document concentrates on the impact on the design and construction of substations. It is pointed out that in some cases this may fundamentally affect the basic concept of the design and construction of the substation.

The WG report that a simple process was adopted to consider the varying degree of impact from the network level, single-line diagram or bay perspective. It was considered that this helps the designer to quickly establish where risks may be evident and offers some remedial actions that could be taken to manage the issue.

Table 10.1 provides a very brief, highly abridged (less than two pages in English) summary of TB 380 'findings', now also reproduced above in a highly abridged format by this writer from the brief content of text (in *Electra*. 2009;**244** (June):44–9). This material has been reproduced and included herein mainly to provide an example, to the reader of this book, of the protracted processes that have to be adhered to, typically by anything up to 80+ worldwide experts contributing to a theme, before a CIGRE TB document is finalised and issued for sale to the industry. This appendix is for guidance only.

CIGRE currently has many specialist Technical Study Committees (SCs), as reported later in this book (see Chapter 23, Tables 23.13 and 23.14), covering studies into a wide and diverse range of strategic technical activities/themes. The regular *Electra* magazine publications/CIGRE websites, SC-AGM Reports and numerous international conferences/tutorial publications provide a rich ongoing CIGRE resource for the reader on strategic information covering many technical themes. In addition, the following footnote[1] provides the reader with an indication and useful perspective of the diversity of CIGRE committee/group structures and specialist publications.

---

[1] CIGRE key: SC, Study Committee; WG, Working Group and JWG, Joint Working Group; TF, Task Force; TB, Technical Brochure; TR, Technical Report (Sometimes referred to as a Technical Committee Report); SC, Scientific Paper.

Chapter 11

Life management of electrical plant: a distribution perspective

John Steed

11.1 Introduction

The effective management of assets to ensure that the user obtains the optimum life for the plant is becoming more vital as electricity distribution systems are worked harder. Systems and equipment need to be reliable.

The life management of plant is concerned with the life of the plant from preconception to final dismantlement (and recycling). In that journey the plant may encounter harsh operating and/or environmental conditions, which may test the limits of the original design. Figure 11.1 shows pictorially the whole operating lifecycle of plant from inception, design, test, commissioning and operating life to refurbishment. During the useful operating life, maintenance testing and monitoring perform an essential element for the user to decide whether refurbishment is an economic option to extend the life of the plant.

Plant reliability is essential to the overall reliability of the power system. Therefore, this chapter starts by considering sources of information on reliability and how this will help the user and manufacturer. Examination of system reliability data will enable the user to make further improvements on system reliability, to look for requirements for new plant and to help towards planning and prioritising an asset replacement strategy for the existing plant. In the majority of the equipment's operating life it will be passive – not necessarily actively moving, but nevertheless operating quite satisfactorily within the whole range of external environmental factors as shown in Figure 11.1. An important part of life management is to consider how long the plant can be left alone – to plan preventative maintenance before severe deterioration leads to failure and to consider what type of condition assessment should be done – whether it should be simply limited to inspection, or be intrusive, and if so, to what extent and how frequently. Condition assessment, monitoring and maintenance, therefore, feature largely in this chapter. A generous list of references is produced for the reader [1–35]. Finally, some comments are given on the modern risk-based approach to asset management and the need to work plant harder.

426 High-voltage engineering and testing

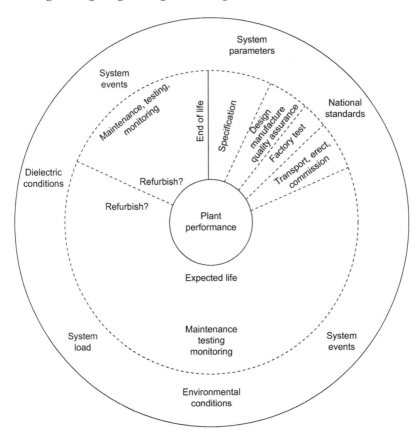

Figure 11.1 The operating life-cycle of plant
Source: After D. Allan & B. Corderoy, CIGRE 1992

There is no easy meaning to the 'typical life' of plant. Much depends on understanding the capability of the plant in all circumstances – being nosey about the design, being nosey about maintenance: knowing what to measure, having the ability to understand it, analyse it and then correctly interpret the result.

11.2 Reliability

11.2.1 Sources of data

There are four sources of data: failure data systems and reports, manufacturers, academia and, not necessarily as a last resort, legislative bodies.

11.2.1.1 Failure data

Within the UK electricity industry, the failure data is held as the National Fault and Interruption Reporting Scheme (NaFIRS), which has been in existence since 1965. This is the most comprehensive database in the world, enabling UK power engineers to have access to both customer interruption data and detailed plant reliability data. Used properly this is one of the most useful sources of system information for asset managers. Information on the general performance of the distribution system is produced [1], enabling benchmarking and comparisons to be made between companies. Plant performance is sometimes reported in general terms in professional papers [2].

Closely allied to the NaFIRS database is the 'fixed' or system data – the lengths of lines and cables and numbers of switches and transformers. This enables basic reliability data to be derived in terms of fault rates (e.g. typical fault rate per 100 km of cable, etc.). Basic population data can be investigated, but age of equipment is not collected nationally.

Apart from the failure database, defects are also collected within the electricity industry and managed by the Energy Networks Association. In this way, a formal network of company contacts is briefed on the details of defects found and the remedial measures agreed with the manufacturer. Allied to this may be notice of any important changes in operational practice for items of plant, which have a direct safety implication.

11.2.1.2 Manufacturers

The manufacturers are obviously the best source of information for the installation, commissioning, operation, maintenance and decommissioning of the plant – especially for the user with a limited plant and operational experience. Close co-operation with them at the pre-tender stage will ensure a good relationship during the operating lifetime of the plant.

11.2.1.3 Academia

Universities and institutes of higher education are a rich source of information on the lifetime of materials and components, and may have the facilities needed for forensic-type examinations of plant. With their current business-like approach, universities are forging close links with users.

11.2.1.4 Other sources

Other sources such as legislative bodies include, e.g., the Health and Safety Executive (HSE) in the UK.

In 1995, the HSE mounted a campaign to identify and deal with the problems of certain older oil-filled electrical distribution switchgear. All users of such equipment were required to complete a simple one-page form showing all their oil-filled switchgears manufactured prior to 1970. Accompanying the survey was an information document, 'Oil-Filled Electrical Distribution and Other Switchgear', Ref. 483/27. This document outlined the legal duties for the owners of premises where such equipment is operated. It is written in terms that the non-specialist manager can readily understand, and includes the following topics:

- knowledge of equipment, definitions
- overstressing of switchgear
- equipment modifications
- dependent manually operated switchgear

- maintenance
- provision of anti-reflex operating handles.

Advice is also given in the publication on the precautions, which should be taken to eliminate or control any risks in operation of electrical switchgear.

11.2.2 Typical distribution company requirements for data

There are three natural levels of managing the information from failure data; staff at each stage have very important duties towards the life management of plant.

11.2.2.1 The strategic or corporate planning level

This is the highest level where senior managers will look to the operation of the system to meet their top-level 'visionary' requirements. For example to seek real improvements in customer service, the vision may be to reduce the minutes lost per connected customer by 5% year on year (irrespective of weather conditions and other factors).

11.2.2.2 The managerial level

This is the second level, where the policy and procedures are written to enable the strategic vision to be achieved. For example there may be a change in emphasis towards more live-line working to ensure continuity of supply for customers during maintenance work. This level also conducts the more detailed search of a wide range of circuits' data to ensure conformity to the strategic goals.

11.2.2.3 The operational level

This is the 'sharp end', where the distribution system is managed. Part of this management will, e.g., include the identification and management of rogue circuits – those circuits that have a fault performance outside the acceptable range. Policy and procedures, e.g. on the life management of plant, have to be implemented. The management of resources – labour and materials – forms an important part of the operational level.

11.2.3 Case studies using data

11.2.3.1 Underlying assumptions

Reference 3 contains several examples illustrating the use of data. Analysis by direct cause, especially those attributed to ageing/wear, is examined in detail. A number of assumptions, however, have to be made when analysing the failure data, these being as follows:

1. Each failure is replaced, i.e., it is a non-repairable system.
2. The year of commissioning is the same as the year of manufacture.
3. The equipment is continuously energised *in situ*.
4. There have been no significant changes in equipment design between any periods considered.
5. Each item of equipment is treated as being a single component for the reliability analysis.

11.2.3.2 11 kV underground cables

A very simple example of information that can be obtained from fault statistics is of underground cables.

The move towards corrugated aluminium sheathed (CAS) cables in the 1970s enabled the replacement of the traditional paper-insulated lead-covered (PILC) cables.

Figure 11.2 reproduced from Reference 3 illustrates the failure rates of two types of 11 kV underground cable. Third-party damages were excluded so that only latent defects remain. The data was necessarily smoothed in three-year moving averages to show the trend. It can be seen that there was an increased failure rate soon after a new design of cable was introduced. This failure rate then decreased to match that of the long-established design (see comments on the bath tub curve, section 11.2.4).

Some care needs to be taken analysing reliability data, as will be illustrated for power transformers.

11.2.3.3 Comments on transformer reliability

Transformers per se are generally extremely reliable. Those 'failures' that do occur are often related to associated equipment (e.g. on-load tap-changers). Failures that cause sustained outages of transformers where a component has to be replaced are, for one Distribution Network Operator (DNO), equal to or below the national five-year average trends. For example the failure rate for 11/0.433 kV ground-mounted transformers is below 0.2% of a population of around 25,000.

Failures of 33/11 and 132/33 kV transformers due to the plant themselves (i.e. windings) are quite rare. For example for that DNO, only four winding failures of

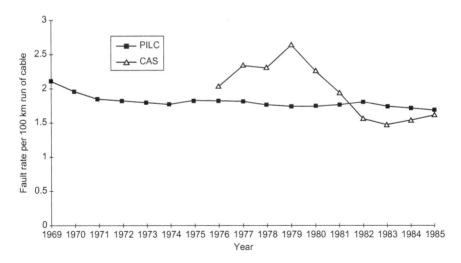

Figure 11.2 Failure rates of two types of 11 kV underground cables – three-year moving averages (all causes except third party)

430   High-voltage engineering and testing

33/11 kV transformers have occurred in over 6,300 transformer-years of operation. There were no inherent failures of 132/33 kV transformers in that period.

With high levels of reliability, there may not seem to be a problem, but with such large populations, the user has three main concerns. First that a full economic life is obtained for the transformer, second that a knowledge of the health of units in service is obtained bearing in mind the high capital cost and long manufacturing lead time for the larger transformers, and finally a reassurance that cascade failures will not occur for those in service of a common age range or type.

11.2.4 The bath tub curve

11.2.4.1 Theory

Without going into a detailed discourse on reliability theory, certain concepts in reliability are useful to understand the life management of electrical plant.

The reliability analyst should be concerned not only with failure rates (where the failures are conveniently expressed in terms of the numbers of the item in service, e.g. faults per 100 km of line, failures per 100 switches, etc.) but also with a related quantity called the *hazard rate*. The hazard rate, or 'instantaneous failure rate', is a more specific quantity and is defined as the *rate at which the remaining items fail*. The hazard rate, therefore, takes account of the *surviving* equipment rather than a number based on failures alone. It is for this reason that reliability analysts prefer the hazard rate for detailed studies rather than the simpler failure rate.

Conceptually, there are three phases that can describe the expected life of a component. The curve shown in Figure 11.3, the traditional bath tub curve, is drawn by plotting the hazard rate against time. The three phases in the life of the component are:

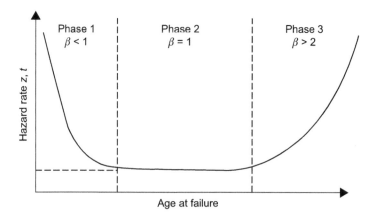

Figure 11.3   'Bathtub' curve

Phase 1: the early life failures with decreasing rate of failure (e.g. installation error including overstress, design inadequate to cope with extended duty, etc.)

Phase 2: the constant rate of failure (comprising random failures during the prime of life, where the operating stress occasionally exceeds the design stress)

Phase 3: the wear-out period characterised by an increasing rate of failure (e.g. due to fatigue and degradation).

Much has been written about the statistical analysis and interpretation of reliability data. There have been very many attempts at modelling different distributions (e.g. Lognormal, Rayleigh, Gumbel, etc.) to the data to predict future trends. One of the most commonly used distributions of the 'extreme value' type is the Weibull distribution. The Weibull distribution has become popular as it can be used to model the data in the form of a bath tub curve. The mathematics can get quite complicated, but in simple terms, if $R(t)$ denotes the reliability of a component at time t

$$R(t) = \exp\{-(\alpha t)^\beta\}$$

where α is known as the scale parameter and β is the shape parameter. Referring to Figure 11.3, the shape parameter values vary as:

- $\beta < 1$ phase 1
- $\beta = 1$ phase 2
- $\beta > 2$ phase 3:

Hence if β values can be found, then this will represent where on the bath tub curve the equipment (or groups of equipment) lie. β values can be obtained graphically by summing the hazard rates, plotting them on log-log paper and then determining the slope of the line to give the shape parameter β.

11.2.5 Practical example – distribution transformers

11.2.5.1 Background

Up until 1985, the then area electricity boards in the UK collected failure data (NaFIRS) and plant data – simply based on numbers of plant items or lengths of cable added each year. In 1985, a major survey was undertaken among the Boards in England and Wales to determine the age profile of distribution and transmission equipment on the system at that time. A considerable amount of detailed plant data showing the age distribution of plant in commission in 1985 was collected. Reference 4 gives examples from the information returned and makes suggestions on plant replacement strategies. Part of this data was combined with NaFIRS data to provide information on transformer lifetime. It enabled, e.g., the provision of the numbers of transformers still in service to be plotted against each year of manufacture. The techniques shown can be applied to individual or groups of individual circuits where detailed information is available for the period that the equipment has been in service.

11.2.5.2 Pole mounted transformers (pmt's)

The pole mounted transformer, unlike overhead lines or underground cables, is a replacement item (i.e. mostly non-repairable on site). So long as accurate age data are available, reporting is as accurate as possible and the user accepts the underlying assumptions (section 11.2.3.1), the pmt lends itself quite well to statistical analysis.

Figure 11.4 shows the cumulative failures of pmts by age, and illustrates, e.g. that in 1985/86, 50% of the failures attributed to deterioration due to ageing/wear involved units less than 25 years old. Also shown is the cumulative population age profile. The difference between the two curves and the crossover at 28 years is significant in that an age-related phenomenon is indicated. Had the curves been parallel (e.g. like the 'all causes except weather'), then the failures would not be age-related (as is the case for the 0–6-year-old failures). For pmt's less than 28 years old there were fewer failures for ageing reasons alone than might be expected, those surviving longer than 28 years exhibit more failures than expected from the population age profile.

Figure 11.5 shows the cumulative hazard rate plotted for the same data. Three points emerge:

1. The Weibull plot appears 'bimodal', i.e. there are two age-related phases shown with β values of 2.65 and 4.45, respectively. This would indicate two separate failure modes (possibly involving failures of different components): one for transformers up to about 10 years old and another for those 10–30 years old.
2. Most of the transformer failures in service occur within the age range of 20+ years.
3. Despite a β value of 4.45 (quite high) for the older transformers, the simple overall fault rate of 0.12% (irrespective of age) is 'acceptable'.

The third point is a conundrum: why should an acceptable failure rate coexist with a high Weibull shape parameter? Intuitively, from such a high β value, one might expect a failure rate at least 10 times of that observed. One explanation is that this result arises from a sensitive combination of low hazard rates and relatively few overall numbers of failures.

Reference 5 considers this example further and also looks at a similar study for ground-mounted transformers.

11.2.6 Human factors in plant reliability

This is the psychology of fault reporting! To derive the most benefit in terms of being able to improve the operation of the distribution system from operating a fault reporting system, there are four human-related aspects to consider:

Motivation: The user should have access to a user-friendly operating system to input and edit information.

Accurate reporting: The user must be encouraged to report accurately: the use of 'unknown' material, component or cause codes must be discouraged.

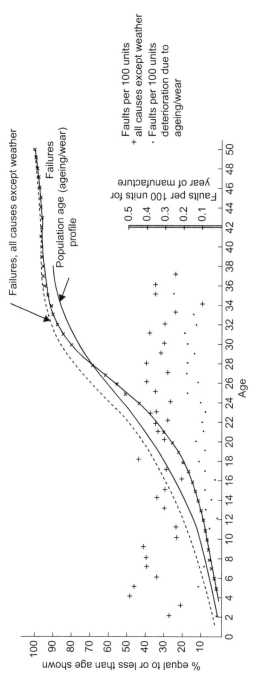

Figure 11.4 Pole mounted transformers: cumulative failures and age profile (1985/86)

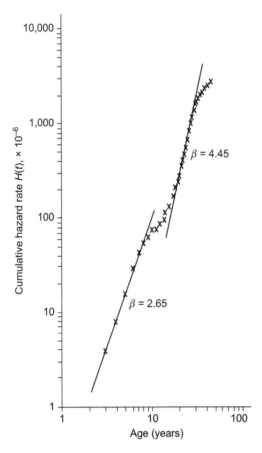

Figure 11.5 Pole mounted transformers 1985/86: cumulative hazard (Weibull) plot

Sophisticated systems with inter-field checks may give a very accurate database but it will more than likely discourage the busy reporter and lead to poor motivation and reporting.

Training: Users should have sufficient training to operate the database system efficiently. It should be the operators who input the data (not a remote or uninformed clerk!)

Feedback: Lastly, it is vital that feedback on the results from database searches is given to users in order that they can readily see the use to which data are put. The illustration of how the information is and can be used will have a positive effect on the motivation and accuracy of future reporting.

11.2.7 Conclusions on reliability

Failure and in-service defect data can provide a vital source of information on plant life management provided that it is correctly reported and properly interpreted.

11.3 Condition monitoring

11.3.1 Definitions

The normally quiescent state of electrical transmission and distribution system plant does not draw attention to incipient faults, which may develop from the gradual deterioration of the equipment. These faults may be detected during routine maintenance, but the ability to have detailed information on the state-of-health of transmission and distribution system equipment prior to carrying out maintenance work or alterations becomes a significant asset and adds an element of *preventative maintenance* to the operation of such assets.

Condition monitoring may be formally defined as:

> ... a predictive method making use of the fact that most equipment will have a useful life before maintenance is required. It embraces the life mechanism of individual parts of or the whole equipment, the application and development of special purpose equipment, the means of acquiring the data and the analysis of that data to predict the trends.

Certain key words are useful to recall from this definition:
predictive, useful life, maintenance, application and development of special purpose equipment, acquiring the data, analysis of that data, prediction of the trends.

In more practical terms, the initial stage of a condition monitoring programme consists of establishing the *baseline* parameters (what to measure) and then recording the *actual* baseline (or 'fingerprint') values. The next stage is the establishment of routine tours of plant and equipment observing the running condition and assessing the parameters previously determined for the baseline. These readings are then compared with the fingerprint, and the trends are determined.

The *state* of the present plant condition can be determined from the *absolute figures*.

The *rate of degradation* and an assessment of the likely time to failure can be estimated from the *trend*.

The resources committed to monitoring the condition of plant will depend on the numbers and on the service experience and reliability of the plant concerned. Readers should note that the widespread application of condition monitoring strategies for transmission equipment such as SF_6 switchgear substations, transformers and other equipment is considered in more detail in other chapters (e.g. see Chapters 20–22).

11.3.2 Benefits of condition monitoring

The benefits of condition monitoring can be summarised as:

- reduced maintenance costs
- the results provide a quality control feature
- limiting the probability of destructive failures, which lead to improvements in operator safety and quality of supply

- limiting the severity of any damage incurred, eliminating consequential repair activities and identifying the root causes of failures
- information is provided on the plant operating life, enabling business decisions to be made on either plant refurbishment or replacement.

11.3.3 Application to equipment

11.3.3.1 Switchgear

There are four aspects relating to possible failure modes. These are the deterioration of the dielectric medium (e.g. oil), the deterioration/failure of solid insulation, the degradation in performance of the mechanism and partial discharge.

The dielectric medium

On oil-filled equipment requiring maintenance, the condition of the dielectric oil will provide much information on the general condition of the plant. Basic oil-tests particular to oil-filled switchgear are shown in Table 11.1.

Testing of solid insulation

This is most often the insulation resistance test, but other tests are done as well (see section 11.3.3.5). Some solid insulation takes the form of tank linings. Testing of these should usually be restricted to a thorough visual inspection. However, in certain cases, samples of the material may be taken for moisture content analysis. In extreme cases where free water is present in the switchtank, the fibrous tank lining absorbs the water and expands by bulging out. The presence of moisture can be confirmed by taking a sample of the lining and immersing it in a dish of hot oil. Free water is exhibited by frothing.

Table 11.1 Basic oil-tests for oil-filled switchgear

Test	Comment/method
Appearance	Water
	Suspended matter
	Colour – cloudy – moisture/suspended particles
	dark yellow – overheating
	black-arcing
	green – copper salt, dispersion of carbon
	dark brown – cellulose/resin products
Odour	Acrid smell – volatile acid
	Petrol smell – reduced flashpoint due to fault
Moisture	'Crackle test' (>150 ppm)
	Karl Fischer (>5 ppm)
Electric strength	Withstand test 22 kV 60 s 2.5 mm gap
Acidity	Proprietary solution
	Titration method
	Potentiometric titration method

Life management of electrical plant: a distribution perspective 437

Tripping circuit timing tests

It is well known that the tripping mechanism of some circuit-breakers can become sluggish after a long period of inactivity. There have been instances where the slow operation of circuit-breakers has led to more serious faults when called on to operate. Digital online timers have been developed to enable the operating time of an oil circuit-breaker to be measured reliably pre- and post-maintenance. Timers can also be used to test newly purchased equipment as part of the pre-commissioning tests. Modern timing equipment can print and display sophisticated traces of mechanism movement ('coil current profiling') clearly showing the areas where particular attention should be given.

Partial discharge activity

The availability of partial discharge measurement and location equipment (e.g. the Transient Earth Voltage, TEV^{TM} instrument) for the estimation of discharge activity within high-voltage plant offers many advantages in that it is a measurement that can be made in 1 min or 2 min while the plant is in normal operation. It has been found that the actual measured values of the discharge may not be typical of the state of the insulation because, as has been known for many years, the discharge activity in synthetic resin varnished paper (SRVP) insulation, used almost exclusively 50 years ago, is intermittent, due to the self-quenching nature of discharges in this type of insulation. Continuous recording of discharge in 33 kV metal-clad switchgear has shown that the discharge activity varies cyclically over each 24 h and seasonally over the year with most activity in the summer. These variations make single quick measurements appear less useful. TEV^{TM} measurements are more appropriate to the location of discharge sites than for instantaneous evaluation of plant health.

Current counters

Time-honoured prescriptive rules dictated that after, say, three fault breaking clearances oil circuit-breakers should be maintained before further operation. It was very probable that the three fault clearances were much less than the 250 MVA rating of the circuit-breaker. It is possible to use a current counting instrument to measure the magnitude of each fault clearance, assign an appropriate count and sum those currents to give an alarm when the remaining circuit-breaker capability falls to, say, half that deemed available. Use of a current counter ensures that inspection and maintenance effort is diverted to the circuit-breaker when it is only necessary. Early versions of the instrument, however, only measured current and took no account of power factor on operation, which can have a significant effect on the condition of the circuit-breaker contacts.

11.3.3.2 Busbar systems

The inspection and maintenance of busbar systems of any voltage are facilitated by the use of non-contact optical thermal imaging equipment, which will quickly identify poor connections. Hot-spots identified in this way need to be considered in terms of the load being carried and the vulnerability or criticality or otherwise of the component and circuit.

The ability to detect hot-spots by using thermovision cameras has been available for more than two decades, but it has only been in recent years with advances in technology that the equipment has been made more portable for substation use.

11.3.3.3 Surge arresters

Surge arresters are used on more complex interconnected distribution systems to protect large expensive items of plant (transformers, cables, etc.) from the effects of high-voltage surges caused either by direct or by indirect lightning strikes, or certain switching transients. The traditional test to prove the integrity of the gapped type arrester is a 5 kV direct current (DC) insulation resistance test.

11.3.3.4 Cables

General

All cables will give satisfactory reliability provided that they are installed and operated correctly. For low- and high-voltage (less than 11 kV) cables there is little monitoring on the cables themselves that is necessary. Regular inspection and monitoring of the terminal equipment will indicate any problems with the cables. There are two techniques of condition monitoring available for EHV (>33 kV) cables. These will be outlined with two other techniques for all cables.

EHV cables – oil leak location

Higher voltage cables often have oil under pressure as part of the dielectric system. This is principally to fill any voids that may occur, and assists with the thermal characteristics of the cable. Regular monitoring of the oil gauges at tank reservoir positions will show whether there is any leak and indicate the likely scale of any leak problem. Accurately locating sites of leaks is an expensive business. 'Normal' monitoring of gauge pressures en route is unlikely to give a location for the site of the leak due to the differing contour levels. However, an intelligent computer-based system has been developed, which monitors rates of oil flow from various tanks to give the approximate location for the leak.

The days of 'just keep on pumping' (adding cable oil) to maintain the pressure against an unidentified leak have passed and to be environmentally creditable, owners of oil-filled cables should adopt a more pro-active strategy for leak location. A risk-based repair and replacement strategy should be adopted based on (i) environmental sensitivity, (ii) circuit criticality, (iii) severity or condition of the leak and (iv) leak detection techniques.

Although the computer-based prediction of the leak position system is based on the precise measurements of the flow of fluid into the cable to feed the leak, and the pressure difference across the leaking section (based on the relative frictional resistance of the cable to the flow of fluid between each end of the leaking section and the leak) is still available, new techniques involving 'tracers' are being used [24]. In this method, a volatile tracer is added to the oil and then, by walking along the cable route, seek its presence above ground. Per Fluro Carbon (PFC) is such a tracer and is a stable and compatible compound with oil and paper insulation and offers easy detection at low mixtures. More general information and practical examples are provided by Goodwin [25].

EHV cables – cable sheath resistance
Some EHV cables are designed with PVC over an aluminium sheath. At 132 kV, e.g., this design is used with single-core cables. The single-core cables are laid trefoil and the sheaths crossbonded en route to minimise sheath current. To monitor the integrity of PVC sheaths, it is necessary to decommission the circuit, break the cross-bonding connections where applicable and apply a DC voltage to the sheath system and test for zero leakage current. Any punctures to the sheath can be located by normal cable location techniques and verified by a 'pool of potential' instrument above the cable route.

Partial discharge mapping
There is a relationship between paper cable ageing and partial discharges. Long-term failure mechanisms of paper insulation (pinholes, waxing, treeing, etc.) will give rise to partial discharge activity. Test equipment is available using very low-frequency techniques to determine the location of the discharge through the 'time of flight' principle. Cables impregnated with oil under pressure will not discharge; the techniques of partial discharge detection are described in Reference 6. Defective polymeric cables caused by voids or by external damage will discharge. Electrical and water trees in polymeric cable will not be detected. If they discharge, this will lead to breakdown within a short period of time.

The discharge technique will also locate defective accessories, caused by, e.g., water ingress, insulation voids, imploded or exploded joints.

A paper by Walton [26] describes the use and experience of partial discharge monitoring of 11 kV paper-insulated cables in the London network.

Samples and ageing tests
A paper by Harrison [7] considered some tests that could be performed on samples from cables that had a poor service history. It is, of course, difficult to decide whether the sample taken is representative of the whole cable length, which could be many kilometres. Harrison's work showed that, for his example, the failures were occurring discretely and likely to be due to original manufacturing faults. Partial discharge mapping techniques were reckoned to offer most help in assessing the overall condition. Seventy-three per cent of failures on one rogue circuit occurred within 40 m of a joint, one failure in four being a failure very close to a repair joint.

11.3.3.5 Insulation systems and bushings
Insulation resistance tests
This is the basic test of insulation colloquially known as 'Megger^{TM}' testing, after the registered trade name of an insulation resistance tester. It is a non-destructive test employing a relatively high test voltage to perform the resistance measurement. The high test voltage is designed to reproduce conduction paths, which may occur under normal operating conditions. Test voltages of up to 5 kV or 10 kV may be used.

Polarisation index tests

The characteristics of the dielectric material are not just resistive but can be more complex. Measurement of the resistive losses, although simple and quick, will not necessarily give the complete picture. There are two losses which exist in the dielectric:

- frequency-dependent loss
- resistive conduction loss

The frequency-dependent losses result from the capacitance of the dielectric under test. When applying a DC test voltage, the capacitive loss will be overcome relatively quickly (about 1 min). As well as the capacitive loss, a second loss called dielectric absorption loss gives more information about the dielectric condition as it varies depending on the chemical structure of the dielectric, which is affected by contamination.

The polarisation index test, which is done using a standard insulation resistance tester, is designed to measure the dielectric absorption. It consists of simply taking a reading of the insulation resistance after 1 min (after which the capacitive effects will be over), and again at 10 min.

$$\text{Polarisation index} = \frac{\text{Resistance at 10 min}}{\text{Resistance at 1 min}}$$

The higher the level of dielectric absorption, the better the condition of the dielectric, so the higher the polarisation index, the better the dielectric. Typical good values may be 1.5–2.0 (poor values 1–1.2); values of polarisation index can be as high as 5.0.

The polarisation index is used as a measure of dielectric deterioration with time and can be a useful part of a preventative maintenance programme.

Partial discharge tests

Partial discharge occurs in gaseous voids in insulation and is a significant source of dielectric damage. Partial discharges are localised events occurring when the local electric strength is exceeded, and generally occur near the positive and negative peaks.

The discharge pattern produced by individual discharge events shows the number of discharge pulses per power cycle and the distribution of these pulses within the power cycle, i.e. their phase relationship. In addition to an individual record of the partial discharges, the user is more interested in any changes that occur in the pattern as a function of the applied electrical stress and time of application. Pattern interpretation is best left to those with expert knowledge. An expert can usually detect a particular type of cavity-type discharge and determine whether, e.g., it is insulation-bound or formed between insulation and metal.

In some circumstances, if the equipment represents a good radiative loop, detection of general partial discharges can be made with a simple aerial and very high frequency (VHF, 30–300 MHz) and ultra high frequency (UHF, 300 MHz to 3 GHz) radio receiver.

Step voltage testing
This is another useful test that can be done with an insulation resistance test set – here, a series of tests of 1 min duration at a steadily increasing test voltage, typically 500 V steps from 0 to 2.5 kV. For good insulation, the conflicting effects of falling resistance due to surface effects and increasing resistance due to dielectric absorption should lead to a relatively stable resistance reading. If the resistance starts to drop at higher values of test voltage, this is an indication of partial discharge activity taking place.

Dissipation factor
The dissipation factor test applies a high voltage at power frequency to the dielectric, and the reactive component of the losses is measured. It is an indicator of the dielectric deterioration or contamination in the same way as the polarisation index, but because it employs power frequency alternating current (AC) and because it more faithfully reproduces normal operating conditions it is preferred. The instrument used is essentially a bridge for the measurement of capacitance and resistance. The dissipation factor is the tangent of the phase angle and can be used comparatively between different pieces of equipment.

Recovery voltage method (interfacial polarisation)
This method consists of applying a DC charging voltage to the insulator (or insulation system, such as a transformer) for a predetermined charging time, then short-circuiting for half the charging time to partially dissipate the charge, and then leaving open circuit when a return voltage is built up on the electrodes.

The maximum value of the return voltage is directly proportional to the polarisability of the dielectric material, and the initial slope is directly proportional to the polarisation conductivity, i.e. a higher initial slope of the response as the material ages and degrades. Put simply, a new transformer with very dry insulation will exhibit a dominant time constant of several hundreds or thousands of seconds, whereas a transformer with fairly high moisture content will have a dominant time constant of the order of seconds. Using software to analyse the results, it is possible to give a qualitative statement on the percentage of moisture content of the paper, and therefore an indication of the maximum operating temperature above which paper ageing is likely to occur.

11.3.3.6 Overhead lines
Traditionally, overhead lines and their associated equipment being visible and frequently patrolled either by foot and/or by helicopter, including for vegetation management surveys, have received little attention from the high-technology condition monitoring technologies. However, with the adoption of more risk-based techniques, operators are keen to make use of new methods. Some of these are described below.

Helicopter and foot patrols – enhanced visual inspection
Made with varying frequencies, this facilitates not only normal visual inspection but also the use of high-definition photography and thermal imaging. Taking

Figure 11.6 High-resolution photograph taken from a helicopter

high-definition photographs (e.g. as many as 20–40 per tower) has the advantage of later ground-based analysis of the pictures. It enables a consistent approach to assessing the health of the apparatus that will enable repairs and/or an assessment of future asset life to be prioritised. Figure 11.6 shows an example of such a photograph.

Conductor corrosion

The carrying out of impedance measurements is often done by using a conductor-mounted trolley detection system. Internal corrosion is a major factor limiting the life of steel-reinforced aluminium conductors (ACSR) and a crucial stage in the corrosion process is the loss of zinc from the central galvanised steel strands. Once this galvanising is lost, the aluminium strands are subject to galvanic corrosion and the conductor deteriorates rapidly. The effects of this form of internal corrosion may not be visible or detectable by infrared methods until the conductor is nearer to failure and then more especially during times of high load. Changes in conductor impedance section by section is a good indication of corrosion.

Tower foundation corrosion measurements

The lightning performance of transmission lines is strongly related to the tower footing resistance since it has an effect on the reliability following back flashovers where during a lightning storm, the current flowing through the tower ground resistance raises the voltage of the tower relative to the phase conductors. The existence of a few towers in a soil with high resistivity can degrade the overall line performance to lightning. Tower foundations may themselves provide a sufficiently good earthing system, but it needs to be periodically checked. The degree of lightning protection depends on the impulse impedance and not the power frequency grounding resistance. A paper by Brookman *et al.* [22] shows that the combined use of polarisation resistance measurement and transient dynamic response has proved to be an accurate way of assessing the integrity of tower foundations in a non-intrusive manner.

Measurements of network earthing: Measurements to be taken, e.g., of the steelwork and neutral earths at a pole supporting equipment.

Verification of conductor clearances: This is done to ensure the conductors meet minimum statutory clearances given that ground conditions (as well as use of the land) may have changed over time.

Partial discharge detection: More recently, vehicle-mounted radio-frequency equipment [23] has shown that it is a viable technique for identifying defective overhead line insulator strings.

Wood pole decay: Wood pole overhead lines can have the poles tested for decay on a regular basis either by simply listening to the report from a hammer blow or by using more sophisticated instruments such as an ultrasonic rot detector, or from the mechanical resistance found by taking small bore samples.

11.3.3.7 Batteries

Batteries form an essential link in a power system, providing vital supplies for both protection systems and main plant operating mechanisms (e.g. switchgear tripping and spring charging). In the last ten years newer valve-regulated lead-acid (VRLA) or 'sealed' lead-acid batteries have been introduced on the market as maintenance-free units to replace the traditional vented plante type cells. The reliability of battery systems may be considered by not only monitoring charging current (continuously) but also carrying out regular tests. Such tests include:

- online or off-line impedance monitoring
- battery earth fault monitoring.

The impedance measured between the cell terminals is inversely proportional to the cell capacity. Impedance data for healthy cells is normally available from manufacturers. Data can therefore be collected over time and compared with the reference value. The reduction in cell capacity and an increase in cell resistance can occur by the process of either 'drying out' or sulphation. (Sulphation is where the lead sulphate builds up on the negative plates; this process can be reversed by quickly recharging.)

Earth faults on battery systems can be detected by injecting a 25 Hz signal between system and earth. The resulting current is tracked with a clamp meter.

11.3.3.8 Transformers

Transformers have had most effort directed on the techniques of condition monitoring. In general, transformer condition monitoring techniques can be grouped into five headings:

- load readings
- diagnostic tests on insulation
- thermal imaging – connections/bushings

- mechanical strength
- winding displacement.

Of the five, apart from load readings, diagnostic tests on insulation are the most common. They can be subdivided into tests on the liquid (oil) and tests on the solid insulation. The tests themselves range from simple and inexpensive tests, which can give early warnings of problems, to those that require sophisticated analytical techniques the results of which can be subject to some degree of interpretation. The user, therefore, has a whole range of tests available, the particular ones are used depending on the cost/benefit including any outage costs associated with a more invasive investigation.

Load readings taken from a winding temperature indicator (WTI) are at best a heavily averaged figure, which *may* be representative of the general thermal conditions inside the transformer. They will necessarily be subject to a time delay and would not represent particular hot spot conditions in the transformer windings. The ability to detect hot spots has been available for some, while advances in technology have made portable versions available for substation use. One new technique for very large transformers is to embed an optical fibre into the windings to directly measure the real-time temperature values and to use this to further improve transformer operating performance (and improve predictive modelling techniques).

The winding displacement technique comprises applying a variable swept frequency signal (from 50 Hz to 10 MHz) to the isolated primary and secondary windings and measuring the response. Any disturbance of the winding caused perhaps by the passage of fault current in an inherently weak part will be detected by a difference in measured frequency response due to the difference in the representational values of distributed inductance and capacitance for each winding. The technique in the early stages of development required the complete electrical isolation of the transformer, so the preparation time in a substation environment was considerably longer than the test itself. The results yield a lot of information about the transformer, but the resources required to carry out the test rule out its application to routine primary (33/11 kV) transformer condition monitoring. The test is more appropriate to one-off problem-solving tests as a preliminary measure to justify more expensive de-tanking. The technique can, however, be used effectively for larger power transformers to provide fingerprint values representing winding displacement at manufacture and when commissioned on site. This then becomes very useful data for future reference.

11.3.3.9 On-load transformer tap-changers

The purpose of a tap-changer is to transfer power from one transformer tapping to another without interruption. There are two types in use: slow-speed reactor and high-speed resistance. The former has fully rated reactors, which become part of the winding during the tap-change. This type was built from about 1940 to 1960. The selector switches and the reactor are in a common oil compartment with the oil in the transformer windings. Diverter switches are in separate compartment(s). Because of the reactive current breaking, this type of tap-changer needs more frequent maintenance – typically at least every two years (as against more than four years for resistor type).

All modern tap-changers are of the high-speed resistance type where short-time rated resistors are inserted during the tap-change. Therefore, some form of stored energy device must be employed – usually a spring wheel or falling weight/spring mechanism. Operation is completed typically within five cycles, arc interruption within one cycle. The advantage of using resistor transition is that operation takes place close to unity power factor. When arc interruption occurs at the current zero, the voltage is also at zero and the restriking voltage does not build up to a maximum for another quarter-cycle. There is therefore a reduction in contact wear when resistive rather than reactive loads are broken. Some types of resistance tap-changers have combined selector/diverter tanks (but separate from the main tank oil), while others have separate selector and diverter compartments.

Clearly, for the resistance type, the mechanism plays a vital role. Mechanism timing tests are possible, but these would only be done after a major overhaul, and then be done by the manufacturer. This would traditionally involve a range of sophisticated equipment. The Queensland Electricity Company has developed and manufactured a portable open circuit indicator for tap-changers. Their device will detect an open circuit due to faulty or misaligned contacts in a star point connected tap-changer. It uses a range of reference delay settings from 1 to 500 ms as a means of determining any open circuit.

Reference 8 describes the use of optical fibre techniques to measure the position of components in the selector mechanism and the duration of current pulses in the diverter mechanism. The system is controlled by some programmed decision-making software, and one prototype system has already worked 'in anger' correctly and avoided a catastrophic failure of a 400 kV transformer tap-changer.

Research is taking place on methods of acoustic monitoring as well as vibration response analysis of tap-changer mechanisms.

Two papers [27, 28] describe the use of a vibration-based condition monitoring technique developed for on-load tap-changers in Queensland, Australia. Field trials demonstrated that the system can provide reliable indication regarding the actual condition of the tap-changer contacts.

A paper in 2000 [29] describes the experience Hydro-Quebec has had in developing a portable acoustic monitoring system. The method, based on the distinctive vibration and noise patterns of an on-load tap-changer, is stated as simple to operate and run, requiring several signatures of normal operation both on- and off-load. The method is claimed to be reliable and very sensitive for wear, synchronism and arcing diagnosis.

11.3.4 What condition monitoring information can tell us about asset management

11.3.4.1 Testing of solid insulation

In the 1980s there was concern among users regarding the destructive failures of a number of types of oil-filled switchgear, which had solid laminated wood insulation components. The breakdowns were caused by poor condition of the insulation system (absorption of water from the oil leading to thermal breakdown of the laminated insulators). The thermal breakdown occurred due to the negative coefficient of resistance of the laminated component. That is, passing leakage current gave rise to a small heating effect, which reduced the resistance of the component which in turn allowed more current to flow. Failure of the component is characterised by the laminations splitting and burnt carbon deposits forming along the length of the laminations. Research showed that the quality of the insulator could be assessed by a simple insulation resistance test and that a value of resistance above a certain threshold would guarantee continued good service provided that the rest of the switchgear was in good condition and properly maintained. The insulation resistance test provided an initial 'fingerprint' on the health of the switch, against which future values could be compared. This was probably the first link between condition assessment and guaranteeing an additional service of six to ten years for carrying out a simple insulation test.

Other solid materials can be removed for inspection and testing, which can range from simple inspection to more sophisticated tests involving specialised equipment.

11.3.4.2 Testing of liquid insulation

Insulating oil can be a rich source of information on the state of health of the main plant, and testing its condition will help determine its useful life.

Typical tests include:

Appearance/odour
Fibre
Dissolved gas analysis (DGA)
Electric strength
Resistivity
Chemical analysis (ICP, PCB, FFA)
Moisture
Dielectric loss angle
Interfacial tension
Acidity

The experienced operator can determine much from the appearance and odour of the oil as to whether it has come into contact with various contaminants and the suitability of carrying out further tests. The electric strength or breakdown voltage test particularly searches for combinations of moisture and other contaminants that reduce the insulation strength. Testing for moisture contamination where the results are expressed in parts per million requires a particularly stringent discipline in properly taking the representative sample. Acidity testing, which can be done crudely on site, to be followed up by laboratory analysis, is likewise a good indicator of the ageing of the oil, as well as ageing of the total insulation system of the main equipment.

The amount of dissolved furfuraldehyde (FFA) found in oil from transformers can be an indication of the quality of the paper insulation.

The work of Schroff and Stannett [9] reinforced by Carballeira [10] clearly shows the importance of measuring the FFA levels in insulating oils and how this can be a prime indicator of ageing. Experimental tests have shown that the degree of polymerisation, DP ('strength'), of the paper insulation is inversely proportional to the FFA level, and the results from experiments indicate that an FFA level of 5 ppm corresponds to a DP of 250, which is reckoned to be towards the lower limit for insulation strength.

Other traditional 'tools' for assisting with the interpretation of oil analysis results from dissolved gas analysis (DGA) are the Rogers Ratio [11] and the Duval Triangle [12]. In the former, faults can be assessed on being high, medium or low temperature, and ratio codes indicate if the fault is due to general overheating or arcing discharges. In the case of the triangle method, gas concentrations are related to a position in a triangle whose area is divided into cause segments. In Duval's paper [12], examples are given of expected times to failure with certain concentrations of dissolved gases.

A more recent paper by Allen *et al.* [13] shows that a general indicator of age of a transformer is given by an increasing rate of dissolved CO_2 production with time. This is then used to compare the relative 'insulation age' with the 'nameplate' age for similarly loaded transformers.

DGA is therefore a powerful tool for condition monitoring and asset management (see also section 11.3.5.3).

11.3.5 Condition assessment leading to asset replacement

11.3.5.1 General

Age will not necessarily be the principal criterion by which judgements are made for the replacement of equipment. In practice, a host of issues will need to be addressed such as reliability, maintenance history, availability of spares, type defects, environmental performance and operational performance. How these issues are dealt with in terms of dealing effectively with the information input is critical to the efficient management of the plant assets. Examples are given in the following sections for switchgear and transformers.

11.3.5.2 Switchgear

Primary distribution switchgear comprising 6.6–66 kV indoor metal-clad switchgear and 22–132 kV open terminal switchgear in outdoor substations can be assessed on a qualitative basis on:

- spares' situation
- performance history
- action taken to carry forward persistent problems
- other relevant documentation.

Secondary distribution switchgear comprising all cable-connected indoor and outdoor fused switches, switches and ring main equipment, being more populous, is subject to an as-found condition assessment. The system is based on a 'cost to restore' the equipment to an acceptable level criterion to obtain a continued satisfactory life-cycle until the next service date. Equipment design and location are also taken into account. The condition assessment is based on a simple point-scoring system designed to take account of good design attributes, places high scores against high cost to repair areas of deterioration and gives relatively low points against deterioration that can be repaired easily and at low cost during normal servicing. The list typically, for oil-filled switchgear, comprises those items shown in Table 11.2.

448   High-voltage engineering and testing

Table 11.2   Oil-filled switchgear assessment – top-level items

Environment	Mechanism
Rust treatment (cost to restore)	Oil chamber (internal)
Compound leaks	Past defect information
Plinth	Oil chamber (external)
Metal condition (external)	Cable boxes

Table 11.3   Transformer assessment

Corrosion	General condition		Insulation system	History
Tank exterior	Overall	Oil pumps	PCB content	
Tank cover ext.	T/C equipt.	Fans	Acidity	
Radiator	Bushings HV/LV	Aux. trans.	Moisture	(Free format)
T/C equipment	Oil leaks	Cable boxes HV	Elect, strength	
Pipework	Gaskets	Cable boxes LV	FFA*	
Fans	Oil valves		DGA*	
Aux. trans.			Insul. resistance	
Outdoor marshalling cubicle				

Each item will have up to nine further subcategories for particular details to be assessed e.g. for 'mechanism': metal condition above oil, insulation condition, contacts, drive links, etc.

In this way, by adding up the points and taking into consideration any type-related problems, the condition of switchgear can be assessed for remaining operation and scheduled for replacement into a suitable planned programme.

11.3.5.3 Transformers

Although the condition of power transformers is regularly inspected, because of the large numbers involved for a typical distribution network operator (DNO), their condition is only more thoroughly examined as transformers approach maturity. In one DNO, the assessment of condition is done on a points-based qualitative approach (as above). Items considered are shown in Table 11.3.

A simple algorithm can effectively combine the qualitative data on general condition and corrosion with quantitative results from the analysis of the insulation system to give an overall point score.

The work establishing the insulation age against nameplate age (section 11.3.4.2) assists in this. For example of the seven values for the 'insulation system', only those items marked '*' will contribute to the assessment of the insulation ageing. Other items, e.g. electric strength, etc., can be improved by on-site reprocessing and therefore, for a revenue expenditure, the transformer could be taken out of the replacement list. Hence, by identifying initial values and rising trends for FFA and DGA (mainly CO_2), equivalent insulation ages can be assessed with reference to Figure 11.7, which is based on the results from tests on 220 132 kV transformers in one

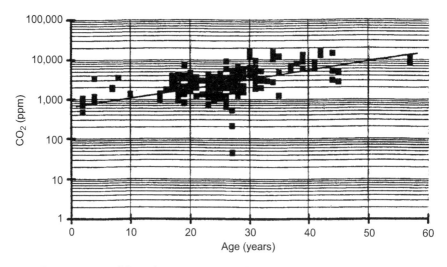

Figure 11.7 CO_2 values against age from 132 kV transformer survey

DNO in the UK. For example a 25-year-old transformer may have a DGA CO_2 value of 6,000 ppm (instead of around 2,000 ppm), which represents an *insulation age* of over 40 years. The CO_2 value needs to be tested and confirmed again on absolute as well as rising trend figures. Likewise the FFA values need to be confirmed to judge whether there has been any degradation to the paper insulation. These results can then be suitably 'weighted' on a point-based system with the scores for the general condition, etc., to give an overall score. With this approach, the transformer can then be scheduled against other units for replacement in an established order of priority.

Although experience has shown that FFA in the oil can be reduced by on-site reprocessing and/or oil replacement, such oil changes will not, of course, alter the irreversible cellulose degradation, which has given rise to the FFA. In these circumstances, oil treatment will mask the underlying cause, and the original FFA values must be kept and added to the new ones.

Allen [14] explains the technique in more detail. He also uses the dissolved content of CO as a measure of ageing. Experiments with power transformers in Queensland, Australia, are being conducted to assign a probabilistic value to the remaining life of a transformer from the CO and CO_2 values.

The techniques of condition assessment (which may form the basis for a condition monitoring programme for particular units) will assist the user *to plan* and *prioritise* an asset replacement strategy.

Reference 15 explains in more detail the practical experiences with using a whole range of condition assessment and condition monitoring tests on transformers.

11.3.6 *The new working environment – users' requirements*

In view of the organisational and cultural changes that have taken place and still are taking place within the electricity industry world-wide, the user now demands:

1. moving away from fixed-period maintenance regimes – adopting a condition-based strategy and risk appraisals
2. simple and cost-effective techniques suitable for multi-skilled labour (fewer available specialists)
3. the information and results with some high degree of confidence on their interpretation to be on-hand
4. the confidence of using the results to ensure that the plant is operated to its optimum economic life and to be able to replace the unit just before failure.

Condition monitoring must not be a purely scientific activity driven by technology; rather, it should be a maintenance approach that is driven by financial, operational and safety requirements. It must produce information on plant condition to allow maintenance resources to be optimised and assist with the optimum economic replacement of the asset.

11.3.7 Condition monitoring – the future

11.3.7.1 Research – transformers

It has been shown in earlier papers [3, 5] how fault statistics can be used to monitor equipment failure rates and predict lifetimes. In the author's opinion, it would be useful to carry out more thorough postmortems on failed equipment to determine the mechanism of failure and to show whether there is more than one mode of failure as predicted. The information from such investigations is also vital as a feedback mechanism to staff to ensure more accurate reporting in the future.

As mentioned earlier, the work by Allen *et al.* [13] and Allen [14] on the distinction between nameplate age and insulation age is significant in assessing the remaining life of transformers. Their papers, and also that of Darveniza *et al.* [16], show how microsamples of paper insulation from transformers in service can be used to yield information on the ageing of the insulation. The results from the techniques of gel permeation chromatography (GPC) [16] to measure molecular weight distributions are interesting, particularly as this incorporates the earlier work on DP measurement (i.e. GPC is the fundamental measurement of which DP is a subset).

11.4 Plant maintenance

11.4.1 General techniques

Regular inspections as part of maintenance of plant are very important and form the first link in the chain of planned maintenance, preventative maintenance, repair and replacement. Figure 11.8 shows the relationships between each in simplified flow diagram form. The distinctions between these various functions are also shown. Maintenance is made necessary by legislation (e.g. in the UK, The Electricity Safety, Quality and Continuity Regulations 2002, as amended, The Health and Safety at Work Act, 1974, and the Electricity at Work Regulations, 1989).

In practical terms, Figure 11.9 illustrates the terms used – particularly preventative maintenance and corrective maintenance when the condition of plant falls

Life management of electrical plant: a distribution perspective 451

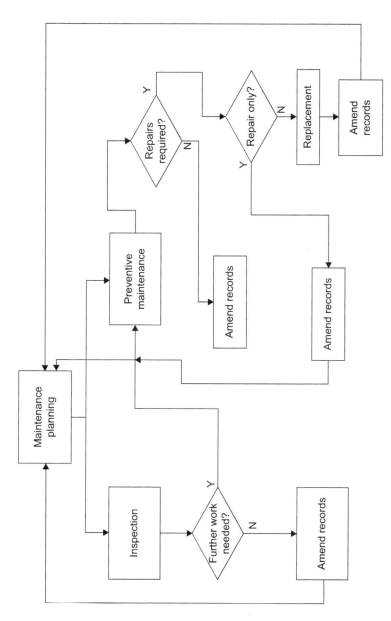

Figure 11.8 *Simplified maintenance flow diagram*

below an acceptable condition. Section 11.2.1.2 referred to establishing a good working relationship or 'partnership' with the manufacturer at the pre-tender stage. This relationship is useful for assistance with maintenance. Indeed, for large and small users, the manufacturer will give best advice for particular maintenance procedures and instructions.

It is important to have a maintenance strategy for plant, which must address both what work needs to be done and the method of getting it done. Staff can be assigned responsibility for plant maintenance – they have the ownership and can be accountable for its performance. Maintenance instructions should be presented in a form that will be clear to staff. This includes routine cleaning and lubrication, contact inspection and replacement, mechanism adjustment, checking of mechanical performance characteristics, checking of insulation characteristics and tightening procedures for oil and pressure gas seals and for porcelain/epoxy to metal joints. Instructions should identify areas where adjustments can be made to compensate for wear and – most important of all – guidance on after-fault maintenance for switchgear contacts. The correct means of handling the dielectric medium (oil/gas/vacuum) with special reference to cleanliness must be considered. Special precautions in dealing with stored energy devices must be observed.

11.4.2 Enhanced maintenance

Table 11.4 is a summary of some recommendations from an ad hoc working group, which looked at enhancing 11 kV switchgear maintenance in the 1990s. It gives good examples of the range and application of preventive and corrective techniques.

Operations and maintenance strategies have been traditionally developed around fixed-period or prescriptive timescales, e.g. 'plant will be overhauled every ten years'

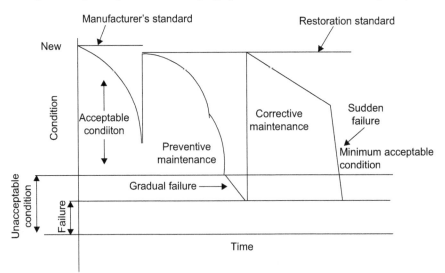

Figure 11.9 Graphical representation of equipment monitoring and maintenance
Source: BS6626: 1985

Life management of electrical plant: a distribution perspective 453

Table 11.4 Enhanced maintenance techniques – 11 kV switchgear

Reference	Action	Comments/benefits
1	Go to site to remove and replace the oil	Enables testing of insulation materials
		Detailed inspection of mechanism
2	Crackle test, only, of oil on site	No time wasted on site on further sophisticated tests on oil
3	Mandatory tenting up	Avoids debris from landing in open tanks
		Gives a controlled working atmosphere
4	Oil sampling by thief tube	Standardised thief tube available
		General improvements in oil-testing and -handling procedures
5	High standards of cleanliness at all times	Switchtanks to be washed down with oil spray and residue removed by aquavac
		Elimination of foreign bodies in tank
6	Oil-handling improvements	Specialised vehicle to reduce manual handling, unnecessary decanting and improve cleanliness
7	Diagnostic testing of solid insulation	Condition monitoring – refurbish or replace?
8	More attention to regular inspection of test equipment	Carriages, test prods and cases checked for cleanliness
9	Improved defect reporting procedures	Better feedback from field to improve future maintenance policy, report on results from as-found condition, etc.
10	Improvements and reviews of existing maintenance practices	Gasket changing each time a gasket seal is broken, for example

regardless of whether it is really necessary for that particular item. As will be explained in section 11.5, users are expecting more by working the plant harder. As part of this, the user is being forced to consider more flexible risk-based approaches.

11.4.3 Reliability-centred maintenance (RCM)

Reliability-centred maintenance (RCM) can be defined [17] as:

> A process used to determine what must be done to ensure that any physical asset continues to fulfill its intended functions in its present operating context.

In essence, RCM builds on the risk-based approach to maintenance referred to later in section 11.5.2. It consists of an appraisal of the functions of the asset, failure modes and effects analysis (FMEA), an evaluation of the consequences of failure and a structured approach to determine the appropriate maintenance for managing each failure mode. Reference 19 refers to using small dedicated teams of staff to engineer the process 'from scratch'. This leads to several advantages, not the least of which is accountability and in dealing with the motivational management of the work.

The process of RCM involves answering seven critical questions [17]:

- What are the functions and associated performance standards of the assets in its present operating context?
- In what ways does it fail to fulfill its function?
- What causes each functional failure?
- What happens when each failure occurs?
- In what way does each failure matter?
- What can be done to predict or prevent each failure?
- What should be done if a suitable pro-active task cannot be found?

By answering each of the seven questions above, one can identify the failure modes of the equipments, the causes of the failure, the criticality of each failure modes and the corresponding action to prevent the failure modes.

References 18 and 19 give practical examples of how RCM has been successfully used.

11.4.4 Condition-based maintenance (CBM)

During the last 20 years the whole philosophy of equipment maintenance programmes has shifted. The first steps towards this change came from the aircraft industry as it tried to reduce the number of in-flight failures being experienced by its equipment. Their approach is a strategy that started from the premise that many equipment failures are maintenance-induced. They realised that to minimise equipment failure they had to minimise invasive maintenance by ensuring that equipment was only dismantled when it really needed it.

The technique of CBM is therefore based on the premise that work is only carried out when the condition of the plant is such that maintenance is needed rather than at fixed time intervals. Adopting CBM requires to have some value for the condition to be measurable (i) online or by (ii) non-intrusive scheduled condition assessment or by (iii) intrusive condition assessment.

CBM employs a risk-based maintenance planning approach. To manage maintenance effectively we need to take a systematic look at the types of equipment and identify all the possible failure modes, and then work out the chance of each failure actually happening and the cost of the event if it does actually happen.

Then look at the maintenance – ask how the present activities mitigate the real risk of failure and therefore whether it is actually worth doing them.

11.5 Working plant harder

11.5.1 Towards a risk-based strategy – the reasons why

As we move into the twenty-first century, managers in the power industry will be judged even more on their ability to control and cut costs, maintain the value and increase the utilisation of existing plant and equipment and grasp strategic opportunities to provide a high-quality service, which the customer wants. In all respects, this is about *working plant harder*. Two aspects are: (i) management of the plant and equipment and (ii) management techniques themselves. Under each aspect are the topics in Table 11.5.

It will be shown that, in the process of working plant harder, a number of risk-based decisions need to be made. The following examples are given, which illustrate this from a designer's viewpoint and then the practical implications for system operators.

As stated earlier, operations and maintenance strategies have been traditionally developed around fixed-period or prescriptive timescales regardless of whether this is too long or too short for that particular type of equipment. To work systems and plant harder, the user is being forced to consider more flexible approaches, e.g., by carrying out a quantitative risk assessment. The technique uses a systematic approach to identify all hazards and estimate the likelihood and consequence of each to arrive at a total risk. The technique can be used to evaluate alternative designs during optimisation, or a single design can be assessed and judged according to a set of acceptability criteria.

11.5.2 Risk assessment – FMEA and FMECA

FMEA and failure modes, effects and criticality analysis (FMECA) are methods of reliability analysis intended to identify failures, which have consequences of

Table 11.5 Aspects of working plant harder

Management of plant and equipment	Management techniques
Innovative applications	Risk and reliability management
Performance under stress	Value management
Condition monitoring	Maintenance management
Knowledge of the ageing processes	Quality management
Refurbishment strategies	Multi-skilling and teamwork
Design techniques for higher reliability and less maintenance	Small fast computer systems with co-ordinated communications
Environmental effects	Training

affecting the function of a system within the limits of a given application, thus enabling priorities for action on the design and construction to be set.

Techniques used to identify hazards include comparative methods (comparing the current situation with known examples or standards) such as checklists and hazard indices and fundamental methods employing systematic examination of all hazardous scenarios.

FMEA is a method of performing a qualitative reliability analysis on a system from a low level to a high level. In practical terms, a system or an item of plant is examined in terms of all the ways it can fail as well as in terms of the consequences of each failure on the system. It is important to remember that the 'system' includes not only the direct electrical system effects (and any other direct effects, e.g. mechanical), but also the secondary effects to personnel, innocent bystanders and the environment. The sequence of failure or events, which leads to a final failure, needs to be systematically analysed using, e.g., a fault tree method. Simultaneous failures of two or more components/systems need to be considered. The report on the process will be made on any compensating or mitigation measures and the severity of the effects of the failure.

FMEA begins at the lowest component level with a consideration of the basic failure criteria and how such failures impact on the 'next level up' as well as the whole system performance. Often the sequence of events in time may also need to be considered.

The analysis is qualitative; e.g., for a transformer, the consequences of two possible failures of one component are shown in Table 11.6.

Reference 20, British Standard BS 5760: Part 5: 1991, defines the terms and gives guidance on the application of these techniques. The BS lists 17 benefits of FMEA, and to illustrate the importance placed on this subject these days in relation to overall management of assets, three of the benefits are highlighted below:

1. to identify failures which, when they occur alone or in combination, have unacceptable or significant effects, and to determine the failure modes, which may seriously affect the expected or required operation
2. to identify serious failure consequences and hence the need for changes in design and/or operational rules
3. to assist in defining various aspects of the general maintenance strategy, such as:
    i. establishing the need for data recording and condition monitoring during testing
    ii. provision of information for development of trouble-shooting guides
    iii. establishing maintenance cycles, which anticipate and avoid wear-out failures
    iv. the selection of preventative or corrective maintenance schedule, facilities, equipment and staff
    v. selection of built-in test equipment and suitable test points.

Further information and examples on the total cost of maintenance employing Commercial Criticality Analysis (allied to FMECA) can be found in Reference 21.

FMECA is a more quantitative analysis and can only be done from an FMEA. In introducing an element of criticality, the concept of severity of consequences of failure is combined with the rate of occurrence or probability of occurrence of failure in a defined period. For example the FMECA will require an FMEA, plus:

- a determination of failure effect severity for each failure mode
- the evaluation of event frequency.

The criticality element is assessed by constructing a criticality matrix, ranking failure events by their contribution to the total failure frequency of each severity.

Again, Reference 20 provides more detailed information and examples of how to perform an FMECA.

11.5.3 Working switchgear harder

The requirement to keep switchgear running with frequent overhauls/repairs against the costly job of replacement should be based on:

- safety
- cost benefits in reducing maintenance/down-time
- operational and other factors
- reliability.

The operational factors include the need to keep spare parts, as well as environmental considerations (e.g. the hazards posed by gases/liquids, compatibility with existing and new substances).

New switchgear is likely to be 'maintenance free', and will be built using fewer components. The expected reliability will be higher.

In some cases, as well as doing switchgear changes with new remotely controlled switches, some 'old' switchgear is being retro-fitted with remotely controlled actuators in order that supplies can be restored quicker following a fault.

In an effort to also reduce switching time on faults, fault passage indicators (FPIs) can be fitted to the 'outgoing' cable leg of some distribution ring main units. This should enable only the minimum number of switches to be operated on fault conditions. This improves the safety of the operator as well as extending the life of the switchgear.

11.5.4 Working transformers harder

A transformer may be loaded above its nominal rating if the user takes measures to monitor the temperature and follows adequate inspection and maintenance procedures. In larger units, the formation of gas bubbles is the main risk. In general terms, running with a hot spot temperature of 98 °C (with an ambient temperature of 20 °C) should correspond to running the transformer continuously at its nameplate rating. If the transformer hot spot temperature is allowed to rise above 98 °C, then the winding insulation will age more rapidly. With hot spot temperatures of up to 140 °C, the rate at which transformer insulation deteriorates is such that it doubles for every temperature rise of 6 °C. The upper limit of 140 °C is set by the insulation undergoing rapid deterioration above this level. IEC 60354, 'Loading Guide for Oil-Immersed Transformers', gives formulas for load and load-cycle capabilities, as well as loss of life calculations. To get a more real-time evaluation of hot spot temperature, investigative work is taking place with optical fibres embedded in the transformer windings. This will enable actual hot-spots to be monitored and the transformer to be worked harder.

While the golden rule for extending the life of transformers is to reduce operating temperatures, there are occasions when operation at elevated temperatures will be necessary. In some cases, especially thermally upgraded paper can be used for the conductors. Whereas the electrical properties of the paper remain roughly constant even though degradation may be taking place, thermally upgraded paper effectively delays the onset of loss of mechanical strength.

Depending on the economic criteria, which the transformers are operated in, it may be possible to consider replacement on energy conservation grounds, i.e. newer units employing lower loss core steels.

The above arguments assume that the transformer insulation system is in a good state and 'dry'. 'Wet' transformers must be operated at a lower temperature than very dry transformers if accelerated paper ageing is to be avoided. The use of the recovery voltage method test (section 11.3.3.5) will help to judge the state of the total insulation system.

11.5.4.1 An 'obvious' word of caution!

It is important to consider fully the effects of any life management improvement to plant in case this affects other aspects in which the plant is operating. For example, in some cases, improved installation techniques may extend the life of the transformer. An enclosure, e.g., will provide greater environmental protection, but it is important to provide adequate ventilation to counter the effects of overheating. More frequent inspection and maintenance of the transformer may be necessary, e.g. tests of the neutralisation value of the oil should be conducted more often as this will indicate the effects of overheating. For a distribution transformer, a simple overhead canopy may be beneficial.

11.6 Future trends in maintenance

A survey was carried out by CIGRE WG 13.08 (life management of circuit-breakers) [30], in which 42 utilities were asked about their maintenance practices for switchgear. The results are shown in Table 11.6 together with what the utilities' future strategy (Table 11.7) was going to be. At that time, it was clear that there was an increase towards CBM and RCM. As risk-based techniques have been further developed and greater confidence in their use is obtained, it is likely that RCM-based techniques have gained further ground.

Table 11.6   Two modes of failure for one component

Component description	Failure mode	Possible causes	Effects on system	Detection method
Oil	Low level	• Leak • Incorrect level	• Flashover in tank • Environmental damage • Fire	• Routine inspection
Oil	Sludge	• Overheating • Lack of maintenance	• Overstress transformer	• Load readings • Temperature readings • Oil samples

Table 11.7   CIGRE WG 13.08 survey (2000)

Strategy	Now (%) (2000)	Future (%)
Time-based (TBM)	41	24
Condition-based (CBM)	38	47
Reliability-centred (RCM)	15	24
Other	6	5

11.7   A holistic approach to substation condition assessment

One risk-based technique that looks at the whole substation involves assigning a risk rating that can be used for asset replacement and/or future inspection strategies by adopting a holistic approach to the whole substation site. In this simple technique, all the substation equipment is looked at separately from the site itself.

For example the equipment is given scores of between 1 (good) and 10 (bad) for issues such as estimated remaining life, breaker condition, switch condition, transformer condition, busbar support condition, fault history, fault ratings, etc. These are then added to provide a 'probability' value.

The site is similarly scored on issues such as visual risk, number of customers, number of critical circuits, etc. These are then added to provide a 'consequence' value.

The holistic site (risk) condition assessment is then the probability multiplied by the consequence. For a number of sites, the holistic risks can be plotted cumulatively to show an overall picture of the heath of the utility's substations. A typical example is shown in Figure 11.10.

In this example, most of the substations have scores lower than around 30, but those sites with scores peaking around the 60–72 range as well as those sites with scores above around say, 100, need scheduling for further work.

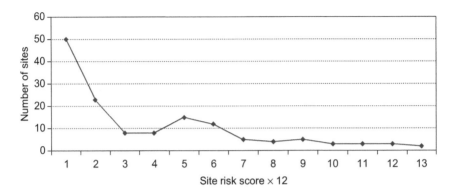

Figure 11.10 Example of holistic substation site scores

11.8 Retrofit, refurbish or replace?

This is the asset manager's dilemma.

A paper by Chan *et al.* [31] describes facing this dilemma for circuit-breakers (and looking to the implementation of RCM techniques). They describe ranking circuit breakers, for example, based on two criteria. The first is the technical condition index C (the results from different physical components of the device used to evaluate the condition based on a set of criteria). The second is the importance index (I) of the equipment being a numerical representation of the impact due to the failure of the device. Chan *et al.* then describe the technical condition index and the importance index on a two-dimensional decision map with the choices limited to either replacement, maintenance or no further action with a fourth category – 'corrective action' applying where the importance of the circuit-breaker to the whole system is lower than some threshold value regardless of the technical condition of the device.

In practice, however, the author sees more fundamental choices for the asset manager – corrective action, retrofit, refurbish or replacement and the natural boundaries between each are more blurred.

Figure 11.11 illustrates the decision-making process. For example equipment in the replacement region has higher importance (almost) irrespective of the technical condition – although that does play an important part. It is suggested that increasing the 'importance index' should trigger more frequent use of condition monitoring. At the other end of the scale, the less important equipment would probably require less frequent condition assessment.

11.9 Current challenges

As has already been indicated, the electrical power industry faces many challenges. An earlier paper by the author [32] listed many of the challenges and considered how skills and competence underpin safety in the industry. The HSE in Great Britain regulates not only the safety of the workforce but also the public safety from

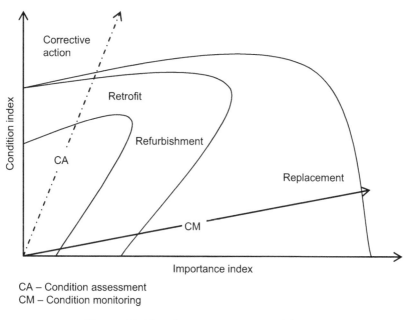

CA – Condition assessment
CM – Condition monitoring

Figure 11.11   The asset manager's dilemma

the workings of the licensed (and certain unlicensed) network operators, suppliers, meter operators and generators, via regulations [33]. These regulations require duty holders to ensure inter-alia that their equipment is constructed, installed, protected (both electrically and mechanically), used and maintained as to prevent danger, interference with or interruption of supply, so far as is reasonably practicable. So far as is reasonably practical, there is a duty to inspect networks with sufficient frequency to ensure awareness of any action needed to ensure compliance. Therefore, in many respects, the 'stewardship' of the assets is risk-based and asset owners have to factor this into their whole business process.

The author sees this stewardship shown in diagrammatic form in Figure 11.12. Here the asset database is populated with information from new assets as well as data from the condition of in-service assets. The database itself drives the inspection and preventive maintenance processes, both of which have a policy foundation. For example, as mentioned above, many asset owners have moved away from traditional time-based preventive maintenance regimes in favour of condition-based or reliability-centred approaches. The results from such fieldwork, including from day-to-day operations and failures in service, are then fed into the decision-making process labelled 'Operations Management' in which various decisions are needed to be made regarding the remaining life of the asset. Underpinning this is the policy that should assist in the decision-making process, e.g. in simple terms whether to do more work at the time, or defer further work and organise more testing. Above all, the central management function must provide guidance on prioritising remedial work based on the criticality of the asset and/or network. Apart from the asset database routinely issuing work orders for inspection and maintenance, much of the rest of the process is human-driven.

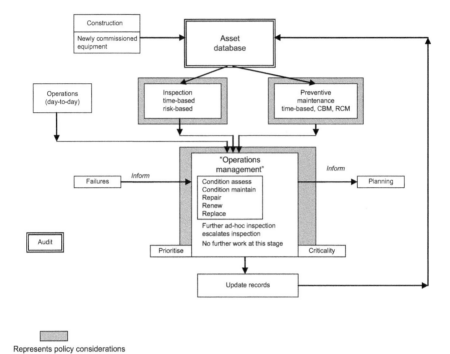

Figure 11.12   The asset management process

As indicated earlier [32], those involved in this process need to have the necessary skills to drive the process and competence to understand and act on the results.

11.10   A standard for asset management

The British Standards Institution published in 2008 a Publically Available Standard (PAS) 55 on Asset Management [34]. Part 1 is the specification for the optimised management of physical assets, and Part 2 contains guidelines for the application of Part 1. The standard was developed over six years by more than 50 public and private organisations in 10 countries and 15 sectors – incorporating feedback from users. The standard offers an internationally recognised framework for asset management and governance. Particularly in the utility industry with lots of high-value assets and a range of professional and technical staff, stakeholders will be looking for PAS 55 compliance as a recognised measure of good management practice. In essence, the PAS 55 framework provides a structure for better decision-making and execution on a day-to-day basis for all matters relating to a utility's operations in order to demonstrate optimal asset management. By utilising a co-ordinated asset management plan, a utility business can begin to guarantee asset availability, forecast production rates and realise costs savings. Compliance would be viewed by senior management as a tool for driving continual organisational improvement.

PAS 55-1 requires organisations to establish, implement and maintain an asset management policy, strategy, objectives and plans. It then describes in some detail the typical elements that are expected within each category. However, the scope of PAS 55 is primarily focused on the management of physical assets and asset systems. It recognises that the management of physical assets is at the 'heart' of the business and is inextricably linked with human assets, information assets, intangible assets and financial assets. As far as human assets are concerned, PAS 55 recognises that human factors such as leadership, motivation and culture are not directly addressed within the document, but that they are critical to the successful achievement of reliable asset management and require due consideration. The interface between human and physical assets includes such items as motivation, communication, roles and responsibilities, knowledge, experience, leadership and teamwork [35]. As explained above and illustrated in Figure 11.12, all of these factors have a critical role to play towards the successful outcome of an asset management system.

11.11 The impact of smart grids on asset management

In the early part of the twenty-first century, much has been talked about 'smart grids', i.e. the use of new technologies that will (a) facilitate the transition to a low-carbon electricity supply system and (b) enable an increase in security of supply. As far as (b) is concerned, this relates to the challenges of being able to integrate inflexible and/or intermittent generation into the system. The essential elements for this will include a wider use of automation and intelligent systems, distributed and centralised intelligence, real-time monitoring and diagnostics and an integrated 'cyber security' data protection and data privacy safeguards. Thus condition monitoring systems will become more and more important and are likely to be applied to more equipment especially those identified as system-critical – informing the users of impending failures in the system.

11.12 Information management

Users will have their own plant databases, which should be robust enough to run from a PC system and enable all users to access and input data, and do searches, sorts and reports on a range of fields. It must be appreciated that the database will hold a vast range of information – some quantitative and some qualitative. Ideally, the more critical the plant, and the easier to ascribe values, the greater the amount of quantitative data held. In this respect, the cautionary notes in section 11.2.6 are relevant.

With access to a range of cohort data, other professionals and professional publications and operating experience, the user should not have to resort to new sophisticated or leviathan computer systems, which will not easily or simply be adapted to the flexible nature of risk-based asset management. The ever changing technology in systems and in asset management will make demands of information management systems. The process should be one-way, in the direction that the information management system should have the 'servant' role.

11.13 Conclusions

The performance of new plant in service very much depends on developing a good working relationship with the manufacturer at the first stages – making sure that both supplier and user understand the specification, and in what environment the equipment will be expected to be operated. Its future life in service, like that for units that already may be approaching maturity, will very much depend on the skill of maintenance staff. Their task will not only be to take measurements and record data (and know how to analyse the data), but also to be aware of 'best practice' techniques such as have been described in this chapter. To get the most life out of plant, the techniques of condition monitoring will be combined with risk assessment methods to enable maintenance resources to be optimised, and then assist with the economic and planned replacement of the asset. Management have a great responsibility to ensure that today's hard-pressed multi-skilled workforce have sufficient knowledge, training and resources to enable them to manage effectively the equipment for which they are responsible. Equally the challenges to operators can be very rewarding, especially when decisions have to be made in the face of uncertainty.

> Ageing is not about how old your equipment is: it's about what you know of its condition and how that's changing over time.
> *Plant ageing, management of equipment containing hazardous fluids or pressure*, HSE Research Report, RR509, HSE Books 2006'

References

1. National Fault and Interruption Reporting Scheme, Annual Report. Originally published by The Electricity Council, London; now the Energy Networks Association, London. http://www.energynetworks.org
   Note that the reports are currently available from Quality of electricity supply report(s) available from Ofgem – the Office of Gas and Electricity markets. http://www.ofgem.gov.uk
2. Steed J.C. 'Using fault statistics to monitor equipment failure rates and lifetimes'. *IEE International Conference on Revitalising transmission and distribution systems. IEE Conference Publications*. 1987;**273**(February):15–20
3. Steed J.C. 'Using fault statistics to monitor equipment failure rates and lifetimes – 1'. *Distribution Development*. The Electricity Council. 1987 (June):9–13
4. Atkinson W.C., Ellis F.E. 'IEE international conference on revitalising transmission and distribution systems'. *IEE Conference Publications*. 1987;**273** (February):1–5

5. Steed J.C. 'Using fault statistics to monitor equipment failure rates and lifetimes – 2'. *Distribution Development*. 1987(September):30–6
6. Mackinlay R. 'Measurement of remaining life of HV cables: Part 1. Experience in the field'. *IEE Cables Conference*. 1986;**270**(November):1–6
7. Harrison B.J. 'Some aspects of failure of 33 kV H-type oil-rosin impregnated cables', *IEE cables Conference*. 1986;**270**(November):16–20
8. Lewis K.G., Jones R.E., Jones G.R. 'A tap-changer monitoring system incorporating optical sensors'. *IEE Conference on the Reliability of Distribution and Transmission Equipment*. 1995;**406**:97–102
9. Schroff D.H., Stannett A.W. 'A review of paper ageing in power transformers', *IEE Proceedings C*. 1985;**132**(6):312–9
10. Carballeira M. 'HPLC contribution to transformer survey during service or heat run tests'. *IEE Colloquium Digest – Assessment of Degradation Within Transformer Insulation Systems*. IEE; 6 December 1991, pp. 5/1–5/14
11. Rogers R.R. 'IEEE and IEC codes to interpret incipient faults in transformers using gas-in-oil analysis', *IEEE Transactions on Electrical Insulations*. 1978; **EI-13**(5):348–54
12. Duval M. 'Dissolved gas analysis: it can save your transformer', *IEEE Electrical Insulation of Magazine*. 1989; **15**(November/December(6)):22–7
13. Allen D.M., Jones C., Sharp B. 'Studies of the condition of insulation in aged power transformers – part 1: insulation condition and remanent life assessment for in-service units'. *ICPADM – 91 Conference*; Tokyo, July 1991
14. Allen D.M. 'Practical life assessment techniques for aged transformer insulation', *IEE Proceedings A*. 1993;**140**(5):404–8
15. Steed J.C. 'Condition monitoring applied to power transformers – an REC view'. IEE international conference on the reliability of transmission and distribution equipment. *IEE Conference Publications*. 1995;**406**(March):109–14
16. Darveniza M., Saha T.K., Hill D.J.T., Le, T.T. 'Studies of the condition of insulation in aged transformers and predicting its remaining life'. *CIGRE Symposium*; Berlin, 1993, Paper 110–22
17. Moubray, J. *Reliability centred maintenance*. Butterworth Heinemann; 1991
18. Basille C., Aupied J., Sanchis G. 'Application of RCM to high voltage substations'. IEE international conference on the reliability of transmission and distribution equipment. *IEE Conference Publications*. 1995;**406**(March):186–91
19. Hamman X. 'Experience with the use of RCM in a transmission maintenance environment'. IEE international conference on reliability of transmission and distribution equipment. *IEE Conference Publications*. 1995;**406**(March): 192–7
20. British Standard 5760: Reliability of systems, equipment and components, Part 5 – Guide to failure modes, effects and criticality analysis (FMEA and FMECA). BSI – British Standards Institute, 1991
21. Upshall P. 'Performance, cost and service, a trial balance'. IEE international conference on the reliability of transmission and distribution equipment. *IEE Conference Publications*. 1995;**406**(March)

22. Brookman P., Bryce M., Hughes D. 'Non-intrusive assessment of steel-tower foundations'. *IET 20th CIRED Conference*, June 2009;**550**, paper 0525
23. Moore P., Portugues I., Glover I. 'Remote diagnosis of overhead line insulation defects'. *IEEE Power Engineering Society, General Meeting*. 2004;**2**(June):1831–5
24. Landucci L., Lanzarone L., Meurice D. 'Leak location in oil paper cables', Paper C7212, *7th International conference on insulated power cables*; Paris, 24–28 June. JICABLE, 2007
25. Goodwin R. 'An innovative method for finding leaks in oil-filled high-voltage cables', *Conference on Power Cables. Cable Tech*. 2010; **S IV**: 254–66
26. Walton C. 'Detecting and locating MV failure before it occurs'. *IET 16th CIRED Conference*. June 2001;**482**, paper 1.40
27. Kang P., Birtwhistle D., Daley J., McCulloch D. 'Non-invasive on-line condition monitoring of on-load tapchangers', *IEEE Power Engineering Society, Winter Meeting*. 2000;**3**(ISS 23–27):2223–6
28. Kang P., Birtwhistle D. 'On-line condition monitoring of tapchangers – field experience', *IET CIRED Conference.*, June 2001, publication 482, paper 1.52
29. Foata M., Beauchemin R., Rajotte C. 'On-line testing of on-load tapchangers with a portable acoustic system'. *IEEE Transmission and Distribution Construction, Operation and Live Line Maintenance 9th International Conference*, October 2008, pp. 293–8
30. The International council on large electric systems (CIGRE), Working Group 13.08. *CIGRE WG 13.08 Brochure 165: Life management of circuit breakers*, 2000
31. Chan T., Chen-Ching L., Jong-Woong C. 'Implementation of reliability centred maintenance for circuit breakers'. *IEEE 2005, Power Engineering Society General Meeting*. 2005;**1**(Pt1):684–90
32. Steed J. 'Skills and competence underpin safety in the UK electrical power industry'. *IET 4th International Conference on System Safety*. IET Publication **555**, 2009
33. The Electricity Safety, Quality and Continuity Regulations 2002, as amended, UK Statutory Instrument No. 2665
34. British Standard PAS 55: Part 1, Specification for the optimized management of physical assets; Part 2, Guidelines for the application of PAS 55-1. September 2008
35. Steed J. 'Safety is critical to asset management in the GB electricity industry'. *IET Reliability of Transmission and Distribution Networks Conference*. IET Publication **580**, 2011

Chapter 12

High-voltage bushings

John S. Graham

12.1 Introduction

A bushing is a device for carrying one or more high-voltage conductors through an earthed barrier such as a wall or a metal tank. It must provide electrical insulation for the rated voltage and for service overvoltages and also serve as mechanical support for the conductor and external connections. The requirements for bushings are specified in IEC 60137: 2008 [1].

12.2 Types of bushings

Bushings are used to carry conductors into all types of electrical apparatus, for example transformers, switchgear and through building walls. Their form depends on the rated voltage, insulating materials and surrounding medium. Bushings can be broadly grouped into two types: non-condenser and condenser graded bushings.

12.2.1 Non-condenser bushings

In its simplest form a bushing would consist of a conductor surrounded by a cylinder of insulating material, porcelain, glass, cast resin, paper, etc., as shown in Figure 12.1. The radial thickness a is governed by the electric strength of the insulation and the axial clearance b, by that of the surrounding medium.

As shown in Figure 12.2, the electric stress distribution in such a bushing is not linear through the insulation or along its surface. Concentration of stress in the insulation may give rise to partial discharge and a reduction in service life. High axial stress may result in tracking and surface flashover. As the rated voltage increases, the dimensions required become so large that this form of bushing is not a practical proposition.

Partial discharge inception voltage can be increased by including in the bushing design a stress control mechanism. With cast resin insulation a control electrode, electrically connected to the mounting flange, can be embedded in the insulation reducing the stress at the flange/insulation interface.

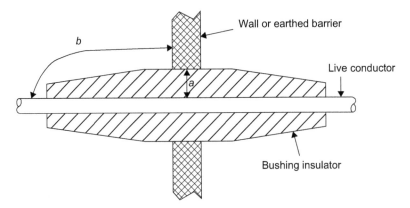

Figure 12.1   Non-condenser bushing

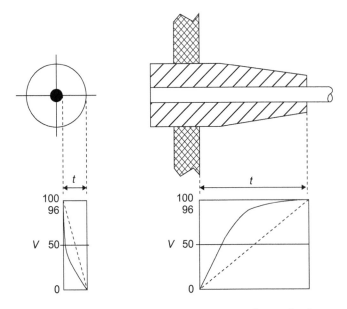

Figure 12.2   Stress distribution in non-condenser bushing

Stress control methods have been developed for power cable terminations using heat-shrinkable stress control tubing which can also be applied to bushings. The heat-shrinkable stress grading tube is installed over the exposed solid insulation and overlapping the flange. The tube reduces the voltage gradient at the flange and along the surface of the bushing (Figure 12.3). It is important that air is eliminated from the interface using a void filling mastic to prevent partial discharge.

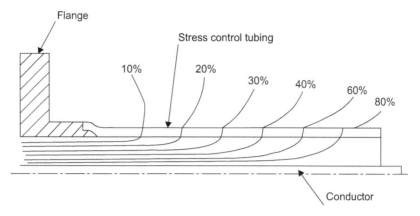

Figure 12.3   Stress control using heat-shrinkable stress control layer

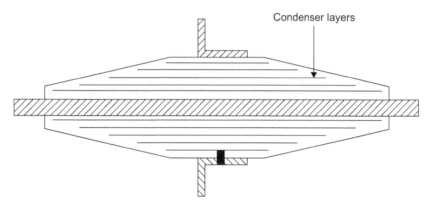

Figure 12.4   Condenser bushing

12.2.2   Condenser bushings

At rated voltages over 52 kV, the condenser or capacitance graded bushing principle is generally used, as shown in Figure 12.4. The insulation material of such a bushing is usually treated paper with the following being the most common:

- resin bonded paper (RBP)
- oil-impregnated paper (OIP)
- resin-impregnated paper (RIP).

As the paper is wound onto the central tube, conducting layers are inserted to form a series of concentric capacitors between the tube and the mounting flange. The diameter and length of each layer are designed so that the partial capacitances give a uniform axial stress distribution and control radial stress within the limits of the insulation material (Figure 12.5).

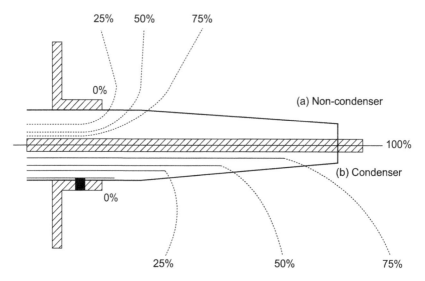

Figure 12.5 Field distribution in non-condenser and condenser bushings

12.2.2.1 Resin bonded paper bushings

RBP bushings were previously used extensively, up to 420 kV, for transformer applications but are now limited to low voltage use, particularly in switchgear, due to technical limitations. In RBP bushings, the paper is first coated with a phenolic or epoxy resin and then wound into a cylindrical form under heat and pressure, inserting conducting layers at appropriate intervals. The use of RBP bushings is limited by the width of paper available and by the danger of thermal instability of the insulation due to the dielectric losses of the material. RBP bushings are designed to operate in service at a maximum radial stress of approximately 20 kV/cm.

The RBP insulation is essentially a laminate of resin and paper. The bushing, therefore, contains a considerable amount of air distributed between the fibres of the paper and at the edges of the grading layers.

Where internal faults exist, as the voltage applied to the bushing is raised, partial discharge inception can occur at the layer ends where the stress is greater than the radial stress between the layers. During manufacture, incorrect winding conditions may result in circumferential cracking produced by shrinkage or weak resin bonding. Here, in the voids produced, the electric stress is enhanced and partial discharge may occur at low levels of stress.

In service, ingress of moisture into RBP bushings can cause delamination and increased and unstable dissipation factor or tangent delta. Discharges at the layer end produce carbon treeing, extending axially, while discharges in voids may produce breakdown between layers. Both forms of discharge are progressive and ultimately lead to failure over a long period by overstress or thermal instability in the residual material [2, 3].

Overvoltages that occur in service are usually surges produced by switching or lightning. Breakdown of a bushing under this type of stress would normally be initiated axially from the ends of the layers due to breakdown of air in the winding. Complete failure may be a combination of axial and radial breakdown.

12.2.2.2 Oil-impregnated paper bushings

OIP insulation is widely used in bushings and instrument transformers up to the highest service voltages. OIP bushings are made by winding untreated paper, inserting conducting layers at the appropriate positions and impregnating with oil after vacuum drying.

The paper used is generally an unbleached Kraft which is available in widths of up to 5 m. This width is adequate for most applications but, for ultra high-voltage bushings, various methods of extending the condenser length by multi-piece construction or paper tape winding have been used. It is important that the paper be sufficiently porous to allow efficient drying and impregnation while maintaining adequate electric breakdown strength.

The oil used is generally a mineral oil as used in power transformers and switchgear [4]. Prior to impregnation, processing is carried out to ensure low moisture and gas content and high breakdown strength [5]. In certain applications, other properties may be important, for example low pour point for low-temperature installations and resistivity and fibre content for direct current (DC) bushings.

Processing of the bushing may be carried out by placing several condenser cores or individual whole bushing assemblies into an autoclave or by applying vacuum directly to the bushing assembly before impregnation. Manufacturing defects are generally detected in routine tests. In the case of properly processed OIP bushings, there are no gaseous inclusions in the material. Internal discharge inception, therefore, occurs at much higher stress levels than with the RBP bushings. OIP bushings are therefore being designed to operate at radial stresses of typically 45 kV/cm. Discharges can occur at the layer ends (due to misalignment) of the layers at the high stress levels associated with lightning impulse and power frequency tests. If this stress is maintained, gassing of the oil and dryness in the paper can be produced, and eventually carbonisation at the layer ends may occur which, due to the high radial component of the stress, tends to propagate radially leading to breakdown.

12.2.2.3 Resin-impregnated paper

RIP insulation was developed in the 1960s for use in distribution switchgear and insulated busbar systems. In recent years, development has increased its utilisation to 800 kV.

In the manufacturing process, creped paper tape or sheet is wound onto a conductor. Conducting layers are inserted at predetermined positions to build up a stress-controlling condenser insulator. The raw paper insulator is dried in an autoclave under a strictly controlled heat and vacuum process. Epoxy resin is then admitted to fill the winding. As a 525 kV bushing may have a core greater than 6 m in length, it is important that the resin has low viscosity and long pot life to ensure total impregnation. During the curing cycle of the resin, shrinkage must also be controlled to avoid the production of cracks due to internal stresses. The resulting insulation is dry, gas tight and void free giving a bushing with low dielectric losses and good partial discharge performance.

During manufacture, the conducting layer follows the shape of the creped paper. The spacing between individual layers varies between the peaks and troughs of the creping. The layer spacing with RIP is therefore coarser than with RBP and OIP bushings and full advantage of the high intrinsic strength of the resin cannot be taken. RIP bushings are designed to operate with a radial stress of about 36 kV/cm.

The advent of UHV DC transmission schemes, particularly in China, has given fresh impetus to the development of resin-impregnated paper bushings.

12.3 Bushing design

It is essential that a bushing be designed to withstand the stresses imposed in test and service. These are summarised in Table 12.1.

Condenser-type bushings are used predominantly at high rated voltages and their design and application will now be considered.

Electrical stresses act both radially through the insulation and axially along its surface. The maximum allowable stresses for each material have been determined by experience and test to give a minimum service life of 40 years.

The condenser design controls these stresses to a safe level. The stress distribution is dependent on five principal factors (Figure 12.6):

r_0 – radius of conductor
r_n – radius of outer layer
l_1 – length of first condenser
l_n – length of last condenser
U_n – rated voltage

Table 12.1 Likely stresses for bushings

Electrical	Lightning impulse voltage (BIL)
	Overvoltages caused by switching operation (e.g. SIL)
	Power frequency voltage withstand
Thermal	Conductor losses
	Dielectric losses
	Solar radiation
Mechanical	Loads due to external connections
	Self-loads due to mounting angle
	Earthquake forces
	Short-circuit forces
Environment	Wind loads
	Nature of surrounding medium (air, oil, gas)
	Pollution

High-voltage bushings   473

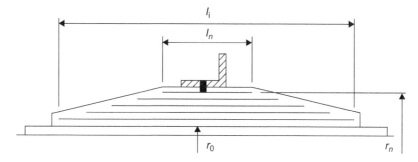

Figure 12.6   Major dimensions of condenser bushings

A typical OIP bushing, one end operating in air and the other in oil, as shown in Figure 12.7, will be considered to describe the procedures of bushing design [6]. Figure 12.8(a) shows 550 kV OIP bushings installed on a three-phase autotransformer supplied by Hyundai Heavy Industries, Korea for a project in USA.

The condenser winding (1) is enclosed in air-side (2) and oil-side (3) insulators which are pressed against the mounting flange (4), the underside of which may be extended to provide current transformer accommodation by springs contained in the head of the bushing (5) acting to tension the central conductor or tube (6). The assembly is sealed by gasket to prevent oil leakage.

12.3.1   Air end clearance

For indoor use with moderate pollution and humidity, resin-based insulating materials need no further protection from the environment. Oil- or gas-filled or impregnated bushings always require an insulating enclosure. This is commonly porcelain but modern glass-reinforced plastic with rubber coatings (polymer insulators) are also used.

The length of the insulator is governed by the lightning impulse and switching impulse requirements. The design of the bushings produces uniform axial stress along the surface of the insulator and the length can therefore be less than for a simple air gap. The length of the insulator is also affected by the service environment. In polluted atmospheres, resistance to flashover under wet conditions even at working voltage is dependent on the surface creepage distance, that is the length of the insulating surface between high voltage and earth, and the proportion of it protected from rain. IEC 60815 [7] gives guidelines on the design of insulator profiles for use in polluted atmospheres. In the latest edition IEC 60815 has been split into five parts covering definitions and general principles and ceramic and polymer insulators for AC and DC applications.

For example for ceramic insulators for AC systems the insulator creepage can be determined as follows. From information on site pollution conditions, taken

474 High-voltage engineering and testing

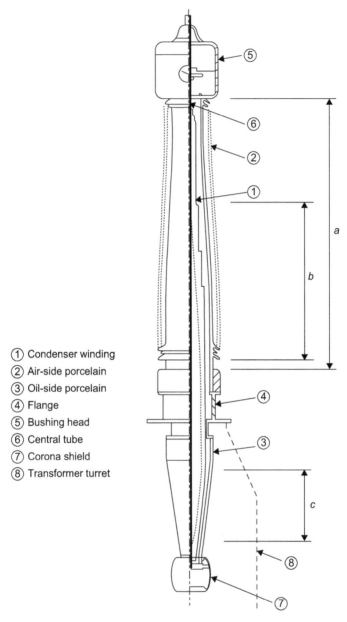

① Condenser winding
② Air-side porcelain
③ Oil-side porcelain
④ Flange
⑤ Bushing head
⑥ Central tube
⑦ Corona shield
⑧ Transformer turret

Figure 12.7 Section of typical transformer bushing

from data of pollution accumulation on sample insulators, a reference unified specific creepage distance (RUSCD) and a site pollution severity (SPS) are determined. The SPS is graded in five classifications from very light to very heavy. Where specific site data is not available the RUSCD IEC 60815 part 2 gives the approximation given in Table 12.2.

*Figure 12.8 (a) 550 kV oil-impregnated paper transformer/air bushing;
(b) 1,200 kV testing transformer with oil-impregnated paper bushing*

Table 12.2 RUSCD as a function of SPS class

SPS class		RUSCD (mm/kV)
a	Very light	22.0
b	Light	27.8
c	Medium	34.7
d	Heavy	43.3
e	Very heavy	53.7

The classification and performance of insulators under polluted conditions is affected by:

- shed profile, overhang and difference in adjacent shed overhang
  - larger values of difference in shed overhang is beneficial under ice, snow and rain conditions
- spacing vs shed overhang
  - important to avoid creepage distance bridging by shed-to-shed arcing
- minimum distance between sheds
- creepage distance vs clearance
  - large ratios can cause localised build-up of pollution in deep and narrow sections of the profile
- shed angle
  - an open profile to allow efficient natural washing without impeding water run-off

- creepage factor
  - an overall check on creepage packing. High ratios of creepage to arcing distance can reduce insulator performance. If the above factors are met the creepage factor requirement is usually automatically respected.

The minimum total creepage distance L is given by

$$L = \frac{\text{RUSCD} \times K_a \times K_{ad} \times U_m}{\sqrt{3}}$$

where

$K_a =$	Altitude correction factor, usually $= 1$
$K_{ad} =$	Diameter correction factor to increase creepage distance with average diameter of insulator D_m
$D_a = <300$ mm	$K_{ad} = 1$
$D_a = 500$ mm	$K_{ad} = 1.1$
$D_a = 700$ mm	$K_{ad} = 1.2$
For other diameters the values can be interpolated linearly.	
$U_m =$ Highest voltage for equipment (kV)	

In certain desert areas, a combination of adverse climatic conditions, long-periods without rain, frequent fog, sand storms, etc., lead to the accumulation of conductive pollutants and insulator flashover. To combat these conditions, enhanced creepage distances of 40 mm/kV or more have been specified.

Modern porcelain insulator designs generally use an alternate long/short (ALS) shed profile which gives superior performance to the previous anti-fog type. ALS profiles allow easier cleaning under natural rain and wind conditions and can be produced economically by modern turning techniques. Typical profiles are shown in Figure 12.9.

The manufacture of porcelain insulators is limited by the tendency to bend during firing if the piece has a high ratio (typically > 6) of height to bore. This is overcome by bushing manufacturers using epoxy adhesive to bond together several sections. The outdoor porcelain of a 420 kV bushing may typically be assembled from three sections to give an overall height of 3.5 m with a height-to-bore ratio of approximately 10. The adhesive produces a high strength, oil-tight joint, and the assembly is considered to be a single piece.

IEC 60815 and IEC 60507 [8] are based primarily on the experience of overhead line insulators and limited testing of bushings and hollow porcelains. The performance of vertical insulators can be restricted due to flashover caused by water cascading over the shed; IEC 60815, therefore, recommends a shed spacing of 30 mm and a creepage-to-clearance ratio (l_d/d) of less than 5. Insulators are normally designed for use in the vertical position; both specifications suggest horizontal use would improve performance. Recent experience has shown this not to be the case [9] and it is now considered that an increase in creepage of 50% is necessary for horizontal use.

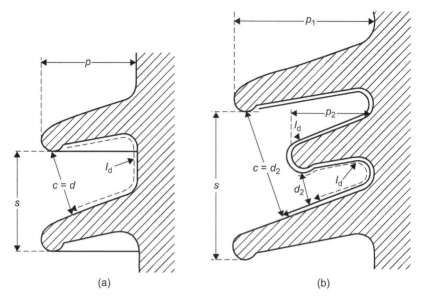

*Figure 12.9 Typical porcelain insulator shed profiles
(a) Normal sheds; (b) alternate long/short (ALS) sheds*

The application of polymer insulators has increased rapidly due to advantages over porcelain being lighter, explosion proof and having a greater hydrophobicity due to the nature of the silicone rubber-moulded shed profile. Improvement in the manufacturing processes for filament wound fibre glass tubes has allowed extension of application to UHV class bushings. The selection of the type of rubber and the shed form is important to achieve good performance under polluted and heavy wetting conditions. Different methods are used in the manufacture of the polymer insulator which use room-temperature or high-temperature vulcanised silicone rubbers applied by moulding or continuous extrusion techniques. The performance of each is different and must be carefully assessed in relation to the site condition.

Having determined the height of the air-side insulator (dimension a), the length of air-side grading (dimension b) can be determined. It is not necessary to grade 100% of the air-side insulator length; in practice 60% internal grading or less gives adequate surface grading for large bushings.

12.3.2 Oil-end clearance

For OIP bushings, the oil-side insulator [3] is usually a conical porcelain or cast resin shell. The internal axial grading over the condenser is dependent on the power frequency test voltage at a stress of approximately 12 kV/cm; this determines dimension c. It is normal that the oil side of a transformer bushing be conservatively stressed due to the consequences of a flashover within the transformer. Dimensions b and c together with the physical requirements of the mounting flange and current transformer determine the lengths l_1 and l_n of the condenser.

12.3.3 Radial gradients

While it is possible to design a bushing with a constant radial gradient, this can only be achieved at the expense of a variable axial gradient. In most cases a constant axial gradient is desirable and the radial gradient may be allowed to vary and will be a maximum at either the conductor or the earth layer.

The values are given as follows:

$$E_0 = \frac{V(a+1)}{2ar_0 \log b}$$

Earth layer stress

$$E_n = \frac{V(a+1)}{2ar_{(n-1)} \log b}$$

where $a = l_1/l_n$ and $b = r_n/r_0$.

E_0 is maximum, when $a < b$ and E_n is maximum when $a > b$. A minimum insulation thickness is achieved when radial gradients at the HT layer and the earth layer are approximately equal, that is when $a = b$; however, this cannot always be achieved.

The radius r_0 is dependent on the current rating, the method of connection between the bushing and the transformer winding and on the bushing construction. An optimum value for r_n can then be calculated.

Having determined the limiting dimensions, the positions of the intermediate layers can be calculated. The detailed calculation method may vary but the object is to achieve acceptable radial stress on each partial capacitor and uniform axial stress, with the minimum number of layers.

Since, as stated, the axial gradient varies throughout the insulation thickness, the layer spacing for constant voltage per partial capacitor will also vary. In this way, a 420 kV bushing may contain about 70 layers.

12.4 Bushing applications

12.4.1 Transformer bushings

Transformers require terminal bushings for both primary and secondary windings. Depending on the system configuration the outer part may operate in air, oil or gas.

At distribution voltages up to 52 kV, non condenser-type bushings are generally used. In the case of dry-type transformers, the bushings form an integral part of the cast resin winding. With liquid-insulated transformers, porcelain-insulated bushings are commonly used for outdoor applications and cast resin for connections inside cable boxes or with separable connectors. These types of bushing are covered by the European Standard EN 50180 [10].

Condenser-type bushings have been developed for rated voltages up to 1,600 kV [11]. With the establishment of UHV AC transmission systems in China and India, IEC and CIGRE have held joint symposia to collect and share information on the state-of-the-art in technologies and to progress standardisation [12]. Figure 12.8(b) shows a 1,200 kV testing transformer manufactured by TBEA Hengyang, China. Transformer bushings are not exclusively of the OIP type. RIP has some advantages over OIP for certain applications particularly where the bushing may be subject to vibration or shock loading. Fully dry-type bushings using RIP condenser and polymer insulators are explosion and seismic resistant reducing maintenance requirements and life-cycle costs.

In many cases, the flexible cables from the transformer winding are drawn through the bushing and terminated at the head of the bushing within the bushing tube. This 'draw lead' type of connection is limited to approximately 1,250 A rating due to the dimensions of the flexible cable required. In cases of higher current ratings, connections may be made at the lower end of the bushing and the bushing tube is itself used as the conductor as shown in Figure 12.10.

With RIP insulation the layers are embedded in a solid material of sufficient strength that there is no need for a supporting tube. In the case of draw lead connection using paper-insulated cables, a transformer end stress shield is unnecessary which allows a reduction in the transformer turret diameter.

The oil end of the bushing may take two forms: conventional type and re-entrant type. The conventional type is considered above. From a comparison of the two forms shown in Figure 12.11, it can be seen that the re-entrant type is shorter and, as no stress shield is required, the transformer turret diameter can be reduced. The re-entrant form has been used extensively in the UK on power transformers but has been replaced by the conventional type. Re-entrant bushings cause difficulties with their installation. The transformer lead must be insulated with paper to

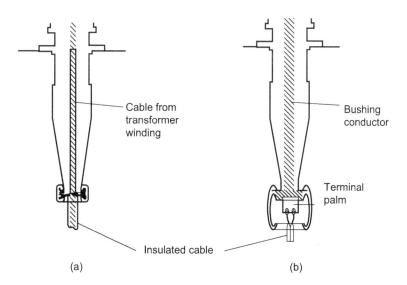

Figure 12.10 Transformer bushing connections
(a) Draw lead type; (b) bottom connection type

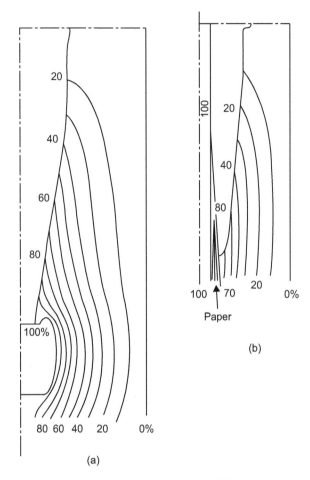

*Figure 12.11 Field plots of the transformer side of HV bushing connections
(a) Conventional type; (b) re-entrant type*

approximately 30% of the service voltage and it is possible for gases to become trapped on the inner surface.

At the mounting flange of the bushing a connection to the last layer of the condenser is brought through a test tapping. This tapping is used during partial discharge and capacitance measurement of the bushing and the transformer. As the capacitance of the tapping is low it is essential that it is connected directly to earth when in service to prevent generation of high voltage and sparking at the tapping terminal. In certain cases, particularly in North America, a potential tapping may be required. In this case, extra layers are included in the winding to provide a capacitance voltage divider. This type of tapping has high capacitance compared to the main bushing and can be used in service, connected to a bushing potential device, to provide a voltage source of up to 5 kV and an output power, typically 100 VA. This power may be used to supply relays and measuring equipment.

12.4.2 High-current bushings

Bushings used on the low-voltage side of generator transformers require special consideration due to their operating condition. Bushings of this type are often required to operate with their outdoor side enclosed in a phase-isolated bus duct. This arrangement can produce ambient air temperatures as high as 90 °C around the bushing, differing greatly from the standard conditions. It is essential that the bushing and the connections are designed to reduce conductor losses and dissipate heat efficiently. At service currents of up to 40 kA, local heating due to poor connections can cause serious damage. To facilitate cooling, a multi-palm configuration is often used at the end terminals. Figure 12.12 shows an RIP condenser-type bushing having an aluminium conductor. However, copper conductors and non-condenser types are also widely used.

Where low-voltage, high-current bushings are mounted in close proximity, consideration should be made of distortion of the current path in the bushing due to magnetic effects.

12.4.3 Direct connection to switchgear

Due to advantages of space saving given by gas-insulated switchgear (GIS) operating with gas (usually sulphur hexafluoride SF_6) at a pressure of about 4 bar (g), it is increasingly common for transformers and switchgear to be directly connected. Direct connection also reduces pollution problems in coastal and industrial areas giving increased system reliability. Oil-to-gas bushings, used to provide the interconnection, generally have a double flange arrangement for connection to the transformer turret and the GIS duct. The bushing design, therefore, needs to be

Figure 12.12   36 kV 31,500 A resin-impregnated paper transformer/air bushing

flexible to cater for the requirements of different equipment manufacturers. To reduce problems of interchangeability IEC 61639 [13] gives dimensions for the gas side of the bushing, in particular the flange fixing and gas end terminal dimensions.

It is important that escape of gas from the GIS is minimised. Precautions must be taken with the bushing design to effectively seal the conductor and flange interfaces to prevent leakage of gas into the transformer. Figure 12.13 shows typical arrangements where double seals are provided at each position, the effectiveness of which can be tested by applying high pressure between the seals [14].

RIP bushings provide an ideal solution (Figure 12.14), and are available up to 800 kV. Due to the gas-tight nature of the insulation an additional porcelain shell is unnecessary. The dry insulation can be mounted at any angle without any need for

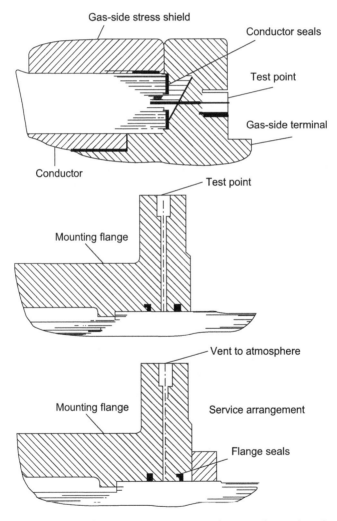

Figure 12.13 Typical sealing arrangements for transformer/gas bushings

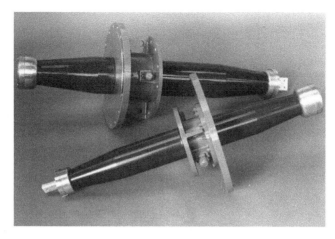

Figure 12.14   Resin-impregnated paper transformer/gas bushings

oil expansion devices as would be required with OIP. The high electric strength of the resin also allows reduced axial dimensions, particularly of the gas part.

Electrical tests for oil-to-gas bushings require special arrangements. The gas side is tested in gas instead of oil to prevent contamination of the bushing seals and the gas duct in service. Draw lead-type bushings are generally not used as the risk of leakage of gas through this site-made joint would be undesirable.

12.4.4   Switchgear bushings

Entrance bushings for high-voltage GIS often utilise pressurised porcelain. The gas within the bushing is common with the duct. Stress control is achieved by profiled electrode screens between the flange and the conductor. The porcelain must be dimensioned to withstand the full pressure of the system and presents an obvious danger if damaged in service. An improvement of this technique was the so-called 'double-pressure' bushing where a glass-reinforced plastic tube is used as a liner and the gap between the tube and porcelain is at reduced gas pressure.

RIP gas-to-air bushings have been manufactured up to 525 kV. The RIP condenser seals the GIS and the porcelain may be filled with a compound material or gas at low pressure. In this case, the low internal pressure enable a lightweight porcelain to be used and operation at any angle without modification. As the gas side may be used directly into, or close to, a circuit-breaker, the components of the bushing must exhibit resistance to the decomposition products of SF_6, particularly hydrogen fluoride (HF). This can be achieved by coating the RIP with a special alumina-rich varnish.

Gas-insulated bushings with polymer weather shell are now commonly used in switchgear applications. The gas filling of this type of bushing is generally common with the GIS (Figure 12.15). Stress control is achieved by internal profiled electrodes which screen the mounting flange and improve stress distribution over the insulator.

In GIS, very fast transients (VFT) generated by disconnector switching are recognised to present a problem to the internal connection bushings. Due to the

Figure 12.15 Polymer gas-insulated GIS/air bushing

speed of propagation of the VFT, it is possible to develop a high voltage between the conductor and the first layer of the condenser. Studies have been made of the high-frequency properties of bushings in this application [15] and the interaction of external and internal resonances. At present, no test exists within IEC 60137 to demonstrate acceptability of the bushing design. However, tests have been proposed which apply lightning impulse chopped within the gas duct at approximately 70% of the rated BIL. The distance between the spark gap and the bushing affects the VFT incident at the bushing.

12.4.5 Direct current bushings

DC bushings require special consideration. HVDC schemes are becoming increasingly popular for the transmission of power over long distances and also for the connection of separate AC networks. These so-called back-to-back schemes may be used on systems of different frequency and asynchronous operation or to increase operational stability. Such systems operate at typically ±80 kV DC while long-distance transmission occurs at up to ±800 kV DC.

The design of a DC bushing is influenced by the resistivities of the various materials used as opposed to their permittivity in the AC case. While permittivities of paper, oil, porcelain, etc., are of a similar order, their resistivities vary by up to 10,000:1. It is therefore important to study the voltage distribution in the core and the surrounding area [16]. The effect on field distribution is shown in Figure 12.16. The upper plot shows the AC field in a typical oil-impregnated paper transformer bushing. A concentric field is produced in the oil between the paper-insulated stress shield and the transformer turret wall. In the centre plot, the DC field of the same arrangement is illustrated, where the high resistivity of the paper, compared to that of oil, concentrates the stress in the shield insulation. To reduce this stress concentration and the stress of the surface of the porcelain, concentric cylinders of pressboard are placed around the bushing. This technique is illustrated in the lower plot. In the practical case a greater number of cylinders and conical barriers may be required to achieve suitable stress control. As the resistivity ratios vary with temperature, studies of the field are made across the operating range of the transformer. It is important therefore that the transformer and bushing manufacturers co-operate in this detail of the design.

In a DC scheme, pollution and fire risks are a major concern. To reduce both, bushings have been developed to operate horizontally to project directly into the converter building. Alternative solutions are contrasted in Figure 12.17, which are comparable to the situations at EDF Les Mandarin and at NGC Sellindge, respectively, at either end of the 280 kV DC cross-channel link. In France, more

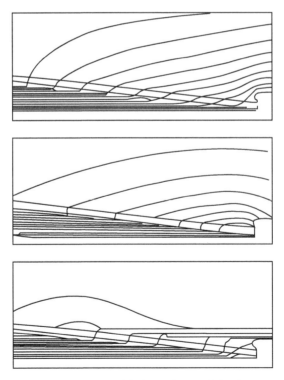

Figure 12.16   Field plots of HVDC bushings

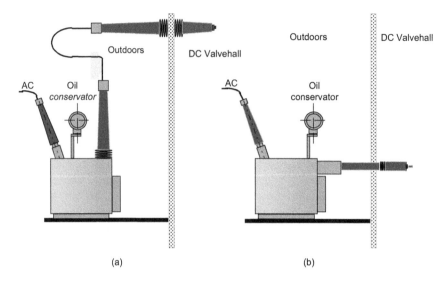

Figure 12.17 Alternative arrangements for converter transformers
(a) Outdoor bushings; (b) through-wall transformer bushings

conventional OIP bushings are used in the transformer connected to the converter equipment by dry-type RIP wall bushings. At Sellindge, due mainly to space restrictions, horizontally mounted through-wall bushings were used on the converter transformer.

Dry-type bushings are now almost exclusively specified in modern HVDC schemes, and to meet the needs of the increasing use of long-distance HVDC transmission in China and India 800 kV DC RIP bushings have been developed.

Figure 12.18 shows an 800 kV DC, 321 MVA converter transformer manufactured by Siemens Transformers, Nuremburg for the Xiangjiaba–Shanghai link. This was the world's largest and most powerful DC connector, at 2,000 km and 640 MW, when it entered service in 2009. The project required special dry-type bushings approximately 16 m in length developed by HSP, Cologne. The through-wall bushings are designed to plug into the transformer insulation system.

In DC schemes pollution-induced flashovers [17] have occurred and where mounted close to a building, non-uniform wetting of bushing insulator has caused problems. Where a polluted insulator is partially protected from rainfall by the building, the difference in surface resistivity in the dry and wet parts reduces the flashover voltage. Various methods of improvement using booster sheds [18] and improving hydrophobicity of the insulator [19] surface have been examined.

12.5 Testing

Adequate testing is essential to ensure reliable operation over the required service life. Routine and type tests are performed in accordance with IEC 60137 and

Figure 12.18   800 kV DC resin-impregnated paper bushings

IEC 60060 [20]. Figure 12.19 shows a transformer bushing installed in an oil-filled tank being prepared for test.

12.5.1   Capacitance and dielectric dissipation factor measurement

This test is probably the most universally applied of all tests on high-voltage bushings and insulation systems. Measurements are made by Schering Bridge, or similar equipment, and give an indication of the quality of the bushing processing. Dissipation factor, or tangent delta, is identical in value to power factor in the range of values obtained. Tangent delta is a measure of the losses in the insulation and can indicate the degree of cure of resinous materials or the moisture content of RBP and OIP. The typical tangent delta/voltage curve for a correctly processed bushing is flat up to at least rated voltage. An increase in tangent delta, particularly below working voltage, would almost certainly cause deterioration in service due to increased dielectric losses or internal partial discharge.

12.5.2   Power frequency withstand and partial discharge measurement

Although classed as separate tests, power frequency withstand and partial discharge are often combined. Partial discharge is a major cause of failure in bushings and, as discussed earlier, may occur in voids, cavities or inclusions in solid- or liquid-impregnated insulation and surface discharges at material boundaries.

Figure 12.19   Bushing tests at HSP Laboratory, Cologne

In earlier times, an audible 'hissing' test was used to assess the quality of RBP insulation. A trained ear could detect discharge of about 100 pC. RBP bushings with this limit of discharge at 1.05 $U_r/\sqrt{3}$ have given years of satisfactory service and this limit has been retained in recent specifications.

Discharges have a more damaging effect on OIP and a limit of 10 pC at 1.5 $U_r/\sqrt{3}$ is agreed. In general, well-processed OIP and RIP insulation are free from detectable discharge at this level.

Modern discharge detection equipment has been developed to improve the sensitivity of the measurement. Most systems monitor current from the bushing test tapping and display on an oscilloscope in the form of an ellipse. Partial discharges appear as pulses, the magnitude of which can be compared to a calibrated pulse. Much work has been published on partial discharge interpretation [21, 22]. From the position of the discharge pulse on the ellipse, it is possible to recognise certain types of fault. When combined with the routine power frequency withstand test, normally applied for 1 min, the stress dependence of any discharge can provide further information on the discharge site. Researchers have shown that potentially damaging voids in GIS spacer insulator may be undetected by conventional discharge measuring techniques. A system has been developed [23] that uses X-ray scanning to reduce the discharge inception voltage, increase sensitivity and locate the discharge source. This technique known as X-ray induced partial discharge (XIPD) detection may find use in the testing of RIP bushings where void location is important; however, safety in the industrial environment is essential.

For transformer bushings, changes have been made in the latest revision of IEC 60137 so that bushing withstand tests should be carried out at 10% above that of the transformer for which they are destined. In some cases, transformer manufacturers specify bushings co-ordinated one level above the transformer to avoid the possibility of internal fault during transformer test and the ensuing cost of rebuild.

12.5.3 Impulse voltage tests

Lightning and switching impulses represent transients occurring naturally in a high-voltage system under operation. Tests with impulse voltage are designed to demonstrate the response of equipment to transients over a wide frequency range. Dry lightning impulses are applied to all types of bushing as a type test and wet switching impulse tests on bushings above 300 kV rating. Fifteen impulses of positive polarity and 15 impulses of negative polarity are applied to the bushing during type test. As the internal insulation is considered non-self restoring, a maximum of two flashovers are allowed in air external to the bushing insulator with no internal fault permitted. To ensure that transformers, complete with bushings, can be safely subjected to impulse test, it is increasingly common for dry switching and chopped lightning impulses to be applied to bushings as a routine test. Negative lightning impulse tests have been carried out routinely by some bushing manufacturers for a number of years and this has now been incorporated in IEC 60137 for bushings rated 300 kV and above. During impulse testing it is common to measure the current flowing through the bushing by a low-resistance shunt in the tapping earth connection; this gives greater sensitivity in the detection of partial breakdown of the insulation.

12.5.4 Thermal stability test

This test is particularly applicable to bushings for transformers of rated voltage above 300 kV and is intended to demonstrate that the dielectric losses do not become unstable at the operating temperature. The test is carried out with the bushing immersed in oil heated to 90 °C. A voltage is applied equal to the maximum temporary overvoltage (usually 0.8 U_r) seen by the bushing in service. By continuously measuring the capacitance and tangent delta of the bushing, the dielectric losses are calculated. Should the bushing be incapable of dissipating these losses, the tangent delta would increase and thermal runaway would occur, resulting in breakdown of the insulation.

Due to the inherently low value of tangent delta of OIP and RIP bushings, thermal stability is not normally a problem. In certain applications such as oil-to-gas bushings where the cooling is restricted, special attention should be paid to thermal stability.

It is the intention of the specification IEC 60137 that dielectric and conductor losses be applied to the bushing simultaneously. This is not always possible due to design restrictions of the bushing, and conductor losses are considered separately during the temperature rise test.

12.5.5 Temperature rise test

This test is intended to demonstrate the ability of the bushing to carry rated current without exceeding the thermal limitations of the insulation. OIP and RIP are restricted to a maximum temperature of 105 °C and 120 °C respectively. The higher thermal rating of the RIP material does not necessarily mean that smaller conductors can be used. RIP is a good thermal insulant and the design of OIP bushings more readily allows cooling of the conductor by convection within the oil of the bushing. The service condition of different types of bushing, particularly high-current bushings used in phase-isolated bus ducts, must be carefully considered. In a typical test, a bushing achieved a rating of 10 kA under the standard test conditions laid out by the specification, while with an increased ambient air temperature, equivalent to the duct, the maximum current was reduced to 7 kA. This causes obvious difficulty in the specification and use of this type of bushing.

12.5.6 Other tests

In addition to the major electrical tests discussed, tests or calculations are usually required to demonstrate the suitability of the bushing. These include:

1. leakage tests – resistance to leakage by internal or external pressure of oil or gas.
2. cantilever test – demonstration of the ability of the bushing to withstand forces imposed by connections, short circuit, self-loads, etc.
3. seismic withstand, usually demonstrated by static calculations – the effect of the stiffness of the equipment to which the bushing is mounted is also important. Guidance on the seismic qualification of bushings is given in IEC 61463 [24]. This report proposes methods of calculation and test. In parts of the USA utilities have prohibited or restricted the use of porcelain insulators due to potentially hazardous seismic damage and have imposed the use of polymer insulation with strict testing. Figure 12.20 shows such a bushing on a shaker table
4. short circuit, again usually demonstrated by calculations, given in IEC 60137, to prove adequate thermal capacity to prevent overheating and insulation damage during short-circuit events.

12.6 Maintenance and diagnosis

Bushings are hermetically sealed devices generally operating in service under low electrical and mechanical stress. Ingress of moisture, however, due to gasket defects is a major cause of insulation deterioration. Internal partial discharge can result from moisture ingress, system overvoltages or inadequate stress control. External contamination build-up and the risk of pollution flashover can be reduced by periodic washing or the use of silicone rubber or grease coatings.

Dielectric diagnosis techniques can be applied to installed bushings [25]. Online infra-red scanning and radio influence voltage (RIV) measurement can

Figure 12.20 Seismic withstand test on polymer transformer bushing

detect thermal problems and corona. Off-line measurement of capacitance and tangent delta can be made on bushings and compared with factory results and other similar equipment. Information on bushing insulation should include ageing, moisture content and condenser breakdown.

A continuous tangent delta monitor for online transformer bushings has been developed [26–28]. Signals derived from test tappings on all bushings within a substation are compared and abnormal changes activate an alarm. The system is claimed to be cost-effective in high-risk situations. Transformer manufacturers now offer integrated packages for online monitoring of power transformers; these use the bushing tapping as a voltage source and partial discharge sensor [29].

With OIP insulation, dissolved gas analysis (DGA) offers an established technique for the assessment of insulation condition. Most published work refers to transformer insulation [30, 31] comparing the relative concentrations of fault gases. DGA information specific to OIP-type bushings is given in IEC 61464 [31]. This report was compiled by bushing experts based on information from bushings in service. Guidance is given on the significant level of fault gases, interpretation of typical faults and recommendations for action [32] (see Tables 12.3 and 12.4).

Table 12.3 Typical faults occurring in bushings

Case no.	Key gases generated	Typical examples	Characteristic faults
1	H_2, CH_4	Discharge in gas-filled cavities resulting from incomplete impregnation or high humidity	Partial discharge
2	C_2H_4, C_2H_2	Continuous sparking in oil between bad connections of different potentials	Discharge of high energy
3	H_2, C_2H_2	Intermittent sparking due to floating potentials or transient discharges	Discharge of low energy
4	C_2H_4, C_2H_6	Conductor overheating in oil	Thermal fault in oil
5	CO, CO_2	Overheating of conductor in contact with paper: overheating due to dielectric losses	Thermal fault in paper

Table 12.4 Normal gas concentrations

Type of gas	Concentration (µl gas/l oil)
Hydrogen (H_2)	140
Methane (CH_4)	40
Ethylene (C_2H_4)	30
Ethane (C_2H_6)	70
Acetylene (C_2H_2)	2
Carbon monoxide (CO)	1,000
Carbon dioxide (CO_2)	3,400

Today the trend is for the development of online diagnosis techniques to minimise the need for periodic line diagnosis in assessing insulation condition and thereby predicting fault development and extending equipment life.

References

1. IEC 60137 2008. 'Bushings for alternating voltages above 1000 V'. Geneva: International Electrotechnical Commission
2. Douglas J.L., Stannett A.W. 'Laboratory and field tests on 132kV synthetic resin bonded paper condenser bushings'. *IEE Proceedings*. 1958;**105**(Part A): 278–94
3. Bradwell A., Bates G.A. 'Analysis of dielectric measurements on switchgear bushings in British Rail 25 kV switching substations', *IEE Proceedings*, 1985;**132**(Part B):1–17
4. IEC 60296 2003. 'Unused mineral oils for transformers and switchgear' (equivalent B5148: 1984). Geneva: International Electrotechnical Commission
5. Kallinikos A. *Electrical insulation*. London: Peter Peregrinus; 1983 (Chapter 1.1)

6. Barker H. 'High voltage bushings'. *Harwell high voltage technology course*, 1968
7. IEC 60815 2008. 'Guide for the selection of insulators in respect of polluted conditions'. Geneva: International Electrotechnical Commission
8. IEC 60507 1991. 'Artificial pollution tests on high voltage insulators to be used on A.C. systems'. Geneva: International Electrotechnical Commission
9. Gilbert R., Gillespie A., Eklund A., Eriksson J., Hartings R., Jacobson B. 'Pollution flashover on wall bushings at a coastal site'. Report no. 15-102, CIGRE, 1998
10. EN 50180 2010. 'Bushings for liquid filled transformers above 1 kV up to 36 kV', Brussels: European Committee for Standardisation (CEN)
11. Bossi A., Yakov S. 'Bushings and connections for large power transformers.' Report no. 12-15, CIGRE, 1984
12. *2nd international IEC/cigré symposium on standards for ultra high voltage (UHV) transmission*; New Delhi, 2009. Available from http://www.cigre.org
13. IEC 61639 1996. 'Direct connection between power transformers and gas-insulated metal-enclosed switchgear for rated voltages of 72.5 kV and above'. Geneva: International Electrotechnical Commission
14. Gallay M., Fournier J., Senes J. 'Associated problems, inspection and maintenance of two types of interface between EHV transformer and metalclad electric link insulated with SF_6'. Report no. 12-01, CIGRE, 1984
15. Johansson K., Gafvert U., Erikson G., Johansson L. 'Modelling and measurement of VFT properties of a transformer to GIS bushing'. Report A2-302, CIGRE, 2010
16. Moser H.P., Dahinden V. *Transformer board H*. Switzerland: H. Weidmann AG, Rapperswil
17. Lampe W., Eriksson K.A., Peixoto C.A. 'Operating experience of HVDC stations with regard to natural pollution'. Report no. 33-01, CIGRE, 1984
18. Lambeth P.J. 'Laboratory tests to evaluate HVDC wall bushing performance in wet weather'. *IEEE Transactions on Power Delivery*. 1990;**5**(4):1782–93
19. Lampe W., Wikstrom D., Jacobson B. 'Field distribution on an HVDC wall bushing during laboratory rain tests'. *IEEE Transactions on Power Delivery*. 1991;**6**(4):1531–40
20. IEC 60060-1 2010. 'High voltage testing techniques'. Geneva: International Electrotechnical Commission
21. Natrrass D.A. 'Partial discharge measurement and interpretation'. *IEEE Electrical Insulation Magazine*. 1988;**4**(3):10–23
22. Kreuger F.H. *Partial discharge detection in HV equipment*. London: Butterworth; 1989
23. Rizetto S., Fujimoto N., Stone G.C. *A system for detection and location of partial discharges using X-rays*. EPRI Project RP2669-1
24. IEC 61463 1998. 'Guide for the interpretation of dissolved gas analysis (DGA) in bushings where oil is the impregnating medium of the main insulation (generally paper)'. Geneva: International Electrotechnical Commission

25. Fruth H., Fuhr J. 'Partial discharge pattern recognition – A tool for diagnosis and monitoring of ageing'. Report No. 15/33-12 CIGRE, 1990
26. Allan D., Blackburn T., Cotton M., Finlay B. 'Recent advances in automated insulation monitoring systems, diagnostic techniques on sensor technology in Australia'. Report no. 15-101, CIGRE, 1998
27. Polovick G.S., Jaeger N.A.F., Kato H., Cherukupalli S.E. 'On-line dissipation factor monitoring of high voltage current transformers and bushings'. Report no. 12-102, CIGRE, 1998
28. Krump R. Haberecht P., Titze J., Kuechler A., Liebschner M. *Bushings for highest voltages: Development, testing, diagnostics and monitoring*. Germany: Highvolt Kolloquium 07, University of Schweinfurt
29. Knorr W., Liebfried T., Breitenbauch B., Viereck K., Dohnal D., Sundermann U., et al. *On-line monitoring of power transformers – Trends, new developments and first experiences*. Report no. 12-211, CIGRE, 1998
30. Barraclough B., Bayley E., Davies I., Robinson K., Rogers R.R., Shanks C. 'CEGB experience of the analysis of dissolved gas in transformer oil for the detection of incipient faults'. *IEE Conference Publication* 1973; **94**:178–92
31. IEC 60599 2007. 'Interpretation of the analysis of gases in transformers and other oil filled electrical equipment in service' (equivalent BS5800: 1979). Geneva: International Electrotechnical Commission
32. IEC 61464 1998. Insulated bushings – Guide for the interpretation of dissolved gas analysis (DGA) in bushings where oil is the impregnating medium of the main insulation (generally paper). Geneva: International Electrotechnical Commission

Chapter 13
Design of high-voltage power transformers
A. White

13.1 Introduction

Energy, and particularly electrical energy, is a commodity that mankind in general tends to take for granted. We switch on a light or our computer and expect an immediate power flow to energise it. Yet, the steady development and rapid progress that has been made in the transmission and distribution of electrical energy during the past 120 years or more may not have been possible but for the capability of linking generators, transmission lines, the secondary distribution systems operating at a variety of loads, with each at its optimum voltage [1]. This linking of systems at different voltages has relied upon a simple, convenient and reliable device – the power transformer. The unique ability of the transformer to adapt the voltage to the individual requirements of the different parts of the system is derived from the simple fact that it is possible to couple primary and secondary windings of the transformer in such a way that their turns ratio will determine very closely their voltage ratio as well as the inverse of their current ratio, resulting in the output and input volt-amperes and the output and input energies being approximately equal.

Coverage in this chapter includes a little of the theory of transformers, the major components involved in their manufacture, and the many different applications of the transformer technology in that are in everyday use.

13.2 Transformer action

A transformer essentially comprises at least two conducting coils having a mutual inductance. The primary is the winding that receives electric power and the secondary is the one that may deliver the power as shown in Figure 13.1.

The coils are usually wound on a core of laminated magnetic material and the transformer is then known as an iron-cored transformer. The modern iron-cored transformer has so nearly approached perfection that for many calculations it may be considered as a perfect transforming device. In the simplest form of the theory of the transformer, it is assumed that the resistances of the windings and the core losses are negligible, that the entire magnetic flux links all of the turns of the

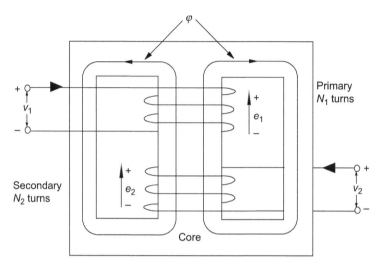

Figure 13.1 Schematic diagram of a transformer

windings, that the permeability of the core is so high that a negligible magnetomotive force produces the required flux and that the capacitances of the winding are negligible. That is, the transformer is assumed to have the characteristics approximating to those of an ideal transformer with no losses, no magnetic leakage and no exciting current. Thus, the instantaneous terminal voltage v_1 is numerically identical to the instantaneous voltage e_1, induced by the time-varying flux linkages, which in turn is equal to the number of turns in the coil N_1, multiplied by the rate of change of flux linkages. Thus, for the primary circuit

$$v_1 = e_1 = N_1 \frac{d\varphi}{dt} \qquad (13.1)$$

For the secondary circuit, similar criteria apply

$$v_2 = e_2 = N_2 \frac{d\varphi}{dt} \qquad (13.2)$$

The parameter φ is the resultant flux produced by the simultaneous actions of the primary and secondary currents, and therefore

$$\frac{v_1}{v_2} = \frac{N_1}{N_2} \qquad (13.3)$$

In an ideal transformer, the instantaneous terminal voltages are proportional to the number of turns in the winding, and their respective input and output waveforms are identical. The net magnetomotive force required to produce the resultant flux is zero and the net magnetomotive force is the resultant of the primary and secondary ampere-turns; hence, if the positive direction of both primary and secondary currents are taken in the same direction about the core, then

$$N_1 I_1 + N_2 N_2 = 0 \tag{13.4}$$

So, for this ideal transformer

$$\frac{I_1}{I_2} = -\frac{N_2}{N_1} \tag{13.5}$$

The minus sign indicates that the currents produce opposing magnetomotive forces.

The performance of the transformer depends upon time-varying flux and therefore in the steady state a transformer operates on alternating voltage only. The transformer is therefore a device which transforms alternating voltage or alternating current or even impedance. It may also serve to insulate one circuit from another or to isolate direct current (DC) while at the same time maintaining alternating current continuity between circuits.

13.3 The transformer as a circuit parameter

As stated previously in this chapter, the simple theory assumes that the transformer is electrically perfect. In a more comprehensive theory of its electrical characteristics, however, account must be taken of some aspects that are not entirely ideal, which occur in iron-cored transformers. First of all, the windings do have resistance; secondly it is not possible for the windings to physically occupy precisely the same space meaning that there is some magnetic flux leakage between the coils of the transformer; thirdly, an exciting current, albeit small, is required to produce the flux; and fourthly, there are hysteresis and eddy current losses produced in the core. Furthermore, whenever rapid rates of voltage change occur, then the capacitances between the windings and to earth potential can no longer be neglected.

When a transformer is supplied with power through a transmission circuit, whose impedance is relatively high, the primary terminal voltage may vary with changes in load over an undesirably large range, because of changes in impedance drops in the various parts of the transmission circuit. In order to maintain the secondary voltage at its desired range of values under varying conditions of load and power factor, it is often necessary to provide voltage tappings to permit compensatory turns variation in one of the windings (Figure 13.2).

Large tap-changing power transformers are also frequently used to control the flow of reactive power between two interconnected power systems or between component parts of the same system, while at the same time permitting the voltages at specified points to be maintained at desired levels. Often the tap-changing apparatus is designed so that the ratio of transformation can be adjusted while the transformer is carrying load without interruption of the current. Two basic design styles of on-load tap-changer are available, these being the high-speed resistor and the reactor types. The most commonly used type is the high-speed resistor

498 High-voltage engineering and testing

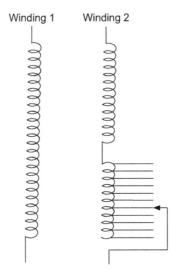

Figure 13.2   Voltage regulation diagram

equipment, but the reactor type has its advantages in some applications. More will be related on this topic later in this chapter.

The voltage transformation affected by a transformer is accomplished at the expense of an exciting current to create the magnetic flux. This has a power component which corresponds to the no-load losses which are practically all confined within the core, and there is also a reactive component which corresponds to the reversible magnetisation of the core and is wattless. At no-load, there are losses due to the current travelling through the copper conductors. These are usually of small magnitude. There are also losses in the insulation, which are usually of minute proportions.

The in-phase and reactive components of the excitation current depend upon the quality and thickness of the steel used to manufacture the core, the flux density, the frequency of power supply, the quality of the joints in the core and the overall length of the magnetic path. They are totally independent of the windings design. The desirable qualities in a core are minimum values of both the power and the reactive components. Of these two, the former is by far the more important consideration.

13.4   Core- and shell-form constructions and components

Two formats of power transformer design are manufactured. These are commonly known as core form and shell form. The core-form transformers usually have concentric windings around a circular section core leg, while the shell-form core is of rectangular cross-section encapsulating the windings that are often of rectangular pancake format (with rounded corners to simplify manufacture) and sandwiched between one another. These descriptions are general and are not intended to suggest exclusivity of any combination of characteristics.

Both core forms have merits and each possesses some advantages over the other. Nevertheless, shell-form core manufacturers are nowadays accepted as being in the minority worldwide, but have a vital role to play in transformer production. Unless otherwise stated, descriptions in this chapter will hereafter refer to core-form transformers.

13.4.1 The core

The core constitutes the heaviest component of the transformer, and as such deserves continuing developments to obtain the very best performances at the lowest cost to the purchaser of the transformer.

One of the most significant historical improvements in transformer manufacture has been the necessity for ongoing development and introduction of higher grade core steels [2]. Prior to the twentieth century, soft magnetic iron materials were used in transformer manufacture (Table 13.1). Early in that century, however, silicon steels were developed, which gave very much reduced no-load losses. Prior to the Second World War, metallurgists were able to produce steels in which the steel grains were oriented in the direction of the rolling operation, resulting in much easier magnetisation along this direction and thus reducing losses. By the early 1970s, Japan began to lead the rest of the world in bulk steel production. This superiority included the area of electrical steels. By clever metallurgical methods they were able to introduce a variety of low-loss steels that are known universally as 'Hi-B'. The technology was licensed to many other countries, and today 'Hi-B' steels are commonly used throughout the world. In 1980 various steel companies, led primarily by the Japanese, were producing ever-thinner core steels. Whereas in 1960 most core steel used in the UK was 0.35 mm thick, the Japanese have been able to achieve 0.05 mm thick materials. Such small thicknesses are extremely difficult to mechanically handle in power transformer manufacture even if expensive and specialist cutting and handling equipment is employed. Bonding thin laminations to marry the magnetic advantages of very thin steels with the mechanical advantages of thicker steels is a strong possibility for future production, when the cost, inevitably, is no longer prohibitive. To date, alternative methods of

Table 13.1 Core material development

Date	Material
1885	Soft magnetic iron
1900	Non-oriented silicon steel
1935	Grain-oriented silicon steel
1970	Hi-B
1980	Thin Hi-B
1983	Laser scribed Hi-B
1990	Very thin magnetic steels

loss reduction have proved to be more economic. These methods include the use of a laser or other scribing methods to break up the long grains in the 'Hi-B' steels, thus allowing easier rotation of those grains in a magnetic field making the steel much easier to magnetise.

There was a further benefit of the higher grade steels noted – in respect of noise level produced by the operating transformer, typically yielding a 2–3 dBA reduction in noise compared with a core of 'conventional' cold rolled grain-oriented silicon steel. Such a reduction is becoming increasingly important where transformers are installed near populated areas and noise pollution is an undesirable feature.

As a consequence of producing core materials that make it easier for core materials to be magnetised along the rolling direction of the core steel, it usually means that it is more difficult to magnetise these materials perpendicular to the rolling direction. Inevitably, in any power transformer design, the flux has to be persuaded to turn through angles, usually of 90°. Adjacent to joints between plates, the flux must further change direction. This introduces localised power and reactive components that can be substantially higher than those that appear in the main body of the laminations remote from the joint. The method used to reduce the effects of change of flux direction is to mitre the joint at an angle (Figure 13.3). The optimum mitring angle for conventional grain-oriented steel from a loss viewpoint is normally recognised as being about 55°. However, cutting plates at this angle increases the amount of wasted material. The economic optimum angle of mitre is 45°, giving minimum waste, but with only minor loss difference between 55° and 45° cuts.

Modern cores generally preclude bolted constructions to hold the plates together. Even cores of masses of 250 tons or more may be held together with bands or straps while still maintaining mechanical integrity in handling, transportation and under the action of short-circuit forces. The move from bolted to banded core construction has yielded improvements in noise levels in addition to a 3% or more improvement gained in no-load loss levels.

The transformer designer is faced with a large number of core steel grades (Table 13.2) but the better the grade, the higher the price that must be paid for the privilege. The selection of the most appropriate grade for each application depends purely upon economics (Table 13.3).

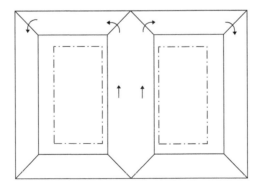

Figure 13.3   Mitred core construction

Table 13.2 Grades and losses of electrical steels

Type	Grade	Loss, W/kg 1.7 T, 50 Hz	Relative loss % of that of M103-27P
Conventional	M111-35N	1.41	144
	M097-30N	1.30	133
	M089-27N	1.23	126
	M120-23S	1.11	113
Hi-B	M117-30P	1.12	114
	M105-30P	1.00	102
	M103-27P	0.98	100
	M100-23P	0.92	94
Etched Hi-B	27ZDKH	0.92	94
	23ZDKH	0.84	86

Table 13.3 Typical net worth of core steels for a single-phase, 50 Hz generator step-up transformer operating at a flux density of 1.7 T. Base material M103-27P

Type	Grade	Net worth with capitalisation at £3,000/kW
Conventional	M111-35N	−1230
	M097-30N	−720
	M089-27N	−460
	M120-23S	−330
Hi-B	M117-30P	−180
	M105-30P	−65
	M103-27P	0
	M100-23P	+40
Etched Hi-B	27ZDKH	+70
	23ZDKH	+180

Transformers are frequently of three-phase construction, although trackside transformers and those for which transport or all-up site mass or dimensions are critical features of the specification may require single-phase designs to be utilised. For three-phase cores, a three-limb design with each limb carrying one phase of windings may be used (Figure 13.4). Core parts are characteristically referred to as 'legs', 'yokes' and 'joints'. Legs are those over which the windings are assembled; yokes are the necessary joining pieces between the legs of the magnetic circuit; and joints are the overlap regions where leg and yoke plates meet. The leg and yoke sections are usually of identical areas and shapes (commonly circular). However, there may be an advantage in mass if the yoke shape is altered to have a 'D' outline, albeit with some added complexity in respect of the yoke securing arrangements.

502　*High-voltage engineering and testing*

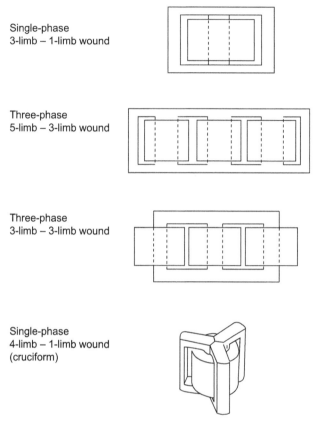

Single-phase
3-limb – 1-limb wound

Three-phase
5-limb – 3-limb wound

Three-phase
3-limb – 3-limb wound

Single-phase
4-limb – 1-limb wound
(cruciform)

Figure 13.4　A variety of core types

Frequently, the transportation height of the transformer is critical or perhaps for certain system design applications in which the zero phase sequence impedance of the transformer is required to be equal to the positive sequence impedance, a five-limb construction is regularly employed. In this case, the three inner wound legs are of the full section area, while the outer limbs and the yokes may have reduced area, since some of the flux which passes through the wound legs is circulated through the outer legs. The flux distributions in this case are complex and the fluxes in the centre yokes and in the outer limbs are not sinusoidal. Depending upon the sectional area chosen for centre yokes and outer limbs, the peak flux densities in these parts may be higher or lower than the nominal flux density that applies to the wound legs. In the event of overvoltages on the system, it may be necessary to reduce the nominal flux density in order to prevent some parts of the core from saturating under conditions of overvoltage or under-frequency. Some designs have yoke and outer leg sectional areas both equal to 50% of that of the main legs. Others may have centre yoke sections of 58% of that of the main legs, while the outer leg section is reduced to 44%; these values generally mean that the peak flux density in any part of the core does not exceed that in the main legs.

There are limitations on the widths of core plates that are commercially available, and consequently assemblage of pairs of plates side by side makes up the larger sectional areas. This may be done with two different plate widths alternating to give an overlapping effect or with equal plate widths that create a deliberate duct between them. In this latter instance, the peak flux densities can be different in each loop of the core.

For single-phase design applications, several variants are possible. Cores may have two wound legs only, or may have a centre wound legs with half-section outer legs, or even two wound legs plus two half-section outer legs. Whenever, the footprint of a transformer is of paramount concern, it is also feasible to use three or four cores in a cruciform arrangement so that one set of windings surrounds one limb of each part-core.

The component parts of the total losses in the core materials are subdivided into two principal categories. First there is the hysteresis loss, which is inherent in the material and is principally a function of metallurgy, frequency and flux density. The second component category is eddy losses that result from circulating currents within the laminations due to flux passing through them. This is a function of the resistivity of the material, the dimensions (thickness and width), and the square of the frequency, as well as the flux density. Electrical steel manufacturers have cleverly researched the metallurgy to reduce the hysteresis component and have steadily reduced the thickness of the material to improve the eddy component. The ultimate aim is to achieve higher permeabilities and lower losses at higher and higher flux densities.

Core plates are cut on high-speed, high-accuracy, automatic machines using regularly sharpened tools to ensure that burrs and slivers are eliminated or minimised. The plates are built layer by layer, usually no more than two plates at a time, with alternate layers displaced to impart a sufficiently rigid mechanical structure when the whole core is fully clamped and secured. Multiple displacements of successive layers, by means of a process known as step-lapping, can result in reductions of up to 8 dBA in sound pressure levels and of between 3% and 8% in no-load losses.

13.4.2 The windings

Windings constitute an essential feature for all transformers and similar devices. A number of different winding constructions are in common usage. Throughout a large part of the world core-type transformers predominate. The type of winding for each particular application depends upon the current and voltage ratings of the transformer and a variety of types are shown in Figure 13.5.

A very high-current winding will normally be associated with low voltage and therefore have few turns of conductor. The principal design intent is to accommodate as much of the conducting medium into the smallest possible physical space, within the constraints of satisfactory dielectric performance of the local

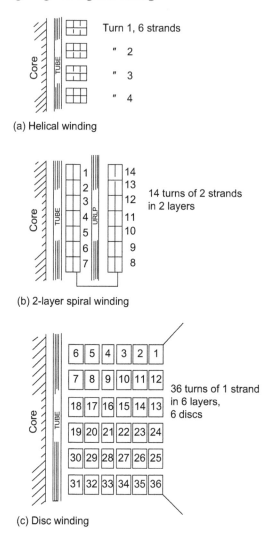

Figure 13.5 Various types of winding

insulation structure. The simplest and straightforward method of achieving this is to use a spiral winding in which each successive turn is wound tightly against its neighbour. As the winding becomes larger, it becomes increasingly difficult to preserve temperatures that are generated by the losses in the coils, such that the strict and specified limits pertaining to the insulation materials are maintained. Inevitably, vertical and/or horizontal cooling ducts must be introduced into the design. Multiple-layer spiral coils provide vertical ducts and helical coils provide horizontal ducts. As the voltage increases, the current tends to decrease and therefore the windings tend to be of smaller cross-section conductors but of many more turns. In due course, the economics of high space utilisation demands a change from the simple helical or spiral construction to a disc arrangement. As the voltage increases further, the insulation between the conductors in the discs adjacent to the line end of the winding reaches a point where simply adding more and more insulation becomes an uneconomical proposition. The dielectric control may then be achieved by the introduction of electrodes of controlled potential which are physically inserted between the turns of the winding. This latter construction is known as the intershielded disc type of winding.

Ultimately further complications have to be introduced into the design and manufacture of disc windings, and a feature known as the interleaved disc winding, invented by George Stern of English Electric in the early 1950s, now has worldwide appeal. Some control methods are illustrated in Figure 13.6. It must be noted that not all core-type transformer manufacturers utilise disc winding constructions. A number of companies have continued the principle, once used by several UK manufacturers, of multiple-layer construction.

Turning next to the conductor material, it is recognised that the lowest resistivity conductor is silver. However, the use of such material in transformers would be totally uneconomic, and copper is the next best alternative. There are also certain financial (and sometimes technical reasons) for the use of aluminium winding conductors instead.

The conductors may be circular in section, particularly where a very large number of turns of low sectional area are involved. Power transformer applications, however, tend to use rectangular conductors for better space utilisation reasons. Such conductors are found in a variety of forms, either individually as strip conductors, or in paired or tripled format with reduced insulation between the individual conductors of the bundle, so as to maintain high dielectric capability between adjacent turns of the winding while providing only the insulation strength that is absolutely necessary between parallel strips. The conductors may be laid side by side or end to end as dictated by the manufacturer's preference. Another alternative conductor form is known as continuously transposed cable, which comprises a number of rectangular conductors each coated with suitable acetate enamel, transposed with one another at regular intervals along the cable length, and the whole bundle wrapped with an insulating material to provide the inter-turn insulation (Figure 13.7).

For low-voltage windings, the insulation value of the enamel alone may be sufficient to withstand the voltage stresses between turns and the outer insulation wrap may then be omitted.

Where very few turns are involved, the use of copper or aluminium foils may prove advantageous.

Fundamentally a transformer requires at least two windings but may have many more than this. It is sometimes feasible to arrange for the regulating coils to have tapping leads simply brought out from the main high-voltage winding. Frequently, regulating windings that are usually associated with the high-voltage winding are wound and assembled as separate coils. A separate tapping winding may be wound concentrically outside the high-voltage winding or inside the low-voltage winding (Figure 13.8). In addition the high-voltage winding may be split

506 High-voltage engineering and testing

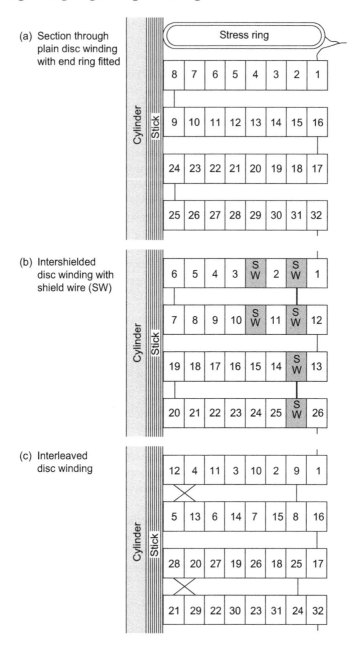

Figure 13.6 Methods of increasing series capacitances in disc windings

into two parallel components with the regulating winding also split to maintain sufficient dielectric clearance to allow the high-voltage lead to pass between the two half-coils. There are sometimes advantages to be had from splitting the HV winding into two series components to achieve a specific impedance requirement or

Design of high-voltage power transformers 507

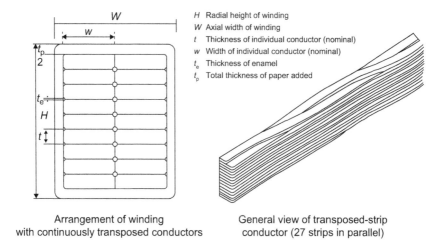

H Radial height of winding
W Axial width of winding
t Thickness of individual conductor (nominal)
w Width of individual conductor (nominal)
t_e Thickness of enamel
t_p Total thickness of paper added

Arrangement of winding
with continuously transposed conductors

General view of transposed-strip
conductor (27 strips in parallel)

Figure 13.7 Continuously transposed conductors

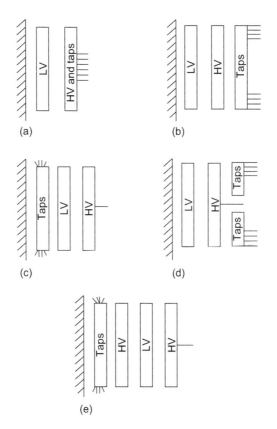

Figure 13.8 Winding arrangements

to reduce some of the force components that would otherwise result from short-circuit current conditions. This arrangement is commonly known as 'double concentric'.

The arrangements may be further complicated by the addition of other windings. A locomotive transformer, for example, may have four primary HV windings, four secondary low-voltage windings and five tertiary windings, all having specified impedance requirements between each pair of windings. These impedance constraints control the physical location of each of the many coils with respect to any other [3].

13.4.3 Cooling

A winding that carries current is subject to the losses in the conductors due to the electrical resistances of those conductors. This component of the so-called load losses is frequently known as the I^2R or galvanic loss. The alternating current passing through the coil also creates an electromagnetic field, and this in turn creates eddy currents that flow within the conductors, giving rise to further losses known as eddy losses. These eddy losses are not uniformly distributed throughout the winding since the electromagnetic field is itself not uniform over the complete winding. Furthermore, since it would be very difficult to arrange each conductor to physically occupy a space in an identical field to that of any other conductor to which it is ultimately connected, the voltage in each interconnected conductor is different from its neighbour and in consequence circulating currents will result. All of these load and parasitic currents create losses and the heat generated by these losses must be extracted and carried away from the windings to a position in which the heat can be dissipated to the atmosphere or to another substantial heat sink such as a river. Within the transformer the commonly used cooling medium is mineral transformer oil, although other natural or synthetic fluids or gases, including air, are to be found in power and distribution transformers.

In fluid systems there is a choice of flow conditions. Flow may be natural, that is the oil flow results from convection. Alternatively a pump may be used to induce higher flow rates than simple convection would permit. This forced oil-flow system provides greater efficiency of heat transfer, but the pump consumes running energy.

Within the winding there is the possibility of having non-directed flow or directed flow. In the former case the flow is determined solely by convection, and in the latter case the flow paths are predetermined by the designer in order to improve the efficiency of heat transfer particularly from the horizontal surfaces of the coil. Generally, the selection of natural or forced oil flow lies with the purchaser, who selects flow type according to local conditions and preferences. The choice of directed or non-directed flow usually remains within the domain of the designer.

The choice of actual oil velocities is also an option that is left to the designer. There are no mandatory limits on oil velocities, although each manufacturer will impose rules that will include some velocity limitations. Oil velocities that are above 1,000 mm/s, for example, might induce surface charge that may be concentrated on local areas of insulation surface especially under moisture-free oil conditions. This surface charge may, in turn, cause partial discharges that would lead to tracking and ultimately to complete dielectric failure.

13.4.4 Insulation

The highest temperature that appears anywhere in any of the windings will generally determine the life expectancy of transformer insulation. On windings having vertical axes, this hot-spot temperature usually appears near the top and is probably located in the first or second disc counted from the top. Of course, the actual hot-spot position is totally a function of the design and manufacture. Only the transformer manufacturer has access to all information that is required to obtain the best mathematical estimate of the exact position of the hot-spot temperature. It is then possible, but extremely difficult, to fit a fibre optic cable to run to that location through the winding and measure temperatures at this and other places of interest, with a resolution of about 1 m of the fibre length. The difficulties of fitting fibre optic cables, with high levels of safety, coupled with their maintenance and monitoring economics must be considered very carefully by the purchaser before selecting this option. Analogue hot-spot temperature measurement is a well-proven channel and is sufficient for many users.

Transformer oil is not only the most common cooling fluid used in transformer manufacture; it is also an excellent insulation material although dielectrically inefficient in regions of large unbroken volume. Cellulosic kraft paper (or paper and enamel) is the most common conductor insulation material. When insulation is used in structural components the cellulose-based material is produced in board form, known as pressboard. Major inroads have been made in recent years in relation to the quality of insulation materials permitting continuously improving space utilisation for the active materials, namely copper and steel. The most universally used insulating board is made completely from kraft. There are a number of niche suppliers that supply material that is a mix of kraft with cotton. The cotton fibres are longer than the kraft fibres permitting oil impregnation to be accomplished within a shorter period of time. The conservationists also argue that cotton is an annual product and its use is therefore environmentally friendly, whereas use of wood kraft involves the destruction of the plant. Unfortunately, harvesting the cotton is a very labour-intensive operation and costs have risen dramatically over the years.

Other insulation systems at lower ratings may use cast resin as an insulant and air as a coolant. This approach provides a low-flammability transformer suitable for use inside large buildings and tower blocks.

High-temperature systems are also available using synthetic fluids such as silicone in conjunction with Nomex and polyester glass boards [4, 5]. These materials are much more expensive than mineral oil, and cellulose that is paper and pressboard.

Hybrid insulation systems may also be economically used in certain applications. Such systems use the expensive high-temperature materials only in the immediate vicinity of the hotter components, principally the conductors and leads, and cellulose materials elsewhere. The cooling fluid in this case is mineral oil. Typical applications of this type of insulation system are mobile transformers and higher rated rewinds of existing transformers.

13.4.5 Tank

Another principal feature of the transformer is the container that is used to encapsulate the fluid. As well as being called upon to withstand the rigours of nature in its allocated environment for periods of typically 50 or even more years, this tank is required to sustain the vacuum necessary for oil impregnation and filling, to withstand the pressure that is exerted by the oil in normal operation and to withstand the forces imposed by transportation at least twice in its lifetime. In addition, many of the forces that are imposed on the transformer under short-circuit current conditions are effectively restrained by the tank in reaction. Tanks are usually of mild steel, but it may be necessary to introduce non-magnetic steels or other non-magnetic metals in order to reduce stray losses in the metallic components or the temperatures resulting from them.

13.4.6 Bushings

Connections that pass through the cover or walls of the tank are required to link up the external power supply system with the internal windings of the transformer. The interconnection devices, which are called bushings, permit the high-voltage leads to pass through the tank while ensuring adequate clearances to be maintained to prevent flashover in normal operation and under fault conditions. The tank is normally held at earth potential for safety reasons.

Lower voltage bushings comprise a current-carrying stem that is centrally located within a solid insulator having an external pollution tolerant surface, which is coupled to a mounting flange affixed to an aperture on the tank. Simple bushings may utilise as insulation medium either porcelain glazed on its external surface or epoxy resin. As the voltage rises, the external surface of the insulator is provided with sheds in order to preserve a high enough creepage distance while presenting a suitable flashover distance.

For higher voltages, the major insulation generally comprises either oil-impregnated paper or epoxy resin-impregnated paper. The impregnation with oil or resin is carried out under vacuum.

In the past, bushings have been manufactured from resin-coated paper which was then cured to bond the paper layers. Unfortunately this method of manufacture did not prove entirely effective in precluding air bubbles within the resin. After long period of time, partial discharge occasionally occurred leading to catastrophic failure after some 10–20 years in service. Subsequently many of these synthetic resin-bonded paper bushings above 66 kV class have been replaced with vacuum-impregnated bushing types.

Higher-voltage bushings additionally require control of the voltage distribution within their insulation systems. A common control application is the condenser arrangement. Very thin metallic foils are inserted at regular intervals within the layers of paper that makes up the major insulation. The inner foil is the longest and is electrically bonded to the current-carrying conductor. The outer foil is often the shortest and is bonded to the fixing flange of the bushing. One or two of the intermediate foils may be electrically connected to externally accessed bushing tappings. These tappings may be used for either checking power factor and capacitance during service or driving low-power circuits at the substation. During normal service these bushing tappings are earthed.

A number of termination possibilities are available – oil on one side/air on other, oil on one side/high-pressure SF_6 on the other, oil on both sides, oil on one side/semi-fluid compound on the other and moulded semi-conducting bushings. Condenser bushings may employ oil to fill the space between the condenser and the outer shell or may make use of a synthetic compound or may even include pressurised SF_6.

This topic is covered in more detail in Chapter 12.

13.4.7 On-load tap-changer

On-load voltage regulation has already been referred to in this chapter. The device which performs regulation on a transformer while it is carrying load is known as an on-load tap-changer. In the USA, reactor-type tap-changing is still available and is frequently used for low-voltage, high-current applications, but the rest of the world generally favours high-speed resistor-type tap-changing.

The on-load tap-changer includes a series of switches which operate in a preset sequence to permit the load current to flow throughout the operation [6]. The tapping which is adjacent to the one which is in service is preselected by the tap selector. The service-tapping selector switch carries the load current. A switching device transfers the current flow from a direct connection to one which permits the current to be bypassed through a resistor. The switching movement then continues such that the selected and the preselected tapping are both permitted to share the load current, and an additional circulating current created by the voltage difference between the selected and preselected taps is restricted in magnitude by two resistors connected in the circuit. The switch then continues its motion and breaks contact with that resistor which is connected to the previously selected tapping. Finally, by further switch action, the second resistor is taken out of the circuit. The switch is designed to operate very quickly and the resistors are therefore only required to have short time ratings. It is therefore essential that the operation is not interrupted in order that the tap-changing equipment is fully protected. The complete sequence is therefore actuated by a motor drive mechanism which operates through geneva gears to operate the tap-selector switches and simultaneously to wind up an energy spring mechanism which actuates the load current breaking or diverter switch. The diverter switch operating cycle is carried out only when the spring is released, and this is done independently of the motor drive mechanism. The actual current transfer time is of the order of 40–60 ms, but the total time for one tap-change between initiation of the motor drive mechanism and the completion of the tap-change operation amounts to between 3 and 10 s, depending on the type and manufacture of the on-load tap-changing equipment.

High-speed resistor tap-changers are classified into two groups, the first being the separate tank tap-changer and the second the in-tank tap-changer. As the name implies, the separate tank tap-changer divorces oil of both the selector and diverter tank from the oil in the main tank of the transformer. For the in-tank type tap-changer the selector switch operates within the main tank oil but the diverter oil is maintained in a separate system. While historically the UK has had a general preference for the separate tank tap-changer, it must be recognised that the in-tank type tap-changer has a greater population worldwide. The principle of in-tank tap-changer design offers greater flexibility in matching the tap-changing equipment to the transformer than does the separate tank tap-changer. Furthermore, the numbers of solid insulation surfaces which are under high dielectric stress in the in-tank design are limited, with the critical surface being associated with the insulating drive tube. The major insulation is thus oil whose quality is easier to maintain and check. With the in-tank tap-changer, however, a certain amount of gas is produced by capacitance current sparking, and many believe that this will disguise any gases which may emerge from the transformer core and windings. Analysis of hydrocarbon gases dissolved in the transformer oil is an important part of condition monitoring.

13.5 Design features

Transformer design is a complex process that must take into account dielectric, electromagnetic, thermal and mechanical performances, many of which are interrelated. The design engineer is, therefore, equipped with a number of calculation tools that facilitate confirmation of the capability of the transformer.

13.5.1 Dielectric design

The winding is designed to have a number of dielectric duties. It must be satisfactory for the voltage conditions which appear in normal operation, but must also be capable of meeting any overvoltage conditions which are impressed on it. This capability is demonstrated in the final test of the transformer as an induced overvoltage test at power frequency. Normally, the voltage between parts and to earth, which is induced during this test, is at least twice the normal rated voltage between these elements. In service, the transformer also will have imposed on it transient voltages in the form of lightning and switching impulses. The lightning impulse, in particular, is of very short duration with very high rates of change of voltage (Figure 13.9).

Because of the high rate of change, the capacitances of the windings have a much more predominant effect on the voltage distribution than they do under power frequencies. The voltage distributions within the winding are therefore determined by capacitance networks, at least for the first 2 or 3 µs of the impulse wave arriving at the terminal. The capacitance network of the transformer winding

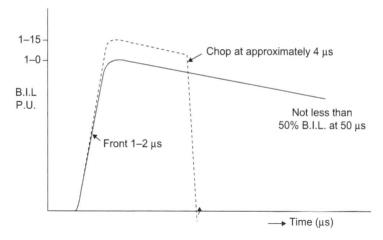

Figure 13.9 Standard impulse voltage wave shape

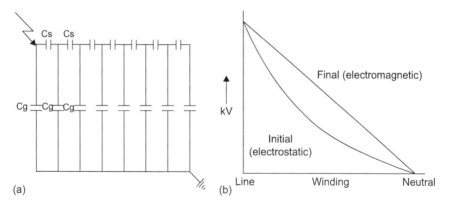

Figure 13.10 Equivalent capacitance network (a) and impulse voltage distribution (b)

assembly gives a non-uniform voltage distribution throughout the coils (as shown in Figure 13.10). The highest voltage drops appear at the turns of the winding which are closest to the impinging lightning impulse wave. If the voltage wave was a step function then ultimately a linear voltage distribution would occur. Between the initial distribution and the final distribution, therefore, considerable variations can appear at various parts of the winding. The effect is similar to that which would be achieved by holding a piece of elastic between the live and neutral terminals on a graph representation of winding length and then stretching the elastic to follow the curve of the initial distribution. Upon releasing the elastic, oscillations are produced which eventually die out to give the linear stretch between the two fixed points. The voltage distributions within the winding under stepped wave application give similar effects.

Usually, control of the amount of insulation on the conductors is determined by the capacitive distribution that results from lightning impulse. Thus, over-insulation of the line end turns may be a cheap and viable solution. However, the feasibility of this method decreases with increasing voltage level, and other solutions then become necessary. The first is the utilisation of capacitive shields within the winding. These may be either of the form of covered rectangular conductors inserted between turns of the winding or alternatively as toroids having a metallic coating which are inserted between the discs. By interconnection of the shields with a part of the coil closer to the line end, the capacitance network may be modified and by intelligent manipulation a more linear initial distribution may be obtained. The distribution is characterised by the value of the square root of the ratio of the shunt to series capacitances in the network. The closer this value is to zero, the more linear is the voltage distribution.

In the detailed winding design, variation of the series capacitance is often easier to accomplish than variation of the shunt capacitance. In the intershielded construction, the shields effectively increase the series capacitance, especially at the line end. There are other methods of increasing the series capacitance, the most notable of which is the interleaved disc construction to which reference has already been made. The interleaving is accomplished by reconnecting the conductors so that the capacitive charging current circulates through each disc twice. The inter-conductor capacitances, which thereby act in parallel, give a very high series capacitance and so reduce the degree of non-uniformity of the initial voltage distribution.

The enhancement may be taken further if there are a number of conductors electrically in parallel. It is then possible to achieve various combinations of capacitance throughout the coil. This is known as grading and by careful grading from fully interleaved through partial interleaving to no interleaving at all, that is continuous disc, the initial distribution can be even further improved.

The windings are required to be separated electrically and often mechanically. This separation is essentially by solid insulation materials but, to introduce cooling oil to the windings, it is common for the insulation to be provided as a combination of solid and oil. In this respect, advantage is taken of the phenomenon whereby a small oil volume is capable of withstanding a greater stress than a large volume. In most major insulation systems, therefore, the solid insulations serve as barriers to break up the oil spaces into smaller volumes. Nevertheless, it is normally the oil which provides the weakest link. Under power frequency and also lightning impulse conditions, the voltage distribution within the major insulations is controlled by capacitive effects and hence by the relative permittivities of the insulating materials. For transformers which are connected to DC equipment and are therefore subject to DC voltages, it is the relative resistivities of the materials which provide the voltage control. Whereas the permittivity ratio of paper to oil is of the order of 2 to 1, the ratio of resistivities can achieve 100 to 1. The effect of this is that the DC stress is concentrated within the solid insulation with generally very low stresses appearing in the oil. However, where the solid insulation terminates abruptly, very high stresses may appear at the boundary between the solid and the oil [7]. This could lead to discharge inception and breakdown in the localised region. The DC field is thus very different from the AC field within a transformer.

13.5.2 Electromagnetic design

A transformer is an electromagnetic device and in operation, electromagnetic fields are induced. The effects of these fields are many. Eddy losses are induced in the conductors of the windings when carrying current. Electromagnetic fields exist outside the windings as well as within them and flux is attracted to, and perhaps concentrated in, magnetic materials, whether these are the core structural steelwork, the tank or devices which are deliberately installed to deliberately influence the field.

The leakage flux produced by the equal and opposite primary and secondary ampere-turns spreads out at the ends of the windings and circulates, some through the core and some between the outer winding and the tank and some through the tank and other metalwork (Figure 13.11).

The flux which impinges on, or travels within, metallic components induces eddy currents and gives rise to loss and temperature increase. To control either temperature rise or the level of loss or both, detailed knowledge is required of the electromagnetic field. Analogue methods of field solution are possible but are extremely limited in application and inaccurate in operation. Rapid developments in computational analysis have resulted in considerable flexibility, high accuracy and considerable improvements in response time. Mathematical solutions may be derived either by finite difference or by finite element methods [8]. A number of

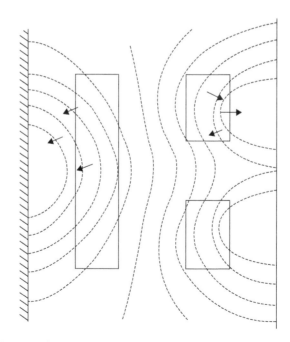

Figure 13.11 Electromagnetic flux distribution and directions of forces

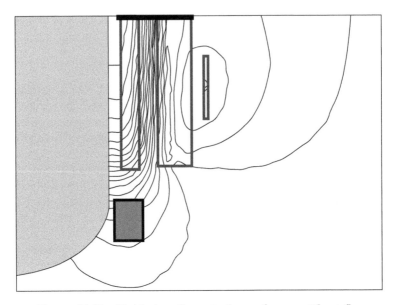

Figure 13.12 Field plot of a typical transformer without flux

finite element analysis software packages are currently available for the solution of two- or three-dimensional fields (Figure 13.12).

Information on the exact nature of the electromagnetic field forms only part of the requirement; the capability of controlling such a field is also extremely important. Control of electromagnetic fields takes two forms. The first is flux attraction and the second is flux rejection. Each plays its part in the manufacture of large power transformers.

It has already been mentioned that the leakage field associated with the load currents spreads out at the ends of the winding. The introduction of magnetic shields above and below the windings will effectively straighten the field at the ends (Figure 13.13) and often gives some reduction in the eddy losses within the windings themselves. In addition, if these magnetic flux shunts are located between the windings and the metallic structural framework, they effectively shield the framework from impinging flux resulting in a reduction of stray loss in the structural materials (Figure 13.13). As the shunts provide an attractive sink for the flux, less flux impinges on the tank, with resulting reduction in tank wall loss and tank heating. It is also possible to provide magnetic shunts on the tank wall. This has the purpose of directing the flux through laminated material, giving lower losses than would result from direct impingement of this flux on the tank wall (Figure 13.14).

The second method of electromagnetic flux control is by rejection of the flux. The provision of conducting shields, usually of aluminium or perhaps copper, has the effect of preventing flux penetration, provided a sufficient thickness of material is available. Before the advent of sophisticated finite element modelling software, the use of rejectors presented the difficulty of determining where the rejected flux was actually transferred and consequential effects.

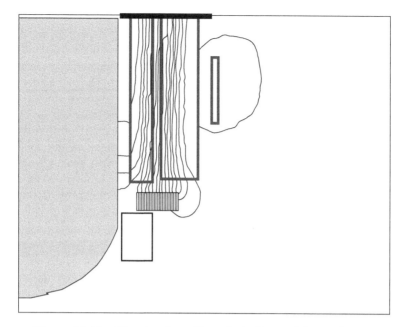

Figure 13.13 Plot showing effect of winding end flux collectors

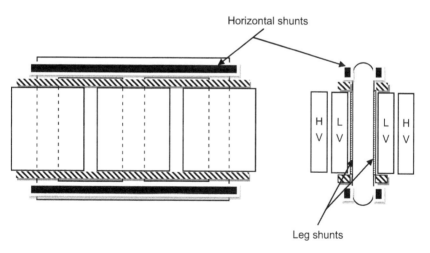

Figure 13.14 Leakage flux shunts

13.5.3 Short-circuit forces

When a conductor that is carrying current lies within a changing magnetic field, it experiences a mechanical force upon it. This force is proportional to the square of the current that flows through the conductor. The transformer is required to be capable of withstanding through fault currents and hence the forces resulting from the peak value of those currents must be resisted and the design prepared accordingly. The action of the forces is complex but, for simplicity, they are often resolved into axial and radial components. The axial component of the force results from the radial component of the flux and vice versa. It may be shown that for a simple two-winding transformer the axial force on a conductor is greatest at the ends of the coil and acts towards the centre of that coil. The summation of the forces on each turn of conductor gives rise to a compressive stress which is usually maximised at the coil centre. The radial component of the force acts on the inner winding radially inwards and is greatest at the centre of the coil. On the outer winding, it acts outwards, thereby imposing a hoop stress on the conductors. Furthermore, any imbalance of the ampere-turns of the two windings, either by design intent or by manufacturing tolerances, will give rise to forces on the ends of the windings. Mechanical engineering takes over from electrical engineering in the design of the transformer in these respects.

13.5.4 Winding thermal design

The heat generated by the currents passing through the winding conductors and the fields they set up must be extracted and transferred to some other part of the cooling system for eventual dissipation. To achieve this, the winding is equipped with a labyrinth of channels which permit oil to come in contact with the covered conductors, extract the heat and, by induced convection or by forced flow, carry it from the winding assembly (Figure 13.15). The heat flow is affected by the proximity or

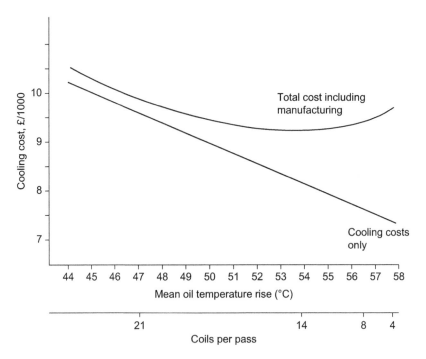

Figure 13.15   Cost of cooling vs number of coils between oil restriction washers

Table 13.4 Temperature effects of varying number of coils per pass (typical)

Coils per pass	9	18
Winding hot spot rise above ambient (°C)	55.4	59.6
Winding average rise above ambient (°C)	45.7	47.0
Hot spot to average (K)	9.7	12.6
Oil rise in winding (K)	4.4	4.4
Winding over local oil (K)	5.3	8.2

Table 13.5 Temperature effects of varying oil flow rates (typical)

Oil flow	+20%	Nominal	−20%	−50%
Average winding rise above ambient (K)	45.4	45.7	46.1	47.1
Hot spot rise above ambient (K)	54.1	55.4	56.9	51.5

otherwise of conductors to a cooling oil duct, the thermal conductivity through the paper and the oil velocity over the exposed surfaces of the winding. For disc-type winding constructions, the bulk of the exposed surface of a coil usually lies in the horizontal plane. To achieve greater efficiency of heat removal, it is necessary, especially under forced oil conditions, to persuade the oil to flow past these horizontal surfaces. This is achieved by inserting oil restriction washers alternately at the inside and outside diameters of the coil. The number of horizontal ducts which are located between these restriction washers has an effect on the winding to local oil gradient (Table 13.4).

For forced oil flow conditions, the rate of flow also has an effect on the gradient. Increasing the flow beyond a certain level (which depends on the particular design) has little effect on the gradient (Table 13.5).

High oil-flow velocities have been found to give rise to charge built up by a phenomenon known as streaming electrification or electrostatic charging tendency [9].

13.6 Transformer applications

Having considered transformer theory and discussed some of the key components, the next stage is to review some of the applications of power transformer technology. Each of the many applications requires a level of specialist knowledge and experience in order to ensure that the particular transformer type is fit for the intended purpose. A number of transformer types together with their particular characteristic requirements and crucial features are deliberated in the following paragraphs [10, 11].

The starting point with respect to electrical power is generation. Nowadays the variety of commercially available energy sources is substantial, ranging from fossil fuels, through nuclear fission, water power, wind wave and solar energy to fuel cells. Getting the power onto the transmission system necessitates some form of generator. The optimum range of sizes of typical generators is highly dependent upon the energy source and may be measured in a few kilowatts in the case of solar power to several hundred megawatts for a fossil fuel fed power station.

13.6.1 Power station transformers

Base load power stations are equipped with very large generators, which in turn demand very large generator transformers to connect them to the transmission systems. The generator transformer is actually designated 'unit transformer' in IEC Standards, and is a step-up transformer that is directly connected to the generator output terminals on the low voltage (LV) side and to the high voltage (HV) network's transmission voltage on the other side. Most generator transformer LV windings are designed to operate at a voltage of between 11 and 36 kV, but the output transmission voltage can currently reach at least 800 kV, and 1200 kV systems are currently being operated for evaluation purposes. The inherently severe loading conditions call for high security factors with respect to their thermal performances. Unlike transmission interconnection transformers, base load generator transformers are fully loaded for about 50 weeks of the year and so necessitate low total loss levels for optimum efficiency. The high LV currents require finely tuned control of the magnetic fields both inside and outside the tank in order to avoid localised overheating of this and associated steel structural parts. The very highest levels of reliability and availability are required for these transformers since they are vital to virtually continuous supply of power to the network.

Also located in the power station, a unit (auxiliary) transformer supplies some of the generated power to the station auxiliaries. These transformers must withstand the same degree of continually elevated loading that is characteristic of the associated generator transformers.

Starting or station service transformers are called upon to supply the externally derived power to the station auxiliaries when such supply is not available from the power station's own generators. These transformers are usually unloaded, other than when the unit auxiliary transformer is not available since logically the external power is more expensive than that produced by the station itself.

Running up and running down a very large base load generator is too slow to meet rapid load fluctuation demands that are encountered on a daily basis. Fast-acting gas turbine generating stations provide much of the fine tuning to permit the system to meet these power demand swings. The step-up transformers for this application must be designed not only to permit rapid load changes, but additionally to be capable of exploiting the insulation life usage to the utmost degree by balancing periods of low load with substantial overload durations.

Hydroelectric stations, such as that at Dinorwig in North Wales, are equipped with rotating plant that is capable of motoring as well as generating. In periods of high demand, the plant can be set to generate and assist the base load stations in producing and supplying power. At times of low demand, the plant may be set to motor, absorbing energy by lifting water back into the upper reservoir and thus eliminating the need to switch out base load power plant. This type of equipment can change over from generation to motoring and vice versa within half a minute. The transformer must permit this high-speed changeover of power flow.

13.6.2 Transmission system transformers

Turning attention to the transmission system, the most common system network transformer is the interconnection transformer, which provides a step-up or step-down facility and interconnects two systems that operate at different voltages, and so permit the exchange of power between the two systems. Interconnection transformers can be manufactured with two separate main windings, which offer galvanic isolation between the two interconnected networks. This is a common solution chosen when the voltage transformation ratio is above 1.5.

For a voltage transformation ratio that is closer to 1.0 (generally between 1.0 and 2.5) the auto-transformer solution is often chosen, as the closer the ratio is to 1.0, the greater the economic advantages of this solution.

The main drawback of the auto-transformer solution is that it offers no galvanic isolation between the two interconnected networks and consequently a disruption in one voltage system will directly affect the other. The auto-transformer usually provides an on-load voltage regulation range to keep the voltage level output constant and to influence the reactive power exchange between the interconnected networks. It is typically specified to have star connection for both main windings, often with the neutral directly earthed, and sometimes with a tertiary winding connected in delta. By implication, in an interconnection transformer, power can as a rule flow in either direction through the transformer.

13.6.3 HVDC convertor transformers

High-voltage DC convertor transformers are used whenever it is economically and technically advantageous to introduce DC links into the electrical supply system. In some situations, this may be the only feasible method of power transmission. When two AC systems cannot be synchronised, as the distance by land or cable is too long for stable or economic AC transmission, DC transmission is likely to be a useful alternative. At one rectifier station AC is converted to DC, which is then transmitted to a second rectifier station where it is inverted to AC to be fed into a dissimilar electrical network. In many instances, power may be required to flow in either direction, in which case, the convertor and inverter modes of the rectifiers reversed.

By general consensus, in HVDC systems regardless of whether converting or inverting, the stations are referred to as 'convertor stations' (see Figure 13.16).

HVDC transmission applications fall into three broad categories and any scheme usually involves a combination of two or more of these. First, there is transmission of bulk power where AC is impracticable, uneconomic (for long distance transmission, the cost of AC transmission and associated losses combined with the necessary reactive compensation plant may prove to be uneconomical

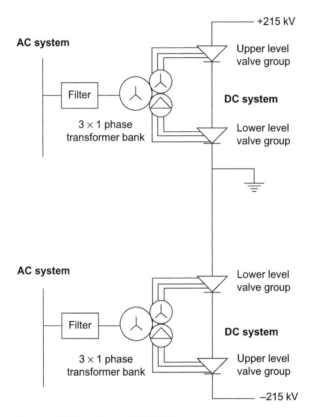

Figure 13.16 Typical HVDC convertor arrangement

compared to DC transmission) or is subject to environmental restrictions. Second DC is often more economical for submarine cable schemes.

Convertors generally utilise 12-pulse bridges that are connected to transformer windings that have a phase-angle difference between them. Usually one winding is connected in star formation, while the other is in delta (see Figure 13.17).

Interconnection between systems which operate at different frequencies or between non-synchronised or isolated AC systems will often benefit from 'back-to-back' HVDC schemes: here the convertor and the inverter have safe working clearances between them and the DC voltage levels are often lower than for a bulk transmission scheme. As a bonus, by adding faster-acting and more accurate control of HVDC power than an AC system could possibly achieve, a connected AC system's performance can be enhanced.

DC transformer components, usually superimposed on the AC element, appear both during service and during factory testing, and generate high levels of stress on the valve windings that are connected to the rectifier bridge. For that reason the design and manufacture of HVDC converter transformers require a fine mastery of insulation structures. Furthermore, the high harmonic content of the load current

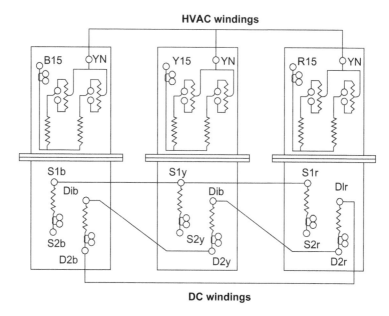

Figure 13.17 Three-phase bank connections

stipulates a very specific and detailed knowledge of the thermal design of each and every component.

13.6.4 Phase-shifting transformers

Improved transmission network reliability and the options available under deregulation have called for strengthening of the large power grids. Reinforcement of these interconnections often means that steady-state power flow has to be controlled within certain parts of these systems to ensure that the power flow through existing equipment on the systems remains within its capability. The main factors requiring careful control are due to different impedances of parallel paths, power factor mismatch and variations between generation and consumption.

When transmission lines are paralleled, the power flow in each line is distributed in accordance with the respective impedances of each line, as shown in Figure 13.18. This means that the highest power flows along the path of least impedance. However, this may be contrary to efficient operation of every part of the grid.

Phase-shifting transformers (PSTs) are one possible solution for provision of power flow control. They permit the insertion of variable or constant voltage with an adjustable phase angle into the line in which they are series connected. By varying the phase angle the real power flow through the line can be controlled (refer to Figure 13.19 for an example of a typical circuit). Variation of the in-phase voltage magnitude can only control the reactive power flow.

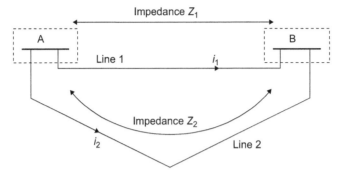

Flow of load current from A to B determined by the impedance

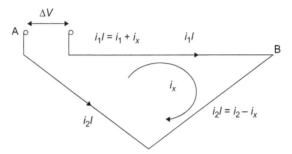

Additional imposed voltage, ΔV, at substation A gives rise to circulating current i_x

Figure 13.18 *A typical application of phase-shifting transformer*

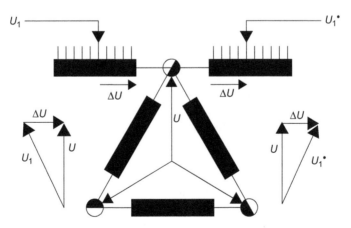

Figure 13.19 *An example of a phase-shifting transformer circuit*

They are used in two main situations: namely, interconnection between two independent networks to allow the coupling and control of power transfer in a condition that is acceptable to both networks, and in already interconnected networks to increase the efficiency of existing systems by optimising the power flow according to capacity of each transmission line and so relieving flow restrictions.

Phase shifters are categorised into two principal families; either having one active part (an active part is a single core and windings entity and may be single-phase or three-phase) or with two active parts. A single active part permits independent phase-angle and voltage regulation and is suitable for a limited voltage range, relatively narrow phase-angle range and restricted power level. For higher power phase-angle variation and voltage, a PST with two active parts becomes economically attractive. Quadrature boosters are special cases of PSTs which offer economical solutions when limited phase-shifting angles are required when a degree of voltage variation is permitted. For larger phase-angle regulation, constant modulus phase-shifting transformers offer voltage modulation independent of phase-angle values.

The design criteria of phase shifters are tailored to the individual network specifications and requirements. These transformers generally have large ratings up to 3,000 MVA. They are therefore heavy units, and hence manufacturing limits as well as transport constraints must be taken into consideration in their design.

13.6.5 Industrial transformers

Static VAr compensators (SVCs) are flexible solutions for transmission networks and grid connections for industrial power supply applications, should improvements of power system efficiency and reactive power balance and control are of paramount importance. SVCs are used to lever existing transmission networks by injection of reactive power in order to maintain voltage levels in systems even when they are in a choked state. They are also employed to manage and control network disturbances at the grid connection immediately before industrial applications, where they provide dynamic control of voltage through absorption/injection of reactive power to maintain the voltage within the contractual range.

SVC transformers are used to connect SVC equipment to transmission lines. These transformers combine the complexity that is intrinsic when DC components are involved, giving high levels of harmonic currents.

All industrial activities require electrical power. Large industrial sites often entail dedicated substations with step-down power transformers connected directly to the HV network. These transformers are conventional transformers albeit with some special considerations pertaining to the industrial application – such as frequent short-circuits, high harmonics, etc.

Further downstream, there is a large range of special applications that require specific power supplies for variable speed drive (VSD) systems or for frequency conversion such as in large ships, rolling mills for metal industries, mining, pumping substations and rolling stock. Large drive rectifier transformers are combined with frequency converters to supply these applications.

According to the particular application, the load and duty cycles will vary and transformer ratings and capacities must be carefully matched.

Large drive rectifier transformers are connected on their secondary sides to thyristor 12-pulse rectifiers that use a Graëtz bridge scheme. The thyristor control covers the system voltage regulation needs and hence eliminates the need for separate regulation of the transformers. Large drive convertor transformers are most frequently of the double secondary winding type. They commonly have a tertiary winding that is connected to a filter bank for harmonic filtering and for power factor compensation. These transformers must possess tightly toleranced inter-winding impedance relationships for correct converter group operation thus optimising the costs of the filter bank.

An electrolysis process is used to produce metals such as aluminium, magnesium, copper and zinc, or chemicals such as chlorine. Electrolysis is a continuous, stable but very energy-intensive process that consumes enormous quantities of high-quality DC current. Electric power is a very large part of the overall production cost, which means that an efficient and safe supply is essential to the process. For example, in aluminium production, the energy consumption is 14,000 kW/h per ton of aluminium produced from alumina. This is three times as much as for smelting 1 ton of steel in an arc furnace application.

To supply electrolysis applications with the very high magnitudes of DC currents several groups of rectifier power transformers combined with rectifiers must be connected in parallel operation. These rectifier transformers are designed and manufactured to support the high currents together with the high harmonic contents created by the thyristor or diode rectifiers that they are connected to. The technical challenge is in mastering high magnetic fields for internal and surrounding components, the high losses and the thermal performance of current exit connections and the tank walls.

Arc furnaces are typically used for production of steel and ferrous alloys. They utilise the heat generated by an electrical arc to melt scrap metal in a furnace and then refine the molten metal in a ladle furnace. These are electro-intensive applications, requiring very high currents and operating with cyclical loads that are determined from the mass of metal and the metallurgic requirements. Arc furnaces are supplied by AC arc furnace transformers, which are connected on the HV side to the grid, usually at a voltage level of 36 kV or 72.5 kV, and are connected on the LV side directly to the arc furnace in the case of AC arc furnaces, or via thyristor rectifiers for DC arc furnaces. These furnace transformers are especially designed and manufactured to carry very high currents on their LV sides, frequent overcurrents and overvoltages generated by short-circuits in the furnace; so they must withstand the associated high thermal, mechanical and dielectric stresses. Case-by-case studies are required to determine the continuous rating for which they must be designed, and for load cycle conditions which include current excursions of a few times the nominal furnace rating.

13.6.6 Railway transformers

Railway trackside transformers supply power to electric railway locomotives and vehicles that use either DC or single-phase AC networks. The trackside transformer is normally connected on primary side to two phases of an HV transmission network, and on the secondary side it is directly connected to the catenary and rails. To improve transmission efficiency and system regulation while reducing earth current and electromagnetic interferences, catenary feeding systems also use booster transformers or auto-transformers at regular intervals along the track. These schemes are extensively used for high-speed trains.

When combined with auto-transformers, the trackside transformer primary voltage is supplied directly from two phases of a high-voltage transmission network, while the secondary one consists of a 50 kV centre-tapped winding giving effectively 25-0-25 kV at its terminals. One of the output terminals is connected to the catenary and the other to a feeder circuit, while the mid-point is connected to the rail (at earth potential). The advantage of this is that the distance interval between feeder stations can be increased.

These transformers must be designed to withstand the mechanical stresses due to the inherent fluctuating load current characteristics and frequent short-circuit conditions created by passing trains.

13.7 A few predictions of the future

What does the near future hold in store for transformer development? Development trends are triggered by major events or circumstances.

The world guzzles ever growing quantities of oil, and this is an irreplaceable resource. Most power transformers use mineral oil as their cooling and bulk insulation medium. There are limits on the number of times that transformer oil may be refined without unacceptable loss of performance. The major players in the transformer industry are introducing ranges of transformers using vegetable oils or synthetic liquids in the interests of environmental friendliness. Perhaps there will be another major breakthrough in this area.

In 1986, sun spot activity brought about the near total collapse of Quebec's electric power grid. Solar eruptions of enormous magnitude are frequently observed giving rise to fears of further events similar to that experienced in Quebec. The sun spot activity induces very low, quasi-DC currents in the surface of the Earth. These currents flow through the neutral earth connections of the star connected transformers and send the cores of those transformers into saturation resulting in high temperatures and possible catastrophic failures. As systems grow and equipment is pushed closer to its limit, the effects of sun spot activity will probably become increasingly problematic and important. System and transformer design must consider and take steps to mitigate the consequences.

Earthquakes and ensuing tsunamis during 2011 in New Zealand and Japan have had very disastrous impacts upon life and commerce, both directly and indirectly. Multiple explosions in nuclear power plants have highlighted the vulnerability, not only the nuclear reactor itself, but of the complete of electrical power system to this type of calamitous natural occurrence. With up to 10,000 of people losing their lives and hundreds of thousands their livelihoods, the odds of occurrence may not have changed but the degree of severity most certainly has been revised upwards. It would appear appropriate to expend time and money in ensuring that transformers (and the systems in which they are connected) remain safe and secure for the benefit of future generations.

Demands are not always sparked off by disastrous events. So, for example, as faster-acting power systems are developed, the need will eventually arise in transformers for speedier voltage variation. In this event, mechanical tap-changing will be unable to provide the essential swiftness of operation, and electronic switching will have to be considered.

As we continue well into a second century of transformer technology, old problems have been solved but new ones and re-activated old ones continue to arise that demand ingenuity rational thinking from the transformer designer. Let us hope that we equip them for the weighty task in hand and train them well.

References

1. Allan D.J. 'Power transformers – the second century'. *Power Engineering Journal*. 1991;**5**(1):5–14
2. Daniels M.R. 'Modern transformer core materials'. *GEC Review*. 1990;**5**
3. Bennett P.C. 'Aspects of the design, construction and testing of BRB's class 90 and 91 locomotive transformers', International conference on main line railway electrification. *IEE Conference Publication 312*, September 1989
4. Sadullah S., Willrich M. 'Performance characteristics and advantages of insulation systems in GEC cast resin transformers'. *Proceedings of 6th BEAMA international electrical insulation conference*, Brighton, UK, May 1990
5. White A., Howitt E.L. 'Low flammability insulation systems for fluid filled transformers'. *Proceedings of 6th BEAMA international electrical insulation conference*; Brighton, UK, May 1990
6. Breuer W., Stenzel K. 'On-load tap-changers for power and industrial transformers – a vital apparatus on today's transformers'. *TRAFOTECH 82: International conference on transformers*; New Delhi, India, February 1982
7. Darwin A.W., Harrison T.H., White A. 'HVDC converter transformers', *GEC ALSTHOM Technical Review*. 1996;**19**
8. Darley V. 'The practical application of FEM techniques in transformer design and development'. *ISEF 1991 proceedings, international symposium on electromagnetic fields*; Southampton, September 1991
9. Davies K. 'Static charging phenomena in transformers'. *Proceedings of 6th BEAMA international electrical insulation conference*; Brighton, UK, May 1990
10. Preston T.W., Reece A.B.J. *Finite element methods in electric power engineering*. UK: Oxford University Press; 2000
11. AREVA Authors, *Power Transformers*, Vol. 2 – Expertise. AREVA T&D; Paris, 2008

Chapter 14
Transformer user requirements, specifications and testing

S. Ryder and J.A. Lapworth

14.1 Introduction

A transformer is usually considered to be a non-mechanical apparatus for changing the voltage of an alternating current (AC) supply. Transformers are built in a very wide range of sizes from small devices used inside consumer products to very large devices used to connect generators to the national transmission network. This chapter will focus on medium and large transformers. In this context these are considered to be any transformers which are so large or else so complex as not to be manufactured in bulk. This generally includes all transformers rated at more than 2,500 kVA and all transformers which do not include a winding intended for connection to the low-voltage distribution network, but practice varies between countries and between users.

Medium and large transformers are by their nature large and complex. They are expensive and have become increasingly difficult to source, with both prices and lead times for new transformers rising sharply in the mid-2000s. Fortunately medium and large transformers are not usually subject to rapid obsolescence. They also contain few if any moving parts and are generally very reliable, partly as a consequence. When failures do occur they can sometimes be catastrophic, especially for large transformers. Detecting incipient failure in the field is notoriously difficult, with only a limited range of tests available online.

One of the main aims for users from the procurement process is to ensure that any medium or large transformers purchased will be reliable in service. The direct costs of unreliability may be comparable with the cost of the transformer itself, and the indirect costs may be very much higher if this involves disruption to other activities. The human consequences of a catastrophic failure may also be very severe – in what is believed to be the worst single incident of this kind in Great Britain, three people were killed by a transformer failure at Teeside power station in August 2001.

The transformer procurement process will be described in more detail during much of the rest of this chapter. Special emphasis will be placed on the specification of user requirements and on acceptance testing, to ensure that these have been met. In an important change from the first and second editions of this book, this chapter will now also include some material on supplier selection to reflect increased difficulties in transformer sourcing since the second edition was published.

Finally some material on transformer operation and maintenance will be presented. This will necessarily brief, but will include some discussion of life-limiting processes for transformers together with on- and off-line diagnostic testing. A bibliography of important CIGRE and IEC publications is included listed 1 to 23, and these may be consulted for further guidance.

14.2 Specification of user requirements

14.2.1 Need for user specifications

Medium transformers are very often purchased against national or industry specifications, which cover all of the important technical requirements. There is frequently no need for a user specification. Where there is a need for a user specification, this can be very simple, covering only the areas not covered by the national or industry specification.

Large transformers are in general not covered by national or industry specifications and it is necessary for users to produce their own. Users who regularly purchase large transformers may wish to split their specifications into two parts, with one part giving general requirements and another giving requirements for the specific project. The general part of the user's specification would thus be similar to a national or industry specification. Project specific requirements are often communicated using schedules.

14.2.2 Functional and design specifications

Specifications are usually classed as follows:

- *Functional specifications*: Functional specifications state the functions required from the transformer, but not how the specified requirements are to be achieved. This is left to the manufacturer. It is considered to be good practice for every function to be associated with a test or other method of verification, although this is not always possible.
- *Design specifications*: Design specifications state how the specified requirements are to be achieved. Most users do not favour design specifications for large transformers, as they lack the necessary design expertise. Users may wish to include some design requirements in an otherwise functional specification, either as certain requirements are difficult or impossible to put into functional terms (e.g. dimensional compatibility) or as a means of communicating poor service experience with certain design features to manufacturers. Users should exercise a degree of restraint in using specifications to communicate poor service experience and should also recognise that certain design features may be problematic when used by one manufacturer, but not by others. Design review can provide a better means for communicating this information, and allows manufacturers to make their own representations.

14.2.3 Specifications and standards

Specifications, including national and industry specifications, are based on standards. Worldwide there are two systems of standards in widespread use. IEEE Standards are used in the USA and certain other countries. Some other countries have national standards which are very similar to or based on IEEE Standards. IEC Standards are used in most other countries worldwide and form the basis of European Union Standards. Various attempts have been made in recent years to converge the two systems of standards. So far these have been only modestly successful. Users should also note that some large emerging market countries, notably China, have their own standards which differ somewhat from both IEC and IEEE Standards.

For most users worldwide, IEC Standards are the best choice. IEEE Standards are a good choice for users in the USA and certain other countries, where they are in general use. Occasionally users may wish to refer to a mix of IEC and IEEE Standards, where one or other system of standards is lacking in a particular area or for compatibility purposes. National standards are more problematic, as many manufacturers may not be familiar with foreign national standards and this may give rise to misunderstandings. For certain large emerging market countries, there is a large choice of domestic manufacturers and less need to facilitate understanding by foreign manufacturers. In these circumstances, national standards may also be a good choice. In general, the opposite is the case and users are advised against using national standards.

The aim of the specification should be to communicate all of the user's particular requirements to the manufacturer as clearly as possible. As was suggested above, users who regularly purchase large transformers may wish to split their specifications into two parts, with one part giving general requirements and another giving requirements for the specific project. Project-specific requirements are often given in tabular form in so-called schedules of requirements. Many users find it convenient to ask manufacturers to complete certain parts of the schedules of requirements to confirm that they can comply with the specific requirements for the project. Certain users also use the schedules of requirements to ask for more general performance or other information.

Users are reminded that keeping specifications short aids understanding. Users should avoid re-stating the standards on which the specification is based, and should concentrate on their own requirements. Users should try to separate their general requirements from project-specific requirements, even where these form part of the same documents. Users should also try to present their requirements in a coherent manner.

14.2.4 Specification content

In general, users are advised that the following content is required:

- applicable standards for transformer and components or materials if necessary; order of precedence for standards if necessary (usually part of the general requirements)

- environment conditions (usually part of the general requirements)
  - temperature rise limits: Alternatively, maximum and minimum ambient temperatures; annual and monthly average ambient temperatures
  - altitude above sea-level
  - seismic requirements, if any; manufacturers will, in general, assume no seismic requirement.

In Great Britain, users are fortunate that ambient temperatures are rather low compared with the limits for normal operation given in IEC Standards 60076-1 and 60076-2, viz. an annual average ambient temperature of 20 °C with a monthly average temperature for the hottest month of 30 °C. This leads to winding hot spot temperatures in service rather lower than the values for normal ageing given in IEC Standards 60076-2 and 60076-7, viz. 98 °C. Consequently ageing of solid insulation tends to be rather slower, even for heavily loaded transformers.

In countries with warmer climates, this was found not to be the case. This led to the widespread practice of rounding temperature rise limits owing to high ambient temperatures down to the nearest 5 K. This practice is recommended in the latest edition of IEC 60076-2.

- Rating data (usually part of the project-specific requirements)
  - indoor or outdoor type, etc.
  - insulating medium, for example oil, natural ester; insulation temperature class if necessary; in the case of liquid-immersed transformers, type of liquid preservation system
  - number of phases for transformer and system
  - rated voltage for each winding
  - for any tapped winding, tapping range and number of steps, on-load/de-energised taps and any other particular requirements
  - rated power for each winding associated with each cooling mode
  - type or types of cooling
  - connection symbol
  - whether any windings are to be auto-connected
  - impedance for each pair of windings

At the time the first edition of this book was published, substantially all medium and large transformers produced worldwide were oil-immersed. This situation has now changed markedly, with natural and synthetic ester-immersed transformers now being widely available up to approximately 100 MVA and 245 kV class. A further change has been the more widespread use of dry transformers, mainly but not exclusively resin encapsulated, in larger sizes. These are now widely available up to 20 MVA and 36 kV class. Pressurised gas-filled transformers have been widely used in Japan and Korea for many years, without restrictions on rated power or voltage. These are now starting to appear in other markets, mainly for use inside buildings where an explosion risk cannot be tolerated.

Concerning taps, usual practice in Great Britain and most foreign countries is for taps in the high-voltage winding. In most cases these are on-load, but

Transformer user requirements, specifications and testing 533

de-energised taps are useful for some applications where an occasional change of configuration is all that it is required. The tap range is selected bearing in mind possible variations in system voltage and load, especially load power factor. Users are reminded that it is not possible to change the tap range of a transformer in service, so it is important to be conservative in specifying requirements. It is possible to limit the impact this has on costs by specifying a maximum current on any tapping, which effectively reduces the rated power of the transformer over part of the tap range. This practice is favoured in the USA.

Specifying rated power is easiest for generator transformers, where the rated power of the transformer is required to match that of the generator and the associated prime mover. Some allowance must be made for operation at reduced power factor. In Great Britain the minimum power factor at full load is usually assumed to be 80%. Similar considerations may apply to process transformers in industrial plants, although it may be necessary to make some allowance for future expansion of the plant. Specifying rated power is more difficult for transmission and distribution transformers. These do not normally operate at high loads, except when other transformers are out of service. Rated power for such transformers is often specified so as to avoid excessive loading under these circumstances. Equally rated power may be selected so as to match that of other equipment, for example cables, switchgear.

Impedance is an important parameter in transformer specifications. Transformer impedance is beneficial to the power system as it limits fault currents and promotes correct current sharing between parallel circuits. Transformer impedance is detrimental as it results in voltage regulation. For a transformer to be added to an existing installation, the impedance is usually selected to match existing transformers. For new installations, the user may specify either a target value or an allowable range. The latter practice is preferable, and is widespread in Great Britain.

Recommended minimum values for impedance are given in IEC Standard 60076-5.

- Dielectric requirements (usually part of the general requirements)
  - method of earthing for each winding and the system to which is connected
  - voltage class for each winding
  - dielectric test requirements for each winding
  - dielectric test requirements for auxiliary wiring and equipment
  - dielectric test requirements for core/frame insulation

Dielectric test requirements are usually specified in accordance with the national insulation co-ordination standard, and users have little discretion.

- Short-circuit withstand requirements (usually part of the general requirements)
  - expected frequency and duration of short-circuits, especially if these are unusually frequent or unusually long
  - system impedance, if any
  - system X/R ratio, if not assumed to be infinite
  - whether a short-circuit withstand test is required

System impedance can be specified in a number of ways. In the author's opinion the clearest method is to specify the breaking capacity of the switchgear to which the transformer will be connected.

- Transport and installation requirements (usually part of the project-specific requirements)
    - any limits on transport dimensions or mass from the user
    - details of transport method for delivery to site
    - details of any lifting or transport lugs required
    - dimensional compatibility requirements
    - auxiliary supply details (voltage, type of system)
    - number and type of terminals required for each winding
    - list of required fittings
    - details of required surface finish
    - language(s) for labels, drawings and manual
    - limits on sound level
    - vacuum withstand level for main tank

Users are advised that large transformers are not, in general, designed to be transported while fully assembled and are also not, in general, designed to be transported regularly. Important exceptions to this are transformers which form part of mobile sub-stations and mobile transformers. These require special design considerations. It may be advisable for users to purchase such transformers from manufacturers with special knowledge, built up through previous experience.

Note that, as a safety measure during transport, all medium and large transformers should have the approximate position of the centre of gravity, as arranged for transport, permanently marked on at least two adjacent sides and preferable all four sides. To ensure the transformer is correctly positioned on installation it is also helpful for the centre of gravity, as arranged to service, to be similarly marked.

Users are advised that it is possible for transformers to be directly connected to gas-insulated switchgear, via trunking and oil/SF_6 bushings. This makes future maintenance of both the transformer and the gas-insulated switchgear rather complicated, and this practice can only be recommended where it is essential to save space. Users are advised that it is possible for transformers to be directly connected to cables. This practice is very widespread for medium transformers. It is less problematic than direct connection to gas-insulated switchgear. It is advisable to make provision for the transformer to be disconnected from the cable for maintenance.

- Design requirements, usually including some of the following (and usually part of the general requirements)
    - any design features which must not be used
    - specific limits for design parameters

- list of approved suppliers for components and materials
- loss capitalisation formula

As was noted above, users should exercise a degree of restraint in using specifications to communicate poor service experience. Design review can provide a better means for communicating this information.

14.2.5 Guidance on specifications

For further guidance users and manufacturers may wish to refer either to appendix A of IEC Standard 60076-1 or to new CIGRE brochure 528.

14.3 Supplier selection

14.3.1 Industry changes

There have a number of major changes in the medium- and large-transformer industry since the first edition of this book was published. In industrialised countries there has been a reduction in manufacturing capacity, for example three of the four manufacturers of large transformers based in Great Britain closed in 2002–3. In certain emerging markets there has been an increase in manufacturing capacity. As a result transformer users in industrialised countries are now having to import transformers from emerging markets, rather than the other way around as was previously the case.

To ensure that unfamiliar manufacturers are capable of meeting their expectations, many users now have pre-qualification procedures. For further guidance users and manufacturers may wish to refer to new CIGRE brochure 530.

14.3.2 Timing

Different users have different approaches to the procurement process, and thus have different requirements regarding manufacturer pre-qualification. Where the user favours open tenders for each new project, a large number of manufacturers may respond each time and the user may wish to pre-qualify any unfamiliar manufacturers before evaluating their tenders. However, where users purchase large numbers of transformers an open-tender for each project may be burdensome, and the user may instead decide to make frame-work agreements to purchase transformers from selected manufacturers. Users may wish to pre-qualify any manufacturers with whom they intend to make frame-work agreements.

14.3.3 Format

The pre-qualification process normally takes the form of a review of written submissions from the manufacturer by the user, followed by a face-to-face meeting with the manufacturer's staff and an inspection of the manufacturer's factory. The process should be proportionate both to the size and to the complexity of the transformers the user wishes to purchase and to their importance to the user.

Users should recognise that the process might be burdensome for the manufacturer and should therefore avoid putting manufacturers who have no realistic chance of being pre-qualified through the process.

Users should recognise that there is a limit to what they can find out during a pre-qualification process, even a very thorough process. Users may therefore wish to place trial orders with new factories to assess their actual performance during design, manufacture and test. The performance of manufacturers may change over time, and users may wish to wholly or partly re-assess manufacturers at regular intervals. In particular it might be hoped that manufacturers would listen to adverse comments from users and improve their practices.

Many transformer factories are wholly or partly owned by large international companies, for example ABB, Alstom Grid, Crompton Greaves and Siemens. Some of these companies have chosen to implement a common design technology at all or more of their factories (e.g. Trafo-Star from ABB and UniPower from Crompton Greaves). This should ensure a common minimum standard for all of their designs. There is sometimes a tendency for certain factories to implement the technology differently from others. Available manufacturing and test facilities may also differ. Users should therefore not assume that once they have pre-qualified one factory from a large international company, all of the others are automatically pre-qualified.

14.3.4 Aims

The aim of the pre-qualification process should be to decide whether a particular factory would be a suitable partner for the user for a given range of transformers. Users should recognise that no particular transformer factory is perfect and that different factories have different strengths and weaknesses. Certain factories may have weaknesses which can be addressed by taking action short of refusing to buy their transformers, for example enhanced design review or enhanced manufacturing inspections. Certain factories may have weaknesses which are acceptable to the user in certain circumstances, but not others. Users may therefore decide that they wish to purchase transformers up to a certain limit of rated power or rated voltage from a particular factory. Conversely they may decide that they do not wish to purchase certain special product lines from a particular factory.

14.3.5 Main elements of process

A thorough pre-evaluation process is likely to include the following elements:

- quality management
  - ISO 9000 certification
  - quality policies
  - effectiveness and independence of quality management
  - adequacy of documented procedures
  - internal/external audits

Users are reminded that while possession of a valid ISO 9000 certificate is a necessary condition for purchasing transformers from a particular manufacturer, it is not a sufficient condition. Users might reasonably expect the quality manager to be a professional engineer with suitable experience in the industry who sits on the main management board.

- basic technology
  - current and recent technology partners
  - technical reference list
  - main design and construction features
  - experience with special products
  - adequate number of short-circuit type tests or equivalent commitment to R&D
- design
  - effectiveness and independence of engineering management
  - adequate design information (manuals, standard drawings, previous designs, etc.)
  - adequate design software
  - adequate design review procedure, internal and external

Users might reasonably expect the engineering manager to be a professional engineer with suitable experience in the industry who sits on the main management board. For large transformers, users might expect the manufacturer to have the capability to model both the electric and magnetic field within the transformer and to make transient voltage calculations.

- purchasing
  - adequate material and component specifications
  - adequate programme of supplier audits
  - adequate checks on materials and components received
  - main suppliers
- manufacturing quality
  - use of GANTT charts or similar to plan production
  - use of manufacturing quality plans at point-of-work
  - adequate labelling of materials and components
- manufacturing facilities
  - general condition of factory and grounds
  - general condition of main workshops
  - access control for critical workshops
  - adequate measures to ensure cleanliness and prevent contamination
  - drying method
  - lifting capacity

Users might reasonably expect significant access controls to be in place for the most critical workshops (insulation components, winding, assembly), especially for large transformers. Users might reasonably expect final dry-out by vapour phase, especially for large transformers. Heat and vacuum might be acceptable for medium transformers.

- test
  - separate test laboratory
  - capabilities for main works tests
  - measuring equipment calibrated
  - measuring equipment correctly stored

  For large transformers it is preferable for tests to be carried out in a separate laboratory, as this avoids disruption to both production and testing.

- environment
  - ISO 14000 certification
  - environmental policies
  - effectiveness and independence of environmental management
  - reduce, reuse, recycle programme
  - dangerous waste disposed of responsibly
- human resource management
  - human resource manager
  - industrial relations
  - equal opportunities/non-discrimination policy
  - adequate training programmes for specialist employees
  - protection for vulnerable workers
- health, safety and working environment
  - reasonable limits on working hours
  - OHSAS 18000 certification
  - health and safety policies
- transport
- installation and commissioning
- warranty

14.4 Testing

14.4.1 Classification of tests

Final acceptance testing should provide proof for both the manufacturer and the user that the transformer as built meets their requirements. As was discussed above it is therefore important that every required function of the transformer is associated with a test or tests. Each test or tests should be associated with some pass/fail criteria. It is preferable if these have been agreed in advance by the manufacturer and the user, at the design review meeting or otherwise. For certain tests this might be considered unnecessary as the pass/fail criteria are included in standards.

Tests can be classified in a number of different ways. For the purposes of this chapter, the following classification will be used:

- performance tests
- thermal tests
- dielectric tests.

Short-circuit withstand is not usually verified by direct testing, but might be regarded as an additional category. Certain further tests are frequently carried out on transformers in the field to check for problems (e.g. winding frequency response), and these are sometimes also performed during works test. These might be regarded as a fifth category.

14.4.2 Performance tests

Performance tests are usually considered to include the following:

- winding resistance
- transformation ratio
- vector group/polarity
- load loss and impedance
- no-load loss and current
- over-excitation (might also be regarded as a dielectric test)
- zero sequence impedance
- (no-load) sound level
- auxiliary power consumption.

Leak/pressure testing of the tank is sometimes regarded as a performance test, but is perhaps best regarded as a quality control check.

The order of tests given above is taken from IEC Standard 60076-1. The manufacturer, in consultation with the user or otherwise, might decide on performing the tests in a different order.

There is sometimes a tendency to regard performance tests of being of lesser importance compared with other tests, especially dielectric tests. This is most unfortunate as poor results in performance tests can point towards fundamental problems with the transformer that will cause problems in service. For the manufacturer the results can provide an important check on their design methods.

The winding resistance, transformation ratio and vector group/polarity tests, taken together, are an important check on the quality of the transformer, and in particular of the winding connections. These three tests should find any problems with leads being exchanged or connected incorrectly and any poor-quality joints. For this reason it is best if the manufacturer can provide the user with expected values for the results of the winding resistance measurements. As the winding resistance measurements are used to correct the load-loss measurements for temperature, it is important that the transformer is in thermal equilibrium when the measurements are made and that the temperature is recorded accurately.

Both load losses and impedances are usually the subject of contractual agreements and failure to meet the guaranteed or expected values can lead to serious consequences, up to or including rejection of the transformer. Certain serious problems with the transformer can also be detected by this test including circulating currents in the windings or connections or excessive losses induced by the leakage flux in structural parts of the transformer. As the guaranteed and expected values of load loss and impedance are not always the same, it is helpful if the manufacturer can provide the user with both. Where unexpected results are obtained, the manufacturer and the user should consider carefully whether the transformer should be rejected, accepted subject to agreement over changes to the contract (e.g. a reduction in price) or accepted anyway. In some cases further investigation might be required, for example an infra-red scan of the tank during the temperature rise test. Where unexpected results are obtained from only one transformer in a long series this requires explanation.

No-load losses are usually also the subject of contractual agreement. The test is usually performed early in the test sequence and additional measurements are often made at 90% and 110% of rated excitation. Sound-level measurements are usually, but not always, made at the same time. As the guaranteed and expected values of no-load loss and current or (no-load) sound level are not always the same, it is helpful if the manufacturer is to provide the user with both. Results are an important quality check on the material used to build the core and the core build quality. Where unexpected results are obtained, the manufacturer and the user should consider carefully whether the transformer should be rejected, accepted subject to agreement over changes to the contract (e.g. a reduction in price) or accepted anyway. Where unexpected results are obtained from only one transformer in a long series this requires explanation. Where the core material is suspected of being the cause of the problem, this should be communicated to the supplier.

Certain transformers may be subject to prolonged over-excitation in service and users of such transformers may wish to specify a test to check the ability to withstand these conditions. As such tests are not included in IEC Standard 60076-1; the user must specify both the over-excitations conditions and the pass/fail criteria. Where specified the test is usually performed directly after the no-load loss and current measurements as it requires the same configuration and uses the same test equipment. It could be regarded as either a performance test or a dielectric test.

Where information about the zero-sequence impedance of a particular transformer is required, then it is useful to measure it directly. As results do not vary very much from one transformer to another in any particular series, it is usually specified as a type test. Note that as it is not required in many cases, it is included IEC Standard 60076-1 as a special test rather than as a type test. Expected values depend largely on the positive sequence impedance and the transformer configuration, and methods for their estimation are given in IEC Standard 60076-8. If measurements are required, it is preferable if they are made following the load-loss and impedance measurements so that the manufacturer and the user are better able to evaluate the results.

14.4.3 Thermal tests

Users should note that a temperature rise test in accordance with IEC Standard 60076-2 includes checks on the temperature of the oil and on the average temperature of the windings. It does not necessarily include any checks on the winding hottest spot temperature, although the standard includes some guidance on methods for determining winding hottest spot temperature. It also does not include any checks on the temperature of the tank, although many users specify an infra-red scan of the tank in addition. It further does not include any checks on the temperature of the core, although this seems to be of little interest to most users and is very rarely measured.

Users should also note that certain thermal problems which might affect transformers in service can be detected using other tests altogether. Circulating currents in the core/frame/tank usually require the failure of the core/frame/tank insulation and so can be detected by checking the core and frame insulation. As is noted above, certain other problems including circulating currents in the windings or connections or excessive losses induced by the leakage flux in structural parts of the transformer can sometimes result in unexpected values of load loss or impedance.

While there are a number of possible methods for carrying out the temperature rise test, medium and large transformers are usually tested by the so-called short-circuit or total loss injection method. This involves first injecting losses into the transformer by injecting current into one winding with one or more other windings short circuited. The loss injection is continued until the transformer reaches thermal equilibrium and then the oil temperatures are determined, usually by direct measurement. Once thermal equilibrium is reached, winding temperatures are then determined by injecting rated current into each winding for 1 h, disconnecting the test source and then making winding resistance measurements while the winding cools. Winding resistance results are extrapolated back to the point where the test source was disconnected and the winding temperature is estimated from the resistance value.

The thermal time constant for the oil is typically several hours, which means that it may take more than one day for a large transformer to reach thermal equilibrium. The long duration of the test leads to high costs, partly as the cost of the electricity consumed supplying the losses is significant and partly as it occupies the test area for such a long time. For this reason the temperature rise is included as a type test only in IEC 60076-1. Not all users agree with this, and some make temperature rise a routine test. There have also been attempts to develop a shortened version of the test, to be carried out as routine on later members of a long series. This is quite a new development and is not included in IEC Standards 60076-1 and 60076-2.

Given the duration of the test it is highly likely that some diurnal variation in ambient temperature will take place while thermal equilibrium is being reached. The manufacturer should take care to eliminate any variation in temperature owing to the losses in the transformer and the test equipment and to minimise any variation owing to other avoidable causes. Particular attention is required as the transformer reaches thermal equilibrium, as rises in ambient temperature at this time may lead to oil temperatures being under-estimated.

The thermal time constants for the windings are typically several minutes, so it is important to begin winding-resistance measurements as soon as possible after the test source is disconnected. The electrical time constant for the windings may be sufficiently long as to cause a delay of some minutes in making the first credible winding-resistance measurement. This is particularly likely in the case of windings having a high magnetising inductance and a low resistance, for example low-voltage windings on generator, arc furnace or rectifier transformers. In these circumstances the user and the manufacturer may wish to agree on alternative or additional methods to determine winding temperature rise, for example the use of fibre optic temperature sensors.

The manufacturer must supply a considerable amount of reactive power during the temperature rise test, typically using a capacitor bank. These impose limits on the rating and impedance of transformers which can be tested by the manufacturer. In some cases it may be possible to 'bend' these limits by choosing to reduce the losses injected and then adjusting the results afterwards. Formulae for making the adjustments are given in IEC Standard 60076-2. For transformers with natural oil circulation, these formulae are somewhat different from the formulae given in IEC 60076-7 for estimating the temperature of transformers in service.

As an additional criterion, many users ask for oil samples to be taken before and after the temperature rise test to check for overheating during the test. The samples are usually analysed for dissolved gases and occasionally also for furans. Without wishing to discourage this practice, users should note that the temperature rise test rarely lasts for more than one day and even severe overheating faults may not, during this time, produce sufficient dissolved gas or furans to be detectable. Certain users have asked for the test to be extended to allow more time for dissolved gas to accumulate in the event of problems.

As was mentioned above, in addition to the requirements of IEC 60076-2 many users also require checks on either winding hot spot temperatures or tank temperatures or both. Winding hot spot temperatures can be monitored directly using fibre optic temperature sensors. These have now been in use for more than 20 years and most manufacturers are familiar with them. The user and the manufacturer must agree both the number and the location of the sensors, usually during the design review. As the sensors are not particularly reliable it is good practice for them to be either duplicated or triplicated. They can be monitored continuously during the test and a hot spot temperature gradient can be estimated at its conclusion. Tank temperatures are most conveniently checked using an infra-red camera, although alternative methods are also possible. Limits on tank temperature are not included in IEC Standards 60076-1 and 60076-2. As the purpose of the limits is usually to prevent injury to persons coming into contact with the tank, users are advised to base them on relevant national or international safety standards, for example ISO Standard 14122.

14.4.4 Dielectric tests

Dielectric tests are usually considered to include the following:

- insulation resistance measurement (might also be regarded as a condition assessment test)
- winding capacitance and power factor measurement (might also be regarded as a condition assessment test)

Transformer user requirements, specifications and testing 543

- applied AC voltage
- induced AC voltage
- lightning impulse
- switching impulse
- applied DC voltage (for HVDC transformers)
- DC polarity reversal (for HVDC transformers).

Over-excitation is sometimes regarded as a performance test, but is perhaps best regarded as a performance test. For further information, please see above.

The order of tests given above is taken from IEC Standards 60076-1, 60076-3 and 61378-2. It is usually considered to be good practice to conduct the AC-induced test after the lightning and switching impulse tests, so that any dielectric damage caused during these tests which might give rise to partial discharge in service can be detected. In the case of HVDC transformers, it is normal for the DC tests to be conducted between the lightning and switching impulse tests and the AC-induced test for similar reasons.

Where a temperature rise test is to be carried out, it is usually considered to be good practice for this to be made before the dielectric tests. This is so that any residual moisture in the solid insulation has migrated into the oil and any contamination from the coolers has been flushed into the transformer.

The usual order for tests on a non-HVDC transformer is shown in Figure 14.1.

Insulation resistance measurements are usually made on each separate winding, between each pair of separate windings and on the core and frame insulation in all possible combinations. During the test non-tested windings or components are usually earthed via the tank. The manufacturer and the user should agree on the voltage to be used and the time for which it is to be applied. Some users and some manufacturers may wish to measure the polarisation index of the insulation in addition, and the method for doing so must also be agreed. IEC 60076-1 and IEC 60076-3 do not give any pass/fail criteria. In general one might expect the insulation resistance to be more than 1 Gohm once capacitive effects had become insignificant. If a significantly lower value were measured then this would require explanation. Likewise, if a significantly lower value were measured on one transformer in a long series this would also require explanation.

As a check on the overall dielectric condition of a transformer, winding capacitance and power factor measurements are generally required as being

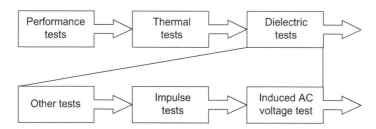

Figure 14.1 Usual order for tests

superior to insulation resistance measurements as the insulation is subject to similar stresses, albeit at lower levels, as in service. It is widely performed on large transformers and is used as a quality check by some users and manufacturers. Measurements are usually made on each separate winding and between each pair of separate windings. IEC Standards 60076-1 and 60076-3 do not give any pass/fail criteria. Capacitance results are highly characteristic of a particular winding arrangement, and should therefore agree well between different transformers in the same series. Where unexpected results are obtained this requires explanation. Power factor results are a measure of insulation quality. Unexpectedly high results require explanation, perhaps rectification in some cases.

Voltage levels for the AC applied and induced tests and lightning and switching impulse tests are specified by the user, usually with reference to national practice with regard to insulation co-ordination. The user must also decide where the lightning impulse test should include chopped waves. The user, perhaps in conjunction with the manufacturer, should decide on the method for induced AC voltage testing and whether a switching impulse test is required in addition. In the case of HVDC transformers the user must also specify the voltage levels for the applied DC voltage and polarity reversal tests, usually with reference to IEC Standard 61378-2. Note that it certain cases the polarity reversal test may not be required. In all cases pass/fail criteria are given in IEC Standards 60076-3 and 61378-2.

The aim of all lightning impulse testing is to ensure that the transformer is capable of withstanding any lightning transients which may occur during service. On overhead transmission lines lightning transients often result in one of the insulators flashing over. This effect is simulated by suddenly collapsing the lightning impulse test voltage to zero, usually by means of a controlled chopping gap. Current practice is to do so shortly after the peak of the applied voltage. The former practice of doing so while the applied voltage was still increasing has now largely ceased. Chopped wave lightning impulse is a special test according to IEC 60076-3. It is widely specified as a routine test by users, especially for windings which will be connected to overhead transmission lines. According to IEC 60076-3 the peak voltage for chopped waves should be 110% of the peak voltage for full waves. Some users have their own requirements, arising from the characteristics of their surge arrestors.

Examples of different lightning impulse waveforms are shown in Figure 14.2(a–c).

The standard lightning impulse waveform according to IEC Standards 60060-1 and 60076-3 has a virtual front time of 1.2 μs ±30% and a time to half-value of 50 μs ±20%. Note that the neither the virtual front time nor the time to half-value should change during an impulse test sequence. The standard time to chop according to IEC Standard 60076-3 is 2–6 μs. This may vary somewhat during an impulse test sequence.

Slower voltage transients can arise owing to a number of causes including switching long lines, single or two phase earth faults and resonances. There are two main tests which have been developed to ensure that transformers are capable of

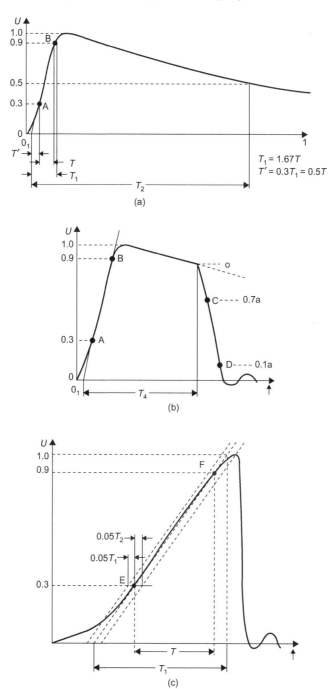

Figure 14.2 (a) Full wave lightning impulse; (b) chopped wave lightning impulse; (c) front-of-wave lightning impulse

withstanding these transients – induced AC voltage and switching impulse. At lower voltages the induced AC voltage test is usually considered to be sufficient. At higher voltages a switching impulse test is usually also required, and the induced AC voltage test is modified and extended to allow more time for measurement of partial discharge. The voltage at which this becomes advisable depends on the users' requirements and in particular the length of the transmission lines to which the transformer will be connected. According to IEC 60076-3 switching impulse testing is routine for 170 kV class and higher.

An example of a switching impulse waveform is shown in Figure 14.3.

The switching lightning impulse waveform according to IEC Standard 60076-3 has a time to peak of at least 100 μs, but usually less than 250 μs, and a time to first zero of at least 500 μs, but preferably at least 1,000 μs. The dwell time at >90% of the specified value should be least 200 μs. Note that this waveform is intentionally different from that according to IEC Standard 60060-1, which is intended for use with other types of equipment.

The core may become saturated during switching impulse testing, which will change the waveform somewhat. This effect can be minimised by demagnetising the core between applications, usually by applying one or more reduced switching impulses of the opposite polarity.

With impulse testing users should be aware that transformers represent a difficult test object and that they may have to accept less than ideal waveforms. Particular problems can arise with controlled chopping gaps if there are oscillations on the peak. Users and manufacturers should also be aware that impulse test generators and controlled chopping gaps operate using the breakdown of air gaps and can malfunction under certain circumstances, for example very high relative humidity, insufficient interval between tests. Users and manufacturers should further be aware that the waveforms used for comparison are measurement results and may differ from the actual waveforms. A poor waveform may not mean that a

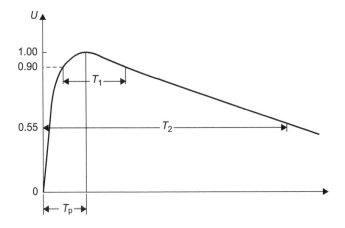

Figure 14.3 Switching impulse

transformer has failed and may point to a problem with the measurement circuit, for example a flash-over on a measurement shunt.

Where non-linear resistors have been used, these may operate during the lightning impulse test, resulting in a change to some of the waveforms. The correct operation of the non-linear resistors is checked in two ways: first by making additional voltage applications at gradually increasing voltages up to the full wave level to determine the voltage at which the non-linear resistors first operate, and second by repeating these additional voltage applications in reverse sequence after the test to ensure that the voltage at which the surge arrestors operate has not changed.

Induced ac voltage testing on medium and large transformers is usually combined with partial discharge measurements, and at higher voltages the test is usually extended to allow more time for these measurements. IEC Standard 60076-3 includes a number of different methods for making the test, depending on the voltage class and winding configuration. Three-phase transformers may require either three-phase or single-phase tests or a combination of both. Given the complications, it is preferable if the method to be used has been agreed in advance by the manufacturer and the user, at the design review meeting or otherwise.

With partial discharge testing it is important to ensure the lowest possible background level during the test. It is preferable if the manufacturer has a separate test laboratory with doors and windows that can be closed to make a Faraday cage. Where poor partial discharge test results are obtained this requires explanation and probably also investigation. Various methods can be used to locate sources of partial discharge including multi-terminal electrical measurements, acoustic measurements, UHF measurements and, in the case of external partial discharge, ultraviolet photography. Users are advised to think carefully before deciding to accept a transformer with high partial discharge as this could seriously compromise its future reliability.

As with temperature rise testing, many users ask for oil samples to be taken before and after the dielectric tests to check for problems of the tests. The samples are usually analysed for dissolved gases only.

14.4.5 Short-circuit withstand

To the best of the authors' knowledge and belief there is no manufacturer of transformers anywhere in the world who is able to short-circuit test a large power transformer in their own test laboratory. The practice of short-circuit testing transformers directly on the transmission network has now largely ceased, so when a short-circuit test is required the transformer must be transported from the manufacturers' works to a suitable short-circuit testing station and then returned to the manufacturer's works for further inspection and testing. There are very few short-circuit testing stations anywhere in the world which are capable of testing a large transformer. Transformer manufacturers in certain countries (e.g. Australia, South Africa) have poor access to short-circuit testing stations. This introduces a further complication as requirements for short-circuit testing should, in general, be drawn up in such a way as not to have a distorting effect on international trade. Users are referred to the WTO agreement on Technical Barriers to Trade. They may wish to seek legal or regulatory guidance on this point.

In most circumstances the cost of performing a short-circuit test is prohibitive, and users chose verify the short-circuit withstand capability of the transformer in other ways. Manufacturers can prove their designs through a combination of calculations and testing of either prototype transformers or models. Note that manufacturers who have poor access to short-circuit testing stations may prefer to use models for verification.

Users may wish to consider a short-circuit test where the transformer is likely to be subject to frequent short circuits in service, for example trackside transformers, or where they intend to purchase a very large number of transformers of the design. Users should also consider that while manufacturers might be expected to take great care with transformers that are short-circuit tested, a high proportion fail. The findings of a recent study by Smeets *et al.* of KEMA are summarised in Table 14.1.

Table 14.1 Results of short-circuit withstand tests at KEMA over 13 years

Transformers	Number	
Presented for testing	102	
Passed at first attempt	73	
No problems reported from visual inspection		69
Problems reported from visual inspection		4
Failed at first attempt	29	
Passed at second attempt		12
Others		17

Note that where a short-circuit test is required, according to IEC Standards 60076-1 and 60076-5 the user is required to inform the manufacturer at the tender stage.

14.4.6 Condition assessment testing

Some users make condition assessment tests on transformers in service and require that the manufacturer make similar tests at their works to provide reference results. Some of the more popular condition assessment tests are regularly performed by manufacturers as work tests, for example insulation resistance, transformation ratio, winding resistance, winding capacitance and power factor. Other condition assessment tests are of more use for diagnosing problems in the field than for proving that the transformer meets the specified requirements. Standards from some of these tests have been developed, e.g. IEC 60076-18 for winding frequency response measurements.

14.5 Operation and maintenance

14.5.1 Limitations on transformer life

As was noted above, medium and large transformers are not usually subject to rapid obsolescence. As a consequence some users now have very long life expectancies,

Transformer user requirements, specifications and testing 549

especially for transmission and distribution transformers. The life of such transformers is limited by a wide range of processes, of which the best known and most widely studied is ageing of the solid insulation.

Life-limiting processes in medium and large transformers include:

- oil ageing
- solid insulation ageing
- ageing of accessories, especially fans and pumps
- ageing of gaskets and seals, leading to oil leaks
- ageing and obsolescence of indicative and protective devices
- ageing and embrittlement of cabling
- external corrosion and weathering
- moisture ingress
- tap-changer contact wear, in some cases tap-changer mechanism wear.

Users should also note that medium and large transformers or their accessories may be damaged as a result of system events, for example through faults, lightning strikes, remote switching. The damage may result in immediate failure, or increase the vulnerability of the transformer to further damage in future events. Accumulation of damage in this way could be regarded as another life-limiting process.

14.5.2 Preventive and corrective maintenance

- *Preventive maintenance*: Maintenance activities made with the intention of preventing the condition of the transformer from deteriorating or deteriorating further.
- *Corrective maintenance*: Maintenance activities made with the intention of improving the condition of the transformer, usually but not necessarily to its original condition.

14.5.3 Time- and condition-based maintenance

- *Time-based maintenance*: Maintenance activities carried out at regular planned intervals. Time-based maintenance is especially applicable to generator and process transformers, which are directly connected to other equipment requiring time-based maintenance.

    Many users carry out some basic and non-invasive activities at regular planned intervals, for example oil sampling, inspections. The information gained can be used to decide whether any more major intervention is required.
- *Condition-based maintenance*: Maintenance activities carried out at irregular intervals, depending on the condition of the equipment to be maintained.

    Many users carry out major interventions on a condition basis, using information collected during time-based maintenance. Examples of major overhauls carried out on a condition basis might include rebuilding the tap-changer, replacing accessories, changing or re-generating oil, and repainting. These address many of the life-limiting processes identified previously, with the important exception of solid insulation ageing. However, users should that solid insulation ageing is promoted by oil ageing and moisture ingress, so even it cannot be reversed by corrective maintenance it can be controlled to some extent.

14.5.4 Oil tests

As was noted above, detecting incipient failure of medium and large transformers in the field as notoriously difficult, with only a limited range of tests available online. Most of the available tests involve taking an oil sample for analysis. Users are therefore advised to take oil samples from medium and large transformers at regular intervals. The optimum time between oil samples depends on both the condition of the transformer and its size and importance.

The author suggests that oil samples should be taken from generator and process transformers every three months for dissolved gases and moisture and samples should be taken every year for a wide range of quality checks (colour/appearance, acidity, breakdown voltage, dielectric dissipation factor, inter-facial tension, resistivity and inhibitor content if necessary) and furans. The author suggests that oil samples should be taken from transmission and distribution transformers every year for dissolved gases and moisture and samples should be taken every second year for oil quality and furans.

Dissolved gas results are sensitive to thermal and dielectric faults. Unfortunately there is not sufficient space in this chapter to discuss interpretation of results in detail. Users are instead referred to IEC Standard 60599.

Oil-quality results are sensitive to a number of problems, especially oil ageing and moisture ingress. Unfortunately there is not sufficient space in this chapter to discuss interpretation of results in detail. Users are instead referred to IEC Standard 60422.

Furans are produced by the degradation of solid insulation, and can thus be used to assess solid insulation ageing. Users in the USA and other countries where so-called thermally upgraded insulation is in widespread use are advised that deterioration of such insulation produces substantially lower levels of furans than conventional insulation. Interpretation of results is further complicated by differences between different transformer designs – some designs age uniformly, producing moderate-to-large quantities of furans long before the solid insulation reaches the end of its useful life; conversely other designs suffer from severe ageing at hot spots, producing only small quantities of furans before failure. It is largely for this reason that IEC have so far been unable to produce an interpretation guide for furans. CIGRE are to begin work on this subject soon, and it must be hoped that they are able to overcome these difficulties.

14.5.5 Electrical tests

Where oil tests cannot give sufficient information on the condition of a transformer, it may be necessary to make electrical condition assessments to investigate. These require the transformer to be removed from service and disconnected.

The following tests are now in widespread use in the field:

- winding resistance, to check electrical continuity
- transformation ratio, to check for short-circuited turns and tap-changer problems
- no-load loss and current at low voltage, to check for short-circuited turns
- impedance
- winding frequency response, to check for mechanical damage to the windings
- insulation resistance measurement, to check dielectric condition (especially applicable to core/frame insulation)
- winding capacitance and power factor measurement, to check dielectric condition
- bushing capacitance and power factor measurement, to check bushing dielectric condition
- dielectric frequency response, to check insulation ageing and dryness.

Depending on the size of the transformer, it may be possible to make other more complex tests, for example induced voltage with partial discharge measurement.

14.6 Concluding remarks

There have been a number of major changes in the medium- and large-transformer industry since the first edition of this book was published. In industrialised countries there has been a reduction in manufacturing capacity, with much of the gap being filled by manufacturers in emerging markets. As a result transformer users in industrialised countries are now having to import transformers from emerging markets, rather than the other way around as was previously the case. This has increased the need for clear specifications and has created a need for supplier-selection procedures.

An important development in transformer design and construction since the first edition of this book has been the emergence of viable alternative technologies to conventional oil-immersed transformers. It is now possible to purchase natural or synthetic ester-immersed medium transformers, and it is increasingly possible to purchase resin encapsulated medium transformers. In Japan and Korea gas-filled transformers have emerged as practical alternative to oil-immersed large transformers. It remains to be seen whether they will achieve similar success in other parts of the world.

Testing techniques for medium and especially large transformers have improved, with better methods now available for both dielectric and thermal tests. Users are encouraged to take advantage of these developments, especially fibre optic temperature probes. Field testing techniques have also improved substantially since the first edition of this book was published. Winding frequency response measurements are now well established as a useful diagnostic test and are being joined by dielectric frequency response measurements. Another exciting emerging technology is the use of dynamic resistance measurements for tap-changer diagnostics.

An emerging technology which has not been possible to discuss in detail in this chapter is transformer monitoring systems. Dissolved gas monitors are currently available which are capable of producing near laboratory quality results at intervals of minutes-hours. These monitors are beginning to be used not merely for fault investigation but increasingly for protection. Development of these monitoring systems is likely to continue at a rapid pace.

Bibliography

Three new CIGRE Technical Brochures have now been produced by WG A2.36:
1. TB 528, 'Guide to Preparation of Specifications for Power Transformers', 2013 [This updates and replaces TB 156]
   TB 529, 'Guide to Design Review for Power Transformers', 2013 [This updates and replaces TB 204]
   TB 530, 'Guide to the Assessment of the Capability of a Transformer Manufacturer', 2013 [This is a new document]
   Note: Readers should be aware that the Transformer Procurement Process is strategically very important; the purchase of a large power transformer is a difficult and complex exercise since the purchase of a new transformer represents one of the biggest single investments that any major utility or industrial user is likely to make.

IEC Standards for transformers are now being consolidated as part of two series IEC 60076 for power and distribution transformers and IEC 61278 for converter transformers. At the time of writing, the following versions of the main standards in the two series were current:

2. IEC 60076-1, 'Power Transformers – General', 3rd edn, April 2011
3. IEC 60076-2, 'Power Transformers – Temperature Rise for Liquid Immersed Transformers', 3rd edn, February 2011
4. IEC 60076-3, 'Power Transformers – Insulation Levels, Dielectric Tests and External Clearances in Air', 2nd edn, March 2000
5. IEC 60076-5, 'Power Transformers – Ability to Withstand Short-circuit', 3rd edn, February 2006
6. IEC 60076-7, 'Power Transformers – Loading Guide for Oil-immersed Power Transformers', 1st edn, December 2005
7. IEC 60076-7, 'Power Transformers – Application Guide', 1st edn, October 1997
8. IEC 60076-10, 'Power Transformers – Determination of Sound Levels', 1st edn, January 2005
9. IEC 61278-2, 'Converter Transformers – Transformers for HVDC Applications', 1st edn, February 2001

IEC Standards for oil are applicable to oil types of oil-filled or oil-impregnated equipment, and not just transformers. As a result they form a separate series. The main standards are as follows:

10. IEC 60422, 'Mineral Insulating Oils in Electrical Equipment – Supervision and Maintenance Guidance', 3rd edn, October 2005
11. IEC 60599, 'Mineral Oil-impregnated Electrical Equipment in Service – Guide to the Interpretation of Dissolved and free Gases', 2nd edn, March 1999

Similarly, IEC Standards for test methods are also applicable to other types of equipment and do not form part of the consolidated series:

12. IEC 60060-1, 'High Voltage Test Techniques – General Definitions and Test Requirements', 3rd edn, September 2010

The following ISO and OHSAS Standards are useful references for supplier evaluation:

13. ISO 9000, 'Quality Management Systems – Requirements', 3rd edn, December 2000
14. ISO 14000, 'Environmental Management Systems – Requirements with Guidance for Use', 2nd edn, November 2004
15. OHSAS 18001, 'Occupational Health and Safety Management Systems – Requirements', 2nd edn, July 2007

Users may also find the following CIGRE brochures interesting:

16. CIGRE brochure 227 (final report of working group A2.18), 'Life Management Techniques for Power Transformers', June 2003
17. CIGRE brochure 248 (final report of working group A2.20), 'Guide on Economics of Transformer Management', April 2004
18. CIGRE brochure 343 (final report of working group A2.27), 'Recommendations for Condition Monitoring and Condition Assessment Facilities for Transformers', April 2008
19. CIGRE brochure 445 (final report of working group A2.34), 'Guide for Transformer Maintenance', February 2011
20. CIGRE brochure 528 (final report of working group A2.36), 'Guide for Preparation of Specifications for Power Transformers', April 2013
21. CIGRE brochure 529 (final report of working group A2.36), 'Guidelines for Conducting Design Reviews for Power Transformers', April 2013
22. CIGRE brochure 530 (final report of working group A2.36), 'Guide for Conducting Factory Capability Assessment for Power Transformers, April 2013
23. Smeets R.P.P., Te Paske L.H., Lzeufkens P.P., Fogelberg T. 'Thirteen Years' Test Experience with Short-circuit Withstand Capability of Large Power Transformers', Paper presented at 6th CIGRE Southern Africa regional conference, Somerset West (South Africa), 18–20 August 2009

Chapter 15
Basic measuring techniques
Ernst Gockenbach

15.1 Introduction

The measurements of voltage and current in high-voltage tests are difficult, because the amplitudes are high and they cannot be measured directly with conventional measuring and recording systems. Furthermore, not only the peak value but also the shape of the signal, particularly at impulse voltage and current, should be measured and evaluated and this requires an adequate recording system.

15.2 Measuring system

A high-voltage or high-current measuring system consists generally of a converting device, a transmission device and a recording device. A measuring system is as good as the weakest part of the system, and therefore it is not necessary to require an exceptional performance on one of the components.

The converting device should reduce the amplitude of the signal to be measured to a value which is suitable for the transmission and recording device. The output signal of the converting device should be an exact replica of the input signal concerning the wave shape and the time parameters. This requirement is very strong in many cases and in addition the measurement uncertainty should be estimated and evaluated carefully.

The converting device is usually the critical component due to its physical size and transfer behaviour. The preferred calibration method is the comparison with an approved measuring system under normal operating conditions. Figure 15.1 shows the required arrangement during calibration by comparison.

The uncertainty can be evaluated by the comparison between the output of the measuring system to be calibrated and the approved measuring system. The calibration should be done at least at 20% of the rated operating voltage of the converting device. Reference measuring dividers, for example, for impulse voltage, are available up to 500 kV which can be used to calibrate a converting device up to 2,500 kV. However, it is necessary to check the linearity of the converting device between the calibration voltage level and the rated voltage of the converting device.

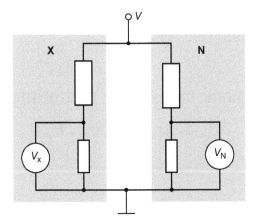

Figure 15.1 Arrangement for comparison measurements
V = applied voltage; N = reference system; X = system to be calibrated [1]

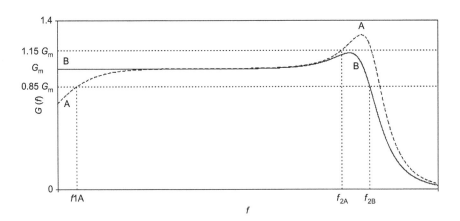

Figure 15.2 Amplitude-frequency response with examples for limit frequencies [1]
G_m = normalized amplitude

The dynamic behaviour of a converting device should be evaluated preferably by an amplitude/frequency response measurement or by a unit step response measurement. Figure 15.2 shows the results of an amplitude/frequency measurement with a lower frequency f_{1A} and the upper frequencies f_{2A} and f_{2B} for the relevant responses A and B. Example B has a constant frequency response up to DC voltage.

Assuming the input voltage or current is an ideal unit step, the output signal is deformed by the converting device. Figure 15.3 shows a simplified output signal of a voltage divider [1] with the reference level epoch.

The reference level epoch should be adequate to the type of voltage to be measured, for example for a standard lightning impulse $t_{min} = 0.84$ µs and $t_{max} = 1.56$ µs.

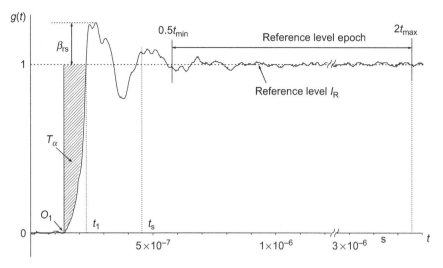

Figure 15.3 Output signal of a voltage divider with a unit step voltage input [1]

The output signal can be described by the partial response time T_α, the response time T, the settling time t_s and the overshoot β_{rs}, whereby the time parameters represent a time-voltage area.

The response time T is the integral from O_1 to t minus the unit step response $g(t)$

$$T = \int_{O_1}^{t} (1 - g(t)) \times dt \tag{15.1}$$

The experimental response time T_N is the value of the step response integral at $2t_{max}$

$$T_N = T(2t_{max}) \tag{15.2}$$

The residual response time $T_R(t_i)$ is the experimental response time T_N minus the value of the step response integral at some specific time t_i, where $t_i < 2t_{max}$

$$T_R(t_i) = T_N - T(t_i) \tag{15.3}$$

The settling time t_s is the shortest time for which the residual response time $T_R(t)$ becomes and remains less than 2% of t

$$|T_N - T(t)| < 0.02t \tag{15.4}$$

for all values of t in the epoch from O_1 to the longest time-to-half value $T_{2,max}$ of the impulse voltage to be measured. Figure 15.4 shows clearly the requirement on the settling time.

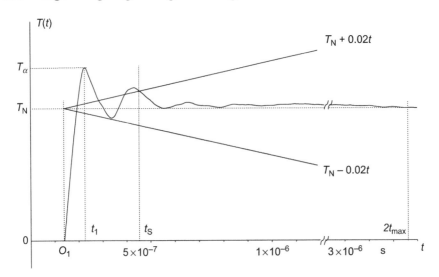

Figure 15.4 Settling time t_s as function of the time [1]

The overshoot β_{rs} (see Figure 15.3) is the difference between the maximum $g_{max}(t)$ and unity as a percentage of unity

$$\beta_{rs} = 100\% \, (g_{max}(t) - 1)$$

The performance of an approved measuring system is given by its uncertainty budget according to IEC 60060-2 [1]. The uncertainty budget comprises the individual uncertainty of the scale factor measurement, the linearity check, the dynamic check, the short- and long-time behaviour, the ambient temperature effect and the proximity effect. If software is used for the evaluation of the uncertainties then also the uncertainty of the software should be taken into account. Within the calculation of the total uncertainty budget, the type of uncertainty should be checked. Depending on the voltage shape, the above-mentioned uncertainties may be different regarding their contribution to the total uncertainty budget. IEC 60060-2 describes in detail the evaluation procedure for the uncertainty budget for different types of voltages like DC, AC and impulse voltages.

The scale factor between the measuring system under test and the reference system for DC and AC voltage should be <3%, otherwise the scale factor should be assigned again. For impulse measuring system the scale factor should be constant within 1% over the reference level epoch, which is given by 50% of shortest front time and 200% of the longest front time according to the relevant tolerances. The time parameter should be measured with an uncertainty $\leq 10\%$.

For a reference measuring system the total uncertainty is lower compared to an approved measuring system. The uncertainty budget for the test voltage value should be $\leq 1\%$ and for the time parameter $\leq 5\%$.

The transmission system is generally a coaxial cable, which does not influence the amplitude but in particular cases does influence the transfer behaviour of the system.

The recording device has very often an additional converting device on its input in order to reduce again the amplitude, but this converting device is small and has normally sufficiently good transfer behaviour. The transformation characteristic is normally not critical but it should be checked and included in the evaluation of the whole system.

An important problem for all measuring systems, particularly at high voltage or high current, is the sensitivity to electromagnetic interferences. Some measures to prevent or suppress these disturbances are described here.

- The most critical tests are impulse tests, where the impulse generation is at the same time a radiation source of an electromagnetic wave. This wave penetrates the whole system – the converter, the transmission and the recording device – and not all components can be shielded. Therefore it is necessary to reduce the amplitude of the radiation and to increase the signal level in order to get a very high signal-to-noise ratio.
- The voltages and currents flowing within the measuring cable are another source of electromagnetic interferences, because they can be induced by capacitive or inductive coupling or by loops of the measuring cable. The measures against these effects are a proper shielding, or in special cases, a double shielding of the measuring and control cables and the prevention of loops by a star connection of all measuring and control cables. Figure 15.5 shows an example of a bad cable connection and Figure 15.6 shows an example of a good cable connection for a simple impulse generator circuit.

There are some cases where a cable loop cannot be avoided. A typical example is the residual voltage measurement of a metal oxide arrester. In this case the high magnetic field induces in the loop, consisting of the test object, the divider connection, the voltage divider and the earth potential, a voltage which is superimposed on the residual voltage. Figure 15.7 shows the schematic diagram [2].

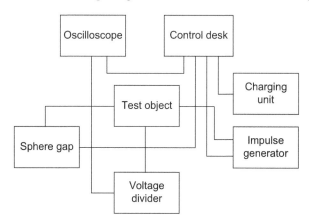

Figure 15.5 Bad cable connection of an impulse voltage set-up

560 High-voltage engineering and testing

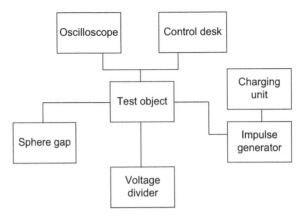

Figure 15.6 Good cable connection of an impulse voltage set-up

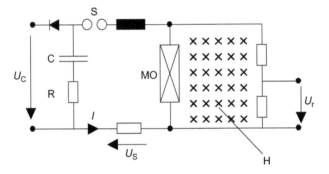

Figure 15.7 Schematic diagram of a metal oxide residual voltage measuring circuit
U_C = charging voltage of capacitor C; C = charging capacitor; R = damping resistor; S = spark gap; MO = metal oxide surge arrester; U_r = residual voltage across the surge arrester; U_s = voltage across the shunt, representing the current I; I = current through the surge arrester; H = magnetic field in the residual voltage measuring loop

The earth connection of the divider cannot be changed for some reasons and therefore a compensation of the magnetic field H by a suitable connection of the voltage divider is necessary, because the magnetic field induces a reasonable voltage in the residual voltage measuring loop. A simple check is the replacement of the surge arrester by a metallic tube with the same geometry and the application of a current impulse. In this case the measured residual voltage across the metallic tube should be zero, if not the measuring circuit is not sufficiently compensated against the induced voltage.

The measuring system should be designed according to the requirements of the measuring uncertainty concerning the transformation behaviour or amplitude measurement and the transfer behaviour or time parameter measurement.

15.3 Amplitude measurements

The amplitude of a supplied voltage or current is the main parameter of the high-voltage or high-current stress of the object to be tested, and therefore it should be measured carefully and with the required uncertainty. Furthermore the amplitude could change very fast depending on the applied voltage or current.

15.3.1 Direct voltage

The amplitude of a direct voltage can be measured with a high-ohmic resistor. The current through the resistor is proportional to the voltage if it is assumed that the resistance of the measuring instrument is negligible. Figure 15.8 shows the equivalent circuit diagram.

The surge protective device (SPD) in parallel to the measuring instrument is very important, because in case of disconnection of the measuring cable or flashover of the high-voltage resistor the full voltage will be applied to the measuring instrument and/or to the operator.

The high-voltage resistor R consists normally of many elements which are connected in series. The whole chain can be fixed on an insulating tube in order to get a reasonable mechanical strength and a certain distance between high potential and ground. To prevent a large measuring error, the minimum recommended current through the high-voltage resistor is 0.5 mA according to IEC 60060-2 [1].

A resistance of the insulating material of 10^{12} Ω leads to a current of 1 µA at a DC voltage of 1,000 kV, which is only 0.2% of the measured current and fulfil the requirement of IEC 60060-2, where a leakage current on the external insulating surfaces is required with a negligible influence on the measuring uncertainty.

Under normal conditions the voltage and temperature coefficients do not play an important role. However, for very small measuring uncertainties the temperature coefficient has to be taken into account because it has a greater influence than the voltage coefficient. Figure 15.9 shows an example of the relative resistance change as a function of the temperature for preselected sets of resistors [3].

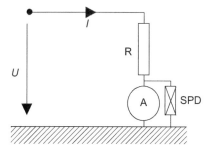

Figure 15.8 DC measuring system
$U = $ DC voltage; $I = $ DC current; $R = $ high-voltage resistor; SPD = surge protective device; $A = $ current measuring device with negligible resistance

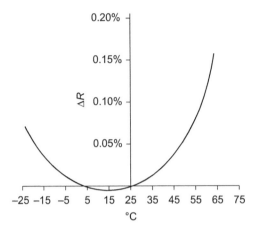

Figure 15.9   Relative resistance change as function of the temperature

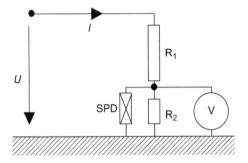

Figure 15.10   DC voltage divider
U = DC voltage; I = DC current; R_1 = high-voltage arm resistor; R_2 = low-voltage arm resistor; SPD = surge protective device; V = voltage measuring device with very high resistance

The relative change in total resistance is less than 0.05% in the temperature range between −20 °C and +50 °C, but such a small deviation can be reached only if the resistor elements are carefully selected, according to its temperature coefficient, and combined in such a way that the temperature coefficient for one stack of resistors fulfils the requirement. Each resistor element may have a larger coefficient than that allowed for the complete chain.

Another possibility to compensate the influence of the voltage and temperature coefficient is the use of a voltage divider, where the elements in the high-voltage and low-voltage arm are under the same voltage stress and temperature. This type of voltage measuring device is commonly used in the high-voltage measuring technique. The simplified equivalent circuit diagram is shown in Figure 15.10.

The direct voltage has normally a certain content of ripple. Depending on the measuring instrument the reading is the mean arithmetic value (moving-coil instrument) or the rms values (static voltmeter). This means that the instantaneous value of the high voltage can be much higher than the measured mean arithmetic value as shown in Figure 15.11, depending on the instrument to be used.

An important point is the voltage distribution in case of fast transient voltages because even in DC test systems a breakdown with a very fast voltage change may happen. The voltage distribution of a resistive divider, shown in Figure 15.12, is linear as long as the frequency is near zero, which means DC or AC at very low frequency. As soon as the stray capacitances C_e influence the voltage distribution, which is the case for voltage contents of higher frequencies, the linear voltage distribution is not any longer valid.

A flashover cannot be always prevented in high voltage tests and therefore the voltage divider should be at least protected against non-linear voltage distribution. The measuring uncertainty may be much higher for fast transient voltages. Figure 15.12 shows the equivalent circuit diagram with stray capacitance and distributed resistor elements.

The ratio of the stray capacitances influences strongly the voltage distribution. Starting from a linear ideal voltage distribution for large parallel or low earth stray capacitances the voltage distribution is getting more and more non-linear if the parallel stray capacitance decreases or the earth stray capacitance increases. This phenomenon should be taken into account by the design of the pure resistive voltage dividers.

15.3.2 Alternating voltage

A resistor or resistive divider can also be used for alternating voltage measurements if the power loss is not too high, but an additional error comes from the phase shift due to the influence of the capacitance and inductance. Therefore a capacitor

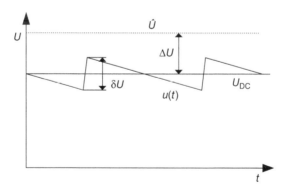

Figure 15.11   *DĊ voltage*
   $\hat{U}$ = *peak value of the AC supply voltage;* U_{DC} = *arithmetic mean value of the voltage u(t);* ΔU = *difference between peak AC supply voltage and DC voltage;* δU = *difference between highest and lowest value of u(t)*

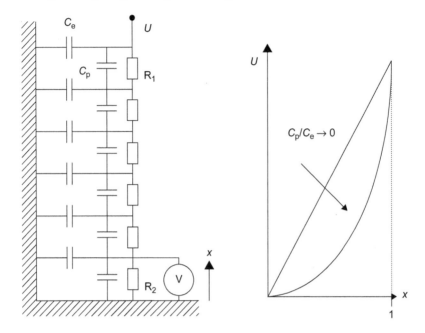

Figure 15.12 Resistive divider with stray capacitance and voltage distribution
$U = DC$ voltage; $R_1 =$ high-voltage resistor; $R_2 =$ low-voltage arm resistor; $C_p =$ parallel stray capacitance; $C_e =$ earth stray capacitance; $V =$ voltage measuring device with very high resistance; $x =$ distance from earthed floor

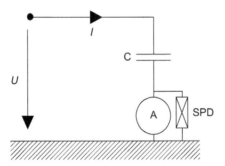

Figure 15.13 AC measuring system
$U = AC$ voltage; $I = AC$ current; $C =$ high-voltage capacitor; $A =$ current measuring device with negligible impedance; $SPD =$ surge protective device

instead of a resistor will be normally used for alternating voltage measurements. Figure 15.13 shows the circuit diagram.

The high-voltage U can be determined by measuring the current I according to the following equation

$$I = U\omega C \tag{15.5}$$

with ω as the frequency of the measured voltage. Assuming that the measuring instrument indicates the rms value, the true rms value of the voltage is given by the equation

$$U = \sqrt{U_1^2 + U_3^2 + U_5^2 + \cdots} \qquad (15.6)$$

but the current, driven by the harmonics, is higher than the current of the basic frequency and this increases the uncertainty of the measurement. Therefore a capacitive voltage divider will be normally used for AC measurements, which is also not influenced by the temperature and voltage coefficient of the capacitor. Figure 15.14 shows the simplified equivalent circuit diagram.

The impedance of the measuring system should be as high as possible, similar to the resistive divider, in order not to influence the load and the frequency independent transformation ratio. The divider output voltage is given by the ratio of the capacitors according to the equation

$$U_2 = U \frac{C_1}{C_1 + C_2 + C_i} \qquad (15.7)$$

with C_i as capacitance of the measuring instrument.

The dielectric strength of insulating material depends on the peak value of voltage, if the voltage stress is short. Therefore, the recommendations require normally the peak value, which can be calculated from rms by multiplication with the factor $\sqrt{2}$, if the wave shape is pure sinusoidal. In all other cases the evaluation of the peak value from the rms measurement is not possible. Figure 15.15 shows a measuring circuit for peak voltage measurement according Chubb and Fortescue [4].

Assuming a symmetrical shape of the voltage and no harmonic content the current through the measuring instrument is given by the integral of the first half positive period representing the arithmetic mean value peak by the equation

$$i_m = \frac{1}{T}\int_0^{T/2} i(t)dt = \frac{1}{T}\int_0^{T/2} C\frac{du(t)}{dt} dt = 2fC\hat{U} \qquad (15.8)$$

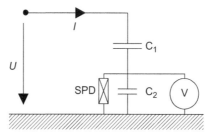

Figure 15.14   AC voltage divider
   $U = AC\ voltage;\ I = AC\ current;\ C_1 = $ high-voltage arm capacitor; $C_2 = $ low-voltage arm capacitor; $V = $ voltage measuring device with very high impedance; $SPD = $ surge protective device

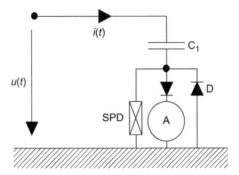

Figure 15.15 AC peak value measuring circuit
$u(t)$ = AC voltage; I = AC current; C_1 = high-voltage arm capacitor; A = current measuring device with negligible impedance; D = diode; SPD = surge protective device

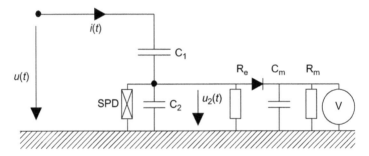

Figure 15.16 AC peak value measuring circuit with a capacitive divider
$u(t)$ = AC voltage; I = AC current; C_1 = high-voltage arm capacitor; C_2 – low-voltage arm capacitor; R_e = earthing resistor; R_m = measuring resistor; C_m = measuring capacitor; $u_2(t)$ = measured voltage; V = voltage measuring device with very high impedance; SPD = surge protective device

From (15.8) the relationship between the measured current and the peak value can be determined by the following equation

$$\hat{U} = \frac{i_m}{2fC} \tag{15.9}$$

In this measuring circuit the frequency influences directly the measuring uncertainty, and therefore the measurement of the frequency is required in addition.

The peak value of a voltage can also be measured with a capacitive voltage divider according to the circuit diagram shown in Figure 15.16.

The measuring capacitor C_m will be charged up to the peak value of $u_2(t)$ and the measuring instrument shows the peak value of this voltage, which is only a part of the high-voltage $u(t)$ given by the divider ratio. The series resistance of the diode D must be small in order to have a negligible voltage drop across the diode.

There are three parameters which influence mainly the measuring uncertainty of this peak voltage measuring circuit. The resistor R_e influences the transformation ratio, the resistor R_m discharges the measuring capacitance C_m and the capacitor C_m is in parallel with the capacitor C_2 during the charging time, which changes the capacitive transformation ratio. The last two factors are in addition frequency dependent, but all these influences on the measuring uncertainty can be compensated.

Another possibility to measure the peak value is the use of an analogue digital converter as a recording device. The evaluation of the recorded signal can be done by built-in computer or by a host computer. Furthermore the computer can make more calculations, for example Fast Fourier Transformation and can evaluate the amplitudes of the relevant harmonics. With a 12-bit resolution of a commercially available digitizer, the measurement uncertainty of the recording device is negligible compared with the other components of the measuring system. There are actually two IEC Working Groups dealing with the requirements of the hardware of AC and DC voltage and current measurement and with the software for the evaluation of the measured parameters [5, 6].

Finally the sphere gap is also a device for peak voltage measurements. Disadvantages are the relative high number of tests, because in the tables of the IEC 60052 [7], the mean value of the disruptive breakdown voltage is given, and also the voltage collapse due to the breakdown of the sphere gaps during the voltage tests. The device is very simple and the gap can be checked with a simple scale but nevertheless the measurement uncertainty is $\pm 3\%$.

15.3.3 Impulse voltage

The measurement of impulse voltages requires a system which has a known scale factor and an adequate dynamic behaviour, because not only the peak value but also the wave shape should be recorded and evaluated.

A resistive divider can be used for impulse measurements, if the resistance is low enough, that the transfer behaviour is not influenced by the stray capacitances. A good estimation of the time constant T of the unit step response of a resistive divider, which characterise the transfer behaviour, is given by the equation

$$T = \frac{RC_e}{6} \tag{15.10}$$

where C_e is the stray capacitance and R the resistance of the divider. From (15.10) it can be deduced that the stray capacitance and the resistance should be as small as possible. Due to a low resistance of the measuring system, the wave shape of the impulse may be influenced in such a way that no switching impulses can be measured with resistive dividers. Therefore resistive–capacitive dividers are often used for the measurement of impulse voltages. Figure 15.17 shows the schematic diagram which is the same as for DC measurements (see Figure 15.12) regarding fast voltage changes.

If the time constants in the high-voltage arm $R_1 C_p$ and the low-voltage arm $R_2 C_p$ are identical, the divider is frequency independent, but the inductance of the

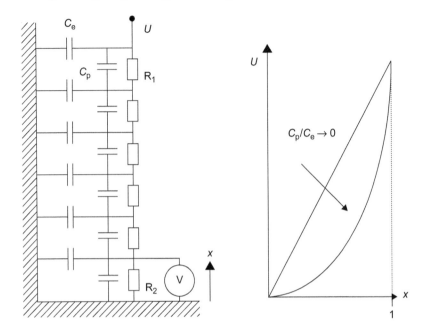

Figure 15.17 Resistive–capacitive voltage divider and voltage distribution
U = Impulse voltage; R_1 = high-voltage arm resistor;
R_2 = low-voltage arm resistor; C_p = parallel stray capacitance;
C_e = earth stray capacitance; V = voltage measuring device with very high impedance; x = distance from earthed floor

high-voltage lead and the capacitance of the divider build up a series resonance circuit, which may cause oscillations.

To prevent these oscillations, in particular for high-voltage dividers with large dimensions, a combination of resistors and capacitors in series is a very often used solution. The so-called damped capacitive divider is shown in Figure 15.18.

It is clear that the time constant R_1C_1 should be equal to R_2C_2 in order to make the divider frequency independent. Two types of damped capacitive dividers exist depending on the value of the damping resistor, the optimum damped and the slightly damped divider. The total resistance R of an optimum damped divider can be estimated by the following equation

$$R \approx 3\text{--}4\sqrt{\frac{L}{C_e}} \tag{15.11}$$

and leads to a resistance of 400–800 Ω for a divider in the MV range.

The slightly damped divider has a smaller resistance, which does not depend on the stray capacitance but only on the inductance of the measuring circuit.

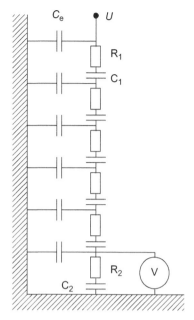

Figure 15.18 Damped capacitive voltage divider
 $U = DC$ *voltage;* $R_1 =$ *high-voltage resistor;* $R_2 =$ *low-voltage arm resistor;* $C_1 =$ *high-voltage capacitor of the n element;* $C_2 =$ *low-voltage arm capacitor;* $C_e =$ *earth stray capacitor;* $V =$ *voltage measuring device with very high impedance*

Therefore it is possible to remove the resistor in the secondary part [8]. The resistance for such a divider can be estimated according the equation

$$R \approx 0.2 \sqrt{\frac{L}{C_1/n}} \qquad (15.12)$$

and leads to a resistance of 60–100 Ω for a divider in the MV range.

For impulse voltage measurements it is necessary that the measuring cable will be matched with its characteristic impedance. With an oscilloscope or an analogue digital converter as recording device the input impedance is very high if the cable is directly connected to the deflecting system or the input divider. Therefore the standard transmission system has the characteristic impedance at the beginning of the measuring cable. Figure 15.19 shows the equivalent circuit diagram for the low-voltage side of a capacitive divider.

The signal $u_2(t)$ across the capacitor C_2 will travel to the recording device via measuring cable, but the amplitude of the signal arriving at the recording device will be half of $u_2(t)$ due to the voltage division between the impedance Z and the characteristic cable impedance Z. Due to the high impedance of the recording device the amplitude of the incoming signal will be doubled and this results in the

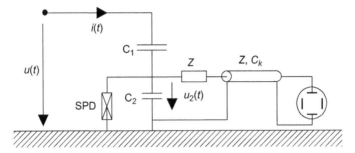

Figure 15.19 Equivalent circuit of a capacitive divider
$u(t)$ = impulse voltage; $i(t)$ = impulse current; $u_2(t)$ = impulse voltage across the low-voltage capacitor; C_1 = high-voltage arm capacitor; C_2 = low-voltage arm capacitor; C_k = measuring cable capacitance; Z = characteristic impedance; SPD = surge protective device

effect, that the amplitude of the recorded signal is the same as the amplitude of the signal $u_2(t)$ across the capacitor C_2. The wave travelling back has no reflection at the voltage divider side because the cable is matched with the impedance Z.

For fast transients the transformation ratio is given by the equation

$$n = \frac{u(t)}{u_2(t)} = \frac{C_1 + C_2}{C_1} \tag{15.13}$$

After twice the travelling time of the signal through the measuring cable the ratio changes to

$$n = \frac{u(t)}{u_2(t)} = \frac{C_1 + C_2 + C_k}{C_1} \tag{15.14}$$

with the cable capacitance C_k.

Depending on the cable length the measuring error can be neglected or should be compensated. Figure 15.20 shows a compensated measuring circuit with a constant ratio for high and low frequencies with the requirement that

$$C_1 + C_2 = C_k + C_3 \tag{15.15}$$

The evaluation or measurement of the peak value can be done by an oscilloscope, a peak voltmeter or a digital recorder. The oscillogram or digital record evaluation permits a determination of the mean curve through oscillations, if necessary, according to the relevant recommendations. A peak voltmeter gives only the highest amplitude recorded, but this is in many cases sufficient, particularly if the wave shape is checked by an oscilloscope or a digital recorder. With an analogue digital converter as recording device the data can be evaluated, stored and treated depending on the need of the test.

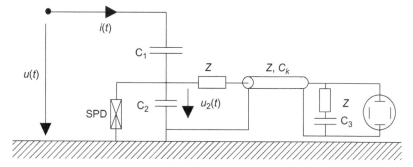

Figure 15.20 Equivalent circuit of a capacitive divider with compensation
$u(t)$ = impulse voltage; $i(t)$ = impulse current; $u_2(t)$ = impulse voltage across the low-voltage capacitor; C_1 = high-voltage arm capacitor; C_2 = low-voltage arm capacitor; C_3 = compensation capacitor; C_k = measuring cable capacitance; Z = characteristic impedance; SPD = surge protective device

15.3.4 Impulse current

The measurement of an impulse current can be done by the measurement of a voltage across a defined resistor, but it is very important that the influence of the high magnetic field will be considered. IEC 62475 [9] describes the current testing and measurement. Figure 15.21 shows a simplified measuring arrangement taking into account the voltage drop across a resistor or shunt and the voltage induced by the magnetic field of the current flowing through the shunt.

It can be clearly seen that with increasing current an inductive part of the measured current is superimposed on the resistive part and this should be compensated by the design of the shunt or by other measures. Furthermore, the power loss and the mechanical forces have to be taken into account at the proper design of a high-impulse current shunt.

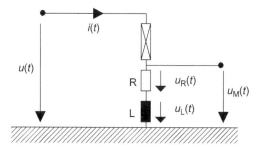

Figure 15.21 Impulse current measurement with a shunt
$u(t)$ = impulse voltage; $i(t)$ = impulse current; R = shunt resistor; L = shunt inductor; $u_M(t)$ = measured impulse voltage representing the impulse current; $u_R(t)$ = resistive part of the measured impulse voltage; $u_L(t)$ = inductive part of the measured impulse voltage

572 High-voltage engineering and testing

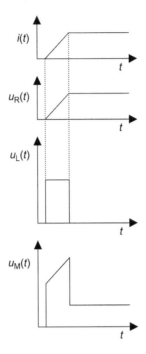

Figure 15.22 Shape of the measured impulse current and voltages

Another technique for impulse current measurement is the use of a so-called Rogowski coil. The measuring principle is very simple and based on the induction of a voltage in a winding with changing magnetic field. The voltage is proportional to the change of the current and therefore integration is necessary to get the current which can be done by an RC circuit. The advantage of the Rogowski coil is the potential-free measurement of the current due to its separation from the ground.

Figure 15.22 shows the single component of the measured signal.

15.4 Time parameter

The measurement of time parameters is mainly limited to impulse measurement. It is necessary to record the shape of the impulse and then to evaluate the required parameter. The evaluation method depends on the recording device. With an oscilloscope as recording device the evaluation can be done only manually using an oscillogram. With analogue digital converter the evaluation can be carried out automatically by a computer. The requirements are defined in IEC 61083-1 [10] for the analogue digital converter and in IEC 61083-2 [11] for the evaluation software.

15.5 Measuring purposes

The measurement has two main reasons:

1. check of the applied voltage and current in order to ensure that the required voltage and current waveform and amplitude have been applied
2. decision on the test results for compliance with a test requirement.

15.5.1 Dielectric tests

A dielectric test consists of applying a voltage up to a certain level and proving that no partial discharges or flashover occur. The measurement during these tests confirms the required waveform and amplitude. In addition the test criterion, flashover or no flashover, will be recorded at the same time, but for that the measuring system is not necessary in any case. For chopped impulse the record of the impulse should show the applied wave shape and the time to chopping, which are determined within certain limits in the relevant recommendations. A switching impulse on a transformer is also determined by the test object itself and the test prehistory. Therefore the measurement should not only detect a flashover but also record the wave shape for the evaluation of the different time parameters.

15.5.2 Linearity tests

Besides pure dielectric tests, linearity tests will be carried out particularly on transformers. The reason for such tests is to check the linear behaviour of the test object. The transformer will be stressed first with 50% of the voltage and later with 100% of the voltage, and the neutral current will be measured. The test will be judged as successful if the comparison between the current and voltage of 50% and 100% levels shows no differences. The result of these measurements can also be the basis for a monitoring or diagnostic procedure. The digital recorded data can be used for further evaluation and improvement in the diagnostic procedure.

15.6 Summary

- The performance of a measuring system is given by the weakest part of the measuring system chain.
- The transformation ratio as well as the transfer behaviour of the complete measuring system should be sufficient for the requirements on the signal to be measured.
- All factors concerning the stability, reproducibility and uncertainty of the measuring system should be taken into account by a so-called uncertainty budget.
- Depending on the signal to be measured the recording device should be selected.
- Even for measurements at DC or low frequencies the behaviour of the measuring system under transient stress should be considered at least for safety reason.

References

1. IEC 60060-2 2010. 'High voltage testing techniques. Part 2: Measuring systems'
2. Modrusan M., Gockenbach E. 'Eine kombinierte Prüfanlage für die Arbeitsprüfung von Metalloxidableitern'. 19. Intern. Blitzschutzkonferenz (ICLP), Graz 1988, paper 6.10
3. Gockenbach E. Modrusan M., Zinnburg E., *et al.* 'DC voltage divider with high precision for HVDC transmission'. *4th international conference on AC and DC power transmission.* London: *IEE Conference Publication.* 1985;**255**: 223–8
4. Schwab A. *Hochspannungsmeßtechnik.* Berlin: Springer Verlag; 1981
5. IEC 61083-3. 'Instruments and software used for measurements in high-voltage and high-current tests. Part 3: Requirements for instruments for tests with alternating and direct currents and voltages'. To be published, for further details see IEC TC 42 WG 21
6. IEC 61083-4. 'Instruments and software used for measurements in high-voltage and high-current tests. Part 4: Requirements for software for tests with alternating and direct currents and voltages'. To be published, for further details see IEC TC 42 WG 20
7. IEC 60052 2002. 'Voltage measurements by means of standard air gaps'
8. Feser K. 'Heutiger Stand der Messung hoher Stoßspannungen'. PTB Seminar 1982, Physikalisch-technische Bundesanstalt, Braunschweig
9. IEC 62475 2010. 'High-current test techniques – definitions and requirements for test currents and measuring systems'
10. IEC 61083-1 2001. 'Digital recorder for measurements in high-voltage impulse tests. Part 1: Requirements for instruments'
11. IEC 61083-2 1996. 'Digital recorder for measurements in high-voltage impulse tests. Part 2: Evaluation of software used for the determination of the parameters of impulse waveforms'

Chapter 16
Basic testing techniques
Ernst Gockenbach

16.1 Introduction

Tests are generally necessary to demonstrate that the equipment under test fulfils the specified requirements and quality standards. The tests may have different purposes, a type test as check and quality assurance for the design of the equipment and a routine test as check and quality assurance for the manufacturing processes. Further tests could be an acceptance test in the factory or on site to demonstrate the quality of the components under test or to check the integrity of the components after transportation and installation. With the introduction of new technologies or the use of higher voltages so-called prequalification test could be required from the customer in order to check the quality of a complete system under on-site conditions but with higher stresses. Such a prequalification test allows the change of the design and adaption of the system as consequences of the test results, but the test takes time in the range of one year or more. An example is the use of extruded cables up to 250 kV for DC transmission systems [1].

Above all a routine test should show that the equipment is able to withstand the test conditions, which are selected according to the stress during the whole time period of service. That means the test stress should be high enough concerning the sensitivity but low enough to prevent an initiation of undetected defects during the test procedure, which may lead to damage after a certain time of service. Therefore, the test requirements are based on experiences concerning the stress and the behaviour of the tested material during normal operation conditions. Furthermore the following parameters have to be taken into account for the type and/or routine tests:

- regulations by law
- requirements
- recommendations
- mutual agreements on technical specifications
- economy.

The following chapters are only related to high-voltage testing requirements and recommendations for type and routine tests, without any consideration of regulations by law, mutual agreement on technical specifications and economic factors.

16.2 Recommendations and definitions

IEC 60071 'Insulation Co-ordination' [2] the relationship between the different kinds of test voltages like:

- AC voltage at power frequency
- lightning impulse voltage
- switching impulse voltage.

IEC 60060 Parts 1 and 2 'High Voltage Test Technique' [3] gives the general definitions and requirements on the voltage shape, the tolerances and the measuring uncertainty.

For some high-voltage apparatus, specific IEC publications exist due to the strong influence of the test object on the generally defined test voltages and due to the specific test conditions. IEC 60076 [4, 5] describes, for example, a switching impulse shape for transformer tests, which is related to the behaviour of the transformer and the strong influence on the wave shape by the transformer as test object. This recommendation has to be taken into account during the transformer test in addition to the general recommendations given in IEC 60060. Also for high-voltage cables, an additional recommendation concerning impulse voltages exist. IEC 60230 [6] allows for high-voltage cable tests with lightning impulses a front time up to 5 μs, because the capacitance of the cable under test is normally so high that a standard front time of 1 μs cannot be reached without oscillations.

Therefore the technical compromise, defined in the recommendations, is to enlarge the front time in order to have a double exponential wave shape not only without any oscillations but also without any influence on the test result. Furthermore a clear definition of all the relevant parameters regarding the test technique and procedure is necessary to ensure comparable and reproducible test results.

The most important definitions, given in [2], should be named and explained here for understanding of the following chapters.

Insulation co-ordination: The selection of the dielectric strength of equipment in relation to the voltages which appear on the system for which the equipment is intended and taking into account the service environment and the characteristics of the available protective devices. Note that the 'dielectric strength' means the rated or standard insulation level.

External insulation: Distances in atmospheric air and the surface in contact with atmospheric air of solid insulation of the equipment which are subject to dielectric stresses and to the effects of atmospheric and other external conditions such as pollution, humidity, etc.

Internal insulation: Internal solid, liquid or gaseous parts of the insulation of equipment which are protected from the effects of atmospheric and other external conditions.

Self-restoring insulation: Insulation which completely recovers its insulating properties after a disruptive discharge.

Non self-restoring insulation: Insulation which loses its insulating properties or does not recover them completely after a disruptive discharge.

Nominal voltage of a system: A suitable approximate value of voltage used to designate or identify a system.

Highest voltage of a system: The highest value of operating voltage which occurs under normal operating conditions at any time and at any point in the system.

Highest voltage for equipment (U_m): The highest rms value of phase-to-phase voltage for which the equipment is designed in respect of its insulation as well as other characteristics which relate to this voltage in the relevant equipment standards.

Overvoltage: Any voltage between one phase conductor and earth or between phase conductors having a peak value exceeding the corresponding peak of the highest voltage for equipment. It is expressed unless otherwise clearly indicated, such as for surge arresters, in pu referred to $U_m \sqrt{2}/\sqrt{3}$.

Classification of voltages

- continuous power frequency voltage
- temporary overvoltage (power frequency)
- transient overvoltage (short-duration overvoltage of ms or µs)
  - slow front overvoltage, 20 µs $\leq T_p \leq$ 5,000 µs, $T_2 \leq$ 20 ms
  - fast front overvoltage, 0.1 µs $\leq T_1 \leq$ 20 µs, $T_2 <$ 300 µs
  - very fast front overvoltage, time to peak $\leq$0.1 µs, total duration <3 ms, superimposed oscillations 30 kHz $< f <$ 100 MHz.

Standard voltage shapes

- power frequency voltage between 48 and 62 Hz, duration of 60 s
- standard switching impulse voltage with $T_p/T_2 =$ 250/2,500 µs
- standard lightning impulse voltage with $T_1/T_2 =$ 1.2/50 µs.

Withstand voltage: The value of test voltage to be applied under specified conditions in a withstand test during which a specified number of disruptive discharges is tolerated.

- Conventional assumed withstand voltage with a withstand probability P_w of 100%
- Statistical withstand voltage with a withstand probability P_w of 90%.

The conventional assumed withstand voltage is specified for non self-restoring insulation and the statistical withstand voltage is specified for self-restoring insulation, both in [2].

Co-ordination withstand voltage (U_{cw}): The value of the withstand voltage of the insulation configuration that meets the performance criterion in the actual service conditions.

Standard withstand voltage (U_w): The standard value of the test voltage applied in a standard withstand test. It is a rated value of the insulation and proves that the insulation complies with one or more required withstand voltages.

Rated insulation level: A set of standard withstand voltages which characterise the dielectric strength of the insulation.

Standard withstand voltage tests: Dielectric test performed in specified conditions to prove that the insulation complies with a standard withstand voltage. These tests are:

- short-duration power frequency voltage tests
- switching impulse voltage tests
- lightning impulse voltage tests
- combined voltage tests.

More details are given in IEC 60060 [3].

16.3 Test voltages

A test voltage is defined by its amplitude, frequency and/or shape within specified tolerances. In [3] the standard and preferred test voltages are described, but as already mentioned it may be sometimes necessary depending on the test object or test circuit to modify the frequency or shape of the test voltage in order to enable the tests.

16.3.1 DC voltage

A direct current (DC) voltage is defined as the mean value between the highest and lowest level within a time period. The duration of the period depends on the generating system. Figure 16.1 shows a typical example of a DC voltage generated by rectification of an AC voltage.

The DC voltage U_{DC} is the arithmetic mean value of the voltage $u(t)$. The voltage drop ΔU is the difference between the peak voltage of the AC power supply

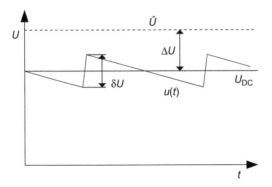

Figure 16.1 Parameter of a DC voltage

and the DC voltage, which is given by the so-called 'internal impedance' of a multi-stage rectifier. The difference between the highest and lowest value of $u(t)$ is called ripple and represents the charging of the capacitors by the AC source during the conductive period of the diodes and the discharging of the capacitors by the load during the nonconductive period of the diodes. It is obvious that the charging time is much shorter than the discharging time and this should be taken into account for the design of the diodes, because the charging current is much higher than the discharge current. According to IEC 60060 the value of the ripple should be less than ±3% of U_{DC}. The main parameters influencing the ripple are the frequency of the AC supply voltage, the value of the smoothing capacitance and the load current. For small load currents the ripple can be estimated by the following equation:

$$\delta U \approx \frac{I}{fC} \frac{N}{4} (N+1) \tag{16.1}$$

with the load current I, the frequency of the AC power supply f, the smoothing capacitance C and the number of stages of a multi-stage rectifier N. Equation (16.1) depicts that the ripple increases linear with increasing load current and quadratic with the number of stages and decreases with increasing frequency and smoothing capacitance.

The voltage drop ΔU influences only the mean value U_{DC} of a DC generator and depends also on the design of the generator. For a multi-stage generator the voltage drop can be estimated according to the equation:

$$\Delta U \approx \frac{I}{fC} \frac{N}{3} (2N^2 + 1) \tag{16.2}$$

with the same parameters as in (16.1).

The most important point is the increasing of the voltage drop by the cube power of the number of stages, but this can be taken into account by choosing the adequate no load voltage of the generator and a fast voltage regulating system. The strongest requirements are necessary for DC pollution tests, where the load current can change very rapidly within a large range from some milliamperes to some amperes. For these cases the voltage should be kept stable within given limits and that requires a strong AC power source, a large smoothing capacitance and a fast regulation system.

Figure 16.2 shows the diagram of a multi-stage rectifier with its main components. The measures to reduce the ripple δU can be deduced from (16.1). The voltage drop ΔU is normally no problem for the test generator because it can be taken into account by the design and the no-load output voltage of the generator. For DC voltage tests on polluted insulators the voltage drop may influence the test results due to fast change of the load. For these particular tests the transient behaviour of the voltage regulator should be taken into account because in the case of high load current the energy supply should be regulated very fast by feeding the current from the smoothing capacitor or from the main power supply or from both, which is normally the most economical solution. Figure 16.3 shows the relationship between the number of stages, the maximum load current and the output voltage for multi-stage rectifiers with a no-load voltage of 400 kV per stage.

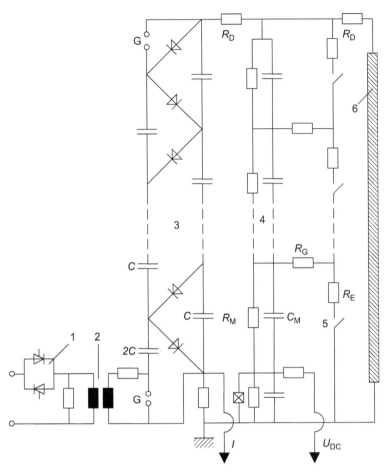

Figure 16.2 Equivalent diagram of a multi-stage rectifier
1 = thyristor-controlled voltage regulation; 2 = high-voltage transformer; 3 = rectifier with diodes, charging and smoothing capacitors; 4 = voltage divider with grading capacitors; 5 = discharging and grounding device; 6 = test object; C = smoothing capacitor; C_M = grading capacitor of the measuring resistor; R_M = measuring resistor; R_D = damping resistor; R_E = earthing resistor; R_G = grading resistor of the earthing switch; U_{DC} = DC output voltage; I = charging current; G = protection gap

16.3.2 AC voltage

The alternating voltage is defined by its rms value and/or by its peak value depending on the purpose. For a pure sinusoidal waveform the relation between the peak value and the rms value is given by the square root of 2. This relation is also

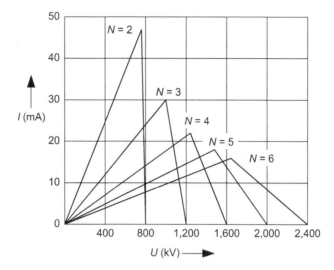

Figure 16.3 Output current as a function of the output voltage at 3% ripple
$N =$ number of stages (400 kV rated voltage)

used in the recommendations [3] to define the tolerance of the sinusoidal wave shape by the following equation:

$$\frac{\hat{U}}{U} = \sqrt{2} \pm 5\% \tag{16.3}$$

The flashover or breakdown behaviour of insulating material depends on the peak value of the supplied voltage for short-time stress. At longer stress time or under service stress very often the rms value is the most important parameter for the breakdown depending on the loss characteristic and the thermal behaviour of the insulating material.

The waveform of an AC voltage can be defined by the ratio of the basic frequency and the harmonics of different order. The harmonics should be taken into account for the measuring devices, which may not be able to measure the true rms value due to its frequency behaviour. Very often rms measuring systems use the relation between the peak value and the rms value and they are in reality peak value measuring systems. With modern digital measuring devices a Fourier analysis of the measured AC voltage is possible and the harmonics can be simply determined.

The generation of an AC voltage is much simpler than a DC voltage because the voltage can be transformed by a very simple circuit. Figure 16.4 shows a simplified equivalent diagram of a voltage transformer.

The ratio between the two voltages U_1 and U_2 is given by the ratio of the number of windings and therefore theoretically every voltage ratio can be reached by a single-stage transformer. Due to the non-linear behaviour of the insulating material it is very often useful to generate high AC voltage by means of a transformer cascade in order to reduce the cost for the insulating material and to have a

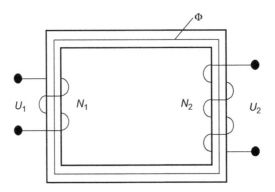

Figure 16.4 Equivalent diagram of a voltage transformer
Φ = magnetic flux; N = number of windings; U_1 = input voltage; U_2 = output voltage

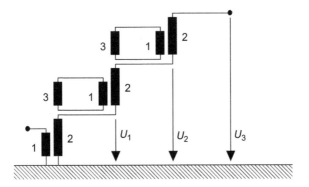

Figure 16.5 Equivalent diagram of a three-stage transformer cascade
1 = primary winding; 2 = secondary winding; 3 = tertiary winding; U_1 = output voltage of the first stage; U_2 = output voltage of the second stage; U_3 = output voltage of the third stage

flexible combination of a number of single transformer units. Figure 16.5 shows a typical three-stage cascade arrangement with identical units. Each unit needs only the insulation for a single-stage transformer.

Important parameters for an AC voltage generator are the short-circuit impedance and the content of harmonics as a function of the load. The short-circuit impedance, given normally in percentage, represents the voltage drop under full-load conditions assuming the resistive part is negligible in comparison to the inductive part. For a cascade, the short-circuit impedance is more than the sum of the short-circuit impedance of the single units and increases more than linear with the number of stages.

The combination shown in Figure 16.5 can also be used with two transformers at the bottom stage and one transformer in the second stage in order to get a higher output current, but at a lower output voltage. Therefore the use of each unit as a single transformer or the combination of the units offers a number of possibilities. Concerning the power distribution within a cascade it is important that the primary winding of

Basic testing techniques 583

the first stage should be designed to carry the full output power of the cascade. The secondary winding in all stages are loaded with the same power, given by the output power divided by the number of stages assuming that all transformers are identical. The tertiary winding of the first stage is loaded with two-thirds of the output power. The windings in all other stages are loaded with the relevant load distribution within the transformer.

Another important factor for AC voltage generation by a transformer is the voltage increase due to the capacitive load, which is normally the case for AC test objects. Figure 16.6 shows the vector diagram for a simplified transformer circuit, whereby all elements are referred to the high-voltage side.

The output voltage U_2 can normally be calculated by the ratio of the windings, but due to the capacitive load the voltage U_2 is higher than calculated. This can be demonstrated by the vector diagram, shown in Figure 16.7.

The increasing of the output voltage due to the capacitive load is normally undesired, but for AC tests the increasing voltage at the secondary side can be used in

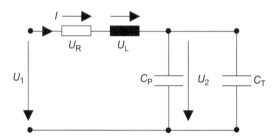

Figure 16.6 Equivalent diagram of a transformer with capacitive load
U_1 = input voltage, referred to the high-voltage side; U_2 = output voltage, across the capacitive load; I = transformer current; C_T = capacitor, representing the transformer capacitance; C_P = capacitor, representing the load capacitance; U_R = voltage across the resistor; U_L = voltage across the inductor

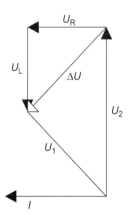

Figure 16.7 Vector diagram of a transformer with capacitive load

order to reduce the input voltage or the size of the test equipment. If the capacitance of the load, including the transformer capacitance, and the inductance of the transformer, including the circuit inductance, are equal the circuit is resonance and theoretically the output increases up to infinity. Due to the existing resistance in the circuit, the voltage will be limited, but the ratio between input voltage and output voltage can be very high (up to 100) and the necessary reactive power is very low.

The resonance conditions can be used for voltage test purposes by changing the inductance of the circuit, resulting in resonance conditions at constant frequency or by changing the frequency resulting in resonance condition at different frequencies, depending on the circuit inductance and capacitance. With this type of test equipment test objects with very high capacitance can be tested with reasonable size and power of the test equipment. In all cases it should be taken into account that the regulation system of the test system should be adapted very carefully to the required voltage level in order to prevent an overshoot of the test voltage.

16.3.3 Impulse voltage

An impulse voltage or current is defined by its peak value and its time parameter. In order to reproduce results at impulse voltage tests the wave shapes are defined in general recommendations [3] or apparatus-related recommendations [4, 5]. The standard lightning impulse voltage has a peak value $\hat{U}$ at the maximum and a front time T_1 and a time to half value T_2 according to Figure 16.8.

The front time T_1 is given by the following equation:

$$T_1 = 1.67\,T \tag{16.4}$$

with T as the measured time between the 30% and 90% level of $\hat{U}$. The straight line through the 30% and 90% level gives the virtual origin point O_1 and the time T' as difference between the 30% and the zero level.

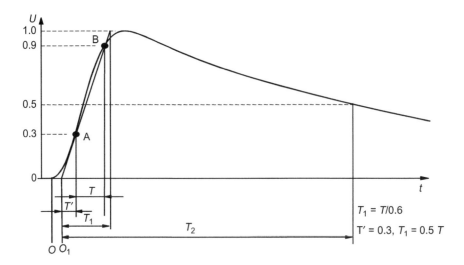

Figure 16.8 Standard lightning impulse voltage

The time to half value T_2 is the time difference between the virtual origin O_1 and the 50% level in the tail of the wave shape. The tolerances of the time parameter are quite large, because the test results are not strongly influenced by the variation in the time parameter and the generation of the required impulse shape is time consuming for different test objects. Therefore, the recommended front time for a standard lightning impulse according to IEC 60060 is 1.2 µs ± 30% and 50 µs ± 20%.

The generation of an impulse voltage or current is generally done by charging and discharging of capacitors through resistor which gives an impulse, which could be described by two exponential functions. Figure 16.9 shows a simplified typical equivalent circuit for impulse voltage generation without any inductance. Figure 16.10 is a similar circuit but with a different arrangement of the resistors.

In both circuits a wave shape, representing more or less two exponential functions, will be generated, but the ratio between the output voltage or the voltage across the load capacitor C_1 and the input charging voltage U_0 is different. In circuit

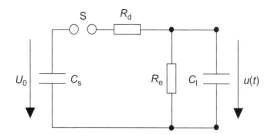

Figure 16.9 Simplified equivalent circuit of an impulse generating circuit type a
C_s = charging or impulse capacitor ($C_s \gg C_1$); R_d = damping resistor; R_e = earthing resistor ($R_e \gg R_d$); C_1 = load capacitor (test object, divider, etc.); S = spark gap; U_0 = DC charging voltage of capacitor C_s; $u(t)$ = voltage across the load capacitor (test object; divider etc.)

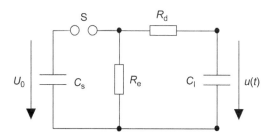

Figure 16.10 Simplified equivalent circuit of an impulse generating circuit type b
C_s = charging or impulse capacitor ($C_s \gg C_1$); R_d = damping resistor; R_e = earthing resistor ($R_e \gg R_d$); C_1 = load capacitor (test object, divider; etc.); S = spark gap; U_0 = DC charging voltage of capacitor C_s; $u(t)$ = voltage across the load capacitor (test object; divider etc.)

type a, the voltage across the load capacitor can be calculated according to the following equation neglecting the influence of the inductance of the circuit and the discharge through the resistor R_e during the charging time of C_1

$$u(t) = U_0 \frac{C_1}{C_s + C_1} \times \frac{R_e}{R_d + R_e} \tag{16.5}$$

The output voltage is given not only by the ratio of the capacitors but also by the ratio of the resistors.

In circuit type b, the output voltage is only determined by the ratio of the capacitors, and is therefore higher as in the circuit a.

$$u(t) = U_0 \frac{C_1}{C_s + C_1} \tag{16.6}$$

An empirically based estimation of the output voltage $u(t)$ of the circuit type b is given by the equation

$$u(t) = U_0 \frac{C_1}{C_s + C_1} \times 0.95 \tag{16.7}$$

which takes into account the damping effect of the resistors.

For a given capacitance of the capacitor C_1 the front time is determined by the load capacitance and the damping resistance. Assuming that the load cannot be changed, the front time can be adjusted by variation in the damping resistance. Under the same conditions, the time to half value is mainly determined by the earthing resistance and the impulse capacitance, assuming that the impulse capacitance is much higher than the load capacitance in order to reach a high-output voltage.

For tests with standard lightning impulse, the value of the resistance should be calculated depending on the test object and the generator to be used. The solutions of the differential equations are not consistent and therefore assumptions are given in the following equations

$$T_t \approx R_e(C_s + C_1) \tag{16.8}$$

$$T_f \approx R_d \frac{C_s \times C_1}{C_s + C_1} \tag{16.9}$$

with T_f as time constant of the front and T_t as time constant of the tail. The relation between these two time constants and the front time T_1 and time to half value T_2 is given for the circuit type b in the following equations assuming a standard lightning impulse 1.2/50 µs:

$$T_1 = 2.96 T_f \tag{16.10}$$

$$T_2 = 0.73 T_t \tag{16.11}$$

The generation of impulse voltages with high amplitude is normally done with a multi-stage impulse generator, known as Marx generator, named after its inventor

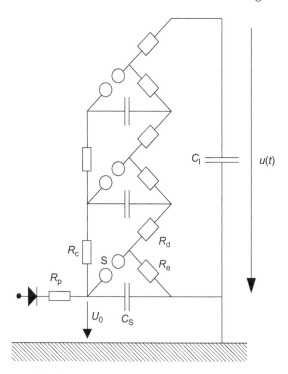

Figure 16.11 Simplified equivalent circuit of a multi-stage impulse generator
C_s = charging or impulse capacitor per stage $(C_s \gg C_1)$;
R_d = damping resistor per stage; R_e = earthing resistor per stage $(R_e \gg R_d)$; S = spark gap per stage; C_1 = load capacitor (test object, divider, etc.), R_p = protection resistor; R_c = charging resistor per stage

Erwin Otto Marx. The principle of such a generator is the parallel charging of a number of impulse capacitors and the discharging in series of these capacitors. The equivalent simplified circuit, again for the type b circuit, is shown in Figure 16.11.

The behaviour of the spark gap S is the most important part for a high reproducibility of the impulse shape and amplitude. The breakdown process and the breakdown time of a spark gap depend on the value and the uniformity of the electrical field. Therefore sphere gaps are generally used for spark gaps, with an electrode material of tungsten in order to reduce the damage of the surface by the arc. Furthermore it is necessary that the breakdown of the spark gap occurs at the same charging voltage in order to generate impulses with identical amplitude and time parameters. This behaviour can be reached by triggering the spark gap.

Two methods may be used for that purposes, the triggering at constant distance of the spark gap or the triggering at constant voltage. For the first method, the spark gap should have a distance, which is greater than the breakdown distance at the desired voltage. Then the impulse capacitors will be charged in parallel up to the required charging voltage. Due to different time constants in the stages the impulse capacitors reach the full charging voltage at different times which should be taken into account. When all capacitors are fully charged the distance of the spark gap will be reduced and the breakdown takes place at the flashover distance, which is constant for a given charging voltage. Then all capacitors are connected in series through the spark gaps and the damping resistors R_d and they charge the load capacitance C_l and generate the impulse by discharging later on through the earthing resistors R_e.

The second method is more often used and based on a triggering device within the bottom spark gap. The spark gap distance is slightly higher than the required breakdown distance at a preselected charging voltage. Then the impulse capacitors will be charged up to the preselected voltage and be kept constant. A trigger impulse at the bottom spark gap, normally generated by a simple spark plug, generates a breakdown of the bottom spark gap and due to natural overvoltage at the spark gaps in the other stages all spark gaps flash over and the impulse capacitors are connected in series for generating the impulse voltage. This method has a high reproducibility and reliability which is necessary for impulse tests, particularly for transformer impulse tests, where a comparison between impulses at different charging voltage is required.

The simulation of switching operations within a network can be done by the same type of generator by generating switching impulse voltages. The standard switching impulse voltage is shown in Figure 16.12 and has a time to peak T_p, time interval from the true origin to the time of maximum value of a switching-impulse voltage, of 250 µs and a time to half value T_2 of 2,500 µs. Furthermore a time T_d is defined as the time at which the value of the switching impulse is above 90% of the peak value $\hat{U}$.

The evaluation of the time to peak has changed compared to the former recommendations. The procedure is similar to the evaluation of the front time of a lightning impulse, but the large tolerance of the time to peak requires a procedure which takes the time to half value into account. The time T_{AB} is the measured time

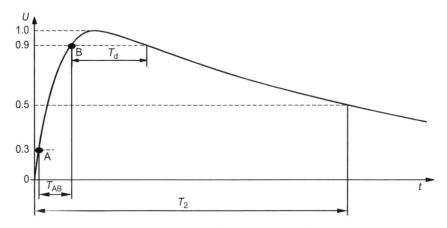

Figure 16.12 Standard switching impulse voltage

(time difference between 90% and 30% of $\hat{U}$) and the time to peak T_p is calculated according to the equation

$$T_p = K T_{AB} \tag{16.12}$$

where K is a dimensionless constant given by

$$K = 2.42 - 3.08 \times 10^{-3}\, T_{AB} + 1.51 \times 10^{-4}\, T_2$$

and where T_{AB} and T_2 are in microseconds.

For on-site tests the evaluation is simpler using the value of 2.4 for K.

The tolerances are much higher compared to the lightning impulse due to the smaller influence of the breakdown behaviour of the tested insulating material. The time to peak has a tolerance of 20%, and the time to half value of 60%. Similar to the estimation for lightning impulse the time to peak and the time to half value can be calculated by the simplified equations

$$T_p = \frac{T_f \times T_t}{T_t - T_f} \ln\left(\frac{T_r}{T_f}\right) \tag{16.13}$$

$$T_2 = T_t \times \ln\left(\frac{2}{\eta}\right) \tag{16.14}$$

with η as efficiency factor given by the ratio of the output voltage to the charging voltage.

This efficiency factor is mainly determined by the value of the impulse capacitors and the load capacitor. For a multi-stage generator with n stages the efficiency for lightning impulse can be estimated by the following equations whereby the factor 0.95 takes the influence of the resistors into account

$$\eta \approx \frac{C_s/n}{C_s/n + C_l} \times 0.95 \tag{16.15}$$

For switching impulses the efficiency is much lower and the factor can be as low as 0.50.

The inductance of the test circuit is also important and should be explained in detail. The circuit shown in Figures 16.9 and 16.10 are without any inductance because normally the circuit is damped in such a way, that no oscillations happen (aperiodic damping) and that the impulse follows more or less a double exponential function. The minimum value of the resistance is given by the equation

$$R_d \geq 2\sqrt{L \frac{C_s/n + C_l}{C_s/n \times C_l}} \tag{16.16}$$

with L as inductance of the complete circuit. It is clear that with given impulse capacitors and load capacitor the damping resistor value for an aperiodic damped

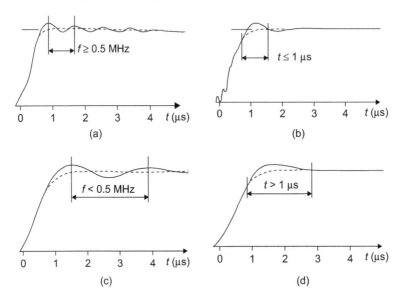

Figure 16.13   Examples of lightning impulses with overshoot or oscillations

impulse is determined by the inductance. At a given front time the required damping resistor may be lower than the value for an aperiodic damped shape and then the impulse may show some oscillations if the inductance cannot be reduced due to the size of the test circuit and the inner inductance of the capacitors. In Figure 16.13 typical lightning impulses are shown as examples. These examples were used in former recommendations to evaluate the test voltage and the overshoot. The actual recommendation [3] describes the evaluation of the base curve which is closed to the former mean curve shown in Figure 16.13 as dotted line. Then the difference between the recorded curve and the calculated curve (residual curve) is filtered by a so-called test voltage function, and the sum of the filtered residual curve and the base curve built the test voltage curve from which the test voltage and the time parameter could be evaluated. The overshoot, the difference between the recorded curve and the base curve, is extended to 10% instead of former 5%.

More details concerning the determination of the peak value, the mean curve and the test voltage are described in the IEC recommendation [3]. Due to these limits the capacitive load of a given circuit is determined by the inductance of the circuit and by resistance given by the tolerance of the front time. A load diagram for an impulse generator is shown as an example in Figure 16.14 under the assumption of the tolerances of the time parameter and only 5% overshoot.

The influence of the load capacitance on the tail is much weaker so that only one resistor element is normally necessary to cover the whole capacitive load range.

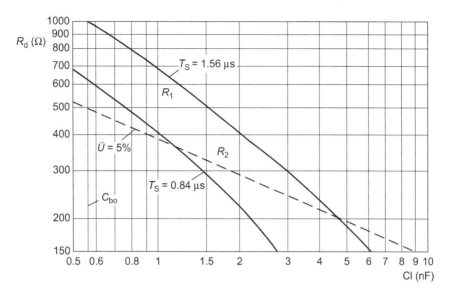

Figure 16.14 Load diagram for standard lightning impulse

16.4 Impulse current

The current testing and measuring technique becomes more important in the last years and therefore all the definitions and requirements concerning current testing and measurement are described in a new IEC recommendation 62475 'High current test techniques – Definitions and requirements for test currents and measuring systems' [7]. The structure and some parts of this recommendation are similar or identical with the recommendations for voltage test and measurement and therefore only an example of an impulse current will be described here.

Analogue to the impulse voltage the impulse currents are defined by its peak value and its time parameter like front time T_1 and time to half value T_2. Figure 16.15 shows a standard impulse current.

The front time T_1 is given by the following equation:

$$T_1 = 1.25\,T \tag{16.17}$$

with T as the measured time between the 10% and 90% level of $\hat{U}$. The time T_2 is the difference between the virtual zero point O_1 and the 50% level in the tail of the impulse. The undershoot or reversal peak is typically for an impulse current and therefore a requirement depending on the equipment under test is necessary. In general, a reverse peak less than 30% is required.

The generation of an impulse current is similar to the generation of an impulse voltage. The main difference is that normally the capacitors are all connected in parallel in order to increase the capacitance and to reduce the inductance of the circuit. Current impulses with short front times need a special arrangement of

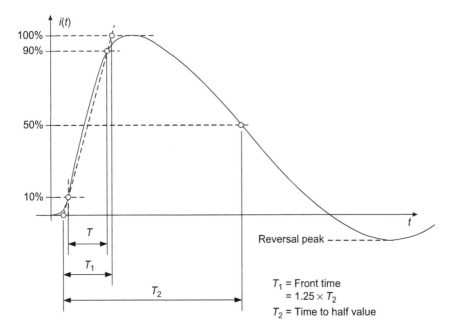

Figure 16.15 Standard impulse current

the capacitors due to the fact that the front time or steepness of the current impulse depends on the derivative of the current according to the equation

$$\frac{di}{dt} = \frac{U}{L} \qquad (16.18)$$

with U as charging voltage and L as inductance of the complete test circuit. It is clear that a lower inductance leads to a higher steepness of the current, but in many cases the inductance is given and the steepness can only be reached by increasing the charging voltage. This should be realised without a remarkable increase in the inductance too.

16.5 Test conditions

The insulation co-ordination comprises the selection of electrical strength of the equipment and its application. The dielectric stress may be divided into the following classes during the service:

- power frequency voltage (AC or DC)
- temporary overvoltage
- switching overvoltage
- lightning overvoltage.

For a given stress the behaviour of the internal insulation may be influenced by its degree of ageing and the behaviour of external insulation and in addition by the atmospheric conditions in the vicinity. A correction of the test voltage due to ageing processes is not possible because the test value takes already into account the reduction of the electrical strength during the life time of the insulation. The influence of the atmospheric conditions, however, can be considered by using correction factors. This allows tests under different laboratory conditions but with the same stress of the external insulation.

The standard atmospheric conditions are an air temperature of $t_0 = 20\,°C$, an air pressure of $p_0 = 1013$ mbar and an absolute humidity of $h_0 = 11$ g/m^3 according to [3]. The atmospheric correction factor K_t has two parts, the air density factor k_1 and the humidity factor k_2. The disruptive discharge voltage of air is proportional to the atmospheric correction factor K_t, which results from the product of k_1 and k_2.

The air density factor k_1 depends on the relative air density δ and can be expressed by the following equation

$$k_1 = \delta^m \tag{16.19}$$

where δ is given by the equation

$$\delta = \frac{p}{p_0} \frac{(273 + t_0)}{(273 + t)} \tag{16.20}$$

The temperature t and t_0 are given in degree Celsius and the pressure p and p_0 in mbar. The parameter m is a function of g which depends on the type of predischarges and g is defined according to the equation

$$g = \frac{U_B}{500 L \delta k} \tag{16.21}$$

U_B is the 50% disruptive discharge voltage at the actual atmospheric conditions in kV, L the minimum discharge path in m at the actual relative air density δ and with the value of k according to Figure 16.16.

The correction is considered reliable for $0.8 < k_1 < 1.05$.

Figure 16.17 shows the value of m as a function of the factor g.

The humidity correction factor k_2 is expressed as

$$k_2 = k^w$$

and the values of k and w are given in the Figures 16.16 and 16.18 as a function of the humidity h related to the air density δ and the factor g described by (16.20).

The selection of the test voltage according to the recommendations is different for voltages up to 300 kV and above. Tables 16.1 and 16.2 show the relation between the rated power frequency voltage and the rated withstand voltages.

Equipment with rated voltages between 52 and 300 kV are tested in general by a short duration (1 min) power frequency test and by a lightning impulse voltage test (15 impulses).

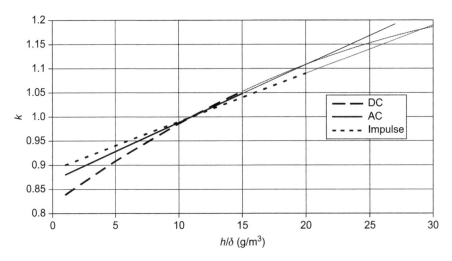

Figure 16.16 Value of k as a function of the humidity h related to the air density δ

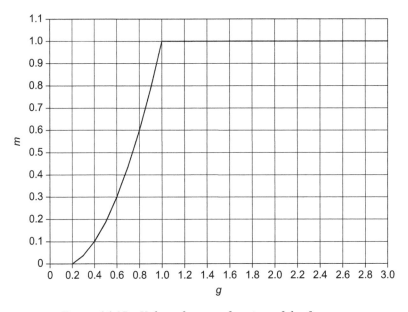

Figure 16.17 Value of m as a function of the factor g

For equipment with a rated voltage above 300 kV the performance under operating or temporary overvoltage at power frequency can be checked by an extension of the test time in order to demonstrate the suitable design regarding ageing and pollution. The performance under impulse voltage can be checked by the relevant impulse tests, lightning and switching impulse. It should be taken into account that the ratio between the rated switching impulse withstand voltage and

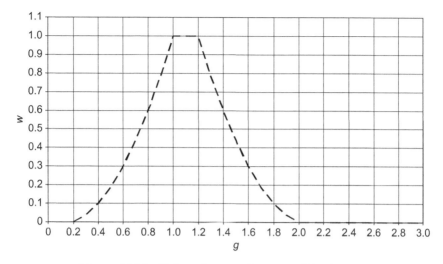

Figure 16.18   Value of w as a function of the factor g

Table 16.1   Standard insulation levels for 52 kV ≤ U_m ≤ 300 kV

Highest voltage for equipment U_m (rms)	Base for pu values $U_m \sqrt{2}/\sqrt{3}$ (peak)	Rated lightning impulse withstand voltage (peak)	Rated power-frequency short duration withstand voltage (rms)
kV	kV	kV	kV
52	42.5	250	95
72.5	59	325	140
123	100	450	185
145	118	550	230
170	139	650	275
	200	750	325
245	200	850	360
	200	950	395
	200	1,050	460

the rated power frequency voltage decreases with increasing rated power frequency voltage from 3.47 to 2.08 in extreme cases. The ratio between the lightning and switching voltage is more or less stable and independent of the voltage level if the lowest lightning impulse voltage is taken into account. The ratio increases with increasing lightning impulse level at the same switching impulse level. The lowest ratio is 1.09 and the highest 1.55.

The use of higher voltage above 800 kV, usually named ultra high-voltage (UHV), requires the extension of the insulation coordination up to 1200 kV level. The recommended voltage levels are shown in Table 16.3.

Table 16.2 Standard insulation levels for $U_m \leq 300$ kV

Highest voltage for equipment U_m (rms)	Base for pu values $U_m \sqrt{2}/\sqrt{3}$ (peak)	Rated switching impulse withstand voltage (peak)		Ratio between rated lightning and switching impulse withstand voltage	Rated lightning impulse withstand voltage (peak)
kV	kV	pu	kV		kV
300	245	3.06	750	1.13	850
				1.27	950
	245	3.47	850	1.12	950
				1.24	1,050
362	296	2.86	850	1.12	950
				1.24	1,050
	296	3.21	950	1.11	1,050
				1.24	1,175
420	343	2.76	950	1.11	1,050
				1.24	1,175
	343	3.06	1,050	1.12	1,175
				1.24	1,300
				1.36	1,425
525	429	2.74	1,175	1.11	1,300
				1.21	1,425
				1.32	1,555
	625	2.08	1,300	1.10	1,425
				1.19	1,550
				1.38	1,880
765	625	2.28	1,425	1.09	1,550
				1.26	1,880
				1.47	2,100
	625	2.48	1,550	1.16	1,880
				1.26	1,950
				1.55	2,400

It should be taken into account that the ratio between the rated switching impulse withstand voltage and the rated power frequency voltage is more or less stable and do not decrease with increasing rated power frequency voltage as for voltage levels from 300 to 765 kV. Also the ratio between the lightning and switching voltage is more or less stable and independent of the voltage. The lowest ratio is 1.25 and the highest 1.47.

There are many activities within CIGRE and IEC to check the relevant recommendations for use at UHV. The most important points from the test technique point of view are the requirements of the overshoot and front time due to the

Table 16.3 Standard insulation levels for $U_m \leq 800$ kV

Highest voltage for equipment U_m (rms)	Base for pu values $U_m \sqrt{2}/\sqrt{3}$ (peak)	Rated switching impulse withstand voltage (peak)		Ratio between rated lightning and switching impulse withstand voltage	Rated lightning impulse withstand voltage (peak)
kV	kV	pu	kV		kV
800	653	2.0	1,300	1.29 1.38	1,675 1,800
		2.18	1,425	1.26 1.37	1,800 1,950
		2.37	1,550	1.26 1.35	1,950 2,100
1,100	900	1.58	1,425	1.37 1.47	1,950 2,100
		1.72	1,550	1.35 1.45	2,100 2,250
		1.86	1,675	1.34 1.43	2,250 2,400
		2.0	1,800	1.33 1.42	2,400 2,550
1,200	980	1.71	1,675	1.25 1.34	2,100 2,250
		1.84	1,800	1.25 1.33	2,250 2,400
		1.99	1,950	1.31 1.38	2,550 2,700

large test circuits, the wet tests, the time to peak of the switching impulse voltage due to the breakdown voltage minimum as a function of the time to peak for large air gap distances and the atmospheric and altitude correction factors due to the limitation of the existing one up to a certain altitude. Furthermore the linearity check of the measuring devices for AC, DC and impulse voltage becomes more important due to the increasing ratio between the rated voltage of the measuring device under test and the available reference measuring devices.

16.6 Summary

- Insulation co-ordination describes the relationship between the different kinds of test voltages at different voltage levels.
- DC voltages are generated by a multi-stage rectifier. Voltage drop and ripple can be calculated with simple equations.
- AC voltages are generated by transformers or by a transformer cascade. Capacitive loads lead to a voltage increase, but this phenomenon can be used to generate high peak values with a resonance test set-up.

- Impulse voltages are generated by a multi-stage Marx generator with variable resistors to form the shape of the impulse. The same generator can produce lightning as well as switching impulses.
- Impulse currents are generated also by capacitors, but the inductance of the circuit should be kept as small as possible.
- Test conditions should take into account the correction factors and their influence on the disruptive discharge voltage.

References

1. 'Testing DC Extruded Cable Systems for Power Transmission up to 250 kV', CIGRE Brochure 219, 2003
2. IEC 60071-1 1996. 'Insulation Co-ordination. Part 1: Definitions, Principles and Rules'. IEC 60071-2 1996. 'Insulation Co-ordination. Part 2: Application Guide' (see also references 14, 15 of Chapter 2 of this book)
3. IEC 60060-1 2010. 'High Voltage Test Technique. Part 1: General Definitions and Test Requirements'. IEC 60060-2 2010. 'High Voltage Test Techniques. Part 2: Measuring Systems'
4. IEC 60076-3 2000. 'Power Transformers. Part 3: Insulation Levels, Dielectric Tests and External Clearances in Air'
5. IEC 60076-4 2002. 'Power Transformers. Part 4: Guide to the Lightning Impulse and Switching Impulse Testing – Power Transformers and Reactors'
6. IEC 60230 1966. 'Impulse Tests on Cables and their Accessories'
7. IEC 62475 2010. 'High-Current Test Techniques: Definitions and Requirements for Test Currents and Measuring Systems'

Chapter 17
Partial discharge measuring techniques
Ernst Gockenbach

17.1 Introduction

A partial discharge (PD) is a localised electrical discharge that only bridges the insulation between conductors and which may or may not occur adjacent to a conductor. This definition is given in the IEC Standard 60270 [1]. Partial discharges are in general a consequence of local electrical stress in the insulation or on the surface of the insulation. The discharges appear normally as pulses with duration of less than 1 µs. Also so-called continuous pulse-less discharges exist in gaseous dielectrics, but this kind of discharges will not be handled in this contribution. Furthermore, the term 'corona' is often used for the partial discharges that occur in gaseous media around conductors which are remote from solid or liquid insulation and therefore corona is a particular kind of partial discharges.

Partial discharges are a sensitive measure of local electrical stress and therefore the measurement is very often used as quality check of the insulation. The inception of partial discharges gives information on the limit of the electrical strength of the insulating material before a complete discharge between the conductors takes place. Therefore, the insulating material can be tested with a high stress but without damaging or reducing the performance of the insulation. Also at partial discharge measurements it should be taken into account that every stress of the insulation will have an influence of the life expectancy of the material, but a reasonable compromise between the stress during the measurement in order to get reliable results and the influence of the life time should be found and fixed in the relevant standards for the particular equipment, for example transformers, cables, switchgears, etc.

The partial discharge measurement is a typical non-destructive test and it can be used to judge the insulation performance at the beginning of its service time taking into account the reduction of the performance during the service time by the ageing whereby the ageing depends on numerous parameters like electrical stress, thermal stress and mechanical stress. Depending on the kind of insulating material different limits for the allowed partial discharge value at a given stress are defined in the relevant recommendations. In particular for solid insulation where a complete breakdown damages seriously the test object the partial discharge measurement is a tool for the quality assessment.

17.2 Physical background of partial discharges

Partial discharges occur at locations where the electrical stress exceeds the limit of the insulating material. These limits depend on different parameters like kind of material, temperature, pressure, duration of stress, purity, etc. In any case the inception of partial discharges at a given applied voltage demonstrates that at least locally a strength limit has been reached. Depending on the insulating material the voltage level at which partial discharges occur and the amount of charges allow a judgement of the performance and quality of the insulation to be tested based on experience values. The most critical insulation is the solid material and therefore the examples are based on that. The partial discharge measurement on a solid insulation should demonstrates that under a given stress the insulation will withstand the service conditions for, for example, 30 years. It is therefore necessary to know the change of performance due to ageing under the service conditions. Figure 17.1 shows a simplified diagram of the electrical performance of a solid insulating material as function of the service stress and time, i.e. in the format U_B versus stress duration [time, t].

The electrical breakdown takes place at the beginning of the stress duration, followed by the so-called thermal breakdown and the later erosion breakdown, mainly caused by partial discharges which needs a long time until the complete discharge between the two conductors occurs. It should be noted that the time axis in Figure 17.2 is not linear due to the different time ranges for the different types of breakdown. For high-voltage equipment very often insulating material is used where the area of thermal breakdown does not exist, because the power losses are very small. For these materials, the breakdown voltage behaviour as a function of U_B versus stress duration, [time t] is shown in Figure 17.2.

There is no clear separation between the electrical and erosion breakdown and the experiences show that this curve can be described by the so-called life time law

$$E^N t = \text{const} \tag{17.1}$$

where E is the electrical field stress, N the life time coefficient and t the stress duration.

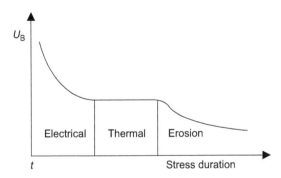

Figure 17.1 Breakdown voltage as function of the stress duration

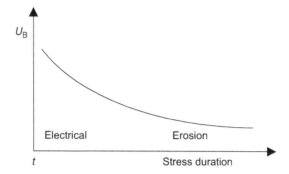

Figure 17.2 Breakdown voltage as function of stress duration for insulating material with very low power losses

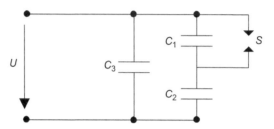

Figure 17.3 Equivalent circuit for the partial discharges
U = applied voltage at power frequency; C_1 = capacitor representing the cavity; C_2 = capacitor representing the insulating material around the cavity; C_3 = capacitor representing the remaining insulating material; S = spark gap representing the discharge of the capacitor C_1

The effect of the volume on the breakdown strength of solid material can also be taken into account by modification of this equation. The strong advantage of this equation is the fact, that with a reasonable voltage stress duration an estimation of the expected life time can be made and that a partial discharge measurement at the beginning of the stress duration gives a reliable information concerning the behaviour of the material in the time range where the complete breakdown is caused by erosion.

The detection and measurement of partial discharges are based on the assumption, that in the insulating material small cavities exist in which the discharge takes place. This discharge causes a charge transfer in the whole circuit by an impulse-shaped current which can be detected and measured. The relation of the discharge event and the phase angle of the applied voltage at power frequency are also important for the partial discharge measurement, because the nature of the defect can be determined. In Figure 17.3 a simplified equivalent circuit for the explanation of the partial discharges is shown.

The voltage across the capacitor C_1 is given by the following equation:

$$U_1 = U \frac{C_1}{C_1 + C_2} \tag{17.2}$$

The discharge of the spark gap S represents the partial discharge and this takes place when the voltage U_1 has reached the so-called inception level of partial discharges. Under ideal conditions the discharge process of the spark gap S needs no time, but for real arrangements the discharge process is still very fast and can be neglected in regards to the timescale of the applied voltage. Due to the discharge of the capacitor C_1 the capacitor C_2 is now stressed with the voltage U and this causes a change of the charges. Before the breakdown of the spark gap S the charge Q_2 of the capacitor C_2 is

$$Q_2 = U_2 C_2 = U \frac{C_1 \times C_2}{C_1 + C_2} \tag{17.3}$$

after the breakdown the charge Q_2^* is

$$Q_2^* = U C_2 \tag{17.4}$$

The charge difference should be delivered by the capacitor C_3 and or by the voltage source. Due to the different time constants the charge will be delivered by the capacitor C_3 in a very short current pulse. This current pulse causes a reduction of the voltage across the capacitors C_2 and C_3 and then the power supply reacts and the voltage will be increased up to the original value U by charging all the capacitors.

More or less at the same time the spark gap S has been recovered and the capacitor C_1 will be charged again until it reaches the breakdown voltage of the spark gap S. The charge difference can be calculated according to the equation

$$\Delta Q = U \frac{C_1^2}{C_1 + C_2} \tag{17.5}$$

Due to the fact, that neither C_1 nor C_2 is known, only the short current pulse through a coupling device or impedance can be measured. Therefore, the capacitor C_3 is normally replaced by a coupling capacitor C_k as shown in Figure 17.4 where the capacitor C_a represents the whole capacitor arrangement of Figure 17.3.

The partial discharge created as a result of the discharge of capacitor C_1 causes an impulse current flowing in the circuit given by the test object C_a, the coupling capacitor C_k and the coupling device CD respectively the impedance Z_{mi}. The integration of this current gives the measured charge. This circuit is normally known as series coupling circuit because the coupling device and the capacitors are connected in series regarding the current flow due to the partial discharge. From the

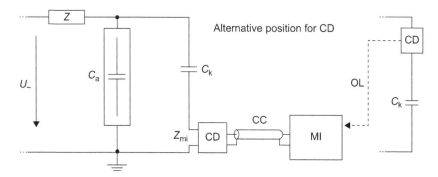

Figure 17.4 Basic partial discharge measurement circuit
$U_\sim$ = high-voltage supply; CC = connecting cables; C_k = coupling capacitor; MI = measuring instrument; Z_{mi} = input impedance of measuring system; C_a = test object (C_1, C_2 and C_3 of Figure 17.3); CD = coupling device; Z = high-voltage filter

measuring point of view it makes no difference if the coupling device is in series with the test object C_a, but in this case the test object should be isolated from ground and this is very often not possible.

The most important point of the partial discharge measurement is the calibration. Normally only the terminals of the test object are accessible and the charge which can be measured there is defined as the apparent charge q according to IEC Standard [1]. A partial discharge pulse with a unipolar charge q, injected at the terminals of the test object in the specified test circuit within a very short time, would give the same reading on the measuring instrument as the partial discharge pulse in the insulation material itself. The apparent charge is usually expressed in Pico-Coulombs (pC). It is clear that the measured apparent charge is not equal to the amount of charge locally involved at the discharge of the capacitor C_1 in Figure 17.3 and there is no more information on the amount of charges available as it can be measured on the accessible terminals of the test object.

The relation between the occurrence of partial discharges and the phase of the applied voltage depends on the type of defect. Assuming a sharp point at the high-voltage electrode of an air-insulated arrangement the partial discharges start when the voltage at the sharp point reaches the negative peak value due to the electron behaviour. With a voltage increase the partial discharges occur also at positive peak value. Therefore, the location of a sharp point generating partial discharges can be easily done. More complicated is the situation in solid materials where small cavities (capacitor C_1 in Figure 17.3) discharge (breakdown of spark gap S in Figure 17.3) and the pulses are measured in a test circuit according to Figure 17.4.

The following diagram (Figure 17.5) shows the typical partial discharge behaviour of a solid material assuming that the discharge voltage is constant and

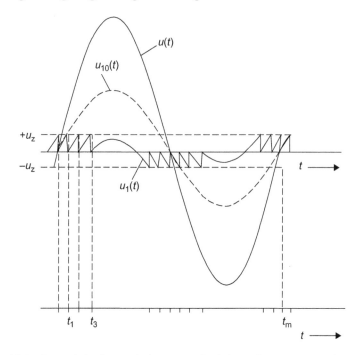

Figure 17.5 Partial discharge behaviour of solid insulating material
$u(t)$ = applied voltage at power frequency; $u_{10}(t)$ = voltage across the capacitor C_1 without partial discharges; $u_1(t)$ = voltage across the capacitor C_1 with partial discharges; u_z = partial discharge onset voltage; t_i = partial discharge occurrence time

polarity independent and that the former breakdowns do not influence the breakdown voltage of the cavity.

It can be seen that in this case the partial discharge occurs around the zero crossing point of the applied voltage which is completely different from partial discharges in air. The number of pulses depends on the breakdown voltage of the cavity and may be different for the two polarities in real insulating arrangements. Furthermore, the amount of charge is also not constant for all pulses, and therefore some other important parameters should be measured or calculated.

17.3 Requirements on a partial discharge measuring system

The most important parameter of the partial discharge measurements is the apparent charge q, because this value gives a reference for the specified test circuit and test object after calibration. Due to the sporadic appearance of the partial discharges an averaging of the measured pulses by the recording device is necessary in order to prevent a wrong measurement only due to a single event. This is in particular important under real measuring conditions where a number of different noise sources influence the partial discharge measurement. The pulse repetition rate n is given by the number of partial discharge pulses recorded in a selected time interval and the duration of this time interval. The recorded pulses should be above a limit, depending on the measuring system as well as on the noise level during the measurement. The pulse repetition frequency N is the number of partial discharge pulses per second in the case of equidistant pulses. Furthermore, the phase angle Φ_i and the time of occurrence t_i are information on the partial discharge pulse appearance in relation to the phase angle or time of the applied voltage with the period T.

$$\Phi_i = 360\left(\frac{t_i}{T}\right) \tag{17.6}$$

The reference point is the positive going transition of the applied voltage.

The interpretation of the partial discharge measurements requires very often further derived quantities like discharge power P, quadratic rate D and average discharge current I, which are described in detail in [1].

For measurement purposes the largest repeatedly occurring partial discharge magnitude and the specified partial discharge magnitude are also important. For partial discharge measurements with voltage at power frequency the specified magnitude of the apparent charge q is the largest repeatedly occurring partial discharge magnitude.

The performance of an insulating material can be evaluated also by the partial discharge inception voltage U_i and the extinction voltage U_e. The inception voltage is the applied voltage at which repetitive partial discharges at first are observed when the voltage is gradually increased from a lower level where no partial discharges occur. The extinction voltage is the applied voltage at which repetitive partial discharges cease to occur when the voltage is gradually reduced from a higher level at which partial discharges are observed. This value is in practice the voltage level at which the specified magnitude of partial discharges is below the specified low value or the limit of the recording system.

For reference measurements a partial discharge test voltage level is specified as a voltage level at which the test object should not exhibit partial discharges exceeding the specified partial discharge magnitude applied in a specified partial discharge test procedure.

Due to the numerous noise sources the partial discharge measurement should be carried out in a frequency range where the noise is as low as possible. The short and small current impulses should not be disturbed by the power frequency or its harmonics. Therefore, lower and upper frequency limits should be defined at which the transfer characteristique of the coupling device has fallen by 6 dB from the peak pass-band value. The measurement of a partial discharge pulse is an integration of the current over the time and the transformation from the time to the frequency domain means that the charge of a pulse is given by the amplitude content at the frequency zero. As long as the frequency response of the coupling device and

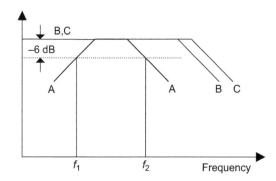

*Figure 17.6 Frequency response of a measuring system and pulses
A = band pass of the measuring system; B = frequency spectrum of the partial discharge pulse; C = frequency spectrum of a calibration pulse; f_1 = lower limit frequency; f_2 = upper frequency limit*

moreover of the complete measuring system is flat, the information concerning the charge of a pulse can be taken at any frequency. However, the noise reduction may be more efficient in a certain frequency range and therefore frequency response of the measuring system should be appropriate. Figure 17.6 shows an example of a characteristique frequency response.

The midband frequency f_m and the bandwidth Δf are given by the following equations:

$$f_m = \frac{f_1 + f_2}{2} \tag{17.7}$$

$$\Delta f = f_2 - f_1 \tag{17.8}$$

The characterisation of the measuring circuit concerning the frequency range is done by its transfer impedance $Z(f)$ which gives the ratio of the output signal amplitude to a constant input signal amplitude as function of the frequency assuming that the input signal is sinusoidal.

The pulse resolution time T_r is the shortest time interval between two consecutive pulses of short duration, same shape, polarity and charge magnitude for which the response will not change by more than 10% of that for a single pulse. The pulse resolution is in general inversely proportional to the bandwidth Δf.

17.4 Measuring systems for apparent charge

It is recommended in [1] that the relevant technical committees should use the apparent charge as the quantity to be measured and that the test circuit and the measuring system shall be calibrated. In order to measure the possible partial discharges within the test object it is necessary that the complete test circuit is free of partial discharges which means that the level of partial discharges within the test circuit is below the required partial discharge magnitude at the relevant applied voltage level. This needs a proper design of the test equipment, the high voltage connections and a proper grounding of all apparatus in the vicinity of the test circuit. Furthermore, very sensitive measurements need also a proper shielding of the whole test arrangement against electromagnetic interferences.

The coupling device is normally a passive four-terminal network which converts the current pulse to a voltage signal and its frequency response is chosen to prevent the power frequency and its harmonics from the recording instrument.

Generally the response of the recording system is a voltage pulse where the peak value is proportional to the charge of the current impulse. The shape, duration and peak value of the voltage pulse are determined by the transfer behaviour of the measuring system and shape and duration may be completely different from the original current pulse. In order to get a reasonable measurement an integration of the recorded pulses should be done in such a way that the indication of the largest repeatedly occurring partial discharge magnitude is according to Table 17.1 for equally large equidistant pulses of a given charge.

This characteristic is necessary in order to get the same readings with different types of instruments including digital recording systems and to establish compatibility of partial discharge measurements.

Regarding the frequency range two types of systems exist, the wide-band and the narrow-band partial discharge measuring systems. The recommended values for a wide-band system according to [1] are:

$30 \text{ kHz} \leq f_1 \leq 100 \text{ kHz}$
$f_2 \leq 500 \text{ kHz}$
$100 \text{ kHz} \leq \Delta f \leq 400 \text{ kHz}.$

The response of such a system to a partial discharge current pulse is in general a well-damped oscillation. The apparent charge q and the polarity of the partial discharge pulse can be determined and the pulse resolution time T_r is small, typically in the range of 5–20 μs.

The narrow-band measuring is characterised by a small bandwidth Δf and a midband frequency f_m. Recommended values are:

$9 \text{ kHz} \leq \Delta f \leq 30 \text{ kHz}$
$50 \text{ kHz} \leq f_m \leq 1 \text{ MHz}$

Table 17.1 Reading R in % as function of the pulse repetition rate N

N (1/s)	1	2	5	10	50	≥100
R_{min} (%)	35	55	76	85	94	95
R_{max} (%)	45	65	86	95	104	105

The response of such a system to a partial discharge current pulse is a transient oscillation with the negative and positive peak values of its envelope proportional to the apparent charge q independent of the polarity of this charge. The pulse resolution time T_r is large and typically above 80 μs.

17.5 Calibration of a partial discharge measuring system

The aim of the calibration is to verify that the measuring system is able to measure the specified partial discharge magnitude correctly. With the calibration of the measuring system within the complete test circuit the scale factor for the measurement of the apparent charge q will be determined. Due to the influence of the test object the calibration shall be made with each new test object.

The calibration is carried out by injecting a short-duration pulse of known charge magnitude q_0 into the terminals of the test object. The reading of the instrument at this given charge can then be adjusted in order to have a reasonable scale factor for the later partial discharge measurements. The calibration should be done for each used measuring range.

The calibration pulses are generally derived from a calibrator producing a voltage step of the amplitude U_0 in series to a capacitor C_0 so that the repetitive charges have the magnitude

$$q_0 = U_0 C_0$$

The rise time of the voltage step shall be very small and not exceed 60 ns. The detailed requirements on the calibrator as well as on the procedure are described in [1].

17.6 Examples of partial discharge measurements

The relevant technical committees should specify the minimum measurable magnitude required for the apparatus [2–4]. In order to obtain reproducible results in partial discharge measurements careful control of all relevant factors is necessary, for example the surface of the external insulation of the test object shall be clean and dry because moisture or contamination on insulating surfaces may cause partial discharge at the required test voltage and then internal partial discharge cannot be detected. Normally the partial discharge magnitude is determined at a specified test voltage. If this magnitude is below the required value the test object has passed the test. Very often the partial discharge inception and extinction voltage are of interest. Then the voltage has to be increased and decreased according to the recommendations. These tests are simple if the background noise is low enough to permit a sufficient and accurate measurement which is the case in well-shielded high-voltage laboratories. If the measurements should be carried out on-site under normally very noisy conditions, then more sophisticated measuring systems should be used and also more effort in data processing may be necessary.

17.6.1 Partial discharge measurement on high-voltage transformers

The partial discharge measurement on a transformer on-site needs in any case a suppression of different kinds of noises. The partial discharge sensors may be capacitors or Rogowski coils depending on the actual situation and the measuring purpose. Figure 17.7 sows the original measured signal during a partial discharge measurement on a transformer. Only after the filtering of sinusoidal noise the partial discharge pulses can be recognised including the calibration signal at about 60 ms.

The filtered signal has still a too high noise level, and therefore further filtering methods reducing the synchronous noise are necessary. The impulse-shaped noise remains because the signals differs not from the true partial discharge signal, and therefore the signal direction may be used for the separation between noise and partial discharge signal as shown in Figure 17.8.

17.6.2 Partial discharge measurement and location on high-voltage cables

The partial discharge measurement on high-voltage cables requires a low background noise, because the specified partial discharge magnitude for XLPE cables is very low. Already the partial discharge measurements within the factory are not simple because special cable terminations are necessary which should be partial discharge 'free' at the test voltage. With the appropriate equipment and test conditions (shielded laboratory) the required low background noise can be reached. On-site partial discharge measurements are more critical and the lowest detectable partial discharge magnitude in the test object is higher than in the factory.

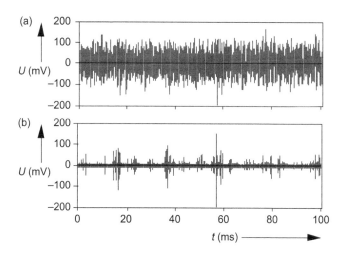

Figure 17.7 Partial discharge measuring signal without (a) and with filtering of sinusoidal noise (b)

610 High-voltage engineering and testing

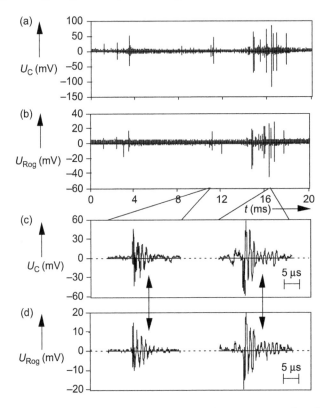

Figure 17.8 Example of more advanced partial discharge signal techniques
U_c = output of the capacitive coupling unit, see (a) and (c)
U_{Rog} = output of the Rogowski coil, see (b) and (d)
(a,b) Partial discharge measuring signal after filtering of sinusoidal and synchronous noise
(c,d) Partial discharge measuring signal after filtering of sinusoidal and synchronous noise and with additional filtering of impulse shaped noise by the detection of signal direction, which is used for the separation between noise and partial discharge signals

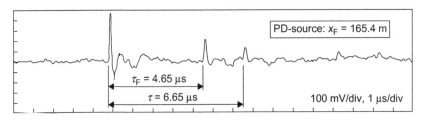

Figure 17.9 Partial discharge measurement and location on a cable on-site

Data processing can reduce the noise level and the particular behaviour of the cable (travelling waves) allows the location of the partial discharge within the cable length. Figure 17.9 shows the result of a partial discharge measurement on a high-voltage cable on-site. The sensor is a Rogowski coil mounted on one cable termination. From the digital record the travelling time of the original signal τ as well as of the reflected signal at the opposite end of the cable τ_F of a single partial discharge can be evaluated. With the known length of the cable the location of the partial discharge can be calculated.

17.6.3 Partial discharge measurement on high-voltage gas-insulated substations

Gas-insulated substations and their components are tested within the factory and this includes partial discharge measurements on the insulating parts and the complete equipment. Anyhow some assembling work is necessary on-site, and therefore a voltage test with AC including partial discharge measurement is the most reliable test to detect weak points concerning the electrical behaviour. Due to the problem of the high noise level on site the partial discharge measurement at ultra-high frequencies (UHF) seems to be the most sensitive partial discharge measurement method. The already described conventional partial discharge method can also be used, but the very low value of the apparent charge for a typical failure in a gas-insulated substation, a needle on the high-voltage electrode, may be not detectable with this method. Therefore, the example shows the result of a partial discharge measurement of a typical gas-insulated substation arrangement [5]. Figure 17.10 shows the test set-up with the position of the failure (free moving particle) and the position of the different UHF partial discharge sensors.

Figure 17.11 shows the measured signal at different positions in order to show the strong dependence of the signal amplitude on the distance between the failure and the sensor position and to demonstrate the high sensitivity of this method.

The partial discharge signal can be still detected with sensor 8 which is about 20 m from the failure position. The only disadvantage of this method is the fact, that the measurement cannot be calibrated according to the relevant IEC Standard [2], but the sensitivity can be easily checked according to a proposal of the CIGRE [6]. A disadvantage of this method is the fact that the measurement can not be calibrated according to the relevant IEC Publ. [2]. However, this is not a problem as the sensitivity can be easily checked in accordance with a CIGRE [6] proposal. With this sensitivity check, it can readily be established whether or not defects producing a partial discharge level above the given value exist within the gas-insulated substation section(s) being tested. Consequently, this partial discharge site testing technique provides strategic information to make the judgement to determine whether the substation equipment can continue in service or whether appropriate remedial action is required.[1]

---

[1] The reader should note that Chapters 20, 21 and 22 also consider condition monitoring in service for a variety of commercial equipment.

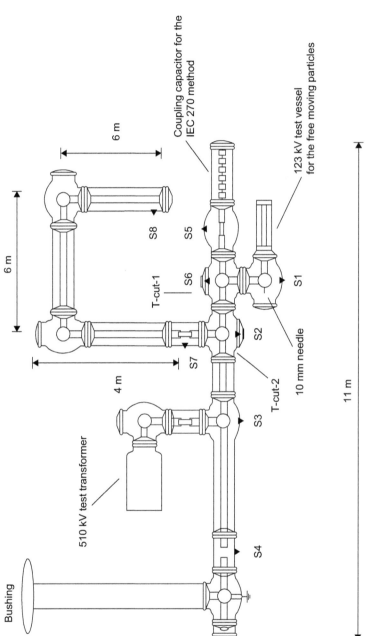

Figure 17.10 Set-up for a UHF partial discharge measurement on a gas-insulated substation

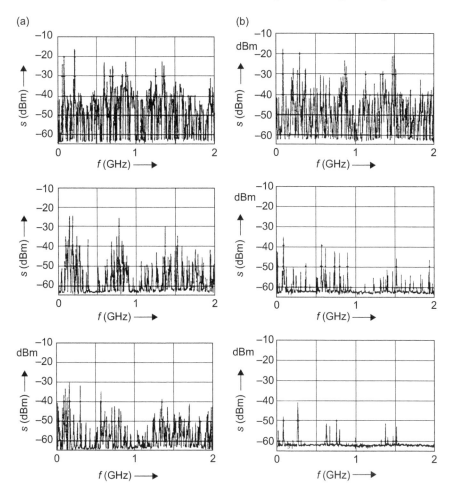

Figure 17.11 Partial discharge signals at different sensors
(a) Partial discharge of a 10 mm free moving particle measured at S1, S3 and S4 (top to bottom)
(b) Partial discharge of a 5 mm free moving particle measured at S1, S4 and S8 (top to bottom)

17.6.4 Development of recommendation

There are activities within IEC Technical Committee 42 'High voltage and current test techniques' to integrate the digital measuring technique in the partial discharge measurement. A maintenance team of TC 42 MT 17 is working on an addendum to the IEC 60270 High-voltage test techniques – partial discharge measurements [2] regarding the implementation of the evaluation of digital recorded partial discharge signal. The digital measuring technique allows the record of every pulse, and therefore the value of the apparent charge as an average value should be more clearly defined.

Furthermore a Working Group within IEC TC 42 is dealing with new methods like UHF measurement and acoustic measurement of partial discharges. Both methods suffer from the situation that no relation exists with the charge amount or/and the measured signal. Here only sensitivity checks are possible, but the methods are still helpful and efficient. The outcome of this work will be an IEC Technical Report IEC 62478 'High-Voltage Test Techniques – Measurement of Partial Discharge by Electromagnetic and Acoustic Methods'.

17.7 Summary

- Partial discharge measurements are a very sensitive tool for the quality check of insulating materials.
- The measurement of the apparent charge is a relative measurement and requires the references values given by the standards and based on practical experiences.
- The main problem of partial discharge measurements is the separation between noise and partial discharge signals, but with a proper selection of the bandwidth and the use of intelligent procedures like filtering the measurements can be carried out also under on-site conditions.
- The examples of partial discharge measurements on high-voltage cables, transformers and gas-insulated substations demonstrate the benefit of the partial discharge measuring technique.

References

1. IEC 60270 2000, 'High-Voltage Test Techniques: Partial Discharge Measurement'
2. IEC 60076-3 2000 'Power Transformers – Part 3: Insulation Levels, Dielectric Tests and External Clearances in Air'
3. IEC 60840 2011, 'Power Cables with Extruded Insulation and their Accessories for Rated Voltages above 30 kV ($U_m = 36$ kV) up to 150 kV ($U_m = 170$ kV) – Test Methods and Requirements'
4. IEC 62271 2011, 'High-Voltage Switchgear and Controlgear – Part 203: Gas-insulated Metal-enclosed Switchgear for Rated Voltages above 52 kV'
5. Kurrer R., Feser K. 'Attenuation measurements of ultra-high-frequency partial discharge signals in gas-insulated substations', *10th international symposium on high voltage engineering*; Montreal, 1997. pp. 161–4
6. Task Force 15/33.03.05 of Working Group 15.03 (Insulating Gases) on behalf of SC 15 (Materials for Electrotechnology) and SC 33 (Power System Insulation Coordination). 'Partial discharge detection system for GIS: Sensitivity verification for the UHF method and the acoustic method'. *Electra*. 1999;**183**: 75–87

Chapter 18
Digital measuring techniques and evaluation procedures
Ernst Gockenbach

18.1 Introduction

The introduction of digital recording system in the high-voltage measuring technique has a great influence on the measuring technique and the evaluation procedures. The formerly used analogue recording instruments were developed for high-voltage measurements and in particular for impulse measurements under very noisy conditions or in other word for high electromagnetic interferences. One of the best measures against the electromagnetic interference was the high signal-to-noise ratio, reached by a high signal voltage level up to 1,500 V for impulse voltage measurements. This high signal level leads also to a high deflection level of about 100 V/cm and requires no amplifier within the analogue oscilloscope.

The use of digital recording devices in the high-voltage measurement requires in particular a number of measures against electromagnetic interferences. The recording device should be inside a Farady cage to prevent the influence of the electric field. The high signal level should be kept by using a so-called input divider inside the shielded cabin in order to reduce the voltage level from some hundred volts up to 2 V, which is the normally used input level of a digital instrument. This voltage divider could be used at the same time for the remote controlled change of the total impulse voltage divider ratio. Furthermore, the requirements on the digital recorder concerning resolution in amplitude and time should be equivalent or better to those of the analogue oscilloscope.

The digital recording devices offer a number of advantages compared to analogue systems. Due to the recording technique the problem of an exact triggering of the analogue oscilloscope does not any more exist, because the digital recorder records continuously the data and stops at a given and preselected time. Therefore, the history before the event takes place can be also recorded. The recorded data is available in digital form and zooming and compressing are easily possible. This means that a change of the timescale can be done by evaluation of the data and does not require another measurement as it was necessary with analogue recording devices assuming that the recorded data allows this procedure. The data can be evaluated automatically with a computer or with a built-in processor. The data can be directly used for documentation, for example for a figure in a test report. It is also possible to compare two measurements with different amplitudes by scaling the amplitudes. The use of mathematical procedures like Fourier transformations, filtering, neuronal networks, genetic algorithms, fuzzy logic, etc., allows further evaluation of the recorded data for a better judgement of the test results and for diagnostic.

18.2 Requirements on the recording device

It is obvious that the performance of a digital recording device should be at least equivalent to the performance of an analogue recording system. The measurement of direct current (DC) and alternating current (AC) voltage is simple and the requirements are very easy to fulfil; therefore, only the requirements for impulse measurements are described in the following part.

It should be taken into account that the recording technique of a digital recorder is different from the analogue oscilloscope and that no information is available between two samples. This means that high-frequency oscillations may be not recorded by a digital recording device, but at least indicated by an analogue oscilloscope, because the signal is continuously recorded and the brightness of the beam changes depending on the deflection velocity. The requirements on the digital recording device are given in the IEC Standard 61083 [1].

The requirements for digital recorders are different for the use in approved measuring systems or in reference measuring systems. Furthermore, the requirements are expressed as an overall uncertainty and individual requirements. A digital recorder used in an approved measuring system shall have an overall uncertainty of less than 2% in the peak voltage or current measurement of full and standard-chopped lightning impulses, switching impulses and rectangular impulses. The overall uncertainty for the peak voltage of front-chopped lightning impulses should be less than 3%. All time parameters (front time, time to chopping, time to peak, time to half value, etc.) shall be measured with an overall uncertainty of less than 4%. These overall uncertainties are usually reached if the individual requirements are fulfilled, but in some cases the individual requirements may be exceeded without exceeding the overall uncertainties. Therefore, the most important individual requirements are described.

The amplitude resolution is given by the rated resolution r, the nominal minimum increment of the input voltage which can be detected in the output voltage. The rated resolution is expressed by the reciprocal of 2 to the power of the rated number of bits N of the digital/analogue converter (digital recorder), namely

$$r = 2^{-N} \qquad (18.1)$$

For approved measuring systems a rated resolution of 2^{-8} (0.4% of the full-scale deflection) or better is required for tests where only the impulse parameter is to be evaluated. For tests which involve signal processing a rated resolution of 2^{-9} (0.2% of the full-scale deflection) or better is required. The full-scale deflection is the minimum input voltage which produces the nominal maximum output voltage in the specified range.

The time resolution is expressed by the number of the samples taken per unit. The sampling time is the time interval between two subsequent samples and the reciprocal of the sampling rate. The sampling rate shall not be less than $30/T_x$, where T_x is the time interval to be measured. For a standard lightning impulse 1.2/50 µs the measured time T_x is $0.6T_1$, the time interval between 30% and 90% of the peak value of the measured lightning impulse. With the lower tolerance of the front time $T_1 = 0.84$ µs the measured time T_x is about 500 ns which leads to a sampling rate of at least

$$\text{Sampling rate} > \frac{30}{500^{-9}\text{s}} = 60 \times 10^6 \frac{1}{\text{s}} \qquad (18.2)$$

For measurement of oscillations in the front the sampling rate should be in accordance with the maximum frequency reproduced by the measuring system.

The scale factor described the factor by which the output is multiplied to get the measured value of the input quantity. Digital recorders may have different amplitude scale factors for different input voltages, for example the static scale factor F_s for DC voltage and the impulse scale factor F_i for impulse voltages representing the shape of the relevant impulse. The timescale factor is defined as the factor by which the recorded time interval is multiplied to get the measured value of the time interval.

The rise time of the digital recorder shall be not more than 3% of the measured time interval T_x. For lightning impulse measurement this value shall be less than 15 ns in order to record the superimposed oscillations.

The interference shall be less than 1% of the amplitude of the expected deflection, the internal noise should not be less than 0.4% of the full-scale deflection for waveform parameter measurements and less than 0.1% of the full-scale deflection for measurements involving signal processing.

The record length shall be sufficiently long to allow the evaluation of the required parameter.

The non-linearity of amplitude and time base are given as integral and differential non-linearity. The static integral non-linearity of the amplitude shall be within ±0.5% of the full-scale deflection, and the differential non-linearity within ±0.8w_0 for static and dynamic tests, with w_0 as average code bin width (see Figure 18.2). The integral non-linearity of the time base shall be not more than 0.5% of the measured time T_x. Figure 18.1 shows the integral non-linearity, and Figure 18.2 shows the differential non-linearity and the code bin width $w(k)$.

The use of a 3-bit digital recorder in Figure 18.2 is only for clarification of the definitions. The differential non-linearity $d(k)$ can be calculated by the equation

$$d(k) = \frac{w(k) - w_0}{w_0} \qquad (18.3)$$

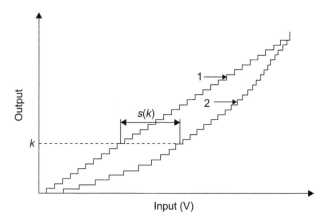

Figure 18.1 Integral non-linearity
1 = ideal 6-bit digital recorder; 2 = non-linear 6-bit digital recorder; k = relevant output level of the digital recorder; s(k) = difference between the input and output at the relevant level k

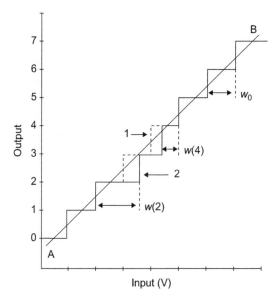

Figure 18.2 Differential non-linearity and code bin width w(k) under DC conditions
1 = ideal 3-bit digital recorder; 2 = large differential non-linearity at codes 2, 3 and 4; w_0 = average code bin width; AB = line between the midpoints of the code bins of an ideal digital recorder; w(2) = difference in code bin width at this level (longer as the average w_0); w(4) = difference in code bin width at this level (shorter as the average w_0)

Digital measuring techniques and evaluation procedures 619

The operating range should be not less than $4/N$ of the full-scale deflection, where N is the number of bits, and this means for 8-bit digital recorder that the peak amplitude should not be less than 50% of the full-scale deflection. For tests which require comparison of records the operating range of not less than $6/N$ is recommended.

For digital recorders used in reference measuring systems for the calibration of approved measuring systems the requirements are more stringent. The overall uncertainty for the peak voltage or current measurement of full and standard-chopped lightning impulses, switching impulses and rectangular impulses shall be not more than 0.7%. The overall uncertainty for the peak voltage of front-chopped lightning impulses should be not more than 2%. All time parameters (front time, time to chopping, time to peak, time to half value, etc.) shall be measured with an overall uncertainty of not more than 3%. The individual requirements are also adapted to the higher performance of the digital recorder.

To establish the impulse scale factor and to check the time parameter determination of an approved digital recorder a pulse calibration by a reference pulse generator is recommended, even if a step calibration is also mentioned in the relevant IEC Standard [1]. The calibration should be done with impulses of the same shape to be measured, and therefore a number of impulses are listed in the Table 18.1.

18.3 Requirements on the evaluation software

The replacement of the analogue oscilloscopes by digital recording devices is worthwhile only when the digital recorded data are used for an automatic

Table 18.1 Requirements for reference pulse generators

Impulse type	Parameter being measured	Value	Uncertainty (%)[a]	Short-time stability (%)[b]
Full and standard-chopped lightning impulse	Front time	0.8–0.9 μs	≤2	≤0.5
	Time to half value	55–65 μs	≤2	≤0.2
	Peak voltage	Within operating range	≤0.7	≤0.2
Front-chopped lightning impulse	Time-to-chopping	0.45–0.55 μs	≤2	≤1
	Peak voltage	Within operating range	≤1	≤0.2
Switching impulse	Time-to-peak	15–300 μs	≤2	≤0.2
	Time to half value	2,600–4,200 μs	≤2	≤0.2
	Peak voltage	Within operating range	≤0.7	≤0.2
Rectangular impulse	Duration	0.5–3.5 ms	≤2	≤0.5
	Peak value	Within operating range	≤2	≤1

[a]The uncertainty is determined in accordance with Annex H of IEC Standard 60060-2 [2] by a traceable calibration where the mean of a sequence of at least 10 pulses is evaluated.
[b]The short-time stability is the standard deviation of a sequence of at least 10 pulses.

evaluation by a computer. Therefore, a standard for the evaluation software was developed and in the IEC Standard 61083 Part 2 the requirements on the software used for the determination of the impulse parameters are described in detail [3]. The raw data are the original record of sampled and quantised information normally corrected by the off-set and multiplied by a constant factor (divider ratio) in order to get the output for example directly in kilovolts. All other data are called processed data. A test data generator (TDG) is available as computer program which generates reference test waveforms with the specified parameters of the digital recorder to be used for example sampling time, full-scale deflection, off-set, record length, polarity, trigger delay and noise. The output of the TDG has to be evaluated and the value of the parameters shall be within the given limits.

The existing impulses of the TDG are divided in groups of waveforms in order to check the software only for the relevant type of waveform. The software passes the test, if for all impulses within the relevant group all parameters to be evaluated are within the given limits. The waveforms are grouped in:

- full lightning impulse voltage (LI)
- front-chopped lightning impulse voltage (LIC)
- tail-chopped lightning impulse voltage (LIC)
- switching impulse voltage (SI)
- oscillating lightning impulse voltage with oscillation in the front (OLI)
- oscillating switching impulse voltage with oscillation in the front (OSI)
- exponential impulse current (IC)
- rectangular current (IC).

The uncertainty of the parameter evaluation of the reference test waveforms is approximately 1% for the peak value and 4% for the time parameter.

The reference test waveforms cover not all possible waveforms of impulse tests. In particular impulse tests on transformers or metal oxide arresters generate waveforms which are different from the reference test waveforms. It could be reasonable to prepare a check of the evaluation procedure for these type of waveforms.

18.4 Application of digital recording systems

18.4.1 DC and AC voltage measurements

The measurement of DC and AC voltage with a digital recording device is very simple and should not be explained in detail here. The advantage of a digital recording system is for example the monitoring of overvoltage superimposed on the DC or AC voltage or the evaluation of harmonics by Fourier analysis. The resolution in time and amplitude depends on the signal to be recorded and requires normally no high performance of the digital recorder. The advantage of a digital recording system is the possibility to trigger the device depending on the event to be recorded. A transient overvoltage superimposed on an AC voltage can be recorded by triggering the digital recording device depending on the frequency or the amplitude and the record includes the event as well as the history before the event.

18.4.2 Impulse voltage or current measurements

The measurement of impulse voltage or current is the most important application for digital recording systems in the field of high-voltage testing. The record of the impulse and the later evaluation of the impulse parameter can be easily done with the support of the digital recording system. The recording device which is in operation during the impulse generation is placed in a shielded cabin in order to prevent the influence of the electromagnetic field, generated by the discharge of the impulse capacitors. The recorded data can be shown with different time and amplitude scales in order to demonstrate details of the shape without any additional measurements. In Figure 18.3 a lightning impulse voltage is shown recorded with an 8-bit digital recorder at a sampling rate of 40×10^6 samples/s and with a record length of 4,096 samples equivalent to about 100 µs.

The same measurement is shown in Figure 18.4 but with another timescale in order to show the front of the measured impulse in detail.

The limits are reached if the zooming goes above the limits of the recording parameters that means for the impulse shown in Figure 18.2 a time resolution below 25 ns which is the time between two consecutive samples.

The digital recorded data allows also a comparison between two impulses. The adaptation of the amplitude can be done by a simple multiplication of the amplitude

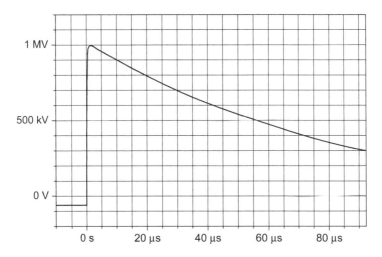

Figure 18.3 Standard lightning impulse

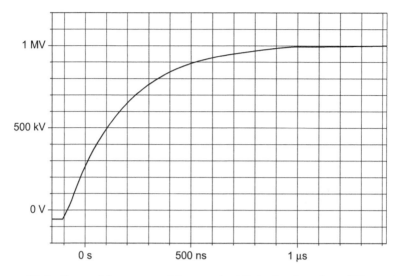

Figure 18.4 Zoomed front part of the standard lightning impulse of Figure 18.3

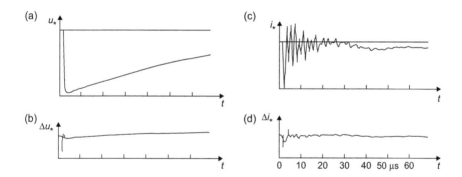

Figure 18.5 Comparison between two impulse voltage and current measurements
(a) $u_* =$ measured voltages; (b) $i_* =$ measured currents;
(c) $\Delta u_* =$ voltage difference between two impulses (zoom factor 8);
(d) $\Delta i_* =$ current difference between two impulses (zoom factor 8)

values with a constant factor. The adaptation of the time can be done by shifting the complete dataset in steps of the sampling time interval or even in parts of sampling time interval. Figure 18.5 shows a comparison between two voltage impulses at 62.5% and 100% together with the current measured at the neutral point of the transformer. The difference between the two measurements of voltage and current is so small that only an enlargement of the difference between the two signals by a factor 8 shows some small deviations at the triggering time.

Besides the comparison of two measurements a two-channel digital recording system can be used for the calculation of the transfer function, which represents the

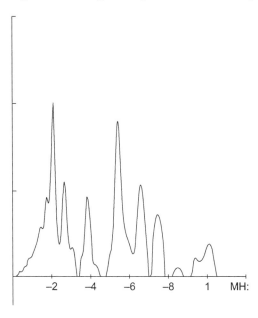

Figure 18.6   Transfer function of a transformer

ratio of the output signal to the input signal as function of the frequency, whereby the current is the output signal and the voltage is the input signal. Figure 18.6 shows the transfer function of a transformer with a standard lightning impulse as input signal.

18.4.3   Partial discharge measurements

A digital recording system can also be used for partial discharge measurements. The monitoring of all pulses allows the determination of the largest partial discharge impulse within a given time, the determination of the number of pulses per time, the determination of the apparent charge per pulse and the relation between the pulses and the phase of the applied power frequency voltage. A digital partial discharge measuring system includes normally filtering procedures in order to exclude noise from the measured signal and to measure only within a small frequency range, if it is necessary. Figure 18.7 shows a measured signal, where (a) the partial discharges cannot be recognised due to the high noise level, but (b) after filtering of the sinusoidal noise. The large impulse at about 60 ms is a calibration pulse of the measuring system.

Figure 18.8 shows in an enlarged timescale in the middle (b) and lower figure (c) the same signal shown in Figure 18.7(b) after filtering of the noise pulses which appear synchronous with the applied power frequency voltage. These procedures are only possible with digital recorded data due to the digital filtering for sinusoidal noise for Figure 18.7 and due to the comparison of signals recorded at the time phase angle for Figure 18.8, because all signals which appears at the same phase angle are declared as noise signals and set to zero.

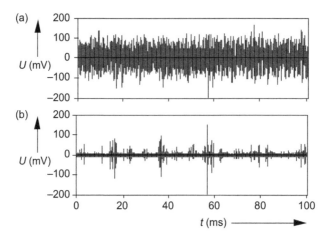

Figure 18.7 Partial discharge measuring signal without (a) and with filtering of sinusoidal noise (b)

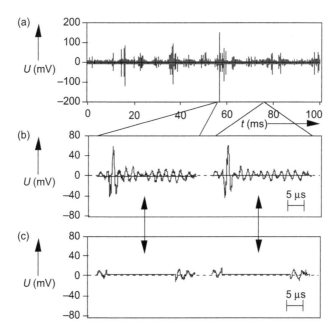

Figure 18.8 Partial discharge measuring signal after filtering of sinusoidal noise (a) and with zooming (b) and with additional filtering of synchronous noise (c)

This procedure is based on the assumption that partial discharge signals occur randomly and never at the exact same phase angle in two consecutive cycles. The most critical noises are impulses which occur randomly with a more or less same shape as the partials discharge signals. These signals can only be detected and removed by additional measures, for example by the detection of the signal travelling direction or by a bridge method. Therefore, it is necessary to measure at least at two different places or with two circuits. An example of a measurement of partial discharges on a transformer is shown in Figure 18.9. The upper two signals (a) and (b) show the already filtered measuring signal of partial discharges measured via a capacitive coupling device u_c and a so-called Rogowski coil u_{Rog}. The output voltage of a Rogowski coil is proportional to the derivative of the current signal and by an integration of this signal the output voltage is proportional to the apparent charge, which is similar to the capacitive coupling method. A zooming of the two signals allows a detection of the signal direction, because signals generated

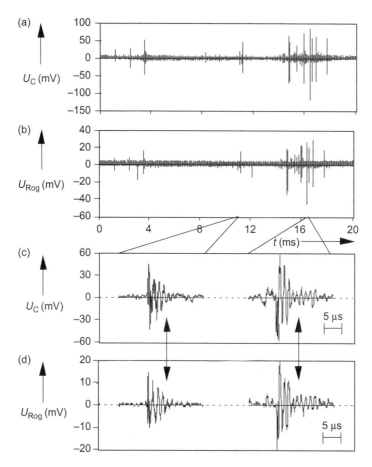

Figure 18.9 Partial discharge measuring signal after filtering of sinusoidal and synchronous noise measured via capacitor (a) or via Rogowski coil (b). Additional filtering of impulse-shaped noise and zooming of time scale depicts difference in the polarity of the measured partial discharge of the two recorded signals via capacitor (c) or via Rogowski coil (d)

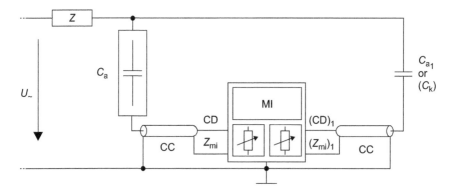

*Figure 18.10 Balanced circuit arrangement for partial discharge measurements
$U_\sim$ = high-voltage supply; Z_{mi} = input impedance of measuring system; CC = connecting cables; C_a = test object; C_k = coupling capacitor; CD = coupling device; MI = measuring instrument; Z = high-voltage filter*

by partial discharges outside the transformer pass the two measuring units in the same direction and signals generated by partial discharges inside the transformer pass the measuring units in the opposite direction. The first signal in Figure 18.9(c) and (d) has a first positive peak for u_c and at the same time a negative peak fur u_{Rog}, and the second signal has a first negative peak for both signals. Depending on the calibration of the measuring device one signal is a partial discharge outside the transformer and the other inside the transformer. With this method a discrimination between signals coming from outside the transformer and interpreted as noise and signals coming from inside the transformer and interpreted as partial discharges is possible.

The same procedure can be used when two more or less identical objects are available and the partial discharge measurement is done in a balanced circuit arrangement. Figure 18.10 shows such circuit, given in [4].

18.5 Application examples of evaluation procedures

The use of digital recorded data in many high-voltage test laboratories leads to extensive research activities regarding evaluation procedure for digital recorded impulse voltages. The main result of these research activities was the implementation of the so-called test voltage factor k, which allows a high reproducibility of the lightning impulse voltage with oscillation or overshoot near the peak. Beside the test voltage function a procedure was determined to evaluate a base curve, which is close to the mean curve of former recommendations. This mean curve was in the past only determined by a drawing in the relevant IEC recommendation (IEC 60060-1) [5]. Now the evaluation of the mean curve is clearly defined. The digital recorded data from 20% of the extreme value (highest recorded sample) in the front up to 40% of the extreme value in the tail should be used for curve fitting following the equation of two exponential functions

$$u(t) = U(e^{-(t/\tau_1)} - e^{-(t/\tau_2)}) \tag{18.4}$$

The off-set of the recorded data should be taken into account for calculating the zero line and the extreme value. There are some programs available for the curve fitting.

With (18.4) the base curve (former mean curve) is now well defined and the evaluation can start. In case of no overshoot or oscillations the smooth curve could be used to evaluate the 30%, 90% and 50% in order to calculate the front time and the time to half value.

In case of oscillation or overshoot near the peak the following procedure is now given and explained in details. Calculate the differences between the mean curve and the recorded curve and filter this residual curve by the test voltage function, developed in a European Research Programme [6], and then add the filtered residual curve to the mean curve. This curve is now representing the test voltage curve from which the test voltage (peak of the test voltage curve) and the time parameters can be evaluated [7, 8].

Figure 18.11 shows the recorded curve, the base curve and the unfiltered mean curve as well as the overshoot β. Figure 18.12 shows the base curve, the filtered residual curve and the sum of both curves, the test voltage curve.

To demonstrate the influence of the filtering both the recorded curve and the test voltage curve should be shown in one diagram. The test voltage factor can also be used for a manual evaluation using (18.5).

$$U_t = U_b + k(f)(U_e - U_b) \tag{18.5}$$

where U_b is the peak of the base curve and U_e the peak of the recorded curve (extreme value).

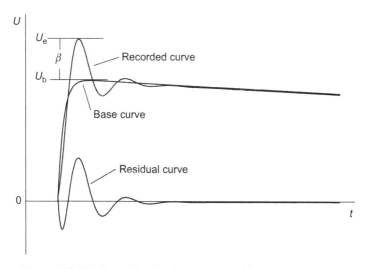

Figure 18.11   Recorded and base curve (former mean curve)

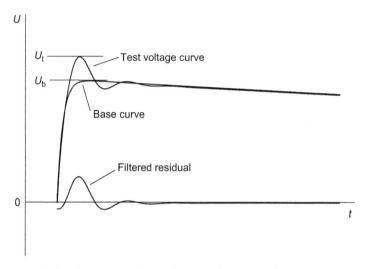

Figure 18.12 Base curve, filtered residual curve and test voltage curve

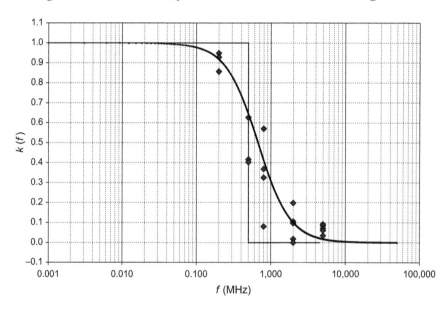

Figure 18.13 Test voltage function with experimental data

The value of $k(f)$ should be taken from Figure 18.13 which shows the experimental values as well as the curve fitting resulting in a simple equation (18.6) for the test voltage factor.

$$k(f) = \frac{1}{1 + 2.2f^2} \tag{18.6}$$

with f in MHz.

18.6 Conclusions

- The available digital recorders can be used for high impulse voltage or current measurements without any problems if the electromagnetic compatibility has been checked.
- The digital recording systems have large advantages in data acquisition and treatment for example impulse or partial discharge measurements.
- The evaluation procedures for smooth impulses are simple, but for impulses with oscillations or overshoot the relevant recommendations show procedures which allow impulse parameter evaluation with high reproducibility and robustness.
- The evaluation of the switching impulse voltage follows now the same procedure as for lightning impulse voltage.

References

1. IEC 61083-1 2001. 'Digital Recorders for Measurements in High-Voltage Impulse Tests, Part 1: Requirements for Instruments'
2. IEC 60060-2 2010. 'High-Voltage Test Techniques, Part 2: Measuring Systems'
3. IEC 61083-2 1996. 'Digital Recorders for Measurements in High-Voltage Impulse Tests, Part 2: Evaluation of Software Used for the Determination of the Parameters of Impulse Waveforms'
4. IEC 60270 2000. 'High-Voltage Test Techniques: Partial Discharge Measurement'
5. IEC 60060-1 2010. 'High-Voltage Test Techniques, Part 1: General Definitions and Test Requirements'
6. Simon P., Garnacho F., Berlijn S., Gockenbach E. 'Determining the voltage factor function for the evaluation of lightning impulses with oscillation and/or overshoot'. *IEEE Transactions on Power Delivery*. 2006;**21**(2):560–6
7. Hinow M., Hauschild W., Gockenbach E. 'Lightning impulse voltage and overshoot evaluation proposed in drafts of IEC 60060-1 and future UHV testing'. *IEEE Transactions on Dielectrics and Electrical Insulation*. 2010; **17**(5):1628–34
8. Garnacho F., Simon P., Gockenbach E., Hackemack K., Berlijn S., Werle P. 'Evaluation of lightning-impulse voltages based on experimental results'. *Electra*. 2002;**204**:31–9

Chapter 19
Fundamental aspects of air breakdown
J. Blackett

19.1 Introduction

Professor Norman Allen was the author of the original chapter entitled 'Fundamental Aspects of Air Breakdown' in editions 1 and 2 of *High Voltage Engineering and Testing* [1, 2] and the accompanying lectures for the IET International Vacation School series. The current author is pleased to acknowledge and incorporate much of his original work including 'Mechanisms of Air Breakdown' [3].

Atmospheric air remains as the main insulant on electricity transmission and distribution systems even though transmission voltages have increased to over 1.2 MV AC and 800 kV DC [+ve, −ve]. Such ultra high voltage (UHV) systems may be prone to more lightning strikes and system overvoltages than ever before. Some transmission systems are being developed that will operate at higher altitudes and in higher humidity conditions, under conditions not yet known, or adequately covered by IEC Standards. Additions to the original Chapter 20 [2] are necessary because, although the fundamental aspects of air breakdown do not change, we continually strive to learn more about the processes that change atmospheric air from being an insulant into a conductor.

The electrical stresses that may lead to insulation failure are from two principal sources:

- lightning surges (natural source) (Figures 19.1 and 19.2)
- switching surges (system operating source).

Direct natural lightning strikes, and also the overvoltages that may be induced from natural lightning, produce the conditions that build up to the breakdown of atmospheric air.

Switching operations on high voltage transmission systems can also cause overvoltages (switching surges). These surges can also cause failure of the air insulation.

Some power frequency overvoltages can occur as a result of certain switching operations. (For details refer to Chapter 2 on insulation coordination for AC transmission and distribution systems.)

632 *High-voltage engineering and testing*

Figure 19.1 Photograph of multiple lightning discharges

Figure 19.2 Cloud-to-cloud lightning in Switzerland

As in the original introduction [2], we are dealing with the dielectric breakdown of air. The term 'disruptive discharge' as defined in IEC Standard 60060-1 [4] was first used by Michael Faraday [5] in 1838.

Because, in practice, the electric fields are highly non-uniform, we will examine, in detail, criteria for breakdown in such situations.

19.1.1 History

In the eighteenth and early nineteenth century, rotating balls of sulphur and Wimshurst machines were used to produce sparks for amusement and for the scientific study of electrical discharges in air. Later in the late nineteenth century, Tesla invented a continuous lightning machine – the Tesla Coil. This resonant air 'transformer' can produce long (~9 m) arc-spark displays. Figure 19.3 shows the discharges produced in 2011 by a Tesla Coil at a 'Teslathon' in Cambridge.

Because transmission and distribution voltages increased from the early years of the twentieth century, higher and higher voltage sources were needed to test the strength of the insulation of the equipment used. The system voltages were 50/60 Hz AC and DC. Thus, voltages were applied in test laboratories to determine the breakdown strength of the equipment under steady conditions. However, transient voltages from lightning and switching surges were much higher than the transmission voltages.

In the twentieth century, Van de Graaff generators produced up to 25 million volt DC and Greinacher circuits used in Cockroft–Walton DC sets could produce 800 kV. Because power frequency AC transmission systems of 735 kV and beyond were being developed, laboratory test-transformers were developed to produce over 2 MV.

Figure 19.3   Photograph of Tesla Coil discharges (Cambridge 'Teslathon' 2011)

Impulse generators (Marx) were made in the mid-twentieth century to produce lightning impulses and switching impulses up to several megavolt. So in addition to be able to test equipment to failure it became possible to study the effects of overvoltages on insulating systems. The electric fields of simple geometric shapes such as rod–plane and spheres could be calculated and the effects of these high electric fields were studied. Thus, the electrical breakdown in air could be investigated from small systems (needle-plane gaps of 1 or 2 mm) to full-scale modelling of 800 kV transmission structures. The laboratories have become larger (and are continuing to do so up to 70 m × 70 m × 100 m buildings) and the applied voltages are over 7 MV.

19.1.2 High-voltage laboratory testing

Tests are made on all high voltage electrical equipment to assure their strength at operating voltages and above. In this chapter, we are only considering simple equipment having 'self-restoring insulation' such as external atmospheric air breakdown or insulator surface flashover. Warning: Different conditions apply when testing equipment containing 'non self-restoring insulation' systems, when 'highest withstand voltages' are determined (such as for transformers and switchgear, or bushings, as considered in other chapters (see Chapters 2, 3, 8–10 and 12 etc.)). The test voltages can be deliberately applied at higher levels than the system voltage to cause a breakdown or flashover in the cases of 'self-restoring insulation' systems. These test voltages determine a 'safety factor' and prove the integrity of the equipment. Other test voltages that are applied in the laboratory can simulate the effects of natural and system overvoltages. Lightning impulse and switching impulse tests are intended to simulate these effects on the insulation and also to help to develop appropriate insulation coordination criteria.

19.1.2.1 Alternating and direct voltage

Rated system voltages will not cause total breakdown of the insulating air in vicinity of power lines in the short term. In air, the fundamental breakdown processes are still at work but the electron avalanches and streamers may only cause localised discharges. Under steady-state conditions, only 'partial discharges' occur. Wet and polluted insulators may eventually suffer 'flashover failure' if partial discharges persist. The detection of partial discharges, particularly in transformers, is now a sophisticated diagnostic tool.

19.1.2.2 Impulse testing

Impulse voltages from laboratory impulse generators may become 'distorted' when they are applied to transformers and other equipment with windings because of interaction with the inductance of the load. Testing large UHV equipment in huge laboratories makes it difficult to avoid overshoots when applying impulse voltages of a few megavolt. Oscillations and overshoots occur near the peak of the applied impulse and the effect on the dielectric becomes complicated. IEC Standard 60060-1 [4] has introduced a factor that takes into account the frequency of these oscillations. Switching impulses in the laboratory simulate the switching surges that can occur on transmission systems. Another phenomenon is that of the so-called U-curve.

Here, the breakdown voltage of a large air gap at UHVs depends on the time-to-peak of the applied switching impulse voltage.

However, it must be remembered that high voltage transformers, DC sets and lightning impulse generators in the test laboratory have very limited energy compared to that of natural lightning.

Because UHV testing is very expensive, some knowledge of the mode of failure of electrical equipment enables designers to estimate breakdown voltages in order to minimise full-scale insulation tests.

We shall now examine, in more detail, the formation of streamers and leaders. An understanding of this may lead to an explanation of how the breakdown voltage in air, for similar gaps, depends upon the wave shape of test voltages.

It is quite remarkable that test laboratories may be over 100 m long, 50 m high and 50 m wide, yet we are looking for processes on an atomic scale that can cause the breakdown air insulation between gaps of tens of metres long.

We emphasise that this discourse is only concerned with such atomic particles as molecules, electrons and ions. Our imagination is still stretched. The simple Bohr image of an atom being a sun with orbiting planetary electrons may suffice. The concept of tiny billiard balls, colliding with one another, is feasible but these tiny particles also have electric charges.

Air is made up of nitrogen, oxygen and other gases but we will regard the processes as a molecular phenomenon. The electronegative nature of oxygen is a property to be taken into account. SF_6 is more highly electronegative than O_2 and has greater breakdown strength.

We shall not be considering any magnetic influences.

19.2 Pre-breakdown discharges

Three fundamental processes are at work.

19.2.1 Electron avalanches

Electrons are accelerated under the influence of an electric field. Some electron collisions with neutral atoms and molecules are energetic enough to release more electrons. Some collisions result in the emission of photons which, in turn, may result in ionisation. These ionisation effects produce an 'ever-increasing' number of electrons and so-called electron avalanches result.

19.2.2 Streamer discharges

If conditions are 'right' then these avalanches develop further to produce bright, conducting, branched streamers.

19.2.3 Leaders

It is characteristic of high voltage electric fields that the final sparkover is preceded by streamers followed in many cases by a 'leader' which develops prior to the breakdown itself.

19.3 Uniform fields

19.3.1 Electron avalanches in uniform fields

'Uniform' electric field electrode systems were initially used when studying the electrical breakdown in air. Flat electrodes, spaced a distance 'd' apart, were used to produce a reasonably uniform electric field.

When direct voltages were applied across the plates, Thompson [6–8] found that in low-pressure air, the current was initially constant but then, as the voltage was increased, the current increased exponentially. We now know that the electric field E produces a force on electrons that accelerate them. When electrons are accelerated through air they may ionise molecules. The electrons released can, themselves, be accelerated in the field to cause further ionisation and release even more electrons, and the electrical breakdown of a gas depends upon these electron and ion swarms.

Figure 19.4 is a simple schematic of such an electron avalanche. The anode is at the top attracting the electrons.

Electrons may be emitted from the cathode or are already 'free' in the interelectrode volume. It is estimated that there are about 12 electrons per cubic centimetre atmospheric air at sea level. These electrons are mainly from decaying radioactive sources with some from cosmic rays. (For the production of lightning at an altitude of about 40 km, the source of electrons is mainly from cosmic rays.)

The theory goes as follows.

Let n_0 electrons be emitted from a cathode per second and increase to n_x as a result of ionisation at a distance x from the cathode. If these electrons move through a lamina dx at x they will generate $\propto n_x \, dx$ new electrons so that

$$dn_x = \propto n_x \, dx$$

The number of electrons to strike the anode at a distance $x = d$ is $n_0 \exp \alpha d$ and the total current I at the anode is

$$I = I_0 \exp \alpha d \qquad (19.1)$$

In a uniform field then the successive avalanches may lead to a breakdown.

A uniform field is only theoretically possible between two infinitely large parallel plates. The electric field E is directly proportional to the voltage V applied and the spacing d of the gap $E = V/d$. Quasi-uniform fields are possible in the laboratory with large, smooth rounded parallel plates. If a high voltage is applied across two electrodes so that there is a reasonably uniform field in the gap, the breakdown strength is about 30 kV/cm or 3 MV/m. For all other configurations, the average applied field to cause breakdown is considerably less. That is the strength of the insulating air appears to be less.

In the 'real' world of electricity systems there are rarely, if ever, any uniform fields, so the study of electrical discharges is usually made using non-uniform fields such as sphere–plane, rod–plane and conductor–plane configurations.

Fundamental aspects of air breakdown 637

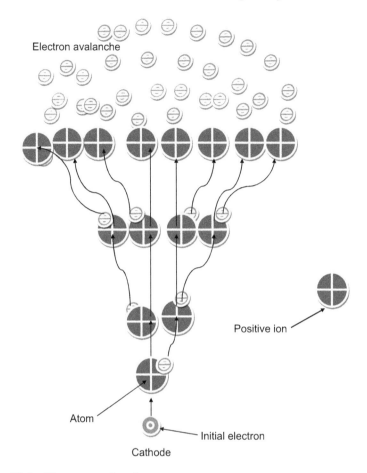

Figure 19.4 Electron avalanche
An initial electron is accelerated under the influence of the electric field. It collides with atoms and ionises them, releasing more electrons leaving positive ions behind. An avalanche of electrons heads towards the anode

19.3.1.1 Photo-ionisation

The Townsend mechanism required several successive avalanches even with over-volted gaps and the theory required further electrons to be emitted from the cathode. The impacting electrons may ionise the air molecules to emit electrons or may excite them to release energetic photons. Even recombination between positive ions and electrons may result in the release of photons. These photons, in turn, may cause photo-ionisation.

If further electrons are released away from the original avalanche then they too may be accelerated in the field to produce further impact-ionisation. This will enhance the development towards breakdown.

Figure 19.5 shows a very simplified schematic picture of such successive electron avalanches. The negative electrons are at the head of the avalanche. It can

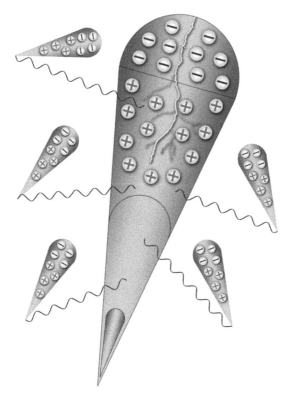

Figure 19.5 Electron avalanches
This schematic drawing shows the effect of a succession of electron avalanches producing streamers and leaders and resulting in breakdown of air

be seen that for every ionised molecule there will be an electron released and a positive ion will be left behind.

The exponential increase as recorded by Townsend is now known to be a result of the increase in numbers of electrons in the electron avalanche. The processes of impact-ionisation and photo-ionisation are complicated.

If the number of electron-pairs created by an electron of a given average energy in a given electric field is proportional to the distance it travels in that field then the number of new electrons dn created in this distance dx is

$dn = \alpha \, dx$

Over a distance d, starting with one electron at the origin, the total number N of ion pairs created becomes $N_a = \exp \alpha d$

$$\int_0^d \alpha \, dx = 20 \qquad (19.2)$$

The criterion for further avalanches to occur, thus leading eventually to breakdown, is that there must be a critical number of electrons in the avalanche head.

$$N_{a\ crit} = 10^8$$

A simple semi-empirical method to predict the DC breakdown of non-uniform gaps was developed by Pedersen [9].

For the critical avalanche number to be reached in a non-uniform field, Pedersen used values of α across the gap between electrodes. He plotted α across a uniform field at breakdown voltage where $N_a = \exp \alpha d$. This was compared to an estimated breakdown voltage applied to the non-uniform gap.

Instead of the number of electrons increasing exponentially with the uniform value of α, Pedersen [9] calculated α across the non-uniform gap of an increasing field (in the case of a positive rod–point or sphere anode).

The critical avalanche number reached at breakdown was compared to a plot of the increasing number of electrons being generated in the non-uniform gap. Breakdown is deemed to occur when the electron number is equal to 10^8.

Empirical methods to predict breakdown voltages may not appear to consider breakdown processes and space charges. The method considers the maximum and average electric fields of non-uniform configurations. However, the critical electron number will be achieved in such estimated fields. (See Reference 10 and Chapter 3 of this edition, 'Applications of Gaseous Insulants', by Ryan.)

Breakdown voltage is given by the relationship

$$V_s = E \eta g \tag{19.3}$$

where V_s is the breakdown voltage, E and g are the appropriate breakdown gradient and gap respectively. η is the utilisation factor (ratio of the average to maximum voltage gradient) E_{av}/E_m.

Actual breakdown records for known configurations have been successfully used to estimate breakdown voltages for other similar configurations with calculable fields.

19.3.1.2 Streamers

When numerous electron avalanches form then streamers may develop. This stage is seen (sometimes literally as the excited atoms emit visible as well as ultraviolet light) as the inception level. The field to produce sufficient electrons and also to accelerate them is E_i. A streamer system is a corona which spreads out like a fan. This streamer or Kanal is fed with ever-increasing electron avalanches.

Even a single electron avalanche will distort the original uniform field. The 'spherical' cloud of positive ions at the head of the avalanche enhances the original field away from the anode. When this space charge field becomes as intense as the externally applied field, the avalanche becomes a streamer and when it reaches the cathode then breakdown occurs. Figure 19.6 shows the development of this field enhancement.

640 High-voltage engineering and testing

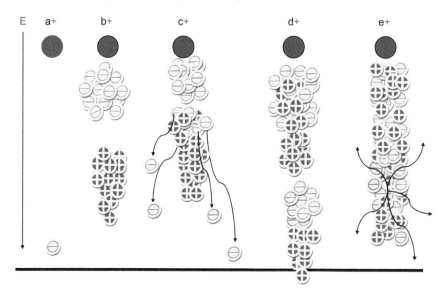

Figure 19.6 Development of streamers
  a = an electron may be produced by cosmic rays, radioactive decay or ionisation of atoms or molecules
  b = the electron avalanche forms a head near the anode leaving a trail of positive ions behind
  c = further electrons are produced by photo-ionisation when excited atoms release energetic photons
  d = further avalanches are formed that distort the original electric field
  e = excitement within the conducting column increases and displays bright streamers

The breakdown voltage of a uniform field is

$$V_s = E_s d \tag{19.4}$$

So far so simple, but, as always, the more we study phenomena, such as breakdown in air, the more complicated it becomes.

We begin with a summary of results obtained with non-uniform fields first with DC applied voltages, then AC voltages and finally impulse voltages.

19.4 Non-uniform fields

In the 'real world of high voltage engineering' there are few, if any, uniform fields. Geometrical non-uniform fields such as point–plane, point-to-point or coaxial electrodes are used in studies of air breakdown. Information from such tests has resulted in empirical formulae. These formulae are used to estimate the probable breakdown of electrical equipment at the design stage. The tests have also produced more information on the processes involved in air breakdown.

The electric field in non-uniform fields is higher at the electrode than in the rest of the gap. There is a logarithmic 'fall' in the value of the field from the maximum.

19.4.1 Direct voltage breakdown

Steady overvoltages on DC transmission systems are very rare. Any surges are of a transient nature. Overvoltage tests in the laboratory are essential to determine the integrity of equipment that has been designed for steady high voltages. However, laboratory studies have revealed some interesting variations in the fundamental processes that lead to sparkover in uniform fields.

When direct voltages are applied to electrodes, such as rods and sphere-ended rods, a high electric field is produced at their tips. The field falls away logarithmically. So for a given gap, the voltage at which inception occurs at the electrode is lower than that for a uniform field gap. The air gap is severely stressed and corona is initiated from the electrodes as the voltage is increased to the inception level.

As shown in the previous section, a criterion for breakdown to be initiated is that the number of electrons in an avalanche reaches 10^8.

Usually, the corona becomes a streamer and even a leader-like extension to the electrode when low breakdown voltage occurs. On other occasions, with the same set-up, a glow forms around the tips of the electrodes. This glow inhibits further corona until a much higher voltage is reached.

The current author encountered this glow in some experiments in a sphere–hemisphere configuration [11]. Although this appeared to be a new phenomenon, Hermstein [12] and Uhlig [13] had already described the glow. Uhlig found that when this glow (called ultra-corona) occurred on smooth wires they could be used as a shield on high voltage equipment under test. Although losses were high, such shields were less expensive than spun-aluminium toroids.

Further references, including Waters and Stark [14], were found and, of course Michael Faraday had already studied the glow on spheres of different sizes [5] in 1855. This 'Hermstein' glow may account for some of the 'dark' periods that occur during impulse breakdown phenomena.

Figure 19.7 shows the development of a Hermstein glow. The sequence of Polaroid photographs shows the gradual increase in applied positive-polarity direct voltage to a sphere–hemisphere gap. Increasing the voltage initially initiates a streamer. The streamer at this stage may cause a disruptive discharge. Sometimes, as the applied voltage is further increased, the streamer is extinguished and a glow develops that enhances the strength of the gap. Final breakdown of the air is ~100% higher than initial streamer that caused the earlier disruptive discharge. For example, the inception of the streamer was recorded at 28.65 kV and the final breakdown voltage was 56.6 kV. The 'glow' had extended nearly over the whole anode sphere.

Figure 19.8 is a schematic diagram of Nasser's theory of the development of Hermstein's glow (after Nasser [15]).

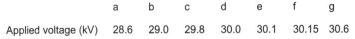

Figure 19.7   Hermstein glow
The sequence of photographs was taken through a port in the outer electrode of a sphere–hemisphere configuration. The inner spheres are seen illuminated at each end of the sequence. The positive DC voltage was increased in steps. Sometimes the breakdown occurred at about 29 kV. The sequence above finally resulted in breakdown at 56.6 kV, the sphere being almost covered in the blue glow

Attempts to predict breakdown voltages of non-uniform field gaps have been made using empirical and semi-empirical methods. Empirical methods are discussed in section 19.5.

19.4.2   Alternating voltage breakdown

Overvoltages on the system are primarily of a transient nature but may be at power frequency. Tests in laboratories produce results to evaluate a 'safety factor' and insulation integrity.

The common problems of power frequencies at rated operating voltages are the partial discharges that occur on the system. These discharges are localised around points of high stress and on polluted insulators. The study of partial discharges in oil and compressed gases is beyond the scope of this chapter. However, some of the fundamental processes that occur in stressed air are also present in gasified oil and compressed gases.

Here, we are concerned with the 'partial discharge' known as corona. The system voltage is rarely, if ever, capable of resulting in a breakdown in air. 'Obviously', flashovers occur when the insulation is contaminated and/or tracking occurs. Corona is more than just a nuisance. It produces radio interference, noise and energy loss. The ultraviolet light that it emits can damage polymeric materials. Worse still, it is known to cause ozone that can attack elastomeric materials. In the presence of humidity, corona also produces acids that may degrade polymers [16].

The initial fundamental processes are the same as for breakdown. Electron avalanches occur and may develop into streamers. The field at sharp points on the live components of an overhead line or structures of substation equipment may be high enough to initiate such corona. Alternating voltage tests in the laboratory are usually made at a level higher than the operating system voltage, e.g. IEC Standard 60060-1 [4]. Alternating tests for 420 kV phase-to-phase voltage are

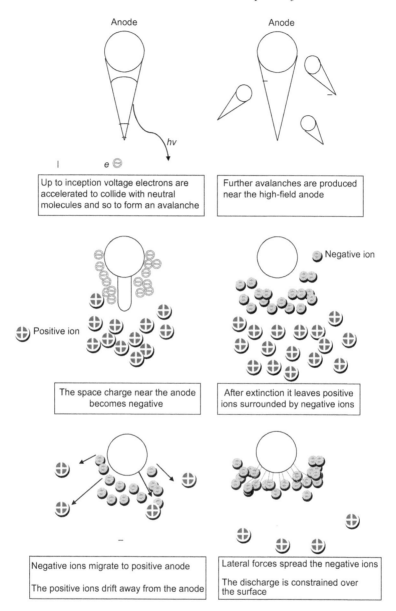

Figure 19.8   Description of Hermstein hypothesis for anode glow

made at 1.1 × system voltage. The voltage to earth is $420/\sqrt{3} = 242.5$ kV. Single phase-to-earth test voltage is $242.5 \times 1.1 = 266.75$ kV. We are concerned with the peak voltage that produces the maximum stress on the insulation. So, the peak applied voltage to earth becomes $266.75 \times \sqrt{2} = 377$ kV. The 'obvious' effect of this is that partial discharges appear at, or around, the positive and negative peaks of alternating voltages.

19.4.3 Impulse breakdown

Overvoltages that occur on an electrical system derive from direct or induced lightning strikes (lightning surges) and those arising from switching activities, such as re-closures on open lines (switching surges).

IEC Standard 60060-1 [4] defines the shape and voltages of test impulses that simulate the above surges. Lightning impulses and switching impulses must be applied in laboratories in order to test electrical equipment.

Lightning impulses rise in 1.2 μs and fall to half value in 50 μs. Switching impulses rise in 250 μs and fall to half value in 2500 μs. Although the fundamental processes are the same in air under stress, the breakdown voltages differ considerably. The time taken for the fundamental processes to develop is an important factor.

Figure 19.9 is a copy of a trace on X-ray photo-sensitive film depicting positive corona. These so-called 'Lichtenberg figures' are obtained by applying a lightning impulse to a pointed electrode on the film surface. The point–plane configuration was subjected to a lightning impulse of peak voltage 180 kV. Breakdown did not occur. The image is an 'integration' of all streamers that developed across the electrode gap during the impulse.

19.4.4 Leaders

In an over-volted gap, the ever-increasing current of the moving electrons may result in the development of so-called leaders.

Figure 19.10 is a schematic of the growth of streamers to leaders. The channel of a streamer heats up because of I^2R and it becomes a leader. The electric field drops and the leader heats up further and effectively results in the extension of the anode.

Figure 19.11 shows a leader developing from the anode and bifurcating. This image was recorded using a photomultiplier by the Renardiers Group (Ed.) [17]. The swarm of excited atoms in the avalanches and streamers ahead of the leaders can be seen clearly.

Figures 19.12 and 19.13 are Lichtenberg figures that were recorded on ordinary colour film. They depict discharges from a point-to-point configuration very close to breakdown (the colours will be false). In Figure 19.12, the positive corona is spider-like or similar to tree roots (dendritic) and bright leaders can be seen in the more tangled negative corona. Figure 19.13 shows a 'final jump' and a number of aborted streamers and leaders even though full sparkover has occurred.

Figure 19.14 is taken from work by Carrara *et al.* in 1976 [18]. This is a simple illustration of breakdown processes at work under the influence of a laboratory high voltage impulse. The illustration is taken from an image converter record of a discharge of length *d*.

This simplified drawing shows some of the more complicated processes that may be at work. Experiments when direct voltages were applied to rod–plane gaps

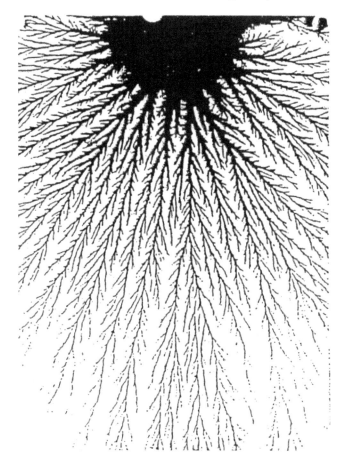

Figure 19.9 Lichtenberg figure of positive streamers (after Hassanzahraee, Ph.D. thesis, University of Leeds, 1989)

revealed some phenomena that may be relevant in impulse breakdown such as dark period as a result of positive ions, glow from anode, streamer to leader transition and upward discharge from cathode.

19.4.5 Sparkover, breakdown, disruptive discharge

The term 'sparkover' is used when air as an insulant fails, usually when the source is of limited energy.

The term 'breakdown' is usually used when the applied voltage falls to zero.

The term 'disruptive discharge' is now used in standards to describe the complete failure of the insulating air and was certainly used by Faraday in 1838.

Breakdown voltages in point-to-point configurations are higher than for point-to-plane.

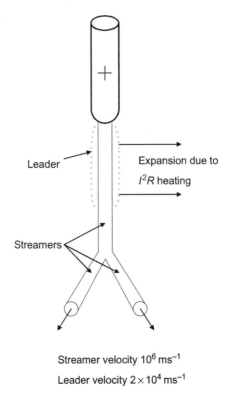

Figure 19.10   Streamer to leader development

The voltage at which breakdown occurs may be quantified in the following way, as depicted in Figure 19.15, which represents the streamers and leaders in a point-to-point gap at breakdown.

Let E_s^+, E_s^-, l_s^+, l_s^- be the gradients and lengths of positive and negative streamers respectively at the instant of breakdown.

The voltage V at breakdown for a gap spacing d is

$$V_s = E_l^+ l_l^+ + E_s^+ l_s^+ + E_l^- l_l^- + E_s^- l_s^- \tag{19.5}$$

Also

$$d = l_l^+ + l_s^+ + l_l^- + l_s^- \tag{19.6}$$

$E_l^+, E_l^-, l_l^+, l_l^-$ are the corresponding quantities for the positive and negative leaders.

In most electrode configurations, the lengths of l are not known although their gradients can be estimated. However, there is a useful simplification with the rod–plane electrode system where, since the negative electrode is only lightly stressed, with a nearly uniform electric field, there are no negative streamers or leaders, i.e. $l_l^- = l_s^- = 0$.

Figure 19.11   Photograph of streamer development into leaders [17]

Hence

$$d = l_l^+ + l_s^+ \tag{19.7}$$

The breakdown voltage becomes

$$V_s = E_s^+ d - l_l^+(E_s^+ - E_l^+) \tag{19.8}$$

Comparison of (19.5) and (19.8) makes it clear that where there is no negative corona, as in the rod–plane, V_s is lower than it would be in a gap where a significant negative corona occurs as in the condition where two highly stressed electrodes are used. Thus, the rod–plane gap exhibits the lowest sparkover gradient of any non-uniform field gap arrangement and this property forms a useful reference with which other gaps are compared.

19.5   The 'U-curve'

Early work on the switching impulse voltage breakdown of long gaps with a point-to-plane configuration revealed a disturbing phenomenon. The breakdown voltage

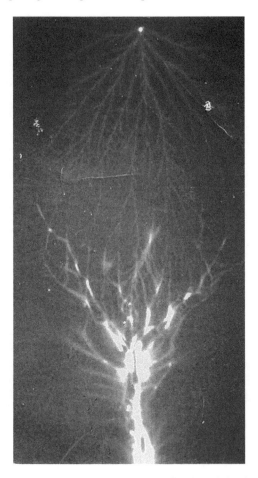

Figure 19.12 Lichtenberg figure of rod–rod discharges

was seen to depend not only on the gap between the point and the plane but also on the time-to-peak of the applied impulse. For a 'fixed electrode spacing', the breakdown voltage was seen to pass through a minimum time-to-peak. There is a different minimum for each electrode gap. In practice, this means that any insulating gap on a system will have a weak breakdown strength that depends on a particular applied overvoltage wave shape. Switching surges on the system produce such wave shapes. Therefore, laboratory studies had to be made and appropriate proving tests had to be devised.

Our simple model of corona inception, streamer development and leader breakdown must be modified. Again, this is important today because of the increasing use of UHV transmission. Large gaps are needed to maintain insulation integrity and large-scale testing is more expensive. The study of the simple geometry of rod-to-plane gaps helps to give valuable practical and theoretical information for designers.

Figure 19.13   Lichtenberg figure of leaders

Figures 19.16 and 19.17 show U-curves of impulses, with impulses of various times-to-crest applied to rod–plane gaps. This phenomenon was initially realised when long-distance transmission systems at voltages greater than 400 kV were being developed and test voltages were >1 MV.

It was found that every gap had a 'minimum breakdown voltage' when tests were made at increasing times-to-crest. For gaps of up to 3 m (Figure 19.16), this time was about 100 μs. At greater gaps (up to 14 m), the time for the minimum breakdown voltage (critical time) was seen to vary in proportion to the gap length.

As humidity increases, the strength of a gap increases and this is true for large gaps under switching impulses (Figure 19.17). The effects of humidity are discussed in section 19.6.4.

How does this phenomenon occur and what effect will it have on the dimensions of UHV transmission systems operating at 1.1 MV and above?

It is obviously a 'time-dependent' phenomenon. We have looked at some of the basic processes leading to breakdown of air-insulating gaps.

We know that streamers travel at about 10^6 m/s. Leaders travel more slowly at about 10^5 m/s.

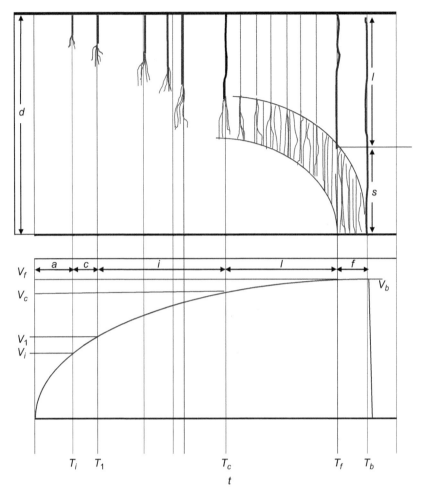

Figure 19.14 Image converter pictures of a discharge in air on a gap of d (after Carrara [18]). Illustration taken from an image converter record of a discharge of length d
l = represents length of leader
s = represents length of streamer
Stage a = absence of phenomena
Stage c = corona, and primary dark periods
Stage i = leader stem, and secondary dark periods
Stage l = leader continues propagation
Stage f = final jump
t = time; V = voltage
T_i, V_i = corona inception
T_1, V_1 = leader inception
T_c, V_c = continuous leader inception
T_f, V_f = final jump inception
T_b, V_b = breakdown

Fundamental aspects of air breakdown 651

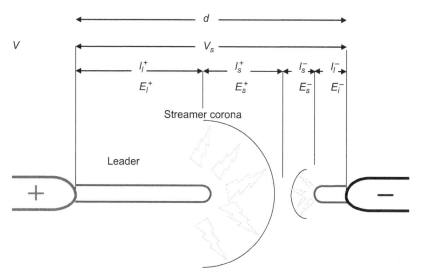

Figure 19.15   Conditions at breakdown in point-to-point configuration

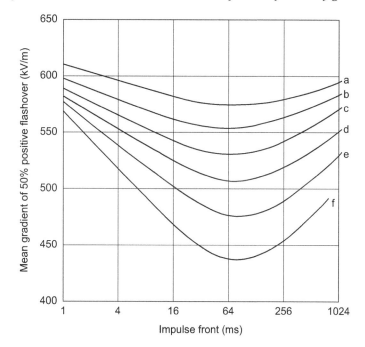

Relation between mean gradient at 50% positive flashover and duration of wavefront

Curve	a	b	c	d	e	f
Gap (m)	0.8	1.2	1.7	2.0	2.5	3.0

Figure 19.16   U-curve from CERL (mean gradient vs. front duration)

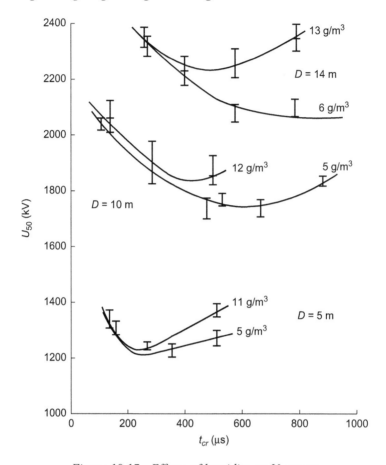

Figure 19.17 Effects of humidity on U-curves

Tests with 'square waves' have shown that there is a 'time lag' between voltage application and breakdown. It consists of two components:

- The time between the application of the voltage and the arrival of an initiatory electron in the stressed space is known as the statistical time t_s.
- The second is the time taken for the 'breakdown processes' to develop from electron avalanche to streamer to leader. This is known as the formative time lag t_f.

For short-term impulses, for example, a lightning impulse rises to crest in the order of 1 μs. Figure 19.18 is the diagram of 1.2/50 μs test curve in IEC Standard 60060-1 [4].

To initiate breakdown, there must be at least one electron available to start the electron avalanche. There are estimated to be about 12 electrons per cubic centimetre of atmospheric air at sea level.

Note: Electrons are produced as a result of radiation. At sea level, the radiation is a result of decaying radioactive material. At higher altitudes, cosmic rays play their part in producing atmospheric electrons.

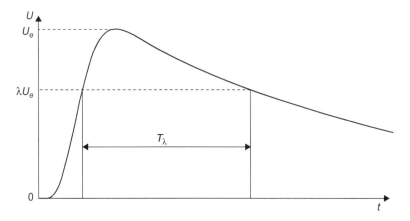

Figure 19.18   Voltage time interval for a full lightning impulse [4]

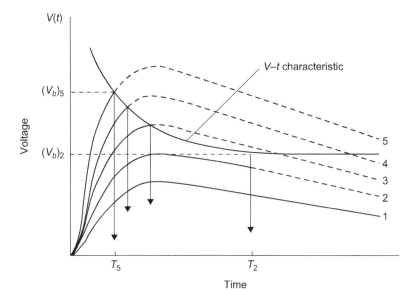

Figure 19.19   Voltage-time characteristics

For the lightning impulse, there is little time for more than a rudimentary leader growth after the streamer corona has occurred (usually on the wavefront). This is because the voltage decreases to half value in the next 50 μs.

For such lightning impulse voltages, there is a certain probability that breakdown may occur. Obviously, the higher the applied voltage the more likely that breakdown will occur.

Figure 19.19 shows that the breakdown under impulse voltage depends on the time that the applied voltage exceeds the static or DC level V_s. The voltage must

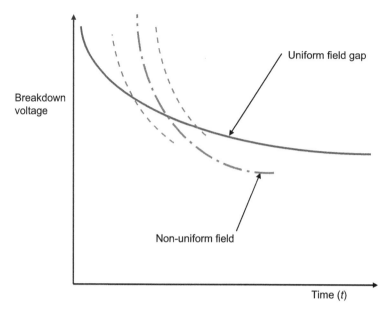

Figure 19.20 Schematic diagram of voltage-time characteristics

remain above V_s long enough for an electron to arrive t_s, and long enough for the breakdown processes to result in a breakdown t_f. These time-to-breakdown values may be plotted as shown in Figure 19.20.

But what happens if the applied voltage has a longer 'front time'?

We have mentioned that for every electron that is released in an avalanche there is a positive ion produced. These ions are 'slow-moving' and can 'choke' the development of streamers. Hermstein found that a glow may develop with DC application, where negative ions may form in regions of low fields where attachment takes place. It is therefore possible that the original electric field may be modified under long switching impulses.

Where a streamer or avalanche is 'choked', the breakdown voltage will rise.

19.5.1 The critical time to breakdown

Note: Energy input comment: The impulse generator has a limited finite energy. A large impulse generator may have a total energy stored in its capacitors of nearly 30 kJ. The capacitors, when charged, must supply the energy to the equipment under test, be it a transformer or an experimental rod–plane gap. Lightning discharges, although having similar fundamental processes such as streamers and leaders, are supplied by more energy that has built up in clouds and can maintain leader growth over vast distances beginning in the rarefied atmosphere at 40 km. The generation of lightning is effectively a direct voltage phenomenon up to the final lightning flash.

If the time that an overvoltage is greater than the corona inception voltage is short, then there will be no chance for streamers and leaders to develop before the

Fundamental aspects of air breakdown 655

energising driving impulse voltage falls. When the voltage (and its applied electric field) falls below the inception level, then no more electrons are generated. Attachment to form ions occurs, reducing the field even further.

If the duration of the overvoltage is too long, then ions and field distortion may result in the choking of streamers. If the time is 'critical', then all the ionising processes are optimised. Electron avalanches and the driving field feed in more electrons. Positive ions cause field distortion that accentuates the applied field. Streamers form and, because of the ever increasing numbers of electrons, leaders develop and make the final jump to total disruptive discharge.

19.6 The gap factor

Although the rod–plane is the 'weakest' and therefore the most pessimistic electrode gap, it does not represent every geometry encountered in the 'real' world of high voltage transmission.

Figure 19.21 shows the breakdown voltages and the average electric field at breakdown for plane–plane (uniform field) and rod–plane and rod–rod configurations for 1 m gaps.

It would be wonderful if transmission systems could operate in uniform fields. The electrode gap 'spacings' would only have to be about a 'foot' (304.8 mm) for 1 MV. The rod–plane represents the weakest configuration (500 kV/m) and this may

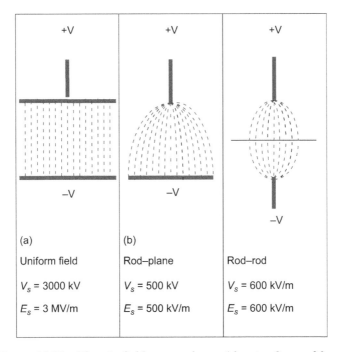

Figure 19.21 Electric fields at sparkover (showing lines of force)

be used, as a 'rule-of-thumb', to estimate breakdown voltages for particular spacings. This implies that a 2 gap could be sufficient to insulate one million volt in air.

We have looked at the electric fields between rod electrodes and the distortion caused by ions. Figure 19.22, taken from Hutzler *et al.* [19], shows values of gap factor K that have been derived by experiment.

19.6.1 Test procedures

The variability in pre-breakdown processes as summarised above also results in a variability in sparkover voltage, particularly when impulse voltages are applied.

Configuration		k
Rod–plane	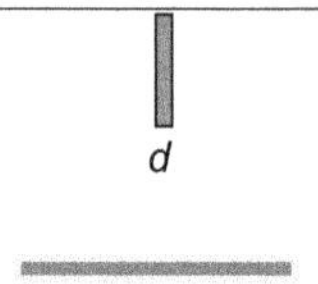	1
Conductor–plane	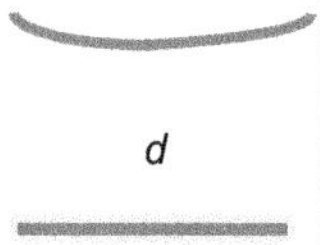	1 to 1.15
Rod–rod	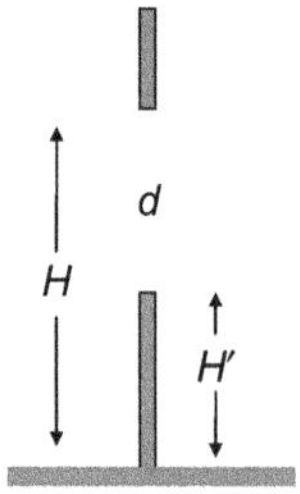	$1 + 0.6\, H'/H$
Conductor–rod	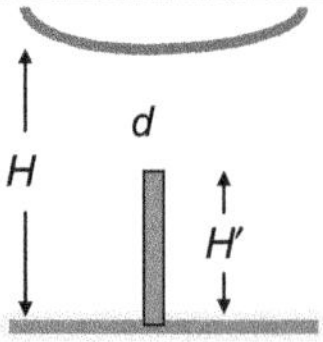	$(1.1 \text{ to } 1.15) \exp(0.7\, H'/H)$
Protuberances	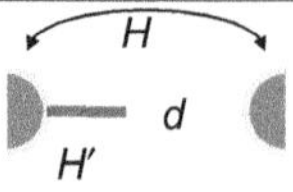	$k_0 \exp\left(\pm 0.7 \frac{H'}{H}\right)^*$ $k \geq 1$

\+ sign for protuberances at the negative electrode
− sign for protuberances at the positive electrode

Figure 19.22 Gap factors

Because of this statistical probability of breakdown, it was necessary to devise the average sparkover to describe the strength of a particular gap. For this purpose, the U_{50} or 50% sparkover voltage is used. This is the impulse 'crest' voltage at which there is a probability of 0.5 that the gap will spark over. With impulse voltages, this can be determined in two ways: (a) the variable voltage method and (b) the 'up and down' method as described in IEC Standard 60060-1 [4].

19.6.1.1 Sparkover voltage characteristics for different gaps
Impulse conditions: Rod–plane gaps
The practical engineer is faced with a variety of possible electrode geometries, but we have seen that the rod–plane gap forms a valuable reference. Jones and Waters [20] have presented a summary showing the mean breakdown gradient U_{50}/d against gap length for this gap. This is reproduced as Figure 19.23. Results for impulse, DC and AC voltages are all shown. A further reference condition becomes apparent. For the positive lightning impulse, the mean gradient remains constant for gap lengths of greater than 1 m at about 500 kV/m. This is approximately the same as the streamer gradient E_s^+. (This reference gradient is utilised in the IEC atmospheric correction procedures in IEC Standard 60060-1 (see section 19.6.4).) Thus, the sparkover voltage increases linearly with gap length. For the negative lightning impulse, the gradient is higher but is not constant with gap. The sparkover voltage increases with electrode spacing, but less rapidly than linearly. No simple

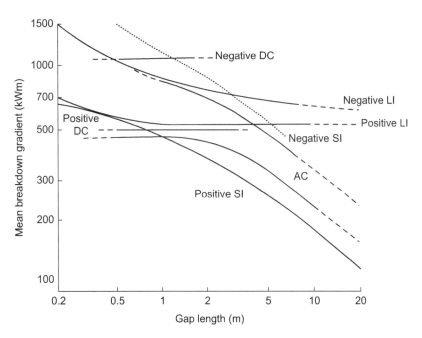

Figure 19.23 Mean breakdown gradient of rod–plane gaps under standard atmospheric conditions as a function of gap length for different testing waveforms

explanation for this can be offered here. Although the 'strength' of the gap weakens, it remains stronger than that of the positive lightning impulse. For this reason, it is less dangerous to power systems than positive impulses.

For positive switching impulses, the gradient declines with increasing gap length due to the increasing growth of the low-gradient leader channel. The breakdown strength depends only on the gap length and is often expressed by the formula [19]

$$U_{50(\text{crit})} = \frac{3400}{\left(1 + \frac{8}{d}\right)} \text{ kV} \tag{19.9}$$

where $U_{50(\text{crit})}$ is the value of U_{50} at the critical time to crest, that is the 'minimum' value. The expression holds for the range of gaps $2 < d < 15$ m.

Figure 19.23 shows that the positive switching impulse characteristic becomes more linear for gap lengths greater than 10 m and a more appropriate relationship is the following

$$U_{50(\text{crit})} = 1400 + 55\, d \text{ kV} \tag{19.10}$$

The following formula has also been proposed by Hutzler et al. [19], as being reasonably accurate for all gaps up to 25 m:

$$U_{50(\text{crit})} = 1080 + \ln(0.46d + 1) \text{ kV} \tag{19.11}$$

Finally, for the particular case of the IEC Standard, positive switching impulse (which is not now at a critical time-to-crest) is well described by

$$U_{50} = 500 d^{0.6} \text{ kV} \tag{19.12}$$

Under negative switching impulses, the following relationship holds with reasonable precision over the range $2\text{ m} < d < 14\text{ m}$ [19]

$$U_{50(\text{crit})} = 1180 d^{0.5} \text{ kV} \tag{19.13}$$

19.6.2 Air gaps of other shapes

The expressions (19.9) and (19.10) can in principle be applied to gaps of more practical interest, provided the RHS of each is multiplied by the gap factor k. This statement must of course be qualified by the reservation already discussed in this section.

For estimation of the flashover voltage of gaps other than the rod–plane, it has become customary to use the gap factor k for switching impulses. This procedure is clearly subject to the same limitations as have been outlined already, but the following equation has been given

$$U_{50} = U_{50(\text{rp})}(13.5k - 0.35k^2) \text{ kV rms} \tag{19.14}$$

Fundamental aspects of air breakdown   659

For gaps in the range 2 m < d < 6 m this formula holds to within ±10% for most of the following gaps: conductor–plane, conductor structure (underneath), conductor structure (lateral), rod–rod, conductor–rod [22]. As noted earlier, the voltages are influenced by the rate of increase of the applied voltage, with a decrease of up to 5% for slow rates of rise towards breakdown of the order of 1 h.

Recent work for CIGRE by Lokhanin [21] is shown in Figure 19.24. This shows sketch drawings of typical air gaps over a wide range of electrode configurations. Results from switching impulse tests are presented in Table 19.1. The 50% disruptive discharge voltages are given for the corresponding inter-electrode clearances of typical air gaps for 1200 kV equipment.

19.6.3   Sparkover under alternating voltages

Here again, the rod–plane exhibits the lowest sparkover voltage and the 50% breakdown condition is given by

$$U_{50} = 750 \ln \left(1 + 0.55 d^{1.2}\right) \text{ kV} \qquad (19.15)$$

for $d > 2$ m.

To give perspective, it may be noted that the peak values of U_{50} are of the order of 20% higher than those for switching impulses of 'critical time-to-crest' for the same gap.

19.6.4   Sparkover under direct voltages

For positive rod–plane gaps in the range 0.5 < d < 5 m, direct voltage sparkover increases linearly with gap length, with a mean gradient of the order of 500 kV/m. Figure 19.23, from Jones and Waters [20], shows the sparkover gradients for this and other gaps and applied voltages.

This characteristic is similar to that of the rod–plane under lightning impulses. In both cases, linearity is due to the dominance of the streamer pre-discharge but, in the positive direct voltage case the reproducibility of sparkover voltage is greater, the standard deviation being less than ±1%.

For the negative rod–plane gap, the gradient is higher being of the order of 1000 kV/m. Gap factors, for gaps other than rod–plane and derived from direct voltage measurements, are given in Table 19.2 [22].

Even UHV laboratories are of a limited size and the proximity of the walls and ceiling within the laboratory may influence electric fields and cause inaccurate breakdown results [23]. Inevitably, some equipment not being used for a particular test may be kept in the laboratory, and these too may have an effect if too close to the test arrangement.

Figure 19.25 (from IEC Standard 60060-1:2010 [4]) gives the recommended minimum clearances of extraneous live or earthed objects to the energised electrode of a test object. NB: These quoted clearances are to ensure 'minimal uncertainty of measurements' and is not supposed to represent the safety margin.

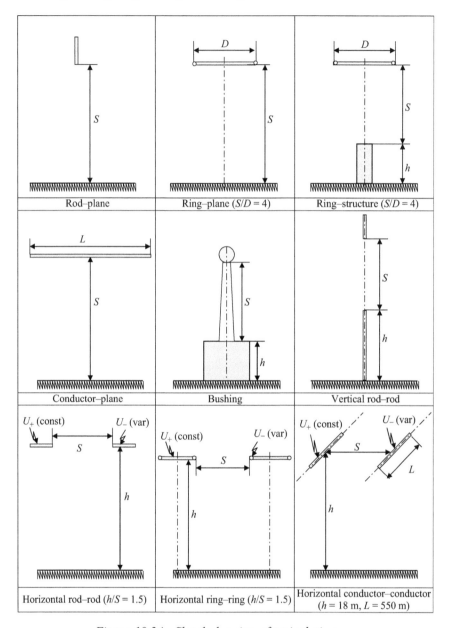

Figure 19.24 Sketch drawing of typical air gaps

However, further work on the effect of walls and objects by Farouk Rizk [23] considered the effects of switching impulses and found that estimation of safety distances should also be considered. A stray switching impulse disruptive discharge of 20 m could affect more than just the instrumentation.

Table 19.1 The 50% probability full discharge voltages and corresponding inter-electrode clearances of typical air gaps for 1200 kV equipment with limiting of switching overvoltages at level 1.6–1.8 Um

Type of air gap		$K_{ov1} = 1.6$ Um			$K_{ov2} = 1.8$ Um		
		U_{50}	Gap length (S)	$\tau_{crit.\,front}^{aperiodic}/\tau_{crit.\,front}^{aperiodic}$	U_{50}	Gap length (S)	$\tau_{crit.\,front}^{aperiodic}/\tau_{crit.\,front}^{aperiodic}$
		kV	m	μs/μs	kV	m	μs/μs
Rod–plane		1950	11.1	780/1530	2270	16.4	1110/2260
Ring–plane		1950	8.8	570/1150	2270	12.0	815/1660
Ring-structure $h = 2.5$ m		1950	7.7	480/990	2270	10.5	700/1420
Conductor–plane		1950	7.1	420/890	2270	9.5	600/1270
Bushing	$L = 10$ M	1950	6.8	400/840	2270	9.2	590/1220
Vertical rod–rod $h = 8$ m	$h = 4.5$ M	1950	5.9	350/720	2270	8.2	520/1060
	$h = 10$ M	1950	6.4	390/800	2270	8.6	550/1110
Horizontal rod–rod $h/S = 1.5$	$h/S = 1.0$						
	$\alpha = 0$	2250	9.0	590/1180	2600	12.3	850/1660
	$\alpha = 0.33$	2250	6.4	390/800	2600	8.4	540/1090
	$\alpha = 0.5$	2250	5.5	320/660	2600	7.2	450/910
Horizontal rod–rod $h/S = 1.5$	$\alpha = 0.33$	3020	11.5	790/1550	3360	14.9	1050/2040
	$\alpha = 0.5$	3020	9.5	630/1250	3360	12.1	830/1630
Horizontal ring–ring $h/S = 1.5$	$\alpha = 0.5$	2250	4.7	480/990	2600	6.0	350/730
	$\alpha = 0.33$	2250	5.5	320/660	2600	7.1	440/900
	$\alpha = 0.5$	2250	4.7	250/520	2600	6.0	350/730
Horizontal ring–ring $h/S = 1.5$	$\alpha = 0$	3020	14.4	1000/2040	3360	—	—
	$\alpha = 0.33$	3020	9.3	600/1230	3360	11.6	780/1600
	$\alpha = 0.5$	3020	7.8	490/1000	3360	9.6	630/1270
Horizontal conductor–conductor $h = 18$ m, $L = 550$ m	$\alpha = 0$	2250	10.4	670/1420	2600	14.7	1020/2130
	$\alpha = 0.33$	2250	6.7	390/830	2600	8.7	540/1140
	$\alpha = 0.5$	2250	5.6	300/660	2600	7.2	420/910
Horizontal conductor–conductor $h = 18$ m, $L = 550$ m	$\alpha = 0.33$	3020	12.1	810/1700	3360	—	—
	$\alpha = 0.5$	3020	9.7	610/1300	3360	12.3	820/1730

Table 19.2 Gap factors under direct voltage (data from Pigini et al. [22])

Configuration	Gap *d* (m)	Sparkover voltage (kV)	Gap factor (*K*)
Rod–plane	1	450	1
	2	910	1
Rod–rod[a]	1	520	1.16
	2	1050	1.15
Single conductor–structure	1	600	1.33
	2	1150	1.26
Bundled (4) conductor–structure	1	460	1.02
	2	950	1.04

Note: The rod/plane has the weakest sparkover voltage and thus has a gap factor of 1 (see column 3) and for a 1 m gap its sparkover is 450 kV (see column 2). A rod–rod gap is stronger and has a sparkover value of 520 kV (column 3). Its gap factor is thus 520/450 = 1.16 (see column 3) etc.
[a]Tip of lower rod is 1.5 m above the ground plane.

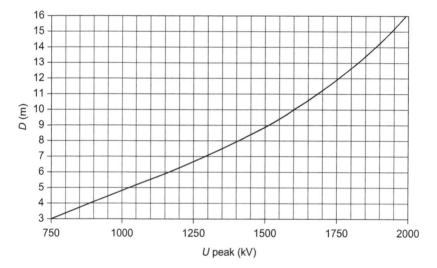

Figure 19.25 Test clearances [4]
Recommended minimum clearance *D* of extraneous live or earthed objects to the energised electrode of a test object during an AC or positive switching impulse test at the maximum voltage *U* applied during the test

19.7 Flashover across insulator surfaces in air

The weakest aspect of the insulation of high voltage systems is frequently the dielectric strength of the solid insulator which is used to separate the conductors. Knowledge of the basic processes involved is still relatively fragmented, but it is known that streamers, once initiated, may travel more rapidly across certain insulator surfaces in air than they do in the air in the absence of the insulator. It is known that charges deposited on the surface by corona may so alter the electric field distribution as to facilitate flashover. However in practice, other factors may likewise cause field distortion and encourage flashover: unequal distribution of capacitance between insulators and insulators; stray capacitances to ground; field concentration at the triple junction between insulator, conductor and air; layers of pollutants; rain and deposition of moisture; all of these can contribute to weakness in the dielectric of a surface. The tendency of a surface to reduce the insulation strength is minimised by the adoption of 'sheds' in order to increase the total path length between conductors. The 'cap-and-pin' insulator takes this concept to the limit and, until recently almost all power lines operating above 100 kV, used strings of insulators of this type. For glass and glazed insulators, the resistance of the surface to the deleterious effects of pollutants may be improved by the use of grease, silicone rubber coatings. 'Nano' surface treatment provides a non-sticky surface that minimises the build-up of pollutants. Figure 19.26 is a photograph of a treated glazed ceramic string insulator being subjected to a wet switching impulse overvoltage test.

Negative impulse voltages may have more effect on the strength of highly stressed polluted insulators. High electric field concentration near the 'triple point' may set up cathode mechanisms that could encourage initiation of discharges.

Non-ceramic insulators are being increasingly used for overhead line and for substation equipment.

Figure 19.26 Wet switching impulse test on a 400 kV insulator string (With kind permission of Alan Edwards)

19.8 Atmospheric effects

19.8.1 Introduction

All that have been written so far must be qualified by the realisation that the temperature, pressure and humidity change in an uncontrolled way in atmospheric air.

It is necessary therefore to establish procedures which demonstrate that

(a) it is possible to estimate breakdown voltages at any atmospheric condition from those measured at another condition, and
(b) measurements can be corrected to a standard atmospheric condition for the purposes of comparison of results obtained in different laboratories at different times.

19.8.2 Density effects

In early experiments in air, it was found that the electrical breakdown of a uniform gap depended on the gap length d and the number density, n, of the gas. The function $V_s = f(nd)_s$ known as Paschen's law [24] holds for a great variety of conditions. It seems obvious today that density has an effect, but we still need to describe the function f and investigate the effects of non-uniform fields.

Temperature and pressure together determine the density of air. The ionisation processes involved in corona and breakdown depend primarily on the air density. Atmospheric temperature has little effect in the range of −40 to +40 °C, 233 K to 314 K. Many experiments have shown that, over a limited range of density, the breakdown voltages vary linearly with density. Thus, it has proved to be a relatively straight forward matter to define a set of conditions that can be regarded as characteristic of a standard atmosphere. It is possible to relate other conditions of pressure and density to this standard atmosphere, assuming the simple gas laws to be correct.

A standard atmosphere is defined for these purposes in IEC Standard 60060-1 [4] as 1013 hPa (1013 mbar) or 760 Torr (760 mm of the mercury column in a mercury barometer at 93 K (20 °C)). Then, at any temperature T and pressure p, the density δ has changed from that at standard atmosphere by

$$\delta = \frac{p\ 293}{1013\ T} \tag{19.16}$$

Clearly, the value of δ is the ratio of the density at (p, T) to that of the standard condition. δ is called the relative air density; $\delta = 1$ is at the standard condition.

If now the value of U_{50} at relative air density $\delta = 1$ is denoted by U_0, then the value of U at (p, T) is

$$U = U_0\ \delta \tag{19.17}$$

It should be observed here that this discussion ignores any effects due to humidity. These effects will be discussed below, but it should be noted that the standard atmospheric condition assumes an absolute humidity of 11 g moisture content per cubic metre of air.

The density correction procedure has proved to be satisfactory over the extreme range of 313 K to 243 K. This procedure has also proved satisfactory for normal variations in pressure at altitudes up to 2000 m. However, as stated in the introduction, a number of electrical transmission systems are being developed at altitudes higher than 2000 m. Further data on the effects of high altitude and humidity will be produced in new high voltage laboratories that are currently being built in regions above 2000 m.

19.8.3 Humidity effects

Atmospheric humidity can change from a moisture content of less than 1 g/m^3 in very cold countries to values of the order of 30 g/m^3 in the tropics.

The effects of humidity on sparkover are quite complex, but the following points may be made.

(a) Humidity has its strongest influence on the positive pre-breakdown discharge. In particular, the streamer gradient E_s (equation (19.5) and Figure 19.18) increases at the rate of roughly 1% per g/m^3 increase in moisture content (i.e. the final sparkover voltage increases).
(b) Humidity has no significant effect on the high-temperature high-conductivity leader gradients (E_l^+ and E_l^-); however, it does have the effect of increasing the leader velocity.
(c) Humidity has no significant effect on the negative sparkover under lightning impulse and under direct voltages.

The disruptive discharge of external insulation depends upon the atmospheric conditions. Usually, the disruptive discharge voltage for a given path in air is increased by an increase in either air density or humidity. However, when the relative humidity exceeds about 80%, the disruptive discharge voltage becomes irregular, especially when the disruptive discharge occurs over an insulating surface.

The disruptive discharge voltage is proportional to the atmospheric correction factor K_t which is determined from the product of two appropriate correction factors

- air density correction factor k_1
- humidity correction factor k_2

$$K_t = k_1 k_2 \tag{19.18}$$

19.8.4 Application of correction factors
19.8.4.1 Standard procedure

If U is a disruptive discharge voltage measured in given test conditions (temperature t, pressure p and humidity h), then it may be converted to the value U_0, which would have been obtained under the standard reference atmospheric conditions.

$$U_0 = U/K_t \tag{19.19}$$

19.8.4.2 Iterative procedure

Conversely, if a test voltage is specified for standard reference conditions (U_0), then the equivalent value at the test conditions (U) must be calculated.

This should simply be

$$U = U_0 K_t \tag{19.20}$$

However, because U enters in the equation to derive K_t, an iterative procedure might have to be used.

The test is to estimate a value for U, and then divide it by K_t. If the result is the specified test voltage U_0, then the estimated value for U has been correct. If the resultant value for U_0 is too high, then the estimated value for U must be increased. Conversely, if the resultant value for U_0 is too low, then the estimated value for U must be decreased.

Successive iterations will produce the correct test voltage U to be applied. For alternating voltages, the peak value must be used because fundamental breakdown processes culminate at the maximum voltage.

19.8.5 Air density correction factor k_1

As seen above (section 19.8.2), density depends on pressure and temperature.

The air density correction factor k_1 depends on the relative air density δ and can be generally expressed as

$$k_1 = \delta^m \tag{19.21}$$

where m is an exponent (Table 19.1).

When temperatures t and t_0 are expressed in degrees Celsius and the atmospheric pressures p and p_0 are both expressed in the same units, then the relative air density is

$$\frac{p}{p_0} \times \frac{273 + t_0}{273 + t} \tag{19.22}$$

The correction is reliable for $0.8 < k_1 < 1.05$.

19.8.6 Humidity factor correction k_2

The humidity factor may be expressed as

$$k_2 = k^w \tag{19.23}$$

where w is an exponent (Table 19.2) and k is a parameter that depends on the type of test voltage, and may be obtained as a function of the ratio of absolute humidity to the relative density δ by using the following equations (for DC, AC and impulse voltages respectively):

DC

$$k = 1 + 0.014\left(\frac{h}{\delta} - 11\right) - 0.00022\left(\frac{h}{\delta} - 11\right)^2 \tag{19.24}$$

for $1 \text{ g/m}^3 < \frac{h}{\delta} < 15 \text{ g/m}^3$

AC

$$k = 1 + 0.012\left(\frac{h}{\delta} - 11\right) \qquad (19.25)$$

for $1 \text{ g/m}^3 < \frac{h}{\delta} 15 \text{ g/m}^3$

Impulse

$$k = 1 + 0.010\left(\frac{h}{\delta} - 11\right) \qquad (19.26)$$

for $1 \text{ g/m}^3 < \frac{h}{\delta} 20 \text{ g/m}^3$

Figure 19.27 shows k as a function of the ratio of absolute humidity h to the relative air density δ.

Impulse equation (19.26) is based on experimental results from positive lightning impulse waveforms. The equation also applies to negative lightning impulse voltages and switching impulse voltages.

19.8.6.1 Exponents *m* and *w*

As mentioned previously, the practical equations to allow for changes in atmospheric conditions were devised from work on the fundamental breakdown processes involved. We have seen that the weakest configuration is a rod–plane gap where the electric field at breakdown is 500 kV/m. Because the dominant breakdown process is that of streamers with a field of 500 kV/m, it is no surprise that the number 500 appears in the equation used to calculate breakdown voltages.

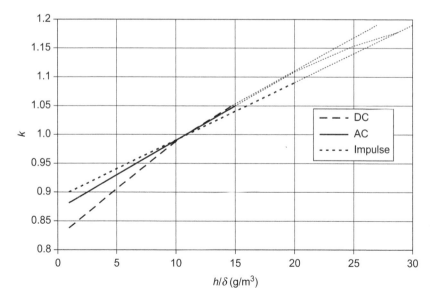

Figure 19.27 Humidity correction factor k as a function of the ratio of the absolute humidity h to the relative air density δ [4]

We consider the parameter

$$g = \frac{U_{50}}{500L\delta k} \qquad (19.27)$$

where U_{50} is the 50% disruptive voltage (measured or estimated) at the actual atmospheric conditions, in kilovolt peak. L is the minimum discharge path in metre δ is the relative air density. k is the dimensionless parameter defined in section 19.6.2, and also in (19.24)–(19.26).

Table 19.3 shows the values of m and w as a function of parameter g[ ].

Table 19.3 Values of m and w as a function of parameter g

g	m	w
>0.2	0	0
0.2 to 1.0	g(g − 0.2)/0.8	g(g − 0.2)/0.8
1.0 to 1.2	1.0	1.0
1.2 to 2.0	1.0	(2.2 − g)(2.0 − g)/0.8
>2.0	1.0	0

Earlier editions of IEC Standard 60060-1 did not have the formulae equation (19.27). It is recommended to use the values in Table 19.2. A small algorithm will suffice. (Using the diagrams is not recommended.)

19.8.6.2 Measurement of atmospheric parameters

Humidity should preferably be determined with an instrument that measures absolute humidity with an expanded uncertainty no larger than 1 g/m^3. The measurement may also be made using a normally aspirated wet and dry bulb hygrometer.

Temperature
The ambient temperature should be measured, with an expanded uncertainty of no larger than 1 °C.

Absolute pressure
The ambient absolute pressure should be measured with an expanded uncertainty of no larger than 2 hPa.

19.8.7 Other atmospheric effects

19.8.7.1 Wet tests

The pre-discharge processes occur over a solid surface in a medium of liquid, vapour and air. The humidity of the surrounding air has little or no effect on the final disruptive discharge. Although density corrections are applied, no correction for humidity is made for wet tests.

19.8.7.2 Artificial pollution tests

The conditions on the surface of insulators under test differ from those in the electrode to electrode air gap. Salt-fog tests involve the spraying of solutions of such pollutants as NaCl directly onto the surface of insulators. Different salinities are applied and both flashover and withstand tests are used. Other pollutants, including kaolin, Kieselguhr, Fuller's Earth, fine silica and cement are applied over insulators. Flashover and withstand tests have been devised to attempt to simulate system conditions. In practice, the flashover is even more complicated than the air breakdown discussed above. As the applied voltage is increased, the leakage current raises the conductivity thermally. Water is lost and local heating occurs, resulting in bands of high resistivity. The so-called 'dry-bands' appear. These dry bands may break down in a similar way to air discharges. The current is limited by the conductivity of the pollutants, but subsequent 'arc-like flashovers' may build up to effect total flashover.

19.9 New developments

19.9.1 UHV at high altitudes

Much of the existing worldwide electrical standards for high voltage testing and measuring were developed from work undertaken in Europe. As discussed above, the most important new developments in UHV are in regions where atmospheric conditions differ considerably from those in temperate and warm areas less than 2000 m above sea level. Consequently, it is essential that other regions of the world will participate strongly in strategic new work directed towards IEC standards in this sector.

The new developments are as follows:

- UHV transmission (DC up to 800 kV (+ve and –ve) and AC up to 1200 kV)
- transmission at high altitudes.

These developments may occur together. Electrical energy from hydroelectric schemes in mountainous areas is transmitted over long distances at UHV and at high altitudes. High humidity does not only occur at sea level.

This is even more important now because most of the standards were written when transmission systems operated at relatively low levels of humidity and at relatively low altitudes. The IEC has recently set up a strategic group on the standardisation of UHV technologies. This includes a proposal to include 'realistic' atmospheric and altitude corrections in future standards.

Further comprehensive work also needs to be done to study the effects of overvoltages on UHV systems. The sometimes strange performance of the electrical breakdown in air under switching impulses needs to be understood. In practice, switching surges may result in disruptive discharges in air but may also cause flashovers on wet and polluted insulators. The subsequent power follow-through is even more dangerous. Clearly, there is much strategic research still to be done and many technical aspects still to be resolved for UHV systems during the next few years.

19.9.2 Testing transformers

When testing transformers with lighting impulses overshoots and oscillatory overvoltages are inevitably produced at the peak of the impulse. The problem is to determine the value of the test voltage U_t that the insulation is subjected to when the applied lightning impulse voltage with an overshoot magnitude β.

The empirical equation is

$$U_t = U_b + k(f)(U_e - U_b) \qquad (19.28)$$

where U_b is the maximum value of the base curve and U_e is the maximum value of the 'noise-free' recorded curve.

Following 10 years of study of IEC Standard 60060-1:2010 [4] has introduced a factor which should help to 'normalise' the handling of test results. The method prescribed in IEC Standard 60060-1:1989 was simply to apply a 'step' filter for oscillations above 500 kHz. The sharp transition from a maximum value to mean curve value led to large errors near the transition frequency.

Five research institutes in Europe were involved in applying overshoots on lightning impulses to uniform field gaps in different media (atmospheric air, oil, oil/paper, SF_6 and polymer sheets). The results had one thing in common: At low frequencies of overshoot the maximum voltage was dominant resulting in breakdown and at higher frequencies the maximum voltage was less dominant. Perhaps, at low frequency the fundamental processes of electron avalanches and development into steamers and leaders have time to develop because the voltage remains above inception level. In the high frequency cases the fundamental processes have less time at the peak developed before the voltage drops below inception level. It does seem to be remarkable that the results were so similar for different media.

The simple filter has the function

$$k(f) = \frac{1}{1 + 2.2f^2} \qquad (19.29)$$

where f is the frequency in megahertz.

Originally, filtering was done 'by hand' on the oscillograms of the test records. The new filter in accord with the digital recording of high voltage impulses can be applied after the raw data has been stored.

The procedure follows the steps shown in Figures 19.28(a–c). In Figure 19.28(a), U_e is the maximum value of the original noise-free recorded curve, U_b is the maximum value of the base curve and β is the overshoot.

A test data generator (TDG) has also been produced following a round-robin study of actual test voltages (and currents) of impulses, with varying 'overvoltages' and frequencies.

The IEC Standard 60183 [25, 26] (hardware and software for recording impulses) will incorporate this TDG to help to incorporate the requirements of IEC standard 60060-1 (parts 1 and 2).

Obviously, transmitting electrical energy at UHV voltages (>1.2 MV alternating voltages at power frequency and also direct voltages 800 kV +ve, −ve) means much larger structures used than of late.

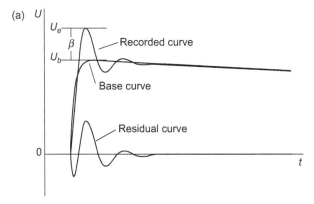

Recorded and base curve showing overshoot and residual curve

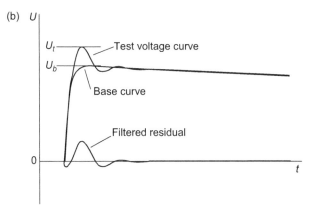

Test voltage curve (addition of base curve and filtered residual curve)

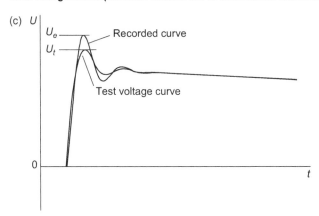

Recorded and test voltage curves

Figure 19.28 (a,b) Evaluation of the test voltage curve from the recorded and base curves (IEC 60060-1 Annex B); (c) comparison of the original recorded curve and the test voltage curve [4]

Figure 19.29   Positive SI breakdown – 9 m clearance [reproduced with permission of Alstom Grid]

Tests on the effect of switching surges on these systems have been simulated in the laboratory with switching impulses applied to well-known electrode geometries as well as actual equipment (Figure 19.24 and Table 19.1).

Thus Figure 19.29 presents disruptive discharges resulting from positive Switching Impulse voltages applied to UHV equipment. DC discharges are not unlike natural lightning discharges. The image of the wayward switching impulse discharge shows how the streamers and then leaders become extensions of the electrode and have time to bifurcate and progress in a more chaotic manner.

19.9.3   Future work

Much work still needs to be done on air breakdown processes and on the flashover characteristics of polymeric and other non-ceramic insulators. The design of transmission towers and equipment for UHV direct voltage will involve the need for even larger high voltage laboratories. More sophisticated computer modelling of the interaction of atoms and molecules with electrons in high electric fields may help designers. High-speed photography and the application of chaos theory may help in the visualisation of the fundamental aspects of air breakdown.

References

1. Allen N.L. 'Fundamental aspects of air breakdown' (Chapter 14), in H.M. Ryan (Ed.), *High voltage engineering and testing* (vol. 17, 1st edn). IEE Power and Energy Series. London: Peter Peregrinus Ltd, 1994
2. Allen N.L. 'Fundamental aspects of air breakdown' (Chapter 20), in H.M. Ryan (Ed.), *High voltage engineering and testing* (vol. 32, 2nd edn).

IEE Power and Energy Series. London: The Institution of Electrical Engineers, 2001
3. Allen N.L. 'Mechanisms of air breakdown' (Chapter 1), in A.M. Haddad, D. Warne (eds.), *Advances in high voltage engineering* (vol. 40). IEE Power and Energy Series. The Institution of Electrical Engineers, London, 2004 [Soft Back, 2009]
4. BS EN 60060-1 2010. 'High-voltage test techniques. Part 1: General definitions and test requirements':
    Permission to reproduce extracts from British Standards is granted by the British Standards Institution (BSI). No other use of this material is permitted. British Standards can be obtained in PDF or hard copy formats from the BSI online shop: http://shop.bsigroup.com or by contacting BSI customer Services for hard copies only: Tel +44 (0)20 8996 9001, Email: cservices@bsigroup.com
5. Faraday M. *Experimental researches in electricity*. 13th series. London: Royal Institution, 1838
6. Townsend J.S. *Phil. Mag.* 1901;**1**:198 [cited in Dutton, J. Spark breakdown in uniform fields. in Meek J.M., Craggs J.D. (eds.), *Electrical breakdown of gases*. Chichester: Wiley, 1978, Chapter 3]
7. Meek J.M., Craggs J.D. *Electrical breakdown of gases*. Chichester: Wiley, 1978
8. Kuffel E., Zaengl W.S., Kuffel J. (eds.). 'Electrical breakdown in gases' (Chapter 5), *High voltage engineering fundamentals* (2nd edn). Butterworth-Heinemann, 2000
9. Pedersen A. 'Calculation of spark breakdown or corona starting voltages on non-uniform fields'. *IEEE PAS-86*. 1967;**2**:200–6
10. Mattingley J.M., Ryan H.M. 'Breakdown voltage estimation in air and nitrogen', *Proceedings of the NRC conference on electrical and dielectric phenomena*. Williamsburgh, VA, November 1971
11. Blackett J. 'High-voltage breakdown and pre-breakdown characteristics of sphere/hemisphere electrode arrangements in air', M.Sc. Thesis, University of Newcastle upon Tyne, 1977
12. Hermstein W. 'The streamer discharge and its transition to a corona or glow discharge'. *Arch. Elektrotech*. 1960;**45**(3):209–24
13. Uhlig C.A.E. The ultra corona discharge; a new phenomenon occurring on thin wires. Paper 15, High Voltage Symposium, National Research Council of Canada, 1956
14. Waters R.T., Stark W.B. 'Characteristics of the stabilised glow discharge in air', *J. Phys. D. Appl. Phys.*, 1975;8
15. Nasser, E., Abou-Seada M.S. *Calculation of streamer thresholds using digital techniques*, NRC Canada Paper, pp. 534–9, 1972
16. Gorur R.S. 'Corona: Silent but deadliest enemy for composite insulation', *INMR*. 2010;**18**(1)
17. Waters R.T. (Ed). (Les Renardiers Group). 'Double impulse tests of long air gaps', *IEE Proceedings A, Physical Science, Measurement and Instrumentation, Management and Education, Reviews*, 1986;**133**:393–483

18. Carrara G. 'Investigation on impulse sparkover characteristics of long rod/rod and rod/plane air gaps', CIGRÉ Report No. 328, 1964
19. Hutzler B., Garbagnati E., Lemke E., Pigini A. 'Strength under switching overvoltages in reference ambient conditions', in CIGRE Working group 33.07, Guidelines for the Evaluation of the Dielectric Strength of External Insulation, CIGRE, Paris, 1993
20. Jones B., Waters R.T. 'Air insulation at large spacings'. *Proc. IEE*. 1978;**125**(11R):1152–76
21. Lokhanin A. 'Sketch drawings of typical air gaps', CIGRE Working Group C4-10 (WG 306)10122
22. Pigini A., Thione L., Rizk F. 'Dielectric strength under AC and DC voltages', in CIGRÉ Working Group 33.07 Guidelines for the Evaluation of Dielectric Strength of External Insulation, CIGRE, Paris, 1993
23. Rizk F.A.M. 'Modeling of proximity effect on positive leader inception and breakdown of long air gaps', *IEEE Trans. on Power Del.* October 2009;**24**(4), 2311–2318
24. Paschen F. *Wied. Ann.* 1889;**37**, 69
25. IEC BS EN 601083-1. 'Instruments and software used for measurements in high voltage impulse tests, Part 1: Requirements for instruments'
26. IEC BS EN 601083-2. 'Digital recorders for measurements in high voltage impulse tests, Part 2: Evaluation of software used for the determination of the parameters of impulse waveforms'

Chapter 20
Condition monitoring of high-voltage transformers
A. White

20.1 Introduction

Transformers and their component parts are critical to the reliable and uninterrupted functioning of all electric power systems. In order to increase availability of critical circuits and to optimise operational management, condition monitoring of power transformers is not only useful, but is fast becoming essential, since utilities are constantly facing the need to reduce costs that are associated with the operation and maintenance of the installed equipment.

Condition monitoring equipment that is used to extend the life of a transformer or to prevent catastrophic failure could pay for itself many times over in the lifetime of the apparatus to which it is fitted. Nevertheless, condition monitoring does not come free of charge, and utilities have tight budgets that have to be met. To determine best value for money, life-cycle costing must be applied to each chosen application. The life cost comes from the summation of the installation, the maintenance, the repair and the lifetime operational costs of the equipment. Several options of monitoring equipment covering a wide spectrum of costs are readily available and others are under development. Monitoring equipment that looks at one or two specific parameters is readily available in the marketplace at modest cost. At the other end of the cost spectrum, not only can many parameters be monitored, but an expert system or artificial intelligence system that is capable of generating an estimate of the overall plant condition may be used to interrogate the data and draw conclusions and recommendations [1]. There are several such integrated systems on the market, such as the MS 3000 supplied by Alstom Grid [2].

A single monitor that costs a relatively small sum of, say, £2000, if fitted to every transformer on the UK National Grid network, would have a total investment cost of more than £1 million. To this expense is added the cost of a communication system to bring the data to a central location. Additionally there is the cost of maintaining the monitoring system itself as well as the individual sensors. A sophisticated expert monitoring system will cost several tens of thousands of pounds, and even then would still not permit a fit-and-forget policy. After spending so much on the monitoring system, one cannot afford to have it fail, and so the additional expense of monitoring the monitoring system must be considered.

In the domestic insurance market, many insurance companies will offer reduced household contents premiums provided that security devices and alarms are fitted to domestic property. With this background, surely it cannot be too long before electrical equipment will be uninsurable unless a minimum degree of condition monitoring is applied.

In recent years there has been a noticeable shift from time-based to condition-based maintenance of electrical plant. Thanks to the evolution of modern monitoring systems, personnel can now concentrate their activities on those tasks that will give the highest added value at any point in time. Utilities are faced with the challenge of exploiting the full capabilities of the equipment, and preventing failures and outages, at the same time as decreasing maintenance expenditure and scheduling investment to when the most cost beneficial opportunity arises.

Condition monitoring can be classified by offline monitoring and online monitoring. Offline measurements have, in the past, been carried out successfully in the determination of the operational state of transformers, and will continue to do so well into the foreseeable future. Modern microprocessor and computer technologies now make it possible to perform many of these procedures online, and as the number of applications grows, inevitably the costs of online monitoring will reduce. Today's technology means that an integrated expert system, with its comprehensive data storage, clever evaluation algorithms and diagnostic functions is capable of providing a significant contribution to the even higher availability of the transformer as well as the whole power grid.

20.2 How do faults develop?

Faults in transformers can develop extremely quickly, in the space of seconds, milliseconds or even in a matter of nanoseconds. In such circumstances, the principal aim is to reduce the degree of damage that might occur on the faulted equipment and protect the remainder of the network from the effects of this failure.

Other faults may develop within periods that are measured in minutes, hours or days. The aim, then, is to prevent catastrophic failure and perhaps even to extend the life of the plant by controlled usage or to achieve a planned isolation of the faulty plant with the minimum of system disruption. Then there are faults that take weeks, months and sometimes years to evolve, so permitting planned refurbishment or replacement.

Field experiences have shown that developing malfunctions and faults in power transformers can be detected much earlier with the deployment of an efficient online monitoring system and as a result the risk to the system is very significantly reduced.

20.3 Which parameters should be monitored?

A CIGRE questionnaire on transformer failures was analysed and reported by Bossi *et al.* in 1983 [3]. This report identified the relative numbers of components involved in transformer failures. Some of the CIGRE report findings identify the principal components involved in transformer failure and are reproduced as Table 20.1.

Table 20.1 Components causing failure of transformers in service

Windings	29%
Bushings	29%
Tank and fluid	13%
On-load tap-changer	13%
Magnetic circuit	11%
Other accessories	5%

While components of the active part of the transformer, viz. the core and winding assembly, are of obvious importance, it is quite significant that bushings and on-load tap-changers also figure prominently in the table.

In fact, each utility will have a unique view on which parameters are most indicative of the dominant type of faults that occur on its transformers and systems and would most benefit from being monitored. There may also be a particular family of transformers which demonstrates a particular failure type and a utility owning such a family is likely to want to monitor more parameters relating to that failure mode than a utility whose prime objective is to reduce maintenance costs. A utility that has suffered a spate of bushing failures will need a different monitor to that of a utility that is seeking to extend the life of transformers on its system. Some parameters, especially those intended to show fast or medium rate fault development will require online monitoring, while others are more economically handled by regular inspection or offline measurement or observation. There is no universal solution for condition monitoring and each utility must make an individual selection of the most appropriate monitoring.

20.4 Continuous or periodic monitoring

Monitoring while in service is sometimes possible on a continuous basis or by periodic checking and trend determination. Some periodic checks, furthermore, may require de-energisation of the transformer, and possibly total disconnection from the system. Each utility must therefore consider the costs and the anticipated value in reaching the decision on which parameters should be monitored and how they should be scrutinised. Intrusive measurements are also a possibility in the extreme, but would generally be adopted only after one or more continuous or periodic monitors have identified a specific problem.

20.5 Online monitoring

In recent years online monitoring systems have been installed on a fairly substantial scale on the most critical power transformers on a network. A combination of different measuring methods plus some relatively sophisticated software tools that are capable of making decisions on the operation of the transformers, has provided enhanced security for the systems in which they are located. A variety of concepts for efficient assessment of the transformer condition can be established, making condition-based maintenance planning a reality instead of just a pipedream. Using the Ethernet or the Intranet, information can be transmitted from the substation to a PC, thus enabling web-based visualisation (Figure 20.1). From the earliest simple monitors, technology has evolved at a rapid rate to the presentation of diagnostics and prognosis with the data.

By means of suitable mathematical models, measurement data can be converted into user friendly information for reliable condition diagnosis. The evaluation of data that is acquired from site located equipment provides the capability of early detection of problems within the active part, the bushings, the on-load tap-changer and the cooling unit before they each develop into major failures. In addition, algorithms for the calculation of overload capacity, life utilisation and remaining life are of increasing importance in getting the very best out of power equipment. The task of online monitoring is then to provide focused and purposeful diagnostic information, so that timely remedial measures can be initiated.

20.6 Degrees of sophistication for transformer monitors

The simplest case of transformer monitoring is the measurement of single parameters up to the conception of a simple model, e.g. measurement of oil temperature, dissolved gas-in-oil levels, moisture-in-oil, partial discharge or thermal imaging derivations. This manner of monitoring makes it possible to advise the user in the event that a pre-set threshold value is reached. With such a simple tool, a correlation with other parameters is generally not possible. The user will merely react to the threshold value being reached; and the possibilities of imminent or eventual failures together with maintenance management proposals are very limited, and the operator takes on the sole responsibility of correctly interpreting the data.

The next level of sophistication comprises a basic monitoring system or the monitoring of a series of separate components of the power transformer, e.g. bushings, on-load tap-changer or cooling units. Such systems scrutinise more than one parameter, but use simple models in order to assess the health of the supervised components. Frequently the correlation of data across the various transformer components is not possible, even if several components are monitored, because the data is sourced from separate devices. The correlation of measured data relevant to an individual component, however, is perfectly feasible, and early detection of abnormal status is possible in many cases. The interpretation of the data imparts

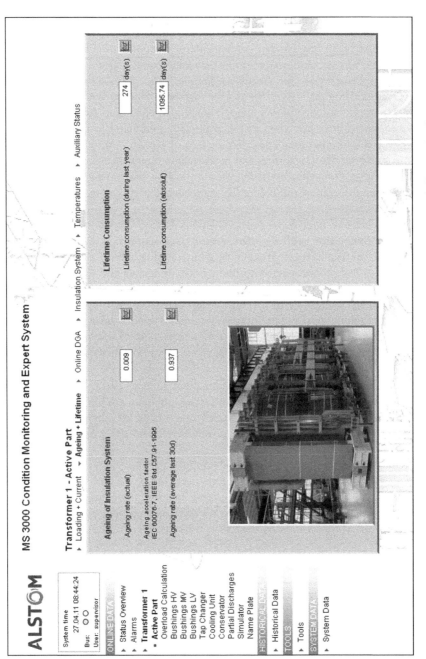

Figure 20.1 Typical condition monitor visualisation (courtesy of Alstom Grid)

potential status information about the monitored transformer component and also about the transformer as a whole.

Moving upward in the degree of refinement, a comprehensive and interactive high-level online condition monitoring, diagnosis and expert system is obtained by the integration of all the relevant components of the power transformer within a single system. The correlation of multiple data concerning a specific component, or even several transformer components is possible, since all acquired data is now located within a common database, and that database together with all calculation algorithms are to be found within a single intelligent electronic device (IED).

Furthermore, the correlation of all of the data pertaining to several transformers is a viable proposition. The interpretation of the data provides health status information about the transformer and of its main components. The expert system and its diagnostic functions assist the user in reaching the most appropriate decisions regarding operation and predictive maintenance plans for the transformer or for the substation.

Comprehensive interactive expert solutions have the benefits of being cost efficient in comparison with a combination of any number of stand-alone monitoring systems. The concept of this type of monitoring system can be employed with field bus and process control technology to implement flexible system architectures. Such a system centres on an IED with server functionality, which makes the simultaneous monitoring of several transformers possible.

Analogue signals emanating from the individual sensors are wired to field bus terminals in a monitoring module on the transformer. Here, the signals are digitised and then transferred to the IED using a suitable field bus protocol. The monitoring IED may be located directly on the transformer (Figure 20.2) or at a remote location in the substation or power plant (Figure 20.3). The monitoring modules and the IED are interconnected with galvanic or fibre optic cables. Connections to a remote location are usually also made via fibre optic cables in order to avoid problems from electromagnetic interferences that might result from wired connections.

Acquisition of the data can often be arranged to take place with millisecond accuracy. Calculated values (e.g. aging rate, overloading condition and change in the bushing capacitance) are derived from the measured quantities, and the data acquisition can be both time-controlled as well as event-controlled. Thus, for example, when a transformer is energised by switching in, the voltages would

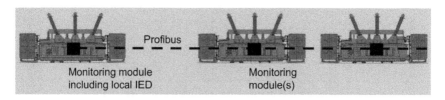

Figure 20.2 Condition monitoring with intelligent electronic device on a transformer (courtesy of Alstom Grid)

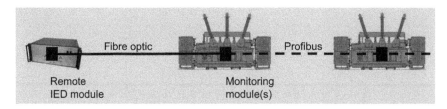

Figure 20.3 Transformers monitored with remote intelligent electronic device (courtesy of Alstom Grid)

typically be recorded over 10 second duration with a resolution of 20 ms. By comparison, the changes in the bushing capacitance would probably be sampled every 20 ms and saved every 15 min. Consequently the size of the database is optimised to a level that is manageable and useful. It is pointless and uneconomic effort to save data that is not going to be used.

Access to the monitoring data is preferably achieved by means of a standardised platform. This eliminates the need for each desktop PC to be installed with individually tailored software. Internet technology provides the underlying foundation for web-based visualisation of the recorded data and the diagnosed information. Browsers such as Internet Explorer, which is a globally proven performer in flexibility and reliability, are standard installations in virtually every computer. In addition, web technology is widely applied as the industrial standard for the visualisation of data in network environments.

The operator expects from a monitoring system, a user friendly and a safe access to all necessary information about the installed electrical equipment. The grid control centre handles information about loading capacity (overload and life utilisation calculations). The operation and maintenance department can perform condition assessment and plan the maintenance procedures with high precision (Figure 20.4). The wide distribution of information can be executed by means of a web-based solution. An additional module installed in the monitoring IED enables the generation of HTML-based web pages, which illustrate both current and historical data. Additionally, if the monitoring IED is connected to the local area network of the utility, every department will be able to access only that information that it is authorised to do so. The number of users directly connected to the monitoring server is virtually unlimited, and password protection gives the appropriate level of data access to specific users. By installing a firewall, it is also possible to have access to the complete substation over the Internet.

20.7 What transformer parameters can be monitored?

Internal and external sensors can be used as sources for monitoring the transformer while it remains in service, but additional data or measurements relating to loading can provide further information on which full assessment of condition may be derived.

682 High-voltage engineering and testing

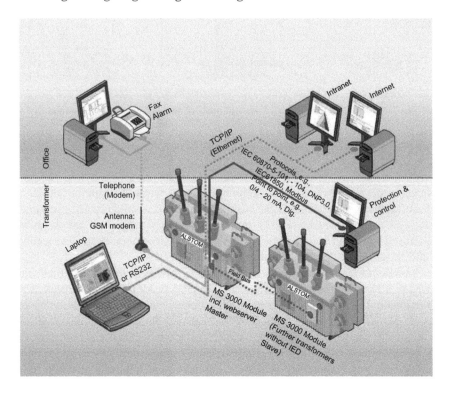

Figure 20.4 Multiple access to condition monitoring database (courtesy of Alstom Grid)

Monitors available in the market place are capable of monitoring such parameters as partial discharge, temperature, dissolved gas-in-oil, fur-furans, moisture, resistance, capacitance, frequency response analysis, polarisation spectrum, tap-changer mechanical position and current. A multitude of measurable variables can be collected for online monitoring. However, coverage of the entire spectrum is rarely cost effective. Therefore, sensor technology should be adjusted to the specific requirements of a particular transformer or transformer bank, depending on age and condition.

20.8 Basic monitors

For the earliest fault detection, monitoring of the active part is frequently of particular importance. Some basic measurements that are valuable include the electrical variables as determined directly at the transformer terminals, viz. the operating voltage and load current. From these, the apparent power and the load factor may be calculated. The load current and top oil temperature are the basic input variables for the calculation of hot-spot temperature and aging rate of the insulation in accordance with IEC 60076-7 (*Loading Guide for Oil-Immersed Power Transformers*).

By the inclusion of ambient temperature monitoring too, these measurements enable the evaluation of life consumption resulting from loading patterns, and also of the short time overload capacity of the transformer. Beyond this, knowledge of the number and amplitude of short-circuit currents may be utilised in the assessment of the mechanical condition of the windings. These events are detected and evaluated using a high sampling rate on the load current signal.

Centred on the aforementioned, basic monitors might therefore include such aspects as follows:

measured values
    operating voltage
    load current
    top oil temperature
    gas-in-oil content
    moisture-in-oil content
    on-load tap-changer position
    operating condition of pumps and fans
    ambient temperature
    inputs from other auxiliary equipment

derived information
    apparent power and load power factor
    oil temperature rise above ambient air or water
    hottest-spot temperature
    aging rate
    number of tap-changer operations completed
    aggregate tap-changer switched load current
    aggregate operating time of pumps and fans
    cooling efficiency.

20.9 On-load tap-changer (OLTC) module

Depending upon a particularly prevalent cause of failure, a utility may wish to maintain a record of tap-changer position and the operating current to assist in the determination of the number of switching operations of the tap-changer and the aggregate switched current, which can be used to predict the level of the wear of diverter switch contacts. This is essential information for a condition based maintenance regime for the diverter switch. Undetected excessive wear might lead to the contacts burning open or else weld them together. Limiting values for the time in service, the number of operations and the aggregate switched load current can be preset in accordance with the maintenance recommendations of the OLTC manufacturer.

Tap-changer failures are often of a mechanical nature, e.g. a broken linkage, a spring fracture, binding contacts, worn gearing or incorrect functioning of the drive mechanism. Mechanical and control variances can often be detected by the measurement of the power consumption of the tap-changer drive, since additional friction, extended changer operation durations and other abnormalities all have a significant influence on the current that is drawn by the drive. An event record of the power consumption is captured during each tap-changing process and analysed by evaluation of typically six characteristic parameters.

The first characteristic parameter might be the duration of the motor inrush current, which normally flows for a period of about 300 ms. The actual duration is related to the static friction and to backlash in the mechanical linkages. The second parameter of interest is the overall duration of the tap-changing process. This typically indicates the correct functioning or otherwise of the timing control. Third, there is the total energy consumed by the motor drive during a tap-changing process divided by the total switching time to give a power consumption index for comparative purposes. The value of this is dependent upon the operation temperature and characterises the average running condition of the drive. During the motion of the selector contacts, the amplitude of the power consumption is monitored. The fourth parameter is the maximum value of power consumed by the drive motor during the opening and subsequent motion of the selector contacts. The fifth characteristic parameter is the maximum motor power consumption during the closing of the selector contacts. Finally, the amplitude of the power consumption is recorded during diverter switch action. The six parameters listed above characterise each tap-changing process, and in the event of significant variations between subsequent operations, warning messages can be triggered by the software.

The oil temperature difference between the OLTC diverter switch compartment and main transformer tank is a function of the severity of contact wear. Excessive contact arcing, misalignment of contacts, loose terminations, locked rotor current of internal tap-changer motors and contact overloading all generate heat within the diverter tank. Serious damage to the transformer is almost certain in the event of the failure of the tap-changer, and hence special monitoring of this mechanically and electrically highly stressed element may be of particular importance.

This high-level tap-changer monitoring module may include, for example:

measured values
  power consumption of the motor drive
  number of switching operations of the selector
  diverter switch current during a switching operation
  OLTC oil temperature
  total switching time
  duration of opening and moving the selector contacts
  duration for closing of selector contacts
  duration of diverter switch action
derived information
  oil temperature difference between OLTC and main tank
  assessment of mechanical quality
  time of inrush current
  switched energy
  contact wear assessment.

20.10 Insulation module

An individual utility may consider insulation condition to be of paramount importance, in which case an insulation model may prove to be a very desirable option.

A capacitive thin film sensor is used for the detection of the amount of moisture in oil. Increasing moisture content in the insulation increases the aging rate. Consequently measurement of humidity of the oil is recommended, in particular for transformers which are already aged or are required to operate for substantial periods with high oil temperatures. This is due to the fact that accurate calculation of the aging rate requires the knowledge of the moisture content of oil and cellulose insulation materials. Under stable conditions the moisture in the liquid is in equilibrium with that in the solid insulation following well established relationship curves that are also functions of the mean oil temperature. From these equilibrium curves, the oil temperature and the measured value of moisture in oil are used to derive the water content of the paper. This value is inputted into the calculation of permissible emergency overload duration. Moisture in paper restricts the loading capacity because of the risk of bubbling or water droplet production as well as increasing the rate of paper aging. The bubbling temperature and its safety margin provide information about the acceptable limit for the hot-spot temperature that is dependent upon the moisture content of the paper.

Typically this module might include:

measured values
    load currents
    overcurrents and short-circuit currents
    number of overcurrent occurrences
    top oil temperature

derived information
    rate of increase of gas-in-oil
    moisture level in paper
    bubbling temperature and safety margin
    hot-spot temperature
    breakdown voltage of insulation system
    life consumption
    overload assessment
    overload capacity
    emergency overloading durations
    overall thermal assessment.

20.11 Bushing module

Another utility may persistently suffer problems with a particular family of bushings. High-voltage condenser bushings, depending upon their constructional features and age, have a relatively high failure rate in comparison with other transformer components. Routine offline measurements of bushing capacitances and dielectric dissipation factor are performed by most utilities, while the transformer is isolated from the system, in order to determine the operational state of its bushings. Modern microprocessor and computer technology make it possible to carry out these procedures online and with the transformer energised, with the help of a monitoring system.

All bushings are subjected to high mechanical, electrical and thermal stresses in normal operation and hence they age mechanically, dielectrically and thermally. Partial breakdowns of the insulation system can affect the operational security to such an extent that continued safe operation can no longer be guaranteed. Dielectric losses in the insulation result in a capacitive loss current that flows through the dielectric material. Depending upon the level of aging, both the bushing capacitance and the tan δ may change. Other reasons for change can also depend upon external environmental influences such as moisture in the insulation system and pollutant depositions on the outside of the porcelain. A change in the bushing capacitance could also be caused by oil ingress into the core of a resin-bonded paper bushing. Apart from a consideration of the values of tan δ (dielectric dissipation factor) and capacitance, analyses of their trends are also of considerable value. An increase in the capacitance for all types of condenser bushing is indicative of partial breakdowns between stress control foils. A short-circuit between two adjacent foils does not result directly in a bushing failure, but the stress across the remaining bushing core layers increases the probability of cascade breakdown.

Measurement of operating voltage is by means of a capacitive voltage sensor that is directly connected to the bushing measurement tap (Figure 20.5). This generally permits reliable measurement with a bandwidth of up to 2 MHz. Thus, overvoltages of both long and short duration can also be detected. Transient

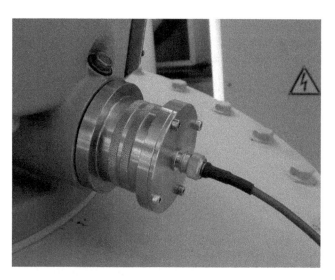

Figure 20.5 Probe connected directly to bushing tap (courtesy of Alstom Grid)

overvoltages cause dielectric aging of the insulation structure of the bushings and of the transformer. Clearly, the detection of these transients is of great significance in the evaluation of the bushing reliability as well as the integrity of the insulation system of the complete transformer.

Transient overvoltages are typically caused by lightning strikes and switching actions. In particular, bushings in GIS or HVDC transformers are subjected to very fast transient overvoltages resulting from pre-strikes in switchgear or in thyristor operation. The magnitudes and waveshapes of such transients are often not determinable with the conventional measuring equipment that is usually present on the network, and hence the potential of failure of operating equipment at the time of transient overvoltage occurrence is higher than for power frequency voltages. The increased risk may often be mitigated by taking specific countermeasures. Information about the overvoltages in combination with other downloaded data is also of considerable importance in root cause analysis following the occurrence of a fault.

Change in the bushing capacitances (ΔC) may be determined by means of a three-phase voltage measurement, whereby the output signal of the voltage sensor of one phase is compared with those of the other two phases. In this way, the influence of temperature is compensated since all three bushings are in close proximity.

Failure of bushings can often be traced to partial breakdowns of stress control layers, or with contact problems in measurement terminals or in mechanical disturbances resulting from external events. These effects can usually be detected reliably by monitoring the change in the bushing capacitance.

The Bushing module could include, for example:

measured values
   operating voltages of all three phases
   overvoltages
   number of overvoltage occurrences
   time stamp of last overvoltage
   capacitances
derived information
   change of capacitances.

20.12 Cooling module

The switching state of the oil pumps and the fans is monitored and also temperature values are measured, the latter with Pt100 probes. The intention of monitoring is to be able to make selective statements regarding the state of the entire cooling plant. The efficiency of the cooling unit is characterised by thermal resistance. For air-cooled power transformers the thermal resistance is calculated by dividing the oil temperature rise by the actual losses. The result has to be averaged over time in order to eliminate variations due to the dynamic behaviour of load factor, the oil and the ambient temperatures. Furthermore, the number of fans and pumps in operation must also be taken into account in order to calculate the nominal thermal resistance. By application of a suitably detailed thermal model, the failure of even a single fan can be identified or partial blockage of the coolers can be identified. Monitoring therefore has the advantage of facilitating the principle of condition-based maintenance.

Overloading is a necessary feature of deregulated electricity markets both for reasons of economy and to ensure continuous energy supply. During an overload cycle, aging must be monitored and controlled and excessive thermal damage has to be avoided. By measurement of environmental and loading conditions the monitoring system can continuously deliver information about maximum continuous and short time overloads considering the actual preload of the transformer in accordance with IEC 60076-7. The thermal model specific to the particular transformer is interrogated and the continuous overload capacity is determined by the monitoring system. The loss of life due to this type of overloading is controlled by ensuring that periods of high load or high ambient and high aging rate are compensated in the long term by periods of low load and resulting reduced aging rate. The module generates warnings in the event of excessive aging or other consequences of overloading, such as tap-changer or bushing limitations.

Intelligent control of the cooling unit by the monitoring system may be applied to increase overload capacity while optimising hot-spot temperature. Depending upon the actual ambient temperature, a specific number of fans can be switched on to dissipate the losses that are generated over that particular duration. By individual switching of the fans, the temperature variations will be lower than those obtained by conventional cooler control that has fixed temperature thresholds. Secondary advantages are reduced variation of oil level within the conservator resulting in reduced transformer breathing activity with consequential drop in water absorption within the conservator, and a reduction in the noise emission from the fans during periods of reduced loading.

Features of the Cooler module would typically include:

measured values
    pump and fan status signals
    cooling equipment power consumption
    inlet and outlet temperatures of cooling equipment
pre-sets
    loading or temperature limitations
derived information
    differences between inlet and outlet temperatures
    control of cooling equipment.

20.13 Advanced features measurements and analyses

The list of measurements that can be made and derived information that can be gleaned is continuously on the rise. A few are listed hereunder:

measured values
    hot-spot by fibre optic
    multi-gas analyser (up to eight gases)

partial discharge
bushing power loss factor (tan δ)
bushing capacitive displacement currents
bottom oil temperature
module cubicle temperature
IED temperature
digital status information from other equipment
moisture level of oil in the OLTC
gas quantity and rate of collection in Buchholz relay
oil pressures
mechanical accelerations (e.g. tank wall, OLTC)
oil level
humidity of air inside the conservator
humidity of ambient air and air pressure

derived information
transformer power factor (cos φ)
transformer efficiency
oil Pressure differences
integrated automatic voltage regulation (AVR).

20.14 Partial discharge monitoring

Dielectric breakdowns in transformers are frequently preceded by partial discharges, i.e. a partial breakdown across a part of the insulation structure. Since partial discharge tends to occur before a complete breakdown, partial discharge monitoring can provide a warning of future, and perhaps catastrophic, failure. Not all partial discharges, however, are potentially damaging in the short term. For example, loose connections, floating metallic objects arcing under oil between metal parts, may not be of immediate concern, but monitoring with a view to eventual correction could save money in the longer term.

Partial discharges can be successfully detected by electrical methods, for example using RIV [4] or a Lemke probe [5]. Other more sophisticated equipment that analyses the discharge signals by magnitude, position and polarity with respect to the voltage wave and compares with a known signature pattern is available or under development.

A highly regarded method for the detection and location of partial discharges relies on data derived from acoustic sensors. Partial discharges generate sound waves in the 100–300 kHz ranges. This sound is at a much higher frequency than the background noise of the transformer, and allows discrimination of the signals. In addition, sound takes a finite, measurable time to travel from source to monitor. The source of a partial discharge can often be located by triangulation from an array of sensors at several positions around the transformer tank.

20.15 Temperature measurement

A variety of temperature measuring devices are used in transformers. Until recently liquid expansion devices have been dominant. Oil temperatures may be read directly from these devices and hot-spot temperature measurement has traditionally used this type of device but with analogue simulation of the thermal effects of the actual load. Should there be a need to produce digitised signals, then transducers are employed.

Extensive use has also been made of resistance temperature detectors, usually Pt100 sensors, again with conversion to digital signals. Here again, analogue simulation of the effects of the load is readily available using the same methodology that is followed with the liquid expansion devices as previously mentioned.

Fibre optic point sensor measurements are fast becoming commonplace as improvements are devised that offer suitable mechanical protection to the fibre as demanded by processes that apply to the manufacture of heavy equipment [6]. Distributed fibre optic thermal sensors are also available with a resolution of about 1 m. These are normally wound throughout the windings in a long loop that also traverses other regions of thermal importance, and have both ends accessible for measurement [7]. This gives the advantage that in the event of breakage of the fibre, measurements can be made from each end up to the point of breakage. The problems faced when using distributed sensors are the heavy engineering environment as applies to the point sensor plus the additional hazard that is posed by the hostile drying procedures that are essential in power transformer production.

Infra-red photography is a fully tried and tested means of measurement of external features on a transformer, such as bushings, bushing connectors and extensive areas of the tank surface. Building this facility into the online condition monitoring equipment is rarely cost effective. However, equipment and operators are hireable on reasonable half-day or daily rates. A wide area scan is often used to determine the optimum locations for point measurements of temperature, whenever more detailed information on thermal performance under various conditions is required.

20.16 Chemical parameters

The first indication of faults inside oil filled transformers is usually obtained from gases that are dissolved in the oil. Oil and cellulose insulation materials degrade under thermal and electrical stresses, and in doing so generate gases in various compositions and concentrations. Hydrogen is a key gas in transformer problem indication, an increase in the output signal of a hydrogen sensor is an indication for irregularities such as partial discharges or hot-spots. The simpler systems of gas-in-oil monitoring rely on detection of this gas. However, much more information can be derived from analysis of other hydrocarbon and carbon oxide gases.

Relative proportions of the gases are an important indication of the type of fault. Rates of gas production allow monitoring of the rate of development of the fault. Regular oil sampling and analysis of the hydrocarbon and carbon oxide gases in the sample are probably the most cost-effective methods of condition monitoring. Sample analysis will identify the longer-term fault development. Some other trigger can identify faults developing in days or months. Devices such as the Hydran Gas Monitor, which provide an alarm when the gas concentration reaches a pre-set level, can be used to initiate a more rigorous and more frequent condition-checking process.

The evaluation of an online measurement signal where it is available, together with the dependency on the temperature of the oil and the load current, provides a reliable basis for the continuous operational assessment of the transformer. In the event of an increase of gas-in-oil content, an immediate reaction can be effected via an offline dissolved gas analysis (DGA) to determine the concentration of the other components dissolved in the oil in order to clarify the cause of the potential damage. Nowadays, online multi-gas analysers that permit monitoring of up to eight gases are available at a reasonable cost premium and may be integrated into a condition monitoring system. In this case, additional analysis in accordance with commonly accepted assessment algorithms and triggers can be applied and the results fed into an expert system. The expert system can then include information derived from other analysis modules to perform failure root cause analysis.

High-performance liquid chromatography can also be used to measure the furfuraldehyde content of the oil, giving an indication of the micro-deterioration of the transformer and therefore the true age of the cellulose insulation [8, 9].

20.17 Dielectric parameters

Frequency response analysis, recovery voltage spectra and dielectric loss angle are often applied to transformers by many utilities throughout the world, but in some areas there still remains polarisation of opinion and therefore there is a need to exchange information and experiences and carry out further work in order to reach agreement on standardised philosophies and procedures.

20.18 Conclusions

Significant inroads into condition monitoring philosophies, equipment and systems have been made in the past few years. The speed of progress has been driven by the need to make the best use of existing and new equipment to give reliable and long service. The means for the possible integration of all substation equipment with modular add-on capability is now readily available. Transformers may constitute one of the most reliable pieces of electrical plant on the grid, but complacency is not an option to progressive utilities and manufacturers. There is still work to be done in the condition monitoring arena in terms of standardisation. New, wider ranging and better equipment will continue to be developed, philosophies will be debated and procedures will be agreed. There is no limit to what can be achieved in the exciting field of condition monitoring.

20.19 Further reading

Further to the specific reference list included hereunder, additional material on condition monitoring, as well as on other power transformer-related topics, may be found in various brochures and technical reports produced by CIGRE Study Committee SC A2 (formerly SC 12). Details of its many activities, together with a comprehensive list of suitable study documents that it has prepared and issued, are reproduced in the *SC A2 Annual Activity Report* that is published in CIGRE *Electra* No. 255, April 2011 (pp. 26–35).

References

1. Jones G.R., Jones C.J., Lewis G., Miller, R. 'Intelligent optical fibre based systems for monitoring the condition of power equipment', *Proceedings of 2nd international conference on the reliability of transmission and distribution equipment.* IEE, Warwick, UK, 1995, vol. 406, pp. 73–78
2. Alstom Grid Brochure. 'MS 3000 on-line condition monitoring and expert system'. Alstom Grid publication, 2010
3. Bossi A., 'An international survey on failures in large power transformers in service'. Final Report of CIGRÉ Working Group 12.05. *Electra.* 1983;**88**:22–48
4. Lapworth J.A. 'Thermal aspects of transformers'. Report of CIGRE Working Group 12.09, *Brochure 096*, 1995
5. Lemke E. 'A new procedure for partial discharge measurements on the basis of an electromagnetic sensor', Paper 41.02. *Proceedings of the 5th ISH*, Braunschweig, 1987
6. Simonson E.A., Lapworth J.A. 'Thermal capability assessment for transformers'. *Proceedings of 2nd international conference on the reliability of transmission and distribution equipment.* IEE, Warwick, UK, 1995, vol. 406; pp. 103–108
7. White A., Daniels M.R., Bibby G., Fisher S. 'Thermal assessment of transformers'. CIGRE Session Paper 12-105, Paris, 1990
8. Darveniza M., Saha T.K., Hill D.J.T., and Le T.T. 'Condition monitoring of transformer insulation by electrical diagnostic methods and comparison with chemical methods'. Paper 110-22 of CIGRE symposium on diagnostic and maintenance techniques, Berlin, 1993
9. Steed J.C. 'Condition monitoring applied to power transformers – an REC view'. *Proceedings of 2nd international conference on the reliability of transmission and distribution equipment.* IEE, Warwick, UK, 1995, vol. 406; pp. 109–114

Chapter 21

Integrated substation condition monitoring

T. Irwin, C. Charlson and M. Schuler

21.1 Introduction

High-voltage equipment such as switchgear, transformers, overhead lines and cables play an important role in the successful operation of the transmission and distribution systems. Managers and operators need access to reliable condition data to know which network components have to be installed, repaired or even replaced. They need to know how the network equipment is performing and with what level of margin in relation to the design specifications. This information must be updated at regular intervals in order to be confident that condition of the equipment is accurately known. It follows from this that the information update period is critical to the successful detection of developing problems in order to ensure correct operation of the equipment. Obtaining the information sometimes requires outages to be taken, to manually fit monitoring equipment and carry out measurements, so there is a cost penalty involved. The data update period, in this case, has to be optimised in order to limit operational costs.

The idea of having online automatic data gathering systems is therefore very appealing, with no need for outages and manual intervention, but it must also be cost-effective. Although the online system would have an initial installation cost, it should be able to significantly reduce the operational cost of monitoring by having automatic, and almost instant, data updates. This instant data would have a major impact on improving the effectiveness of the information and therefore the reliability of detecting potential problems and predicting the outcome.

Over the last 25 years, various discrete monitoring systems have been tried by a range of manufacturers and considerable on-site live operational experience has been obtained. Technology changes and advancement over the last 15 years have progressively allowed more adventurous hardware and software applications. Major improvements in the speed of data processing and the generation of graphical displays followed and the realisation of real-time diagnostics was achieved by these systems.

Although the monitoring systems have been effective in service, the task of handling data and alarms from a number of independent monitoring services has posed an ever increasing burden on the staff responsible for data management. The impact of this is that utilities now require a more coordinated approach by:

1. marshalling the data from each of the individual monitoring systems within the substation into a single substation system
2. global presentation of the data from all substations at the asset management centre.

It has been realised that the value of the monitoring systems can be greatly enhanced by adopting the above approach, and CIGRE working groups [1] are in the process of investigating current monitoring technology and ways of optimising its application along with asset management systems.

The main thrust throughout all of this work is the realisation that condition monitoring (CM) systems can contribute to the network optimisation by providing information for each 'supply-chain' element in terms of efficiency and reliability [2]. Also, by moving to condition-based maintenance, it is possible to minimise maintenance costs and, by using this information, to extend average life cycles.

This chapter reviews a range of monitoring systems that are currently available and the steps taken to provide the coordinated approach by developing an integrated system to present the 'complete substation picture'. An overview to illustrate the structure of the chapter is given in Figure 21.1.

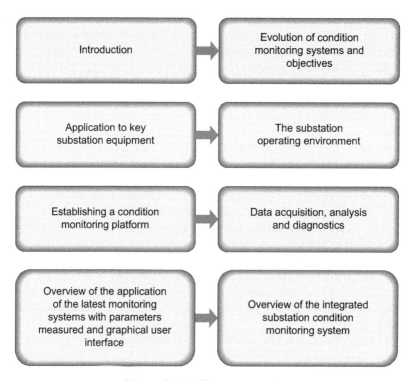

Figure 21.1 Chapter overview

Principal terms used in this chapter are summarised in the list of acronyms provided in Box 21.1.

Box 21.1 Acronyms

3G	Third generation of mobile phone standards and technology
A/D	Analogue to digital
ACM	Arrester condition monitor
ADC	Analogue to digital convertor
ADSL	Asymmetric Digital Subscriber Line (reference, broadband)
AIS	Air-insulated substation
CB	Circuit breaker
CM	Condition monitoring
CT	Current transformer
DEM	Disconnector and earth switch monitor
DGA	Dissolved gas analysis (reference, gas in insulation oil)
DTS	Distributed temperature sensing (reference, along a length of cable)
EHV	Extra high voltage
E1/T1	Port adaptors for wide area links
EMI	Electromagnetic interference
FMES	Fault making earth switch
GDM	Gas density monitoring
GIS	Gas-insulated substation
GPS	Global positioning system
HMI	Human machine interface
HV	High voltage
ISCM	Integrated substation condition monitoring
KM	Knowledge module
LAN	Local area network
LCC	Local control cubicle
NTP	Network Time Protocol
PCI	Peripheral component interconnect electrical bus
PDM	Partial discharge monitoring
POW	Point on wave in relation to power frequency voltage
PXI	PCI eXtensions for instrumentation

(*Continues*)

RTU	Remote terminal unit
SCS	Substation control system
SF_6	Sulphur hexafluoride
TGPR	Transient ground potential rise
UHF	Ultra high frequency
UPS	Uninterruptible power supply
USB	Universal serial bus (PC)
VFT	Very fast transient
VT	Voltage transformer
WAN	Wide area network

21.2 The evolution of condition monitoring systems

The world's first online CM system, designed for ultra high frequency (UHF) partial discharge monitoring (PDM) of gas-insulated substations (GIS), was installed by Reyrolle Ltd in the UK for National Grid Company in 1991.

The system monitored 66 UHF sensors via 22 detector units using a distributed data acquisition system. The data acquisition system and detector unit interface electronics for the sensors were arranged as a network of nine zones distributed around the 420 kV GIS (see Figure 21.2). The control PC was located in the substation control room and used a screened copper communications cable to control

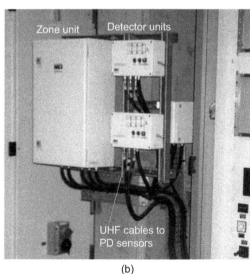

Figure 21.2 UHF PDM system 1991
(a) PDM main control PC; (b) PDM zone and two detector units in the GIS

and retrieve data from the zone units. The system was commissioned during the HV on-site commissioning tests in February 1992 and was in service when the substation was first energised. The system provided a new facility for 'soak' tests while commissioning the substation and recorded the UHF emissions from the circuit breaker and disconnector operations as the feeders were sequentially first energised.

The computer system installed then was very basic, running at 33 MHz with a DOS 5.1 based system, but it did have a modem connection running at 9.6 KB for remote data transfer and remote desktop facilities. Not very impressive by today's standards, but at that time it was pre-Windows 3.1 and also around the same time, Sir Tim Berners-Lee was going through the first trials of the World Wide Web with a few preliminary sites using browser and web server software. Consequently, the idea of being able to access and control PCs using a remote desktop was relatively new.

The monitoring system, although 'in-service' and fully operational, was a first experimental step into assessing the performance and longevity for this application on extra high voltage (EHV) substation equipment. The monitoring system has seen several upgrades for software and PC hardware, over what is approaching a 20 year life, but the original zone units and detector units are still fully operational and have fully served their purpose.

In the early 1990s, the uptake for online monitoring systems was relatively slow and mainly limited to GIS and UHF partial discharge monitoring, but gathered momentum by the mid-1990s and saw expansion into a range of monitoring functions.

21.2.1 Periodic monitoring, 1960–1990

Historically, periodic monitoring, using manually operated measurement systems, was used to gather data for condition assessment.

Typically, the following items of plant would be monitored periodically:

- circuit breakers: taken off-line and timing tests carried out
- transformers: oil samples taken for dissolved gas analysis
- SF_6 (sulphur hexafluoride) gas density: pressure gauges; readings taken
- GIS insulation: UHF and conventional PD measurements, some taken off-line and online [3]
- surge arresters: leakage current and operations counter readings.

21.2.2 Basic discrete online monitoring, 1990–1999

During this period, several different types of monitoring systems would be in service at the substation, each with their own specific set of software and data access points:

- transformer monitoring
- GIS UHF partial discharge monitoring [4, 5]
- circuit breaker trip and close operation monitoring using controlled switching devices [6] and fault recorders, primarily for post-fault analysis.

These systems had varying degrees of remote data access [7, 8] but data could be manually coordinated particularly for fault development analysis. Also, UHF PD monitoring was being adopted by most utilities for GIS HV site testing [9].

21.2.3 More intelligent discrete monitoring, 2000–2009

Moving into the twenty-first century saw a wide range of data acquisition technology become available 'off-the-shelf'. This had the impact of reducing development time for monitoring systems and also increased the scope and capability of the measurements. These mass produced data acquisition systems were very competitively priced and this increased the marketability of the CM product, resulting in an increased uptake in monitoring equipment.

Data acquisition systems, with on-board data handling and storage, coupled with faster PC processing speeds, allowed much improved processing rates and major enhancements for data displays and real-time diagnostics [10, 11].

After a period of some 20 years since the first system went online, CM systems had finally 'come-of-age' with hardware and software sufficiently developed to get quality information with interactive response at an acceptable speed and cost.

Online monitoring systems now cover a range of power system equipment, some of which are listed below:

- GIS Insulation (SF_6 Gas Density [GDM] and UHF Partial Discharge [PDM])
- power transformers [12, 13]
- cables and overhead lines
- surge arresters
- current and voltage transformers
- isolators and earth switches.

Coordinating the data from the many types of monitoring system requires each individual system to provide data files which can be collected periodically using remote data access. In the UK, for example, National Grid is running 'Smart Asset Management' web server applications on the gas density monitoring (GDM) and partial discharge monitoring (PDM) systems, accessing data files directly as a 'new' approach for data centralisation and a first step towards integrating CM systems.

21.2.4 Integrated substation condition monitoring (2010 onwards)

CM systems are supplied by a wide range of manufacturers who have their own specific displays and data handling requirements, and as more systems come online, the task of monitoring and collating the data will become immense, if not unmanageable. This was seen as a barrier to the more general application of CM systems, as reported by CIGRE in TB 343 in 2008 [13], particularly relating to transformer monitoring. Utilities have recognised that, in order to move forward with substation monitoring, the task of data handling needs to be simplified and better coordinated and are now specifying a monitoring system package, rather than individual systems. The package has to cover all of the required aspects of substation CM, with a local area network (LAN) to centralise the substation database, and with access to a wide area network (WAN) to allow centralisation of all substation CM data.

Developments have been put in place to allow the implementation of data integration from the existing individual monitoring systems as currently used at each substation and to provide a global asset monitoring system capable of accessing each substation. Integrated substation condition monitoring (ISCM) is a concept that can be applied as a retrofit option for existing substations as well as for new substation projects.

The ISCM system must provide the user with an overview of the status of the complete substation by continuous monitoring of the substation systems. The aim is to allow the user to have sufficient information to help anticipate any breakdown and save on expensive maintenance.

In general, the ISCM system aims to provide the following benefits:

- precise and actionable information for fast and targeted reaction
- integration of the asset monitoring systems into a unique platform
- reduction of unscheduled downtime through proactive maintenance and planned repair
- effective prediction and prevention of equipment failure
- intelligent consideration of a given normal range of operating characteristics instead of mere 'hard limits'
- complete condition information available in a consistent format
- integrated customised functionality to deliver the right data to the right personnel
- clear rules for all safety-critical and security-relevant situations
- seamless integration by common communication, user interface and diagnostic software
- easy implementation in any given substation automation infrastructure
- use of existing CM systems where technically viable
- a more comprehensive asset management overview
- a co-coordinated approach for CM analysis and visualisation of T&D assets.

Control software has been created which is capable of displaying and handling data and alarms from all of the different monitoring applications. Both hardware and software can now be assembled from a range of devices such as CM modules and together form an ISCM system. This is of great advantage to the user as it provides the same 'look and feel' user interface for all CM modules.

All data and alarms are handled as 'condition monitoring' data rather than as individual data associated with different types of monitoring system. Hence, from one display the user can identify any alarms with any CM modules connected to the system.

A typical ISCM system display is shown in Figure 21.3, which is an example of a multi-CM module system with access to transformer, GIS, cable and surge arrester monitoring. The alarms for all systems are displayed on the main overview screen. The individual CM modules associated with the ISCM system are accessible via menu buttons from the ISCM system main menu screen, which in the case of Figure 21.3 is the GIS PDM system. The ability to display on one screen an overall condition status for the substation is now a reality and from this one screen, access to alarms, 'trending' and diagnostics can be achieved.

In general, the substation hardware for data acquisition will be designed for remote monitoring and control, with very little information displayed directly on the equipment. Consequently, CM systems require a range of software-generated graphical displays to present information to the user and for configuring and

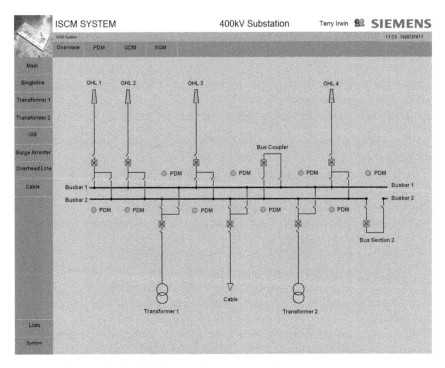

Figure 21.3 ISCM alarms overview and data access screen

controlling the data acquisition systems. Throughout this chapter, examples of the graphical display screens (in greyscale) are given to illustrate the functionality of the user interface, in broad terms, since the image scaling prevents detailed examination. Also, it must be realised that the overview can only include a limited range of examples from the many CM systems that are currently in service.

21.3 Objectives of condition monitoring

The objectives of condition monitoring can be summarised as follows:

- to enable maintenance to be planned in accordance with the condition of the equipment
- to prevent equipment failure and the resultant forced outages
- to prevent catastrophic failures with consequences for safety
- to provide a life assessment for plant items and to predict the rate of deterioration
- forward strategic planning for new substations and extensions.

Ideally, all plant items could be monitored but initially key equipment will form the vanguard for the first generation of integrated systems.

21.4 Application to key substation equipment

Although the cost of monitoring systems has been significantly reduced over the last 10 years, the cost ratio of plant being monitored to the cost of monitoring must be considered. The higher this ratio, the more cost-effective the system will be. Consequently, only the most expensive key items of plant will be monitored for in-service performance, that is:

- circuit breakers (CBs)
- SF_6 gas (CB/GIS) system integrity
- insulation condition of GIS
- cable monitoring
- insulation condition and running status of power transformers.

Monitoring the remaining balance of plant would be considered on a substation-by-substation basis.

Circuit breakers perform the key role of switching plant, such as transformers and compensating reactors, as well as clearing power faults on any item of plant along with faults on overhead lines and cables. Early detection of changes in operating speed, for example, may flag up a maintenance request and prevent a failure to operate.

Falling SF_6 gas density due to leaking seals can occur, for example, where drive shafts enter through the gastight enclosures. These leaks may only occur during the mechanical operation, so could be difficult to identify without continuous monitoring. If sufficient gas escapes then this can cause lock-out for circuit breaker operation and effectively take this key item of plant off-line.

The insulation condition of GIS uses PD as the indicator of possible developing problems and requires sophisticated diagnostics to identify the problem and assess risk of failure.

Cable temperature can be monitored along with the load in the cable using internal optical fibres. This allows the early detection of progressive degradation of the cable insulation and incipient failures, specifically those exhibiting line impedance variations, PD or hot spots in cables.

Power transformers can also be monitored for PD using similar techniques used for GIS. However, a more traditional, dissolved gas in oil analysis, using online equipment, can also provide insulation condition assessment. Monitoring current flow and temperatures at various locations within transformers and cables can also provide valuable information on additional load flow capacity.

21.5 The substation environment

Dealing effectively with electromagnetic interference (EMI) generated by HV equipment operation in the substation is a key feature of the design of the monitoring system. Electronic measurement equipment has been damaged, while being used in substations, by the effects of electromagnetic radiation and by conducted

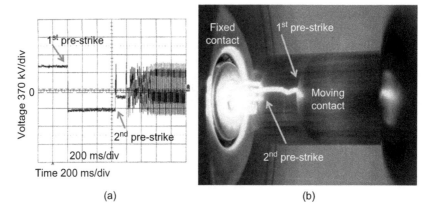

*Figure 21.4 GIS disconnector pre-arcs and busbar voltage transient
(a) GIS busbar voltage; (b) disconnector closing operation with multiple pre-strikes*

disturbances which are generated predominantly by operating circuit breakers, disconnectors and earth switches.

Figure 21.4 shows the pre-arcing between disconnector contacts during a close operation. Multiple pre-arcs can be seen to occur as the contacts move closer together before final contact touch. Also shown is a measurement record of the busbar voltage taken during this operation, which shows a 'trapped charge' (charge stored by the capacitance of the GIS chamber which in this case is equivalent to peak system voltage) prior to starting the close operation.

The 'trapped charge' situation provides a worst case scenario by increasing the voltage across the disconnector contact gap in order to provide a maximum pre-arc length along with a maximum transient overvoltage. This busbar voltage progressively alternates in polarity, by the action of each of the many pre-arcs (>50), and finally shows the power frequency voltage when contact touch occurs. Each pre-arc creates a very fast transient (VFT) voltage, which travels along the busbars throughout the length of the substation. In addition to this travelling wave, bursts of ultra high frequency EMI are also generated which can radiate out from discontinuities (e.g. exposed support barriers and earth switches) in the GIS enclosure.

The connections between measurement equipment and sensors mounted on the switchgear need to be screened, to limit EMI, and also isolated, to prevent earth loop circulating current [14]. Equipment also needs surge protection and have isolated power connections, particularly when used in GIS.

GIS has a very compact construction, and transient ground potential rise [TGPR] [15] is a phenomenon that occurs when operating the switchgear and can locally raise the 'earth-strap' potential to several kV by high-frequency charging currents flowing in the enclosures and earth connections [3 kA peak at frequencies in the range 1–50 MHz] [16].

Figure 21.5 shows the GIS enclosure voltage near to an earth switch during a closing operation to earth the busbar. The busbar had a 'trapped charge' with a value near to peak system voltage (340 kV) prior to the operation.

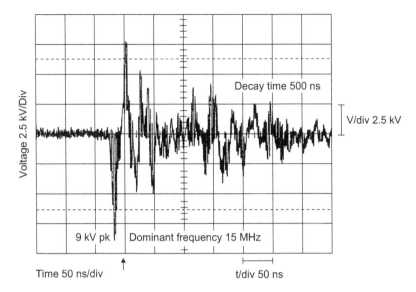

Figure 21.5 GIS enclosure TGPR generated by earth switch operation

The resultant enclosure voltage reaches a peak of 9 kV which is about 3% of the initial busbar voltage. Transient ground rise voltages like this can be conducted along cable screens and enter monitoring equipment.

IEC standard [IEC 61000] [17] specifies a range of tests that simulate the EMI. The tests apply a range of power frequency and pulsed voltages and magnetic fields to the equipment for both conducted and radiated EMI. The equipment is powered and fully operational during the tests and must not be disturbed by the tests. Both the electronic circuitry and software for control and data acquisition must operate correctly during the tests and the hardware must not sustain any damage.

The equipment must be able to work inside the substations with a temperature range of -5 to $+55$ °C and a relative humidity of 5–95% non-condensing. The enclosure protection rating is typically IP55 (dust protected, protection against low pressure jetting water). Equipment mounted outdoor, for example in the overhead line gantry area, must be able to work with a temperature range -30 to $+55$ °C. The enclosure protection rating is typically IP65 (as IP55, but dust tight).

21.6 Condition monitoring platform

Various commercial monitoring systems are available for a range of substation plant, for example power transformer, cable and surge arrester monitoring systems. By comparison, the availability of monitoring system for switchgear and associated equipment was limited and manufacturer specific. A development strategy for a CM platform was therefore established in order to provide, where possible, a flexible common platform that would cover a range of switchgear, whether air-insulated substation (AIS) or SF_6 gas-insulated substation (GIS) systems, and achieve a design with a high degree of 'future proofing'.

Having one platform for all types of GIS monitoring functions was essential in achieving a competitive market price and was a first step to ensure a fully integrated monitoring system for the substation switchgear equipment. Also, utility operational monitoring costs can benefit by achieving a common set of spares along with the same maintenance routines for all GIS systems.

The following sections will review the procedures and background information necessary to develop this common platform along with the performance expectations. Previously existing monitoring systems for transformers, cables and other important additional equipment such as surge arresters will be reviewed before looking at the final arrangement of the 'integrated substation condition monitoring system' developed by the authors.

21.6.1 Data acquisition systems

By carrying out an analysis of the range and type of measurements to be taken by the monitoring systems, it was possible to select a small range of equipment that could be programmed to meet all measurement requirements. In other words, the hardware remains constant and the software determines what the system monitors.

'Future proofing' the design of an integrated monitoring system is required to protect the investment by maximising the life of the installed system. For example, data handling with the data acquisition system requires a formatted text file to be produced and if at some time in the future a data acquisition module becomes obsolete and replaced by another, then the replacement is configured to provide the same formatted text file. This allows the same software to be used for post-processing and display and so the system continues irrespective of the actual data acquisition module in service. The following criteria need to be achieved:

- The system will have to operate in an environment which is very hostile to electronics and therefore needs to be industrially hardened to survive.
- The system needs to be easily maintained and upgraded from time to time.
- The system needs to be of modular construction to minimise any loss of data in the event of a data acquisition circuit fault because this allows the other healthy modules to function without interruption. Modules also allow for ease of manufacture and installation which helps reduce costs.
- A communication system for remote access to the data acquisition system is required for data transmission and to view real-time data. This communication system will also be subject to the hostile substation environment and needs to be industrially hardened.

In order to meet the above criteria, the data acquisition system can be subdivided into the following sections:

- sensor/transducer module to measure the required signal
- conversion module for signal transmission
- signal interface module for signal conditioning and transient protection
- data acquisition module to digitise the signal

- control PC module to control data acquisition, data analysis, data filing and storage and communication system
- communication module.

21.6.2 Sensors and transducers

Sensors and transducers provide the interface with the substation equipment. A range of devices are needed in order to convert a range of monitoring parameters into signals that can be transmitted to a conversion unit if required. The following two chapters (Chapters 20 and 22) will also cover a similar topic and consider sensors and transducers. These devices are discussed further later in this chapter.

21.6.3 Conversion module

The conversion module can use various options for signal transmission in noisy and potentially hazardous environments for electronics. Once the signal has been generated by the sensor, it can be converted to voltage, current, frequency or light. In some instances, the conversion module is included within the transducer and the signal needs no additional conversion.

21.6.4 Interface module

Interface modules provide the signal interface between the conversion modules (or transducers) and the data acquisition modules. Their purpose is to reject the system noise to provide a clean signal and to give protection against possible damaging transient overvoltages. The interface module also allows the optimisation of the signal for scaling with the data acquisition modules. The final signal output stage then usually presents a voltage signal to the data acquisition module.

21.6.5 Data acquisition module

Analysis of signal types from the interface module has to be carried out in order to establish the type of data input and the speed of data capture. Signal parameters which affect the selection of the module are typically:

- signal type: analogue or digital
- signal frequency range
- signal pulse width and repetition.

Also, the required accuracy of the measurement has to be considered when selecting the 'bit resolution' of the analogue to digital convertor (ADC).

21.6.6 Control PC module

The control PC module must be a 'ruggedised' industrial type PC. In most instances, the hard drive will use solid-state technology with no moving parts to provide a more durable solution to traditional hard disk drives. Mechanical shock must be considered when applying the system to circuit breaker monitoring (CBM) or when the equipment is used for other purposes, but is mounted in close proximity (several metres) to circuit breakers.

The operating system normally used will be 'Windows based' and will follow the path of the Windows development. It currently sits at 'Windows XP Professional' but will be moving to 'Windows 7'. A Windows embedded operating system may be used for more compact, remote, stand-alone applications.

The PC module human machine interface (HMI) performs a range of functions:

- control of the conversion and interface modules
- control of the data acquisition module
- reading data from the data acquisition module and creating data files
- analysing the data and performing diagnostics and creating alarms
- graphical data displays for user real-time information
- data and alarm uplink to the substation monitoring control cubicle
- monitoring system performance analysis for maintenance schedules.

21.6.7 Communication module

The communication system utilises the package supplied by the Windows operating system. The substation EMI environment, however, requires that the communication system is to be well shielded for noise-free operation. The best way to achieve EMI noise free data transmission is to use optical hubs and routers to create a fibre optic Ethernet LAN system. The optical LAN provides the link to the substation monitoring control cubicle for data uplink and remote access to the control PC module.

21.7 Data acquisition, analysis and diagnostics

The control PC (HMI) continuously creates a set of formatted text files from the data acquisition module, which is then automatically analysed as each 'new data set' becomes available. The various measured parameters are compared with user settings for alarms. Once an alarm has been generated, diagnostic routines will be automatically performed to establish possible causes for changes in the data, and report the findings. In some applications, diagnostics may be continuously applied, for example, to remove noise from the data and control what data is saved and used for alarms. Having created an alarm, the data will sometimes need to be manually viewed in order to make a more detailed risk assessment or to further clarify the diagnostics, so various graphical display features will be used to examine the data more closely.

For real-time diagnostics the data needs to be pre-processed at the point of measurement, as the data is being collected, particularly where noise discrimination is required. This makes full use of the distributed processing power provided by the individual control PC modules, easing the data processing burden for the overall substation control cubicle PC. There will be additional HMI facilities, used at the overall substation control cubicle PC, for historical data, trending and alarm displays.

21.7.1 Developing a common monitoring platform

When developing a common platform for the monitoring system, all of the requirements for each type of application must be reviewed in order to determine the scope for input/output channel type, frequency response and sampling rates.

The platform is required to accommodate at least, but is not limited to, monitoring:

- gas density
- circuit breakers
- UHF partial discharge
- disconnectors and earth switches.

A cross-section view of a 420 kV GIS is shown in Figure 21.6. This shows the main items of equipment to be monitored along with the SF_6 gas-filled enclosures, gastight barriers and support insulators which require gas density and PDM. In addition to the monitoring application itself, there is also a common requirement for self-monitoring parameters such as power supply status, temperature and detection of faulty transducers.

A preliminary review of hardware requirements revealed a cost benefit of having at least 50 input channels for data acquisition grouped into one monitoring unit.

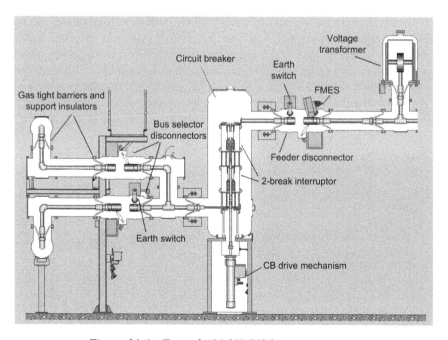

Figure 21.6   Typical 420 kV GIS bay cross section

This unit will form a single node unit within the GIS monitoring scheme. The nearest standard data acquisition module has 64 input channels, so this was adopted as the platform standard. The types of parameters to be monitored for each application were thoroughly investigated and a short summary is given in the appendices and further in this section.

21.7.1.1 Gas density monitoring

The high global warming potential of SF_6 became known in 1995, as a consequence of which SF_6 was put on the list of greenhouse gases in the Kyoto protocol 1997, and explicitly listed as the gas with the highest global warming potential. CIGRE initiated an investigation and in 2002 the result of a survey [18], concerning the environmental implications of SF_6 gas used in electric power equipment, was presented by CIGRE WG 23.02. It used data available at that time on global SF_6 production and emission and critically evaluated and extrapolated to the future and recommended limits that should be observed. Following this work, the impact of recent European legislation on the control of greenhouse gases has resulted in utilities and manufacturers alike being forced to reduce SF_6 emissions to minimal levels, namely 0.5% per annum. In order to achieve this, and also to prove to the regulatory bodies that this is being achieved, a more accurate inventory system for the SF_6 gas in use on substations is required. A recent CIGRE, *SF_6 Gas Tightness Guide* TB N 430-WG B3.18: 2010 presents a valuable and detailed summary of this topic. The 'SF_6 Gas Density Monitoring System' was developed during 2005–2006 with the aim of having a suitable system in place to meet the present requirement.

Typically, each switchgear bay for 420 kV GIS will have 15 gas zones per bay. The GDM node unit is capable of monitoring three bays covering 45 inputs, leaving 19 channels spare, some of which would be used for self-monitoring. The inputs to the GDM node unit are provided by gas density transducers and one node unit will be monitoring typically 3000 kg of SF_6 gas. The gas density transducer is shown in Figure 21.7 and the operating principle, output options and sampling methods are given in Appendix A.

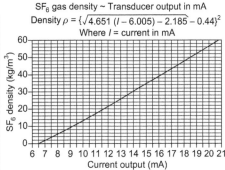

Figure 21.7 Gas density transducer and output characteristic (courtesy of Trafag)

Measuring gas density, as opposed to gas pressure, is very important because gas pressure is affected by temperature, whereas gas density is independent of temperature. Gay–Lussac's law states that 'if a gas's temperature increases then so does its pressure, if the mass and volume of the gas are held constant'. Consequently, when filling busbar enclosures with SF_6 gas, either a density transducer is required or if a pressure gauge is used, a nominal pressure is specified at a reference temperature (reference temperature is 20 °C). The pressure variation with temperature is based on the absolute temperature scale (Kelvin), so a daily 15 °C temperature swing, will result in approximately a 5% change in gas pressure, which would make small gas leak rates very difficult to detect. A clear indication of gas loss will require a measurement system with a stability of better than 0.5% in order to be confident that any changes are not due to ambient effects and ensure the earliest possible warning of leaking gas. The combination of the gas density transducer and associated measurement system in the GDM node unit provides the required stability.

21.7.1.2 Circuit breaker monitoring

Circuit breaker operating time, contact speed and phase spread for both closing and opening operations are very important and, along with other parameters to indicate 'wear and tear', will mean that typically 20 parameters (for 420 kV GIS) are required to be monitored for each switchgear bay. The CBM node unit is therefore capable of monitoring three bays, covering 60 inputs, leaving 4 channels spare which could be used for self-monitoring. Data sampling and timing requirements are given in Appendix B.

A range of transducer types are used for CBM, with some using 4–20 mA loop technology and others with low impedance inputs to give improved noise immunity. The transducers mounted on the actual switchgear must be capable of withstanding the mechanical shock generated by the circuit breaker particularly during trip operations. Open terminal circuit breakers in AIS generate high levels of radiated EMI which additionally requires that the transducers and signal cabling are 'well shielded'.

21.7.1.3 High-frequency PD monitoring

Most manufacturers now install internal UHF sensors on GIS substation equipment, since there is an IEC requirement to use PDM during HV on-site testing. However, external type sensors can also be used if the internal type is not already installed.

Typically, each switchgear bay for 420 kV GIS will require, on average, nine inputs to be monitored per bay (3 × three-phase sets of UHF sensors). The number of sensors relates to a sensitivity requirement as given in Joint CIGRE TF document 15/33.03.05 [19]. This document specifies that a PD magnitude, equivalent to 5 pC from a metal particle, must be able to be detected from any location in the substation. It is known that the UHF signals do attenuate with distance [5] due to the geometry of the GIS and solid insulation used for gas zone barriers within the enclosures. The attenuation is proportional to the sensor spacing so there is a maximum allowable spacing between sensors related to the attenuation between them.

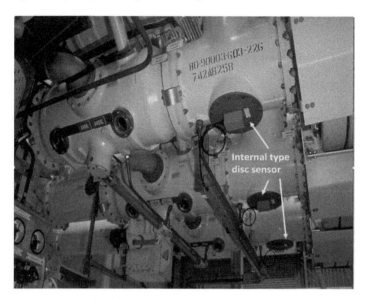

Figure 21.8   UHF sensors installed at Killingholme 420 kV GIS (1991)

Typically, with 9 sensors per bay, the PDM node unit is capable of monitoring 5 bays, covering 45 inputs, leaving 19 channels spare which could be used for noise monitoring and PDM system hardware monitoring. When used with a three-phase GIS enclosure (i.e. 145 kV GIS), one node unit will be capable of monitoring 15–20 bays.

The UHF sensors mounted on the GIS enclosure (see Figure 21.8) are connected to the convertor module via short lengths of low loss cable (<15 m) where the UHF signal pulses are optically converted for transmission as light in order to achieve very high noise immunity during the transmission to the optical receiver modules. The input signal pulses from the sensor are usually in the range of 10–2000 μV for internal partial discharge. Pulse sampling and timing requirements are given in Appendix C.

Measuring tens of microvolts, where the electric field inside the enclosure is driven by hundreds of kilovolts, requires specially designed UHF amplifiers to boost the UHF signals from the sensors, before they are processed. These amplifiers must also withstand switching transients from equipment operation, that is circuit breakers and disconnectors, which will produce sensor outputs of hundreds of volts in the 5 MHz range and hundreds of millivolts above 500 MHz.

21.7.1.4   Disconnector and earth switch monitoring

Typically, each switchgear bay for 420 kV GIS will require the following parameter count:

- three parameters per disconnector
- three parameters per earth switch
- five parameters for fault making earth switch (FMES).

In the main, each bay would require three disconnectors, two earth switches and one FMES. This gives a total parameter count of 20 parameters to be monitored per bay. Each disconnector and earth switch monitor (DEM) node unit is capable of monitoring three bays, similar to the CBM system, covering 60 inputs, with spare channels for self-monitoring.

A range of transducer types are used for disconnector and earth switch monitoring with some using 4–20 mA current loop technology and others with low impedance inputs to give improved noise immunity. Data sampling and timing requirements are given in Appendix D.

21.7.2 Specification of the node unit data acquisition system

The results of the review for the CM platform showed that the most demanding data acquisition application was that to be used for UHF PD monitoring. The data acquisition sampling rate was selected to meet the UHF system criteria with 250,000 samples per second. The A/D resolution was selected to be 16 bit in order to improve the accuracy and stability of the gas density data giving a more clearly defined detection capability for early warning of SF_6 gas leakage.

A compact PCI eXtensions for Instrumentation (PXI) system along with a Windows XP Pro operating system was selected for the data acquisition package since this provided the required EMI and temperature performance. The PXI system also provides modular PC and data acquisition units, along with a modular power supply which meets the specification for ease of maintenance and speed of repair in the event of a failure. There are also facilities for multiple data acquisition modules which allow for hybrid node units to be created for economy, where more than one type of monitoring system can be installed within one node unit.

The final arrangement for the CM platform shown in Figure 21.9 is configured as the PDM option. The PXI control PC module and data acquisition module are shown along with the interface module installed with 11 optical receiver modules ready for service.

The node unit enclosure provides the necessary environmental and EMI protection and houses the interface module along with the power supplies required to drive the interface module. A fibre optic communications module is installed to ensure 'an interference-free communications system'. The EMI filtering equipment and protection circuits are also located within the node unit to make it a self-contained monitoring system requiring only an input power supply usually from a remote uninterruptable power supply (UPS) system.

21.8 GDM node unit overview

The GDM node unit can house up to nine density transducer interface modules. Each interface module contains 5 analogue input channels for density transducers, resulting in a maximum of 45 density transducers per node unit.

712 *High-voltage engineering and testing*

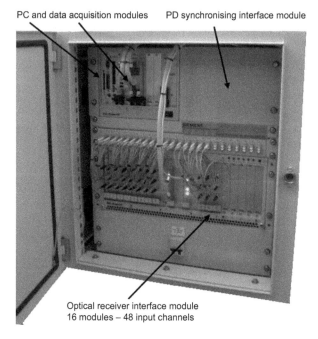

Figure 21.9 GIS PDM node unit

Data acquisition and processing at the node unit is carried out by the integrated measurement system which provides a fully self-contained monitoring system. Access to the measurement system to configure the node unit is provided by an Ethernet connection, and the following parameters are assigned to each channel:

- circuit breaker bay
- phase
- gas zone reference
- working density (g/l)
- first-stage and second-stage alarm levels (g/l)
- installed gas (kg).

The main GDM monitoring screen, shown in Figure 21.10, displays the settings from all 45 transducer inputs. The readings can be displayed as either density values (g/l or lb/ft^3) or pressure values (bar absolute or psi) and are presented as real-time values with a 1 s update.

A calibration facility is included to account for small variations in signal gain and DC offset as the signals pass through the interface module to ensure that a high degree of measurement accuracy is maintained.

21.8.1 Data collection

The node unit data acquisition system acquires data from all transducers every second and displays the real-time values on the main node unit monitoring screen.

Integrated substation condition monitoring 713

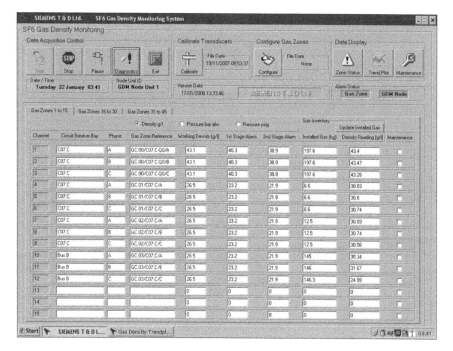

Figure 21.10 Main GDM node unit monitoring screen

The real-time data is checked against the alarm criteria and the alarm situation is shown locally at the node unit and is also sent to the ISCM control cubicle. A range of data display features are available for viewing data trends and alarms which can be accessed from the ISCM control cubicle using remote desktop facilities.

The GDM node unit, shown in Figure 21.11, was part of the first installation (installed in 2010) of an ISCM system employing both PD and SF_6 gas density monitoring. The node units are mounted in the substation close to the transducers and sensors that they monitor. In this case the PDM and GDM nodes can be seen side by side providing a very compact and easily accessible arrangement.

21.8.2 Predictive alarms and SF_6 gas inventory

One of the main advantages of the GDM system is the fact that it can readily identify leaks at a very early stage and predict when alarm criteria will be met. Conventional gauge reading systems, with two-stage gas loss alarms, are unable to provide such early warnings and a significant amount of SF_6 gas can be lost to the atmosphere before the 'Stage 1' alarm is activated.

The current gas loss status for each gas zone is shown in Figure 21.12 on the zone status screen. This information includes, among others, the present gas leak status along with the current gas loss (kg) and uses this leak rate to predict the activation dates for first- and second-stage alarms in order to prioritise

714 High-voltage engineering and testing

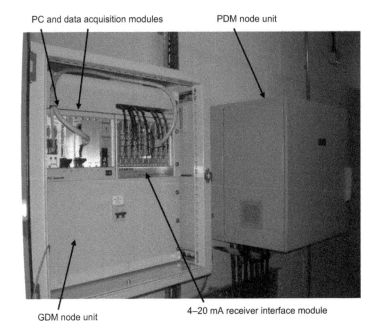

Figure 21.11 GIS GDM node unit installed on site

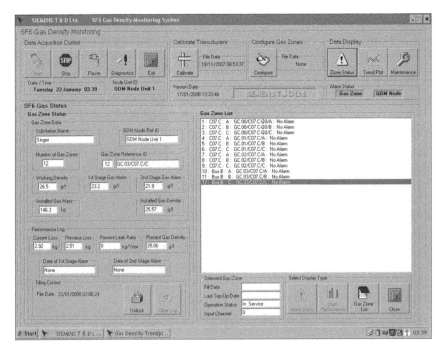

Figure 21.12 Zone status screen

Integrated substation condition monitoring 715

maintenance schedules. A log of the leak performance of the individual gas zones is automatically produced. Those gas zones which have lost the largest amount of gas and those that are leaking at the fastest rate are displayed at the top of the list so that they can be easily identified and 'targeted'.

21.8.3 Maintenance

The maintenance facility, to regulate the data collection when any of the gas zones requires maintenance, is provided by the maintenance settings screen as shown in Figure 21.13.

A range of user selectable monitoring options is provided to allow control of the data logging during substation maintenance. Logging of the following aspects of the monitoring system can be halted as required for the duration of the maintenance period:

- gas inventory logging
- gas alarm logging
- transducer fail logging
- internal flashover/fault logging.

These settings allow for a measure of flexibility to be built into the system, to cope with any gas zone work during the lifetime of the substation, without causing disruptions to the historical data. Consider two typical examples:

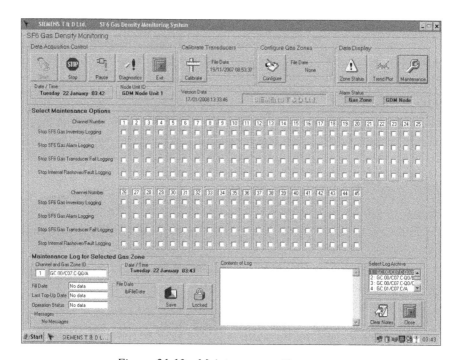

Figure 21.13 Maintenance settings screen

716 *High-voltage engineering and testing*

- This type of facility makes it possible to 'block gas loss logging' so that any gas zone can be de-gassed and re-gassed during maintenance routines for internal maintenance work on disconnectors.
- It also allows for repair work on the actual monitoring system itself, say, transducer calibration or replacement.

21.9 CBM node unit overview

The key function of the circuit breaker is to interrupt current flow during system faults. The trip coil performance during trip initiation is very important, as this can delay the operation of the circuit breaker. The mechanism stored energy gives the interrupter its acceleration power, so it ultimately controls the speed of the contacts during arc interruption. If the speed falls below design levels then failure to interrupt may occur at high fault currents. From this, the key static parameters will be:

- DC control voltage applied to the trip/close coil
- mechanism stored energy
- SF_6 gas density.

The circuit breaker contacts must carry the load current without excessive heating, and must also withstand the heat of the arc produced when interrupting the circuit current. Contacts are made of copper or copper alloys, silver alloys and other materials, and the service life of the contacts is limited usually by erosion during arc interruption.

In order to monitor the performance of the circuit breaker, key dynamic parameters such as the contact separation speed along with operation time and the arc duration along with current magnitude must be recorded and stored.

The CBM software requires a wide range of performance-related alarms and settings in order to cover the range of available circuit breakers. The alarm settings are designated by the user as shown in Figure 21.14, which allows circuit breaker specific settings for CT ratios, trip and close coil resistance values, interrupter stroke length etc. to be configured. These alarm settings can be changed online if the user requires to modify them during the lifetime of the switchgear.

21.9.1 Signals measured and recorded

The following 'dynamic' signals are recorded by the monitoring system when the circuit breaker operates:

- breaker contact travel (if required): 3 inputs
- hydraulic pressure (if applicable): 1 input
- air pressure (if applicable): 1 input
- SF_6 gas density (circuit breaker zone): 1 input
- Trip 1, Trip 2 and close coil currents: 3 inputs
- auxiliary DC supply voltage: 1 input
- Trip/Close commands: 3 inputs

Integrated substation condition monitoring   717

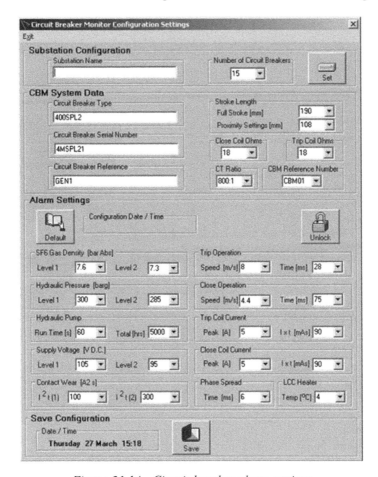

Figure 21.14   Circuit breaker alarm settings

- AC phase currents: 3 inputs
- AC phase voltages: 3 inputs
- mechanism proximity switches (if applicable): 3 inputs
- local control cubicle (LCC) temperature: 1 input
- ambient temperature: 1 input.

The following 'static' signals are recorded on an hourly basis by the monitoring system:

- hydraulic pressure (or equivalent mechanism drive type)
- SF_6 gas density (circuit breaker zone)
- auxiliary DC supply voltage
- AC phase currents
- AC phase voltages

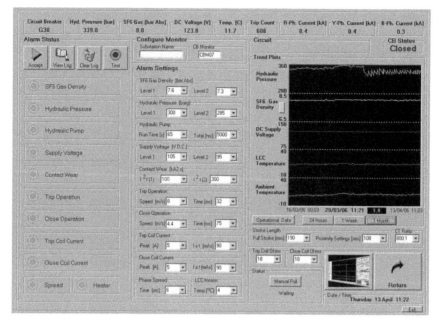

Figure 21.15 Node unit alarm and trend plot screen

- LCC temperature
- ambient temperature
- humidity
- hydraulic pump starts and run times recorded as they occur
- spring charging motor number of starts and duration of run times (if applicable)
- air compressor motor number of starts and duration of run times (if applicable).

The circuit breaker node unit screen, as shown in Figure 21.15, displays detailed information about the circuit breaker. If any of the parameters exceed their 'alarm' parameters, they are clearly displayed in red so the user can see, very quickly and easily, where any problems lie.

The software has been designed to allow the automatic generation of alarms should any of the parameters fall outside of the limits from the user settings.

21.9.2 System alarms

A comprehensive range of alarms is required to provide an early warning of any impending problem with the circuit breaker system. Some of the alarms may be transient, for example a change in trip coil current profile due to a long standstill time without circuit breaker operation, and then subsequent operations show no further alarms. In other situations, the alarms may be consistent and show further excursions into the next alarm level, indicating that an investigation is required on site. Typically, alarms are provided as follows:

- SF_6 gas density: two level alarms
- hydraulic pressure (if applicable): two level alarms
- air pressure (if applicable): two level alarms
- hydraulic pump activity: overrun time and total run time alarms
- spring charge motor: overrun time and total run time alarms
- air compressor motor: overrun time and total run time alarms
- auxiliary DC supply voltage: two level alarms
- main contact wear: two level alarms for I^2t
- trip operation: speed and operate time alarms
- close operation: speed and operate time alarms
- trip coil 1 current: peak current and profile alarm
- trip coil 2 current: peak current and profile alarm
- close coil current: peak current and profile alarm
- closing phase spread: limit alarm
- LCC temperature: limit alarm
- AC phase current: two level alarms
- AC phase voltage: two level alarms.

21.9.3 Data storage and display

Trend facilities for SF_6 gas density, hydraulic pressure, air pressure, Aux DC supply voltage, humidity, LCC and ambient temperature are provided for the circuit breaker 'control' parameters.

During the operation of the circuit breaker, more detailed information regarding the operate times, speeds, pole scatter, switched currents etc. are measured, calculated and stored for alarm purposes and for later viewing. An assessment of contact wear is provided by arcing and pre-arcing times, which are also monitored and stored along with the effective arc energy (I^2t) for each phase.

Figure 21.16 illustrates the record of a 36 kA yellow phase fault. The record has been 'zoomed' via the software's zoom feature to show more detailed information on the analogue traces. The trip coil current pulse, which initiates the mechanism operation, shows that from the start of the current pulse to the start of travel, takes about 20 ms, which is normal. Any problems with the trip coil would show a change in the current pulse profile, along with an increase in the delay to the 'start of travel'.

The mechanism proximity switches, which are mounted close to the interrupter drive shaft, produce the output as shown by the frame under the travel record frame. The gap in the traces indicates that the drive shaft is in motion and marks the 'point of contact separation and the end of linear contact travel' and can be seen to be approximately 11 ms for each phase. From this, the trip speed is calculated and is shown for each phase on the trip statistics panel on the right, and can be seen to be in the range of 9.8–10 m/s. This high speed is required for current interruption,

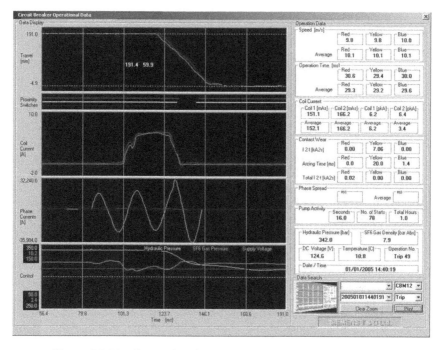

Figure 21.16 Circuit breaker operation screen – trip operation

as explained earlier. During the trip operation, the interrupter moving contact assembly must accelerate from standstill to 10 m/s within 12 ms. To put this in perspective, if this acceleration was maintained, it would be equivalent to getting from 0 to 60 mph in 0.03 s.

The fault current trace shows approximately three full cycles of power frequency current before arc interruption occurs. The arc duration is measured from the point of contact separation to the point of arc interruption, which can be seen to be approximately 20 ms. This time is then used along with the fault current to calculate the I^2t value.

The I^2t values have been incremented automatically based on the amplitude of the fault current and the duration of the arcing time. This is calculated automatically by the software. This figure will accumulate over the lifetime of the substation and provide an indication of the 'work' done by the main contacts. Two alarm limits for the I^2t value are provided to create an early warning (30% of maximum) along with a warning of approaching the recommended maximum level so that an outage can be planned and the main contacts inspected.

In contrast to the trip operation, Figure 21.17 shows a circuit breaker 'close' operation with a closing speed of 5 m/s. This is much slower than the trip operation. Again, as for the trip operation, the travel, proximity switches and coil current records (close coil) can be seen. In this instance, the load current can be seen, with some initial asymmetry, reaching about 700 A peak and finally settling to about

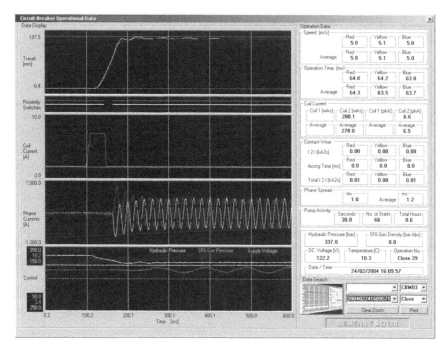

Figure 21.17   Circuit breaker operation screen – close operation

400 A rms. The traces below the load current trace show the hydraulic pressure dropping as the mechanism is triggered and hydraulic fluid moves into the drive mechanism, but this pressure will be restored within a few tens of seconds, when the hydraulic pump detects the pressure drop and restores the pressure. Also shown is the DC supply voltage which remains constant during the close operation. The third trace is that of the SF_6 gas density which can be seen to oscillate with a period of 100 ms. This oscillation is caused by a pressure wave inside the interrupter head caused by the sudden movement of the drive shaft and main contact assembly during closing. The compression of the SF_6 gas will produce localised density increase as the wave passes the density transducer. This will finally settle within approximately 5 s after the interrupter comes to a halt.

All operations are automatically compared with a reference operation (usually the first operation measured by the ISCM system after commissioning). If any parameters are noted to be varying from this 'fingerprint' record, an alarm is raised. This provides the user with an early warning system that something may be failing on the circuit breaker and maintenance should be planned. Similarly, if all records are within accepted limits and not indicating any variations, then this gives the user the vital information that the circuit breaker is in a healthy condition. This eliminates unnecessary time-based maintenance which would otherwise be performed on a healthy circuit breaker, thus saving money and outage time, and reduce the risk of faults being introduced during a maintenance programme.

21.10 PDM node unit overview

The PDM node unit, shown in Figure 21.18, was part of the first installation (in 2010) of an ISCM system employing PD and SF_6 gas density monitoring.

The PDM node unit can house up to 16 UHF optical transducer interface modules. Each module contains three optical input channels. This allows 16 sets of three-phase UHF opto-detector units to be monitored giving a maximum of 48 UHF inputs per node. Data acquisition and processing at the node unit is carried out by the integrated measurement system which provides a fully self-contained monitoring system. Access to the measurement system to configure the node unit is provided by an Ethernet connection which can be accessed by the LAN or by a direct connection to a laptop PC.

The installed node unit is shown in Figure 21.18 along with one of eight of the optical detector units configured with this node unit. The remaining eight spare optical receiver modules can be used to accommodate additional UHF sensors required for any future extension work.

The detector unit is connected to the node unit via a hybrid 'copper and optical fibre' cable which both powers the detector unit and allows optical signal transmission between them. Three UHF cables can also be seen connected to the optical detector unit which provide the signal input from the UHF sensors installed inside the nearby GIS enclosures.

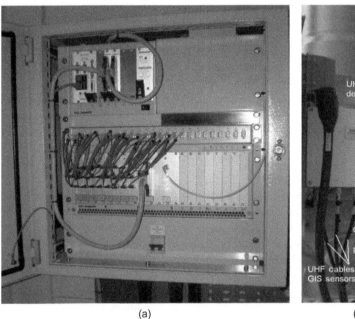

(a) (b)

Figure 21.18 PDM node and optical detector installed at the substation
(a) PDM node unit installed at the substation; (b) UHF optical detector unit

Each node unit is configured independently by loading, for example, PDM sensor reference (Sensor ID), minimum detection level (base) and alarm settings.

In addition to the measurement system and the interface module, the node unit system has a range of power frequency synchronising features. PD diagnostics require the synchronisation of the PD pulses with the power frequency voltage. The option selected will depend on whether a fixed frequency voltage is being used (i.e. normal 50 or 60 Hz power frequency) or a variable frequency voltage, such as that used for HV testing (range 30–300 Hz). The range of reference options, as shown in Figure 21.19 'POW Reference Setting' section, is as follows:

- power supply reference
- GIS VT reference
- internal UHF sensor reference
- digital synchronsing pulse (e.g. from HV test set).

The data status report in Figure 21.19 and Figure and 21.20 show typical low level signals (diagnosed as noise in this case) that may be detected from time to time dependant on the sensitivity and noise floor setting. A range of other controls are also provided for flexibility of operation along with a 'PD Sequencer' system which is used to capture a series of 1 s PD files over periods of tens of minutes to provide PD replay and diagnostic tagging data, which is discussed later.

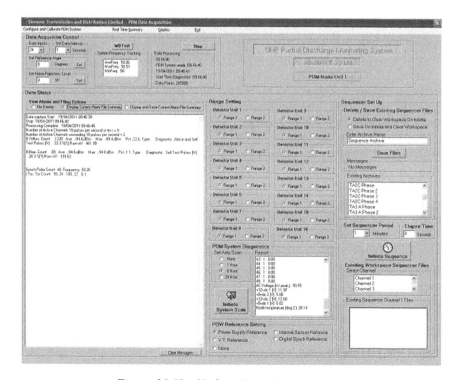

Figure 21.19   Node unit monitoring screen

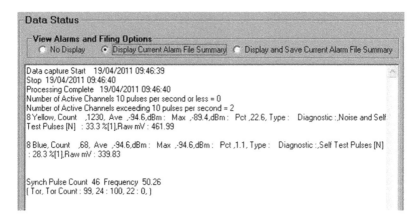

Figure 21.20 Node unit alarm data display

21.10.1 PDM node unit data collection

The signals from all of the opto-detector units connected to the node unit are recorded every second. The readings are displayed as real-time values with their respective ID tags on the main node unit monitoring screen (Figure 21.20).

All data is saved on the local memory store of the node unit. To provide a backup of this data, the ISCM cubicle will periodically poll the node units and transfer any new data to its memory store. This data will then be backed up to an external hard drive located within the ISCM cubicle.

Real Time PD Summary screen can be selected to see the extent of the PD activity as shown in Figure 21.21. Data point-on-wave (POW) information can be displayed as real-time instantaneous and continuously updated data.

The graphical data display shows the signal pulse patterns and is updated every second to show the real-time progress of the signal with PD pulse count and amplitude information. Simultaneously, signal diagnostics are carried out as the signals are detected on this second-by-second basis. This allows for noise and PD segregation, ensuring that spurious alarms are minimised and/or eliminated.

The summary display can provide sufficient capacity to cover simultaneous data for 3–4 bays of switchgear along with the associated feeder circuits.

21.10.2 Node unit sequencer controls

A 'self-learning' facility has been incorporated to allow user designated 'Information Tags' to be created. These tags are produced from the sequence data where analysis of the data files for each active PD provides a signal profile.

The tags are then used for real-time diagnostics at the node units and can help the user to categorise PD and noise types without knowing the precise cause of the signal. This information can also be transferred to other substation PDM systems to check for similar signal types and sources of PD.

Integrated substation condition monitoring 725

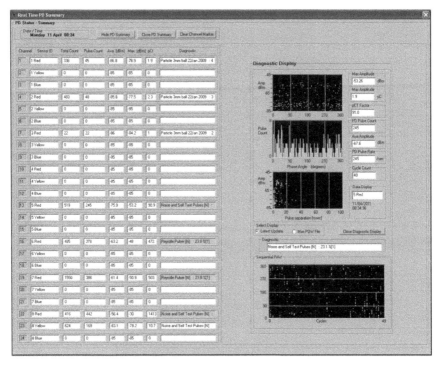

Figure 21.21 Node unit real-time PD summary screen

The sequencer controls shown in Figure 21.22 allows for periods up to 50 min to store a continuous set of 1 s data records for all UHF data being monitored by the node unit. The data can be 'uploaded' to the ISCM control cubicle PC for further analysis.

Once the information is available at the control PC, any new tags (noise or PD) can then be downloaded to all node units on the system to keep diagnostic information up-to-date.

The sequencer system is compliant with the requirement, in accordance with IEC 60517, to provide a continuous record of any PD activity that may occur during the HV on-site test procedure.

21.10.3 Node unit noise monitoring and alarm control

GIS substation PD monitoring systems can experience nuisance alarms created by UHF signals from external corona in air, radar, mobile phones and faulty lighting equipment, among others, that can penetrate the GIS enclosure and be detected by the internal PD sensors. Diagnostic software can be trained to deal with some of these noise sources, but in certain situations this may not be enough and other methods of noise control have to be used, for example, by adopting a noise monitoring configuration using one detector unit to provide three noise monitoring channels. Each of the three sensors, connected to the noise channels, can be located

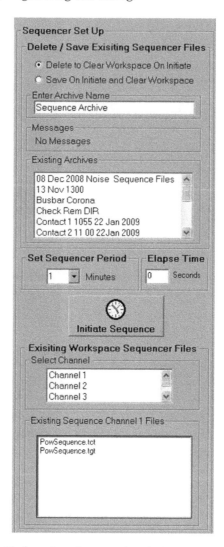

Figure 21.22 Node unit main operational screen – sequencer controls

in the vicinity of the PD detector units to collect noise which is present primarily in those specific areas. The PD detector units will then be assigned a specific noise sensor for that particular location. This allows substation-specific noise and general background noise to be eliminated from the data records on a pulse-by-pulse basis as it occurs, leaving only PD signals for diagnostics and alarms. Typically, the three channels available for each node unit could be allocated, for example, to

- busbar region
- line entry region
- outdoor gantry area.

Integrated substation condition monitoring 727

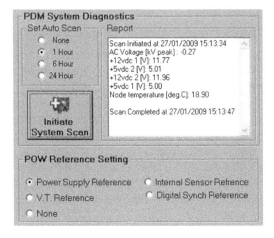

Figure 21.23 Node unit main operational screen hardware diagnostics

Noise control options are as follows:

- monitoring substation background noise for diagnostic information on the cause of the noise and the degree of activity
- noise capture and automatic tagging for second-by-second noise elimination
- noise gating to allow pulse-by-pulse removal during 1 s sweeps.

21.10.4 Node unit system diagnostics

A built-in self-monitoring system is available to flag an alarm for maintenance to minimise any downtime. In the event that the node unit indicates a hardware alarm condition, the nature of the alarm can be identified by looking at the diagnostics report on the main node unit monitoring screen. Figure 21.23 provides an illustration of the diagnostics screen.

The PD data acquisition system is scanned at regular intervals to check for correct operation of the optical transducers, optical convertors and the data acquisition and storage facility. Further diagnostic information is made available to identify possible causes of the alarm in order to advise maintenance requirements.

21.11 Power transformer monitoring

Transformers belong to the most important assets for the transport of electrical energy. They are used from generation to transmission and finally to distribution. Some transformers have to be taken out of service earlier than planned (25 years instead of 40 years) because of several factors:

- ageing of the winding insulation
- degraded bushings

- on-load tap-changer wear
- contaminated or impure oil.

Some of these problems could be prevented or delayed through condition monitoring which would show key performance data with sufficient detail to determine what's happening inside of the transformer.

Insulation deterioration, inadequate maintenance, loose connections, the status of the tap-changer, moisture and overloads can all be monitored and rectified early so as to avoid unplanned outages and its impact on available power supply. As an example of what can occur in service, Figure 21.24 shows the effects of a loose core clamp inside a 400 kV/132 kV interconnector transformer.

The surface of the rear clamp shows 'significant carbon spattering' as evidence of PD between the core clamp foot and the top of the core. There is also a pool of carbon in oil at the bottom of the clamp.

The photograph in Figure 21.24 was taken during an internal inspection of the transformer which was initiated as a result of PD detection in service. The PD signal was detected by the online UHF PDM system installed on the GIS, primarily

Figure 21.24   Transformer internal partial discharge from a core clamp

from the UHF sensor mounted near the transformer SF_6 gas-to-oil bushing, but was of sufficient magnitude to be seen at other sensor locations. This PD was active for some 6 months before the outage was taken for investigation. Initially the gas-in-oil detection system showed no evidence of a problem but after some 3 months, gas-in-oil was just detectable. Finally, the gas-in-oil reached a sufficient level to warrant an internal examination of the transformer after a further 3 months. The PD signals also showed increasing magnitude and pulse count activity during this period, which provided further justification for the outage and resultant cost of the investigation.

PDM is one of many ways [13] to obtain information about the transformer condition. In order to be able to determine the current condition of a transformer accurately, numerous influencing factors must be recorded and evaluated:

- temperature
- measurement of the load of voltage and current
- fan status (cooling)
- DGA: dissolved gas analysis (gas in oil)
- moisture
- tap-changer and bushings
- partial discharge.

A range of intelligent sensors are available to make these measurements (examples of which will be reviewed later) which can transmit the measured values continuously to a central location. Typically, this is used to record the condition of the:

- transformer windings, core and oil
- cooling system
- bushings
- load tap-changer.

Remote terminal units (RTUs) read the sensors and ensure that reliable communication is maintained with the Asset Management Centre via standard interfaces. The acquisition of data about a transformer's condition is nothing new in itself but the feature of data collation at a central location significantly improves the availability of up-to-date information which can be readily assessed by the asset management team. It takes the combination of many individual items of data and their trends to create a comprehensive picture of the actual condition. The data, as shown in Figure 21.25, is analysed automatically by a specific knowledge-based management system using a range of combinable knowledge modules (KMs). These modules are based on the comprehensive knowledge gained from many years of experience by transformer manufacturers. They are held at the central location where the relevant operating and condition parameters are computed on the basis of the available measured data, design parameters and historical events, by means of the KMs.

The KMs communicate with the Asset Management Centre via an interface where maintenance and repair requirements can be deduced from diagnosis of possible faults and forecasting of future operating states.

730 High-voltage engineering and testing

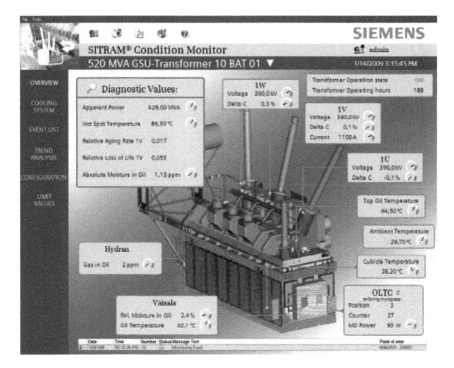

Figure 21.25 Power transformer monitoring

The individual modules are summarised as follows:

- Thermo-hydraulic model (THM): The THM supplies all information needed to operate the transformer with maximum efficiency. Losses, forced and natural oil flows taking into account hydraulic resistances and heat transfer coefficients, local temperatures and other important parameters are calculated reliably on the basis of a specific network model.
- Moisture module: This delivers information about the influence of moisture in the insulation on the dielectric strength, load capability and ageing of the transformer.
- Cooler diagnosis module: This compares the measured cooler oil temperatures with the computed values of the THM. As soon as there are any signs of a variation in the trend the operating personnel is notified.
- Cooler control module: Optimum cooling is crucial for the efficiency of a transformer. Conventional cooling concepts only take the maximum oil temperature into account, whereas the cooler control module also includes the load current. Cooling measures can thus be initiated before the temperature rises.
- Gas-in-oil module: This provides a picture of possible accelerated ageing or faults in the transformer by determining the content of hydrogen, carbon dioxide, carbon monoxide, methane, ethane, ethylene, acetylene and oxygen in the oil.

- Bushing module: This analyses the condition of the insulation system of the bushings and informs the personnel of any changes observed.

21.11.1 Sensors for dissolved gas analysis

Through time, transformers endure stresses that can contribute to a variety of failure mechanisms. Online DGA using a range of diagnostics tools can help utilities avoid unplanned outages due to failures, lower maintenance costs and extend transformer useful life. The DGA diagnostic and maintenance tool can be used to detect most failure mechanisms and indicates the level of any probable damage.

There is a range of DGA sensors supplied by various manufacturers, some of which are shown in Figure 21.26, which have compact package designs allowing them to be mounted on or near the transformer. The sensors measure critical fault gases found in transformer oil and can also measure moisture in the oil, oil temperature and ambient temperature.

The sensor periodically collects oil samples, every 2 h if necessary, and the measurement data is 'date/time stamped' which can be used to correlate with transformer load. The benefits from using these sensors are as follows:

- Field-based fully automated DGA requires no manual oil sampling or remote lab testing.
- Accurate and timely DGA results improve ability to protect transformer against fault damage.
- Data generated can be used to support condition-based maintenance programmes.
- Time-date stamped test results allow correlation between real events and measured conditions.

Monitoring and performing real-time transformer modelling can help reduce the risk of unexpected and sometimes catastrophic failures. This also helps to avoid explosive clean-up, replacement, and unplanned downtime. Early detection of

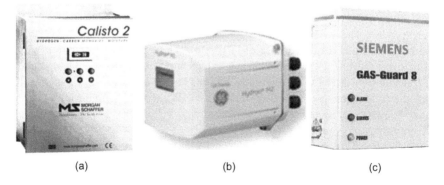

Figure 21.26 DGA sensors
 (a) Calisto 2 (Morgan Schaffer); (b) Hydran (GE); (c) Gas-Guard 8 (Siemens)

732 *High-voltage engineering and testing*

potential transformer problems is vital to the lifespan extension of critical transformers and provides significant business and operational benefits that will

- reduce inspection and maintenance costs by extending the time between routine maintenance activities
- reduce unplanned outages with continuous CM and early detection of primary faults
- provide greater lifespan confidence through the use of online model computations providing real-time transformer condition information.

Additionally, in-service CM can defer major replacement costs by optimizing the transformer's performance and extending its lifespan.

21.11.2 Sensors for tap-changer monitoring

Tap-changer monitors can be used in transformers to perform a range of strategic measurements that can provide essential information to assess the possibility of any damage that could occur to the 'on-load tap-changer and transformer'. This helps to improve operating safety and extended maintenance intervals.

The tap-changer motor drive, which works on the principle of switching in steps, causes a voltage change with each step at the transformer terminals. Each step involves a change of contact position, resulting in a switching transient within the tap-changer. The relevant switching events are recorded during the tap-changer switching operation, along with limiting values such as those for load monitoring. Movement of the tap-changer position will be prevented if any critical limiting value is exceeded or if a failure is detected.

21.12 Cable monitoring

Cables are mainly used in lower voltage levels up to 60 kV where they provide the majority of system interconnections. In contrast at the higher voltage levels, cables are only used in about 4% of the system interconnections. This is due to the expensive investment costs of cables, which are clearly higher than those of the overhead line. However, cables are expected to be used more frequently in future at the higher voltage levels because of environmental issues, with the alternative of using overhead lines. In some instances, cables are used as the only viable method of transmission of electrical energy, for example, for offshore-wind parks to provide the connection to the coastline substations. Wind parks with distances shorter than 50 km are connected to a three-phase power system. For longer distances, HV DC transmission would be used.

Cable monitoring allows utilities to get important information about the temperature and the load in the cable through integral optical fibre. This allows the early detection of progressive degradation of the cable insulation and incipient failures, specifically those exhibiting line impedance variations, PD or cable hot spots. The insulation performance of the cable system is recorded on site, continuously or periodically, by sensors which are specially developed for a simple configuration and installation. Monitoring can also allow performance improvements in the cable transmission capacity. The following benefits can be achieved through cable monitoring:

- cost-effective operation with minimal personnel effort required after installation
- detecting defects occurring shortly before failure
- improvement of cable grid reliability by detection of pending failures
- reduction of unexpected outages
- planned shutdown times in line with maintenance strategy
- reduced repair costs since repair works become planned
- maintaining continuous power supply during measurements
- continuously registering data that captures time-related information for trend analysis
- technical evidence for condition-based maintenance programmes.

21.12.1 Sensors for cable monitoring

During high-load conditions and under emergency circumstances (such as when a failure occurs in a segment of the grid and power must be shifted to other sections to compensate), it often becomes necessary to load cables to the maximum permitted limits by the relevant regulations. In these operational situations, it must be ensured that maximum temperature limits are not exceeded. Due to the increasing complexity of the thermal relationships along cable routes, the ability to continuously measure the temperatures along the cable has proven invaluable. Critical operational data is made available along with the identification of developing cable hot spots, which provides the means to prevent cable failure by early correction.

The measured values from the sensors are analysed through the distributed temperature sensing (DTS). DTS is a powerful tool that allows the accurate rating of HV power cables in real time, and provides the following operating benefits:

- reduce power outages or blackouts
- ensure continuity of supply
- activate hidden capacity reserves of existing assets
- react quickly to overload conditions
- conduct precisely and in real-time load predictions as new sources of energy are added to the grid.

Optical fibres are the most important components in the sensor cable. The optical fibres are preferably encased in a stainless steel tube which significantly increases the mechanical stability of the sensor cable. In addition, the inside of the pipe is lined with gel to ensure that the sensor cable remains permanently waterproof.

The fact that temperatures are measured purely optically produces two major advantages for this technology. First, the high electromagnetic tolerance means that fields of disturbance, such as the HV cable itself, electric motors (e.g. pump) or any kind of transmitter (e.g. mobile phones), do not trigger disturbances. Second, the sensor cable is practically maintenance-free.

Temperature or pressure and tensile forces can affect glass fibres and locally change the characteristics of light transmission in the fibre. Effectively the optical fibre can be employed as a linear sensor, using the resultant changes in the attenuation of the light in the quartz glass fibres through scattering, allowing the location of the physical effect to be determined. This technology provides a system for real-time diagnosis and condition monitoring of installed electric cables and can measure cable length ranges from a few meters to several hundred kilometres, depending on the cable structure and attenuation.

The system can monitor the global, progressive degradation of the cable insulation due to harsh environment conditions and detect local degradation of the insulation material due to mechanical effects or local abnormal environment conditions. Identification of the location of an anomaly in the cable can be made with an estimation error better than 0.5% of the cable length. The following operational benefits can be achieved through cable monitoring:

- connection to energised cables, no need for outages
- online continuous condition monitoring
- early warning of local and global deterioration
- prediction of remaining life time of cable.

21.13 Surge arrester monitoring

The surge arrester is a device used on electrical power systems to protect the system insulation from the damaging effect of lightning and switching surges. A surge arrester protects the insulation by limiting the voltage transient, using the non-linear resistance property of the zinc oxide core. It has a HV terminal and a ground terminal and when a lightning surge or switching surge travels along the transmission system to the arrester, the impedance of the arrester reduces, under the influence of the overvoltage, and draws current from the surge and diverts it to ground through the substation earth mat.

The surge arrester is a very important asset to monitor, and although it is not a costly item, it protects very expensive substation equipment such as transformers. Surge arrester monitors, as shown in Figure 21.27, have been used in various designs over the last 30 years, particularly on gapless metal-oxide surge arresters. The continuous leakage current, under dry, wet and polluted conditions, along with the energy absorbed by various surge currents, is key information that can be used for condition assessment.

The leakage current is composed of two components, one capacitive (typically 10 mA) and one resistive (typically <1 mA). These currents will vary depending on weather conditions and the temperature of the metal oxide core, which will rise following surge current flow. The resistive current component is very important because this will also cause heating of the metal oxide core which must be limited.

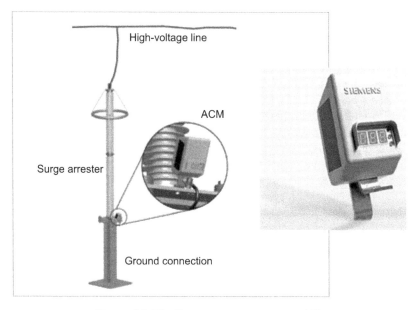

Figure 21.27   Surge arrester monitor ACM

The arrester condition monitor (ACM) was developed to monitor these parameters on gapless metal-oxide surge arresters and performs two basic functions:

- A.C. leakage current measurement: Leakage current measurement determines the total leakage current and the resistive leakage current component by analysing the leakage current's third harmonic component. In addition, leakage current is measured once a day and the results are saved to the long-term memory for the purpose of detecting and evaluating trends.
- Surge current impulse measurement: When registering surge current impulses, the ACM determines both the total number of surge current impulses and the level and duration of the current for the individual surge current impulses and allows analysis of energy converted in the surge arrester.

Communication with the ACM is achieved with the aid of the universal serial bus (USB) wireless module so that data can be sent across site to central control PC at the substation.

Key features for CM data are as follows:

- Automatic long-term trend registration: This function automatically registers relevant signals and provides a detailed history log for performance analysis.
- Measurement of energy absorption: The energy absorbed by the surge arresters is measured in order to predict the point of overheating due to excess energy absorption, so that appropriate action can be taken.
- Solar-powered energy supply: Solar power allows a true 'isolated' measurement system that does not cause interference with the surge arresters' main current path to ground. This gives a more accurate measurement of the resistive component of leakage current.

21.14 ISCM system overview

The previous sections of this chapter have presented an overview of a limited range of the many CM systems that are currently in service. A review of sensors and parameters to be measured has been given, along with examples of the graphical interfaces available directly from these systems. Some in-service examples of data derived from the systems have also been presented to illustrate the range of information that can be produced.

This final section looks at how these individual systems can be brought together by importing and archiving data and alarms, along with the graphical user interface to form an ISCM system. The general functionality of the ISCM system will be shown along with application examples. (The reader should again refer to the list of principal terms, e.g. CM and UPS, referred to in this chapter and summarised in the list of acronyms at the beginning of the chapter.)

Figure 21.28 shows the basic architecture of the ISCM system, where the node units pre-process the acquired data, create alarms and forward the information to the substation gateway control PC via an optical LAN. Similarly, information from transformer, cable and surge arrester monitoring systems can also be forwarded to the control PC via LAN or 'USB wireless' devices. Remote access is achieved by an asset management WAN which will upload all of the latest status alarms and diagnostic data and reports.

The system can 'import' CM data from a wide range of monitoring systems, with communications ranging from Ethernet to IEC 61850. Each monitoring system, connected to the ISCM system, is arranged as a 'CM module', allowing information to be transferred as variables within that CM module. The number and type of CM modules will vary as they are 'project specific'.

As an example project, the salient features of the first installation (in 2010) of an ISCM system employing PD and SF_6 gas density monitoring will be described in the following sections. This was a UK 420 kV GIS project for National Grid at Stella West. It is significant to report that other projects involving only single CM modules have also been installed with the present ISCM system in order to provide the facility for future seamless additions of other CM modules as required.

21.14.1 Substation gateway

The display shown in Figure 21.29 is a two CM module system consisting of PDM and GDM modules where the alarms are displayed for both systems.

The 'Main' operational screen provides access to more detailed information associated with the different CM modules connected to the ISCM system. The control features are the same as that used with the well proven substation control system (SCS) which is used to operate the substation equipment with the appropriate 'User Authorisation' levels for security.

The ISCM system has an integrated alarm and event list so that if any alarm conditions are exceeded, with any aspect of the ISCM system or the parameters that

Integrated substation condition monitoring 737

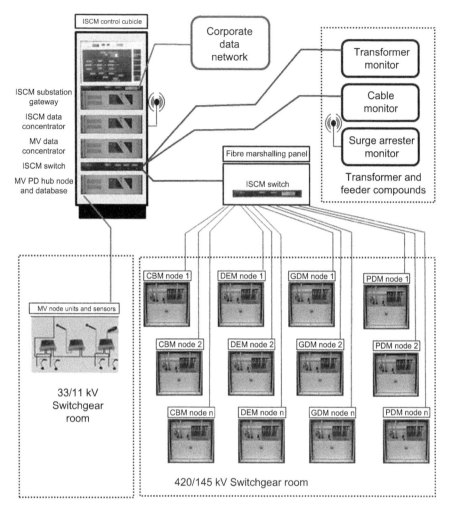

Figure 21.28 ISCM overview of substation monitoring

it is monitoring, then the alarm is automatically registered within this list, along with a text message describing the nature of the alarm or event.

The ISCM control cubicle, as shown in Figure 21.30, forms the hub of the fibre optic Ethernet communication system (LAN) to provide data retrieval and control for each of the monitoring systems. All of the substation CM data is available for display at the control cubicle.

The software installed on the ISCM control cubicle is built using the Siemens SICAM 230 control system platform, which provides pre-processing and support software for each of the monitoring systems. The ISCM system provides a common user interface to a number of CM options, and from one display, the user can identify any alarms with any CM modules connected to the system.

738 High-voltage engineering and testing

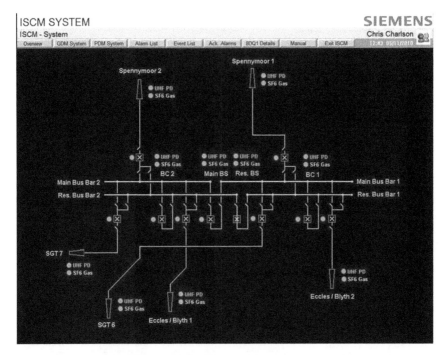

Figure 21.29 ISCM system main operational screen

21.14.1.1 Remote access

The ISCM system gateway (substation gateway) provides a number of facilities for remote access to make best use of the infrastructure available at the substation. For example, via E1/T1 WAN links, ADSL connections, LAN links and 3G wireless links. This provides a great deal of flexibility and allows the fastest possible communications to be established between remote location and the ISCM system.

21.14.1.2 Local alarms

Local potential-free alarm contacts are provided by the ISCM control cubicle for a number of alarms associated with the ISCM system modules. The potential-free contacts can be connected to a local RTU, for example, for transmission to SCADA. The alarms can also be transmitted electronically via a number of available protocols to the local SCS.

21.14.1.3 Uninterruptible power supply

All of the components of the ISCM system are powered via a UPS so that, in the event of a temporary supply failure, the UPS will power the whole of the ISCM system for a period of typically 5–30 min, before 'initiating a controlled shutdown' of the system computers.

Figure 21.30 ISCM control cubicle showing PDM diagnostics display

21.14.1.4 Global positioning system

The ISCM control cubicle uses a GPS timing source to synchronise the main control cubicle computer. This main computer then synchronises all of the other computers on the ISCM system network periodically via NTP. This ensures that all data captured and processed within the ISCM system is synchronised and can be cross-referenced effectively with other systems such as fault recorders for example.

21.14.2 CM module: PD monitoring

The PDM system overview screen is shown in Figure 21.31 and provides the user with the location of the PD sensors, giving a real-time view of the substation PD alarm situation.

If a PD sensor detects PD activity that has exceeded a preset alarm threshold level, then the colour of the PD sensor locator changes accordingly to represent the status of the alarm. There are four alarm status levels ranging form 'green', which indicates no detectable PD, through 'yellow', 'orange' and 'red', the latter being the most severe PD condition level.

21.14.2.1 Detailed bay information

PD interpretation is a complex procedure, with large amounts of information to process, in order to assess the PD performance of the substation. The screen, shown in Figure 21.32, provides PD diagnostics information which has been broken down into alarm levels, text diagnostics and graphs which are simple to understand, even when the user has only a limited knowledge of PD. The complex tools that produce this simplified information are still available to be accessed by users with the required authorisation level.

The single phase representation of the three-phase sensor from Figure 21.31 is now broken out into three individual sensors per measurement point. The colour of these circles is 'alarm level dependent' and enables the user to very simply

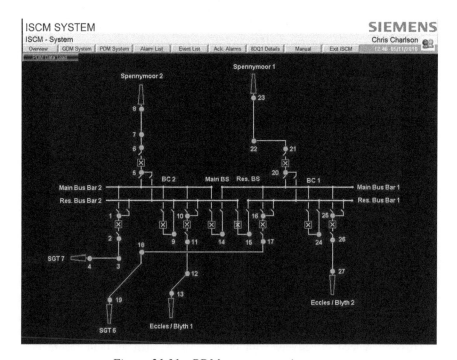

Figure 21.31 PDM system overview screen

Integrated substation condition monitoring 741

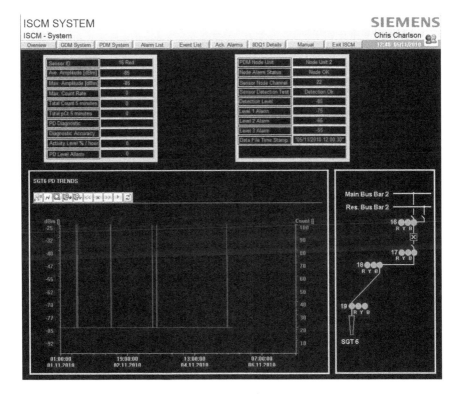

Figure 21.32 PDM system bay overview screen

identify the location and severity of the PD activity, measured within the bay at that time.

The PD measurement details for the bay sensors can be displayed in the tables at the top of the screen (Figure 21.32) for each sensor. The data to be displayed will be as follows:

- sensor ID
- average amplitude (dBm)
- maximum amplitude (dBm)
- maximum count rate
- total count (5 min)
- PD diagnostic
- activity level %/hour
- PD level alarm number
- node unit alarm status
- detection level setting
- level 1 alarm, level 2 alarm and level 3 alarm settings.

The trends have two y-axes, namely a PD amplitude axis and a PD count axis to show the data for each measurement sensor for periods of 24 h or over longer periods of up to 1 year. Any particular time period can be displayed, dating back to the installation date of the monitoring system.

21.14.2.2 PDM system diagnostics

An overview of PDM system status (see Figure 21.33) provides valuable alarm information with regards to each node unit attached to the system, along with any communications system alarms that may be active. It also provides the 'shutdown facility' for the node units in the event of maintenance requirements.

The status and update of the diagnostics tag facility is also shown for each node unit. The details of the individual tag libraries for each node can also be viewed so that the diagnostics library can be updated with the latest information via this user interface.

Coordinating the data from the many types of monitoring systems requires each individual system to provide data files which can be collected periodically using remote data access. In this example, the screen shows the status of National Grid access for the 'Smart Asset Management' web (SAM web) server application which has direct access to PDM status and data.

21.14.2.3 Display and diagnostics

Simultaneous 3-D and 2-D data display for diagnostic assessment are provided to allow the user to view current or historical PD activity as shown in Figure 21.34. Sequencer data for replay is also available to view a sequence, recorded for example, during switching operations or during HV test periods.

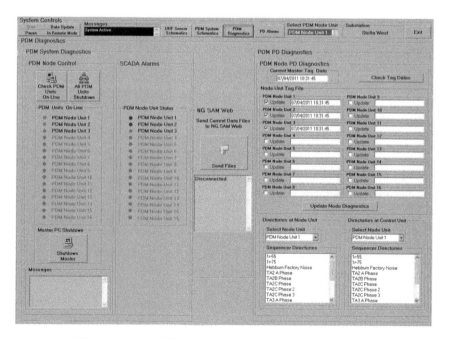

Figure 21.33 PDM system diagnostics overview screen

Integrated substation condition monitoring 743

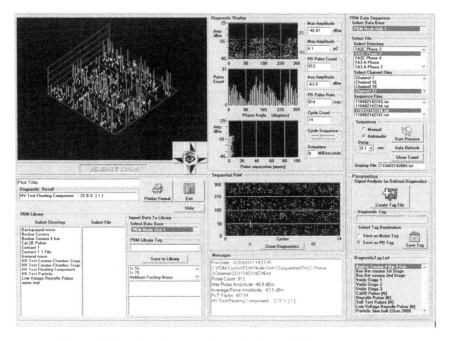

Figure 21.34 PDM system data display and diagnostics screen

The PD signal shown in Figure 21.34 was recorded during an 'on-site' HV commissioning test. The signal was seen to vary depending on voltage level and the diagnostics changed from 'floating component' type to 'particle' type as the voltage increased. The record shown was recorded at a lower voltage level with the diagnostics indicating floating component. The PD source was located to a particular gas zone which was internally examined and a large (5 mm × 5 mm) flat metallic particle was discovered which correlated with the partial discharge POW pattern and diagnostics.

21.14.2.4 Self-learning diagnostics

A 'self-learning' facility has been incorporated to allow 'user designated information tags to be created'. These tags are used for real-time diagnostics at the node units, which allows the software to automatically categorise 'PD' and 'noise' type data. This helps improve the accuracy of determining the nature of any PD source by having substation specific PD data tags and noise filtering. This information can also be transferred to PDM systems, at other substations of a similar type, to ensure that common 'PD' or 'noise' signatures are readily identified and categorised.

The 'Diagnostics Tags' are created by using the data input from a sequence of files recorded at the node unit. The sequence of data is analysed using pattern recognition and statistical methods. The accuracy of the newly created tag is then assessed by using data from other sequences of the same PD source and comparing the calculated 'confidence index' of the diagnostic result. The addition of this tag to

744 *High-voltage engineering and testing*

the diagnostics library means that all future PD data will be compared with this tag as well as all other tags within the library and automatically produce a 'best fit' diagnostic. Noise tags can also be added to the diagnostics library in the same way, so that noise signals can be identified and moved to the noise filing system.

21.14.3 CM module: gas density monitoring

The GDM system access screen, as in Figure 21.35, shows each gas zone within the substation displayed on a single line diagram.

The gas zones are separated by markers which represent the gastight barriers. If all gas zones are in a normal healthy state then the line colour of each gas zone remains green.

If the ISCM system identifies that there is a gas leak in any of the gas zones, then that particular gas zone line colour will be displayed in a yellow. If the gas leak is high enough so that it passes the first-stage alarm level setting, then the gas zone line colour will be displayed in an orange. If the gas zone continues to leak and the gas density level falls below the second-stage gas alarm, then the gas zone line colour will be displayed in a red. Conversely, if there is an inter-gas zone leak and a gas zone passes an overpressure threshold, then the gas zone line colour will be displayed in a blue.

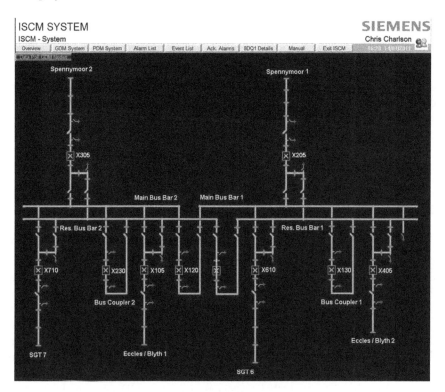

Figure 21.35 GDM system overview screen

21.14.3.1 Detailed bay information

Any alarms shown on the overview screen (see Figure 21.35) can be further investigated by selecting the relevant bay name. Gas zone details for the particular bay will be displayed as in Figure 21.36.

Selecting any gas zone allows the gas zone's data to be displayed in the tables at the top of the page. The current gas density figure is also displayed in the analogue meter at the top of the page. The information displayed in the table for each selected gas zone is as follows:

- gas zone ID
- current density (g/l)
- gas zone temperature (°C)
- circuit name and phase ID
- nominal density setting (g/l)
- Stage 1 alarm setting (g/l), Stage 2 alarm setting (g/l)
- overpressure alarm setting (g/l)
- installed density (g/l)
- installed mass (kg)
- inventory date

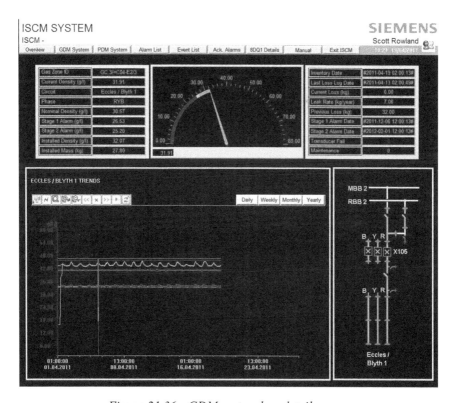

Figure 21.36   GDM system bay detail screen

- last loss log date
- leak rate (kg/year)
- previous loss (kg)
- predicted Stage 1 alarm date, predicted Stage 2 alarm date
- transducer fail alarm flag
- gas zone on maintenance flag.

In addition to this information, the trends for each gas zone within the bay are also displayed in the trend window. The information can be displayed in a 24 h format or over longer periods of up to 1 year format for any period during the lifetime of the substation.

21.14.3.2 GDM system alarms

A range of alarms are required for the general maintenance of the monitoring system in order to highlight any possible problems with the communications system, gas density transducers and power supplies, for example, in order to quickly resolve any loss of monitoring time.

21.14.3.3 Gas inventory

The gas density inventory screen, as shown in Figure 21.37, provides the user with the details of the SF_6 gas inventory within the substation. The 'Substation Status'

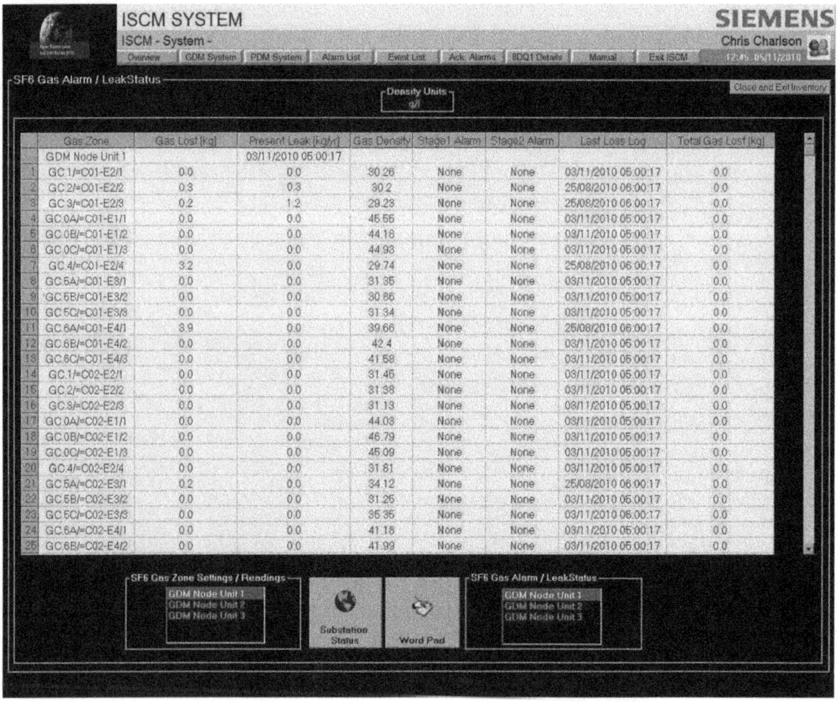

Figure 21.37 Gas density inventory screen

displays the total gas density inventory for the entire substation. It details how much gas has been installed and how much has been lost since the start of the inventory. For example, a 12 bay 420 kV GIS will typically have an installed gas inventory in the range of 15,000–20,000 kg depending on feeder arrangements. A breakdown of the gas loss associated with each node unit can also be provided, which details each individual gas zone and will display the following information:

- gas zone ID
- current gas lost (kg) (since last top-up)
- present leak rate (kg/year)
- gas density value (g/l)
- predicted date of Stage 1 alarm
- predicted date of Stage 2 alarm
- date of the last loss log
- total gas lost (since inventory began).

From this information any gas zone leak can be easily identified. Furthermore, the rate of gas loss and gas zone identity can be very quickly determined, allowing planned outages for essential maintenance. In summary, the GDM system provides:

- proof of performance
- strict accountability of gas use
- visibility of primary leaking zones so that they may be targeted
- effective and efficient planning of maintenance.

21.15 ISCM systems going forward

One of the main features of ISCM is that all monitoring systems can run on the same platform, with no duplication of power, network or IT infrastructure. It also provides the user with a common user interface that is modular and scalable. Another important benefit is that historically the CM systems have always been divorced from the control and protection systems and as such had taken a relatively low profile in the general day-to-day running of the substation. Subsequently, while the CM system was collecting useful data, it was seldom made available to the substation operators and its full potential was never realised. Conversely, the ISCM system allows for the CM systems to be integrated with the control and protection systems in a seamless manner and therefore the data produced is now more visible with easy access, allowing the substation operators better interaction with effective utilisation of the CM information.

This chapter has reviewed key monitoring parameters for each of the individual substation elements. Additionally the number and type of alarms and diagnostics that will need to be handled by the ISCM system has been sufficiently detailed to demonstrate to the reader the extent of the matrix of information that is generated. Any ISCM system must be capable of monitoring many data points and handling an even larger number of variables that are necessary to create and track alarms and diagnostic messages. When the automatic data throughput of the ISCM system is fully realised, then it highlights the enormity of the previously manual task of collating this data and it is understandable that the utilities reported 'an ever increasing burden on the staff responsible for CM data management'.

The need for better integration and management of the substation CM has been further supported by a significant global increase, over the last 3 years, in the uptake of monitoring systems, particularly on new substation projects. In some instances on projects in the USA, the sheer number of gas zones, in just one substation, has made the task of manual gas leak monitoring unfeasible and, without monitoring systems, would be unable to cope with the requirements for greenhouse gas emission control. In this particular substation example, the GDM system had the following transducer and ISCM parameter counts:

substation: 230 kV GIS
monitoring system: ISCM SF_6 gas density monitoring
installed SF_6 gas: >55 imperial tons
number of SF_6 gas zones: 495
number of gas density transducers: 495
number of SICAM parameters: >17,000
number of gas density alarms: >2000
number of GDM system alarms: >550
monitoring update data rate: >600 measurements per minute.

The parameter counts for PDM and CBM systems are similar to the GDM system with the PDM system providing 1 s updates for alarms and 5 min updates for trends. The CBM system updates the static parameters on an hourly basis with the dynamic update period being dependent upon the operation frequency of the circuit breaker.

Retrofit systems have also seen extra activity where a twenty-first century approach to data management and handling is driving the market. The most recent (2011) development for the support of retrofit ISCM systems comes from Hong Kong, where a significant investment is being targeted to provide substation monitoring as a complete package for over 50 existing substations within the 132 and 420 kV transmission systems. This single package approach was adopted to ensure, at both the substation level and at a corporate 'global' level, that a fully integrated monitoring system would be successfully achieved in the shortest time and at the optimum cost.

21.16 Concluding remarks

The ISCM system overviewed in this chapter is one of several systems that are emerging to provide a much needed change in the way that CM data is handled, acted upon and archived. The ISCM system is still a relatively new venture but has included many of the lessons learned over the last 20 years. It is essential to address the cost of monitoring to allow the market to accept the development. Also, the 'future proofing' aspects of the system design are necessary to give the user the confidence that the CM investment is protected for the lifetime of the substation.

To this end, ISCM systems must be able to be extended to incorporate new or updated monitoring systems during the lifetime of the substation. The resident infrastructure for data transport and handling must be able to fully implement the integration of the new system with minimal disturbance to the existing system, and minimal cost impact. Using the approach of a common node platform for substation switchgear also assists in reducing the cost impact both from manufacturing costs and the users costs for training and maintenance.

Establishing a common node platform can also allow the manufacturer to considerably shorten development time for CM products. New applications can be 'realised' by simply adapting the interface module to accept the required range of transducers along with modified software application created using many of the display modules already developed. Data processing can be arranged to allow the data to be imported directly into the ISCM gateway control PC using a new CM module. The infrastructure already created within the existing ISCM system would then allow the user to have full access to the new data using the corporate data network.

The ISCM system can utilise many of the existing monitoring packages already developed by the experts from within the full spectrum of manufactures of substation equipment (and on which a lot of time and money has already been invested on these developments).

It is hoped that these first, but significant steps in fully integrating the CM systems for substations, as reported in this chapter, will provide the solid foundation to achieve complete asset-related condition information in a consistent format and provide a major contribution to the optimal use of resources.

Acknowledgements

The authors wish to thank Siemens Transmission & Distribution Limited and National Grid and Scottish Power for permission to publish this chapter and also to thank Trafag, General Electric, Morgan Schaffer, Reinhaussen, Lios Technology and Wirescan for photographs and product information and specifications as used in this chapter.

References

1. 'CIGRE Technical Brochure No. 462 – WG B3.12: Obtaining value from on-line substation condition monitoring'. *Electra*, June 2011, 256
2. 'CIGRE Technical Brochure No. 420 – WG D12.17: Generic guidelines for lifetime condition assessment of HV assets and related knowledge rules'. *Electra*, June 2010, 250
3. Hampton B.F., Irwin T., Lightle D. 'Monitoring of GIS at ultra high frequency', Paper 23-02, Sixth International Symposium on High Voltage Engineering, New Orleans, August 1989
4. Irwin T., Jones C.J., Headley A. 'Achieving enhanced gas insulated switchgear reliability using diagnostic methods', CIRED 1991, Paper 1.08
5. Irwin T., Jones C.J., Headley A., Dakers B. 'The use of diagnostic based predictive maintenance to minimise life cycle costs' (vol. 5). CEPSI 1992

Proceedings of 9th conference on electric power supply industry, Hong Kong, 1992. pp. 225–234
6. Dalziel I., Foreman P., Irwin T., Jones C.J., Nurse S., Robson A. 'Application of controlled switching in high voltage systems'. CIGRE, 1996, Paper 13–305.
7. Irwin T., Halliday S.P. 'Condition monitoring', APSCOM – Hong Kong, December 1993
8. Irwin T., 'Six years experience with UHF monitoring', *Proceedings of IEEE on Colloquium on Partial Discharges in GIS*, April 1994, 1994/093
9. Bell R., Charlson C., Halliday S., Irwin T., Lopez-Roldan J., Nixon J. 'High voltage on-site commissioning tests for gas insulated substations using UHF partial discharge detection'. *IEEE Transactions on Power Development*. October 2003, **18**(4):1181–91
10. Irwin T., Lopez-Roldan J., Charlson C. 'Partial discharge detection of free moving particles in GIS by the UHF method: Recognition pattern depending on the particle movement and location' (vol. 3). IEEE PES Winter Meeting, Singapore, 2000. pp. 2135–2140
11. Charlson C., Irwin T., Jones C.J., Buchgraber G. 'Substation monitoring systems – recent results and trends in communication architectures'. *Proceedings of the Substation Equipment Diagnostics Conference XII*. EPRI, New Orleans, February 15–18, 2004
12. Lopez-Roldan J., Blundell M., Tang T., Irwin T., Charlson C. 'Benefits and challenges in the application of the UHF method for onsite partial discharge detection in hybrid switchgear and transformers'. *TechCon Asia Pacific*, Australia, 2009
13. 'CIGRE Technical Brochure No. 343 – WG A2.27: Recommendations for condition monitoring and condition assessment facilities for transformers'. *Electra*, April 2008, 237
14. Irwin T., Lopez-Roldan J. 'Substation earthing: Special considerations for GIS substations'. IEEE Publications No. 2000/033. Birmingham, UK: IEEE; June 8, 2002. pp. 5/1–5/5
15. Irwin T., Jones C.J., Nurse S.G. 'Disconnector switching in gas insulated substations'. Power Technology International. Sterling; 1987
16. Irwin T., Lopez-Roldan J., Nurse S.G. 'Design, simulation and testing of an EHV metal enclosed disconnector', IEEE, PE-029PRD (04-2001), 2001
17. IEC 61000-4-1, Ed. 2.0 (2000). Electromagnetic Compatibility (EMC)– Part 4-1: Testing and Measurement Techniques–Overview of IEC 61000-4 Series
18. O'Connel P., Heil F., Henroit J., Mauthe G., Morrison H., Niemeyer L., Pittroff M., Probst R., Taillebois J.P. 'SF_6 in the electrical industry, status 2000 substations'. Paper presented in the name of CIGRE WG 23.02, *Electra*, February 2002, No. 200
19. 'Joint CIGRE TF 15/33.03.05, 1998: Sensitivity verification for partial discharge detection on GIS with the UHF and the acoustic method'. *Electra*, April 1999, No. 183

Appendices

A Gas density monitoring transducer options

The measurement system, within the SF_6 gas density transducer, is based on the oscillating quartz principle in which the transducer utilises two quartz oscillators. The frequency of the oscillation is dependent on the density of the gas surrounding the quartz oscillator. One oscillator is placed in a vacuum chamber within the transducer body and the other is allowed to vent to the SF_6 gas enclosure in order to sample the gas. The consistent resonant frequency of the quartz oscillator under vacuum is then compared with the resonant frequency of an identical quartz oscillator, situated in the sample gas. The difference in the resonant frequency is proportional to the density of the sample gas. This difference is then processed into a digital or analogue output signal. In addition to frequency, the pulse width of the frequency system is temperature dependant and this can allow gas temperature to also be measured. Gas density transducers have two output options:

> Option 1: Transducer with a 4–20 mA output which provides a current that is directly proportional to the SF_6 gas density. The output signal from these transducers is inherently noise free and all 45 channels can be sampled once per second to provide an adequate measurement of gas density. However, some utilities require flashover detection for faults within the GIS gas zones to allow fast identification of a faulted gas zone section of the substation. Detection is achieved by monitoring the 'pressure wave' within the SF_6 gas that results from the power arc when flashover occurs. If this application is additionally required, then the sampling rate would need to be increased to sample all inputs in 100 ms giving a sampling rate of 500 samples per second. From an accuracy point of view, a density reading of better than 0.5% repeatability is required, so an 8 bit digitisation would be the minimum requirement.

> Option 2: transducer with a 16 mA current pulse output with the pulse frequency proportional to gas density and the pulse width proportional to temperature. This option would be used if the utility required, in addition to SF_6 density, that a gas temperature is also recorded at each gas zone. The addition of the temperature is required to detect the internal heating of a gas zone indicating a joint problem ('hot-joint').

SF_6 gas density readings will be derived by frequency analysis and will require a sampling rate of 100,000 samples per second (10 µs acquisition time) in order to give the required accuracy at the highest frequency. All 45 channels polled within 5 s giving a total of 450,000 samples to analyse for all 45 channels within the 5 s polling period.

The temperature readings will be derived by 'Pulse Width' analysis using the same data recorded for gas density. A variation of 800 µs in pulse width creates a temperature span of 130 °C. The sampling rate will give a temperature resolution of ±0.8 °C, which is adequate for purpose.

The most demanding scenario for data acquisition sampling rate is therefore option 2 giving a minimum requirement of an 8 bit resolution with a sampling rate of 100,000 samples per second.

B Circuit breaker monitoring data sampling rates

Timing accuracy forms the main parameter for sampling rate. Circuit breaker trip operating time is typically 20 ms and this is to be measured with 1% accuracy, giving a time error of 200 µs. Therefore, sampling for all 20 channels must be completed within 200 µs, giving a sampling rate of 100,000 samples per second with an 8 bit resolution.

The signals must be digitised at a frequency that will provide sufficient sample points to allow accurate and early assessment of a developing problem. The sampling rate is set to 0.2 ms and a total of 600 ms, including a pre-event record of 100 ms (3000 data points per parameter are recorded with typically 60,000 points per operation).

The 600 ms capture period was selected to allow sufficient data capture prior to the operation of the circuit breaker, for example for pre-fault analysis and for a following break-make-break circuit breaker sequence of operation in the event of a high-speed auto-reclose operation onto a sustained fault.

The system has to be configured to accept a wide range of transducer types so a transducer settings facility is required to configure the system and to allow for the replacement of transducers and any scale changes during service life. Also if some transducers become obsolete during the lifetime of the system, then new generation replacements can be fully accommodated.

C Partial discharge monitoring data sampling rates

Timing accuracy and pulse magnitude form the main parameters for sampling rate. The sampling rate selected for PDM node unit is based on 50 channels being sampled and completed within 200 µs (to give point-on-wave (POW) resolution of 100 points per cycle at 50 Hz, 20 ms period) giving a sampling rate of 250,000 samples per second with a minimum of 8 bit resolution. The provision of large-scale sampling, covering a three-bay area of substation, allows pulse synchronisation checks to identify common signal sources, which is useful for PD source location. Similarly, this synchronisation can also be used for noise elimination if one of the input channels is used for external noise detection.

D Disconnector and earth switch monitoring data sampling rates

Timing accuracy for the FMES forms the main parameter for the maximum sampling rate. FMES operating speed is typically 5 m/s with an operate time of 7 ms.

The specified operation time limit is to be within ±0.25 ms and therefore requires sampling for all 5 FMES channels to be completed within 0.25 ms, giving a sampling rate of 20,000 samples per second with an 8 bit resolution.

Usually disconnectors, earth switches and FMES are operated independently and only one at a time. The operation trigger for the data acquisition can therefore be arranged to recognise what is operating, and then set the data acquisition rate accordingly to minimise unnecessary large data files. A disconnector or earth switch operating speed is much slower than the FMES and the operate time is accordingly longer.

Disconnector and earth switch operate times are typically 5 s, with a timing accuracy of ±5 ms, and therefore requires sampling for all three disconnector or earth switch channels to be completed within 5 ms, which gives a total of 3,000 samples to be analysed for all parameters.

Chapter 22

Intelligent monitoring of high-voltage equipment with optical fibre sensors and chromatic techniques

G.R. Jones and J.W. Spencer

22.1 The nature of intelligent monitoring

An example of the condition monitoring of one component (transformers) of a high-voltage power system has been given in Chapter 20. Chapter 21 has described developments in the integrated monitoring of a combination of high-voltage substation components. The present chapter provides examples of some future monitoring technologies, based upon optical fibre sensors and chromatic techniques, which are being researched and evaluated for extending intelligent monitoring capabilities in a traceable manner.

In relation to the scope of this chapter, it needs to be recognised that 'monitoring' is concerned with providing information about the condition of a system in order to alert about impending faults. As such it differs from 'measurement' which is concerned with ascertaining the magnitude of a parameter by comparison with a standard unit. A third related function is 'diagnosis' which is concerned with ascertaining the condition of a system from symptoms. Thus, 'measurement' is reductional in nature, 'diagnosis' deals with resolving complexities while 'monitoring' may be regarded as interconnecting reductionalism and complexity.

'Intelligent monitoring' is concerned with the provision and interpretation of monitored data to provide a diagnosis. Examples of techniques used for the interpretation function are neural networks, principal components, fuzzy logic etc. [1–4]. These approaches have a number of fundamental limitations. There is a lack of traceability, which prevents the physical cause of a decision from being identified; they require *a priori* trend patterns to be established; the use of conventional sensing (normally used for 'measurement') is not questioned as being the most suitable for monitoring (i.e. the functions of sensing and interpretation are segregated).

An alternative approach for intelligent monitoring is with 'chromatic techniques' [5]. These can be deployed for both sensing and interpretation in either segregated or combined modes. The approach provides a high level of traceability while also respecting the complex nature of the monitored system. An important property of 'complexity' is 'emergence', that is the evolution of a condition from a previously unidentified state. The identification of an emerging condition is facilitated by the cross correlation of monitored information which chromatic techniques inherently employ for both sensing and interpretation. Their deployment respects the need for ultimate decision making to remain with the human operator. In addition, the techniques often accommodate the possibility for the operator to install the monitoring system retrospectively without intrusion to affect the operation of equipment. In addition the monitoring system often has the capability of being transferrable between equipment and sites. Examples of both online and offline monitoring are given in this chapter.

22.2 The basis of chromatic monitoring

Chromatic techniques are based upon non-orthogonal sensing and data processing to extract the required information. They derive from the photic field concepts of Moon and Spencer [6]. As a simple example of the basis of the approach, Figure 22.1(a) shows the response of two sensing elements, R and B, as a function of a measurement parameter P (e.g. optical wavelength). The responses of R and B do not overlap and so are regarded as 'orthogonal'. Figure 22.1(b) shows the response of a third sensing element, G, which overlaps with each of the other two elements R and B. As a result the three elements are said to be 'non-orthogonal'. Consequently the range of the parameter P is more fully covered and values obtained from each of the three sensing elements can be cross-correlated. Conventional measurements are based upon the use of orthogonal sensors in order to segregate the data from each sensor. However, chromatic techniques are based upon the use of non-orthogonal sensing or processing in order to extract correlated information [5].

As an example, consider a signal (also shown in Figure 22.1(b)) which is a complex function of the parameter P (e.g. an optical spectrum). Addressing this complex signal with the three non-orthogonal sensors/processors yields three outputs, Ro, Go, Bo, which define the signal distribution. These outputs may be transformed using various algorithms to produce three parameters, which quantify the complex signal distribution [5] (Appendix A).

One such chromatic transformation of Ro, Go, Bo leads to the signal distinguishing parameters x, y, z (Appendix A). These parameters quantify the relative magnitude of the output from each processor with respect to the total signal strength [5] so that

$$x + y + z = 1 \tag{22.1}$$

A signal may therefore be represented as a point on a chromatic Cartesian diagram with coordinates x, z (Figure 22.1(c)). Significant features of such a diagram are as follows:

- The locus $x + z = 1$ represents signals with no y (i.e. G) component.
- The point 0.33, 0.33 represents a signal with equal contributions from each of the three processors.
- The locus $x = z$ represents a scale for the y component.

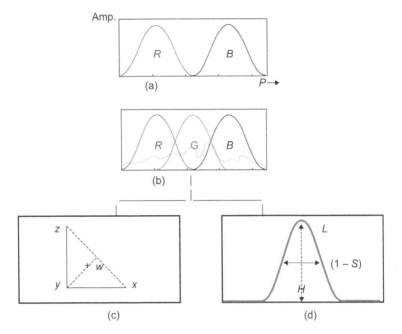

Figure 22.1 The nature of non-orthogonal sensors
(a) Responses of two orthogonal sensors (R, B)
(b) Responses of three non-orthogonal sensors (R, G, B) addressing a distributed parameter
(c) R, G, B transformed chromatic parameters x, y, z
(d) R, G, B transformed chromatic parameters H, L, S

An alternative transformation of Ro, Go, Bo leads to a different set of signal identifying parameters H, L, S (Appendix A) where H quantifies a dominating parameter value, L the effective strength of the signal and $(1 - S)$ the nominal spread of the signal. If the distribution of the signal was truly Gaussian, H, L, S would represent the features shown in Figure 22.1(d). Thus, for a more complex signal, H, L, S may be regarded as representing the three attributes of an equivalent Gaussian signal, that is they define a Gaussian family to which the signal belongs. If additional signal features are required (e.g. symmetry and kurtosis), additional non-orthogonal sensors may be deployed but investigations suggest that little additional information is obtained with more than six such sensors [5].

One particular example of chromatic information acquisition is the manner in which human beings distinguish colour. The human colour vision relies upon three non-orthogonal receptors [5] and H, L, S represent the hue, lightness and saturation of colour science with the algorithms defined mathematically in Appendix A [5]. Chromaticity is more generic than colour in that:

- it extracts the required information (not replicating human vision)
- the responses of detectors or processors are adapted to address the information being sought
- additional information can be obtained by deploying up to six non-orthogonal processors
- the approach is extendable beyond the visible spectrum and in the domains of time, frequency, space, acoustic, mass, discrete data distributions etc.

It has the advantage of inherently identifying unexpected events at an early stage as a result of the cross-correlation produced by the non-orthogonal nature of the sensors.

The chromatic approach may be deployed in several ways for monitoring high-voltage systems and equipment. It may be deployed for the optical fibre sensing of particular parameters, for remote monitoring (e.g. using CCTV cameras), for extracting information from signals obtained with conventional instrumentation, for cross-correlating signals from an array of different sensors etc. Some of these approaches are described in the following sections.

22.3 Online monitoring of high-voltage equipment using optical fibre chromatic techniques

22.3.1 Optical fibre chromatic sensors

Optical fibre sensing is based upon the use of optical fibres for addressing sensors which monitor particular parameters. Use is made of light rather than electric current for undertaking the sensing and for transmission to and from the sensing element. The inherent electrical insulation of optical fibres makes them attractive for installation on high-voltage systems, their optical nature makes them immune to electromagnetic interference, and their silica material base makes them corrosion resistant. The sensing elements themselves may be intrinsic (the fibres serving to transmit and modulate the light in response to a measurand), extrinsic (a separate optical element used for sensing and the fibres used only to transmit the light) or hybrid (the sensor being non-optical but the signal transmission is optical via the fibres). The properties of the light wave, which can be modulated by a sensing element, include its intensity, phase, polarisation state and wavelength. While conventional approaches with these modulation methods are based upon the use of highly monochromatic light (e.g. lasers), chromatic methods utilise broad wavelength band light (e.g. white light emitting diodes). Table 22.1 lists a range of physical parameters, which have been monitored using optical fibre based sensing along with the optical sensing principle involved. These sensing principles include changes in the state of the polarisation of the light due to the magnitude of an electric field (Pockels effect) [7], magnetic field (Faraday effect) [8], mechanical stress (photoelasticity) [9] and temperature (thermochromic) [10]. Mechanical vibration can be monitored by sensing changes in the spatial distribution of the light emerging from a fibre subjected to the vibration which affects the interference

Table 22.1 Chromatic optical fibre sensing

Parameter monitored	Physical effect employed[a]
Voltage	Pockels effect
Current	Faraday effect
Pressure	Fabry–Perot interference
Temperature	Fabry–Perot interference
Micron particles	Mie scattering
Mechanical vibration	Speckle pattern
Liquid samples	Optical absorption
Spatial location	Chromatic scale
Battery cells	Optical emission

[a]Details are given in References 5 and 7–13.

of the fibre transmitted light with itself (speckle pattern) [11, 12]. Micron-sized particles can be monitored by the preferential scattering of polychromatic light [13].

22.3.2 Examples of chromatic optical fibre sensors for high-voltage systems

22.3.2.1 Electric current monitoring

An example of a chromatically based optical fibre sensing unit is for the monitoring of electric current on high-voltage systems. The example presented illustrates the deployment of chromatic optical sensing using optical properties other than wavelength-dependent light absorption but rather, for instance, wavelength-dependent optical polarisation.

One form of optical fibre current transducer is based upon monitoring the B field produced by the current via its effect upon the polarisation state of the light (Figure 22.2(a)). The optical sensing part of the transducer is made of a material which rotates the plane of the polarised light in proportion to the magnetic field experienced (Faraday Rotation) [8]. With polychromatic light, various wavelengths are rotated to different degrees. If the output light is viewed through a polarising filter, the intensities at different wavelengths relative to each other are governed by the magnitude of the B field and hence the current flowing. Such wavelength-dependent intensities can be addressed chromatically via three detectors (R, G, B) (Figure 22.1(b)) and the resulting outputs related to the current flowing via variations in the dominant wavelength, H.

A schematic diagram of such a transducer is shown in Figure 22.2(b). It consists of a ferromagnetic yoke placed around the current carrying cable (Figure 22.2(b)) [8]. There is a gap in the yoke which houses the light polarisation sensitive material and which in turn is attached to input and output optical fibres. A typical result obtained with this type of sensor is shown in Figure 22.2(c) [14] in

760 High-voltage engineering and testing

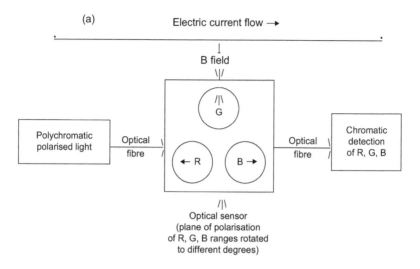

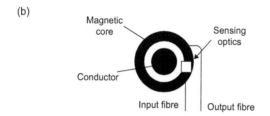

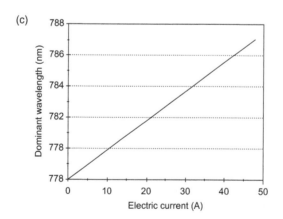

Figure 22.2 Chromatic optical fibre-based current sensor using Faraday rotation
(a) Principle of operation – chromatic sensing of polarised light
(b) Schematic of sensor structure
(c) Typical test result – chromatic H versus current

Intelligent monitoring of high-voltage equipment 761

the form of a calibration curve between chromatic H and the B field producing current. Such transducers have been deployed on high-voltage tap-changers for checking their correct operation.

22.3.2.2 High-voltage circuit breaker monitoring

Optical fibre sensing based upon the various principles described in section 22.3.1 has been deployed for researching and during the development of high-voltage SF_6 circuit breakers. Figure 22.3(a) shows a schematic diagram of a high-voltage gas blast interrupter indicating various types of optical fibre chromatic sensors which have been deployed [15]. These include sensors for monitoring the gas pressures in the interrupter tank and piston chamber, the contact potential, fault current, temperature of the contact stalk, contact travel, mechanical vibration and arc radiation. A few examples of test results obtained with such sensors are shown in Figures 22.3(b–e) [10, 13, 15–17]. The four results shown are for the micron particle concentrations following various fault current interruptions (Mie scattering [13]) (Figure 22.3(b)), the time variation of the piston chamber pressure (with a Fabry Perot pressure sensor [9]) (Figure 22.3(c)), the contact travel (chromatic linear scale [15]) (Figure 22.3(d)) and the mechanical vibration during circuit breaker operation (distributed fibre sensing [16]) (Figure 22.3(e)). Such data taken together provide an insight into the various conditions which occur during the fault current interruption process so that improved understanding of the interrupter operation can be obtained. For example, the particle monitoring has assisted in comprehending the effect of particles upon circuit breaker performance (Chapter 7).

22.3.3 Time and frequency domain chromatic processing of optical fibre sensor data

During the operation of high-voltage systems and equipment, the time variation and frequency components of various sensor outputs can provide valuable information about the condition of the system. Information may be derived from such time varying sensor outputs by deploying chromatic processors in the time or frequency domains to quantify chromatically changes which might occur [5].

22.3.3.1 High-voltage circuit breaker operation – time domain example

One example of such a procedure is for characterising the operation of a high-voltage interrupter unit using the time variation of the various parameters measured with optical fibre sensors as described in section 22.3.2.2. Figure 22.4(a–c) shows examples of the time variation of three interrupter parameters (contact velocity, piston pressure, acoustic vibration) measured with chromatic optical sensors and which can be addressed by a further chromatic transformation procedure [5]. This involves sampling the time varying signal with three non-orthogonal processors (e.g. Figure 22.1(c)) each of whose response varies with time as shown in

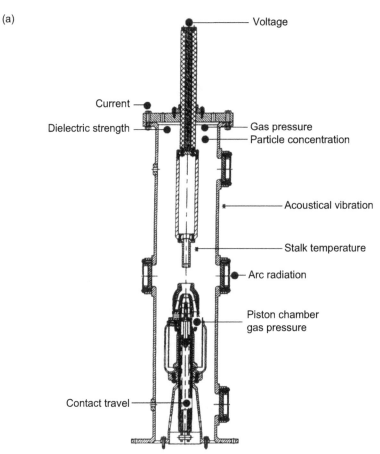
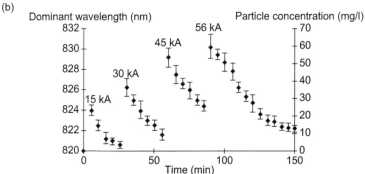

Figure 22.3 Optical fibre sensing of various circuit breaker parameters
 (a) Parameters sensed
 (b) Micro particles concentration post-arcing (light scattering)
 (c) Dynamic piston chamber pressure (Fabry–Perot)
 (d) Linear travel record (dot density modulation)
 (e) Mechanical vibration during interrupter operation (speckle pattern)

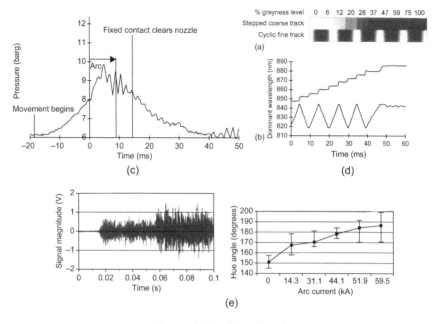

Figure 22.3 (Continued)

Figures 22.4(a–c). The characteristic features of each signal – dominant time (H), effective strength (L), nominal time spread (S) – may then be derived from the outputs of the three processors (R, G, B) using the chromatic algorithms given in Appendix A. Thereafter each signal may be represented as a point on a polar diagram of H versus S as shown in Figure 22.4(d). The interrupter operation is then characterised by the distribution of these points on the chromatic H–S polar diagram.

The outputs from such a procedure for a particular parameter can be used to explore that parameter's variation over a series of interrupter operations in order to check for possible incipient reduction in performance. For example, the effect of a reduction in the hydraulic drive pressure of the interrupter piston can be monitored via the optoacoustic waveform (Figures 22.3(c) and 22.4(c)) and represented as a series of points on a chromatic H, S Cartesian diagram (Figure 22.4(e)). As a result it is possible to identify progressive hydraulic pressure degradation as a locus of the points on the H, S diagram. For example, as the piston pressure decreases from 285 to 189 bar, the operating point shifts from coordinates $S = 0.7$, $H = 250$ to $S = 0.55$, $H = 280$ (Figure 22.4(e)).

22.3.3.2 Transformer tap-changer monitoring – frequency domain example (A.G. Deakin and D.H. Smith)

Optical fibre sensors have also been successfully deployed for monitoring in-service high-voltage transformers. An optical fibre rotary encoder based upon the

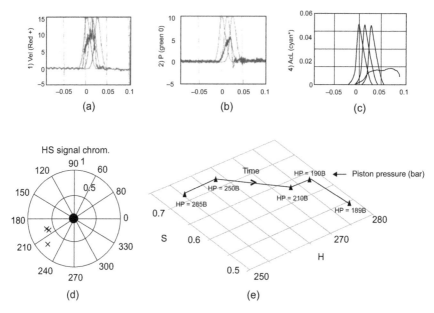

Figure 22.4 Time domain chromatic processing of sensor outputs
(a) Contact velocity; (b) piston pressure; (c) mechanical vibration;
(d) H–S chromatic polar diagram; (e) variation of H, S with
hydraulic drive pressure for a high-voltage interrupter

displacement sensing principle used for monitoring piston movement of puffer circuit breakers (section 22.3.2.2) has been adapted for the in-service monitoring of the rotary action of transformer tap-changer mechanisms (Figure 22.5(a)). Optical fibre current transformers based upon the Faraday Effect principle described in section 22.3.2.1 (Figure 22.2) have been used for the in-service monitoring of tap-changer currents. The optical fibre rotary encoder and current transformer are essentially retrofitable to suit operator needs.

A further evolution of the use of optical fibre sensing is as a monitoring system which is non-invasive and transferrable between different pieces of equipment and various sites. An example of such a system is an intrinsic optical fibre system for monitoring mechanical vibrations [11, 12, 16, 18] as described in section 22.3.1 (Figure 22.3(e)). Such a system is currently being used at a number of substations for monitoring acoustic signals produced during the operation of tap-changers in high-voltage transformers using an optical fibre sensor situated outside the tap-changer housing.

The sensing is based upon the homodyne interferometer principle [11, 18]. Light from a monochromatic source (i.e. laser) is propagated through a length of unjacketed multimode optical fibre. This fibre is attached to the outside wall of a tap-changer tank which houses the drive mechanism (Figure 22.6). Acoustic signals

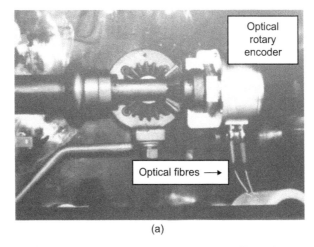

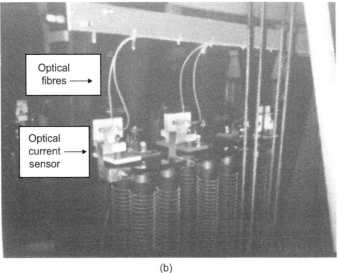

Figure 22.5 Extrinsic optical fibre sensors installed upon a high-voltage transformer tap-changer
(a) Optical fibre-based rotary encoder; (b) optical fibre-based Faraday current transformer

from the transformer and tap-changer mechanism produce small changes in the refractive index of the core and cladding of the fibre which alter the mode propagation of the monochromatic light in the fibre [18]. The output from the fibre is detected and processed using chromatic-based algorithms (Appendix A) [5] which quantify the signal features in terms of the parameters H, L, S.

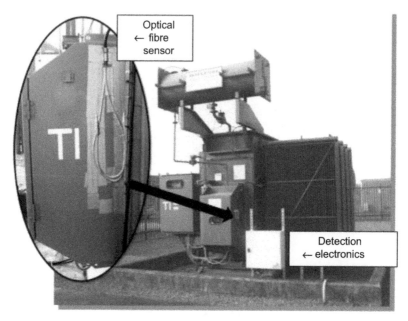

Figure 22.6 Intrinsic optical fibre acoustic monitoring of transformer tap changers showing the optical fibre attached to the outside of a tap changer housing along with the location of the detection electronics

Some typical results from tests on a substation transformer are given in Figure 22.7. Figure 22.7(a) shows examples of four different time varying signals, each of 5 min total duration, detected with the sensor. These illustrate the distinctive nature of such signals. Figure 22.7(a) (i) corresponds to the background 'hum' of the transformer while the others show signals produced by various events. The frequency spectrum of such signals is transformed to yield frequency domain chromatic parameters H, L, S. Figure 22.7(b) shows the H–L chromatic signatures of such signals on a polar diagram (H – azimuthal angle, L – radius). On this diagram, background signals corresponding to the 'normal' hum of the transformer are represented by the two circles, a non-tap-changer event by a square and four tap-change events by the triangles. These results show that the background and non-tap-changing events all have the same H coordinate value ($H = 0$) but different signal strengths, L. In contrast, all the four tap-changing events have not only high values of L but also distinctly higher values of H and so can be discriminated from the non-tap-changer events. Furthermore, two of the tap-changer events have similar values of H (~18 degrees), whereas the fourth tap-changer event has a distinctively different value of H (~110 degrees) which is believed to be due to mechanism stiffness. The results illustrate the discriminating capabilities of the chromatic approach.

Such systems are currently being deployed at various electric power distribution substations. The outputs from the optical fibre sensor are transferred to a

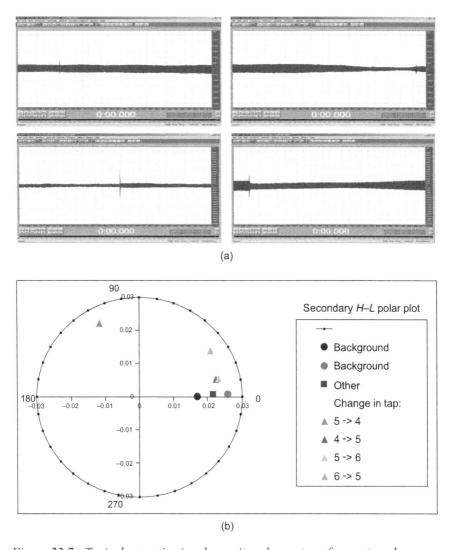

Figure 22.7 Typical acoustic signals monitored on a transformer tap-changer
(a) Examples of time varying signals
(b) Representation of signals on a polar chromatic diagram of H versus L (frequency domain chromatic transformation; triangles denote tap-changes, circles are background and other events)

central server via the electronics module (Figure 22.6) for chromatic processing and display. The optical fibre and supporting electronics (Figure 22.6) are conveniently demountable and can be transferred for use at another site. To date installation times have been of the order of a few hours.

Preliminary investigations with such systems also indicate, as implied by the H–L signatures shown in Figure 22.7, that the optical signals from the sensing fibre can be further interrogated in order to detect other features associated with the transformer operation. For such purposes, different forms of chromatic processing algorithms may be used to extract the additional latent information from the raw time varying signals.

22.4 Chromatic assessment of the degradation of high-voltage insulation materials

22.4.1 *Chromatic characterisation of partial discharge signals (M. Ragaa)*

One symptom of progressive deterioration of a solid dielectric material used for high-voltage electrical insulation is partial discharge activity [2, 19–21]. Such discharge activity may occur due to gas filled voids, cavities and defects within a solid insulator due to ionisation induced by elevated stresses at these locations [1]. Signals produced by such intermittent discharge activity have been explored for diagnosing the electrical insulation condition. The form of the partial discharge signal as a function of the phase angle of a 50 Hz voltage oscillation varies under the influence of at least two factors:

- the type of partial discharge
- the progression with time towards full electrical breakdown.

Figure 22.8(a) shows three examples of different partial discharge signals covering the whole period of a 50 Hz voltage oscillation. The three examples correspond to partial discharge activity within the bulk of a solid dielectric (Figure 22.8(a)(i)), on the surface of a dielectric (Figure 22.8(a)(ii)) and for a floating-on-oil condition (Figure 22.8(a)(iii)) [22]. These show the complex nature of partial discharge signals and how both the amplitude and phase spread of the signals differ for the various discharges.

Figure 22.8(b) shows examples of changes which occur in a signal from partial discharges in a highly contaminated oil as the peak alternating voltage applied to the sample was increased (3, 4, 5 kV) and there was progression towards full breakdown [23].

The complex nature of these discharge signals has led to investigations into the use of neural networks, expert systems, fuzzy logic, pattern recognition, wavelet transforms etc. for diagnosing the cause of various time varying partial discharge patterns [1–4]. However, such methods appear to have several limitations. Neural networks have insufficiently rapid response for real-time application, require training data and have yet be shown to be capable of distinguishing between idealised sources of partial discharges let alone within insulation of real high-voltage power apparatus [1]. Furthermore neural networks classification has no unconditional classification capability [4]. Although expert systems and fuzzy logic approaches have been successfully applied, they require human intervention and there are difficulties in acquiring the necessary knowledge [24]. Pattern recognition techniques have limitations in extracting features [2] and Wavelet Transforms have limitations in providing ideal recognition results [2]. Against this background, chromatic techniques, with their potential to provide a high level of traceability, are now being investigated for monitoring the signals from partial discharges [25, 26].

Partial discharge signals of the form shown in Figure 22.8(a,b) may be pre-processed by normalising the amplitudes of the averaged partial discharges per phase and their standard deviations per unit phase angle and integrating over a three minute period to give signal envelopes. Examples of such envelopes are shown in Figure 22.8(c) for the three signals of Figure 22.8(b). The chromatic approach involves addressing the phase angle varying signal with three-phase domain non-orthogonal processors (R, G, B). The time window covered by the processors can be varied, one preferred window being to cover one quarter of the 50 Hz voltage oscillation period. This enables the discharge activity during the increasing and decreasing voltage parts for each polarity of the voltage oscillation to be addressed separately and for different sectors of the phase domain to be considered. Figure 22.8(c) shows the deployment within such a window of three-phase domain processors (R, G, B) covering the first quarter cycle of the voltage oscillation. The outputs from these processors are transformed into various chromatic parameters x, y, z, and L, $(1 - S)$ using the algorithms given in Appendix A.

The results of such Chromatic Transformations may be displayed using various chromatic diagrams, examples of which are given in Figures 22.9 and 22.10. Figure 22.9 relates to the signal distribution during the first quarter cycle of the 50 Hz voltage oscillation. It not only primarily indicates progress towards full electrical breakdown but also provides some discrimination between different discharges during the approach to breakdown. Figure 22.10 compares the symmetry between signals from adjacent quarter cycles and primarily provides an indication of the type of partial discharge.

Figure 22.9(a) shows a Cartesian diagram of $z(1):y(1)$ which quantifies the dominant phase angle during the first quarter cycle. The various loci shown on this diagram represent trajectories of different signal types towards the point 0.33, 0.33. This point represents equal contributions from all sectors of the quarter cycle and is one indicator of possible full breakdown. Some typical test data [21–23] from various sources are also shown in Figure 22.9(a).

Figure 22.9(b) shows the variation of the parameter $L^*/(1 - S^*)$ for the first quarter cycle with $z(1)$. $L^*/(1 - S^*)$ is a measure of an effective amplitude of the first quarter cycle signal. Increases in the value of this parameter are indicative of possible progression towards full breakdown. Results for various test data from several sources [21–23] are shown to illustrate the typical variations which can occur. Consequently, characteristics of the form shown in Figure 22.9(a,b) may be combined to indicate the likely approach to full breakdown.

Figure 22.10(a,b) shows chromatic diagrams of $z(1):x(2)$ and $z(3):x(4)$ respectively, which involve the other quarter cycles (2, 3, 4) in addition to the first. $z(1)$ corresponds to the first quarter cycle processor adjacent to the second quarter

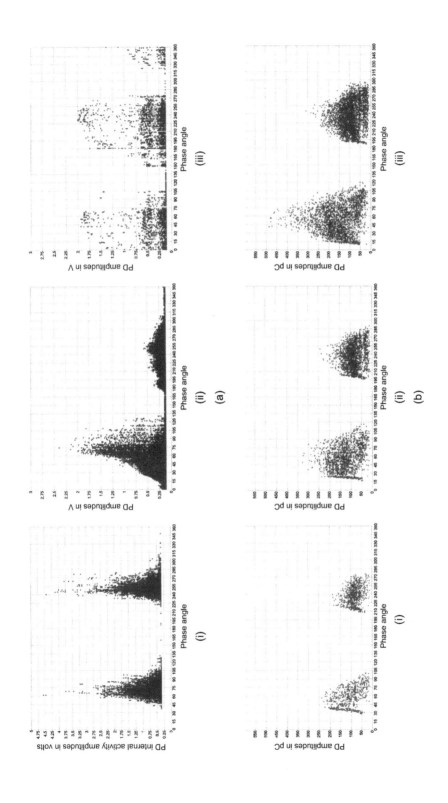

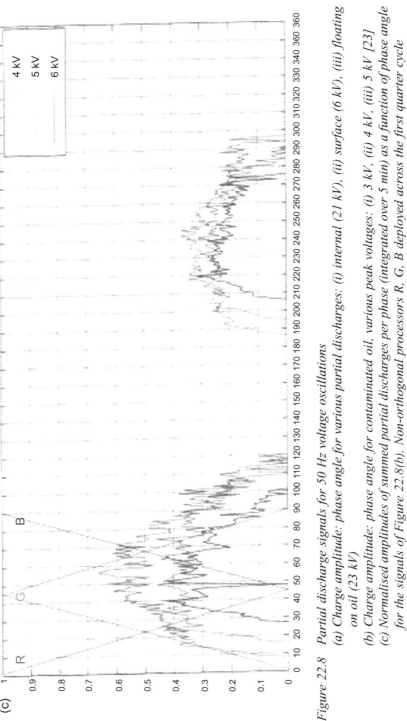

Figure 22.8 Partial discharge signals for 50 Hz voltage oscillations
(a) Charge amplitude: phase angle for various partial discharges: (i) internal (21 kV), (ii) surface (6 kV), (iii) floating on oil (23 kV)
(b) Charge amplitude: phase angle for contaminated oil, various peak voltages: (i) 3 kV, (ii) 4 kV, (iii) 5 kV [23]
(c) Normalised amplitudes of summed partial discharges per phase (integrated over 5 min) as a function of phase angle for the signals of Figure 22.8(b). Non-orthogonal processors R, G, B deployed across the first quarter cycle

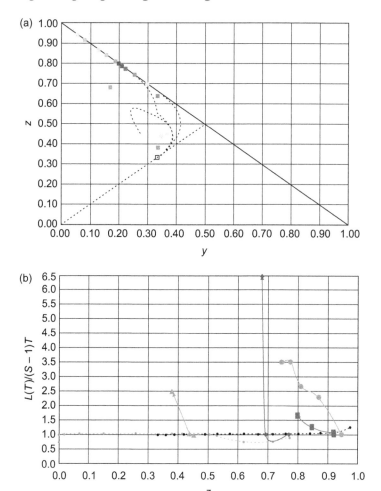

Figure 22.9 Chromatic quantification of various partial discharge conditions
(a) Chromatic Cartesian diagram (z:y) showing the relative contributions from different phase regions (z = last-phase region, y = mid-phase region, x = first-phase region) for various partial discharges
(b) Chromatic Cartesian diagram for the effective amplitude $L^*/(1-S^*)$ of various partial discharges as a function of z

cycle and x(2) corresponds to the second quarter cycle processor adjacent to the first quarter cycle. Similarly, z(3) and x(4) are the adjacent processors for the third and fourth quarter cycles respectively. The loci $z(1)=x(2)$ and $z(3)=x(4)$ correspond to a symmetrical distribution across the boundary between the adjacent quarter cycles. Results for a series of tests with different partial discharges (corona, surface discharge in air, internal discharge and floating in oil [22]) and applied voltages are presented in Figures 22.10(a,b). These show that each discharge type

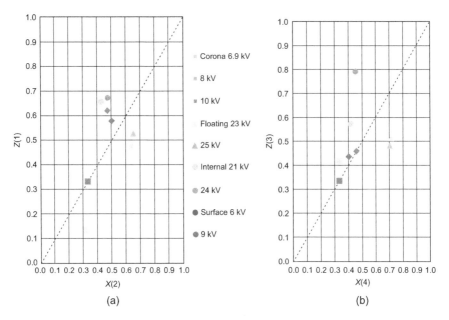

Figure 22.10 Comparison of chromatic parameter values from different quarter cycles for various types of partial discharges
(a) First (z(1)) versus second (x(2)) quarter cycles chromatic parameters
(b) Third (z(3)) versus fourth (x(4)) quarter cycles chromatic parameters (quantification of asymmetry and polarity effects)

clusters at various locations on the $z(1):x(3)$ and $z(3):x(4)$ characteristics, which indicate significant differences in the symmetrical distributions relative to the $z(1) = x(2)$ and the $z(3) = x(4)$ loci. For example, the internal discharge data are biased towards the $z(1)$ and $z(3)$ axes, and the floating oil data are biased towards the $x(2)$ and $x(4)$ axes etc. Hence, in combination with characteristics of the form shown in Figure 22.9(a), this has the potential for distinguishing between different types of discharges at least under ideal test conditions.

The examples given in Figures 22.9 and 22.10 show the potential of chromatic processing for discriminating between different partial discharges and detecting trends towards full breakdown.

22.4.2 Offline assessment of high-voltage transformer oils with chromatic techniques (E. Elzazoug and A.G. Deakin)

An example of the use of chromatic techniques for condition monitoring with data gathered offline is for assessments of the degradation of the electrically insulating oil from a high-voltage transformer. Such monitoring is used for anticipating possible transformer failures [27]. At least three different approaches, each based upon analysing samples of oil offline, are used [28] – colour index, dissolved gases, acidity, water content and electric strength of the oil.

The colour index (CI) is a quantitative assessment of the colour of the oil which can vary as the oil degrades [29]. Dissolved gases are gases produced as a consequence of various oil degradation processes which include severe overheating (CO_2, CO, C_2H_4), severe local heating (C_2H_6) and electrical discharges (C_2H_2, H_2, CH_4) [30]. The acidity level and water content increase are also indicators of degradation as well as decreases in the electric strength of the oil. The results of such offline tests may each be evaluated separately with chromatic processing methods and the ensuing individual quantifications may then be integrated chromatically to provide an overall degradation signature. The procedures provide an example of the high level of traceability and early detection of an incipient fault which is provided by the chromatic technique, in identifying the emergence of a particular dominant aspect, first from the overall signature and traced back to the main indicating feature.

22.4.2.1 Polychromatic light absorption oil signature

The CI approach is based upon quantifying the colour of an oil sample by transmitting white light through a sample and measuring the colour of the emerging light [29]. The chromatic approach also involves transmitting polychromatic light through a sample, but in one manifestation the optical spectrum of a computer screen image of the oil is obtained which is then addressed by suitably chosen R, G, B wavelength domain processors [5]. This allows a laptop computer to be used for illuminating an oil sample, capturing the spectrum or image of the sample and chromatically analysing the data [31]. Figure 22.11(a) shows typical screen spectra for three different oil samples and Figure 22.11(b) shows three non-orthogonal filters deployed upon one such spectrum. As a consequence, chromatic parameters H, L, S or x, y, z (Appendix A (a), (b)) are obtained which quantify the screen spectrum or image associated with an oil sample rather than simply replicating the colour of the oil directly. These parameters may then be combined to produce a scale based upon empirical information which can be made more sensitive to various oil degradations than the CI. Figure 22.11(c) compares the quantification produced by one such chromatic parameter ($Fn(x(O)/z(O))$) (Appendix A (c)) with that from the CI, which shows how the chromatic approach better highlights some highly degraded oil samples. Figure 22.11(d) shows the variation of $Fn(x(O)/z(O))$ for several transformer oil samples arranged in increasing order of magnitude of $Fn(x(O)/z(O))$. T1 is a fresh oil sample and the other samples are from various transformers. Thresholds for oil samples which are becoming suspect and those which are critical can be established empirically on such a characteristic.

22.4.2.2 Chromatic processing of dissolved gas data

Empirical information exists regarding reasons for the production of various gases dissolved in transformer oil along with limits below which the oil may be regarded as normal and limits above which it is critical and action is needed (Table 22.2) [28]. Such data may be combined chromatically to provide an integrated and transparent quantification as already described by Zhang *et al.* [32]. Further

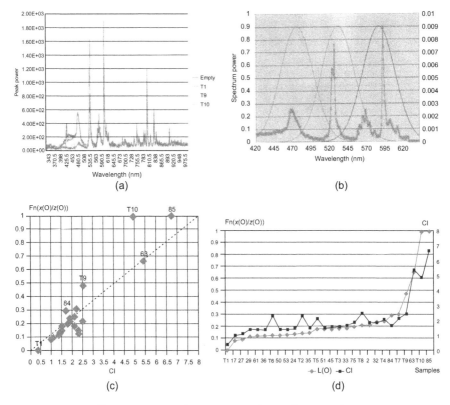

Figure 22.11 Chromatic optical signatures of transformer oils
(a) Optical spectra of computer screen images of various transformer oils
(b) Chromatic processors R, G, B addressing a screen image of an oil sample
(c) Comparison of the chromatic parameter $Fn(x(O)/z(O))$ with the colour index (CI)
(d) Numerical values of $Fn(x(O)/z(O))$ for various oil samples in ascending order

Table 22.2 Dissolved gases in degraded transformer oil

Gas	Normal limit	Action limit	Cause
CO	500	1000	Severe overheating
CO_2	10000	15000	Severe overheating
C_2H_4	20	150	Severe overheating
C_2H_6	10	35	Local overheating
C_2H_2	15	70	Arcing
H_2	150	1000	Corona
CH_4	25	80	Sparking

Source: Reference 30.

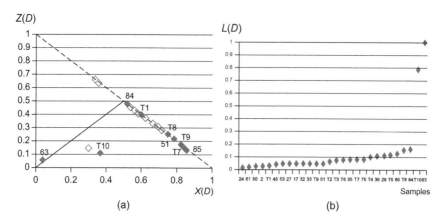

Figure 22.12 Chromatic representation of dissolved gases in transformer oils
(a) z:x Chromatic Cartesian diagram of severe overheating (z(D)) versus electrical discharging (x(D)) with local overheating (y(D)) as parameter (equi-distribution point – 0.33, 0.33)
(b) Chromatic strength of dissolved gases for the various oil samples of Figure 22.6 in ascending order

development of the chromatic processing of such dissolved gas data involves dividing the gases into three groups corresponding to those produced by severe overheating, severe local overheating and electrical discharging (Table 22.2). The range of values for each dissolved gas is normalised with respect to its critical limit for action (Table 22.2) and these normalised values for each member of a group added to provide an overall figure for that group. The three overall figures for the three groups of gases are then regarded as the chromatic inputs R(D), G(D), B(D) from which the chromatic parameters $x(D)$, $y(D)$, $z(D)$ are calculated (Appendix A). An oil sample may then be represented on a Cartesian chromatic diagram of $z(D)$ against $x(D)$ (Figure 22.12(a)) on which $z(D)$ represents the level of severe overheating, $x(D)$ the level of electrical discharging and the origin when only severe local overheating occurs. The point (0.33, 0.33) represents an oil sample with equal contributions from all three groups of gases while the boundary

$$x(D) + z(D) = 1$$

represents oil samples with no local overheating. The oil samples shown on the light absorption results of Figure 22.11 are also shown on this dissolved gas chromatic diagram, most of which lie on the zero local heating boundary.

The $x(D):z(D)$ diagram represents the relative magnitudes of the contributions from the three groups of dissolved gases. The absolute magnitude of the effect is represented by the chromatic $L(D)$ parameter (A.4). The variation of $L(D)$ for the group of oils shown in Figure 22.12(a) is presented in increasing $L(D)$ magnitude order in Figure 22.12(b). This shows that two of the oils (T10, 63) were approaching a highly critical state.

22.4.2.3 Chromatic processing of acidity, water content and electrical strength data

An increase in the acidity and water content and a decrease in electrical strength can be indicative of an increased electrical conduction of the insulating oil [33]. An increase in acidity can also increase sludge formation in the oil [34]. These three factors may be grouped together for chromatic processing in the same manner as described for the dissolved gases (section 22.4.2.2), that is the normalised magnitude of the water content, acidity, electrical strength reduction may be regarded respectively as the R, G, B inputs from which the values of the chromatic parameters $H(A)$, $L(A)$, $S(A)$ are derived (Appendix A). The resulting values of $H(A)$ and $L(A)$ for the series of oils considered in sections 22.4.1.1 and 22.4.1.2 are shown on the $H(A)$–$L(A)$ polar diagram of Figure 22.13(a). Also shown in Figure 22.13(a) are three polar degradation regions. The inner region corresponds to radial $L(A)$ values for oils in a 'normal' condition. The outer region corresponds to radial $L(A)$ values for oils in a critical condition. The intermediate region between these two $L(A)$ values corresponds to oil samples which are degraded but not yet critical. For example, the oil sample 85 (Figure 22.13(a)) is in a critically degraded condition while samples T8, T10, 63, 51 have degradation causing concern. The angular location, $H(A)$, of a sample on the polar diagram indicates the dominant component causing the degradation (for sample 85 both water content

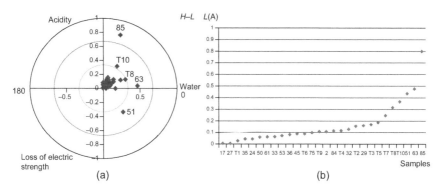

Figure 22.13   Chromatic representation of the water content, acidity and loss of electric strength of transformer oils
(a) H:L Chromatic polar diagram showing the relative levels of water ($H=0$), acidity ($H=120$) and loss of electric strength ($H=240$) of various oil samples of Figure 22.6
(b) Combined effects of water, acidity, loss of electric strength $L(A)$ on oil degradation for the various oil samples of Figure 22.6 in ascending order

and acidity contribute equally to the degradation; for sample 51 both water content and electrical strength reduction contribute equally).

The overall degree of degradation of the various oils, $L(A)$, is shown in Figure 22.13(b) in ascending order of magnitude so that the relative degradation state of each sample becomes clearer.

22.4.2.4 Chromatic representation of the overall oil condition

The overall condition of an oil sample may be represented by combining the most significant chromatic parameter from each of the three groups of degradation indicators, that is $Fn(x(O)/z(O))$ for optical absorption (section 22.4.2.1), $L(D)$ for dissolved gases (section 22.4.2.2), $L(A)$ for electrical conduction etc. (section 22.4.2.3). These three parameters are treated as the R ($\rightarrow Fn(x(O)/z(O))$), G ($\rightarrow L(D)$), B ($\rightarrow L(A)$) chromatic inputs to yield $H(T)$ (dominant degradation group), $L(T)$ (overall degree of the degradation), and $S(T)$ (extent of the degradation). The most significant of these parameters is the overall degree of degradation ($L(T)$) which is compared in ascending order for the various transformer oils in Figure 22.14(a) along with three degradation thresholds corresponding to normal oil, oil of concern and critical oils. (Each threshold corresponds to one group of indicators having reached a maximum level of degradation.)

The relative contribution to the degradation condition of each of the three groups can be displayed on a chromatic $H(T)$–$L(T)$ polar diagram (Figure 22.14(b)). This shows that the sample T10 is regarded as critically degraded due to equal

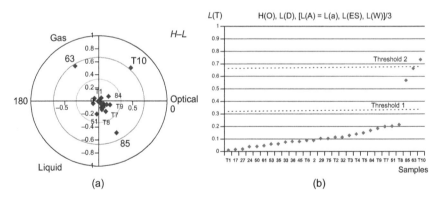

Figure 22.14 Chromatic representation of the overall condition of transformer oils

(a) H:L Chromatic polar diagram showing the relative contributions of the optical absorption ($H=0$), dissolved gases ($H=120$) and water/acidity/loss of electric strength ($H=240$) for various oil samples of Figure 22.6

(b) Overall degradation ($L(T)$) for the various oil samples of Figure 22.6 in ascending order

indications from the optical and dissolved gases data, whereas the sample 63 is critically degraded mainly due to the dissolved gases and the oil of concern 85 is due to both optical and electrical conduction concerns.

22.4.2.5 Condition assessment procedure for a systems operator

The procedure for assessing the condition of an oil sample using the above components is summarised in Figure 22.15(a,b). The operator enters values for each of the thirteen monitored parameters of the three groups of dissolved gases, acidity/liquid and optical parameters (Figure 22.15(a)). These are chromatically processed as described in sections 22.4.2.1 to 22.4.2.4 above to produce a chromatic quantification of the overall oil condition, $L(T)$ (Figure 22.14(b)). Further details about dominating features of any degradation may then be obtained by following the processing flow chart shown in Figure 22.15(b), which is software based in the monitoring computer. This enables checks to be made about any anomalous single component, for example electrical discharging. As such the approach is not over-deterministic but provides the user with a high degree of freedom in decision making as well as a high level of traceability to fundamental physical causes (e.g. details of a particular dissolved gas).

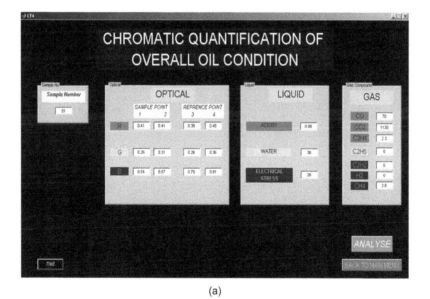

(a)

Figure 22.15 Example of user interface screen display protocol for the chromatic oil condition monitoring process
(a) Data input screen; (b) flow chart for tracing dominant physical parameter threatening the condition of the transformer oil

780 High-voltage engineering and testing

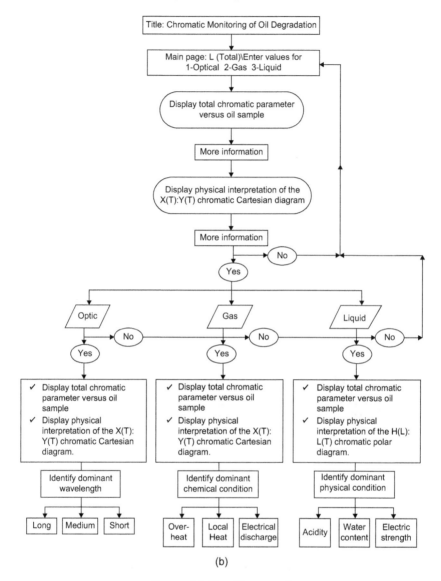

Figure 22.15 (Continued)

22.5 Conclusions

This chapter has described a range of applications of optical fibre–based sensing and chromatic monitoring techniques in relation to high-voltage equipment and systems. It has illustrated how the approach provides the basis for intelligent monitoring with a high degree of traceability and without usurping the decision function of the human operator. Not only can emerging features of complex conditions be detected and discriminated, but their causes may also be traced. The monitoring may be regarded as providing significant information to the operator so that he/she is in a better position to make critical decisions.

There is a growing body of high-voltage applications of the technology being identified, in addition to those outlined in this chapter. Three such further examples are as follows:

- A research prototype battery bank monitor relevant to substation emergency electrical supplies has been demonstrated. This is based upon the use of optical fibre transmission to reduce electrical insulation costs and chromatic transmission of data from each cell of the bank to reduce detection and display costs.
- A single monitoring unit combining optical and infra-red sensing functions for chromatic deployment in the space domain for non-intrusively monitoring personnel activity and incipient changes in plant conditions has been envisaged based upon extensive testing in the social care sector [5].
- Space domain chromaticity has also been demonstrated for a simple, cost-effective monitoring for locating sources of radio frequency emissions.

In the research domain, new chromatically based analyses are emerging and the use of a limited number of additional chromatic processors (3–6) is being investigated for providing additional discriminating information.

Acknowledgements

The provision of partial discharge data by Professor P. L. Lewin of the University of Southampton, UK and Professor T. R. Blackburn and Dr. W. G. Ariastina of the University of New South Wales, Australia is appreciated. The provision of transformer oil data by D. Jones of Electricity North West Ltd. is also acknowledged, as well as Electricity North West, Central Networks and Scottish Power who have been involved in the deployment of tap-changer acoustic sensors.

M. Ragaa (section 22.4.1), E. Elzazoug (section 22.4.2), A. G. Deakin (sections 22.4.2 and 22.3.2.2) and D. H. Smith (section 22.3.2.2) are at the Centre for Intelligent Monitoring Systems, University of Liverpool.

References

1. Salma M.M.A., Bartinikas R. 'Determination of neural network topology for partial discharge pulse recognition'. *IEEE Transactions on Neural Networks*. 2002;**13**(2):446–56
2. Tu Y., Wang Z.D., Crossley P.A. 'Partial discharge pattern recognition based on 2-D wavelet transform and neural network techniques', IEEE Power Engineering Society Summer Meeting, July 21–25. Institute of Electrical and Electronics Engineers, Chicago, IL, p. 411, 2002

3. Abdel-Galil T., Sharkawy R.M., Salma M.M.A., Bartnikas R. 'Partial discharge pulse pattern recognition using an inductive inference algorithm'. *IEEE Transactions on Dielectrics and Electrical Insulation.* 2005;**12**(2): 320–7
4. Kranz H. 'Diagnosis of partial discharge signals using neural networks and minimum distance classification'. *IEEE Transactions on Electrical Insulation.* 1993;**28**(6):1016–24
5. Jones G.R., Deakin A.G., Spencer J.W. *Chromatic monitoring of complex conditions.* FL: Taylor and Francis; 2008. ISBN 13.978-1-58488-988-5
6. Moon P., Spencer D.E. *The photic field.* Cambridge, MA: MIT Press; 1981
7. Jones G.R., Spencer J.W., Yan J. 'Optical measurements and monitoring in high voltage environments' (Chapter 12), in M. Haddad, D. Warne (eds.), *Advances in high voltage engineering* (vol. 40, p. 571). The Institution of Electrical Engineers; 2004. ISBN 0-85296-158-8
8. Jones G.R., Li G., Spencer J.W., Aspey R.A., Kong M.G. 'Faraday current sensing employing chromatic modulation'. *Optics Communications.* 1998;**145**: 203–12
9. Murphy M.M., Jones G.R. 'An extrinsic integrated optical fibre strain sensor'. *Pure and Applied Optics.* 1993;**2**:33–49
10. Messent D.N., Singh P.T., Humphries J.E., Spencer J.W., Jones G.R., Lewis K.G., Hall W.B. 'Optical fibre measurement of contact stalk temperature in an SF6 circuit-breaker following fault current arcing'. *Proceedings of the XII international conference on gas discharges and their applications*; Greifswald, vol. 1, pp. 543–6. Kiebe-Dr.; 1997. ISBN 3-00001-760-7, 978-3-00-001760-5
11. Cosgrave J.A., Vourdas A., Jones G.R., Spencer J.W., Murphy M.M., Wilson A. 'Acoustic monitoring of partial discharges in gas insulated substations using optical sensors'. *IEE Proceedings A.* 1993;**140**(5):369–74
12. Parada S., Spencer J.W., Jones G.R., Cunningham J., Richey K. 'An optical fibre based technique for determining the electrical status of a 3-phase 11 kV belted power cable'. *Proceedings of the 3rd international conference on electrical contacts, arcs, apparatus and their applications (IC-ECAAA)*; Xian, P.R. China; 1997
13. Issac L.T., Spencer J.W., Jones G.R., Hall W.B. 'Monitoring particle concentrations produced by arcing in SF_6 circuit-breakers using a chromatic modulation probe'. *IEE Proceedings on Science, Measurement & Technology.* 1999;**146**(4):199–204
14. Jones G.R., Li G., Spencer R.A., Aspey R.A., Kong M.G. 'Faraday current sensing employing chromatic modulation'. *Optics Communications.* 1998; **145**:203–12
15. Issac L.T. 'Puffer circuit-breaker diagnostics using novel optical fibre sensors'. Ph.D. thesis, University of Liverpool; 1997
16. Cosgrave J., Humphries J.E., Spencer J.W., Jones G.R., Russell P.C., Hall W.B., Lewis K.G. 'Chromaticity characterisation of optoacoustic signals from faiult current arcs in high voltage circuit-breakers'. *Proceedings of the*

XII *international conference on gas discharges and their applications*; Greifswald, pp. 512–5. Kiebe-Dr.; 1997. ISBN 3-00001-760-7, 978-3-00-001760-5

17. Isaac L.T., Spencer J.W., Jones G.R., Hall W.B., Taylor B. 'Live monitoring of contact travel on EHV circuit-breakers using a novel optical fibre technique'. *Proceedings of the XI international conference on gas discharges and their applications*; Tokyo, vol. 1, pp. 238–41; 1995
18. Oraby O.A., Spencer J.W., Jones G.R. 'Monitoring changes in the speckle field from an optical fibre exposed to low frequency acoustical vibrations'. *Journal of Modern Optics*. 2009;**56**(1):55–66
19. Paoletti G.J. 'Partial discharge theory and technologies related to medium voltage electrical equipment'. *IEEE Transactions on Industry Applications*. 2001;**37**(1):90–103
20. International Standard. International Electrotechnical Commission, http://www.thefullwiki.org/International_Electrotechnical_Commission; IEC 60270 2000. http://www.thefullwiki.org/British_Standard; BS EN 60270 2001. High voltage test techniques – partial discharge measurements'. International Electrotechnical Commission
21. Lai K.X., Lohrasby A., Phung B.T., Blackburn T.R. 'Partial discharge of electrical trees prior to breakdown'. *Proceedings of the 2008 international symposium on electrical insulation materials*, 7–11 September. 2008; Yokkaichi, Mie, Japan, pp. 649–52. IEEE, Piscataway, NJ. ISBN 978-4-88-686005-7, 4-88-686005-2
22. Lewin P.L. University of Southampton, UK, Private Communication; 2011
23. Blackburn T.R., Ariastina W.G., University of New South Wales, Australia, Private Communication; 2011
24. Bahadoorsingh S., Rowlands S.M. 'A framework linking knowledge of insulation ageing to asset management'. *IEEE Electrical Insulation Magazine*. 2008;**24**:38–46
25. Zhang J., Jones G.R., Spencer J.W., Jarman P., Kemp I.J., Wang Z., Lewin P.L., Aggarwal R.K. 'Chromatic classification of RF signals produced by electrical discharges in HV transformers'. *IEE Proceedings: Generation, Transmission and Distribution*. 2005;**152**(5):629–34
26. Ragaa M., Spencer J.W., Jones G.R., Deakin A.G. 'Characterisation of partial discharge signals using a chromatic technique'. *Proceedings of the XVIII international conference on gas discharges and their applications*; Greifswald, pp. 556–9; 2010
27. Shoureshi R., Norick T., Swartzendruber R. 'Intelligent transformer monitoring system utilizing neuro-fuzzy technique approach'. Final Project Report, PSERC Publication 04–26, pp. 1–111; 2004
28. Aghaei J., Gholami A., Shayanfor H.A., Dezhankory A. 'Dissolved gas analysis of transformers using fuzzy logic approach'. *European Transactions on Electrical Power*. 2010;**20**(5):630–8
29. Neumann D., Gegenfurtner K.R. 'A color indexing system based on perception (CISBOP)', in H. Bulthoff, M. Fahle, K.R. Gegenfurtner, H.A. Mallot (eds.),

TWK 2000, Beitrage zur 3.Tubinger Wahrnehmungskonefenz. Knirsch, Kirkentellnisfurt 45

30. Lindgren S.R. 'Transformer condition assessment experiences using automated online dissolved gas analysis'. Paper A2-202, 1-8 CIGRE Session 2004 *Proceedings of the International Council on Large Electric Systems*, Paris, 2004, A2-202, http://www.cigre.org

31. Jones G.R., Deakin A.G., Brookes R.J., Spencer J.W. 'A portable liquor monitoring system using a PC-based chromatic technique'. *Measurement Science and Technology*. 2009;**20**:1–8

32. Zhang J., Jones G.R., Deakin A.G., Spencer J.W. 'Chromatic processing of DGA data produced by partial discharges for the prognosis of HV transformer behaviour'. *Measurement Science and Technology*. 2005;**16**(2):556–61

33. Meshkatoddini M.R., Abbaspour S. 'Ageing study and lifetime estimation of transformer mineral oil'. *American Journal of Engineering and Applied Science*, 2008;**1**(4):384–8

34. Van der Zel L. 'Continuous online transformer oil purification, dry-out, degassing and life extension'. Electric Power Research Institute (EPRI), Product ID 1013321 URL; 2006

Appendix A
Algorithms for chromatic transformations

The two sets of chromatic parameters x, y, z and H, L, S are obtained from the outputs of the three chromatic processors R, G, B using the following algorithms [5]:

(a) x, y, z agorithms

$$x = \frac{R}{(R+G+B)}; \quad y = \frac{G}{(R+G+B)}; \quad z = \frac{B}{(R+G+B)} \quad (A.1)$$

so that

$$x + y + z = 1 \quad (A.2)$$

(b) H, L, S algorithms

$$H = 240 - \frac{120.g}{(g+b)} \quad \text{if } r = 0$$

$$= 360 - \frac{120.b}{(b-r)} \quad \text{if } g = 0 \quad (A.3)$$

$$= 120 - \frac{120.r}{(r+G)} \quad \text{if } b = 0$$

where

$$r = R - \min(R, G, B)$$
$$g = G - \min(R, G, B)$$
$$b = B - \min(R, G, B)$$

$$L = \frac{(R+G+B)}{3} \quad (A.4)$$

$$S = \frac{(\text{MAX} - \text{MIN})}{(\text{MAX} + \text{MIN})} \quad (A.5)$$

where MAX and MIN refer to the maximum and minimum values respectively of the parameters R, G, B.

(c) *secondary chromatic parameters*

Secondary chromatic parameters can be derived from the primary parameters in various ways.

For providing a linear calibration in monitoring samples of transformer oils via polychromatic light absorption, the parameter $Fn(x(O)/z(O))$ has been utilised (section 22.4.1.1, Figure 22.8(c)) where

$$Fn(x(O)/z(O)) = \frac{1}{2}(x(O)/z(O)) \quad \text{for } (x(O)/z(O)) < 1 \tag{A.6}$$

$$= 1 - \frac{1}{2}(z(O)/x(O)) \quad \text{for } (x(O)/z(O)) > 1 \tag{A.7}$$

$z(O)$, $x(O)$ are the primary parameters x, z from the optical (O) oil monitoring.

The overall oil chromatic parameters (e.g. $L(T)$) (section 22.4.2.4, Figure 22.11) derived from the chromatic parameters from the optical $(Fn(x(O)/z(O)))$, Dissolved Gases $(L(D))$, Acidity $(L(A))$ etc. may also be regarded as secondary chromatic parameters since

$$L(T) = [Fn(x(O)/z(O)) + L(D) + L(A)]/3 \tag{A.8}$$

Chapter 23

Some recent ESI developments: environmental, state of art, nuclear, renewables, future trends, smart grids and cyber issues

H.M. Ryan

Preface

Currently, we are living at a time of challenging worldwide economic downturn during which there has been much public concern and economic, environmental, technical and political debate in the UK relating to high consumer energy costs which are directly linked to 'high-cost' controversial government decisions on the future directions for energy development (and subsidies) in Britain.

Additionally, there have been perhaps unparalleled public concerns expressed in the UK press regarding, for example, the future of 'new-build nuclear plant' in Britain following on from the Fukushima nuclear accident in Japan, March 2011. The selected locations and the *huge number of wind farms recently being proposed by the UK government*, with still more being installed, developed or planned onshore and offshore; certain operational/environmental/logistical difficulties with wind farms and availability, or otherwise, of appropriate energy-storage technology for back-up; the magnitude of generous public subsidies to various energy technology sectors; and the serious worries regarding the already high household fuel bills and the anticipated further price increases on the horizon, which sadly, seem to be sending some members of the UK public into 'fuel poverty' situations.

Consequently, energy and environmental topics – with strong political connotations/elements – have remained regularly 'on the radar' in the UK 'quality' national press during 2010–12 and it still continues! Many valuable and thought-provoking 'positioning or technically updating press articles' appear almost weekly, as energy-related debates run on and on, covering various *nuances* of current energy sector issues.

Note: The UK Energy Bill had its second reading in parliament in December 2012, and is due to be given royal assent during 2013. Recently, an influential cross-industry group, business owners and consumer groups called the Powerline, 'via a leaked internal memo' to the press, has intimated its intention to launch scathing attacks on this bill via advertising and social media. The business leaders are furious at what they see as the Energy Bill's over-reliance on costly renewable energy at the expense of traditional and lower-cost energy sources such as coal-fired and gas power stations; the press has commented that against the backdrop of a potential triple-dip recession, the Powerline group is concerned that the extra costs the bill will impose on individuals and business energy users will lead to further decline in UK competitiveness. Apparently a leaked memo from the Powerline group calls for a 'dispassionate debate' over *green energy*. It indicated, 'our ambition is to stop Britain sleepwalking into the blackouts, soaring energy bills and a desert of lost jobs'. The group pose the question as to whether the government should rethink its policy to reflect the huge changes that have occurred since the bill was first conceived. They include the financial crisis that has hit the big utilities and the USA's shale gas revolution – which, as Danny Fortson (*The Sunday Times*) very recently informed his readers, 'has unlocked more than a centuries worth of reserves and sent fuel prices plummeting'. The Powerline campaign will be seen as a further blow to the UK government's energy policy as it continues to struggle to put private sector funding in place to ensure that the proposed new-build nuclear generation facility at Hinkley Point in Somerset becomes a reality.

Because of these strategic issues, the writer has adopted the unusual and perhaps unique approach in this chapter of including, and selectively quoting from, several of the more relevant press coverage 'positioning energy articles' with strategic quotes from senior figures in the industry/government. In addition, many valuable references to technical papers and CIGRE Technical Brochures are included, covering important and strategic overview aspects. In this way, it is intended to provide the reader with a useful 'balanced overview' of the topic; the 'rapid transient switches' in government energy policies and of the wide-ranging debates; and the invaluable published technical work of IEC, CIGRE, IET and IEEE, etc., currently available, as the industry moves at considerable investment costs towards further 'lifetime' and other 'enhancement' improvements culminating in intelligent or 'smart-grid' networks in the near future [1–61]. With the widespread plans to increase offshore wind farms around UK and Europe, there has also been a recent initiative to plan a North Sea Super Grid (NSSG) offshore in an effort to improve [27, 28] the integration of the European Energy Market. There must also be ongoing vigilance and development, to ensure *cyber-security* is maintained worldwide, against all intrusion challenges referred to as cyber-crime, cyber-hacking, etc. in the energy, business and private sectors. The energy industry fully recognises this as a rapidly growing problem area.

23.1 Introduction

The environment within which electricity power supply systems operate and develop worldwide has continued to change significantly over the past two decades due to many strategic factors and can be *influenced by many economic, environmental and technical issues* and sometimes *political*, as was the case in the recent past when a multinational energy company operating in the UK was 'dissuaded' from contract negotiations because of 'political' interventions from a foreign government. They include:

- electricity market liberalisations; financial crisis/economic downturn (government); electricity market reforms
- electricity industry deregulation and increased multinational ownership of energy companies (e.g. EDF and GDF Suez (French); RWE and E.ON (German) in the UK)

- 'environmental-protection' and carbon issues, 'increased mix' of energy sources
  Despite numerous disappointments, ascribed by pundits to the 'self-interests' of certain 'big-polluting' countries such as China, USA and India who have not agreed to participate, recently the European Union (EU) re-dedicated itself to a 30% reduction in greenhouse gas emissions by 2020 over the 1990 levels, with similar targets set in most parts of the developed and developing world.
- future energy-generation changes in Britain and strategic changes in power flow patterns that the transmission networks of the twenty-first century will require to handle the challenges facing the system operator, and network developments required to meet the challenges
- recent increased penetration of new technologies and the increasing application of information technology, diagnostic and other in service-monitoring strategies. This condition monitoring (CM) aspect has been covered in detail in other chapters (including Chapters 7 and 20–22), with further themes deemed to be of strategic importance by CIGRE 'touched on' in appendices to Chapters 5, 8–10, and with the 'attendant risk' of cyber-crime and security attacks and the *'commitment'* by utilities to build towards 'smart-grid status' in the next one or two decades also considered briefly (see Appendix B and section 23.13, respectively, later in this chapter)
- global environmental disasters, events, government security issues and possible conflicts with neighbouring 'arms-race-warring Middle-East States' can also affect various strategic government policies and cyber-security or dealing with cyber-hacking activities in strategic regions of the world.

According to experts (e.g. CIGRE, IEEE, IET), the energy industry has to move forward in several strategic directions as will be 'touched on' in this chapter. These include:

1. To carefully prepare and ensure that a strong, robust and 'smart power system network of the future' is planned and achieved [30, 35], which will be capable of transporting power from *non-carbon sources* nationally and transnationally, over great distances – *onshore and soon offshore* – yet with the flexibility to engage with local grids and local energy networks facilitating optimised operation of the network, microgrids, dispersed generation, effective storage for wind farms countrywide, intelligent loads, using vast sophisticated levels of integrated condition and load monitoring techniques and probably even more while still being capable of providing security against the evolving and even more sophisticated cyber-threats! To achieve all this is indeed 'a major ask'; it will be extremely costly (necessitating special funding scenarios) and will require effective integration of the complex network, which obviously will depend on utilising and relying heavily on new technologies, strategies, utilisation of UHV, alternating current (AC) and direct current (DC) schemes, advanced insulation coordination strategies, increased power electronics and ICT-integrated condition and load monitoring techniques.
2. Because of the significant under-investment in the UK energy sector for many years, attention has been focused on making the best use of existing electricity

networks and equipment [e.g. switchgear, substations, cable, transformers] and transmission/distribution network systems, working systems harder [e.g. for overhead lines (OHLs) working nearer to their thermal limits], operating assets up to the end of their effective/useful life, regularly assess their condition, maintain, refurbish, extend their life and replace. The reader can follow up and examine the many publications, e.g. IET, IEEE, CIGRE, on these and related themes [1–61]. Many phased changes and enhancements to energy network system layouts are currently being evaluated or planned, including the possible construction of a North Sea super grid (NSSG) concept within the next 10 years, to support strategic offshore wind farm developments [27, 28].

3. *Environmental concerns and issues* have been at the forefront in the industry for some time and attention to this aspect must continue and efforts will continue as manufacturers, utilities and standards bodies attempt to answer all environmental concerns in many areas including:

   - using existing and new assets effectively
   - environmental friendly materials (or revisit the dash for gas-mark 1 scenario!)
   - reduction of carbon footprint of energy sector
   - less intrusive location of wind farms, transmission line cables onshore and offshore [27, 28], taking cognisance of areas of natural beauty
   - develop knowledge systematically, utilising efficient/effective collaboration processes involving, wherever possible, cooperation and collaboration of worldwide experts, for example the CIGRE model.

It has been predicted that to attain the desired 'smart-grid status' within a time of one to two decades, it will be necessary to successfully achieve the above changes [30, 35], and, equally importantly, ensure that training avenues are in place to produce the next-generation technical staff capable of handling all current and emerging technologies and techniques and the seemingly 'never-ending new challenges' which will most certainly appear [1–61].

In this chapter, some of these aspects, themes and concerns will now be briefly considered – in no particular technical order. This provides the reader with a broader perspective and appreciation of several of the current strategic energy aspects presently under review, development, or public debate, coupled with guidance to help readers achieve 'empowerment' by systematically accessing appropriate important reference sources including strategic CIGRE papers and Technical Brochures (covering many varied energy themes) ranging from A1, Rotating electrical machines (SC A1) through to D2, Information systems and telecommunication (SC D2) – see table below – including renewable energy and other green developments, opportunities and challenges.

The recent UK media coverage of these and customer concerns at the rapidly escalating energy costs and the eventful journey towards smart-grid Networks of the Future [30, 35] will be bumpy and further changes/challenges are likely along the way. These will be discussed.

Some recent ESI developments   791

The CIGRE Technical Study Committees (SCs) and contact address information are summarised below.

All CIGRE Study Committees (SCs) since the 2002 reform [after, A History of CIGRE, 2011]

[*Note*: Full version, 'Fields of activities of the CIGRE[1] Study Committees', is reproduced later in Table 23.13 of this chapter]

A1 Rotating electrical machines [www.cigre-a1.org]	C1 System development and economics [www.cigre-c1.org]
A2 Transformers [www.cigre-a2.org]	C2 System operation and control [www.cigre-c2.org]
A3 High-voltage equipment [www.cigre-a3.org]	C3 System environmental performance [www.cigre-c3.org]
	C4 System technical performance [www.cigre-c4.org]
B1 Insulated cables [www.cigre-b1.org]	C5 Electricity markets and regulation [www.cigre-c5.org]
B2 Overhead lines [www.cigre-b2.org]	C6 Distribution systems and dispersed generation [www.cigre-c6.org]
B3 Substations [www.cigre-b3.org]	
B4 HVDC and power electronics [www.cigre-b4.org]	D1 Materials and emerging test techniques [www.cigre-d1.org]
B5 Protection and automation [www.cigre-b5.org]	D2 Information systems and telecommunication [www.cigre-d2.org]

Note: 'Since 1921, CIGRE's priority role and "added value" have always been to facilitate mutual exchanges between all its components: network operators, manufacturers, universities and laboratories. It has continually led to the search for a good balance between the handling of daily problems encountered by its members in doing their jobs and the reflections on the future changes and their equipment. Its role in exchanging information, synthesising state of the art, and serving members and industry has constantly been met by CIGRE throughout events and through publications resulting from the work of its Study Committees' (www.cigre.org; www.e-cigre.org).

23.2 International takeovers in UK power sector and possible impacts

In Britain, major changes have occurred in the energy sector in the last two decades, as a result of the continuing strategic 'downsizing' of the UK power sector manufacturing base and the major strategic international takeovers of several energy companies since the privatisation in the 1990s. The recent financial crisis/economic downturn has also presented problems and made it difficult for utilities to obtain finance for renewable and fossil fuel plants and also newly built nuclear projects. There have also been uncertainties/apprehensions surrounding the government's electricity market reform and local government bill:

1. As a result, some of the UK power industry and energy companies have changed dramatically in recent years and are 'now playing by different rules' significantly controlled by international rivals. Consequently, strategic technical decisions, for example for 'UK new-build nuclear plant contracts', can sometimes be influenced by the policy 'swings' of foreign governments – RWE/E.ON (German owned) were recently 'encouraged to withdraw from a proposed new nuclear project after being 'dissuaded' by a non-UK government – according to a company executive quoted in the UK press recently'. ('New-build nuclear developments' in the UK will be discussed again later in this chapter.)

---

[1] CIGRE key: SC, Study Committee; WG, Working Group and JWG, Joint Working Group; TF, Task Force; TB, Technical Brochure; TR, Technical Report; SC, Scientific Paper; IP, Invited Paper.

2. The balance of power has 'swung' towards international company ownership control, as explained recently by Webb and Thompson. The material has been 'edited' and reproduced in Table 23.1, with a few additional key points discussed in the text to follow. (The reader is encouraged to read the original article, *The Times*, 17/04/12.)
3. The main thrust of the article mentioned in point (2) centred around the fact that yet another British energy company, *International Power* (IP), appeared set to fall into 'foreign-hands' after IP agreed to a £6.4 billion 'takeover offer' from the world's largest utility *GDF Suez*. This *The Times* article also covers some historical background and much valuable technical data, and observations (see Table 23.1). Some of this energy information is reproduced or commented upon below:

Table 23.1 *Further UK balance of power tilt towards international rivals [source material mainly after T. Webb and S. Thompson,* The Times, *17/04/12]*

International Power Today		**GDF Suez** Today	
Global presence		Global position in different markets	
Net Capacity Ownership		**Power N⁰1***	**Gas N⁰1***
North America	13,261 MW	*Independent Power Producer	*Purchaser in Europe
Latin America	6,521 MW	[IPP] in the world	*Importer of LNG in Europe
Australia	3,068 MW	*Producer of non-nuclear	*Storage operator in Europe
Asia	3,948 MW	energy in the world	*Transmission and Distribution
Middle East, Turkey and		*Wind producer in France	networks in Europe.
Africa (META)	7,139 MW	11.4.5 GW of installed power	
Europe	9,351 MW	production; 17.5 GW of	
		capacity under construction.	
		Services N⁰1 supplier of energy and environmental efficiency services in Europe.	
		N⁰2 supplier of water and waste services in the world.	
		180 district heating and cooling networks in the world are operated worldwide.	
Sales of Leading Utilities	[Euros bn]		[Euros bn]
GDF Suez +IP	84[a]	Tepco	44
E.ON	82	Iberdrola	25
GDF Suez	80	Kansai Electric Power	23
EDF	66	Kepco	K 22
Enel	64	IP	14
RWE	46		

[a]Takes into account **GDF Suez's** 70% ownership of **IP**.
EDF is a **French company**; **GDF Suez** is 35% owned by the **French State**.
E.ON and **RWE** are **German companies**.
Iberdrola is a Spanish company.

  [Source: Tim Webb, Susan Thompson, *The Times*, 17/04/12, business briefing]
  See original for fuller coverage plus:
- **IPs** – Net % capacity geographically worldwide 'today'; plus, its UK assets 'today'; and % Capacity under construction: [e.g. oil (6%), hydro (38%), renewable (7%), coal (9%), gas (39%) and pumped storage (1%).]
- **GDF Suez** – Its oil and gas exploration licences and profit breakdown – 'today'; and its annual oil and gas production figures 42–53 million barrels of oil equivalent, 2006–2010.

Some recent ESI developments   793

(i) Webb/Thompson state, 'of the so-called "*Big Six*" energy companies that dominate the British market *only two – Centrica, the owner of British Gas, and Scottish and Southern Energy – remain independent, with their own listing*'. 'The others were snapped up by giant German, Spanish and French utilities in an orgy of deals in the decades after the UK privatisation in the 1990s'.

(ii) 'Historically (*as this writer clearly recalls*), few UK politicians or regulators complained at the time, "content to let the market hold sway". But the European owners of British power industry play by different rules. They enjoy dominant positions in their home markets, which remain largely closed to foreign entrants'.

(iii) Table 23.1 're-formats' and reproduces much of the valuable strategic information quoted in this *The Times* article, such as size of global presence of GDF Suez, IP and other companies; the percentage scale of IP's current assets, capacity under construction; the global high status positions of GDF Suez in various markets.

(iv) The *dominant positions* of French and German companies, as evidenced by sales performance data, are clearly shown. This is rather worrying from a UK perspective. The combined sales totals for the two French companies (EDF + GDF Suez) and for the two German companies (RWE + E. ON) are, respectively, >*146 and 128 billion Euros* (as quoted in *The Times* article of 17/04/12, by Webb and Thompson).

(v) '*GDF Suez* (35% French State owned) wants full control of *International Power* (IP) to increase its presence in emerging markets. IP apparently owns, or partly owns, dozens of power plants worldwide, including 11 in the UK. *GDF Suez* is part of a consortium currently considering building nuclear plants in the UK' [Later, they pulled out of this initiative!]

23.2.1   Warning: kid gloves treatment

'*The big six energy* companies could be driven out of Britain if they were treated too harshly in any pricing investigation', a senior UK government minister, Charles Hendry, has informed *The Times* (Ref. *The Times*, Industrial Editor, Robert Lea (on 7/2/11)). Lea states:

> Minister Hendry said, 'that suppliers might be too important to future UK energy security to be exposed to the full rigours of a competitive investigation'. This intervention was likely to put pressure on regulators. The warning from the Energy and Climate Change Minister came amid speculation that the Competition Commission could be called in to rule on whether the suppliers were profiteering. *Ofgem recently launched its second pricing investigation in two years and is examining whether profit margins are too big*. It is reckoned that every £100 of an average £1200 dual fuel bill is pure profit and could be rising.

Lea goes on to state: 'Centrica, the country's biggest household supplier through its subsidiary British Gas, is likely to receive much adverse media comment this month when it publishes its profits', *predicted beforehand by City analysts – to be up 25% on last year*.

However, Charles Hendry has warned Ofgem of the implications of referring the industry to the Competition Commission. He said, 'we have to be aware that these are the same companies which we are asking to help find much of the £200 bn of investment in the energy market that this country needs'. 'If we drive these companies away, we are storing up problems in five to ten years time. This is not an issue just about prices. We have to recognise the important role that the *big six* have to play in our energy security.'

Note: In essence, as reflected by the title of Lea's article, the minister appears to be shielding the power giants from a profiteering enquiry!

Lea reports: 'The big six are made up of British Gas, the Southern Electric Group SSE, the German companies E.ON and RWE npower, EDF, of France, and the Spanish owned Scottish Power. All have been putting up prices.'

Together they control >99% of the household energy supply market in the UK. Market share of gas and electricity markets, price rises in winter 2011/12 and 'dual-fuel rises' in winter 2011/12, are listed below. The high average rises in gas and electricity bills for 2008–11 are respectively, 28% and 43%.

Lea said:

- 'The Government's Electricity market reform White paper due in spring envisages/anticipates big investments in nuclear power and renewable energy projects, including thousands of wind turbines, because of the widespread closures of coal and oil plants and old nuclear reactors'.
- '*Consumer Focus* is challenging Ofgem to provide the transparency that will provide evidence of whether households and businesses are getting a raw deal'.
- He further reported that Audrey Gallacher, the Watchdog's head of Energy, said: 'There are major questions over whether wholesale price charges have been fairly passed on by suppliers'.
- 'If Ofgem's review shows systemic problems, and the regulator does not have the powers to make the changes needed a (commission) enquiry cannot be put off. Customers will foot the £200 billion bill for investment. It is more important than ever for customers to be able to see whether they are being charged a fair price,' she said – according to Lea's article.

Impact of UK financial crisis downturn: Table 23.2 provides main aspects, reproduced from a recent article by Lea, dealing with the fact that *clean energy targets* are 'drifting out of reach' as the current financial *downturn* has made it difficult to obtain project finance. It also provides valuable strategic information regarding spending by the 'big six' *energy companies* and can be examined in conjunction with Table 23.1. Most of the information which has been 'reproduced and re-formatted' requires no further comment – other than, once again, the reader is encouraged to read the full version in the original article.

Table 23.2   *'Clean energy targets drift out of reach as cash dries up'*
             [after The Times, *23/4/12, Robert Lea]*

* 'In this article Robert Lea reports, 'a worrying slowdown in investments in renewable *sources means that* Britain is on course to miss its 2020 green energy generation commitments by as much as 15%', according to research by Bloomberg New Energy Finance which was commissioned by Greenpeace.
* 'Britain has made a commitment that by 2020 it will produce 30% of its electricity generation from renewable sources'. Lea informs the reader, 'The Bloomberg research study indicates that at the present rates of investment the UK will fall far short of this'.
* 'The report says that investment in the past decade was encouraging – with wind farms sprouting up off the English and Welsh coasts and on Scottish, while families and industry have embraced solar panels – but this has tailed off'.

Lea informs readers that the report says, 'Investment in "new capacity" is slowing down as the recession-induced fall in power demand and new renewable and gas capacity has led to low prospective returns for fossil-fuel-build.'

* 'The financial crisis has made obtaining project finance a challenge and for both fossil plants and renewables this has been exacerbated by uncertainty around the Government's Electricity Market Reform'.
* The Bloomberg report 'suggests that Britain is likely to achieve a level of only 25% renewable energy by 2020. While the industry could overshoot this pessimistic projection if more coal plants *shift* to co-firing biomass, *The Times* paper argues that its forecast is more probable because of the slowness of final investment decisions for offshore "super" wind farms expected later in the decade'.

Lea includes much strategic information from Bloomberg report and his comments on this research document are reproduced below *quoting Lea, or the report, as appropriate*:

* Since 2006 half of all new generating capacity has been in renewables.
* Lea says, 'Of the six big UK energy companies, the German giant RWE npower, the Spanish owned Scottish Power and SSE, the quoted company behind Southern Electric, account for more than a third of all renewable generation'.
* 'That compares with, a figure of 2.3% for Centrica, and 3.2% for EDF Energy, of France'.

The Bloomberg report

* 'It found that RWE npower, SSE and Scottish Power had invested £9.7 bn in green energy, compared with the £4.5 bn invested by Centrica, EDF and E.ON, of Germany'.
* 'It accepts that Centrica and EDF are committed to a £20 bn joint nuclear power venture, but points out that the most reluctant green investor, Centrica also happens to be Britain's most consistently profitable energy company'.
* 'While the leading players control 70% of Britain's power generation market, with all the legacy gas, coal and nuclear plants, they account for only 47% of more newly installed renewable energy installations'.
* 'Figures also show that the aggregate of small solar photo voltaic installations, typically put up by households, accounts for more than 14% of the renewable markets, outstripping the installed total renewables of RWE npower, the single biggest green investor of the Big Six energy companies'.

Lea comments that the report 'has found a deep divide between Britain's Big Six energy companies. While all have "talked a good green game," analysis of their investment shows that whereas SSE, Scottish Power and RWE npower are living up to their promises, the British Gas Group Centrica is lagging far behind'.

Finally, on the same p. 18 of *The Times* is an article, accompanied by a picture,* informing the reader that *Asia's mightiest solar power farm has been switched on – in the Indian desert state of Gujarat*. Seemingly, it is part of a drive to transform the region into one of the world's leading green energy producers and to provide electricity to some of the 400 million people who have no access to electricity [The Gujarat Solar Park which spans a remote 3,000 acres near the border with Pakistan started generating 214 MW of electricity.] *Final output capacity is predicted to be 1,000 MW, making it the biggest single existing solar power plant in the world,* according to an official [*The Times* reports].

    [*Caption: Solar giant turns the desert green (and blue) in Gujarat]
    (Previously, the largest Solar Park in Asia was China's 200 MW Golmud Solar Park.)
Note: The reader is encouraged to examine the fuller original article.

Share of UK domestic market and price rises in the winter of 2010–11
[after **The Times**, *7/2/11, Robert Lea]*

	Gas market		Electricity market		2010–11 (Due fuel rises) (This winter)
	%	(Rise)	%	(Rise)	
British Gas/Centrica	44	(7.0%[a])	22	(7.0%[a])	(£82)
RWE npower	12	(5.1%[a])	15	(5.1%[a])	(£54)
SSE	15	(–)	19	(9.4%[a])	(£67)
E.ON	13	(3.0%[a])	18	(9.0%[a])	(£62)
EDF Energy	7	(6.5%[a])	13	(7.5%[a])	(£72)
Scottish Power	9	(2.0%[a])	12	(8.9%[a])	(£54)
Average rise in gas bills, 2008–11		28%			
Average rise in electricity bills, 2008–11		43%			

[a]Expected profits for British Gas for 2010, £2.1 billion, a 25% rise on previous year.

23.3 Some aspects of renewable energy development in the UK

23.3.1 Energy-mix and perceived renewable energy costs (2004–10)

Chapter 5, by Adrian Wilson, provides the reader with an overview of renewable energy developments. Some additional perspectives and experiences will be included in the present chapter. First, it is appropriate to provide some brief 'introductory observations' regarding certain UK energy-mix aspects and related issues, considered and widely debated over the period 2004–10:

- *The UK energy mix in 2006* was approximately 35% supplied by coal-fired power stations, 38% from gas turbines, 22% from nuclear and 3% (approx) from renewable sources including hydroelectric. (From a total energy generation capacity of 75 GW (gigawatt) the selling price of this power was approximately £20 billion at that time.)
- UK government had a commitment for renewable energy sources to contribute 10% of UK generation by 2010, rising to double that figure, that is to 20% by 2020. This provided an opportunity for the 20% energy gap to be filled by suppliers of renewable/alternative energy sources, representing an income to the 'successful' suppliers > £4 billion/year by selling electrical power (at 2006 prices).
- During this period (up to 2010), there was much debate regarding provisional costs associated with various renewable technologies: wind, solar, wave, others – against concerns, fears, etc. regarding the continuing availability of gas/oil supplies worldwide. A strategic question – how would renewable energy generation compare on cost with fossil fuel generation? We will now consider some early estimates.

In March 2004, the RAE London estimated:

Gas-fired generation cost	2.3p per kWh (2p)
Nuclear generation cost	2.3p per kWh
On-shore wind generation	3.7p per kWh (3p)
Off-shore wind generation	5.5p per kWh (5p)
Embryonic wave/tidal generation	(7p)
Solar generation	(24p)

For comparison, figures within brackets are costs given by the Independent Carbon Trust in 2003. The reader could follow up this aspect and research representative costs, over time, associated with all energy sources if he/she so wishes. However, as A. Mc Crone in his *The Sunday Times* article of 29/01/2006 stated:

- 'It is difficult to cover/compare all the variables, for example, wind-power generation is intermittent, adequate wind does not blow all the time, so additional back-up generation is required'.
- On the other hand, it should be noted that the market price of gas rose sharply, since 2003 from 3$ a unit to 8.7$ via a peak of 15$. Despite this, it is recognised that if a 'UK Generator' still wanted to burn more gas (for generation) he may be forced to purchase extra Carbon Emission Certificates (We will return to this aspect later – such certificates have been widely used in Europe and elsewhere.)

In 2006, some experts considered that the world was on the brink of a big switch from gas to coal as the preferred fuel for power stations, predicting 40% of orders for turbines for the next decade would be coal-fired units – with the share of gas-fired plants reducing to less than 1/3. Naturally, the rapidly rising price of gas and the encouraging trends towards renewable subsidies were to have a strategic impact on final outcomes and areas of expansion in a direction other than gas!

Returning briefly to Lea (Table 23.2), the reader will recall that between 2006 and 2012 half of all new generating capacity in the UK has been in renewables; of the six big UK energy companies, the German giant RWE npower, the Spanish owned Scottish Power and SSE, the quoted company behind Southern Electric, account for more than a third of all renewable generation; and the corresponding figures for Centrica, and EDF Energy, have only 2.3% and 3.2% respectively.

Another option considered was the mixing of 'biomass fuels' with coal. This process is sometimes referred to as 'co-firing'. Biomass 'wood' or exotic bamboo-like plants in the UK have been used, and can be mixed with coal and fired in existing giant coal-fired stations (e.g. Drax, Yorkshire, UK). The UK government has encouraged novel energy sources with extremely attractive subsidies.

(This could be considered as a big topic on its own right, but meantime a prime strategic question is 'how long the subsidy is to be available?' A significant 'subsidy timeframe' is essential for this to be a viable option!)

By 2010, the landscape and seascape had changed greatly with (i) many onshore wind farms developed, and more recently (ii) offshore farms have been planned around Britain and abroad (see Bowden, *The Sunday Times*, 31/10/10). Renewable energy sources such as sun and wind represent intermittent sources, being weather dependent. The amount of energy produced is influenced by, and dictated by, the weather. This can result in large variations or 'swings' in the amount of electricity entering the grid. Consequently, it is difficult to predict when these swings will occur. Unexpected 'surges' or sudden 'lulls' could lead to power interruptions. As Bowden observed, unless a solution is found, temporary shortages of power will occur and will become commonplace as renewable energy – probably wind energy – makes up an ever-increasing proportion of UK power generation.

Note: On 30/5/2010, the National Grid (NG) paid Scottish Power £13K to *shut down two of its wind farms* to assess how the grid would react/cope with such a new situation. Bowden also reported on research work currently being carried out to offer a solution to this situation, or dilemma:

> 'A company is piloting a new storage technology in the UK that could store electricity during periods when a surplus is being produced, then release it during a "lull" period. The pilot-plant uses energy from a "biomass plant" to freeze air from a temp of $-196\,°C$ so transferring air from a gas to a liquid – it shrinks to 0.14% volume. The liquid is stored until the grid needs the power again – it is then heated up expanding back to gas 700 times larger volume'. 'The sudden expansion of air provides a force to drive a turbine to produce energy in electrical form, which would feed back into the grid!' (A suggested efficiency figure of about 50% has been mentioned.)

Numerous diverse storage technology concepts have been discussed in the literature in recent years. Many informed sources still consider that additional new modern hydro or biomass plants, to support storage needs for wind energy systems, could offer to be the best solution for the UK.

Prediction of installed wind capacity by 2020

- Although renewable energy technologies vary widely in their technical and economic maturity, they all share a common feature, namely they produce little or no greenhouse gases. They rely on a virtually inexhaustible supply of natural sources for their 'fuel'.
- Excluding hydroelectric generation, wind generation has experienced the greatest growth among the sources of renewable generation.
- At the end of 2009, wind capacity worldwide was approximately 159,213 MW with 38,312 MW being added in 2009. This is the highest volume achieved in a single year.

- The installed wind capacity is more than doubling every third year. (*Source*: CIGRE/*Electra*. 2010;**253**(December))
- If this trend continues, the prediction of a global wind energy capacity of 1,900,000 MW is possible by the year 2020! ('*Putting the wind in their sails*', M. Hanlon and J. Leake, *The Sunday Times*, 29/7/12).

As a subset of a bigger article, these reporters commented accurately that developing new sources of energy is expensive and in early wind, wave and tidal sources cannot compete with coal gas and other established sources of power. For example, price per megawatt hour of generating electricity, before subsidies, was as follows:

Nuclear	£74	Hanlon/Leake comment that wave/tidal prices are unknown, for now, but should fall below offshore wind by 2020 (*The Sunday Times*, 29/7/12)
Gas	£77	
Coal	£95	
Onshore wind	£101	
Hydroelectric	£101–£128	
Offshore wind	£126–£146	

The previous Labour government recognised this and introduced the 'renewables' obligation. As Hanlon/Leake point out this work by 'obliging' electricity suppliers to source some of their energy from low-carbon renewable sources, such as wind, biomass, etc. The suppliers pay the generators a premium, called a renewables obligation certificate (ROC), on top of the going rate for electricity – this can be called a subsidy. The extra costs of this are then passed on to the consumers in the form of higher bills. (As Hanlon/Leake points out this is in addition to the feed-in tariff scheme, under which private householders are paid for solar, wind or other energy they generate and feed into the NG.) Conventional power sources such as gas and coal plus nuclear receive no subsidies under the renewables obligation.

- Onshore wind receives 1 ROC worth about £38 per megawatt hour.
- Offshore wind gets 2 ROC.
- Tide and wave power are paid 5 ROC, providing a substantial boost for the fledgling marine energy industry.

Note: On 30/7/2012 a new marine energy park reported as the largest in the world was officially opened. It is called the Pentland Firth and Orkney Marine Energy Park. UK funding of £20 million is being allocated to marine energy with a generous subsidy (5 ROC as stated above) to encourage development in tidal or wave sea power.

23.3.2 Renewable energy vs landscape calculations (The Sunday Times *article of 20/11/11*)

In a brief *The Sunday Times* article (of 20/11/11), Jonathan Leake reported on 'some thought-provoking' energy calculations, by Professor David MacKay, then chief science advisor at the Department of Energy and Climate Change (DECC), as revealed to the Royal Society previous week, when discussing the future of solar power. Professor MacKay presented the following revealing illustration headed *March of the Turbines* and provided the following details as to how relying on renewable sources would transform landscapes:

March of the turbines (by D. MacKay)

Wind
    1 GW power station = 1,400 onshore wind turbines[a]

Solar[b]
    Land that would need to be covered by solar panels to supply all Europe's
       power = 140,000 square miles [1.5 × area of UK]

Biofuels
    Area needed to grow crops to fuel a car for 1 year = 158 acres

[*Source*: The Sunday Times, 20/11/2011, Jonathan Leake]

---

[a]**Britain** already has 3,400 onshore and offshore wind turbines which in 2010 only supplied about 2.5% of the nation's power.

[b]*Note*: Gujarat Solar Park, Pakistan, spans remote 3,000 acres, was recently switched on (214 MW output). Eventual rating of 1,000 MW was predicted (*The Times*, 23/04/12).

So, the message is simple and clear. The above figures provide a stark indication that reliance on renewable sources can have a huge impact and would radically transform the landscapes of the UK and elsewhere.

23.3.3 UK energy storage: call to build a series of dams to store power from wind turbines (after D. MacKay) (Jonathan Leake, The Sunday Times, 18/3/12)

In another article, with main facts extensively reproduced below, Jonathan Leake, reports on a suggestion by Professor David MacKay, the chief scientist at DECC, to build a series of dams in the UK to store power from wind turbines:

- This could involve a network of huge hydroelectric schemes designed to store green energy from wind farms when power is plentiful and release it when the wind fails.
- The scheme, which would see dams built in mountainous regions of Wales and Scotland, is being proposed by David MacKay. It is reported that he spoke at the Institution of Mechanical Engineers (IME) last week. He suggested that several such 'pumped storage' systems could be built around Snowdonia in Wales and up to 13 in Scotland. Most of the energy generated would be supplied to England. Jonathan Leake comments that 'the idea will infuriate environmentalists'. MacKay has suggested some of the schemes could be built in national parks such as Loch Lomond.

Note: At this point, the writer draws the reader's attention to a DTI Report entitled *Review of electricity energy storage technologies and systems and of their potential for the UK* (see Reference 34, Chapter 5).

- 'The proposal's attraction is that it would involve well-proven technology. Britain already has four pumped storage systems, of which Dinorwig, in Snowdonia, is perhaps the best known. It works by pumping 247 mft^3 of water from one reservoir into a second, 1,600 feet higher up. When demand surges, this is released to generate hydroelectric power'.

Although this meeting between MacKay and IME was private, Leake goes on to comment that Professor MacKay has also set out his ideas in print.

He said, 'the new schemes should be much bigger than Dinorwig'. 'We are interested in making much bigger storage systems ....'; 'We have to imagine creating roughly 12 new sites, each storing 100 gigawatt hours – roughly 10 times the energy stored in Dinorwig'.

Leake posed the question, 'why might Britain need so many new dams?'

- The answer lies partly in the unreliability of wind and also in the scale of Britain's commitment to green energy. The government has said 'that by 2030 Britain should have about 8,000 wind turbines with a maximum power output of about 10 gigawatt hours – roughly an eighth of what the country currently needs at any one time'.
- As Leake states, 'the problem is that if wind becomes such a big part of Britain's power supply, it will have to be backed up for times when winds fail. One answer would be to keep lots of fossil fuel power stations on standby. However, a much greener and cheaper alternative would be to store energy from low-carbon sources, such as wind or nuclear power, when they are producing a surplus, in pumped storage systems'.
- In the latest edition of his book *Sustainable Energy Without the Hot Air*, MacKay says, 'Certainly we could build several more sites like Dinorwig alone'.
- In Scotland he suggests that a huge scheme could be built using Loch Sloy and Loch Lomond, which are already linked by a hydroelectric power system. Apparently, this would involve raising Loch Sloy's existing dam by 130 ft. He suggests the mountains could easily provide 10 sites for similar projects.

MacKay also proposes even more ambitious schemes. One would see dams constructed across the mouths of 'hanging valleys' around Britain's sea cliffs. These could then be filled with sea water. Another would see a huge chamber constructed three-quarters of a mile beneath London, with water generating power as it pours in from a ground-level reservoir then pumped out when power is in surplus.

Tim Fox, head of energy and environment at IME, said MacKay was forcing Ed. Davey, the new energy secretary, to confront uncomfortable issues surrounding the green energy agenda. 'There has been a step change in the DECC understanding of engineering since MacKay arrived', he said.

Craig Dyke, strategy development manager at NG, said the demand for power would increase sharply in 2030 under government plans to replace most petrol and diesel cars with electric cars by 2025, and to heat buildings with low-carbon electricity rather than gas. 'We are going to see the demand for power varying a lot more than it does now, so energy storage systems will be important', he said. (*Note*: The idea that parts of Britain's remaining countryside could be sacrificed to UK's power industry will certainly anger some environmentalists.)

The following text provides a brief overview of the operating concept of this 'MacKay plan', together with possible new site locations. (See the original article for full information.)

Dam the countryside?

Professor David MacKay, the chief scientist at DECC, has said Britain must find a way to store energy as a buffer against the intermittency of wind farms. *Giant hydro power for England electric schemes across Wales and Scotland could provide backup power for England when the wind farms are 'becalmed'.*

---

1. When low-carbon energy is in surplus it is used to pump water into a high-level reservoir.
2. When wind turbines cannot operate, or when demand rises, water is released from the upper reservoir into giant pipes.
3. Turbines in the pipes generate hydroelectric power to compensate for decreases in wind farm output (and the water level in lower reservoir rises).

Britain already has four 'pumped storage' schemes, plus one under construction. Possible new sites:

- *Scotland*: Loch Awe, Blackwater, Loch Ericht, Loch Katrine, Loch Leven, Loch Morar, Loch Maree, Loch Ness, Loch Rannoch, Loch Shiel, Loch Shin, Loch Sloy and Loch Tay
- *Wales*: Bowydd, Croesor.
  (*Source*: Mainly *The Sunday Times*, 18/03/12, Jonathan Leake)

Note: As Tim Fox of IME commented (see above), 'this latest input from Professor MacKay certainly forces Ed. Davey, the new UK Energy Secretary to confront several uncomfortable strategic issues urgently, relating to the green energy agenda'.

Although there is still time for the UK to produce a robust, effective and coordinated system with effective energy-storage capabilities to support UK wind farms, for years to come, it appears to this writer that progress in new commercial scale and storage concepts has been minimal in recent years.

The reader should review recent published IET, IEEE and CIGRE work in the area of energy storage (e.g. CIGRE Technical Brochure(s)); the DTI, Reference 34 of Wilson, Chapter 5 also provides a valuable UK viewpoint.[1]

23.3.4 Press articles: Some very public energy discussions (a few commentaries on articles by D. Fortson et al., The Times, 2010–12)

Chapter 5 provides the reader with a brief overview of renewable energy developments. As already indicated, much useful additional technical information can be accessed from a variety of sources:

(a) numerous quality papers on this theme available in refereed publications of IET, IEEE, CIGRE, etc., or from international conference events, manufacturer or utility websites
(b) the many excellent energy-related 'snippets' or articles which appear very regularly in the UK quality press, for example in *The Times* or *The Sunday Times*, *The Guardian*, etc.

These can 'empower' and provide readers with rich and varied resource materials on energy-related themes from all over the world (e.g. technical/ environmental/political/legal/economic), each offering widely differing viewpoints by senior people from within the industry, professions or government, together with regular reports/comments and strategic features on latest trends, developments and concerns, etc.! To illustrate this, consider three energy articles in (a), (b) and (c) below:

(a) '*£32 bn for a light switch*'[2] *(The Sunday Times* (10/10/10, p. 13) by D. Fortson)

This lead-in was used by Fortson in his article, when discussing the fact that the NG will have to be 'rebuilt' to handle all the renewable energy scheduled to 'come online'. Some significant cost comparison figures were quoted:

---

[2] Historically, it should be noted that in 1933 when energy was coming to the masses, a Special Edition of *The Times* was sponsored by The British Energy Development Association to explain the benefits of the technology, including the wizard on the wall – a light switch! (*The Sunday Times*, 10/10/10). In this edition, in an article by Fortson Nick Winser of National Grid is quoted as predicting, 'by 2020 "micro-generation"' – from rooftop solar panels, fuel cell boilers and the like – 'could meet up to 10% of national needs', up from virtually nothing today. In that sense, 'we are going back to the future', but the power being produced would obviously be much cleaner.

Money spent to build first network in 1935	£26 million
Investment required to 'update it', over next 20 years (i.e. to improve the network to handle offshore wind, solar, tidal power, etc., and the new generation of nuclear reactors that will replace the country's ageing fossil-fuel plants)	£32 billion
Peak demand in winter in 1935	4 GW
Peak demand in winter now	60 GW
Old Battersea Power Station, 1935 (initial capacity – the largest in Europe)	105 GW
Modern new nuclear plants	3,500 MW
1935, Extent of transmission lines, km	4,800 km
2010, Extent of transmission lines, km	7,000 km

Over the next decade, 'plant' accounting for 25% of UK 78 GW of generating capacity will close, leaving about 60 GW – equal to peak *'tea time demand'* on a cold winter's day [much has already changed by 2013].

A 2013 update: Tim Webb, 20/2/2013 (*The Times*), reports that 'next month alone, about 10% of the UK's total energy generation capacity will shut down as old coal- and oil-fired power plants will close under a European Union environmental directive'. 'As a result within three years the country's spare cushion of power generation will slump from 14% to less than 5%; that is uncomfortably tight', said Mr A. Buchanan, Ofgem's chief executive who is due to step down in June 2013. He went on to say, *'To keep the lights on*, the proportion of Britain's electricity generated by gas plants will have to double from 30% today to about 60%'. Webb goes on to report that the Ofgem chief is also speaking of the 'horror' of trying to import gas at a time when the world faces a global supply crunch: delays in completing giant liquefied natural gas projects in Australia and the America's inability to export its huge shale gas anytime soon are expected to drive up global gas prices in the short term. Regarding the horror element for the UK, Mr Buchanan said that it's happening 'just at the time when it looks like we have to turn towards gas-fired power stations to see us through this period'. Webb adds that 'the government admitted last night that Britain may face a "looming energy gap" but it insisted that its electricity market reforms which will fix subsidies for new wind farms, nuclear reactors and gas plants would solve the problem'.

(b) *'Ditching wind farms will save £34 bn'* peak demand in winter *(The Sunday Times,* 06/11/11, by Danny Fortson)

- Reviewing a recent 'controversial' unpublished report study by KPMG – it is reported that Britain can reach the 2020 target on reduced pollution by a third less cost than predicted, representing a potential saving of £34 billion.
- To do so, it is suggested that the proportion of wind power envisaged in the current plan would need to be slashed and the energy shortfall made up by new gas-powered stations and nuclear reactors (cost of building an 800 MW plant capable of powering 800,000 homes by gas-fired power

station – £400 m; a nuclear plant would notionally cost £2.4 billion *(although some other very recent articles have quoted cost of certain new third-generation reactors as £4 billion and even >£7 billion)*, while an offshore wind farm could cost >£2.4 billion).

A note in passing: An earlier article by Fortson, 'Carbon Plan could double householder's energy bills' (*The Sunday Times*, 13/3/11) starts with, 'A government scheme to rid Britain of Carbon Emissions will send household bills soaring, hasten the closure of old power-stations and "accelerate" the introduction of electric cars' – still a worrying prospect for energy consumers and commuters in the UK. However, recent projects in the North East of England (reported by Ruth Lognonne, *The Journal*, 4/4/2013) aim to develop a technology that will help electric cars stay powered for longer. 'So-called *range anxiety*, i.e. the fear of running out of power before you have reached your destination, means that some motorists are still to be convinced about converting to an electric vehicle'.

The article in (b) by Fortson (of 06/11/11) created quite a stir, as explained by Fortson in his follow-up article in *The Sunday Times*, of 04/03/12:

(c) *'Blown away: we can ditch solar and wind power, and still meet our targets for emissions, an expert report concluded. So why was it buried, Danny Fortson asks?'* (*The Sunday Times*, 04/03/12)

Only a brief summary of this 'follow-up Fortson article' is necessary here. Fortson reports:

- 'John Griffith-Jones, the chairman of KPMG, was apparently bombarded by angry e-mails and phone calls (these lasted for weeks) having started in November 2011, when *The Sunday Times* (of 06/11/11) ran the story based on the above unpublished KPMG Report' (see (b)) outlined above.
- Fortson claims that his earlier article (see (b)) of 06/11/11, based on this unpublished report, had explosive consequences for the UK Government's energy policy.

Note: There are many 'nuances' and the reader is encouraged to carefully examine the various stances of 'pro' and 'anti' wind-energy protagonists reported in both the articles in (b) and (c) before making up his/her mind.

AF Consult was the firm KPMG commissioned to do the work, and Fortson reports in his article (see (c)) that *AF Consult* were due to publish a report on 07/3/12 in the interests of presenting an 'independent' perspective in a debate 'led by groups with vested interests!' One feels, the constant 'drip-drip' of quotes from senior spokespersons from either 'camp' represent a strong 'Nudge effect' on the UK public!

Unfortunately, this writer has not yet seen this full report, so can't yet give a strong view on economic case scenarios presented 'by the various protagonists' but, 'cost uncertainty' regarding energy oil, gas and coal over the next two or three decades is a big worry and one would wish to see realistic/robust, full life-cycle cost comparisons.

Fortunately, assuming wind generators to be reliable, and have long service life, financial estimation as to ongoing life-cycle costs of wind farms should be 'a positive' as wind farms should not attract ongoing fuel costs providing the wind remains free to all!

It is enough to reproduce data here from Fortson's article[3] (*The Sunday Times*, 04/3/12):

	Unit cost estimate	Total cost estimate
If the UK goes for 'CO_2 reduction target alone'	7.2 p/kWh	£960 billion
Whereas, if in future, 'UK sticks to renewable energy goals as well'	8.4 p/kWh	£1,100 billion

In other words, from these projected estimates, Fortson argues:

- It should be possible for the UK to 'hit the 2050 projected targets' without renewable energy and save £150 billion in the process.
- Referring to the 2020 power 'struggle' Fortson observes:
  – 'For energy executives the year 2020 looms large.
  – That is the year by which Britain must 'hit' key targets.
  – UK needs to cut pollution by a third over 1990 levels while producing four times as much power from renewable resources as we do today.
  – So eight years from D-Day, how are we going? (In terms of carbon dioxide emissions, we should be fine.)
  – A lot of the worst offenders – coal-fired power stations – will be mothballed by then.
  – Even if they are not succeeded by wind farms or solar panels, their replacements will be a lot cleaner.
  – One big unknown is how much electricity we will need by then.
  – Since 2001, British energy consumption has fallen by 15%. (*The current recession is partly to blame.*)
  – *The manufacturing industry, a big energy consumer, has shrunk.*
  – Demand would 'spike' again, however, if we end up 'making the move' towards electric cars.
  – And *our renewables target?*
  – Today about 7% of our power is green.
  – If we are to honour our 2020 'target', or goal, at least 30% of our power will have to be green' (generated by the sun, wind, sea).

23.3.4.1 Wind energy: extensive UK involvement by a Danish company

The Danish company Dong Energy, one of the biggest energy producers, has recently been rocked by an alleged pay scandal, which *The Sunday Times* newspaper (on 8/04/12, by Bojan Pancevski) states, 'could endanger its plans in Britain':

---

[3] Article also provides two illustrations showing energy mix (i) if Britain goes for the CO_2 reduction target alone and (ii) in future, if Britain sticks to renewable energy goals as well.

- 'Last week, the chief executive Anders *Eldrup C.E.* of Dong Energy was forced to resign after it emerged that he had given unauthorised pay packages worth millions to a group of top executives – including the head of its lucrative British wind energy arm'.
- There are some initial concerns that 'the dismissal of Eldrup, who was the driving force behind Dong's push into Britain, raises questions over whether the Company will pull back', Pancevski reports.

In this section, we merely note the above recent news and move on to deal only with outlining for the reader the technical projects this company is involved with in the UK and note any aspects of particular interest. Dong Energy is the Danish energy company behind the huge London Array wind-farm complex located in the Thames Estuary, UK. This is due to be completed in 2013.

The company has already invested more than £3.9 billion in the UK since 2004. It currently employs about 200 people in Britain and the number is expected to rise. 'Dong Energy also owns a gas fired power station near Newport in Wales'.

Apparently, 'the Chief executive of the company, Anders Eldrup had planned to build up to 20 huge offshore farms around Europe' (*The Sunday Times* article states).

Dong Energy is a backer of the London Array, the world's largest offshore wind farm. Indeed, it has been outstanding in taking advantage of the UK's wind subsidy scheme, deemed to be one of the most generous in the world. 'It has eight wind-farm projects (see below) either in operation or being built', *The Sunday Times* reports:

- Walney 1 & 2 (10 miles west of Barrow-in-Furness, Cumbria)
- Barrow
- Burbo Bank (south of Barrow-in-Furness, Cumbria)
- Lincolnshire (off Skegness)
- Gunsfleet 1 & 2 (north of London Array)
- London Array (175 turbines, a £2 billion wind farm; 12 miles off the Kent coast)

London Array: An offshore area of 100 km^2, with 175 turbines capable of producing enough power for about 480,000 homes a year in Kent – saving 925,000 tonnes of CO_2 year.

When first phase is completed in December 2012, it will be the largest in the world and a showcase of the UK government's low-carbon revolution. Dong Energy has invested more than £3.9 billion in the UK since 2004. It has previously been reported that Dong Energy is main backer of the London Array and also owns stakes in the Lincolnshire farm nearing completion – off Skegness and the Walney farm, west of Barrow in Furness, Cumbria.

Pancevski goes on to state:

'Once the wind turbines are turned on, their owners are paid millions on top of the normal electricity price. These subsidies are one reason, along with the rising wholesale price of gas, *for the soaring household energy bills*', in Britain.

Readers should watch the technical, economic and in-service performance of these wind farms, and others installed across the country nationally with interest, taking due note of the contributions to this topic already made by CIGRE, IET, IEEE and other professional bodies.

Note: They may also have read in the press that planning applications for another 190+ wind farms are pending in north east of England; reports in *The Journal Newspaper* Ref., p. 6, article, 09/04/12, 'Wind farms: coming soon to a field near you'.

Tim Webb, of *The Times* (20/7/12), moves this story on a pace in his article, 'Wind farm deal raises hopes for new plant'. He reports that provisional plans to build a turbine factory in Hull and create more than 800 jobs have moved closer after *an estimated 2.5 billion euros contract to supply North Sea wind farms had been agreed.* Siemens, the German Engineering group behind the project, has intimated that it had agreed to provide 300 of the largest giant offshore turbines to Dong Energy Wind Farm Developer. Webb reports, 'The company said that it planned to build the *6 MW turbines,* powerful enough to supply 60,000 homes with electricity, when the wind blows, at plants in Denmark and Hull'.

The acting chief executive of Dong Energy, C.K. Thomson, told *The Times* that 'offshore wind farms will need smaller subsidies in future'. 'We believe that the costs of energy from offshore wind should go down. We do not believe that in the long run, the current level of subsidy can be maintained. Offshore wind needs to become competitive with other forms of generation'.

Webb also reminds his readers that (i) Gamesa, Siemens, General Electric and Mitsubishi have drawn up similar plans for new wind turbine plants in Britain but most are dependent on receiving sufficient orders and (ii) previous month, Vesta another Danish manufacturer, had abandoned plans to build a turbine factory in Kent in response to European governments reducing their renewable subsidies. For the community on Humberside, it is to be hoped that the proposed 'turbine-build plant' for the Hull region does not suffer the same fate as the Kent initiative (as many fear) because of ongoing uncertainties regarding subsidies and potential orders for offshore wind farms.

23.4 UK government's recent wind of change

When the writer was in the early stages of setting out the 'scope' of this chapter, late in 2011, Chris Huhme, the 'then' UK energy secretary, published a report in December 2011 calling for a vast escalation of wind farms in the UK. Huhme – an extremely enthusiastic supporter – was a very strong proponent of wind turbines! He proposed the installation of up to 32,000 new wind turbines in Britain, 10,000 of these to be additional onshore turbines. Huhme resigned in February 2012, soon after the publication of this report. During this period, a significant level of concern had been expressed by public and politicians alike regarding the direction the UK energy policy planning was taking at this time.

Two months later, in April 2012, the public were surprised to learn of major strategic changes regarding UK energy policy. An interview with Greg Barker, the UK Climate Change Minister, was reported in *The Sunday Times* of 15/4/12, by Isabel Oakeshott, Political Editor, under the headlines 'Energy minister vows: No more wind farms' (p. 1) and 'Wind farm expansion axed' (p. 2), and the reader learnt of significant new developments on this topic.

Oakeshott reports that according to Barker, *the UK do not need more onshore wind farms*, in what will be seen as a distinct government 'shift,' or change, in strategy. *The Climate Change Minister has declared there will be no significant expansion in the number of onshore wind turbines beyond those already in the pipeline.*

As indicated above, this move comes just five months after his department unveiled plans for, new or in the planning stage, 10,000 extra onshore turbines. This had prompted an outcry from more than 100 Conservative MPs who had written to the prime minister expressing strong concerns describing the onshore wind farms as inefficient and attacking the scale of government subsidies to the industry.

In this later interview with *The Sunday Times*, Greg Barker claimed that *the DECC had adopted an 'unbalanced' approach to wind farms in the past and must now look to other options.*

Isabel Oakeshott reports that Greg Barker said:

Far from wanting thousands more, actually for most of the wind we need ... they are either built, being developed or in planning. *The notion that that there is spectre of a new wave of wind farms is somewhat exaggerated.*

Note: The report of December 2011, referred to by Huhme, had called for 32,000 new wind turbines, 10,000 of which could be onshore and several hundred offshore, with many more being planned for offshore.

As *The Sunday Times* and many of the public have commented, the plan would have transformed Britain's landscapes, alarming local MPs and the public. Oakeshott continues by saying Huhme's resignation in February 2012 appears to have provided an opportunity for 'a retreat' and the introduction of a revised strategy.

- Barker dismissed the 10,000 figure, saying, 'it's about being balanced and sensible. We inherited a policy from the last government which was unbalanced in favour of onshore wind'.
- Barker wants to 'focus on offshore farms' and admitted that some onshore farms' locations had been misguided. There have been some installations in insensitive or unsuitable locations – too close to houses, or in an area of outstanding beauty.

The Sunday Times reports that previous week Barker had also announced new details of the government's 'green deal', a scheme to enable householders to invest in energy-efficient installations such *as double glazing and under-floor heating without 'having to pay upfront'.* He went on to say:

- 'The economic "downturn" had forced the government to change its approach to "green issues" to deliver better value for money'.
- 'There is a requirement to rethink the economics of green. *We have to have a more nuanced and sophisticated policy*'.
- Basically, that means reducing costs quicker, looking to commercialise sooner and thinking more carefully about the use of public subsidy.

810 High-voltage engineering and testing

23.5 Nuclear power plants: recent events and future prospects

Background notes: Initially, the main intention of this brief section was to look at the nuclear power prospects mainly in the UK, post the Fukishima nuclear incident in Japan in March 2011, following on from the tragic Tsunami. However, because of the vast and diverse UK media coverage following this event, and over the next 15 months, covering not only nuclear energy but additionally the unprecedented debates on energy costs, the future direction of energy development in the UK, environmental aspects and the very high domestic fuel bills in Britain, much valuable information emerged via the UK press with additional information being made available via high-level government sources. During this research, the following aspects of nuclear power accident/subsequent developments and aggressive arms race implications were identified and will be touched on first.

23.5.1 Fukushima nuclear accident: short-term impact on global developments

Major natural disasters such as the earthquake (9.1 magnitude) in Japan, 2/3/2011, are now reported 'almost immediately' worldwide, via TV and other media sources, as the horror unfolds. To date, this was the most powerful earthquake ever to strike Japan. It erupted off this island's north-east coast at 2.46 pm, triggering a tsunami (causing an estimated >16,000 deaths and destroying >115,000 buildings).

- After the devastation, the cost of the rebuilding programme was estimated to be in the region of £150 billion. It soon emerged that the tsunami had damaged the emergency generators that provided the power for the cooling systems of the 40-year-old Fukushima nuclear plant. As a consequence, within a few hours of the earthquake/tsunami, some nuclear fuel had melted, severely damaging the plant and producing a major nuclear accident.

    Clearly, energy planners for nuclear plant should always thoroughly evaluate the preferred location for the plant and all of the risks throughout the lifetime of the plant, including severe nuclear accident(s) and in the case of Japan, potential earthquake/tsunami situations.

- Millions of TV viewers worldwide watched as the destructive powers of the Tsunami unfolded. The initial reaction of leading economists was that the earthquake in Japan threatened the global recovery because of the profound knock-on effects around the world as supply chains were seriously disrupted.

    The Times/The Sunday Times and other quality media sources produced numerous regular technical/economic reports and features on this topic for several months after the Fukushima nuclear accident, and a careful review of all of these will provide the reader with a thorough appreciation of this event, its global impact, etc.

- This includes a late feature 'in graphical detail' entitled 'Eight days that shook the world', by *The Times*, 19/3/11. Valuable strategic 'world news listings' and energy-related articles can also be found monthly in the IET *Engineering and Technology Magazine* (*E&T*) (www.EandTmagazine.com) and elsewhere. In this manner, the reader has a choice of quickly looking through such sources,

or carrying out a more thorough review on this, or other particular themes within the energy sector.
- Additionally, as and if appropriate, his/her supplementary review could be broadened to include major resource materials, that is (i) IET, IEEE, CIGRE or (ii) other appropriate international media or technical publications. (Subscribers to *The Times* and *The Sunday Times* can *access* further resource materials.)

Historically, levels of radiation at several nuclear incidents in recent years can be 'designated' in range 1–7:

1. Anomaly	
2. Incident	Attucha, Argentina, 2005
3. Serious incident	Sellafield, England, 2005
4. Accident with local consequences	Tokaimura, Japan, 1999
5. Accident with wider consequences	Three Mile Island, USA, 1979
6. Serious accident	Kryshtym, Soviet Union, 1957
7. Major accidents	Chernobyl, Soviet Union, 1986
	Fukushima, Japan, 2011

EDF, the French company which also runs UK nuclear plants, ordered immediate safety checks on all British nuclear plants and procedures as a result of the failures at the Fukushima accident in Japan. The company runs more nuclear plants (e.g. in France and the UK) than any other operator. At this point of time, *it was planning at least four new reactors in Somerset and Suffolk, UK*, at a cost of £20 billion. Germany's two big utilities RWE and Eon also have plans to build new nuclear plants in the UK.[4]

Six months after the explosions and release of radiation at the Fukushima plant in Japan, an excellent BBC *Horizon* TV programme, with Professor Jim Al-Khalili, investigated the safety of nuclear power. This programme, dealing mainly with the Fukushima and Chernobyl nuclear accidents, was very reassuring (see BBC 2 viewing guide, 3733, transmitted 9 pm, 14/9/11) from a 'human-life perspective' and illustrated, by comparison, just how much progress had been made 'medically' since the Chernobyl accident in 1986. As far as this writer is aware, no one has died or, according to the government, been seriously injured at the 'Fukushima nuclear' accident. The evacuation of the area and the effective use of iodine tablets were considered adequate to contain health effects on the wider population.

23.5.2 Future nuclear developments

This TV programme also touched on the point that thorium nuclear power is again being 'touted' as the safe green alternative to uranium nuclear power. Despite significant prototype development in USA at Oak Ridge Laboratories, thorium

---

[4] Germany is, and has been, very strong in the energy sector, designing, developing and building nuclear plant and a variety of renewable energy technologies, for example wind, solar, as well as traditional and modern 'cleaner' fossil-fuelled stations.

plant designs were 'dropped in the mid-1970s' in favour of uranium plants. Some environmental experts have claimed that thorium is more efficient, is cheaper and produces less waste. It has been suggested that thorium dissolved in a liquid fluoride salt should be considered again by US scientists as the fuel for a new type of safer, more efficient nuclear reactor.

(An international conference of thorium advocates was scheduled for New York, USA, in October 2011.)

The reader may recall in 2010, seeing in the press an article stating that the French company Areva, the world's largest nuclear energy company, is developing a new type of nuclear reactor that could permanently 'destroy' atomic waste. The CEO of Areva informed *The Times* that her company was developing a technology to burn up 'actinides' – highly radioactive uranium isotopes that are the waste products of nuclear fission inside a reactor. The technology could clearly be critical in achieving a much greater global support for nuclear energy. (Obviously this could also cut carbon dioxide emissions.)

'In the future, we will be able to destroy the *actinides* by making them disappear in a special reactor', Anne Lauvergeon, the company's CEO, told R. Pagnamenta Spence of *The Times* (22/3/10).

Swadesh Mahajan, Institute of Fusion Studies (IFS), University of Texas in Austin, USA, commented that his group is also working in a similar area. 'It is felt this invention could in time hugely reduce the need for geological repositories for waste: "we want to make nuclear energy as socially and environmentally acceptable as possible". The volume and toxicity of waste material could be cut by 99% for civic nuclear reactors, it is claimed.

About 440 nuclear plants were operating in 31 countries worldwide with a collective generating capacity of 370 GW of electrical power, or 15% of the global total. However, 'electricity produced from nuclear fission also produces 12,000 tonnes of high-level radioactive waste per year, including plutonium that can be used to manufacture weapons'.

Ms Lauvergeon said that 'the volume of high level nuclear waste produced by all of France's 58 reactors over the past 40 years could fit into one Olympic size swimming pool. Of course it would be better to have nothing but this is fully managed and we have to view this issue in a balanced way compared to other solutions'. Nuclear power produces 80% of France's electricity, it was reported.

The concept of a hybrid fission–fusion reactor was first developed in the 1950s but limited research was conducted for several decades, it was reported. Meantime, Sellafield high-level nuclear waste plant in Cumbria – a temporary facility – will continue to store Britain's nuclear waste.

Negative impact: Immediately following the Fukushima nuclear accident (post-Tsunami), there were initially some knee-jerk 'negative reaction' statements in the UK media relating to the future for nuclear energy. Two initial examples:

1. Lord Paddy Ashdown (a UK politician) in an interview argued that the 'disaster' at the Fukushima nuclear power plant in Japan has made nuclear energy 'unsellable'.

2. *Angela Merkel*, the German Chancellor, announced a strategic change/switch in the Germany's energy policy from nuclear power to renewable energy.[4]

Despite Japan's troubles following the Fukushima nuclear accident (post-Tsunami):

- 'informed-opinion' still continued to 'strongly support' further development of nuclear energy in the UK and elsewhere. The last reactor built in the UK was at Sizewell, on the Suffolk Coast. This was commissioned in 1995. (*Note*: Typically, the lifespan of a nuclear plant is 40–60 years. A labour force of up to 4,000 is required to build each new plant.)
- Britain's nuclear industry has recently (late 2011) been given a clean bill of health by an official 'six-month-long safety review' set up after the Fukushima accident. 'The outcome of the UK Government's "commissioned review" gave a green light to *faltering plans* to build up to eight nuclear reactors over the next 15 years to avert an energy crisis' (Tim Webb wrote in *The Times*, 12/10/11).

The UK is now committed 'stutteringly' to developing further nuclear plant and future installations have been provisionally approved. Thus, nuclear energy could possibly continue to play a significant strategic role in the UK 'energy-mix' for years to come. But there are several challenges.

(By 2025, new nuclear plant could probably be built at Sellafield, Heysham, Wylfa, Oldbury, Hinkley Point, Bradwell, Sizewell and Hartlepool – if press reports at the time were to be believed!)

Obviously, all new plants will have many modern improved operational and safety features which were absent in the 40-year-old Fukushima plant, and of greater importance, *low seismic exposure conditions exist in the UK*. In passing, it should be noted that a local council in Somerset is demanding an unprecedented share of the annual income from a new nuclear reactor planned to be built locally!

However, there have been some further 'late negative' nuclear developments in the UK.

1. In late 2011, Scottish and Southern Energy announced that it was *pulling out* of the *NuGeneration joint venture* to build a new nuclear plant at a site near Sellafield, West Cumbria, UK.
2. In late 2011, NuGen, the joint venture established between *GDF Suez* of France and *Iberdrola* of Spain (Scottish Power's parent), did not expect to complete a new nuclear plant in the UK before 2023, even if approval is even agreed! This looks rather doubtful after the decision of Scottish and Southern Energy to pull out of the proposed Sellafield, a newly built nuclear plant project site. (Worse UK nuclear news was to follow, and we will return to this theme shortly.)

Very recently, it has been announced that Japan's future nuclear reactors will be limited to a 40-year life, with exceptions to this limit being rare, under planned legislation soon to be submitted to parliament in Japan. Not surprisingly, the draft plans will make it mandatory for utilities to prepare for severe nuclear accidents! In April 2012, Japan gave the go-ahead for two idled nuclear reactors at the Ohi nuclear plant in Fukui to be restarted. They were the first nuclear reactors to do so since the Fukushima accident. Government agreed that the restart was necessary to avoid power shortages in the summer. (Later, on 9/6/12, Japan PM Yasuhiko Noda was to state that the Ohi nuclear reactors, located in western Japan, needed to be restated to protect the economy. Noda urged this action to avoid an 'energy crunch' in the summer.)

Postscript: Since this section was written, ironically, now – some 13 months after the Fukushima accident – a 'snippet' item in *The Times* (p. 32, 26/4/12) by Richard Lloyd Parry states that scientists have said that the oldest nuclear reactor in Japan – the Tsuraga plant, 350 km west of Tokyo – is built on an active fault line. This revelation does nothing to generate confidence in Japan, or indeed worldwide, and is clearly a 'further blow' to the industry still recovering from last year's nuclear accident (2/3/11) at the old Fukushima plant.

Historical note: In February 2012, the existing nuclear power station at Oldbury, Severn Estuary, UK, was shut down permanently, after 44 years. At this time, it was the oldest active nuclear power station in the world. 'Horizon', a joint venture between Germany's utilities RWE and E.ON, plans to build a new station at Oldbury and another at Wylfa on Anglesey, UK (also the home of an earlier nuclear plant). (*Source*: D.O.'Connell, *The Sunday Times*, 11/3/2012.)

There will be great interest to see who the supplier(s) will be for the new breed of nuclear stations in the UK (e.g. Areva, the French nuclear specialist or Westinghouse, now part of Toshiba of Japan!).

However, by the end of March 2012, further dramatic negative developments were widely reported in the UK press, along the lines:

The UK government's plans for nuclear power were now 'in disarray', 'in tatters' or 'leaves Britain's nuclear future hanging by a thread' (30/3/12) or 'in meltdown', when it was announced that RWE and E.ON had 'pulled out of a venture', or 'scrapped plans' to build up to six new nuclear reactors in Britain. Clearly, a coherent, 'new-build nuclear programme', with a robust technical, economic and security infrastructure, is considered to be essential to a successful UK energy strategy.

The companies blamed the decision to pull out of this £15 billion joint venture on (i) the recent *Merkel decision* to phase out nuclear power in Germany (post-Fukushima, 2011), and (ii) they had also come under *political pressure from Berlin*. A spokesman for E.ON, UK (T. Crocker, CEO) is reported (*The Times*, 30/3/2012, T. Webb) as commenting: '*E.ON* has decided to focus its investments in the UK on other strategic projects that will allow us to deliver earlier benefit for customers and our company, rather than the very long-term and large investment new nuclear calls for'. EDF Energy, the large French company, is still anticipated to build new nuclear reactors in the UK in the near future, but estimated completion date has 'gone-out' from 2017 to probably 2019. However, here again, 'political' problems could arise. *In France, the world's most nuclear-dependent nation, about 75% of electricity is supplied via 58 nuclear reactors.* For the 2012 French presidential election, nuclear power was a major political issue, as *Francois Hollande*, the Socialist candidate standing against President Sarkozy, had *suggested a cut by >50%*, the proportion of electricity supplied by atomic power by 2025.

Hollande won the election. This had become a relatively high-profile debate in France in 2012! EDF Energy was expected to make its final investment decision on the Hinkley Point nuclear plant by the end of 2012 but all indicators confirmed that the decision would probably be delayed and negotiations would 'drag on' for several months well into mid-2013 as EDF sought its best possible 'deal'. Despite the delays it is still anticipated that there will be a *'favourable' income guarantee* extracted by EDF Energy from the UK government.

- In yet another *The Times* article, 30/3/12, Tim Webb reports that the Russian state-controlled nuclear energy group, *Rosatom*, is 'considering stepping into the gap' regarding the Horizon venture (as E.ON and RWE are now looking for new owners) and building new nuclear reactors in Britain.
- Webb states, 'any involvement by Rosatom in Britain's faltering nuclear new-build programme would be controversial and extremely political sensitive'. So we must await any new developments!
- One strategic aspect to be settled at outset would be whether Russian nuclear design will comply with stringent UK safety regulations.

These 'negative decisions' clearly show the disadvantages of the present UK energy market which was fully liberated following privatisation, but foreign companies bought up some UK utilities.

Now, UK requires vast sums of money for investment. But, to date, no investors have been found.

Another concern: it is now 'demonstrably obvious' that no longer can the UK honestly consider that it has a robust independent energy policy. The UK now has to respond to the whims/decisions of German and French governments, or others, because of their strategic control of certain of the UK's energy utilities.

Subsidies can't just be thrown at each technical problem! Effective, long-term planning and R&D are essential! Most of these difficulties were readily apparent from the outset, some time ago! They may also apply to other sectors such as the UK gas and water industries! Oh yes the oil/gas industry has been very quiet of late! Hopefully, the present strategic problems will not unduly delay moves in the UK towards full strategic developments relating to state-of-the-art 'smart grids' and robust 'cyber-security' in our energy and other networks. This aspect is discussed again in section 23.13 and Appendices A–E.

23.5.3 Future prospects of 'new-build' nuclear plants overseas

China, India and the oil-rich countries in the Middle East are also committed to future developments in nuclear power in the next few years. Unconfirmed source indicate that China has plans for >10 new nuclear power stations. In contrast, there have been recent press statements, indicative of moves away from nuclear power in mainland Europe, by Germany, Belgium and Switzerland.

Turning the clock back to 2010 for a moment, an alarming report by Hugh Tomlinson (*The Times*, 8/10/10) entitled 'Revealed: the 15 nuclear reactors that may spark an arms race in Middle East' stated: 'The UAE's first South Korean-built reactor is to "come online" in 2017, but uranium enrichment may be ruled out'. *The Times* revealed that:

- United Arab Emirates (UAE) had gone out of its way 'to assuage international concerns' about its intentions, resolving to forgo uranium enrichment and reprocessing before signing its cooperation agreement with the USA. Abu Dhabi has signed or will sign all conventions on safety, security and non-proliferation.
- The Middle East is poised for a dramatic surge of nuclear energy development over the next two decades. *The Times* reported that by 2025, at least 15 new reactors will be built across the region from Egypt and Turkey to the UAE.
- *Iran's centrifuges*, in its uranium enrichment facilities, at *Natanz, are still spinning uranium* in defiance of international conventions. Early in 2012, Iran announced its first domestically produced nuclear fuel rods for its research reactor. It claimed to have added 3,000 'state-of-art centrifuges' to its Natanz facility (making 9,000 in total), tripling Iran's ability to enrich uranium to 20% – just 1% short of the weapon-grade material (21%)!

Fears will mount that the plans now being announced by the Middle East governments mark the first steps in a nuclear arms' race and international disputes across this volatile region. Reactor plans are:

Egypt	2 Reactors, open 2023
Turkey	4 Reactors, open 2019
Kuwait	4 Reactors (site not selected), open early 2018
Iran	1 Reactor, August 2010, 4 major nuclear weapon sites[a]
UAE	4 Reactors, open 2017
Saudi Arabia	4 Reactors expected. Site not selected. Decision expected in 2011–2013
Israel	1[b] Major reactor and weapon site (Israel is believed to have more than 200 nuclear warheads.)
Jordan	1 Reactor, open 2019

[a]If Iran's nuclear programme goes unchecked, Tomlinson reports that there are growing concerns of a surge in proliferation among its neighbours.
[b]*Israel*: Recent natural gas finds. The US Geological Survey has recently estimated that the Levant Basin, which includes the territorial waters of Lebanon, Israel, Syria and Cyprus, could hold some 122 trillion cubic feet (tcf) of gas and 1.7 billion barrels of oil.

The excellent news for Israel is that the largest 'finds' are in its territorial waters. This has been reported as the largest natural gas field discovery in the world for 10 years with some 16 tcf of natural gas, while neighbouring Tamar has some 8.7 tcf. These and other smaller 'finds' give Israel great confidence that it is 'at the dawn' of a new energy bonanza.

For many years, *Israel had often to look far and wide for imports but it now appears that these new natural gas 'finds' have transformed its energy profile*. These new vast resources will certainly change the geopolitical climate in the Middle East. Israel is currently forging links with Cyprus and has major new investments in power plants, etc., enabling the country to move towards strengthening and growing its economy. (*Source*: *The Times*, Advertising Supplement on Israel.)

Postscript 1: Report by Fukushima Accident Independent Investigation Commission (FAIIC) findings

The report and findings of the above commission effectively brings closure on the Fukushima nuclear disaster in Japan. This was reported in the UK press by *The Times*, in an article written by Richard Lloyd Parry, from Tokyo (6/7/12, p. 27), who starts with eye-catching dramatic '*bullet*' headings:

- *National character of obedience blamed by report into Fukushima*
- A very Japanese disaster
- Company failed to take precautions that could have prevented meltdown
- Government regulator and power bosses accused of betraying the country.

The commission findings are particularly damming, states Lloyd Parry in his report:

The Fukushima nuclear disaster was a catastrophe 'made in Japan', a consequence of the nation's conformity and failure to ask awkward questions, according to the devastating report by expert investigators. The report calls for fundamental reforms to the energy industry in Japan, its regulators and Japanese Government to prevent any further disasters.

The Tokyo Electric Power Company (TEPCO) has been blamed 'for wilfully failing to take precautions that could have prevented the meltdown of three reactors after last year's earthquake, Tsunami and nuclear accident, for its incompetent response'.

The report also reproaches the government for failing to organise a swift evacuation of contaminated and for its continuing failure to screen people for long-term radiation damage.

The nuclear regulators were accused of letting themselves be manipulated by the industry they were supposed to oversee.

The TEPCO Fukushima nuclear accident, according to the report by the Fukushima Nuclear Accident Independent Investigation Commission, 'was the result of collusion between the Government, the regulators and Tepco'.

Lloyd Parry goes on to quote, 'They effectively betrayed the nation's right to be safe from nuclear accidents'; 'Therefore, we conclude that the accident was clearly man made.' (See in full the excellent report by Lloyd Parry which contains much additional 'worrying' information.)

Just one final point from the findings of this report by the Commission: It states that TEPCO were too quick to blame the meltdown on the Tsunami alone, suggesting that the extensive damage was caused by the preceding earthquake. 'This is an attempt to avoid responsibility by putting all the blame on the unexpected (the Tsunami) and not on the more forseeable earthquake', it states.

Finally, a reminder of some key dates from this disaster, via *The Times* article:

- *11 March 2011*: Magnitude 9.0 earthquake and ensuing flood at Fukushima Dai-ichi Power Plant, Japan.
- *17 April 2011*: Progress is made stabilising conditions, at the plant.
- *5 May 2011*: All 50 of Japan's nuclear plants are shut down.

- *2 July 2011*: Reopening of the Oi nuclear plant leads to angry public protests.
- *5 July 2012*: Japanese panel (FAIIC) highlights the poor government and company response as being pivotal in the disaster.
- *4 March 2013*: Recent world news, www.EandTmagazine.com – *Areva, the French energy company, hinted that it believes there could be six nuclear reactors in Japan which would restart at the end of 2013. Areva planned its first nuclear fuel shipment of mixed oxide fuel (MOX) to Japan since the Fukushima accident.*
- *13 June 2013*: Tim Webb reports in *The Times*, 'after the nuclear disaster at Fukushima, Japan in 2011, the share of the world's electricity generated by nuclear power dwindled in 2012 to the lowest since 1984, before the Chernobyl disaster 2 years later' (see Table 23.15).

Postscript 2

Mid-2012, Martin Fletcher, *The Times* (6/7/12), reporting from *Iran*, 'Nuclear opinion poll withdrawn as votes go "wrong" way', commented: 'Iranian leaders seize every chance to proclaim that nuclear power is the inalienable right of their citizens, and that people's quest to obtain it will never be halted by western sanctions'. However, Fletcher reports that 'an on-line survey conducted by a state TV channel this week suggests that the opposite is true. It showed that 63% of the respondents wanted the regime to stop enriching uranium so that the crippling sanctions imposed by several countries could be lifted – or at least did so until the IRINN news network in Iran abruptly removed the survey from its website. Various reasons were given for this farce. Blame was also being pointed at Britain and the BBC claiming that the website had been *hacked* from IP addresses registered in Britain!' There is still no closure on this sad 'nuclear initiated' issue. Press reports tell of economic hardship being inflicted on Iranians: high inflation, food prices rising daily, huge drop in value of Iranian currency, factories closing, productivity plummeting and unemploymrnt soaring – the latest round of draconian Western sanctions will accelerate this downward spiral, says Martin Fletcher in his *The Times* article.

2013 Updates:

1. Despite the concerted international action to 'derail', or 'stop', Iran's nuclear ambitions, as recently as early 2013, Iran was still driving on, defiantly, with its nuclear plans.
2. Iran said it planned to build more nuclear power reactors in the earthquake-prone coastal area, just a day after a 6.3 magnitude quake that killed 37 people. The tremor hit 89 km south-east of Bushehr, but Iranian officials said the nuclear power station 18 km south of the port area was unaffected and the head of the Islamic state's Atomic Energy Organisation said more would be built there (10/4/13, www.EandTMagazine.com/news).
3. In April an AP report from Vienna stated diplomats as saying that Iran has trippled the number of machines at its main uranium enrichment facility to

600 in three months, suggesting that Iran has both the technology to mass-produce advanced centrifuges and the ability to evade sanctions (18/4/13).
4. *Ukraine* received a €600 million loan to upgrade its nuclear power plants and to bring them into compliance with international safety standards. The European Bank for Reconstruction and Development (EBRD) will provide a €300 million loan, with the European Atomic Energy Community (Euratom) making up the other €300 million. The total upgrade will cost €1.4 billion and is scheduled for completion by late 2017 (13/3/13, www.EandTMagazine.com/news).
5. *North Korea* said it would restart a nuclear reactor closed since 2007, a blow to China's stated aim of restarting de-nuclearisation talks on the Korean peninsula. The restart of the 5 MW reactor at Yongbyon could produce more plutonium for nuclear weapons, and its uranium enrichment plant would also be put back into operation, a move that could give it a second path to the bomb (2/4/13, www.EandTMagazine.com/news).
6. *Centrica* announced a $10 billion deal to import gas from the United States. The 20-year contract with Chieniere Energy Partners will deliver 89 billion cubic feet of annual liquefied natural gas (LNG) volumes from the Sabine pass liquefaction plant in Louisiana. This is the first time the UK has entered into a formal gas import agreement with the USA (23/3/13, www.EandTMagazine.com/news).

23.6 Some aspects of carbon trading

As with other commodities, carbon is now traded in the city and represents a rapidly growing global market. Investment banks trade carbon on behalf of their clients – owners of power stations and other heavy polluters operating within Europe's trading scheme. Briefly, this operates as follows:

1. Companies require a permit for each carbon tonne emitted, currently £12 per tonne.[5] (More permits are required if a company pollutes with more carbon than its allocation.)
2. European Union Emission Trading Scheme (EUETS) awards free carbon permits to energy companies.
3. Each energy company can switch on electricity generating plant and use its permits – the 'opportunity cost' – or it can sell its permits.
4. Energy companies pass on 'cost' to consumers. The total cost of this carbon trading scheme passed on to each UK household amounted to £32 per year in 2011. (This includes an 'opportunity cost', per household, of £20 per year.) For further details see article 'British bills pay for "dirty" power plants in developing countries' (Tim Webb, *The Times*, 27/8/2011).

---

[5] It is of interest to examine the complete correspondence on these items (*The Times*, 07/3/12 and 09/3/12 and Fortson's article of 04/03/12 in (c)) (see also a contribution in Table 23.3).

5. The system, although flawed with some associated scandal, is big business and the market has mushroomed in size. It has been estimated that the value of carbon permits trading in 2011 will total about 106 billion euro, 10% more than the 2010 figure. Supporting industries have also sprung up, which help energy companies to identify suitable 'offsetting' projects.

(This United Nations-led global carbon scheme allows the 'rich world' to offset its carbon emissions by investing in green energy projects in the 'developing world', for example Brazil. Carbon reduction targets in the UK are legally binding and consequently they will be difficult to change.)

The reader should be aware that the creation of a market in pollution 'permits' a rich vein for organised crime in Europe (e.g. *The Times*, Adam Sage, Paris, 14/1/12, 'Carbon trade leaves a "mafia-trail" of fraud and murder'). Sage reports on comments made in the Paris Criminal Court by Fabrice Sakoun, a French merchant: 'it was the lady Gaga of business'. *Sakoun was sentenced recently to 5 years in prison, plus a one million euro fine for his role in a 43 million euro VAT fraud in 2009, involving this EU scheme,* while in separate cases (i) an 'alleged' *50 million carbon fraud* is being processed and (ii) in another case in Frankfurt, Germany, six men were recently sent to prison for a 300 million euro fraud.

Further, when referring to Europe's Emission Trading Scheme which was launched in 2006 to reduce greenhouse gases by giving pollution permits to companies, Sage said: 'there was a loophole. Some countries charged top rate VAT on the permits, while others, including Britain, did not'.

'Fraudsters bought the permit (in a country) where there was little or no VAT, sold them where it was high and "pocketed" the profits. It seems European gangs used the market to launder drugs money. French police suspect recent unsolved killings are connected to a multi-billion-euro fraud, targeting the EU's carbon trading scheme. From press-information sources, it appears many court cases across Europe involve gangs who have been exploiting Brussels' environmental policy and have achieved lucrative financial gains'. Perhaps reassuringly, a spokesman informed Sage 'that most European countries had stopped charging VAT on the Permits' and said, 'Controls have been reinforced and globally, the market is now safe'. Time will tell.

23.6.1 Coal-fired to co-fired stations in the UK to avoid paying rising climate taxes (after Danny Fortson, The Sunday Times, *Energy Environment, 26/2/12*)

In 2011, the UK government proposed critical changes to biomass energy payouts:

1. remove the cap on amount of renewable obligation credits or ROCs worth about £45 for each kilowatt-hour of electricity paid to generators who burn biomass alongside other fuels
2. 'ramp-up' incentives for old plants that generate at least 15% of their power by burning organic material.

The UK government has now decided that it is much more 'cost-effective' to incentivise co-firing (at existing coal plants) rather than build big new ones, Fortson has reported (*The Sunday Times*, 26/2/12). Briefly, he comments:

- 'With emission targets tightening, Britain is facing the closure of old coal-fired power stations. *Some will be mothballed, but others may be given a new lease of life – by converting them to biomass burners*; consider the situation at Drax coal-fired power station, Selby, Yorkshire, UK. Every day the plant conveyors tip 3500 tonnes of biomass-wood chips, grass, olive stones, and other organic materials into its boilers. This produces enough electrical power to supply 350,000 homes'. *This is called 'a co-firing scheme' because the biomass is mixed with coal.* It is currently the largest such scheme in the world.
- Today nearly 50% of the carbon dioxide permits Drax buys under the European emissions trading are supplied to the company at zero cost. 'Next year, together with the rest of the country's energy firms, Drax will for the first time have to pay all the carbon permits'.

Fortson observes that this will be painful! He informs the reader that:

- In 2011, Drax released 21.5 million tonnes of carbon dioxide into the atmosphere. Carbon permits trade at about £8 each. *This equates to a bill of £170 million in 2013 for Drax!!* Under the current trading regime, 9.5 million of these permits was given at zero cost to Drax (representing a saving of £114 million last year) – when the company paid an average of £12 for permits.
- *Drax Power Station is the biggest and most polluting coal-fired station in Western Europe* and coincidentally *is also one of the UK's largest sources of green power*.
- Currently Drax is pressing ahead with plans to increase its co-firing operation. By the end of 2012, it is hoped that Drax will be able to produce 20% of its power from renewable sources with an ultimate goal of producing more than 50% of its power from biomass.
- The goal it has set for biomass electricity power, based on the new subsidies proposals, is 50% (i.e. Drax will burn more trees in its existing coal-fired plant *to avoid paying rising climate taxes*).

The reader should check out further literature on other biomass initiatives in this sector:

- For example, npower recently explored a scheme to convert the old Tilbury coal-fired plant (near London) into the world's largest wood burning station.
- In 2012, a huge 'logistical issue' still to be overcome was to ensure an adequate, 'ongoing' supply of wood-chips 'via cargo ships' to keep the plant operational. Some predict wood-chips could be transported vast distances, from as far away as North America to the UK. This has become a reality (see recent biomass update of 2013).

- The City was recently concerned with the strategic issue of the sustainability of fuel supplies with such proposals. It is recognised that the environmental benefits of modern biomass are, in some cases, reduced when it has to be transported to the plant over distances exceeding 30–40 miles.
- Clearly, there is a need for critical requirement for a 'full-cycle assessment' in this case. We will probe this aspect further in the next item.

A recent biomass update [2013]:

Drax 'decided to go down the biomass route'. Tim Webb, *The Times*, reported in March 2013, 'this radical plan to convert Britain's largest coal plant to biomass took a large "bite" out of Drax's profits in 2012, which fell by 11% to £298 million last year. Drax, which generates 7% of UK's electricity, spent £180 million last year (2012) building two giant silos to store the wood-chip biomass material on site as well as a conveyor belt'. The current £700 million plan to convert Drax to biomass involves converting three of its six boilers to burn biomass instead of coal.

This will earn the company green energy subsidies and avoid paying a carbon tax. The first of the boilers would be converted by April 2013. The chief executive of Drax recently reassured investors that its biomass technology was 'tried and tested', giving encouragement to Drax shareholders.

Some strategic aspects for the reader to ponder over regarding present UK energy initiatives relating to carbon tax, domestic fuel prices and environmental issues for the application of biomass technology, for example, at the Drax, the giant coal-fired power station in Yorkshire, UK (touched on earlier) are as follows:

1. **Carbon tax will lift domestic fuel prices:** *The Times*, 20/6/13, p. 41, report that 'wholesale energy prices in Britain may almost double those in Germany within three years because of the Government's new carbon tax, according to a report from Credit Suisse. The bank forecasts that by 2016–17, prices will be 85% higher than those in Germany and will remain higher in general for the next seven to ten years. It blames the fivefold increase in the carbon tax on coal and gas plants that was introduced in the UK in April 2013'.
2. **Biomass usage: is it environmentally friendly?** Matt Ridley, *The Times*, 20/6/13, p. 29, writes a compelling 'opinion' article arguing that 'replacing coal with wood pellets in our power stations is bad for the climate, for health and for our pockets'. He observes, 'In the UK Energy Bill currently going through Parliament there is allowance for generous subsidy for a huge push towards burning wood (instead of coal) to produce electricity as is already happening at Drax power station. The UK Government currently estimate that by 2020 up to 11% of our generating capacity will be from burning wood'. The reader is encouraged to read the full transcript of Ridley's

excellent article. Consider some further strategic points made by him, reproduced in the table below:

He states that reporters from the *Wall Street Journal* recently found that the two wood pelleting plants established in the southern USA specifically to supply Drax are not only taking waste or logs from thinned forest, but also taking logs from a cleared forest, including swamp woodlands in North Carolina cleared by shovel-logging with giant bulldozers (running on diesel) with objections from local environmentalists.

The logs are taken to the pelleting plants where they are dried chopped and pelleted in an industrial process that emits lots of CO_2 and pollutants.

The pellets are then trucked (more diesel) to ports, loaded on ships in the USA (diesel again), offloaded at Humberside, UK, on to 40 trains (yet more diesel trains) which arrive at Drax each day.

Until recently the UK government was in denial about this diesel usage: 'No net emissions during production are assumed', it stated in its 2007 Biomass Strategy. More recently it has admitted that the energy costs of transporting biomass can be up to 46% of the energy generated by combustion at the power station if shipped from afar.

- Storage of pellets is an issue too dry and they burst into flame as has happened at Tilbury power station.
- In 2012; storage of pellets also produce CO which can result in suffocation in a confined space; burning the pellets produces smoke.
- Airborne pollutants kill more people than road traffic accidents.

In summary, Matt Ridley reports:

1. Over 20 or 40 years various studies have shown that wood burning is far worse even than coal, in terms of greenhouse gas emissions.
2. Despite the evidence, the UK government persists in regarding biomass burning as zero-carbon and therefore deserving of subsidy.
3. Moreover unlike gas or coal, you are pinching nature's lunch when you cut down trees.
4. Unfelled, the trees would feed beetles, woodpeckers, fungi and all sorts of other animal wildlife when they died, let alone when they lived. Nothing eats coal.
5. Compared with gas the biomass dash is bad for the climate, bad for energy security and dependence on imports, bad for human health, bad for wildlife and very bad for the UK economy. Apart from that, what is there not to like?

23.6.2 Frying note: storage energy back-up

Frying note: Using the heading 'Power cut? I'll put the chip fryer on!' Oliver Shah wrote an article, describing the following aspects (in *The Sunday Times* of 25/3/12):

> When a blaze broke out at Tilbury power station in Feb, 2012, the National Grid, NG, required back-up-fast; the emergency at the plant in Essex, which runs on wood chips, took out 1/10th of Britain's renewable energy capacity at a stroke (100 MW).

One of the sources 'tapped' to plug the electricity gap was certainly unusual: *recycled chip fat*.

Apparently, NG has a 'standby arrangement' with Renewable Energy Generation (REG) which has two small generating plants, with <8.5 MW (total maximum reserve power). These two plants use recycled 'discarded' cooking oil which REG collects and burns.

(This was a small proportion of the 100 MW *shortfall* left by the Tilbury power station blaze or the 5,000 MW reserve power NG can draw on, *but it was a step in the right direction*, wrote Oliver Shah.)

Note: The reader may be unaware of the fact that vast amounts of cooking fat from London kitchens finish up by choking up the large sewers in the city and have to be 'dug out' regularly. As far as is known this waste has not been used yet for power generation purposes!

Table 23.3 Some comments on energy issues
[after Tony Lodge, Letter to the Editor, The Times, *9/3/12]*

'Sir, Alice Thomson rightly calls on the Chancellor to play a bigger role in energy policy in the interests of industry and the consumer. The present debate over wind energy warrants examination of the fuel sources that have provided Britain's electricity over the past month, a period which traditionally necessitates high winter energy demand. It is also important to understand how the Government is looking to further tax those fuel sources which are meeting electricity demand *and the effects this will have on household and industry costs*.

- Official statistics show that *during February 2012, Britain relied heavily on coal to generate electricity, averaging 47% of supply. Gas came second with 26%, nuclear at nearly 19% and wind supplied just over 3.5%.*
- *Wind's 'intermittency' also means it cannot meet important peaks in demand.* In the light of the UK's ongoing heavy reliance on coal – and increasingly gas – for the generation of electricity, *the Chancellor should review and delay his intention to impose a unilateral UK carbon price floor on electricity generators from 2013.*

This will see UK emissions taxed at £16 per tonne of carbon dioxide emitted, while the price for carbon on the Continent looks set to remain roughly at 8 euro per tonne.

- As fuel poverty rises and new nuclear investors hesitate to provide an in-service delivery timetable for new nuclear power plants, *the introduction of UK carbon price floor trajectory from 2013 will significantly increase energy bills for consumers and industry while delivering no environmental benefit as emitters in Europe will 'soak-up' any carbon savings made in the UK.*
- The *carbon price floor is a tax-raising measure* which should be reconsidered and instead effort put behind strengthening the EU Emission Trading Scheme and supporting a future EU-wide carbon price floor so as not to place the UK at huge economic and social disadvantage with the rest of Europe'.

In a separate 'Letter to the Editor' (*The Times*, 9/3/12), one of three replying to and addressing issues on themes of affordable energy, fuel poverty and profits from wind farms, *Tony Lodge,* research fellow, Centre for Policy Studies, made several strong points in his letter, reproduced now in Table 23.3.

His views 'resonate' with the opinion of many in the UK!

23.7 A new green technology: carbon capture and storage (CCS) (Tim Webb, *The Times Business Dashboard*, 29/3/12)

A set of interesting comparative energy cost figures has become available via Tim Webb.

His energy article starts, 'Britain's flagging efforts to clean up its dirty power stations have received a boost after Samsung took a stake in an experimental project in South Yorkshire, UK'. Industry executives warned the government that Britain was 'in the last-chance saloon' as ministers start a second push to establish a new green technology-carbon capture and storage (CCS).

Webb reports that:

- The South Korean company has bought a 15% stake in the *Don Valley, UK* project to build a 650 MW coal power plant and also increase oil production from the North Sea. The project, developed by *2Co Energy*, is backed by the private equity group TPG and will cost £5 billion. Samsung will contribute to the investment.
- Using CCS Technology, the plant's carbon emissions will be captured instead of being released into the atmosphere. The gas will be piped out to sea to be stored in depleted oilfields. It is claimed by its developers that the pressure from the CO_2 will 'squeeze out' 150 million barrels of 'hard-to-reach-oil', which will be pumped via a different pipeline back to shore.
- Developers of UK schemes, including 2Co Energy, are also applying for about *300 million euro (£250 million) in European funding*.
- This technology has not yet been deployed on a large scale and needs subsidies to be put into practice. 'The Don Valley scheme in Yorkshire is one of several hoping to secure a share of £1 billion from UK Government's latest funding round, due to be launched in April 2012, after an earlier round was a fiasco and collapsed ....' (See original article for other interesting background facts.)

23.7.1 Some committed CCS developments worldwide

It appears that Britain has failed to live up to the ambitions of the Labour administration in 2008 to be a world leader in developing carbon capture technology. Webb comments that this failure by the UK to award CCS funds in earlier years has allowed other countries to 'steal a march'.

According to Bloomberg New Energy Finance, Carbon Capture and Storage Association:

(i) The amounts committed to CCS projects around the world are:

USA	$5,595 m	Netherlands	$267 m
Canada	$3,643 m	China	$134 m
Norway	$1,234 m	UK	$145 m
South Korea	$1,100 m	France	$36 m
Australia	$641 m	Germany	$21 m

(ii) CCS is much cheaper than most renewables in 2011, figures quoted in terms of:

Electricity generating costs	£ per MW/h
Onshore wind	9.5
Offshore wind	15.5
Solar PV	46
Tidal stream	26.5
Wave	36.5
Coal (CO_2) CCS	15

(iii) **Estimated proportion of global power generated from fossil fuel by 2030**   **80%**
UK government funding for 2008 programme   **£1 billion**
How much UK government has spent on CCS to date   **£91 million**
Generated by the UK's only CCS pilot project   **5 MW**

(iv) **How much carbon capture and storage can save the UK**   **20 GW**
(i.e. UK target by 2030 is 20 GW) and by 2030, the 100% figure for carbon emissions abated is equivalent to over one sixth of the UK's total emissions in 2010)

Summary: How carbon capture and storage works

(1) Instead of being released via chimney into the atmosphere, it is captured at power stations and compressed

(2) CO_2 is piped to recover 'hard-to-reach-oil' in depleted fields in the North Sea. (This will be pumped via a different pipeline back to shore.)

(3) The CO_2 is also pumped down into the porous rock deep beneath the sea bed which formerly held gas.

(4) The CO_2 filters into the porous sandstone reservoir, filling tiny 'spaces' that once held natural gas. It is trapped by the layers of solid rock above

Notes:

1. *CCS has been used for decades by the oil industry to boost production.*
2. Piping carbon dioxide into depleted oilfields allows operators to squeeze out the last drops in a process called enhanced oil recovery.
3. The technology has emerged only in the past decade as a way of tackling climate change.
4. According to the International Energy Agency, CCS could contribute up to one fifth of the required reduction in global carbon emissions by 2050. (*Source*: Bloomberg New Energy Finance, Carbon Capture and Storage Association.)
5. The equipment can be attached to coal and gas plants as well as factories.

Possible storage problems: A worrying aspect of this carbon capture and storage scheme from UK perspective is that although £1 billion public funding was announced recently (3/4/12), Britain appears to be powerless to stop the North Sea being turned into a giant carbon waste dump.

Tim Webb reports, 4/4/12 (*see Table 23.4 for details*) that *The Times* has learnt that 'the UK Government has admitted that European Law will force it to allow member states to store the waste gas using an experimental green technology'. 'The relevant EU Directive on CCS forces countries to give other member access to their storage capacity or face a fine'; Webb states that the government's recent policy document admitted: 'The CCS Directive requires us not to discriminate against other EU Member States when permitting access to the UK's storage capacity'.

Later, a government spokesman reported safeguards would be in place to protect UK taxpayer in the event of storage gas leakage. Table 23.4 reproduces most of the information from Webb's article.

In summary, it is anticipated that the UK government's carbon reduction plan will:

- deliver significant economic benefits
- reduce CO_2 emissions
- enhance security of electricity supply.

***Notes*:**

1. At the end of March 2012, it was revealed in the press that, during February 2012, more than 100 Conservative MPs had written to PM David Cameron to demand cuts in the £400 million/year subsidies paid to the industry.[6]
2. Recent public statements by the UK government indicate that the number of wind farms that have been agreed, or are passing through planning procedures, is now sufficient to meet the UK renewable targets set for 2020.

Because of the many government public-funded energy-related initiatives, one overall strategic concern that this writer has is how to safeguard and ensure thorough and effective implementation and ongoing management, monitoring and evaluation of each 'initiative-strand' throughout plant life cycle, together with comparative evaluations of each strand. From a UK taxpayer perspective – is he/she getting 'value for money'?

In summary, the present writer considers there is still much to re-examine and critically evaluate on the above issue, from technical, economic and environmental viewpoints.

- One must not forget the era of the 'smart grid' will be with us in the next decade or two. Clearly, there are strong policy reasons to continue along the 'renewable route' but much still needs to be resolved, with thorough evaluations made for each option including gas.

---

[6] *Solar panels update note*: A recent press feature, in 2012, described a new design of solar panel being developed by *Naked Energy, UK,* which mixes traditional photovoltaic technology with thermal energy production. According to the company and research carried out at Imperial College London, *this panel produces up to 46% more energy than the typical solar panel on the market today.*

Table 23.4 *'Britain may be forced to store Europe's carbon dioxide emissions in North Sea': Government sets aside £1 billion to fund (CCS) technology [after Tim Webb,* The Times, *Business, p. 39, 4/4/12]*

Ed Davey, Secretary of State for Energy and Climate Change, said 'the potential rewards from Carbon Capture and Storage (CCS) are immense: a technology that can de-carbonise coal- and gas-fired power stations and large industrial emitters, allowing them to play a crucial part in the UK's low-carbon future'. Some background facts presented:

- *jobs* that the CCS industry could support in Britain by 2030 – *100,000*
- how much the industry could be worth to Britain by 2030 – *£6.5 billion*
- public funding announced 3/4/12 for CCS power plant projects in Britain – *£1 billion*
- deadline for when they must be operational – *2020*
- duration of the last UK government funding round – *4 years*
  (*It failed to award any money in 2008.*)

Webb wrote:

- Britain is powerless to stop the North Sea being turned into a giant carbon dump, *The Times* has learnt. The government has admitted that EU law will force it to allow member states to store the waste gas using an experimental green technology. But experts have warned that the UK risked being 'lumbered' with huge financial liabilities and of sustaining unknown environmental damage if the gas was to leak.
- The revelation came as the government unveiled its plan to develop the experimental carbon capture and storage technology yesterday. Ministers formally invited energy companies to bid for £1 billion of public funding to build coal or gas plants in Britain equipped with CCS, which will stop carbon dioxide from being emitted into the atmosphere, the gas being piped instead into old oil and gas fields in the North Sea for permanent storage under the seabed.
- If the technology works, power plants and factories will have to find an area that is geologically suitable – and publicly acceptable – to store the unwanted carbon dioxide. Pumping it underground onshore is deeply unpopular and has been banned in countries including Germany after safety fears.
- For much of Europe, this leaves the North Sea as the only alternative. It has enough capacity to store 78 giga tonnes of CO_2. More than five times the capacity that the UK needs.

The EU has issued a directive on CCS which forces countries to give other members access to their storage capacity or face a fine. The government's policy document, published yesterday, admitted: 'The CCS Directive requires us not to discriminate against other EU Member States when permitting access to the UK's storage capacity'.

It warned of the 'financial implications' involved when ownership of CO_2 is transferred from one country to another and said there was 'considerable concern' over the liabilities.

The Green Alliance think-tank warned 'that broad public support for CCS would be lost if Britain had to shoulder the liabilities should storage facilities in the North Sea leak'.

But a government spokesman insisted that an agreement with countries exporting the gas would be reached to ensure that taxpayers were 'fully protected'.

Note: The original article also presents a very useful map showing location in and around Britain of gas power stations, coal power stations, saline aquifers and depleted hydrocarbon fields (also size bands of 10, 100, 1000 (CO_2 mtonnes)).

We must aim to achieve a robust, strongly integrated, intelligent system with good back-up storage technology plant (e.g. new hydro power or biomass) to be 'assigned' to work in tandem with specified groups of 'local wind farms' if UK decides to re-commit to renewables.

- The politicians have got 'us' to the present position and it seems unlikely that they 'will-about-face' again! One becomes somewhat cynical regarding the present energy situation: the conflicts, the widely varying provisional estimated costs and indecision as to the best way forward, etc. We could even see another 'rush for gas' scenario played out once more in the UK; not many people would be surprised! (This reminds the writer of the 'Definition of a Cynic' (Oscar Wilde, 1854–1900)):

A man who knows the PRICE of everything and the VALUE of nothing

(Currently in the UK energy sector, 'price' and 'value' are both causing confusion!)
Postscript: But the British public will certainly pick up the bills!
The UK public electricity consumers will 'hope' that politicians and all 'strategic decision makers' can, quickly, provide a good, accurate, prudent and robust 'steer' on both *value* and *price* (cost) regarding our alternative energy 'future development options,' otherwise UK energy consumers may suffer very high energy bills for many years to come (and we will wistfully think of 'smart grids', wow)! Whatever is eventually decided, it will take some considerable time for the finalised plans to be fully implemented and operational.

Strategically, it should be noted that Edward Davey, the new, *replacement*, secretary of state for energy and climate change, also in 'Letters to the Editor' in *The Times* (9/3/12, p. 27), ended his piece with these two strategic observations in his last two paragraphs:

1. 'Onshore wind is getting cheaper: the gap between onshore wind and gas-generated electricity costs has halved in just five years. It could be cost-competitive with fossil fuels in four. Hence we propose to cut the support it receives'.[6]
2. In the final paragraph, reference is made by Davey to a piece in an earlier letter commenting on the AF report (discussed by Fortson, in (b) and (c) above).

'Finally, the piece quotes a DECC spokesman describing the AF Consultant report as "shoddy nonsense". That's rather an understatement'.[5]

In certain rural regions, there have been widespread feelings expressed recently – that it was now timely to consider halting the further consents for wind-turbines meantime. Some individuals feel that certain counties are contributing more than their fair share to this sector. There are clear concerns emerging regarding the limited benefits communities receive from wind farms and the impact on the countryside and tourism.

General notes: The reader is again encouraged to review the numerous and diverse current energy sector-related discussions and issues, such as in the three items considered above, IET/IEC/IEEE/CIGRE TBs, web pages, *The Times*, etc., and to form his/her own judgement, as to the preferred way ahead for the energy industry, both globally and nationally.

In this way, the technical material in this book can be 'augmented' with these latest incremental supplementary reference sources to maintain the body of knowledge in this book 'current' for many years to come, and to 'continue to empower the reader with the very latest incremental developments and trends!'

For example, from the IET, visit www.EandTmagazine.com/news, volume 7, issue 1, February 2012:

- On 13/1/2012, Canada announced it was withdrawing from the Kyoto Protocol, becoming the first country to pull out of the climate change treaty. The Canadian Prime Minister Stephen Harper's Government has close ties to the energy sector said that the treaty could penalise Canada by $14 bn for not cutting emissions to the required amount by 2012.

Some other examples: visit www.EandTmagazine.com/news, volume 7, issue 5, June 2012:

- The UK DECC announced it would work with the USA to develop 'floating' *wind turbines* by collaborating in the development of wind technology to generate power *in deep waters* that were currently off-limits to conventional turbines.
- South Korea approved an emission trading scheme, which would cap carbon pollution across the economy. The scheme, due to start January 2015, has attracted opposition from the nation's top industry body, who said it would add unnecessary costs. However, the government said the scheme was crucial to reining in emissions.
- Iran investigated a suspected cyber-attack after a virus was detected inside the control systems of Kharg Island – the country's largest crude oil export facility. The virus hit the Internet and communications systems of Iran's Oil Ministry and its national oil company.
- Reports of numerous international activities/research into *shale fracking gas* with a cautionary statement from Durham University that shale gas fracking should only take place at least 600 m down from aquifers used for water supplies. The study found that fracking caused fractures running upwards and downwards through the ground of up to 588 m from their source.[7]
- Japan shut down its last working nuclear reactor on 4/5/12, leaving the country without nuclear power for the first time since 1970. The shutdown came just over a year after the earthquake and Tsunami damaged the TEPCO Dai-ichi plant in Fukushima, causing the worst civil nuclear disaster since Chernobyl.

(Many more NEWS items are detailed monthly; see also the complete range of IET publications/activities!)

---

[7] *India's shale gas boon*: *The Times* recently reported rising demand from American oil companies that need to produce shale gas – from guar beans – is providing a boom for farmers in the deserts of Rajasthan in India. This is a multi-million dollar industry in the USA. Guar gum powder is used to make hydraulic drilling fluids that are pumped underground during 'fracking', the process used to extract natural gas from subterranean rocks.

(Guar Beans produce a gum used in oil extraction. *The USA relies heavily on guar gum powder – a thickener also found in ice cream, toothpaste, pet food and sausage skin – using it to make specialist hydraulic drilling fluids that are pumped underground at high pressure during 'fracking'*, the process used to extract natural gas from subterranean rocks. Naturally, oil companies are researching for cheaper alternatives to guar gum, materials that can be used as 'viscosifiers', which, when mixed with sand and water, produce a gel-like substance that forces out gas when it is pumped underground, says Pragnamenta – reference below.) (See a full article by Robin Pagnamenta, Mumbai, in *The Times*, 11/06/12, p. 40, '*Indian farmers strike it rich amid America's new oil boom*'.)

One recent article by Damian Carrington in *The Guardian* (19/4/13) reports on strategic work published today by Stern and Carbon Tracker. Carrington starts with dual banner titles of (i) *carbon bubble 'creates global economic risk'* and (ii) *economists fear massive oil, coal and gas reserve warrants further consideration.* Carrington starts, 'The world could be heading for a major economic crisis as stock markets inflate an investment bubble in fossil fuels to the tune of trillions of dollars', according to leading economists. 'The financial crisis has shown what happens when risks accumulate unnoticed', said Lord Nicholas Stern, a professor at the London School of Economics. He went on to say that 'the risk was very big indeed and that almost all investors and regulators were failing to address it'. The so-called 'carbon bubble' is the result of an overvaluation of oil, coal, gas reserves held by fossil fuel companies. *Carrington refers to a stark report by Stern and Carbon Tracker published today*, which comments that 'at least two-thirds of these reserves will have to remain underground if the world is to meet existing internationally agreed targets to avoid the threshold for "dangerous" climate change'.

'If the agreements hold, these reserves will be in effect unburnable and thus worthless, leading to massive market losses. But the stock markets are betting on countries' inaction on climate change'. Carrington informs his readership. This report is supported by several organisations, including HSBC, Citibank, Standard & Poor's and the International Energy Agency. These, together with the Bank of England, all recognise that a collapse in the value of oil, gas and coal assets, as nations tackle global warming, is a systemic risk to the economy, with London being particularly at risk owing to its huge listings of coal. *Stern reports that far from reducing efforts to develop fossil fuels, the top two hundred companies spent $674 billion (£441 billion) in 2012 to find and exploit even more new resources, a sum equivalent to 1% of global GDP.*

This amount could end up as 'stranded' or valueless assets. HSBC warned that 40–60% of the market capitalisation of oil and gas companies was at risk from the carbon bubble with the top 200 fuel companies alone having a current value of $4 trillion, with $1.5tn debt.

The report calculates that the world's currently indicated fossil fuel reserves equate to 2,860 billion tonnes of carbon dioxide, but that just 31% could be burned for an 80% chance of keeping below a 2 °C temperature rise. For a 50% chance of a 2 °C or less, just 38% could be burned.

Duncan Clark in a companion analysis item: 'What's in it for me? Fossil fuels and vested interests', *The Guardian*, 19/4/13, p. 2, says, 'the report released by Lord Stern and the think-tank Carbon Tracker paints a picture of society in denial. Clark says, 'it shows we're pumping almost $700 billion (£458 billion) of hard-earned savings and pensions annually into finding new reserves of fossil fuels, even though it is clear that almost all of these reserves will have to be written off *to provide a decent chance of keeping the planet safe*'.

'The ever-inflating "carbon bubble" is only part of the bigger picture, because most of the world's fuel – about 75% in total and almost all the oil and gas – is owned not by listed companies but by governments'.

'And we don't need only to stop expanding the world's fossil fuel reserves; we also need to get used to the idea that we can't burn most of what we already have'. Clark goes on to say 'that this is a much trickier problem, because with Carbon Tracker's detailed analysis and growing awareness of the carbon bubble, investors will surely soon start waking up to the madness of putting capital into expanding fuel reserves. But there is little self-interest – only planetary interest – in leaving existing fuel assets in the ground.

'The need to write off existing reserves shines a revealing light on global climate politics, because when you map out the world's fossil fuel reserves, a striking correlation emerges between the amount of carbon a country has in the ground and its keenness for – or resistance to – a global climate deal'.

Take Britain, for example. Clark reports. 'sure, there's lot wrong with our green policies but nonetheless the UK climate laws are world leading'. Why? Clark asks. 'Partly because we are good campaigners, perhaps, but also perhaps because we have virtually no fossil fuels – and therefore nothing much to lose. According to BP, even the UK's economically viable reserves would run out at current production rates in just 7 years for oil, 4 years for gas and 12 years for coal'.

Clark states that a 'similar situation is true for Europe as a whole and Africa', and he goes on to comment, 'so its perhaps no surprise that these two continents – along with the low-lying island nations – have pushed hardest for a global climate deal. They collectively own less than tenth of carbon'. Clark further comments, 'even if you factor in all the nations involved in the Cartegena Dialogue, a loose-knit body of countries proactively engaged in efforts to push for a global deal, you get only a fifth of the total. By contrast, those countries with the biggest fossil fuel reserves – such as the USA, China, Saudi Arabia and Canada – tend to be recalcitrant (not susceptible to control or authority) when it comes to climate politics'.

The USA has 18% of the total fuel reserves and thus plenty of assets that might need to be written off if the world agreed to tackle climate change.

Finally, Clark points out 'that it was surely no coincidence that of all the South American nations it was oil-rich Venezuela that did its best to block the last minute progress at the 2011 talks in Durban'.

Note: The reader is encouraged to examine the complete articles by Carrington and Clark, and also this valuable report by Stern and Carbon Tracker – which certainly makes very clear what many had suspected for many years.

23.8 Recent developments in UK Network/European Grid links

23.8.1 New UK/International DC cable links

UK/Iceland DC link: Iceland's largest energy firm Landsvirkjun, the State-owned giant which produces 75% of the Island's electricity, recently carried out a feasibility study relating to the laying of a 1,170 km (730 miles) *DC cable between Iceland and the North of Scotland (Peterhead)*. This would be capable of transporting up to 6 terrawatt hours (TWh) of energy to Britain per year, with power generated by volcanic heat, sufficient to power 1.5 million households (*Source*: *The Sunday Times*, Energy and Environment, p. 13, 06/03/11). The proposal by Landsvirkjun would cost about $2.3 billion to build, plus a further $2.4 billion being required for construction of several geothermal plants/hydroelectric dams to generate the 6 TWh of energy. It is projected that 5% of the energy will be lost during transmission. Apparently:

1. The longest submarine power line to date in Europe is NorNed, a 580 km *link between Norway and the Netherlands*, capable of carrying 6 TWh a year
2. BritNed, a 260 km *cable between the Isle of Grain in Kent and Rotterdam*, is scheduled to come into service by 2012
3. In passing, it should be noted that another cable project is due to be completed in 2012, and will provide a 500 MW connection via a sub-sea DC *cable link between the UK Grid and the N. Ireland power grid*.

23.8.2 Proposal case for a North Sea Super Grid (NSSG)

Recent UK–Norway government energy agreement: To start this section, it is appropriate to first mention the main aspects of two press announcements of a recent collaborative agreement between the governments of UK and Norway (reported in the press by (i) A. Woodcock, in *The Journal, Newcastle*, 7/6/12 and (ii) T. Webb in *The Times*, 8/6/12) to lend more context to the NSSG contributions, which will follow:

- 'In the past five years British companies have invested £13 billion in Norwegian oil and gas, while Norway now meets more than 25% of the UK's entire energy needs', said D. Cameron (i), the UK Prime Minister, while in *The Times* (ii) Webb reports, 'Norway is to pump £18 bn into North Sea oil fields project' (£6 billion initially). The two fields involved, both operated by *Statoil* (a Norwegian company), *the Marine and Bressay fields off the North coast of Scotland*, are expected to produce 110,00 barrels oil per day when they come on stream in 2017.
- Norwegian companies are major investors not only in oil and gas but also in the £30 billion *Dogger Bank* offshore wind project development which is predicted to be able to supply 10% of UK electricity requirements when fully operational. 'This new deal will take this vital relationship to the next level', said the UK Prime Minister; 'This will mean more collaboration on affordable long-term gas supply, more reciprocal investment in oil and gas and

renewables and – underpinning all of this – a set of new business deals creating thousands of new jobs and adding billions to our economies', reported Woodcock. Mr Cameron continued by stating, 'the UK–Norway energy partnership was a really visionary idea for "the future" which could lead to all sorts of innovatory developments'.

- Under this recent UK–Norway deal a joint business advisory group will be established to allow companies to talk to the government directly and on a regular basis, develop supply chain and encourage new technologies such as energy capture and storage. *Despite the RES-type words just used, one can't help but feel a stuttering rush for gas coming on soon!*

This UK–Norway partnership will focus on affordable long-term gas supply as well as a two-way investment in oil and gas exploration and the development of renewable technologies.

Consider now briefly, plans described in two recent articles [27, 28] regarding the possible construction of the NSSG. It is anticipated that there will be a large deployment of offshore-wind turbines in the North Sea in the next decade. The transmission of electrical power from large offshore wind farms represents considerable technical/strategic challenges for this 'fledgling' technology. This is in contrast to the oil/gas industry, which has successfully operated commercially in the North Sea region for more than 50 years. Undoubtedly, a robust, effective and properly designed NSSG could be an asset in transmitting the large quantities of power to strategic load centres – *if critically assessed to be a viable economic and technical option.*

- Offshore wind farms located near shore (within about 50 km of onshore connection-point) can usually be individually directly integrated into the power system onshore with radial AC cables – somewhat similar to onshore wind farms [27, 28].
- Wind farms located >50 km from shore generally need to be connected *via HVDC voltage source converter (VSC) technology*, for example BARD Offshore Installation [1]. Davies reports [28] that 'Hub' connections generally become viable for distances above 50 km from shore, when the sum of the installed capacity in a small area (about 20 km round the Hub) is relatively large, and standard available HVSC VSC systems can be used.
- When complete, the BARD Offshore 1 wind farm *will comprise 80 wind turbines, each of 5 MW capacity*, the power produced being fed via a 36 kV AC cable system, then transformed to 155 kV AC before reaching the HVDC light converter station (dedicated platform) where the AC power is converted to ± 150 kV DC and fed into two 125 km long sea cables, then 75 km via two land cables to the land-based converter station at Diele, Germany [28].
- With wind farms being built at greater distances from shore, Sean Davies [28] looks at the 'challenges and the available options'. The challenges can sometimes include *hostile environment at sea, equipment-handling problems at sea, and strategically how best to get the electricity produced offshore, back to land.*

We are living at a time when power network designers 'onshore' are looking to developing 'smart grids' in next decade or two, but at huge financial cost [30, 35], (see section 23.13 also). In contrast, offshore grid technology is lagging somewhat and indeed offshore AC wind farm technology has recently even been termed a 'fledgling' technology by CIGRE Working Group on this topic [37].

Davies in his brief yet extremely valuable article [28] provides a strategic summary touching on the requirement for a modern and efficient grid, both onshore and offshore, where the challenge for the latter 'is to more efficiently connect power harvested at sea with the onshore transmission system, while at the same time building a system that can actively contribute to stability and security of supply by enabling further integration of the European power market', he said.

He goes on to comment on an Offshore Grid project (1) co-financed by the European Commission under the Intelligent Energy Europe programme, cost in the range £55.5 billion to £67 billion.

Davies [28] comments on a final 3E report on this Offshore Grid project, published in late 2011, and he adds, 'not surprisingly, the technical solution deemed most viable was a meshed grid consisting of linked hub connections'. 'This would entail connecting projects located in close proximity so that they could share a single transmission line to shore'. 'The report predicted that this would "shave off" £11.5 bn from the original £67 bn'. Interesting figures are also supplied in the article. (*Note*: The reader is encouraged to read the original article which provides valuable locations of offshore international projects in the North Sea region.) Davies also touched on two potential cost-effective grid designs considered in the above 3E report:

- '*The Direct design* where interconnectors are built to promote unconstrained trade between countries and electricity markets as average price differentials are high. Once additional direct interconnectors become non-beneficial, tee-in, hub-to-hub and meshed grid concepts are added to arrive at an overall grid design. Tee-in connections are when a wind farm or hub is connected to a pre-existing or planned transmission line or interconnector between countries, rather than directly to shore, while hub-to-hub involves the interconnection of several wind farm hubs to form transmission corridors'.
- *The split design* is essentially designing an offshore grid around the planned offshore wind farms. As a starting point Davies explains 'interconnections by splitting the connections of some of the larger offshore wind farms between countries. These split connections establish a path for trade, and the offshore wind farm nodes are further interconnected to establish an overall meshed design where beneficial'.

Davies also observes that 'which connection concept best depends on several factors, such as the distribution of the offshore wind farms, the distance to shore, and in the case of interconnecting several wind farms and/or countries, the distance of the farms to each other'. Apparently:

DolWin2 power link will connect offshore wind farms in the North Sea to the German mainland grid.

Davies reports that DolWin2, the largest power transmission project in ABB's history, will feature the world's largest offshore HVDC system with a rating of over 900 MW, with losses constrained to less than 1% per converter station. When completed in 2015 Davies observes that 'it is estimated that the link will be capable of supplying 0.5 m houses with *clean wind generated energy*'.

DolWin2 is ABB's third offshore wind connection order for TenneT, following the 800 MW DolWin1 link awarded in 2010 and the BorWin1 project and the DolWin1.

Notes:

1. Both the DolWin1 and BorWin1 projects feature conventional fixed platforms to house the offshore converter stations, whereas the DolWin2 platform will be based on a new GBS Design concept, building on experience gained from semi-submersible floating platforms for the oil/gas sector (for further information see Davies article [28]). The GBS platform offers several attractive features and reduces environmental impact. It reduces the weather dependence of the installation phase, usually being constructed and all platform systems commissioned on shore before it is towed into position, secured to seabed. Offshore commissioning is limited energising and testing trials, after installation of cables, with significantly less impact to wildlife. (See CIGRE guidance documentation TB 333 on offshore wind farm installations.) It can be easily decommissioned and removed at the end of its service life.
2. Davies (28) also reports that the wind farms in the DolWin cluster will be connected by 155 kV AC cables to the HVDC converter station platform, situated in the North Sea. This will then transmit the electricity to at +320 kV DC via 45 km of subsea cable and 90 km of land cable to the HVDC converter station at Dorpen-West on the German mainland grid, where it will be converted to 380 kV AC.
3. Davies [28] also quotes an impressive statistic supplied by Peter Jones, the engineering manager grid systems at ABB UK, who had explained, 'ABB pioneered HVDC technology in the 1950s and is a world leader in the field'. 'In total around 140 GW of HVDC transmission capacity is installed in some 145 projects worldwide'.
4. HVDC light transmission technology is now available, states Davies [28], which deploys a new insulated gate polar transistor converter in combination with cross-linked polyethylene (XLPE) DC cable systems – the innovation complements the traditional bipolar semiconductor-based converter technology to provide a state-of-the-art power system with increased controllability – see original Davies article for more information.
5. The first HVDC link to connect an offshore wind farm with an AC grid is the 400 MW BorWin1. Based on HVDC Light technology, this is 200 km link that connects the BARD Offshire1 wind farm off Germany's North Sea coast to the HVAC grid on the German mainland.

When complete, the *BARD Offshore 1* wind farm will consist of 80 wind generators each with capacity of 5 MW. These will feed their power into a 36 kV AC cable system. This voltage will then be transformed to 155 kV AC before reaching the HVDC light converter station on a dedicated platform. Here the AC power is then converted to ± 150 kV DC and fed into two 125 km sea cables, which then continue into two 75 km land cables to the land-based converter station at Diele in Germany.

Stricter environmental regulation of the oil/gas industry implies the electrification of offshore rigs-gas turbines on platform can be replaced by power cable to shore [1]. The electrical integration of offshore loads and generation is advantageous, since a part of electrical power produced offshore can be consumed locally [27, 1 (p. 6)].

Recent developments suggest that an NSSG may possibly become viable in the near future but it will require a huge investment with many challenges and risks to be assessed. In general, due to the long distances involved, HVAC transmission does not seem viable and HVDC seems a more realistic option in many situations. Some multi-terminal schemes exist (34).

Advocates for an NSSG consider its infrastructure can be divided into four levels (see the sketches provided in original article):

1. generation and loads
2. wind farm collection grids
3. offshore cluster grids
4. long-distance HVDC transmission.

The article [27] then proceeds to outline the following aspects with references in parentheses:

Offshore generation and loads	**Long-distance HVDC transmission** (9, 10)
Active Power Balancing in Onshore Grids (5)	Voltage Source Converter (VSC) HVDC (4, 11–13)
Active Power Balancing in Offshore Grids (6)	Current Source Converter (CSC) HVDC (14–17)
Wind farm collection grids	Hybrid HVDC (15, 18, 19)
AC Collection Grids (7)	**Multiple HVDC links**
DC Collection Grids (8)	Series Connection (3, 4, 20–22)
Offshore cluster grids	Parallel Connection (23, 24)
AC Cluster Grids	Meshed Connection
DC Cluster Grids	**Future NSSG development** (4, 9, 24–26)

It has been suggested in Reference 27 that there are three regions in the North Sea (see Figure 29.1, sketch 8) which will play a major strategic role in the deployment of the NSSG, namely:

1. German Bight (cluster grid)
2. Valhall Area (cluster grid)
3. Doggerbank (cluster grid).

It is not yet clear to this writer whether a robust commercial case can be, or has been, made for a real viable *NSSG, but who knows.* Meantime, we must await the offshore energy developments in the next few years (2014–2017) to see if sufficient offshore wind farms are developed in the general vicinity of the proposed NSSG. Nevertheless, this is a very interesting article [27] well worth reading, together with the supplementary brief postscript comments set out in 1–3 below and the many relevant related reference sources [1–26] provided in Reference 27, together with related contemporaneous CIGRE work.

Postscript 3

1. Apparently, many remote offshore wind farms are planned in the *German Bight* [16]. Construction work has started and the world's first HVDC-connected wind farm is already operating (BARD Offshore, [1]).

    It is reported that intelligent-grid solutions offer benefits compared to radial connections, since the distance between those wind farms will be smaller than the distance to shore.

2. The *Norwegian Valhall* gas field will be connected via HVDC [4] and the link is under construction. This is considered as another step towards the electrification of the North Sea, but it only involves a small power rating (78 MW). There is optimism of major oil/gas developments, and if this is well founded, there could be a need for greater stronger interconnections. It has been pointed out that old oil/gas platforms and their foundations could be taken out of service, due to reservoir depletion, and could eventually be reused for electrical installations of offshore nodes – it is claimed [16].

    A 'big-driver' *for Scandinavian hydro-power expansion* would be the integration of Norway into the NSSG. This is of special strategic importance, not only due to the oil/gas industry, but especially *to market the storage capacity of Scandinavian hydro power* [16, 27].

3. Clearly, Britain has or recently had big plans for offshore wind farms around its shores, but the one at the Doggerbank is still in planning phase, as far as this writer is aware. Nevertheless, the authors [16] are optimistic, stating that several gigawatts of wind farms are to be constructed, indicating that it could become an important node in the NSSG. Based on this argument, they consider that these three regions can more or less be determined, and they have provided the simplified topology. Several topologies have apparently also been examined but studies are still considered to be very exploratory and some unrealistic, it has been claimed.

4. CIGRE TB 447-WG B4.48 (2011), 'Component testing of VSCs system for HVDC applications': Increasing application of VSC technology in HVDC transmission demands in-depth research on the tests of related components in the subsystems in VSC system. This TB gives detailed description of component stresses under steady-state condition and fault condition respectively, followed by the test philosophies. The useful test procedures are also presented.

Finally, the reader can easily keep appraised of these offshore wind farm developments by perusal of CIGRE, IET, IEEE and other professional sources, plus manufacturer/consultant websites, e.g. www.EandTmagazine.com/news. A controversial offshore wind project off the coast of Aberdeen has been approved by the Scottish government. The 11-turbine European Offshore Wind Deployment Centre is bitterly opposed by The US businessman Donald Trump, who has complained that it will spoil the view from his nearby major golf course development.

In a final example, a CIGRE Technical Brochure, TB 483, and other support documentation have been prepared to assist workers in this sector with special problems associated with AC substation design and development. This is briefly touched on in the next section together with an accompanying outline of the scope of TB 483, which has been compiled in tabular format (see Table 23.5).

Additional note: As mentioned earlier, the UK DECC recently announced it would collaborate with the USA to develop 'floating wind turbines'. DECC indicated that this would involve developing technology to generate power in deep waters that were currently still 'off-limits' to conventional turbines (www.EandTmagazine.com/news, volume 7, issue 5, June 2012).

23.8.3 Challenges facing AC offshore substations for wind farms and preliminary guidelines for design and construction

It must be realised that AC offshore substations for wind farms still represent a 'fledgling technology' *with very much still to be learned regarding the challenges ahead.* Two recent CIGRE reports have addressed this situation meantime by the relevant WG (B3.26), and its work represents a valuable introduction/background to this topic for workers in the industry. *A new CIGRE Working Group* has recently been formed (*B3.36*) dealing with 'special considerations for AC collector systems and substations associated with HVDC connected wind power plants'. Anyone wishing to keep abreast with incremental design and construction updates/future developments relating to the work of either of these WGs (B3.26 or B3.36) is advised to refer to the full CIGRE TB 483 documentation or SC B website (regarding WG B3.36) and give consideration as to whether he/she could make effective contributions and would wish to perhaps even participate in these initiatives, over the next few years, as these topics 'mature' and robust IEC Standards are produced. At this point, let us consider briefly CIGRE documents (1) and (2) relating only to AC offshore substations:

1. CIGRE Technical Report WG B3.26: 'The Challenges Facing AC Offshore Substations for Wind Farms' (*Electra*. 2010;**253**(December)).

    This article highlights the challenges and strategic issues in the design, construction and operation of offshore substations for wind farm connections. A CIGRE Working Group, B3.26, has been established to examine these issues, provide interim sector guidelines and advice on 'best practice'.

    This article reports on the early findings, principally the design challenges that these substations present. Advice on managing these issues was reported later in (2).

    TB 483 was issued in 2011 and this document was later outlined in an article in (2) below.

    Illustrations are provided in (1) of AC offshore station, North Sea oil platform, typical configuration of an offshore substation, replacement of a transformer, installation of an offshore platform and challenges of platform access.

Table 23.5 Guidelines for design and construction of AC offshore substations for wind power plants [abridged chapter summary of CIGRE TB No. 483, WG No. B3.26: Compiled from abstracted Electra. 2011;**259**(December)]

Chapters 1 and 2: Before beginning the design process there are certain fundamental policy decisions to be made by developers/utilities involving risk assessment, maintenance policy and certification. These fundamental considerations are all addressed in the opening chapter. Arriving from these considerations, a policy for the total development of the offshore wind power plant should be established which will provide an appropriate framework for the design of the offshore substation. The design work can then begin.

However, it should be recognised that with the unusual conditions normally existing, such as long submarine cables with significant generation of VARs, necessity to comply with Grid Codes, the development of the single-line diagram for the wind power plant usually involves looking at the overall system, up to even the complete system. This involves reliability, availability and maintenance issues as well as system properties such as total substation power, reactive power management, applied voltage and harmonics. Focus is on the electrical system. The purpose is not to provide standards or solutions for the design issues, but rather to provide guidance in the considerations that need to be taken into account when designing an offshore substation.

The chapter concludes by including a list of the studies which will normally be required. The completed studies will lead to the final single-line diagram and provide some of the key parameters for the primary plant to be installed on the offshore substation. In *Chapter 3* guidance is given on the writing of the technical specifications for the main electrical equipment to be located on the offshore substation. When considering the specification aspects for equipment these can generally be divided into the following four main subgroups:

- *Parameters determined from system studies*: These parameters are strategic technical requirements such as the short-circuit level, full-load current, lightning impulse withstand level and transformer impedance.
- *Parameters defined by the operation parameters*: These are the requirements for modularity *and maintenance regime* requirements for condition monitoring, and need for special tools, for example tap-changer removal tools.
- *Parameters specific to the type*: These are items specific to the type of plant itself and could be *of plant itself* – environmental considerations, vibration and transport forces, special technical considerations and physical and interface requirements.
- *Important items to define to the platform. (It may be necessary for the equipment supplier to define supplier associated with regard to the platform supplier, specific requirements for the accommodation for the equipment room in which the equipment is to be accommodated.)*

Chapter 4 attempts to elucidate design considerations with respect to the high-voltage AC substation platforms and their associated substructures including environmental impacts, remote location maintenance issues, access management, etc. Aspects covered include an overview of the platforms and the different technologies used today; a brief discussion of the most important parameters that need to be considered; also, health safety and environment (HSE) – one of the most important subjects when working offshore. Indeed, HSE must be considered in all parts and, from the very beginning, permeate the thinking and be part of the fundamental design strategy. The boundary conditions for the design are then set – typically parameters or inputs that are external to the design and cannot be easily changed, for example local and global legislations, site location and ambient conditions such as temperature, currents, wave heights and wind speed. Unlike boundary conditions that are to be considered as more or less fixed, the next chapter discusses parts or aspects of the transmission system which will have a significant influence on the platform design but may be subject to discussion and/or iteration. (Examples of such equipment or parameters are electrical components and secondary systems, sub-structure interface, cable installations and installation programmes, commissioning tests, etc.)

Having 'set the scene' in the previous sections, the actual design philosophies, design parameters and issues within its own discipline that will have a major influence on the final platform design are discussed.

Aspects related to structural integrity, what to consider for the general arrangement layout, primary access and egress systems emergency response and platform auxiliary systems are considered. Furthermore, a comparison of stressed skin vs clad truss-based design is performed and corrosion protection, operation and installation and commissioning of plant onshore considered. This leads to different types of platform concepts like container deck, semi-enclosed and fully enclosed topsides. To some extent pros and cons of self-installing concepts like floating and jack up solutions are discussed and compared. Having thoroughly dealt with the topside, the next section covers what is underneath, that is the substructure. Different concepts are compared and the pros and cons discussed. Aspects of load out, transportation, installation and hook up and the consequences these may imply on the overall design of the topside and substructure are considered. A brief overview of the available lifting vessels is included for information. Finally, an assessment of fire and explosion design, together with fire detection/alarm and passive/active fire suppression, is presented.

Chapter 5: This chapter deals with the substation secondary systems which are those systems which provide the functionality necessary to (i) ensure safety of personnel engaged in operation of the substation and associated systems, (ii) permit operation of the substation primary circuits, (iii) monitor the performance of the installation, (iv) detect and manage abnormal conditions on the system and in primary equipment and (v) manage the environment in which the equipment operates. The detailed functionality depends on the specific installation and the way in which it is operated. The guidelines set out in this chapter assume that the offshore substation is classified as a normally unmanned (unattended) installation but allows for the use of the substation as a marshalling point for staff involved in the maintenance of the substation and associated systems.

It also looks at how the secondary equipment requirements differ from what we are all familiar with on the onshore substations.

This includes how the normal aspects such as protection, control and metering are addressed as well as these new items such as CCTV, navigation aids and aeronautical aids which are not normally associated with onshore substations.

Chapter 6: The final chapter briefly summarises the future systematic work now required, from a new CIGRE Working Group B3.26, to address the aspects associated with AC collector substations for wind power plants which will be connected by HVDC links, which was expressly excluded from the content of this brochure. It is sincerely hoped by the whole WG team involved in the preparation of this brochure that this document will assist all utilities, developers and contractors to achieve satisfactory solutions for the offshore substations required for wind power plants.

2. CIGRE: 'Guidelines for the Design and Construction of AC Offshore Substations for Wind Power Plants', CIGRE, Technical Brochure TB 483-WG B3.26 (see also *Electra*, **259**, December 2011).
    (i) TB 483 has been prepared by CIGRE, Working Group B3.26. It is intended to assist utilities, developers and contractors to achieve satisfactory solutions for offshore substations associated with AC-connected wind power plants.

(ii) All aspects are addressed including the risk strategy, maintenance identifications, certification, system design, plant maintenance, plant specification, platform structures and installations, control and metering protection.
(iii) The industry has installed many 'onshore' wind farms and has recently moved 'offshore'.

An 'edited abridged' outline summary of the chapter contents, based only on the text of *Electra* article (2), and prepared by this writer, is set out in Table 23.5. Once again, the reader is referred to the original Technical Brochure, TB 483, for fuller precise treatment of the subject area.

Table 23.5 has been included now only to provide the reader with just a 'brief overview' of how the TB process evolves to assist the industry, with many participating 'subject-expert' members (typically >110) making strategic inputs/contributions towards the final documentation.

(Together with illustrations of a typical wind farm arrangement in (1), typical SLD for Offshore S/S, rigging laid on topside prior to sea transportation, plus three foundation examples–Monopile/Jacket/Self-installing–are also shown.) It should be emphasised again that this section relates to AC offshore wind farms.

Note: It is again emphasised that *a new CIGRE Working Group* has recently been formed (B3.36) dealing with 'special considerations for AC collector systems and substations associated with HVDC connected wind power plants'. (Reports on this CIGRE Study will be issued as a Technical Brochure in the near future.)

The reader should recognise that CIGRE, SC B3 activities address a wide range of customer types including management, commercial groups, equipment suppliers, contractors, consultants, equipment maintenance providers, grid planners, operators, utilities, asset/facility managers, science education and public groups (universities, research institutes), young engineers; authorities, media, NGOs, other international organisations like CIRED, Eurelectic, IEC, IEEE, IET, T&D Europe, etc., and standardisation organisations (e.g. BSI).

The introduction to the initial report in (1) takes the reader through climate change and the ambitions of governments of Europe, identified as being the main drivers behind renewable energy. In densely populated regions, it is pointed out that:

- bulk offshore generation is being pursued to exploit the higher wind-availability figures and greater 'likelihood' of gaining planning consent.
- offshore is a relatively new arena to the electricity transmission industry; consequently, experience is very limited. It is a harsh and unforgiving environment. In contrast, the oil and gas industries have 'extensive' experience operating under such conditions.
- offshore AC substations for wind farms can therefore be considered as a 'fledgling industry', with much experience and learning yet to be established.
- early first-generation, offshore AC substations for wind farms will comprise strings of wind turbines connected by sub-sea cable (usually at 33 kV), forming feeder arrays connected to a collector substation with suitable switching facilities to optimise the onshore connection.
- The report states that to justify the investment, the offshore wind farms are in general going to be large-scale projects, typically between 100 and 1,000 MW.

These larger wind farms will operate at transmission voltages (132–220 kV), requiring step-up transformation offshore plus reactor control to manage switching and voltage regulation to transmit the power ashore in a more efficient manner. AC transmission will be considered for projects close to the shore (up to 100 km), with ratings up to approximately one gigawatt. 'However, it must be realised that economic breakeven costs will be very subjective at this early stage in the development of such plants'.

The WG B3.26 has divided the issues considered into:

- those affecting the complete system, which are described as system design issues
- design issues specific to the offshore substation, divided into electrical, physical and secondary equipment issues.

This article (1) continues in some detail but at this point only a few further initial observations will be made here! Some of the issues surrounding the design, construction and operation of offshore AC substations have been discussed:

- 'There are a number of new areas still to be considered which are not traditionally associated with the power industry and experience needs to be established'.
- *The WG indicates that initial indications suggest that offshore transmission is going to be much more expensive than onshore transmission.* Cost increases with weight, size, water depth and distance to shore and losses. It is currently being suggested that it will cost 10 times more to do work offshore than the onshore equivalent cost. This *'drives up the need' to do as much work as possible 'onshore', including commissioning.*
- Impact of weather on the construction and maintenance programme will be significant with any delays likely to cost above thousands of euro per day. It is considered probable that solutions which are based on lowest installation cost may prove more costly in the long run as operational costs are significantly higher when working offshore.

Access offshore is likely to be a major factor on the design as it affects the design life cycle, in particular the options to maintain, replace or upgrade equipment. The CIGRE WG rounded off this initial document (1) with the statements:

This is a new regime and pioneers will make mistakes but we should collectively learn from these rather than repeat them. While it is probably too early to optimise designs, they are constantly evolving as the developers learn lessons from previous projects. Appropriate 'Certification' and 'Standards' will be established to manage these issues.

- This article (1) *has identified some of the key challenges which must be addressed to help future development with their substation designs in this fledgling industry.* The brochure in (2), published in 2011, provides guidelines and possible solutions to the issues discussed in (1).
- However, it is realised that 'in a fast developing environment there will always be more problems to solve and questions to answer, especially as we look at HVDC and the potential of DC Grids'.

Notes:
1. Over the next few years, the reader will be able to follow up, and track – via CIGRE, IET, IEEE, IEC, etc. – the development and in-service performance capabilities of both onshore and offshore wind-turbine plant AC or DC. Offshore AC technology should develop from the present first-generation technology standing hopefully to a mature technology, articulating any specific weather constraints, typical outage periods, etc., and *the eventual issuing of 'robust mature IEC Standard(s)' for this sector.*
2. It will also be interesting to see how the NSSG concept touched on in section 23.8.2 develops. It is suggested that developments in the offshore grid concept must work closely with CIGRE on this initiative. Obviously, upgraded super grid links would certainly allow Europe to better share renewable energy, allowing wind and wave power generated in the North Sea to be used more effectively across the continent in support of the EU 'goal' of achieving 30% of its energy from green sources by 2020.

23.8.4 Network upgrades and some operational experiences

Tough carbon targets in the UK have given NG confidence for a £16 billion update by 2015 of the energy infrastructure (which carries gas/electricity around the UK). On 20/5/11, N. Fletcher, of *The Guardian*, quotes the NG Chief Executive Steve Holliday as saying:

- 'It is a huge task to put in place the kind of changes we need to tackle the twin threats of global warming and security of energy supply, and it is important to get (policy) consistency from government'. NG said anything that gave long-term certainty to the energy sector helped with its own investment decisions.
- New power lines are required to connect the new generation of offshore wind farms and nuclear plant. (*Costs would be cut by buying transformers and switchgear equipment from China.*)
- Holliday also stated that NG was pressing ahead with other projects such as:
  – research on carbon capture
  – a new 1,000 MW electrical 'inter-connection' link with Belgium.
- Over the next decade, 25% of the UK generation capacity will close down and NG will need to make a lot of new connections to 'connect effectively' with these new nuclear power stations and offshore wind farms which will produce more power; 'so much power we can't get it down existing lines', according to an NG spokesman (*The Times*, 02/4/11).
- More than 200 miles of new transmission line cables will be required to connect these new offshore wind farms and more powerful new nuclear power stations.
- Significantly, the DECC has also attempted to 'make it harder' for planning inspectors to ban the erection of new pylons and wind turbines.
- When put to the test by any objectors 'to a proposed new line route', this will be a real challenge, as some of the proposed new lines planned for the next

decade have routes due to pass through the *picturesque* Lake District, Mendip Hills, Snowdonia, Kent Downs and Dedham Vale, the landscape on the Essex/Suffolk border (i.e. Constable Country)! Such information has been 'drip-fed' in UK newspapers for some time.

- A new report[8] into the future UK power lines, published end of January 2012, analyses the whole life cost of installing HV transmission lines: (i) *underground,* (ii) *under-the-sea* and (iii) *over-ground.*

23.9 Some UK operational difficulties with wind farms: Independence issues

Difficulties have been experienced transmitting energy along the present grid system from wind farms in Scotland, as will now be briefly outlined [see published examples (1–3)]. Example (4) outlines a case of millions of pounds being spent to get wind farm operators to shut down. Example (5) considers possible economic implications, regarding renewables, should Scotland elect to 'go' the 'independence route' in 2014. Finally, example (6) considers The Crown Estate assets from the perspective of the Scottish independence debate. These published examples (1–6) are discussed in sections 23.9.1–23.9.6, respectively.

23.9.1 Wind farms paid £900,000 to switch off (1)

Wind farm operators in Scotland were paid approximately £900,000 to keep their turbines idle (i.e. to cease power production) on the night of 5–6 April 2011, 'because National Grid did not need the power', main reference source below.[9]

This article was written by J. Leake/M. Macaskill, entitled 'Too much power, Scottish wind power should have flowed to England but transmission lines could not cope so turbines were switched off'. Compensation payouts were as follows:

1. Whitelee[a]	£312,654 (Near Glasgow)
2. Hadyard Hill	£134,095 (A 52 turbine development)
3. Black Law	£132,263
4. Farr 2	£132,012 (Near Inverness)
5. Farr 1	£131,472 (Farr 1 + 2 represents a 40 turbine development)
6. Millennium	£32,534
Total	£875,030[b]

[a]Scottish Power claim to have invested £1 billion in the UK Renewables and committed to spend a further £400 million by the end of 2012 (*over the period 2007–12*). Whitelee is claimed to be the largest wind farm in Europe.
[b]*Main source*: *The Sunday Times*, 01/05/11, by Jonathan Leake and Mark Macaskill.

---

[8] Research and production is by Parsons Brinckerhoff in association with Cable Consulting International, with the IET providing QA of the report (available from the IET website).
[9] These organisations, including CIGRE – via its specialist Study Committee structure – rely heavily on unpaid assistance from experts worldwide within the energy community; equipment designers, utilities, consultants, testing laboratories, etc., and academia, regarding participation preparation and development of standards, technical papers, Technical Brochures, organisation of meetings/conferences, etc.

It is reported that these payments were in some cases 20 times the value of the power the six wind farms would have produced. This compensation was offered by the NG because 'it urgently needed to reduce the electricity entering the system'. Apparently:

- NG was oversupplied with power on this wet and blustery night of 5–6 April 2011, when the demand for electricity was low
- although the power could have been used in England, the transmission cables (lines) lacked the capacity to carry it south. The Scottish turbines were 'disconnected' and operators received six-figure payments to compensate for the loss of their subsidy for generating green energy. The NG apparently confirmed that it had made the payments. 'On the night of 5–6 April 2011, the demand for power was low but the nuclear power plants in Scotland were running as expected. There was also heavy rainfall which meant hydro-power plant was also operating well', a spokesman said. The English demand was met mainly by fossil fuels
- the disclosure of the subsidy payments called into question the economic logic of the subsidies paid out to wind firms and posed awkward questions, bearing in mind the cost of these payments is eventually paid for by the customers
- Leake/Macaskill reports: 'a typical turbine generates power worth about £150,000 a year but attracts subsidies worth £250,000 – designed to encourage power companies to create more wind farms. The subsidies are added directly to consumer bills. The payments emerged in research by the Renewable Energy Foundation (REF), a green think tank'
- John Constable, its director of policy and research, is reported as blaming the UK government for building wind farms in Northern Britain without ensuring there were enough high-voltage cables (lines) to take the power southwards to cities where it was most needed.

    'Hasty attempts to meet targets for renewable energy mean some Scottish wind farms are now in the extraordinary position of not only printing money when they generate, but printing it even faster when they throw their energy away,' he said
- Again wind turbines 'normally' produce less than 5% of power needs.[8] Britain is committed to a target of producing 20% of its power from its renewable energy sources by 2020.

This provides an example of generation changes in the UK, and the changes in power flow patterns that UK transmission networks will be required to handle and also the challenges facing the system operator. 'Time will tell' how well the planned network developments will be successful in meeting the challenges ahead.

23.9.2 Storm shut-down is blow to the future of wind turbines (2) (Jon Ungoed-Thomas and Jonathan Leake, The Sunday Times, 11/12/11)

These reporters observed:

- 'Britain's plans for a huge expansion in wind energy face a setback after it emerged that the power output from wind turbines "plunged by more than a half"

from 2,000 MW to 708 MW during last week's storms in Scotland; the turbine operators had predicted that they would be operating normally'. It was also reported that one wind-turbine burst into flames during last week's storm.
- The article continues, and mentions a report which, it is claimed, 'casts further doubts about all forms of alternative energy, including wind-turbines'. Moreover, it argues that the UK government's focus on renewable energy sources is misguided.

This is a joint report from the Adam Smith Institute and the Scientific Alliance, entitled *Renewable energy, vision and mirage*. It states that plans for renewable energy are unrealistic and that the technologies cannot provide the secure energy supply the country needs. If current policies are pursued, then Britain faces an energy crisis by the middle of the decade.

The authors warn: 'as renewable energy sources produce power intermittently, they cannot replace gas, coal and nuclear generation, even with further development'. The report also states, 'the storms have raised concerns among energy experts about the reliability and cost of wind turbines which will face another test this week as forecasters warn of more rough weather to come'.

During the storm, 105,000 homes in Scotland lost their electricity supply – *10,000 of these for a significant period of time!*

Energy experts say the unreliability of wind turbines means extra expenditure is needed to ensure they are 'always "backed-up" by other power sources'. John Constable, director of policy and research for the REF said, 'It is a very expensive way of generating power because this (experience) shows you need two systems running in parallel'.

23.9.3 Energy speculators now bet on wind farm failures (3) *(*The Times, *December 2011, by J. Gillespie)*

The failure of wind farms to generate sufficient energy in windy conditions, as considered in example (2), section 23.9.2 above, is apparently not a rare occurrence:

- As anticipated, subsequent failures occurred the very next week when many turbines were forced to shut down in storm winds of 65 mph, resulting in NG spending about £0.25 million to maintain electricity supplies. Added to these two major failures, it should be recalled that in December 2010, the lack of wind caused most of UK turbines to stop producing during a big freeze over Christmas. It was reported in *The Times*, December 2011, by J. Gillespie, that traders specialising in future prices of energy are now 'factoring-in' the possibility of wind farm outages for the first time.
- Traders in the energy market can exploit this volatility, pushing up the 'future prices' for electricity. 'Volatility is great if you can trade it', one market participant is reported as saying. *Apparently, wind speeds of 56 mph are enough for machines to begin switching off automatically to avoid damage.*
- When this happens, NG needs to buy from other sources, such as coal-fired power stations, to make up the shortfall and, as in any market, *if the demand*

increases while supply declines then prices go up. Gillespie comments that at one point the power from wind turbines dropped to 1,771 MW – less than 50% the amount NG was predicting – of 3,760 MW.

Energy Market Analyst ICIS Heren commented that this loss in generation from wind farms was equivalent to three nuclear reactors unexpectedly going off-line within hours of each other. Some 'renewable sources-analysts' see this increased volatility as a good opportunity to speculate to make money (or if they get it wrong, to lose money)! Gillespie comments that 'there are about 3,000 onshore turbines, generating about 3% of the nation's electricity'.

The Energy Secretary (at this time) Chris Huhne wants another 32,000 built, which will introduce more volatility into the market. 'The government is seeking to stabilise the energy market but changes will not come in until 2015, allowing speculators to trade for three years in an increasingly volatile sector', observed Gillespie.

'It's a normal part of our business to be making sure there is sufficient stand-by generation available. We're learning more and more about wind generation, forecasting and cut outs', an NG spokesman observed.

23.9.4 *Millions paid to wind farm operators to shut down (4)*

Now consider a UK strategic summary viewpoint. In another valuable article, T. Webb (*The Times*, 18/1/12) provides interesting, yet alarming, information concerning the millions paid to wind farm operators to shut down when the weather is too windy. Table 23.6 reproduces article title and includes most of its contents. Webb reports:

- For the first time the NG, an FTSE-100 company, has revealed it paid wind farms £25 million 'not to operate' in 2011. The corresponding figure for 2010 was £180,000. In 2011, NG paid wind farms to stop generating for 149,983 MW/h, equivalent to 1.49% of total wind output for year.
- This means NG effectively paid the equivalent of a large 64 MW peak capacity onshore wind farm, consisting of 26 turbines never to operate throughout the year (2011).
- Many of the payments are made to onshore wind farms in remote places, such as the Scottish Highlands where the grid has not been properly upgraded.
- NG argues that it is usually cheaper to off pay wind farms on the occasions when they would be operating at full capacity than spending billions of pounds to strengthen these isolated parts of the UK.

Webb also reports:

(a) The UK has an onshore-wind peak 'capacity' of 4.4 GW.
(b) The average wind speed in autumn 2011 was 13% faster than in autumn 2010.
(c) *On 28 December 2011*, wind turbines provided a record 12.2% of the UK electricity.

Table 23.6 *Millions blown away paying wind farms to close (money shared after particularly blustery year) (4)*
[Abstracted from The Times, *18/1/12, Tim Webb]*

---

*Wind farms receiving millions of pounds to shut down when the weather is too windy. The Times has learnt.
*Dozens of onshore facilities shared £25 million last year, a 13.733% increase on 2010, according to figures released by the NG.
*Payments to stop operating are made by NG because it cannot cope with the amount of power being fed into the system when it is very windy. But experts and consumer groups have accused wind farm operators of abusing the system by demanding excessive payments.
*Ultimately the cost of being shut down is passed on to households because NG charges energy suppliers who add the levy to bills.
*Wind farms already receive large subsidies from consumers because they cost more to operate than coal or gas plants but produce no carbon emissions.
*In total in 2011 NG paid operators to stop generating for 149,983 MWh, equivalent to 1.49% of the total energy generated by Britain's wind farms. This is equivalent to one large onshore farm being paid to be switched off all year.
*It is the first time that NG, a FTSE-100 company, has revealed how much it paid wind farms not to operate. Many of the payments are made to onshore wind farms in remote places, like the Scottish Highlands, where the grid has not been properly upgraded. NG argues that it is usually cheaper to pay off wind farms on the occasions when they would be operating at full capacity than spending billions of pounds to strengthen these isolated parts of the grid.
*On one of the windiest days in October 2011, *NG paid wind farms £1.6 million*, or £361 per MW/h, on average about four times the price that operators would expect to sell their electricity, according to ENDS (the specialist environmental information provider).
**Consumer Focus* said that wind farm operators should not be able to hold NG to ransom by demanding huge payments in return for not generating electricity.
**Richard Hall*, head of energy regulation, said: 'if wind farm generators are asked to cut production, they will clearly expect some compensation. But to keep costs down for customers, we believe this should be at a level which reflects the realistic value of the loss to the company, not an arbitrary level that the firms set themselves'.
**Ofgem*, the energy regulator, said that it had 'long-standing concerns' about the level of payments.
*Since 2007, the amount of these 'constraint payments' to all power generators has doubled, as the amount of renewable sources being built has risen. Wind farms receive a 'disproportionately high amount' of these payments compared to coal and gas plants.

The size of payments will soar further as Britain tries to meet its target of generating a third of its electricity from renewable sources, mostly wind farms, by 2020.

Phil Hare, vice-president for North West Europe for Poyry, Management Consulting, the Energy Consultant, said: 'if wind farms are receiving much more in constraint payments than they would if they sold the electricity, they are making "a-turn" they shouldn't be. By 2020, because of all the wind farms which will (then) be in the system, the ups and downs of power generation will be staggering and very hard to deal with'.

Note: It should be appreciated that other forms of generators are also paid to switch off if requested. The chief executive of Renewable Energy Association observed (*The Times*, 9/3/12) that *ten times more was paid to coal plants not to generate in 2011*.

23.9.5 Clean energy financial support; impact of Scotland leaving the Union after an independence vote in 2014 (5) (Karl West, The Sunday Times, 22/1/2012)

Karl West makes an interesting point in his recent article that if Scotland breaks away and votes for independence in 2014, with Scotland leaving the Union by 2016, its strong and thriving clean energy sector could lose billions in support. The reader is encouraged to examine the original articles (for this item (5) and the following item (6) in section 23.9.6) carefully and look at the possible strategic and financial implications to Scotland. We will touch on only a few aspects at this time. West points out that:

- 'Scotland currently has a flourishing renewable energy sector. It accounts for approximately 50% of the UK's total wind capacity and 25 of the 30 wave power projects around the UK's shores are located off the Scottish coast'. (The Pelamis wave energy device is located at Isle of Hoy, Orkney.)
- However, this leading position (of Scotland) in green power has been subsidised by UK taxpayers. Investors and analysts are currently questioning whether this dominance could be at risk if the Scots vote for independence in 2014, with Scotland leaving the Union by as early as 2016. West comments, according to research by Citigroup: 'the maths just "do not stack up" for the sector if Scotland breaks away. Most of the projects are based north of the border, while 93% of the subsidies to support them come from electricity consumers who live outside Scotland'.

West also reports that Peter Atherton, Utilities and Renewables Analyst at Citi-group, wrote in a 10 January 2012 research note to clients:

> 'This prospect represents a significant political risk for utilities and renewable energy companies operating in Scotland'. 'Given the uncertainty over who would pay to subsidise the sector if Scotland "goes-it-alone", Citi-group warned that 'investors should exercise caution to committing new capital to renewable power in Scotland'.
>
> (Reader should be aware of *counter-views* to this scenario (see original article).)

Meantime, some interesting supplementary points on this topic, as supplied by West, are noteworthy.

In the run up to this 2014 referendum, *Alex Salmond, Scotland's first minister*, will be under some pressure to explain in detail how an independent Scotland will overcome the subsidy problem.

- 'Onshore and offshore wind projects currently derive 50% and 60% respectively of their revenue from the UK wide subsidy mechanism. This financial support is issued through ROCs, which are given by Ofgem, the energy regulator, for supplying green power. In general, one Certificate is issued to generators for each megawatt hr of renewable energy output. Offshore wind installations receive two Certificates for each MWhr of output'.
- 'This means that as only 7% of UK electricity consumers are in Scotland 93% of the green subsidy is funded by consumers who live outside Scotland'.

Finally, some strategic financial, wind capacity and green power data have been supplied by Karl West in his thought-provoking article ('Scots face threat to pull subsidies plug', *The Sunday Times*, 22/1/2012), which is reproduced below:

Financial support [The renewable sector is heavily subsidised by electricity consumers]

By 2012		The subsidies
Cost of doubling renewable		Funded by all UK electricity consumers
Generation capacity in Scotland	£46 billion	7% in Scotland
Subsidies flowing to assets in Scotland	£4 billion	93% outside Scotland
The UK's total renewable electricity generation [37% Scotland, 63% rest of the UK] Scotland's share of the UK population is 8%		
The UK's current wind capacity		
Operational onshore and offshore		Almost 50% in Scotland

Green power [after James Gilllespie, *The Sunday Times*, 11/3/12]
High wind speeds and rough seas make Scotland an ideal location to generate green power:

Onshore wind in operation:	2,595 MW	Under construction:	965 MW
Offshore wind in operation:	190 MW	In 'scoping':	9,930 MW
Total onshore and offshore wind:			2,785 MW
Capacity in operation in Scotland			
Scotland has a huge resource in hydro.			1,396 MW
[Approximately 90% of the UK's total]			

In summary: Clearly, there is still much to be resolved regarding the implications of future Scottish independence (5) and 'how it all might unfold' in the energy sector. There could also be other implications for the business sector in Scotland and elsewhere. Some additional statistics (6) are included in section 23.9.6, which considers the resources of The Crown Estate – there is still much for Scotland – and indeed other regions in the UK to assess and discuss from this data, and when it is linked with other offshore activities and UK international agreements with Iceland, Norway, etc. In particular, the comment by the CE of The Crown Estate, Alison Nimmo, indicated that agreements would give 'much greater control to local communities and give them much more say in how they use the assets'. It will be interesting to see how this 'unfolds with time' – but for all regions in the UK and not just Scotland.

23.9.6 Crown Estate: Scottish assets worth arguing over in independence debate (6) (Deirdre Hipwell, The Times, 21/06/2012, pp. 34–5)

In this article (6) Deirdre Hipwell moves the impact of the possible Scottish independence vote (2014) on from (5), in section 23.9.5, when she deals with the strength of The Crown Estate (6), which, as she says, may add more fuel to Alex Salmond's fight for Scottish independence. The main aspects from her article have been edited and have been reproduced in Table 23.7 together with the portfolio map of the UK and legend.

Table 23.7 Scottish assets worth arguing over in independence debate (6)
[after Deirdre Hipwell, The Times, 21/06/2012, pp. 34–5]

Deirdre Hipwell starts her article with observation that the strength of The Crown Estate may add more fuel to Alex Salmond's, the Scotland's first minister, fight for Scottish independence. She states:

- When he is not arguing why Scotland's future should not be in its own hands, he is advocating the need to reclaim control of The Crown Estate's offshore Scottish assets.
- In March 2012, The Scottish Affairs Select Committee said that the revenue from the seabed around the coast of Scotland should be taken away from The Crown Estate and handed back to local communities.
- This committee's hard hitting report claimed The Crown Estate behaved like an 'absent landlord or tax collector' and did not reinvest in local communities, reports Hipwell – who goes on to write – 'Westminster has refused to agree that power over Scotland's marine resources should be de devolved and its response to the influential Commons committee is still pending'.
- Alison Nimmo, the chief executive of The Crown Estate said: 'Scotland is a very important part of our business. The report has highlighted a number of areas where we can improve around local accountability and transparency and we are working with the local communities and looking at how we can improve the way we do business in Scotland'. She went on to say that local management agreements would give 'much greater control to local communities and give them much more say in how they use the assets'.
- Control of Scotland's seabed, out to 12 nautical miles, includes valuable rights to salmon fishing, wild oysters and mussels and culturally important sites, such as the King's park at Stirling. The Crown Estate also controls the rights to the sites of fish farms, renewable energy developments, ports and marinas.
- In The Crown Estate's results for the year to end of March 2012, Scotland accounted for 3.9% of a total £314.2 million of revenue. The total property value of the Estate's Scottish assets, excluding properties in joint ventures, is £220 million. Revenue from leasing Scotland's seabed could reach £49 million in 2020.
- Ms Nimmo said that 'a lot of the work it had done in Scotland has been praised' and she added: 'we want to keep working constructively with the Scottish Government and the Scottish people'.

Some Crown Estate's statistics

1956		The year The Crown Estate was officially listed as an organisation.
		The four divisions of The Crown Estate are **Urban, Marine** (including energy), **Rural and Windsor**
Urban	£5.5 billion	£1 billion being invested in The Crown Estate's 20 year Regent Street regeneration programme
		14 Retail Parks in Britain are owned by The Crown Estate
Rural	£1.2 billion	263,000 acres of livestock and arable farming land owned by The Crown Estate
		66,500 acres of common land, principally in Wales, owned by The Crown Estate
Marine	£725.6 billion	Revenue rose by 17.3% to £55.6 million
		Demand for construction aggregates, such as sand and gravel, has been boosted by Olympics
The Windsor Estate		Revenue was up 8.8% to £7.4 million
£195.9 billion		15,600 acres size of the windsor estate
£240		Profit in the year to 31/3/2012, reflecting 4% increase on 2011
£2 billion		The amount The Crown Estate has contributed to the Treasury's coffers in the last decade
£7.6 billion		Overall value of The Crown Estate's property portfolio
£8.1 billion		Overall capital value of all The Crown Estate's assets which is up 11% on the prior year
The Portfolio		**See map of the UK and legend** in original article.
Marine		**Energy**
Moorings and marinas		Round 1 + 2 farms
Stewardship projects		Wind farms in Scottish territorial waters
Aquaculture		
Dredge areas		Wave + tidal

23.9.7 Some poor wind farm performance statistics

Finally, Table 23.8 reproduces – without further comment – some recent perceptions by James Gillespie (in *The Sunday Times*, 11/3/12) and a few additional performance statistics on some of the worst UK wind farms, but still receiving very generous government subsidies and some other energy source data.

Table 23.8   Worst wind farms 'puffed up' by subsidy
[after James Gillespie, The Sunday Times, *11/3/12]*

Gillespie states: Generous government subsidies are enabling Britain's 10 worst-performing wind farms to earn a total of £1.3 million a year, despite producing electricity worth only half that, according to new figures.

   Among the poorest performers are (1) two turbines at GlaxoSmithKline Pharmaceuticals (GSK) plant at Barnard Castle, Co Durham, and (2) Ecotricity, a wind turbine at Green Park in Reading, Berkshire, which the company boasts is 'Britain's best-known,' because it is adjacent to junction 11 of the M4.

1.   The Barnard Castle turbines running at just 8.2% of capacity earned £26,000, half of which was paid as subsidy by the government's renewables obligation scheme.
2.   The turbine beside the motorway (as Gillespie refers to it!) runs on average at just over 16% of its capacity. It earned £229,000 in 2010–11, but half of that was paid in subsidy. The electricity generated was worth about £115,000.

The outputs of three other farms were also quoted as a percentage of capacity as

3.   Blyth Harbour wind farm Northumberland   8.7%
4.   Chelker wind farm, North Yorkshire   10%
5.   Castle Pill wind farm, Penbrokeshire   10.9%

Gillespie goes on to state that last week Edward Davey, the new energy secretary, admitted for the first time *that the number of existing wind farms added to those going through the planning process was already enough to meet the target the government had set for 2020.*

   Chris Heaton-Harris, Conservative for Daventry, Northamptonshire, said the speed with which they were being erected showed the subsidy level was too high and that turbines were being put up in inappropriate places to take advantage of the payments. Although the worst performers are relatively small farms, the figures were not much better for some larger ones.

   Whitelee in East Renfrewshire, Britain's biggest onshore wind farm, with 140 turbines, runs at 20% of capacity or load factor – the proportion of power generated compared with the theoretical maximum.

   This was still enough to ensure that, in 2011, it received £31 million in subsidy according to government figures.

   The REF based its analysis on returns submitted to Ofgem, the energy regulator, by the wind farms. According to Gillespie, they show that out of 300 UK onshore wind farms with a capacity of more than 500 kW, only six worked at more than 40% capacity.

- The industry generally quotes average load factor of 30%. In fact, 79% of Britain's wind farms are operating at less than 30% of capacity.
- By contrast, the load factor of nuclear stations in 2010 was 59.4%, a long way from its peak of 80.1% in 1998.
- Gas power plants in 2010 had a load factor of 60.6%.
- Coal-fired plants averaged a load factor of 40.9%.

The renewables industry blames the poor performance in 2010 on 'a shortage of wind'. However, the figures also raise questions about where turbines are being built/installed. All the top 10 performers are in Scotland. The only two that operate at more than 50% of capacity are on the Shetland Isles.

   John Constable, director of REF, said:

   The consumer is paying to make these wind farms artificially viable. Why are our subsidies so generous that it makes sense to build wind farms in places where there is frankly no wind? This is just not smart.

   A spokesman for the Renewable UK, the trade association, said: 'The load factor is only one part of a wide series of considerations, when deciding the best form of generating electricity. To concentrate on low-load factors is to miss the point. Wind is a way of adding benefits into the mix in terms of producing low-carbon energy, which in the long run we believe is a cost-effective form of energy'. Although the government does not pay a subsidy for building wind farms, the returns from state payments for generating electricity are such that a developer on average recoups their investment after just over ten years.

   Dale Vince, founder of Ecotricity [ii], which runs the M4 Turbine, said, 'the REF's claim that capacity factor is the same thing as efficiency, is nonsense and they know better'.

23.9.8 'Flying wind farms pluck energy out of the blue', states Gillespie in recent The Sunday Times *article (08/07/12, p. 7)*

James Gillespie reports:

They have been condemned for scarring the countryside and blighting our coastlines but prepare for a new cloud on our horizon: wind farms are taking to the skies.

Airborne turbines, now being developed by commercial companies, have the advantage that the wind is stronger and more consistent at higher altitudes. Traditional turbines and in particular those 'on land' have often been criticised for poor output because of varying wind speeds.

One of the most advanced airborne turbines has been developed by Altaeros Energies, based in Boston, Massachusetts. Apparently, this company has carried out a successful test flight with a 15-yard-wide prototype which reached an altitude of 350 ft. A circular, helium-filled, inflatable shell with propeller at the centre is used in this Altaeros airborne or flying turbine, which is tethered and sends the electric power down to ground.

Altaeros claim that the airborne turbine can be in operation within days of a suitable site being chosen and a positive feature with almost no environmental or noise impact. Adam Rein of Altaeros said, 'In the long term we see the biggest opportunity in the offshore sector – places like the North Sea off the Coast of Scotland, with a huge potential wind resource'.

NASA, the US space agency, has also been examining airborne turbines but is currently focusing on a kite device. NASA came up with its own 'turbine kite' to generate electricity. This system apparently involves a generator on the ground which collects energy from the tether of the kite as it catches the wind.

Additional note: As mentioned earlier, the UK DECC has recently announced it would collaborate with the USA to develop 'floating wind turbines'. DECC indicated that this would involve developing technology to generate power in deep waters that were currently still 'off-limits' to conventional turbines (www.EandTmagazine.com/news, volume 7, issue 5, June 2012).

23.10 UK air-defence radar challenged by wind turbines

The reader should also be aware that in the past few years, press reports have drawn attention to concerns that UK air-defence radar faced a challenge due to the presence of wind farms. It is therefore pleasing to see (www.EandTmagazine.com/news, volume 7, issue 1, February 2012, p. 11) an article by Lorna Sharpe entitled 'Air defence radar orders clear the way for wind development'.

1. The article reports that SERCO Group has been awarded a £45 million contract by MOD to use a different innovative high technology, L band radar system, which uses high-integrity pulsed Doppler radar processing. This technology, which will be used at two sites in NE England and one in Norfolk (already installed), *should now prevent wind farms interfering with UK air defences.*

2. The deployment of this new technology to radar systems in the UK *allows the development of up to 4 GW of renewable energy.* It clears the way for MOD and DECC towards significant further wind farm developments to help the UK government achieve its target of reducing carbon emissions.

23.11 Noise pollution: wind turbine hum (*The Sunday Times*, 18/12/11)

Another possible concern with wind power relates to turbine hum, as reported by David Derbyshire in a recent article entitled 'Thousands at risk of turbine hum'.

This is a rather worrying article for the public as it conveys *warnings of acoustic scientists who estimate that up to 20 of wind farms generate a low-frequency hum (or pulsating audible sound, or noise), which can be audible for a distance of more than a mile.*

As the UK government is planning a huge expansion of wind turbines, Derbyshire reports that some experts have urged that limits on wind turbine noise of 35 dB during the day and 43 dB at night – the equivalent of a buzzing fridge – should be lowered. Apparently

- Wind-turbines may generate low-frequency.
- Noise penetrates distant buildings.
- Noise emanates from blades, whose tips can move at more than 100 mph.
- Pulsating sound is generated by air passing over a blade and by vortices at trailing edge.

Noise experts refer to the pulsating sound as amplitude modulation (AM) which experts believe to be caused by the turbine blades' striking patches of turbulent air. It seems that M. Stigwood, a noise consultant, has estimated that 20% of wind farms suffer from AM.

- He believes that this is caused by differences in wind speed and turbulence at the top and bottom of the blades as they spin.
- It appears that walls and roofs tend to filter out higher frequency sounds but allow deeper noises such as AM to penetrate.

The charity, *REF*, has called for tougher regulations and its spokesman Dr Lee Moroney, the planning director, is reported as saying, 'Noise of this kind disrupts sleep and so can have a devastating effect on people's lives. The solution is greater separation distances between turbines and dwellings'.

Now a later variation on this noise theme: 'Council tax cut for homes near wind farms', writes J. Leake (*The Sunday Times*, 22/7/12, p. 14). Briefly, The Valuation Office Agency (VOA), which decides council tax valuations, has accepted that having a wind farm built near your home can sharply reduce their value and has as a result moved some into lower tax band. The same picture emerges:

- People living near newly built wind farms say the turbines have slashed property values through noise and visual intrusion.

- *Shadow flicker*: When sun gets behind moving turbine it generates fast moving shadows, experienced as a powerful 'strobing' effect that can penetrate blinds and curtains.
- *Visual*: Turbines tower above surrounding buildings, constantly catch eye when moving.
- *Noise*: Turbulence around moving blades generates swooshing sound that keeps changing in pitch and loudness.
- *Vibration*: Some people living near turbines report vibrations in windows, floorboards and doors.
- Some people consider that the blame lies with the government planning guidelines known as ETSU-R-97 which Jonathan Leak comments was drawn up in 1996 when turbines were smaller, and less was known about their impact. Despite the complaints, the DECC revalidated the rules last year with few changes.
- Jonathan Leak reports that the director of REF, a charity publishing data on the energy sector, commented, 'current policies are making renewables deeply unpopular by creating a few rich and happy winners and a large number of very angry losers'.

In contrast Renewable UK, the industry body, said a review of AM was under way, adding that 'we think that the regulations are robust enough. Once research into AM is finished, we will look into ways we can minimise it'. (*Closing the gate after the horse has bolted, comes to mind here!*)

Further points:

- Current rules governing the noise created by wind turbines are based on continuous background noise and do not take into account the bursts of loud noise also associated with AM.
- To date, one wind farm operator has been taken to court by a married couple who claimed that the noise from nearby wind turbines had forced them to leave their Lincolnshire home. Derbyshire reports that they settled for an undisclosed sum.
- In another problem situation, a householder has apparently been unable to open the windows of his bungalow since a 360 ft turbine some 500 yards from the house was switched on in 2011. This householder is quoted as saying: 'the noise is absolutely horrendous. It is worst in the evenings when everything else is quiet'.
  - Clearly, extensive and thorough noise measurement studies are required on wind turbine farms in proximity to houses and results need to be carefully evaluated before realistic, acceptable robust international standards can be produced.
  - *It would appear that a satisfactory resolution to this AM noise issue is required soon and, if appropriate, tighter and tougher regulations may be necessary.* This should be given high priority.
  - The estimation that 20% of wind farms could suffer from AM is very alarming bearing in mind the planned expansion of wind farms.

(It may not be straightforward to reduce, retrospectively, the 'noise level impact' for existing wind farms.)

In passing, it should also be noted that, in some instances, the presence of wind turbines is also known to slightly 'perturb' local weather patterns downwind of the blades and also ground temperatures can sometimes be affected (up to 0.5 °C in one reference and a 4 °C variation in another have been reported) – close to the turbines, upwind/downwind of turbines. Wind turbines have also been accused of blighting wildlife on land and sea. 'The report in the *Marine Pollution Bulletin*, based on a four year research programme, commissioned by E.ON highlighted the devastating impact of building offshore wind farms close to the habitat of Britain's seabirds in Great Yarmouth, Norfolk (between 2002 and 2006),' said John Simpson. A better understanding of the impact on wildlife of wind farms is clearly required (after J. Simpson, p. 13, *The Times*, 30/4/12).

23.12 Balancing fluctuating wind energy with fossil power stations

In an *Electra* article, 2002;**204**(October), when considering the topic of 'balancing fluctuating wind energy with fossil power plants', the authors Leonhard and Muller posed the question, 'where are the limits?' At the start of their article, some thought-provoking comments were made:

- 'Wind energy fed to the grid to save resources and reduce emissions, requires control power for balancing fluctuations; this causes fuel losses in the thermal power stations and limits the degree of energy substitution.
- *Facilities for energy storage are needed* when greatly extending wind power use offshore, at the same time generating secondary fuel for stationary and mobile applications'.
- In contrast to energy produced from storable hydro and biomass, which can be matched quickly, effectively and closely to the varying energy load demand needs, wind and sun are considered as 'sustainable future resources', and development, that is conversion to electricity, is encouraged via generous government subsidies. Despite this enthusiasm, there are problems when wind or sun power is used to produce electricity, that is that electricity must match consumption in real time, whereas the power from wind (and sun) is complex and both subject to seasonal and climatic variations/changes and also statistical effects.
- 'If the wind energy is not converted to electricity immediately it is lost. This is why large storage facilities are needed to match supply and demand by preserving (i.e. storing) temporary surplus energy for periods of insufficient supply – allowing for unavoidable energy losses'.
- While the development of such storage provisions is feasible, in principle it would require changes in environmental policies, which are at present (i.e. based on a paper written in 2001) focused on direct grid connection.
- *Wind energy to substitute fossil resources* – Leonhard and Muller comment: 'Feeding with priority wind-generated electrical energy into the supply grid is intended to supplant electricity from traditional power stations with the aim of reducing fuel consumption and emissions. When substituting power from hydro stations, emissions are not directly diminished, but natural energy is upgraded into storable and more valuable potential energy, making it available for later uses'.

- For a fuller review of this study, as these authors report, in Northern Germany at that time (2001), where wind farms were concentrated and wind power must predominantly be balanced by coal- or gas-fired generation, the situation is different; the conversion processes are slower and more complicated than with hydro stations and they can be operated only in a restricted power range, for instance above half-rated load.
- Priority wind energy pushes these generating stations towards part-load operation where the output-related fuel consumption and CO_2 emissions per kilowatt-hour rise; hence, part of the expected savings is not realised but evaporates in the form of hidden 'control losses'. Apart from cost issues, this raises questions of how effective present environmental policies really are.

In their paper, Leonhard and Muller considered 'that since the use of renewable energy is aimed at the substitution of fossil fuel resources, they would make an attempt to quantify these effects of energy substitution with the example of a fairly large grid supplied mainly by fossil fuels'. Briefly, in order to quantify the proclaimed savings in a grid control area mainly supplied by coal or gas fossil power stations, the secondary control system with its generating units and associated grid controller was modelled, subject to particular power scenario, observed in April 2001. 'On this occasion, when the wind represented <15% of the load energy, the wind power varied widely unrelated to the grid load'.

Once again, for a fuller review of this paper, the reader is directed to the original publication which also considers the development of grid control algorithms, shows results of many simulation studies and depicts examples of the efficiency vs load curves of fossil generating plants. Efficiency is usually decreasing at part load. Hence, the specific fuel consumption and emissions per kilowatt-hour rise when plants are operated outside the preferred optimum efficiency region. One would expect further publications on this topic in the near future to advance knowledge. It seems that a significant amount of further study is required on this activity. The simulation study results showed that even at this low penetration of wind energy, the 'infeed' caused a hidden increase of the specific fuel consumption in remote fossil fuel generating stations. Expressed differently, they are now processing less electrical energy but with a higher fuel consumption and CO_2 emissions per kilowatt-hour.

23.13 Future developments including smart grids

In many countries, the existing power delivery infrastructures were designed several decades ago and they are not appropriate to meet the needs of the twenty-first century, with the restructured electricity market place and the increased electricity load demands, opportunities and also increasing risks of cyber-threats with the widespread usage of IT. Moreover, examples of the scale and frequency of such threats will be discussed briefly later in this chapter – see Appendices B and C for cyber-crime and treatment of information security for electricity power utilities (EPUs) respectively, providing instances of various malicious international and other cyber-threats to the business sector, major energy equipment manufacturers, energy networks and industrial premises. Appendix C briefly considers a framework for

EPUs on how to manage information security and also the CIGRE treatment of information security for EPUs. (There is a significant body of 'follow-up' articles within CIGRE on this theme.)

Rather worryingly, investment in expansion and maintenance of electrical transmission and distribution infrastructure has generally not kept pace with the load growth, technological developments and dramatic changes to energy mix with the recent vast expansion in renewable energy wind farms in the UK and abroad. Indeed, in the UK investment in the expansion and maintenance of the electrical transmission and distribution networks has been lagging for decades and UK consumers are angry at having increasingly to 'foot increased bills and much of the investment costs' in recent years.

These are unfortunate considerations, since there is now an urgent commitment to modernise and improve the power delivery systems in the UK and worldwide, such that all necessary enhancements can be made to get systems up to a fully functional power delivery system, or 'smart grid' as experts have so named this enhanced status, within the next 10–20 years! Furthermore, the fact that the world is now in a recession and the UK is in 'double-dip' economic recession does not help matters.

A valuable strategic *Electra* article by *Gellings* [30] from EPRI, USA, is discussed further. This indicates the huge expenditures of >$100 billion that could be required to achieve 'smart-grid' status in the USA. Europe is also facing an urgent need for substantial power generation and infrastructure investments over the next decade, estimated at >1 trillion euro by 2020, as EU and other international organisations have developed strategies for a low-carbon economy by 2050, and industry has promised to achieve carbon neutral status by 2050.

So the long-term vision for 'smart grids' is clear, having been widely debated in theory, but as many experts have already commented, turning this into a reality in the USA and EU is quite a different matter and the position has been further complicated by the considerable changes, and also future uncertainties, in energy mix – and the huge switch to wind turbine renewable energy, at very great cost, or not, in several countries including Britain at a time of worldwide economic downturn.

The emergence of renewables, distributed generation and future electric vehicles demand, effective energy-storage capabilities, an infrastructure that actively integrates the actions of generators, consumers and 'smart grids', clearly require smart legislation to realise the full potential.

Obviously, much is still to be done to ensure European distribution grids can take 'upcoming challenges' in the next decade, for the benefit of customers and society as a whole. However, strategically, there is still much to be fully resolved as to what the final energy mix will be 'one or two decades down the road'.

To develop this theme further, and provide the reader with a brief relevant outline of this *'smart-grid' concept and of the scale of costs, benefits and enhancements to achieve this position – some years down the road*, advantage will now be taken to discuss two distinct and contrasting strategic viewpoints, presented in recent CIGRE *Electra* articles:

1. A recent *Electra* article by C.W. Gellings [30], from EPRI, USA, deals with a comprehensive study, 'Estimating the costs and benefits of the "smart grid" in

the United States'. The full study is available online (www.epri.com, 'Estimating the costs and benefits of the smart grid', TR-1022519)
2. The second viewpoint relates to a CIGRE perspective [35] on electricity supply systems of the future, a TC Advisory Group Report entitled CIGRE WG 'Network of the Future'; Electricity supply systems of the future.

Briefly, CIGRE states that its aims [35] are being primarily driven by environmental concerns which are the landmark of this decade and will certainly also shape the energy systems of the future. CIGRE will aim to:

- 'Identify the key parameters, which, in the opinion of participating CIGRE WG experts, will shape the Networks of the Future, in a time horizon of two decades
- Highlight the relevant activities which have been and are still taking place within the various Working Groups of CIGRE
- Identify missing activities, which could be the objectives of future working bodies
- Identify synergies within and outside CIGRE'.

The over-reaching CIGRE aim 'is to contribute to the vision and the development of the Future Energy Supply Systems in a global and co-ordinated way'.

It is now convenient to discuss (1) and (2) separately, with most of the available technical information now reformatted and arranged/set out in detail in tabular style (*see Tables 23.9–23.14*).

For this comparison exercise, the reader has a choice:

1. to quickly scan through this material initially to get a quick and initial 'feel' for forthcoming developments
2. as/if appropriate, look into topic materials/issues again later at greater detail up to the individual level required.

23.13.1 US study by Gellings et al. from EPRI [30] (1)

The full study is available online (www.epri.com, 'Estimating the costs and benefits of the smart grid', TR-1022519). For convenience, the estimated costings quoted by Gellings are reproduced and summarised below as:

(A) Summary of estimated cost and benefits of the smart grid		(B) Total smart-grid costs		
20-Year total ($ billion)		Costs to enable a fully functioning smart grid ($ million)		
			Low	High
Net investment required	338–476			
Net benefit	1,294–2,028	Transmission and substations	82,046	90,413
Benefit-to-cost ratio	2.8–6.0	Distribution	231,960	339,409
		Consumer	23,672	46,368
		Total	337,678	476,190

Over and above the investment to meet electric load growth, the estimated net investment needed for the 'smart grid', that is the estimated net investment needed to realise the envisioned power delivery system of the future in the USA, according to Gellings, '(A) is between $338 and $476 billion. The total value estimate range of between $1294 and $2,028 billion when compared to the future power delivery system cost estimate results in a benefit-to-cost ratio of 2.0 to 6.0'.

- Thus, based on his underlying assumptions, this comparison by Gellings [30] shows that the benefits of the envisioned future power delivery system significantly outweigh the costs.
- This indicates 'an investment level of between $17 and $24 billion per year will be required over the next 20 years. The costs cover a wide variety of enhancements to bring the power delivery system to the performance levels required for a smart grid'.

The costs include the infrastructure to integrate distributed energy resources (DER) and to achieve full customer connectivity, but exclude the cost of generation, the cost of transmission expansion to add renewable sources and to meet load growth, and a category of customer costs for smart grid-ready appliances and devices. A summary of estimated cost and benefits of the smart grid is given in (A) above.

Included in the estimates of the investment required to realise the smart grid are estimated expenditures needed to meet load growth and to enable 'large-scale' renewable power production. ('As part of these expenditures, the components of the expanded power system will need to be compatible with the smart grid', Gellings reports.)

Turning now to the benefits of the 'smart grid', the total smart-grid costs in (B) above show the various types of improvements that correspond to each of the three attribute types used (see Table 23.9), 'in the root study' by Gellings, who points out:

- 'A key aspect of the value estimation process in general is its consideration of improvements benefits to the power delivery system [b] (left-hand column), as well as improvements benefits that consumers directly realise [b] (right-hand column)'.
- Apparently, this was done by Gellings to ensure that emerging and foreseen benefits to consumers in the form of a broad range of value-added services are addressed in the estimation of value.
- The cost of energy attributes is the total cost to deliver electricity to customers including capital costs, O&M costs and the cost of losses on the system.
- Therefore, the value of this attribute derives from any system improvement that lowers the direct cost of supplying this electricity. SQRA (security, quality, reliability and availability) is the sum of the power security, quality and reliability attributes, because the 'availability part' of SQRA is embedded in the power quality and reliability attributes.

Table 23.9 *Attributes and types of improvements assumed in the value estimation of the future power delivery system [After C.W. Gellings] [30]*

(Left: Power Delivery System Improvements)   (Right: Improvements That Consumers Realise)

Power delivery [b] (Improvements Benefits)	<< Attributes >>	Consumer [b] (Improvements Benefits)
O&M cost Capital cost of asset T&D losses	**Cost of energy** **(Net delivered life cycle cost of energy service)**	End use energy efficiency Capital cost, end user infrastructure O&M, end user infrastructure Control/manage use
Increased power flow New infrastructure Demand responsive load	**Capacity**	Improved power factor. Lower end user Infrastructure cost through economies of scale and system streamlining, expand opportunity for growth
Enhanced security Self-healing grid for quick recovery	**Security**	Enhanced security and ability to continue conducting business and everyday functions
Improve power quality and enhance equipment operating window	**Quality**	Improve power quality and enhance equipment operating window
Reduce frequency and duration of outrages	**Reliability and availability**	Enhanced security Self-healing grid for quick recovery Availability included
EMF management Reduction in SF_6 (sulphur hexafluoride) emissions Reduction in clean-up costs Reduction in power plant emissions	**Environment**	Improved aesthetic value Reduced EMF Industrial ecology
Safer work environment for utility employees	**Safety**	Safer work environment for end-use electrical facilities
Value added electric related services	**Quality of life**	Comfort Convenience Accessibility
Increase productivity due to efficient operation of the power delivery infrastructure Real GDP	**Productivity**	Improved consumer productivity Real GDP

Note: In his conclusions, Clark W. Gellings [30] commented, 'Implementation of the Smart Grid will result in Substantial benefits to society including improved reliability, greater comfort and control as well as improved energy efficiency and productivity along with improved environmental performance. These benefits will require a concerted effort across the industry to enable and deploy the technologies mentioned here involving all the stakeholders and policy makers'. Supporting references provided by Gellings:

➤ 'Future inspection of overhead transmission lines', EPRI, Palo Alto, CA: 2008. 1016921
➤ 'Automated demand response tests: An open ADR demonstration project', EPRI, Palo Alto, CA: 2008. 1016062
➤ 'Estimating the costs and benefits of the smart grid', EPRI, Palo Alto, CA: 2011. 1022519.

A further useful reference:

➤ 'Power system infrastructure for a digital society: creating the new frontiers', keynote speech (Montreal Symposium CIGRE/IEEE), Marek Samotyj, Clark Gellings, and Masoud Amin, *Electra*. N°210, October, 2003.

The quality of life attribute, quoted by Gellings, refers to the integration of access to multiple services, including electricity, the Internet, telephone, cable and natural gas. This involves the integration of the power delivery and knowledge networks into a single intelligent electric power/communications system, which 'sets the stage' for a growing variety of products and services designed around energy and communications.

This US study (1) reported by Gellings [30] considers a variety of new technologies as implied by the functionality described above. These technologies are touched on separately as:

(A) transmission systems and substations
(B) distribution
(C) consumers.

and are now abridged slightly, and Gelling's views are reproduced in a tabular format in Tables 23.10–23.12 respectively, which deal separately and at some length with (A) transmission, (B) distribution and (C) consumer sectors.

It is hoped that this material [30] provides the reader with a 'flavour' of (1) Gelling's detailed strategic smart-grid concept and just how much is still to be achieved [30]. Next, it is appropriate to move on and consider the achievements of the CIGRE initiative (2).

(Once again, the reader is referred to the full study available online at www.epri.com, 'Estimating the costs and benefits of the smart grid', TR-1022519, together with future developments worldwide as they are reported in the technical literature CIGRE, IET, IEEE, etc.)

23.13.2 Some CIGRE perspectives of energy activities and future development [35] (2)

First, some CIGRE background information: CIGRE (International Council on Large Electric Systems) (www.cigre.org) has also been a key player, since 1921, in the development of electrical power systems worldwide. This organisation works closely on many strategic energy programmes/themes with IEEE, IET, IEC and other international professional bodies and standards organisations worldwide. As mentioned earlier, CIGRE does not carry out/undertake specific system or equipment development work, 'but is an independent and critical analyser of different solutions and the provider of high-quality, unbiased publications and other contributions to the electrical supply industry'. CIGRE considers [35] it adds value by:

- 'expressing its view on the different conditions/solutions which preparation of standards and are appropriate in different regions of the world
- identifying new issues and challenges to be investigated
- providing information about the development of new techniques, indicating the challenges for new development or for new applications of existing techniques
- supporting and collaborating with associations such as CENELEC, IEC, BSI, IEEE, IET, etc. for the development of new technical standards and best practices' [35].[8]

The 'distinctive published outputs' from CIGRE can provide the reader with 'ongoing strategic information' relating to established or new emerging energy topics/ issues and technical/practical aspects of strategic concern 'garnered-worldwide' over

Table 23.10 Towards the smart grid: cost, components etc. (after Gellings [30])

[A] Transmission Systems and Substations	[B] Distribution	[C] Customers

[A] Transmission Systems and Substations

Gellings informs the reader that:
The total cost for enhancing the US transmission system and substation performance to the level of a 'smart grid' is estimated as between $56 billion and $64 billion. The cost includes several categories of technology whose functionality overlaps significantly between the transmission system and substations and also some elements of the distribution system described later, and also enterprise level functions such as cyber-security and back-office systems.

Cost components of the smart grid: Transmission systems and substations
Core components of cost for the transmission and substation portion of the smart grid are:

- transmission line sensors including dynamic thermal circuit rating
- storage for bulk transmission wholesale services
- FACTS devices and HVDC terminals
- short-circuit current limiters
- communications infrastructure to support transmission lines and substations
- core substation infrastructure for IT
- cyber-security
- intelligent electronic devices (IEDs)
- phasor measurement technology for wide area monitoring
- enterprise back-office system, including GIS, outage management and distribution management
- other system improvements assumed to evolve naturally include:
  - faster than real-time simulation
  - improved load modelling and forecasting tools
  - probabilistic vulnerability assessment
  - enhancement visualisation.
- substation upgrades will enable a number of new functions including, but not limited to:
  - improved emergency operations
  - substation automation
  - reliability-centred and predictive maintenance.

A few of the key technologies are highlighted below:

Dynamic thermal circuit rating (DTCR)
Dynamic rating and real-time monitoring of transmission lines are becoming important tools to maintain system reliability, while optimising power flows. Application of dynamic ratings can benefit system operation in several ways, in particular by increasing power flow through the existing transmission corridors with minimal investments.

Sensors
The smart grid will require a more diverse and wider array of sensors throughout the power system to monitor conditions in real time. In particular, sensors in transmission corridors and in substations can address multiple applications:

- **Safety:** The application of sensors for transmission line or substation components will allow for the monitoring and communication of equipment conditions.
- **Workforce deployment:** If the condition of a component or system is known to be at risk, personnel can be deployed to prevent an outage.
- **Condition-based maintenance:** Knowledge of component condition enables maintenance actions to be initiated at appropriate times rather than relying on interval-based maintenance.
- **Asset management:** Sensor data used together with historic performance information, failure databases and operational data allows better allocation of resources.
- **Increased asset utilisation:** Higher dynamic ratings can be achieved with more precise, real-time knowledge of the asset's condition.
- **Forensic and diagnostic analysis:** After an event occurs, there is limited information to understand the root cause. Sensors allow the capture of pertinent information in real time for a more rigorous analysis.
- **Probabilistic risk assessment:** Increased utilisation of the grid is possible if contingency analyses are performed using a probabilistic, rather than deterministic method.

Table 23.10 Towards the smart grid: cost, components etc. (after Gellings [30])
(*Continued*)

[A] Transmission Systems and Substations	[B] Distribution	[C] Customers

[A] Transmission Systems and Substations (*Continued*)

The transmission system of the future will utilise a synergistic concept for the instrumentation of electric power utility towers with sensor technology designed to increase the efficiency, reliability, safety, and security of electric power transmission. The system concept is fuelled by a list of sensing needs for transmission lines and towers – illustrated in Fig. 2 of the original. [e.g. System Encroachment (vegetation), connector splice, shield wire lightning strike, falling aerial injured boy, shield wire corrosion, avian nesting, insulator cracking, tracking, vandalism (e.g. gunshot damage), system encroachment (structural), structural damage, foundation ageing, fallen line, insulator contamination (pollution build-up), terrorism, phase conductor broken strands, and other sensors]. This system scope is limited to transmission line applications (i.e. 69 kV and above), not distribution, with focus on steel lattice and pole structures, not wooden.

Short-circuit current limiters (SCCL): is a technology that can be applied to utility power delivery systems to address the growing problems associated with fault currents.

Flexible AC transmission systems (FACTS): There are a number of flexible AC transmission systems **(FACTS)** technologies which are critical to the 'smart grid'. These all incorporate power electronics and can be applied to the transmission system. Currently, these include both the control and the operation of the power system and applications that will extend eventually to the transformers themselves.

Storage: Bulk storage is one of the major limitations in today's **just-in-time** electricity delivery system and one of the great opportunities for **smart grid** development in the future.

Communications and IT infrastructure for transmission and substations: Smart substations require new infrastructure capable of supporting the higher level information monitoring, analysis and control required for smart grid operations, as well as the communication infrastructure to support full integration of upstream and downstream operations.

Intelligent electronic devices (IEDs): Intelligent electronic devices (IEDs) encompass a wide array of microprocessor-based controllers of power system equipment, such as circuit-breakers, transformers and capacitor banks.

Phasor measurement technology: Phasor measurement units (PMUs) provide real-time information about the power system's dynamic performance. Specifically, they take measurements of electrical waves (voltage and current) at strategic points in the transmission system (30 times/s).

Cyber-security: Electric utilities have been incorporating cyber-security features into their operations since the early 2000s. In recent years as the smart grid became increasingly popular, cyber-security concerns have increased significantly. Cyber-security is an essential element of the smart grid. It is the protection needed to ensure the confidentiality and integrity of the digital overlay which is part of the smart grid.

Enterprise back-office systems: All large utilities already have enterprise back-office systems which include geographic information systems (GIS), outage management and distribution management systems (DMS). To enable the smart grid [to be achieved], additional features will be required, including a historic data function in conjunction with analytic tools to take in data streams, compare and contrast with historical patterns and look for anomalies in the data.

Impacts on system operators: Independent systems operators (ISOs), transmission system operators (TSOs) and other independent operators (referred to as ISOs here) are making investments in an increasingly robust communications infrastructure as well as an enhanced analytical and forecasting capability.

Power electronic devices for mitigating geo-magnetically induced currents: Geo-magnetically induced currents can cause serious problems to high-voltage equipment and promote blackouts. The future application of power electronics to be applied to the grounded neutral on substation transformers could neutralise these currents.

Table 23.11 Towards the smart grid: cost, components, etc. (after Gellings [30])
(Continued)

[A] Transmission Systems and Substations	[B] Distribution	[C] Customers

[B] Distribution

Gellings informs the reader that investment in the US distribution system has averaged $12–14 billion per year for the past few decades, primarily to meet load growth, which includes both new connects and upgrades for existing customers. An urban utility may have less than 50 feet of distribution circuit per customer, while a rural utility can have more than 300 feet of primary distribution circuit per customer. Assuming a rough average 100 feet of line for each of the 165 million US customers, this indicates the USA has an installed base of more than 3 million miles of distribution line. Upgrading a system to the level of performance required in a smart grid will require substantial investment.

Cost components for the smart grid: Distribution: Smart grid investments in the distribution system entail high bandwidth communications to all substations, intelligent electronic devices (IEDs) that provide adaptable control and protection systems, complete distribution system monitoring that is integrated with larger asset management systems, and fully integrated intelligence to mitigate power quality events and improve reliability and system performance. The key cost components for the distribution portion of the smart grid are as follows:

- communications to feeders for AMI (advanced metering infrastructure) and distributed smart circuits
- distribution automation
- distribution feeder circuit automation
- intelligent re-closures and relays at the head end of feeders
- power electronics, including distribution short-circuit current limiters
- voltage and VAR control on feeders
- intelligent universal transformers
- advanced metering infrastructure (AMI)
- local controllers in buildings, on micro-grids, or on distribution systems for local area networks.

A summary of the key technologies is included below:

Communications: Communications constitute the local backbone for integrating customer demand with utility operations. Detailed, real-time information is a key to effectively managing a system as large and dynamic as the power grid. Each smart meter in the advanced metering infrastructure must be able to communicate with a wide range of user control systems, as well as reliably and securely communicating performance data, price signals, and customer information to and from an electric utility's back-haul system.

Distribution automation: Distribution automation (DA) involves the integration of SCADA systems, advanced distribution sensors, advanced electronic controls and advanced two-way communication systems to optimise system performance.

Intelligent universal transformers: Conventional transformers suffer from poor energy conversion efficiency at partial loads, use liquid dielectrics that can result in costly spill cleanups and provide only one function-stepping voltage. These transformers do not provide real-time voltage regulation nor monitoring capabilities, and do not incorporate a communication link. At the same time, they require costly spare inventories for multiple unit ratings, do not allow supply of three-phase power from a single phase circuit and are not part-wise repairable. Future distribution transformers will also need to be an interface point for distributed resources, from storage to plug in hybrid electric vehicles.

Advanced metering infrastructure (AMI): An advanced metering infrastructure (AMI) involves two-way communications with smart meters, customer and operational databases and various energy management systems. AMI, along with new rate designs, will provide consumers with the ability to use electricity more efficiently and to individualise service and provide utilities with the ability to operate the electricity system more robustly.

Controllers for local energy network: Local energy networks are means by which consumers can get involved in managing electricity by reducing the time and effort required to change how they use electricity. If usage decisions can be categorised so they are implemented based on current information, and that information can be readily collected and processed, then consumers will purchase and operate such a system, install and operate a home area network (HAN), an electronic information network, connected to an energy management system (EMS), a decision processor.

Table 23.12   Towards the smart grid: cost, components, etc. (after Gellings [30])
(Continued)

[A] Transmission Systems and Substations	[B] Distribution	[C] Customers

[C] Customers

Gellings informs the reader that:
There are more than 142 million customers in the USA, of which 13% represent commercial and industrial accounts. Customer base is expected to grow 16% over the next 20 years to more than 165 million. By then most of the consumer appliances and devices will be grid ready (i.e. DR-ready) – these appliances and devices are manufactured with demand response (DR) capabilities already built in. Because of this, the actual number of individual communication-connected end nodes will be more than double.

Cost components of the smart grid: Consumer/Customer technologies
Gellings considers the key components for the customer portion of smart-grid costs to be, as listed below:

- integrated inverter for photo-voltaic (PV) adoption
- consumer EMS portal and panel
- in-home displays
- grid-ready appliances and devices
- vehicle-to-grid two-way power converters
- residential storage for back-up
- industrial and commercial storage for power quality
- commercial building automation.

A summary of the key technologies is included below:

PV Inverters: Inverters are microprocessors-based units used to transform DC to AC power that can be used to connect a photo-voltaic (PV) system with the public grid. The inverter is the single most sophisticated electronic device used in a PV system, and after the PV module itself, represents the second highest cost.

Residential energy management system (EMS): A residential EMS is a system dedicated (at least in part) to managing systems such as building components or products and devices.

In-home displays and access to energy information: Providing real-time feedback on energy consumption holds significant promise to reduce electricity demand.

Grid-ready appliances and devices: Grid-ready appliances do not require truck rolls to retrofit with remote communications and control capabilities. Grid-ready appliances and devices, which are often referred to as 'DR-ready', are manufactured with demand response (DR) capabilities already built in.

Plug-in electric vehicle charging infrastructure and on-vehicle smart-grid communications technologies: Plug-in electric vehicles (PEVs) are defined as any hybrid vehicle with the ability to recharge its batteries from the grid, providing some or all of its driving through electric-only means. Almost all of the major automotive manufacturers have announced demonstration or production programmes in the timeframe 2010–2014. Their announced vehicles feature all-electric, plug-in hybrid-electric and extended-range electric vehicle configurations.

Communication upgrades for building automation: Today, over one-third of the conditioned and institutional buildings in the USA have some form of energy management and control systems installed. Automated demand response (ADR) can be accomplished by communicating to the building energy management system using an Internet-communicated signal or some other form of direct link.

Electric energy storage: Advanced lead-acid batteries represent the most prevalent form of electric energy storage for residential, commercial and industrial customers wanting to maintain an uninterruptible power supply (UPS) system. Commercial and industrial systems can supply power for up to 8 hours at 75% efficiency and maintain performance through more than 5,000 cycles. Residential versions typically involve 2 hour duration at 75% efficiency and 5,000 cycle performance.

relatively short timescales, in extremely diverse fields, each covering a clearly specified wide range of 'relevant technical aspects', within a particular energy sector.

Note: This writer considers that research academics worldwide working on related or relevant engineering studies to those of CIGRE would do well to ensure that their research students, and indeed all engineering undergraduates students, are made fully aware of the vast body of valuable CIGRE work, and how it is 'garnered worldwide', including publications, IEC type activities and TBs, covering many strategic activities. This material can also provide much valuable background for the more 'fundamental studies in the universities' – but each sector should become more aware of the other's work to prevent opportunities being lost in 'both camps'. While there has been good progress in this joint awareness in recent years, it can still be 'patchy in some countries' and efforts should continue to build on and optimise such links.

(This was one of the strategic aims for the long-running HVET International Course, for many years, via the then IEE.)

Table 23.13 (after 'A history of CIGRE', 2011, Appendix 3, pp. 175, 176) provides valuable details of the fields of activities (determined by CIGRE experts) for each of the designated and diverse, specialist CIGRE Study Committees (SCs). SCs, which are occasionally referred to, within generalised subject groupings as A_{1-3}, B_{1-5}, C_{1-6} and D_{1-2} were established following the CIGRE 2002 reforms.

Note for further guidance: From the activities within the scope of each specialist Study Committee (SC), see Table 23.13, identified 'Preferential Subject Themes' for invited papers are also agreed well in advance, for the next CIGRE international conference held in Paris every two years. [The themes selected 'reflect topics of current strategic importance', agreed by CIGRE every two years].

To illustrate this, the reader could find out the identified 'Preferential Subject Themes' for invited papers submitted for example to the 43rd and 44th CIGRE Paris sessions of 2010 and 2012, respectively. These are listed and available from CIGRE and can readily be compared and contrasted over time.

Such retrospective comparisons provide the reader with a historic opportunity to visit CIGRE sites and obtain a clear idea of current worldwide 'strategic technical topics or themes of current strategic interest in each sector over a selected time period'. Session papers address a strictly limited list of topics – referred to as preferential subjects (PSs) (e.g. PSs 1–3 is typical per the SC).

In-depth consideration of each SC can readily be researched by each reader to obtain a better overview of all Table 23.13 activities and more importantly, the scope of each CIGRE SC activities. Study Committees, namely A_{1-3}, B_{1-5}, C_{1-6} and D_{1-2}, were established since its 2002 reforms. Having done this for any CIGRE sector activities, of particular interest, it is then a simple matter to add and insert relevant published references on related themes from IET, IEEE, etc. to enable the reader or research group to quickly familiarise themselves within a 'template' detailing all relevant source materials of interest worldwide. This has proved very successful to this writer.

Consider again the second strategic report (2), *prepared on behalf of the CIGRE Technical Committee*, which can often prove to be of considerable value to the reader. This TC Advisory Group Report was published in *Electra* [35] entitled 'CIGRE WG "Network of the Future"; Electricity supply systems of the future'. Briefly, in the introduction to this report, the stated 'mission of modern power

Table 23.13 Fields of activities of the CIGRE Study Committees, since 2002 reform [after A history of CIGRE, 2011]

SC A1 Rotating electrical machines [www.cigre-a1.org]
Economics, design, construction, test, performance and materials for turbine generators, hydrogenerators, high-power motors and non-conventional machines

SC A2 Transformers [www.cigre-a2.org]
Design, construction, manufacture and operation of all types of power transformers, including industrial power transformers, DC converters and phase-shift transformers, and for all types of reactors and transformer, components (bushings, tap-changers, etc.)

SC A3 High-voltage equipment [www.cigre-a3.org]
Theory, design, construction and operation of devices for switching, interrupting and limitation of currents, lightning arrestors, capacitors, insulators of busbars or switchgear and instrument transformers

SC B1 Insulated cables [www.cigre-b1.org]
Theory, design, applications, manufacture, installation, tests, operation, maintenance and diagnostic techniques for land and submarine AC and DC insulated power cable systems

SC B2 Overhead lines [www.cigre-b2.org]
Design, study of electrical and mechanical characteristics and performance, route selection, construction, operation, management of service life, refurbishment, uprating and upgrading of overhead lines and their component parts, including conductors, earth wires, insulators, pylons, foundations and earthing systems

SC B3 Substations [www.cigre-b3.org]
Design, construction, maintenance and ongoing management of substations and of electrical installations in power stations, excluding generators

SC B4 HVDC and power electronics [www.cigre-b4.org]
Economics, application, planning, design, protection, control, construction and testing of HVDC links and associated equipment. Power electronics for AC systems and power quality improvement and advanced power electronics

SC B5 Protection and automation [www.cigre-b5.org]
Principles, design, application and management of power system protection, substation control, automation, monitoring and recording, including associated internal and external communications, substation metering systems and interfacing for remote control and monitoring

SC C1 System development and economics [www.cigre-c1.org]
Economics and system analysis methods for the development of power systems: methods and tools for static and dynamic analysis, system change issues and study methods in various contexts and asset management strategies

SC C2 System operation and control [www.cigre-c2.org]
Technical and human resource aspects of operation: methods and tools for frequency, voltage and equipment control, operational planning and real-time security assessment, fault and restoration management, performance evaluation, control centre functionalities and operator training

(Continues)

Table 23.13 Fields of activities of the CIGRE Study Committees, since 2002 reform [after A history of CIGRE, 2011] (Continued)

SC C3 System environmental performance [www.cigre-c3.org]
Identification and assessment of the environmental impacts of electric power systems and methods used for assessing and managing the environment impact of system equipment

SC C4 System technical performance [www.cigre-c4.org]
Methods and tools for power system analysis in the following fields: power quality performance, electromagnetic compatibility, lightning characteristics and system interaction, insulation coordination, analytical assessment of system security

SC C5 Electricity markets and regulation [www.cigre-c5.org]
Analysis of different approaches in the organisation of the electrical supply industry: different market structures and products, related techniques and tools, regulation aspects

SC C6 Distribution systems and dispersed generation [www.cigre-c6.org]
Assessment of technical impact and new requirements which new distribution features imposed on the structure and operation of the system: widespread development of dispersed generation, application of energy storage devices, demand side management and rural electrification

SC D1 Materials and emerging test techniques [www.cigre-d1.org]
Monitoring and evaluation of new and existing materials for electro-technology, diagnostic techniques and related knowledge rules and emerging technologies with expected impact on the system in medium to long term

SC D2 Information systems and telecommunication [www.cigre-d2.org]
Principles, economics, design, engineering, performance, operation and maintenance of telecommunication and information networks and services for the electric power industry; monitoring and related technologies

---

Note: 'Since 1921, CIGRE's priority role and "added value" have always been to facilitate mutual exchanges between all its components: network operators, manufacturers, universities and laboratories. It has continually led to the search for a good balance between the handling of daily problems encountered by its members in doing their jobs and the reflections on the future changes and their equipment. Its role in exchanging information, synthesising state of the art, and serving members and industry has constantly been met by CIGRE throughout events and through publications resulting from the work of its Study Committees' (p. 169; www.cigre.org; www.e-cigre.org).

systems' from a CIGRE perspective is to supply electric energy satisfying the following conflicting requirements.

- high reliability and security of supply
- most economic solution
- best environmental protection.

'From their inception, reliability and security of supply of electrical power systems have always been, and will continue to be, of prime importance to CIGRE. Such considerations have shaped the design and operation of power networks from the very beginning of their formation.' Some salient aspects covered broadly in this CIGRE document were:

- '*In recent decades the need for a more efficient operation of the network with the aim to reduce prices and increase the quality of service led to the unbundling of the power system and the liberalisation of the energy markets.*
- More recently, there has been growing concern about global warming/climate change issues, and in particular on any possible effects energy production may have on greenhouse gas emissions.
- This moved attention on to renewable energy options and led to the wide integration of Renewable Energy Sources (*RES*) and Dispersed Generation (*DG*) into electrical networks and, from a global warming viewpoint, a more environmentally friendly public perception of "impacts" power systems may have on our environment.
- The increased share of renewable generation in electrical networks as a result of power system developments moving to comply with a stringent EU commission target figure, now well known in the electrical supply industry – requirement to achieve 20% reductions in greenhouse gas emissions by 2020 over the 1990 levels.
- To obtain efficient integration of large shares of RES and DGs "imposes, or introduces a new flexible mindset" – the need to re-think, "or re-visit", the current ways of thinking regarding the planning, management and control of the power systems, both at transmission and distribution level, with the introduction of "higher intelligence" to help further improve efficiency'.

The CIGRE WG go on to observe that these aims, 'which are primarily driven by the *environmental concerns* are the landmark of this decade, will certainly shape the *Energy Supply Systems of the Future*'. The aims of this document or CIGRE Technical Brochure [35] are:

- to 'identify the key parameters, which, in the opinion of participating CIGRE WG experts, will shape the Networks of the Future in a time horizon of two decades
- to highlight the relevant activities which have been and are still taking place, within the various Working Groups of CIGRE
- to identify missing activities, which could be the objectives of future working bodies
- to identify synergies within and outside CIGRE'.

The over-reaching CIGRE aim (35) 'is to contribute to the vision and the development of the future Energy Supply Systems in a global, co-ordinated way'.

23.13.2.1 Driving factors, network development scenarios and relevant activities

The CIGRE WG considered the driving factors that could be generally identified [35] as:

- 'increased customer participation
- international and national policies encourage low carbon generation, the use of RES and more efficient energy use

- *integration of RES and DG into the grid*
- need for investment in end-of-life grid renewal (ageing assets)
- necessity to handle grid-congestion (with market-based methods)
- progress in technology including information and communication technology (ICT)
- environmental compliance and sustainability of new built infrastructure'.

These factors 'suggested to WG [35] that two models for network development are possible, and not necessarily exclusive:

- an increasing importance of large networks for bulk transmission capable of interconnecting load regions, large centralised renewable generation resources including offshore, and to provide more interconnections between the various countries and energy markets
- the emergence of clusters of small largely self-contained distributed networks, which will include decentralised local generation, energy storage and active customer participation intelligently managed so that they are operated as active networks providing local active and reactive support'.

These two models lead to three scenarios illustrated in the table below.

After CIGRE discussions, it was 'agreed that the most likely shape of the future energy supply systems would include a mix of the above two models'. The WG considered it was not possible to predict the quantitative composition of future systems, but this was not considered to be very important. 'It considers that both models are needed in order to reach the ambitious environmental, economic and security targets sought. CIGRE has therefore decided to deal with all technical issues relevant to both models including the study of isolated systems'. This is now reproduced in a visual-friendly format.

Summary: emphasis for future development (CIGRE) [35]

Large networks	Smaller networks
A shift towards larger and larger aggregated networks and new, major interconnections between large loads generation centres over long distances (including continents)	Greater shifts to distributed generation and localised solutions, a slowing and reversal of greater interconnections between grids and parts of grids, more self-sufficiency and reduced reliance on generation sources, large distances from load centres.
Projected trajectories: system models of future development	
Scenario 1 Predominantly larger networks	- Greater emphasis on large networks
Scenario 2 A mix between large and small networks	- Less emphasis on small networks
	- No favour given to large or small networks
Scenario 3 Predominantly smaller networks	- Less emphasis on large networks
	- Greater emphasis on small networks

The key technical issues
The CIGRE WG considers that the main driver [35] for the power system evolution can be summarised as:

- 'massive integration of RES and DER
- active customer participation
- increasing end use of electricity and non-acceptability for building new infrastructure'.

In their CIGRE *Electra* article, the WG report that the evolution of today's power system towards the new models described above *is based on the following technical issues* (*TIs*) that should be considered in detail (TI.1–TI.10), shown in heavy print at top of each group section of Table 23.14.

These have been re-arranged by this writer, in Table 23.14, to provide 'just a taster' to the reader of some of the huge amount of strategic work ahead for the sector! Particular 'participating Study Committee (SC A1), etc.' are also detailed in this table.

Also it is of strategic interest to be aware of planned technical themes or preferential subjects at the bi-annual Paris sessions. This provides short descriptions of the range of strategic themes being covered by each SC (namely A_{1-3}, B_{1-5}, C_{1-6} and D_{1-2} of Table 23.13). These can readily be obtained from CIGRE websites.

WG notes: relevance of CIGRE SC's key challenges (see Table 23.14)
'It is important to emphasise the scale of work involved in producing state-of-the-art reports, or Technical Brochures (TBs) for each of the technical issues listed in this table, each may require the involvement of >100 experts from different CIGRE SCs, over two-to-three-year period. Since many issues have various/different facets, and involve different technologies and specialisms working together' [35].

CIGRE WG has carried out a preliminary assessment of which SC might be involved in addressing each TI, and, at this time, *only a preliminary list is included in Table 23.14 – so subject to variation!*

It should also be noted that each TI is likely to be handled by a number of separate WGs, some of which will have members from a single SC and others will be joint WGs involving members from two or more SCs [35].

(A particular SC may be singled out with the statement, for example 'activities of SC C6 are mainly driving the TI'.)

The writer has outlined and described in some detail this 2011 CIGRE WG report together with certain CIGRE TBs and *Electra* articles with the 'sole specific aim' of indicating to the reader – *many of whom are largely unaware of the scale and depth of CIGRE activity worldwide* – the vast amount of detailed coordinated work and painstaking 'garnering' *of facts/data*, which lies ahead in these ambitious and wide-ranging CIGRE programmes to achieve this 'thorough worldwide understanding' of electricity supply systems of the future.

Table 23.14 Some key CIGRE technical issues [35]

SCs

TI.1 Active distribution networks resulting in bidirectional flows within distribution level and to the upstream network	C6, C3, C4

Activities of **SC C6** are mainly driving issue TI.1.
Protection issues are an important challenge and are dealt with by **SC B5** in TI.6.
Future work
Modelling and analysis of active networks are important and should be dealt with by **SC C4**.
Environmental issues are not yet covered and will be developed in cooperation with **SC C3**.
Indicative activities
- Smart buildings
- Smart cities
- DC distribution networks
- Potential ancillary services from distribution
- How to protect islanded electronically dominated distribution systems?
- How to address the micro-grid issues?
- High renewable penetration in islanded communities

TI.2 The application of advanced metering and resulting **massive need for exchange of information**	B5, C6, **D2**

Activities of **SC D2** are mainly driving issue TI.2.
Coordinated protection is an outstanding issue dealt with by **SC B5** in TI.6. The operation of active networks with increased DG and customer participation requires increased exchange of information. This is dealt with in **SC C6**.
Future work
Effects on power system operation and control are an important issue to be pursued by **SC C2**.
Indicative activities
- Evolution of existing telecommunications infrastructures to cover the new needs
- Cyber-security
- New techniques and applications for maintenance of communication equipment
- How to cope with and best utilise the information from intelligent monitoring?

TI.3 Massive integration of HCDC and power electronics at all voltage levels and its impact on power quality, system control, and system security, and standardisation	B1, **B4**, B2, C4, C6, C1, D1

Activities of **SC B4** are mainly driving TI.3.
The increased integration of power electronics requires new advanced modelling and analysis techniques.
This is developed in cooperation with **SC C4**. Impacts on system development are dealt with in cooperation with **SC C1**, and the HVDC cable components by **SC B1**. Impacts on materials are dealt with by SC D1.
Future work
The penetration of power electronics at medium and low voltage levels is an important outstanding issue that should be pursued in cooperation with **SC C6**.
Indicative activities
- Power electronics at lower voltages including within generators
- FACTS applications for active and reactive power control
- New HVDC technologies
- Enabling of multi-vendor HVDC grids
- Coping with multiple harmonic sources with the networks.

TI.4 The need for the development and **massive installation of storage systems**, and the impact this can have on the power system development and operation	C6, C4, C1, D1

TI.4 is expected to be an important future activity, which is currently partially covered by **SC C6** at the distribution level only.
Future work
Issues not covered include the developments in storage technologies, in material and devices or methods, which could be developed by **SC D1** and **SC A1** respectively and storage analytical models that could be developed by **SC C4**. Storage solutions which are connected to the network via power electronics will provide opportunities for reactive power management, and potential integration with HVDC grids, and this aspect will be analysed by SC B4. The effect of large-scale storage in the development and operation of the power system is an important future issue to be pursued by **SC C1** and **SC C2** respectively.
Indicative activities: Storage techniques overview and selection, integrating energy storage in HV grids, dual utilisation of power electronic converters associated with energy storage, influence of bulk energy storage, maximise the value of energy efficiency measures by using storage to allow demand reduction to occur when it has the highest value.

Table 23.14 Some key CIGRE technical issues [35] (Continued)

	SCs
TI.5 New concepts for system operation and control to take account of active customer interactions and different generation types	**C1, C2**, B5, C6, C5

Activities of **SC C2** are mainly driving issue TI.5.
SC C6 develops the work on operation of active distribution networks and control of DER.
SC C4 develops models of specific components (gas turbines, wind parks, etc.) and risk-based and probabilistic tools.
Future work
TI.5 is closely linked with the activities of SC C2 and also SC C1; joint work is in progress and should be developed further. Protection is another relevant activity that should lead to joint work with SC B5.
Indicative activities
– Challenges in control centres due to renewable generation
– How much conventional generation is necessary in the future?
– Critical infrastructure protection against cyber-attacks from outside
– ICT convergence
– Providing ancillary services from intermittent generation
– Harmonisation of grid codes for wind farms connecting to the grid
– Harmonisation of grid codes for connection of power stations to the network
– Impact of instrument transformers on substation concept
– Planned and unplanned outage management in a controlled and efficient manner.

TI.6 New concepts for protection to respond to the developing grid and different characteristics of generation	**B5, B4**, C4, C6

Activities of **SC B5** are mainly driving TI.6. **SC B4** develops joint work on the impact of HCDC networks.
Future work
Modelling of protection devices and consideration of protection in analytical tools are an activity pursued by **SC C4**. Protection of active networks including low-voltage networks is an outstanding activity to be pursued jointly with **SC C6**.
Indicative activities
– New protection, automation and communication for distribution networks
– Synchrophasors for system-wide protection
– IEC 61850 Process bus and beyond the substation
– Advanced protection algorithms.

TI.7 New concepts in planning to take into account increasing environmental constraints, and new technology solutions for active and reactive power flow control	**C1**, C3, C4, C6, B2, B4, B5, C5

Activities of **SC C1** are mainly driving TI.7. **SC C4** develops probabilistic models for assessing network capability; SC C1 is exploring how best to integrate these tools in making planning decisions on justifying network augmentations. SC C6 develops work on planning active distribution networks.
Future work
Planning is closely related to environmental effects and the functioning of electricity markets.
These are issues that can be the focus of joint work with **SC C3** and **SC C5** respectively.
The effects of protection should be considered by **SC B5**.
The integration of HVDC grids and AC networks can be the focus of joint work with **SC B4**.
Indicative activities
– Demand forecasting recognising the changing nature of supply and demand
– Manage strong need for reinvestment and new investments for future demand in the context of current and future technologies and growing uncertainty
– Future of reliability in terms of what the customer needs and the future system can deliver
– How can markets give incentives for improving the generation portfolio?
– System flexibility for accommodating variable intermittent generation
– Impact of smart technologies on market designs
– More dynamic coupling between the markets and the infrastructure investments, recognising the changing investment risk
– Integration of renewables potential from remote areas such as the North sea and African desert
– 'Investigate tools' and approaches for planning networks which integrate HVDC networks and HVAC networks solutions
– Future use of new unconventional generation technologies including super conducting machines.

(*Continues*)

Table 23.14 Some key CIGRE technical issues [35] (Continued)

	SCs
TI.8 New tools for system technical performance assessment because of new customer, generator and network characteristics	C4, C6, B3, B4

Activities of **SC C4** are mainly driving issue TI.8.
SC C1 has also developed work on modelling complex networks.
Future work
The development and operation of active networks require new tools for their technical performance, especially their dynamic behaviour, islanding and power-quality effects. Work should be developed in cooperation with **SC C6**. Models will be required for HVDC converter stations, HVDC grids and for FACTS devices, and these can be developed jointly by B4 and C4.
Indicative activities
- Simulation tools for distribution levels
- Models for dynamic performance of PV, and wind mill generation
- Islanding
- Phasor measurement units used at distribution level
- Peer-to-peer approaches with multi-agent techniques
- More transients and higher harmonic content in the network, means to review test programme.

TI.9 Increase of right of way capacity and use of overhead, underground and subsea infrastructure, and its consequence on the technical performance and reliability of the network	B1, B2, B3, B4, C1, C3, C6, C5

Activities of **SC B2 and SC B1** are mainly driving TI.9.
SC B4 has also developed work on HVDC and FACTS, which can increase the right of way capacity.
SC C4 has developed models on long AC cables and HVDC cables jointly with **SC C1**.
The effect of the transmission infrastructure on the environment is an important activity considered by **SC C3**.
Future work
The increased use of interconnections is a key component of the network of the future as far as the supergrid is concerned. This development will have major implications on planning, operation and control and the establishment of electricity markets covering wider areas.
Relevant activities could be developed in cooperation with **SC C1, SC C2** and **SC C5**.
Indicative activities
- Smart components for transmission lines
- Cost estimation in new transmission lines
- Transmission interconnections
- Integration of superconducting transmission and distribution elements/lines
- Integration of power stations to the load centres by superconducting cables
- Offshore platform connection to the grid (reliability use)
- Towards fluid free components (dry type).

TI.10 Need for stakeholders awareness and engagement [An increasing need to keep stakeholders aware of the technical and commercial consequences and keeping them engaged during the development of the network of the future]	B1, B2, B3, B4, C1, C2, C6, C5

Activities of **SC C3** are mainly driving TI.10.
The effects of cables and overhead lines on the environment are considered by SC B1 and SC B2.
Future work
Public opposition to new construction is a key problem impeding power system development.
This issue concerns networks, substations, HVDC converter stations and generating stations.
Relevant activities should be pursued by **SC B1, SC B2, SC B3** and **SC B4**.
Public opposition also affects power system development and should be considered in cooperation with **SC C1**.
The facilitation of distributed generation (including renewable generation) and 'active customer participation' are important relevant activities that should be developed in cooperation with **SC C6**. **SC C6** is also responsible for rural electrification, a subject very relevant to TI.10. Finally, the development of electricity systems of the future in the new context requires extensive engagement with key stakeholders to ensure the required investments receive government and community support and gain access to scarce capital. These investments are required to both enable the new energy solutions and enhance the power system to deal with more extreme weather events.
Indicative activities
- Improve and make more effective the stakeholders engagement in order 'to smooth the public acceptability' of power system infrastructures
- Leveraging CIGRE activities into developing countries
- Developing education tools together with methods to value the true cost of energy, thus strengthening the awareness of stakeholders and general public
- Producing simple public documents that promote key 'customer-focussed' messages.

Note: The writer and colleagues were involved in organising an annual IET HVET International Summer School in the UK for approximately 16 years. During this time delegates were strongly encouraged to follow the latest CIGRE work closely and to participate in the 'wide-ranging' professional activities including standards work of CIGRE/IEEE/IET/IEC/BSI to help them attain an effective/thorough understanding of the electricity supply industry worldwide, in the past, at present and in future.

It is very gratifying that many IET HVET International Summer School attendees moved on to become actively involved in, and committed to, strategic CIGRE/IEC/BSI/relevant IET technical and academic programmes/activities in subsequent years – thanks to the positive and strategic generous contributions made by all school lecturers (i.e. represented a valuable part of each individual's career with CPD).

- Finally, the reader is again reminded that certain additional CIGRE technical studies are also discussed in the appendices to Chapters 8 and 9, while appendices of Chapter 9 focus on certain recent strategic activities of substation SC B3 and also SC B5. The reader could equally well check out himself/herself other SC activities from the CIGRE sites, and list of available TBs, etc., issued by each SC during, say, the past five years.
- This writer has found CIGRE published TBs, articles, magazines, papers, etc. (available in French or English) which are quite distinctive from IEEE or IET publications, have provided invaluable practical resource materials for research projects in degree and post-graduate studies, etc., for many years.

23.14 Discussion and conclusions

23.14.1 Discussion

This chapter has briefly considered a wide range of strategic needs: energy development and delivery challenges; problems; issues; concerns; technical environmental, cyber and economic considerations; changes of energy source mix touching on new-build nuclear and renewables (onshore and offshore); the very large costs essential for the necessary huge UK and worldwide energy network developments/upgrades/enhancements required to incrementally meet the strategic needs in the short term and in the next two decades as the UK, USA, EU, etc., attempt to achieve 'smart grid-enhanced transmission and distribution Networks of the Future' status.

A substantial amount of resource material has also been provided which should be of assistance to the readership.

While preparing this chapter, which also included reviewing a cross section of the UK press energy articles over a period of 2010–2013, there has been a 'heavy barrage of press coverage on energy-related issues', mainly negative.

1. The huge projected energy costs discussed in this chapter, regarding the essential modernisation of energy networks in the UK and elsewhere, have shown the following:

(i) Renewable energy wind farms – initially onshore and – more recently – the offshore wind farms, which are more expensive and many feel should first have costs and performance assessments thoroughly evaluated prior to huge expenditures being incurred.

(ii) Difficulties have been encountered in obtaining project finance for several 'new-build' third-generation nuclear plant projects in the UK, planned for the next few years, and while this writer hopes they are approved, he considers EDF Energy will be in a very strong negotiating 'bargaining position' at the final agreement/contract discussions (at the end of 2012), with the UK government.

(iii) Some of the challenges associated with environmental CCS carbon and green and coupon issues and co-fired biomass usage have been considered.

2. The 'marked' and 'confusing' rapid changes or 'swings' in the UK government energy policy (*sometimes 180 degree turns!*) regarding the desired/preferred final energy mix, without clear explanation of rationale – over very short timescales. Indeed, recently 'breaking news' indicates further swings in the UK government policy, and we will return to this shortly. All proposed new energy projects must be thoroughly evaluated before approval and should comprehensively satisfy robust technical, economic and environmental acceptance criteria.

3. Some of the recent disappointing and poor performance problems experienced with renewable energy. Wind farms within the UK energy supply network have been identified together with occasional inadequate wind farm site approval processes, noise and visual pollution difficulties with wind turbines, and also certain technical and 'still-to-be-resolved' energy storage 'unavailability' issues in certain locations.

4. The strong UK public dissatisfaction expressed in the press regarding (i) the recent uses of public subsidies to fund energy projects with weak economic justification together with (ii) increasingly high domestic energy charges, and as a direct consequence large numbers of consumers moving into fuel poverty, is cause for great concern.

5. *It is considered vital* that all 'fledgling' UK offshore wind farm infrastructures being planned will be first benefited from full, rigorous and relevant technical/economic/projected 'in service' performance evaluation studies, as to location, performance under typical operational/environmental scenarios to establish, and what-ifs, that is, will any upgrades be required to transmission and distribution networks nearby, 'appropriate or matching storage facilities', etc., beforehand, to ensure cost-effective and 'fit-for-purpose' installations are produced. Also it must be demonstrated that the wind farms should represent value for money and be able to deliver reliable and effective service to the UK public.

6. Hopefully, all new schemes will be fully compatible with the UK drive towards achieving the planned 'smart-grid' and enhanced infrastructure network status planned to be in place sometime within the next two decades.

23.14.1.1 Further press observations

It should be emphasised again that the *'concerned' UK press has continued its 'incessant barrage' coverage of energy matters unabated.* Consider the latest examples [1–5] covering some very recent additional aspects:

1. 'Energy "will be unaffordable in three years" as bills soar' (after Cheryl Latham, *Metro* Paper, 30/5/12)

   This was yet another article expressing concern at burgeoning domestic fuel bills in the UK.

   It stated as follows:
   - Millions of families will be unable to pay to heat their homes within three years as energy bills soar past the £1,500 barrier as figures quoted in this article show.
   - One in three households is likely to find that gas and electricity costs become unaffordable by 2015.
   - They have already seen bills more than double in the past eight years, according to consumer groups. Further increases at the same rate could reach the 'tipping point of £1,582 in three years' and £2,766 by 2018.

   Note: On a separate platform, that is in *The Sunday Times*, 1/7/1012, D. Fortson said, 'Household bills have doubled in the past 5 years to an average £1345/year'. See later comments below.

   Ann Robinson from *u.switch.com*, which apparently compiled her figures, said, 'The UK is hurtling towards a cliff beyond which the price of household energy will become unaffordable', reported Cheryl Latham. An estimated 6.5 million people already live in fuel poverty, where they have to choose between paying for heating or other daily essentials, Latham reports.

   As bills climb further, 3 out of 4 families say they will have to ration energy and 6 in 10 are likely to go without adequate heating, u.switch.com estimates. 'Time is running out – if pricing trends continue we will hit 'crunch point' in less than three years', Ms Robinson added.

   - Moreover, it seems that these predictions do not include any extra costs added to bills by power companies investing in technology to make their fuels greener, writes Latham.
   - A final point raised: Maria Wardrobe, from the fuel-poverty charity National Energy Action, said: 'It is time to utilise some of the revenues from carbon taxes and the increased revenue from VAT, on bills to pay for a programme to protect the most vulnerable'.

2. *Also in the UK press recently Chancellor George Osborne is reported as being in favour of* a 25% cut in government subsidy for onshore wind farms which could 'theoretically' reduce or prevent many more being built, but realistically, the local government Finance Bill legislation due to be introduced April 2013 will, it would appear, negate any such subsidy reduction impact – by making it more difficult for councils, etc., to resist new wind farms being built (*The Journal*, 5/6/12).

   Let us just take a quick snapshot at press coverage on 26–29/7/2012 (*The Journal*, 26/7/12). Dave Black reports, 'DECC announced yesterday that the huge

Government subsidy for new onshore wind projects – which are paid for via consumers' energy bills – will be cut by 10% from next year. This is somewhat *less than Chancellor Osborne was predicting (i.e. 25%) following on from recent call by more than 100 Conservative MPs for action to curb wind farm developers'.*

Tim Webb says, 'Drax caught between ROC and hard place' (*The Times*, 26/7/12) (Drax can generate about 7% of UK electricity needs.) Disappointment over the subsidies for burning renewable fuel sent shares in Drax into a tailspin yesterday as investors worried that the government was losing its enthusiasm for green energy. Shares dropped 25% but later rallied closing about 15% lower. Mr Edward Davey, the DECC secretary, informed "that subsidy units known as Renewable Obligation Certificates (ROC) worth about £40 for each megawatt hour of electricity would only be granted for boilers that were converted to burn 90% biomass" – Drax has 6 boilers. Webb reports Davey as saying, "We tried to strike the right balance between on the investment that we need and also getting the best bang for the buck for the consumer". (*Note*: Who said the English language was dead?)

Tim Webb says, 'Gas gets the green light to power Britain' (*The Times*, 26/7/12). Centrica is to create 4,000 jobs in Britain with the development of a huge North Sea gasfield (after the UK government signalled a 'New Dash for Gas'). British Gas said that it would invest £1.4 billion with its French partner GDF Suez. Edward Davey, is reported by Tim Webb as saying, 'gas would be at the heart of Britain's energy strategy and promised that new gas-fired power plants would play a "key role" in keeping the lights on'. He went on to say that the consumer-funded subsidies for onshore wind farms would not be cut by as much as expected. As a UK consumer, one can only hope that adequate gas fired power station options would, in fact, currently be available to play the key, or strategic energy, role just referred to above. The commercial exploitation of shale oil and gas in the USA resulted in American power stations buying up much of the cheap shale gas, and gas-fired generation in the USA jumped by 21% in 2012. Webb has also reported it in another article (see Table 23.15).

Webb states the moves came after intervention by the Treasury. *Note 1*: *Emily Gosden* tells her North *Telegraph* readers on the same day that the site is *Cygnus field* – about 100 miles off the North Norfolk coast. The government's new tax allowance will apply to income from shallow-water field production such as the Cygnus field. The first £500 million of income will be exempted from the 32% tax, saving the companies £160 million.

Note 2: This pledge on this, the second and, surprising to many, the latest: 'dash for gas' appears to increase – with its announcement – the uncertainty over the UK government's commitment to the green agenda and risks scaring off investors. This can be viewed from the two legal obligation perspectives:

(i) *the UK obligation 1*: commitment to the EU to generate 32% of UK electricity by 2020 from renewable sources
(ii) *the UK obligation 2*: Climate Change Act – under this the UK will cut carbon emissions by 80% within 40 years.

Peter Troy (in the *The Journal*, 31/7/12) 'touched' on some small print of Edward Davey's statement of last week:

- That by 2017, the UK hope to generate 79 TWh of electricity a year from renewables, rising by 2020 to 108 TWh.
- The latter figure is necessary to meet the UK renewable energy target.
- To achieve this target 'loads and loads and loads' of wind farms will still be necessary!

(Expect a follow up to this announcement by DECC Secretary Edward Davey!)

Energy statistics via *Tim Webb regarding his contribution above:*

- 40% – the proportion of electricity produced from gas last year (2011)
- The Climate Change Act 2% – the proportion of electricity produced from gas in the early 1990s
- 3.9 GW – combined output-gas-fired power stations (2011) owned and operated by Centrica England/Wales
- 4.9 GW – of new gas-fired electricity generation capacity is projected to come online by (2020) 4.1 GW is projected by 2016
- 0.5 GW – biomass co-firing capacity at Drax plant (2011)
- 7% – the proportion of Britain's electricity needs (2011).

3. *Share of state energy support* (an earlier feature): In May 2012, the Department of Energy and Climate Change Secretary Edward Davey is due to publish a draft energy bill providing details of a package of subsidies that the government will use to inject more than £200 billion into the energy power sector over the next few decades. If this legislation is approved it should become law in 2013 and 'pull the UK into the twenty-first century energy production wise'. If the Energy Market Reform (EMR) is approved in its present form, the UK may be saddled with an inappropriate network system and energy mix, huge energy bills for years to come, and in the view of the writer of this chapter, several aspects need further and broader debate and revision in the best interests of Britain's energy future, encompassing incorporating the smart grid of the future before the EMR is passed.

Meantime, strategically, the UK government will have to determine how this more than £200 billion will be divided/shared for the next few decades between wind and low-carbon technologies. Naturally, each sector will expect a good share of the money available and the reader can expect a barrage of quotes, articles, statistics from the government and 'all vested interest groups', and should make up his/her mind or merits regarding the veracity (formal truthfulness) and cogency (strong, persuasive, convincing) of all arguments! Significant amounts of 'questionable' or 'dodgy data' may be presented in the press and other media to win the minds of the public – by 'nudge factor strategies' – one way or another.

4. *In fact, revisiting the press* after this chapter was effectively written, *The Sunday Times*, 3/6/12, commented upon this aspect, 'Greens' propaganda war', and its subheading read, 'Environmental projects face dirty tricks as factions struggle for a share of state support', by D. Fortson'. Fortson also quotes Dale Vince, Ecotricity, a green energy developer, as saying, 'There are huge sums of money at stake. It's perhaps to be expected that this generates significant lobbying and dirty tricks that often come with it'. So to the reader – watch this space!

(4a) Following on from (4), 'Wind energy developers have recently been accused of deceiving local councils and the public by using computer generated images in planning applications that make the turbines seem smaller than in reality', said J. Gillespie in his article (*The Sunday Times*, 15/7/12, p. 7, 'Photo trickery makes wind farms smaller'). The claim is contained in a new book, *Wind farm visualisation: Perspective or perception*, by A MacDonald, whose company architect specialises in computer-generated images. The article also reports a separate study by University of Stirling found serious flaws in the images that were presented as part of a visual assessment in the planning process. Gillespie's article was supported by photographic evidence (ARCHITECH) under the heading: 'The wind farm con: how images shown to planners can give a false impression of turbine "size"'. Apparently:

1. a 50 mm photograph is joined to others to create a panorama
2. the turbines appear smaller set in a panorama too wide for the human eye
3. an actual photograph is also shown which clearly makes the point!

Certainly comparison of an actual 50 mm image and a wind farm developer's image was very concerning, and it seemed that other readers of the paper – who read this article and made a response – considered the use of words 'trickery' to be *apposite*, if a tad generous!

The University of Stirling report also found the use of the industry standard 50 mm lens to be misleading, said Gillespie, while a response from Renewable UK, the wind energy industry body, is quoted as: 'It is in developers' interests to ensure that their visualisations are accurate'.

5. *Nuclear: As stated earlier*, the UK government faces a tough negotiating 'battle' with EDF Energy regarding the building of a new-build nuclear plant at Hinkley Point, Somerset, UK. Without doubt, EDF and partner Centrica will be holding out for huge subsidies before agreeing to commit to build the proposed new two-reactor £14 billion new nuclear plant at Hinkley Point, as they are the only group left who plan to build new-build nuclear reactors in the UK to replace old reactors soon due to be retired. The UK government is in a weak/poor position (i) following the withdrawal of German companies RWE and E.ON, which put their joint venture Horizon up for sale from building new nuclear plant in the UK and (ii) 25% of British nuclear capacity is due to close by 2020.

Meantime, EDF Energy, the French state-backed energy group, has selected a consortium made up of Bouygues TP and Laing O'Rourke as its preferred bidder to build Britain's first nuclear reactor for decades. *The Times* report that the contract is worth £2 billion to the construction companies if EDF Energy group agrees to proceed with the Somerset project, at the end of 2012 (*The Times*, 19/7/12, p. 31. Col. 1, 'EDF's choice for reactor').

D. Fortson (*The Sunday Times/The Times*) has reported that analysts at Citi-group, have estimated that the UK government will have to pay a minimum price of £166 per megawatt hour to convince EDF Energy to agree to build this strategic nuclear plant at Hinkley Point, by approximately 2020. It seems likely that this forecast minimum power price of £166 MWh (*>3 times current rate*), to cover new nuclear, could be realistic and contribute to getting third-generation nuclear plant installed in the UK but as outlined earlier, the UK households will certainly be faced with soaring energy bills – doubled in 5 years to an average £1345 per year in 2012 – which in turn will result in many households (>4.5 million UK households) in fuel poverty (when more than 10% of income is used to pay energy charges). Moreover, several commentators have commented – including, for example, Fortson (*The Sunday Times*) – along the lines, 'Raising the price floor for nuclear as well as costly low carbon technologies, such as huge offshore wind farms, will add hundreds of pounds more'. Later, in a comment focussed on the nuclear debate, Fortson said, 'Whitehall (London) is keen to stress that the nuclear programme will be subsidy free. The government's position is seen by many as little more than "semantics"; however, because taxpayers will still foot the bill for new reactors, only they will do it through higher energy bills rather than taxes'.

It is appropriate that this is our last comment on this theme, distressing as it may be to UK energy consumers!

No, wait!

In a further nuclear energy 'nudge article', Tim Webb, *The Times*, 16/7/12, restates, 'French demand a high price for rescuing nuclear industry with two new reactors'. He states, 'families and businesses are being asked to find an extra £2.8 bn a year for the next 25 years by EDF Energy as the price for rescuing Britain's faltering nuclear power programme'. Most of the article has been covered earlier except for two other 'gems':

- Gillespie also states, 'The government will offer special subsidies to EDF Energy because its package of electricity market reforms will not be ready for several years'. (Last week the Coalition (UK government) appointed KPMG to represent it when formal negotiations (with EDF Energy) begin.)
- EDF Energy said 'that the subsidy for nuclear will represent a fair and balanced deal for customers. It will show the cost competitiveness of nuclear 'new-build' installations as compared to other types of generating plant. The process will be transparent – and the details published in due course'.

This writer notes the many press energy comments and recalls the words of the losing UK trainer, outside the ring, when an ageing British heavyweight boxer yet again lost a fight and was KO'd, just as his trainer 'threw in the towel!' and said, 'So OK then – we certainly did our best – under the circumstances, that's it now!'

6. *IET – Regional focus around UK* (www.EandTmagazine.com/news, volume 7, issue 6, July 2012) Now for some 'good energy-related news' reported in this edition:

   (A) The UK and Iceland Energy Ministers have signed an energy agreement which could result in the volcanoes in Iceland supplying electricity to the UK (31/5/12). Part of this agreement includes:

   (i) to explore options for building the world's largest sub-sea electricity interconnector
   (ii) pledge to exchange information on the development of the deep geothermal sector in the UK and on the development of oil and gas industries
   (iii) work with their respective ministries for international development on renewable energy projects in developing countries.

   (B) A wind farm project on the island of South Uist in the Western Isles (23/5/12) will generate more than £20 million over the next two decades for residents, it is claimed. Three 2.3 MW turbines will be erected in the community wind farm by August 2012, with money from the renewable energy scheme reinvested in the area to improve tourism, leisure and local business.

   (C) Underwater turbine successfully completed tests (17/5/12) in the fast-flowing tidal waters around Orkney. The power generator, which will generate electricity from tidal power, has been providing electricity for homes and businesses on the island of Eday since it was installed by Scottish Power Renewables (SPR) in December 2011. SPR plans to use this technology in the 'Sound of Islay' as part of the 'world's first' tidal turbine array.

   (D) Offshore wind cost 'can be competitive', says industry

   Two reports were launched jointly at the Global Offshore 2012 Conference in London, organised by Renewable UK.

   (After E&T News report – Reader is encouraged to read the full original text of this IET Magazine renewables item.)

   According to a report by the Offshore Wind Cost Reduction Task Force, it is claimed that the cost of offshore wind energy could be cut by as much as 30%. The report apparently builds on detailed evidence in a study by *The Crown Estate* to show how reductions can be achieved, setting out specific actions to drive costs down. Maria McCaffery, the chief executive, Renewable UK, said: 'By committing to slash the costs of developing offshore wind, the UK has once again shown why it is the world leader in this industry'; 'We are ready to play our part to encourage the industry to follow the recommendations in the report'. This brief E&T renewable energy item goes on to state:

   > Such a cut would see the cost of delivering 18 GW of electricity from offshore wind farms (approximately 20% of the UK's total electricity demand) drop from £140/MWh today to £100/MWh by 2020, saving over £3 bn a year and making offshore wind more cost competitive with other forms of energy generation such as nuclear and carbon-abated fossil fuels.

Note: As just reported above, Citi, the investment bank, has estimated that the UK government will have to pay a minimum price of £166/MWh to convince EDF Energy to agree to build this strategic third-generation nuclear plant at Hinkley Point, Somerset, UK, by approximately 2020.

- *The Task Force report* is quoted as charting 28 recommendations on how the industry can reduce the cost of generation. In particular more efficient contracting *and the concept of alliancing*, used successfully by the North Sea oil and gas industry to reduce risk and bring down costs, have the potential to be transformative in lowering cost and improving working practices.
- 'Offshore wind will be a vital part of a diverse and secure low-carbon energy mix in the decades ahead', said Energy Minister Charles Hendry, 'but we are clear costs must come down'.
- This E&T news item goes on to write about the separate report by The Crown Estate, published in parallel with the Task Force report. The Crown Estate, which manages the seabed around Britain, also supported the view that a major cost reduction by 2020 was possible, stating that producing more efficient turbines was the main driver for cheaper wind farms.
- Overall, the cost of building an offshore wind farm in 2020 could fall by 39%, this report apparently said, largely through the development of larger and more reliable types of wind turbines.
- More intense competition among turbine manufacturers support structure providers and installers in the UK and the rest of Europe could also help cut costs by as much as 6%.
- Rob Hastings, the energy and infrastructure portfolio director of The Crown Estate, said: 'we believe that there is no single solution to reduce the cost of offshore wind and all participants in the sector need to play their part'.

(E) Plans dashed for wind turbine plant in Kent (after Robert Lea, *The Times*, 23/06/12, p. 44)

Robert Lea reports, 'Ambitious plans to build the world's biggest wind turbines at a state-of-the-art factory in Kent have been ditched'. Vestas, a Danish company, had announced a proposal to build the plant at the Port of Sheerness on the Thames Estuary some 13 months ago and hopefully create 2,000 jobs. However, yesterday the company said it would no longer proceed.

Lea comments that this decision comes 'against a backdrop of stalling investment in the British wind energy industry; there are few concrete commitments to begin work on the so-called Round three projects erecting turbines as tall as the Gherkin office tower in the City of London – around the coast of Britain'. (Vestas said it would 'remain active' in the British wind market.)

Lea also observes:

- 'That at the same time, the world's main manufacturers have had their heads turned by the high level of investment being planned in China to develop wind energy;
- Vestas had said that it wanted to construct its V164-7.0 turbine in a purpose-built facility on land owned by Peel Ports. At 164 m in height it would have been the world's tallest wind driven electricity generator, with single turbines capable of producing 7 MW – enough to power 6,500 homes'.

Lea reminds the reader:

1. 'Vestas had warned that its investment decision would be dependent on government support, and the Government subsequently pledged assistance from its Regional Growth Fund. (The plan had been to build a plant on 70 hectares which would be ready by 2015.)
2. Since then, the global slowdown in spending on wind power and stiff competition for work from China have forced Vestas to cut investment in Europe and eliminate 2,300 jobs, about 10% of its worldwide workforce.
3. It is not the first time that the Danish company has disappointed in Britain. It sparked an outcry in 2009 after closing a plant producing onshore turbines on the Isle of Wight with a loss of 450 jobs'.

Postscript to item: B. Marlow, on 1/7/12 in *The Sunday Times*, wrote that Vestas, the world's biggest maker of wind turbines, with 20,000 employees, is considering putting itself up for sale as concerns mount over 'its giant debt pile' and is entering debt restructuring talks with its lenders, which include Royal Bank of Scotland and HSBC.

Ben Marlow states, 'The company's, woes reflect a significant shift in the wind industry since the recession took hold. The sector has been battered by a perfect storm of government belt tightening, lack of project financing and rising cost'.

(F) *'British Sea Power'*: Michael Hanlon and Jonathan Leake posed the question, 'Wind power is controversial and solar unreliable: could the answer to our energy needs lie in the ocean?'. They went on to report (in *The Sunday Times*, 29/7/12, p. 17):

1. the formal opening of the Pentland Firth and Orkeny Marine Energy Park
2. a £20 million Government investment, announced in April for marine energy, 'and as of last week an enticing subsidy package to encourage the development of technologies that can turn sea movement into electricity'. The article provides an illustration of a 1,000 kW underwater tidal generator, by Rolls Royce, together with Pelamis and Oyster wave generator illustrations.

(Tidal and wave power will be paid 5 Rocs, under the government renewables obligation scheme which provides a great boost for the fledgling marine energy industry. The authors speculate that wave/tidal should fall below the cost of offshore wind by 2020.)

Summary: The developments (A)–(C) above are encouraging and positive indicators meantime towards future developments, but clearly, much progress will be necessary if (D) offshore wind farm and substation installation costs can be significantly cut, as anticipated in (D) above – clearly here – *everyone seems to be saying the correct things at present!* but all will have to be, and stay, *'onside' plus many other indicators must be carefully assessed*, if we are to achieve the proposed cost-cutting outcomes. *The final sentence quote by Rob Hastings is particularly apposite!*

Turning to item (E), this clearly demonstrates the hard economic facts of life. It shows the critical dependence of wind energy companies obtaining strong/robust government support guarantees, be it from a government in the UK or China! Now the item (*F*), on a week when the government announced another 'dash for gas': it would be nice if the tidal/wave energy fraternity came out of this initiative with flying colours! So very good luck.

(Many will feel that, as with new-build nuclear initiative, the British public will still have to pay a very high price for the proposed offshore wind farm developments. Let us develop an effective infrastructure that the UK can be proud of – robust, technically effective, flexible, intelligent, linking in/compatible, with the 'smart grid' concept of the future and trusting that it will stand the test of time – and all will marvel at another brilliant technological achievement – in true British/European tradition.)

Meantime, looking into the future via one's 'crystal ball', the immense economic investment issues/difficulties so clearly apparent at present, during this period of world recession, make it very likely that the time frame for the full development of the idealised 'smart-grid' network concepts may take somewhat longer to come to practical fruition.

Final notes

In this chapter, a wide range of technical energy aspects have been considered and extensively documented. They include the following:

1. Numerous tables have been provided (Tables 23.1–23.9) covering many energy and environmental themes and issues. Several tables reproduce strategic energy, environmental, design, operational, cost and logistical issues mainly using resource material/information originally reported in selected press articles, mainly from *The Times* and *The Sunday Times* or *The Guardian*. See Tables 23.1–23.15 which cover many aspects including some renewable energy and nuclear experiences, wind farm issues including operational experiences, cyber issues, coal and biomass generating plant, carbon capture issues, network issues, worryingly high energy subsidies, ever increasing energy bills for consumers, increasing fuel poverty in the UK, and the anticipated huge costings necessary to achieve smart-grid systems of the future (i.e. within the next two decades).
2. Outlining the smart-grid concept proposed by Gellings, EPRI [35] (see Tables 23.10–23.12) explaining the CIGRE SC activity structures post 2002 (see Table 23.13) and a range of Study Committee activities and publications, mainly to *empower* the reader with the technical resource listings available, including a strategic table (Table 23.14) detailing/covering key strategic technical issues to work on worldwide over the next few year.
3. As an engineer and a UK citizen, one has to be somewhat concerned at the confusion in government positions: interview statements; signals; messages supposedly relating to a UK energy policy, which have emerged during the production of this chapter (2010–2013) (e.g. compilation and evaluation of many press articles/interviews/reports). There has been much, indeed a barrage, of energy-related articles in the UK press during 2010–2013 and earlier, reflecting the growing consumer discontent at the high level of fuel bills, with many articles on all aspects of UK energy sector activities and subsidy plans.

Senior members of each sector, government, renewables, nuclear, etc., seem to 'drip-feed' the party line as they get their message across and 'nudge the agenda along' new *appropriate* directions – even though this is often seen in conflict with last week's, and next week's edicts. Strategic aspects of these issues are discussed in this chapter.

Table 23.15   Latest press articles relating to energy situations in the UK, USA and EU (Business articles, [1, 2] after Tim Webb, The Times, June 13, 2013, p. 45)

---

[1] **Ofgem's plan for reform 'no benefit to consumer', Tim Webb:** The largest independent power supplier in Britain has dismissed Ofgem's boast that its flagship reforms announced yesterday will break the stranglehold of the UK Big Six energy companies, Webb reports. **Darren Braham**, chief financial officer of First Utility, said that **the regulator's plan would not make the market 'significantly better' for consumers or small suppliers**. This criticism is a blow to Ofgem's claims that it is championing the cause of the consumer by tackling the dominance of the big energy groups – **EDF Energy, SSE, E.ON, RWE npower, Centrica and Scottish Power**.

'The regulator published its reforms yesterday', which, Webb comments, 'were significantly **"watered down"** from its original proposals unveiled some two years ago. Ofgem had proposed a radical plan to force the biggest energy groups to auction at least 20% of the electricity they generate to the market. These '**Big Companies**' own 85% of Britain's generation capacity and they supply >95% of households and businesses. Webb informs the reader that 'the idea was that an auction would make it easier for independent suppliers to buy electricity for customers on the open market'.

However Webb reports that the regulator [Ofgem] dropped the plan and instead will require the big energy groups to make available to the market limited amounts of electricity for certain periods of the day. Moreover, the Big Six will also be required to publish the wholesale electricity prices, at which they buy and sell, two years in advance.

Ofgem hopes that this will boost liquidity in the power market and make it easier for independent suppliers to trade. A new code of conduct requiring the big companies to trade fairly with the smaller companies is also being introduced. Tim Webb goes on to state **that Andrew Wright, the interim chief executive**, said, 'Our aim is to improve consumer confidence and choice by putting strong pressure on prices through increased competition in the energy market; Ofgem's proposals will break the stranglehold of the Big Six and create a more level playing field for independent suppliers'.

Webb goes on to state that nevertheless, despite these reassurance, '**Mr Braham, (First Utility)**, insisted that the proposals would have only a limited impact, at best.' 'It's encouraging they are doing something, but it's hard to see how it will make the market significantly better. You won't see a step change in liquidity', said Mr Braham, who went on to comment that 'with many of the details still to be worked on, the potential was for the proposal to have somewhere between '**zero impact and or a small impact**' for liquidity.

First Utility have called for energy companies to be prevented from selling any of the power they generate to themselves, which would have been similar to a full scale auction. Mr Braham warned that the Big Six would lobby to limit the impact of the final proposals. 'I'm sure they will be pushing in a certain direction in terms of watering them down'.

[2] **America's new almost accidental oil rush, Tim Webb:** Webb reports that the shale gas revolution in the USA has turned into an oil bonanza after the country recorded its biggest jump in production last year (2012). Apparently, drillers are now targeting more valuable shale oil because the domestic gas '**glut**' has depressed prices. US oil production has rocketed by 13.9% to 8.9 million barrels per day in 2012. The USA is expected to become the world's largest oil and gas producer by 2017. Webb reports that the USA pumps nearly 1 in 10 of the world's barrels and more than Iraq, Kuwait, and Norway combined. This huge jump in US oil production easily outstripped that of any other country last year.

Some recent ESI developments 889

However (almost predictably for long suffering UK consumers), **Bob Dudley**, the chief executive of BP, said 'that it would take a long time before the US shale gas revolution resulted in lower energy bill for British consumers'. Webb reports, 'The US is planning to start exporting some of its surplus energy across the Atlantic as liquefied natural gas, but domestic manufacturers, who have enjoyed a renaissance thanks to rock bottom fuel prices, want shipments to be restricted'. 'That will take a long time', Dudley had commented. He went on to observe, 'There will be some [impact on prices] no doubt; but it's a wide spectrum [of debate] about whether the US will actually export a lot of gas'.

Webb comments that only the USA and Canada are currently producing shale gas in large quantities, but he reports **that 'dozens of countries including Britain are estimated to be sitting on huge untapped deposits'**. This week, a report from the US government Energy Information Administration [EIA] said that shale oil and gas were '**globally abundant**' and estimated that there were 345 billion barrels of recoverable shale oil, equivalent to 10% of the world's crude oil.

Webb goes on to comment that EIA also said that Britain alone held 700 million barrels of oil and 26 trillion cubic feet of gas that can be recovered. While prospective shale gas producers such as in the UK look on, Webb reports that the USA continues to feel the dynamic effects of its shale gas boom.

According to BP, US power stations 'bought-up' much of the cheap surplus gas produced in 2012 by the shale drillers:

- Gas fired generation jumped 21% in 2012, the biggest rise for more than 40 years to produce a record amount of electricity.
- Coal plants could not compete and their output (2012) slumped to the lowest since 1987.
- After the nuclear disaster at Fukushima, Japan in 2011, the share of the world's electricity generated by nuclear power dwindled in 2012 to the lowest since 1984, before the Chernobyl disaster two years later.

Webb ends his article with some material relevant to the USA and also abroad:

- **Christof Ruhl**, the chief economist of BP, warned that the construction of wind farms could slowdown in Europe as taxpayer subsidies have become unsustainable. 'If they scale up too fast, the cost of that support can become unmanageable', he said.
- Ruhl also claimed that the '**business audiences he speaks to have become weary of tackling climate change**. People are disillusioned because they **have been told it's a problem relatively easy to fix with renewables and [carbon dioxide] sequestration**', he said. 'It's not going to be cheap and easy. It [action] will happen only if people are going to feel the need'.

4. Worthy of special mention is the last table in the main chapter (i.e. Table 23.15). This reproduces two articles in *The Times*, dated 13/6/13, by Tim Webb, reporting on [1] and [2]. Some aspects are listed below:

   1. The energy regulator Ofgem published recently its reforms aimed to break the 'stranglehold' of the Big Six energy companies in the UK. Ofgem owns 85% of the UK's generating capacity and supplies 95% of UK households and businesses. It contains some information on reforms which seem to be significantly 'watered down' from the original proposal 'muted' two years ago. This will probably result in further price increases for long suffering UK energy consumers.
   2. Webb reports that the shale gas revolution in the USA has recorded its biggest jump in production last year (2012). US oil production has increased dramatically, up by 13.9% to 8.9 million barrels per day in 2012.

This week, a report from the US government Energy Information Administration (EIA) said that shale oil and gas were 'globally abundant' and estimated that in the USA there were 345 billion barrels of recoverable shale oil, equivalent to 10% of the world's crude oil. Webb also comments:

(i) EIA had also stated that Britain alone held 700 million barrels of oil and 26 trillion cubic feet of gas that can be recovered.
(ii) According to BP, American power stations 'bought-up' much of the cheap surplus gas produced by the shale drillers last year (2012) and
  – Gas fired generation jumped 21% in 2012, the biggest rise for more than 40 years to produce a record amount of electricity.
  – Coal plants could not compete and their output (2012) slumped to the lowest since 1987.
  – After the nuclear disaster at Fukushima, Japan in 2011, the share of the world's electricity generated by nuclear power dwindled in 2012 to the lowest since 1984 before the Chernobyl disaster two years later.
(iii) Christof Ruhl, the chief economist of BP, has warned that the construction of wind farms could slowdown in Europe as taxpayer subsidies have become unsustainable. 'If they scale up too fast, the cost of that support can become unmanageable,' he said.
(iv) Ruhl also claimed that 'the business audiences he speaks to have become weary of tackling climate change.' 'People are disillusioned because they have been told it's a problem relatively easy to fix with renewables and [carbon dioxide] sequestration', he said. 'It's not going to be cheap and easy. It [action] will happen only if people are going to feel the need'.

23.14.2 Conclusions

This chapter has discussed the development of *'smart grids' and enhanced transmission and distribution networks of the future*, anticipated to be in place within two decades [30, 35] in some detail (see section 23.13, Tables 23.9–23.14) and has provided details of additional technical, environmental and economic resource materials. It has also looked carefully into several aspects of renewable energy, wind farms, nuclear energy prospects post Fukushima nuclear accident in March 2011, some worrying cyber issues, future trends and concerns relating to ever-increasing energy consumer bills, energy subsidies being paid for by UK customers and more.

Above all, and of major concern to many, including this writer, however, was the evidence obtained relating to what appears to be the absence of a coherent UK energy policy and robust ownership of it by the UK government. This clearly emerges from the compilation of the press reports and interviews with senior figures over a three year period. This writer has tracked and recorded this and other issues in this chapter and highlighted the latest in the preceding discussion section.

Undoubtedly, the economic downturn may also delay somewhat certain of these huge network development programmes coming to fruition on schedule – but naturally, this will be much clearer once the UK government finalises its energy plans soon.

Clearly, when discussing the smart grids and enhanced transmission and distribution networks of the future, the writer 'has called heavily on' Gellings, EPRI [30] and CIGRE [35] WG document published by *Electra* and many other sources such as technical reports/articles mainly published in the UK press/media.

He gratefully acknowledges this and again reminds the reader always to refer to original and supporting documents for a fuller 'empowered' understanding of the strategic issues briefly touched on in this chapter.

References

1. Fichaux N., Wilkes J. *Oceans of opportunity*. European Wind Energy Association, September, 2009. p. 34
2. Trotscher T., Korpas M., Tande J.O. *Optimal design of subsea grid for offshore wind farms and transnational power exchange*. Sintef Energy Research; 2009
3. Nakajima T., Irokawa S. 'A control system for HVDC transmission by voltage source converters'. *IEEE Power Engineering Society Summer Meeting*, 1999. pp. 1133–19
4. Available from http://www.abb.com/hvdc (Accessed 5 January 2011)
5. Kundur P. *Power system stability and control*. McGraw-Hill; 1994
6. Vrana T.K., Hille C. 'A novel control method for dispersed converters providing dynamic frequency response'. *Electrical Engineering*. 2011 (Springer-Verlag)
7. Hauck M., Spath H. 'Control of a three phase inverter feeding an unbalanced load and operating in parallel with other power sources'. *EPE-PEMC*; Dubrovnik & Cavtat, 2002
8. Max L. *Design and control of a DC collection grid for a wind farm*. Doctoral Thesis, Chalmer's University, 2009
9. Vrana T.K., Torres-Olguin R.E., Liu B., Haileselassie T.M. 'The North Sea Super Grid – a technical perspective'. *IET ACDC Conference*; London, 2010
10. Agelidis V.G., Demetriades G.D., Flourentzou N. 'Recent advances in HVDC power transmission systems'. *Proceedings of IEEE international conference on industrial technology (ICIT)*. pp. 206–13; 2006
11. Lindberg A., Larsson T. 'PWM and control of three level voltage source converters in an HVDC back-to-back station'. *AC and DC Transmission*; London, UK, 1996
12. Available from http://www.siemens.com (Accessed 5 January 2011)
13. Barker C.D., Kirby N.M., MacLeod N.M., Whitehouse R.S. 'Renewable generation: Connecting the generation to a HVDC transmission scheme'. *CIGRE Canada conference on power systems*; Toronto, 4–6 October 2009

14. Nakao H., et al. 'The 1,400 MW Kii-Channel HVDC system'. *Hitashi Review*. 2001;**50**(3)
15. Zhao Z., Iravani M.R. 'Application of GTO voltage source inverter in a hybrid HVDC link'. *IEEE Transactions on Power Delivery*. 1994;**9**(1): 369–77
16. Jonsson T., Holmberg P., Tulkiewicz T. 'Evaluation of classical CCC and TCSC converter schemes for long cable projects'. *EPE 09*; Lausanne, 1999
17. Xu L., Andersen B.R. 'Grid connection of large offshore wind farms using HVDC'. *Wind Energy*. 2006;**9**:371–82
18. Iwata Y., Tanaka S., Sakamoto K., Konishi H., Kawazoe H. 'Simulation study of a hybrid HVDC system composed of self-commutated converter and a line-commutated converter'. *Sixth international conference on AC and DC power transmission*; 1996
19. Torres-Olguin R.E., Molinas M., Undeland T.M. 'A model-based controller in rotating reference frame for hybrid HVDC'. *ECCE*; 2010
20. Haileselassie T., Molinas M., Undeland T. 'Multi-terminal VSC-HVDC system for integration of offshore wind farms and green electrification of platforms in the North Sea'. *Nordic workshop on power and industrial electronics*; June 2008, pp. 9–11
21. Hendriks R.L., Paap G.C., Kling W.L. 'Control of a multi-terminal VSC transmission scheme for connecting offshore wind farms'. *European wind energy conference*; 2007
22. Pan W., Chang Y., Chen H. 'Hybrid multi-terminal HVDC system for large scale wind power'. *Power systems conference and exposition proceedings*, pp. 755–59; 2006
23. Groeman F., Moldovan N., Vaessen P. 'Ocean grids around Europe'. *KEMA Briefing Paper*; November 2008
24. Van Hulle F. *Integrating wind*. Trade Wind project report, 2009
25. De Decker J., Woyte A., Schodwell B., Volker J., Srikandam C. *Directory of offshore grid initiatives, studies and organisations*. Offshore Grid; 2009
26. Cole S., Vrana T.K., Curis J.B., Liu C.C., Karoui K., Fosso O.B., et al. 'A European supergrid: Present state and future challenges'. *PSCC conference*, Stockholm; 2011
27. Vrana T.K., Fosso O.B. 'Technical aspects of the North Sea Super Grid'. *Electra*. 2011;**258**(10):6–19
28. Davis S. 'But where do you plug them in? The challenges of building a grid for offshore wind'. *IET Engineering & Technology*. 2012;**7**(5):80–3
29. Gellings C.W., Zhang P. 'The ElectriNet'. Invited CIGRE Paper. *Electra*. 2010;**250**(6):4–10
30. Gellings C.W. 'Estimating the costs and benefits of the 'smart grid' in the United States'. Invited CIGRE Paper. *Electra*. 2011;**259**(12):6–14. Available from www.epri.com (TR-1022519)
31. *Future inspection of overhead lines*. Palo Alto, CA: EPRI; 2008 (1016921)
32. *Estimating the costs and benefits of the smart grid*. Palo Alto, CA: EPRI; 2011 (1022519)

33. Zomers A., Waddle D., Mutale J., Kooljman Van Dijk A. 'The global electrification challenge – the case of rural and remote areas'. Invited CIGRE Paper. *Electra*. 2011;**259**(12):16–29 (29 references)
34. Ito H., Yoshizumi T., Amano N. 'Compact substation technologies, state-of-the-art and future trends'. Invited CIGRE Paper. *Electra*. 2002;**203**(10):4–11
35. CIGRE WG. 'Network of the future' electricity supply systems of the future. TC Advisory Group Report. *Electra*. 2011;**256**(6):42–9
36. 'Background of Technical Specifications for Substation Equipment Exceeding 800 kV AC', TB 456-WG A3.22 (also CIGRE WG A3.22. *Electra*. 2011; **255**(4):60–9.)* [see also a recent important update, 'Insulation Coordination for UHV AC Systems', TB 542, CIGRE WG C4.306 (also WG C4.306. *Electra*. 2013;**268**(6);72–9); will be proposed for the application guide IEC 60071-2 (1996) and IEC apparatus standards)]
37. 'The challenges facing AC offshore substations for windfarms'. CIGRE Technical Report (10 References) (see also CIGRE WG B3.26. *Electra*. 2010; **253**(12):32–8)*
38. Samotyj M., Gellings C., Amin M. 'Power system infra-structure for a digital society, creating the new frontiers'. Invited CIGRE Paper. *Electra*. 2003; **210**(10):20–30
39. 'The Impact of Renewable Energy Sources and Distributed Generation on Substation Protection and Automation', TB 421-WG B5.34, 2010 (see also WG B5.34. *Electra*. 2010;**251**(8):34–43)*
40. 'Technical and Commercial Standardisation of DER/microGrid Components', CIGRE TB 423-WG C6.10, 2010 (see also WG C6.10. *Electra*. 2010; **251**(8):52–9)*
41. 'Demand Side Integration', CIGRE TB 475-WG C6.09, 2011 (see also WG C6.09. *Electra*. 2011;**257**(8):100–7)*
42. Jenkins N., Allan R., Crossley P., Kirschen D., Strbac G. *Embedded generation*. IEE Power and Energy Series, No. 31. London: IET; 2000. ISBN 0-85296-774-8
43. Jenkins N. 'Impact of dispersed generation on power systems'. Invited CIGRE Paper. *Electra*. 2001;**199**(12):6–13 (9 References)
44. 'Increasing Capacity of Overhead Transmission Lines: Needs and Solution', CIGRE TB 425 (JWG (B2/C1.19)), 2010 (see also JWG (B2/C1.19). *Electra*. 2010;**251**(8):70–7)*
45. 'The Impact of Implementing Cyber Security Requirements using IEC 61850', TB 427-WG B5.38, 2010 (see also WG B5.38. *Electra*. 2010;**251**(8):84–90)*
46. 'Treatment of Information Security for Electric Power Utilities', CIGRE TB 419-WG D2.22, 2010 (see also WG D2.22. *Electra*. 2010;**250**(6):64–71)*
47. 'Security technologies guideline: Practical guidance for deploying cyber security technology within utility data network', CIGRE WG D2.22, Report (by BARTELS A. *et al.*). *Electra*. 2009;**244**(6):11–7 (6 References)
48. 'Guide for Qualifying High Temperature Conductors for Use on Overhead Transmission Lines', TB 426-WG B2.26, 2010 (see also WG B2.26. *Electra*. 2010;**251**(8):78–83)*

49. 'Technical assessment of 800 kV HVDC applications', CIGRE TB 417-WG B4.45, 2010 (see also WG B4.45. *Electra.* 2010;**250**(6):51–9)*
50. 'Status of development and field test experience with high temperature superconducting power equipment', CIGRE TB 418-WG D1.15, 2010 (see also WG D1.15. *Electra.* 2010;**250**(6):60–3)*
51. 'Modelling new forms of generation and storage', CIGRE, Task Force 38.01.10, TB 185, 2001 (see also TF 38.01.10. *Electra.* 2001;**195**(4):54–63)*
52. 'Grid integration of wind generation', CIGRE TB 450-WG C6.08, 2011 (see also WG C6.08. *Electra.* 2011;**254**(2):62–7)*
53. 'EMS for 21st century-system requirements', CIGRE TB 452-WG D2.24, 2011 (see also WG D2.24. *Electra.* 2011;**254**(2):74–9)*
54. 'FACTS technology for open access', JWG 14/37/38/39/39.24, TB 183, 2001 (see also JWG 14/13/37/38/39/39.24. *Electra.* 2001;**195**(4):38–45)*
55. De Almeida S.A.B., Engel Brecht C.S., Pestana R., Barbosa F.P.M. 'Prediction of faults caused by lightning for transmission system operations'. Invited CIGRE Paper. *Electra.* 2001;**254**(2):4–19 (11 References)
56. Obtaining value through optimising withstand design'. CIGRE WG B3.01 Report. *Electra.* 2010;**252**(10):30
57. 'Review of the current status of tools and techniques for risk based and probabilistic planning in power systems', CIGRE TB 434-WG C4.601, 2010 (see also WG C4.601. *Electra.* 2010;**252**:78–85)*
58. 'Integrated management information in utilities', CIGRE TB 341-WG D2.17, 2008 (see also WG D2.17. *Electra.* 2008;**236**(2):52–66)*
59. 'Current practices of electrical utilities towards sustainable development', CIGRE TB 340-WG C3.03, 2008 (see also WG C3.03. *Electra.* 2008;**236**(2):42–51)*
60. 'Consideration relating to the use of high temperature conductors', SC B2 CIGRE, TB 331, 2007 (see also SC B2. *Electra.* 2007;**234**(10):28–37)*
61. 'Modelling and dynamic behaviour of wind generation as it relates to power system control and dynamic performance', CIGRE TB 328-WG C4.601 (see also WG B4.19. *Electra.* 2002;**233**(8):48–56)*

(*Readers are encouraged to refer to websites of organisations involved for latest information.*)

Note: The asterisk (*) identifies any reference (No. 20 onwards) relating to CIGRE Technical Brochures.

Appendices

A Cyber-crime and cyber-security

Perusal of recent articles in the UK press, briefly touched on, for example see items (1)–(5) below, provides a valuable strategic insight into current cyber development issues, problems and 'secure' business. Later in Appendix C we will touch briefly on CIGRE treatment of information security for electric power utilities, certain energy-related cyber issues and reports.

B Cyber-crime

(1) At the 2011 international conference in London on the future of cyberspace (an article by T. Coghlan, *The Times*, 2/11/11) delegates were informed of major changes occurring on the Internet:

- According to this report by Tom Coghlan, these changes also include mounting government censorship with restrictions on Internet access, which is already imposed in 60 countries including some within Europe. Britain and America appear to have issued clear warnings to countries that seek to restrict Internet freedoms. Some strategic points reported by Coghlan are reproduced below. (See original article for full coverage.)

    Mr William Hague, the foreign secretary, UK, said: 'We must aspire to a future for cyberspace which is not stifled by government control or censorship, but where innovation and competition flourish and investment and enterprise are rewarded'. There was a need to curb a "free for all" on the Internet but Hague warned that 'without international action to establish "rules of the road" for cyberspace, a "darker scenario" could prevail'; 'Cyberspace could become "fragmented" and "ghettoised", subject to separate rules and processes in different regions, set by isolated national services, with state imposed barriers to trade, commerce and free-flow of information and ideas'.

- *Mr Joe Binden*, the US vice president, echoed the words of Mr Hague, warning that despite the Internet offering the opportunity for wrong-doing 'on a vast scale', he went on to say, 'In our quest for security, we believe that we cannot sacrifice the openness that makes possible all the benefits and opportunities that the internet brings'; 'These countries that try to have it "both ways" by making the internet closed to freedom of expression but open for business will find that this is no easy task'.

    In the next 20 years, it is anticipated that changes to the Internet also include:

    - an expected growth in Internet users from *2 billion to 5 billion*
    - an exponential rise in cyber-crime – currently costing the UK economy a staggering £27 *billion a year in 2011.*

- Western officials have been critical of Russia and China for suppressing Internet access to their own citizens, on the one hand, and for alleged

government-sponsored cyber-attacks on Western state secrets and private sector intellectual property, on the other.
- Mr Hague also indicated to *The Times* reporter that such attacks on government networks *now number at least 1,000 per month.*

(2) An earlier article in *The Times* (21/9/2011) by Richard Lloyd Parry, from Tokyo, entitled 'Hackers breach the defences of Japan's biggest weapons maker: Reports suggest Chinese origin for attacks' Parry, reports that *foreign cyber-hackers have stolen secret information from the biggest arms (weapons) manufacturer in Japan* (Mitsubishi Heavy Industries), which makes submarines, nuclear power plants and American-designed fighter jets on missile systems. Some of the strategic aspects reported by Parry are set out below:

- Apparently, the company admits that computers and servers at its offices and factories have been infected with Trojan viruses, capable of transmitting information back to the cyber-attackers. To quote Parry: 'neither the Government nor the company has made comment on the origin of the attacks but the Japanese media reported clues that it may have originated in mainland China'.
- 'At least 45 servers and 38 computers were found to have been infected with several different kinds of virus at 11 Mitsubishi sites including its headquarters in Tokyo, and also in its shipyard in Kobe, which produces submarines and nuclear power-plants, and in Nagasaki, the latter being involved in making destroyers in joint strategic Japan/USA defence projects, and where rocket engines/missile interceptors are made'.
- Computers at the Nagoya Guidance and Propulsion Systems Works in Komaki were also hit. Apparently, this facility makes rocket engines for Japan's space programme and missile interceptors for the defence shield that Japan is developing with the USA.
- It is reported that the *attacks were traced back to more than 20 servers in China, Hong Kong, the USA and India, according to the press in Japan* (*The Yomiuri Shimbun newspaper* (YS)).
- 'Parry reports that the attackers appeared to be using the simplified script used on the mainland (i.e. China), suggesting that the attackers originated there. The Japanese government was unaware of some attacks until reported in the press', YS reported.
- In a statement, Mitsubishi said, 'at this stage, it's not confirmed that information on our products has passed outside the company'.

(3) 'The chink in the West's cyber-armour – you' (*The Times*, article by Misha Glenny, 30/10/11), whose latest book, *Dark market: cyber-thieves, cyber-cops and you*, is published by Bodley Head.

Glenny provides another brief yet excellent article with a 'subtitle theme': 'Millions are spent defending vital services from hackers, but they have found a way in – through social networking sites' (e.g. Facebook, Twitter). Glenny observes that, in the past three years, criminals have discovered that one of the quickest ways to spread viruses or attack businesses is by using Facebook and

Twitter. A few strategic aspects of this reporter's article are touched on and reproduced below.

A UK Major General, Jonathan Shaw, the head of cyber-security for Britain's armed forces, was reported as saying: 'the biggest threat to this country by cyber is not military, it is economic. If the moment you come up with a brilliant new idea, it gets "nicked" by the Chinese then you can end up with your company going bust'.

Reporter Misha Glenny comments: 'The Major General was right to be concerned. A large company can spend tens or hundreds of millions of pounds securing its computers but all that cash can be wasted in an instant if a careless employee is duped by one of the increasing number of skilled social engineers, the criminal hacker or the spy, who persuades us to behave unwittingly on our computer in a way that can have a fatal impact on our interests'.

The estimated annual cost of such attacks is now moving into the *trillions*, according to various sources.

Major companies spend huge sums of money to have their networked systems thoroughly checked to prevent such attacks.

This article also reports that one of the largest pharmaceutical companies, because of the vast amounts it channels into research and development, is 'compelled to create a Digital Fort Knox' around its networked systems.

'One slip and a new drug could be everywhere – as a generic version in India, or China. Hundreds of millions of dollars of investment – and potentially thousands of jobs – could go down the drain in a nanosecond of a successful computer breach'.

Glenny also states: 'Yet the fastest growing threat to company's data security comes in the shape of the disgruntled worker'. He goes on to state that one such worker 'is facing a prison sentence of up to ten years after pleading guilty to having wiped out large depositories of data and freezing the networks of his former employers, the US branch of the Japanese pharmaceutical company Shionogi'.

- *Note*: The author of this chapter had a similar but more modest experience (i.e. less costly) many years ago when another disgruntled employee – shortly to be made redundant – erased a vast amount of computer code and other data from a specialised computer system used for switchgear R & D purposes and after it had been incrementally improved year on year, and thus ended, in an instant, all usage of an excellent commercially and internationally used analytical design tool (back-up got 'lost! also)!
- Rex Hughes, a lawyer who specialises in cyber-security and is a visiting fellow at Cambridge University, is quoted as stating: 'this is another area that companies are going to have to deal with'; 'Not only do companies have to protect themselves from external attack, they will need to introduce a more extensive internal monitoring'.

(4) A malicious code: Stuxnet claimed to be the world's first cyber super-weapon (*The Sunday Times* article by Christopher Goodwin, 4/12/11):

- 'Stuxnet' is claimed to have destabilised a nuclear power plant in Iran, and the analyst who deciphered the code believes it now has the potential to wreak havoc across the globe!
- This article provides the reader with a James Bond-type 'wrongdoings in cyberspace' scenario – allegedly that Stuxnet might be 'targeting industrial control systems' via their programmable logic controllers (PLCs) resulting in extensive damage to many centrifuges at Iran's National Nuclear uranium enhancement facilities. (*It was later reported that the Stuxnet worm had caused hundreds of Iran's centrifuges to accelerate to breaking point.*)
- An earlier article (*The Times*, 30/11/11), 'Tehran nuclear plans "are under attack" as blast hits second site' also refers to major incidents.
- Early in 2012, Iran announced its first domestically produced nuclear fuel rods for its research reactor. The government claimed to have added 3,000 state-of-art centrifuges to its Nantanz facility (making 9,000 in total) and tripling Iran's ability to enrich uranium to 20%.

Notes:
(1) 'Weapons grade material' is 21%.
(2) A later report has suggested that Iran is planning further two new nuclear installations.

(5) Chinese steal jet secrets from BAE (*The Sunday Times* article by D. Leppard, 11/3/12)

- It was recently reported by D. Leppard that for the past 18 months 'Chinese spies hacked into computers belonging to BAE Systems, Britain's biggest defence company, to steal details about the design, performance and electronic systems of the West's latest fighter jet, senior security figures have disclosed'.
- Apparently, the Chinese have exploited vulnerabilities in BAE's computer defences to steal vast amounts of data on the £200 billion, F-35 joint strike fighter (JSF), a multinational project to create a plane that will give the West air supremacy for years to come – according to sources. It should be noted that Joel Brenner, a former chief of American counter-intelligence, is quoted as warning, at a 2009 business conference of a potential situation, 'where a fighter pilot can't trust his radar!'. This is a worrying situation.

Note: Additional information is available in the original article. Leppard also reports that, in 2010, BAE had also been targeted by two Chinese nationals 'who tried to smuggle thousands of its microchips out of America, after buying them from BAE in Virginia'. (It was reported that both were convicted after a 'sting' operation.)

(6) 'Flame malware: Israel blamed for new "Flame" virus attack on Tehran' (after James Hider, Middle East correspondent, *The Times*, 30/5/12).

According to a recent follow-up strategic article in *The Times* (James Hider reports from Iran), the Middle East is still in a state of tension. Hider says Iran has recently claimed that Israel is to blame for a new computer virus

'Flame' – even more powerful than the 'Stuxnet' worm that *sabotaged* its nuclear facilities (see item (4) above). Reporting from Iran in *The Times* of 30/5/12, Hider comments, 'the strength and sophistication of the virus described as a super-weapon and nicknamed "Flame", has led to speculation that it was 'state-engineered', possibly by Israel, who was believed to have collaborated with the US Stuxnet several years ago'. Marco Obisco, cyber-security coordinator for the UN International Telecommunication Union, stated, 'This is the most serious warning and alert we have ever "put-out" indicating the threat faced by the virus'. Hider reports that 'the UN Agency is responsible for assisting member countries to secure their national infrastructure'.

Moshe Yaalon, Israel's Deputy Prime Minister, is claimed by Hider to have done little to dispel the speculation, saying that 'whoever sees the Iranian threat as a significant threat is likely to take serious steps including those to hobble it', and 'Israel is blessed with high technology and we boast tools that open all sorts of opportunities to us'. Ironically, the article also states that Israel itself has been 'hit'. This Flame virus is thought to have been in operation 'in the wild' for two years:

- It appears that the main goal with the 'Flame' virus is *espionage* rather than *sabotage*, having the capability of siphoning off documents, 'screen sheets' and audio-recordings, which it sends to servers all over the world. (Look at the similarity of the Japanese case referred to in item (2) above.)
- The *Flame virus* is actually being used as a cyber-weapon in attacking entities in several countries. Hider reports that Eugene Kaspersky, the founder and CEO of Kaspersky Laboratory Company, Moscow (one of the world's biggest producers of anti-virus software), has said, 'the Flame malware looks to be another phase in the war and it's important to understand that such cyber weapon can easily be used against any country'.
- The virus 'is actively being used as a cyber weapon attacking entities in several countries', and the company added, 'the complexity and functionality of this newly discovered malicious programme exceed those of all other cyber menaces known to date'.
- Apparently, 'no security software had detected it because of its extreme complexity', the company is reported as stating. 'Flame was one of the most advanced and complete attacking tool kits ever discovered', the company also said.
- An independent publication has been published within *The Times* of 8/12/11, by Raconteur Media (pp. 1–16) (edited by Bryan Betts): This feature quotes that the spending on cyber-security in the UK is expected to reach nearly £3 billion in 2011, according to the PwC study entitled '*Cyber Security M&A*: Decoding deals in the global cyber-security industry'.

    Betts comments that (i) spending on it may grow 10% 'year on year' for the next three to five years. Earlier, he also emphasised points already made above by others, including (ii) 'we forget at our peril that security is at least as much about people as it is about technology'.

(7) From the above, the reader can see the importance and strategic urgency of being able to combat effectively and work towards eliminating cyber-crimes.

Table B.1 provides additional self-explanatory information which further illustrates the lack of UK preparedness at this time, relating to cyber computer attacks. (*Source*: *The Times* and PwC, 2011)

Finally, the reader may recall just a few other examples of 'noteworthy' historical strategic cyber-attacks:

- *11 June 2011* – the International Monetary Fund stated it had suffered a sophisticated cyber-attack (IMF sources stated that the material would be political dynamite in many countries!).
- *20 April 2010* – Sony was forced to shut down its PlayStation network after the 'biggest cyber hack attack in history' with the details of 70 million users stolen.
- *12 January 2010* – Google was subjected to a highly sophisticated attack on corporate infrastructure originating from China. This resulted in the theft of intellectual property.
- *January 2009* – Data from more than 130 million credit and debits cards stolen in a 'hack-attack' had cost nearly $130 million to Heartland Payment Systems, a credit card-processing company.

Table B.1   PwC states that computer attacks soar with UK PLC unprepared (7) [after The Times *UK Business Report, A. Spence, 29/11/2011]*

Recent international research by PwC shows that cybe-attacks represent the biggest criminal threats to British businesses:

- More than 25% of British companies were victims of computer-related crime (hacking or intellectual property theft) in 2010.
- A recent international economic research crime survey by PwC has shown that cyber-attacks represent the biggest criminal threats to British businesses.
- In this article by Alex Spence, it is reported that PwC consider cyber-crime is ranked as the third most common form of crime aimed at UK businesses, in 2011, after (1) asset misappropriation and (2) accountancy fraud. However, cyber-crime is soaring in contrast to these other forms of attack ((1) and (2)), which are falling.
- T. Parton (a forensic investigations partner at PwC) was quoted as saying: 'This is a dramatic finding and marks the promotion of cyber-crime to the premier league of fraud'.
- 'As well as direct financial costs, there are other commercial consequences of cyber-crime, such as reputational and brand damage, poor employee moral or service disruption'.
- PwC also said that many British companies were still not alive to the extent of the threat and were not taking measures to protect themselves.
- W. Beer, the director of cyber-security services at PwC, said British companies are struggling to cope with other types of economic crime too. Responding to this 2010 survey, more than half of UK companies polled indicated that they had been victims in the past year, compared to 34% of companies globally.
- Incidence of all types of such criminal activity against businesses has risen by 8% since the last time the survey was carried out. (The latest survey polled 3,877 companies in 78 countries.)

- *2007* – The US government suffers an espionage 'Pearl Harbor' in which an unknown foreign power breaks into a number of agencies including the Pentagon, and downloads 'terabytes of data'. (*Source*: *The Times*, London, 2011)

C   CIGRE: Treatment of information security for electric power utilities

- Worldwide, the strategic need for improved cyber-security systems for electricity power utilities (EPUs) has also been recognised, as evidenced, for example, by the following CIGRE International Collaborative Programme contribution to the topic, culminating in the production of a Technical Brochure (TB 419) in 2010, as briefly outlined and reproduced below:

Summary – CIGRE TB 419-WG D2.22: 'Treatment of Information security for Electric Power Utilities (EPUs)': Electric power utilities, as providers of critical products and services, 'need to develop new security systems and procedures that are responsive to the improvements in Information and Communications Technology and also recognise the development of threats and attacks'. There are a great and increasing number of studies that support EPUs and also create a confusing situation when a security framework is to be chosen and the information security issues are to be managed. The Information Security Domain Model and EPU Risk Management and Risk Assessment Models are discussed.

It is claimed by CIGRE that this Technical Brochure TB 419 will answer this question.

(Readers will get only a 'feel for the content' of this TB from the following (present) brief 'abridged text' but *they are urged to carefully examine the complete original CIGRE document(s) for further details and supporting references*.)

Better still, an abridged background to TB 419 was presented in *Electra*. **250**, June 2010 (pp. 64–71). The article based on the content of this TB document was prepared by a Working Group, WG D2.22 and the group of participating CIGRE experts (see original!). – It starts by posing the question: 'Is Information Security Important for an Electric Power Utility (EPU)? Why?' This *Electra* article, written in a very positive and somewhat 'cyber-energy-evangelistic style', goes on to comment:

- 'EPUs are providers of critical products and services for society. Consequently, they need to develop new security systems and procedures that are responsive to the improvements in Information and Communications Technology (ICT) and also recognise the developments of threats and attacks. *Naturally, this implies that utilities not only need to deal with physical intrusion, but also deal with cyber intrusion*'.
- 'Cyber security has become a critical issue. Experts predict that the problem is likely to develop rapidly and this could have serious consequences in the energy sector as we move forward with developments towards the predicted Smart Grid networks of the future'. (*Note*: We have already demonstrated this elsewhere in this chapter.)

- *Numerous cyber-security intrusions (such as those already touched on in this chapter)* have clearly demonstrated cyber vulnerabilities to control systems in the energy sector. The literature has already described events which have occurred in electric power control systems for transmission, distribution, generation, as well as control systems for water/oil/gas/transport/nuclear plants/ chemical plants, paper and agricultural businesses together with hacking of top secret designs of equipment for US/Japanese governments (see, item (5) in the previous section) and many more.
- On the energy sector, some of these events have resulted in damage, for example, quoted in this article: 'intentional opening circuit-breaker switches and shutting down industrial facilities'.
  - *As with similar cases in other business sectors, very few of the reported incidents have been publicly described and attempts/initiatives to set up a database of all reported incidents normally encounter understandable resistance* (see items (1) to (5) in Appendix B).
  - As already reported in this chapter, 'cyber-attacks have a very wide range of capabilities and motives. Threat agents can arise from many diverse groups of people, such as young/teenage/adult hackers, employees, insiders, contractors, competitors, traders, foreign governments or companies, organised crime, extremist groups, terrorists and alliances between some of these groups'.
  - This writer stresses that potential hackers obviously have a wide range of expertise, capabilities, resources, organisational support, motivations and access to advanced hacker programmes such as Stuxnet, which worryingly is now widely available.
  - The purpose (or rationale) for TB 419 covers the efforts of WG D2.22 (with the above stated title), being the successor of Joint Working Group (JWG) D2/B3/C2-01 on 'Security for Information Systems and Intranets in Electrical Power Systems'.

CIGRE WG D2.22 comments that the stated purpose has been to focus and deepen the study on the following three issues:

- 'frameworks for EPUs on how to manage information security
- risk assessment (RA): common models and methods for treating vulnerabilities, threats and attacks
- security technologies for SCADA (Supervisory Control and Data Acquisition)/ control systems including real-time control networks'.

As intermediate results, the WG D2.22 of CIGRE has produced six papers which are included as appendices of TB 419. Highlights of some of these papers are also given in the CIGRE bimonthly magazine *Electra*.

Notes: At this juncture, it is relevant 'to pause' to mention that a companion CIGRE TB 427-WG B5.38 entitled 'The Impact of Implementing Cyber Security Requirements using IEC 61850' has also been produced to better understand the impact of cyber-security on IEC 62850 systems. WG B5.38 surveyed existing, and work-in-progress standards, reports and technical papers to assess the quality of the solution offered. The assessment describes the strengths and weaknesses of each solution from a protection engineer's point of view. A gap analysis shows where further research is needed: It is recognised that substation protection and automation systems are an integral part of the critical infrastructure required to maintain the quality of life in all countries. The substation automation system relies on reliable communications in the substation and many substations use legacy infrastructures and proprietary protocols to achieve this. However, these systems need protection against cyber-attacks that could significantly disrupt the reliability of the electrical grid.

'Until recently, the security was able to rely on obscurity, isolation and locked doors. But with the introduction of IEC 61850, there is concern that these measures are no longer satisfactory. For more details and a complete listing of references please see the full Technical Brochure'. (Secondary source: *Electra*. 2010;**251** (August).)

Frameworks for EPUs on how to manage information security

- Here, the CIGRE WG D2.22 informs the *Electra* reader that 'the objective is to provide some practical guidelines when developing frameworks that include SCADA/control security domains'.
- 'This is done by elaborating the specific security requirements of these types of domains, and also by giving a view of interrelated domains and high level frameworks that are necessary to manage corporate risks'.
- 'The WG D2.22 of CIGRE comments that it has not attempted to select standards on behalf of the electric power industry but rather it has aimed to support the EPUs in the selection process by analysing some of the most important approaches in order to achieve a consensus on the best "pieces"'.

Even though this is a 'fast changing landscape'/'a moving target', an overview of relevant standards that support the EPUs in the selection is presented. Alignments to other types of frameworks are also treated.

An Information Security Domain Model is discussed in the TB work, with good 'visuals' in figures provided. The reader is directed to the TB 419 for further information, including a full listing of references.

'Risk assessments (RA): 'Common models and methods for treating vulnerabilities, threats and attacks': here the CIGRE WG D2.22 informs the *Electra* reader that 'Risk Assessment is an integral part of the risk management process employed by EPUs. The objective is to ensure that all the risks faced by the EPU are appropriately identified, understood and treated. *Thus, risk treatment options range from mitigation to acceptance for any given risk'*.

- 'Risk management needs to be applied on a hierarchical, multi-level basis to work within the hierarchical and multi-level nature of the organisations, business processes and systems that exist in the modern EPU environment, with complex dependencies.
- For the purposes of developing an appropriate Risk Management Framework for EPUs, it is suggested we decompose the organisations, as shown in Figure 2. The four layers illustrate the various hierarchical levels of the operations part of a typical EPU which operates power plants and/or electricity networks'.

Security technologies for SCADA (Supervisory Control and Data Acquisition)/ control systems: Guidance on the application of cyber security controls to electric utility networks is provided in this part of TB 419:

- 'Ideally, assets within the network can be mapped to security domains.
- Appropriate security technologies can then be deployed within the network to meet the requirements of each domain.
- A specific diagram is provided, integrating transmission and distribution (T&D) grid data networks, and data networks associated with generation plants.
- The approach in this section can be applied to other EPU data networks'.
- 'An essential part of any EPU security deployment process is ensuring compatibility with legal systems. The WG of experts comment that this part of the CIGRE TB 419 is mindful of both modern IT and legacy SCADA environments, and the technological approach given herein allows for consistent and effective security measures across both'.
- 'The security domains recovered are logical. Although security controls can be assigned in general to each security domain, a successful security implementation involves adding cyber controls to the data network, that is, appropriate policies, procedures, segmentation, technology devices, and software'.
- 'To ensure that devices and software are placed in the data network to enforce security applicable for each domain, physical network assets must be mapped to logical security domains. Identifying security domains and implementing cyber-security technology within an EPU is a complex process, and specific details vary according to the data network configuration'.

Further works and CIGRE conclusions: The WG D2.22 experts who combined to write CIGRE TB 419 feel strongly that 'the work on information security must continue both within and outside CIGRE'.

One of the key strategic CIGRE questions to be answered here is: what do the EPUs need? Meantime, the CIGRE WG D2.22 has identified the following:

- 'To be informed about how to do risk assessments for electricity industry control system environments.
- To have a way of convincing management to invest in security measures for electricity industry control systems. This could include awareness, quantification of risks, presentation of avoided potential costs and damages etc.
- A security management system for operations which is appropriate and implemented at all levels of the risk management framework.
- Guidance on the embedding of security into the project life cycle for the implementation of electricity industry control systems.
- Deeper knowledge of what kind of information security requirements an EPU should present to a potential vendor of SCADA/control systems.
- Knowledge of what are the most relevant security metrics that can be used by an EPU to quantitatively measure and monitor the effectiveness of the security of the EPU's ICT systems'.

Based on this CIGRE work, it was considered that the following can be concluded:

- 'There is a great and increasing number of works which could both support an EPU but also create a confusing situation, where a Security Framework is to be chosen and the information security issues are to be managed.
- An overall framework should be based on existing standards and "best practices," taking into account legal requirements, culture and specificities of the EPU. (The wheel should not be re-invented!) The Security Framework should "fit" into the landscape of other frameworks adopted by the EPU.
- Risk assessment works show that this area could be more mature and be further developed in order to provide guidance to security management. A survey has shown that IT-based support for risk assessment was only used to a limited extent.
- The technical solution should be based on the domain model and the assigned controls, which are based on the risk assessment work and possible legal requirements imposed by the government. Also limitations related to the already installed base, such as "legacy systems," must be taken into account in a way that does not hamper the organisation security posture.
- The world of standards has been widened. An EPU needs guidance when selecting the appropriate and proper standard(s) for its needs of managing information security'.

Based on these and other experiences faced by the CIGRE WG D2.22, it is evident that information security for an EPU will continue to be an important issue!

Index

accessories 243
  joints 246–9
  for subsea cable systems 255–6
  terminations 243–6
AC offshore substations for wind farms 839–44
  guidelines for design and construction 840–1
AC test voltage generation 580–4
advanced metering infrastructure (AMI) 866
after-laying tests 262–6
age profile 431–2
air breakdown 631
  atmospheric effects 664–9
  flashover across insulator surfaces in air 662–3
  future work 672
  gap factor 655–62
  high voltage electrical equipment, tests made on 634–5
  history 633–4
  non-uniform fields 640–7
  pre-breakdown discharges 635
  testing transformers 669–72
  'U-curve' 647–55
  UHV at high altitudes 669
  uniform fields 636–40
air clearances, atmospheric 96–111
air density 593, 664–5
air density factor 593, 666
air gaps 658–60
air insulated substation 320, 321
air-insulated switchgear (AIS) 244
air-insulating gaps, breakdown of 649
air pressure 593
air/solid interfaces 94
air temperature 593
Allgemeine Elektrische Gesellschaft (AEG) 337

ALSTOM design 157, 180
alternate long/short (ALS) shed profile 476
alternating voltage 634
  breakdown 642–3
  test 264
aluminium conductors 246
amplitude measurement 561
  alternating voltage 563–7
  direct voltage 561–3
  impulse current 571–2
  impulse voltage 567–71
analogue digital converter 567, 570, 572
analogue to digital converter (ADC) 705
annual mean wave power 191
apparent charge 603–4, 606–8, 611
arc chute 372
arc chute interrupter 372
arc control 281–3
arc control devices 370
arc extinction media 116–19
arc interruption 310–16
  gas mixtures 315–16
arc quenching 281–3
arc resistance 416
area electricity boards 21
arrester condition monitor (ACM) 735
artificial pollution tests 668–9
ASEA 337
Ashdown, Lord Paddy 812
asset management (AM) 425, 445–7
  process 462
  Publically Available Standard (PAS) 55 on 462
  smart grids on 463
  standard for 462–3
  themes, CIGRE work on 353–4
Asset Management Centre 729
asset manager
  corrective action 460
  dilemma 461

refurbish 460
replacement 460
retrofit 460
asset replacement 447–8
Association of Edison Illuminating Companies (AEIC) 258
ASTM D5334-08 258
asymmetrical breaking current 365
atmospheric air 631
atmospheric air clearances 96–111
   test areas 96–100
atmospheric conditions 593
atmospheric correction factor 593, 665
   iterative procedure 666
   standard procedure 665
atmospheric effects 664
   air density correction factor 666
   artificial pollution tests 668–9
   atmospheric parameters, measurement of 668
   correction factors, application of 665
   density effects 664–5
   humidity effects 665
   humidity factor correction 666
   wet tests 668
atmospheric pressure 593
auto-expansion circuit-breaker 409–10
automated demand response (ADR) 867
automatic voltage regulator (AVR) 195
axial gradient 478

backflashover 63, 81–4, 87–8
back-to-back converters 5
back-to-back schemes 484
balancing fluctuating wind energy with fossil power stations 857–8
banded pair 157
baseline parameters 435
bath tub curve 430–1
   shape parameter 430
batteries 443
   valve-regulated lead-acid 443
Binden, Joe 895
biomass-fed generators 196
biomass resources 193
Black, Dave 879
breakdown 645–7
breakdown strength 600–1

breakdown voltage prediction 143–7, 639
   empirical 143–6
   semi-empirical 146–7
British Electricity Authority 21, 338
Brown Boveri 337
Brunke, J.H. 340
BS 5760 456
built-in self-monitoring system 727
busbar systems 437–8
bushing applications 478–86
   direct connection to switchgear 481–3
   direct current 484–6
   high current 481
   switchgear 483–4
   transformers 478–80, 510–11
bushing design 472–8
   air end clearance 473–7
   alternate long/short (ALS) shed profile 476, 477
   axial gradient 478
   GIS/air 483–4
   oil end clearance 477
   pollution level 473–6
   radial gradient 478
   reference unified specific creepage distance (RUSCD) 474–6
   stresses 472–3
   transformer/air 475, 481
   transformer/gas 482–3
bushing diagnosis 490–2
bushing maintenance 490–2
bushing module 685–7
bushings 467
   condenser 469–72
   dry-type 486
   non-condenser 467–8
   oil impregnated paper (OIP) 471, 479, 486–91
   resin bonded paper (RBP) 470–1, 487–8
   resin impregnated paper (RIP) 471–2, 479, 481–3, 488–9
   stress control 468
bushing tests 486–90
   cantilever 490
   dissipation factor 487
   impulse voltage 489
   leakage 490
   partial discharge 487–9

Index 909

power frequency withstand 488
seismic withstand 490
short-circuit 490
temperature rise 490
thermal stability 489

cable insulation 89
cable monitoring 732
  sensors for 733–4
cables
  for AC systems 261
  condition monitoring 438–9
  for DC systems 261
  EHV 438
  failure rate 429
  partial discharge measurement 609–11
cable systems 211
  accessories 243–9
  conductor 218–22
  current, magnetic field and inductance 214
  current ratings 234–43
  direct current and subsea cable systems 249–56
  early lighting systems 215
  early telegraph cables 215
  elementary theory 212–14
  features of 218–34
  flexible 215–16
  fluid-filled 217
  future directions 269–71
  historical development 214–18
  impregnated 216
  installation 232–4
  insulation system 222–8
  joints 246–9
  multi-core and multi-function cables 228–9
  outer layers 229–32
  polymeric 217–18
  polypropylene-paper laminate 218
  pressure-assisted 216–17
  renaissance of high voltage 216
  standards 256–8
  terminations 243–6
  testing 258–69
  time-dependent ratings 235–6
  voltage, electric field and capacitance 212–13

cable temperature 701
cantilever test 490
capacitance per unit length 213
capacitive current switching 312–14, 414–15
capacitor commutated converters (CCC) 53
'cap-and-pin' insulator 663
carbon bubble 831, 832
carbon capture and storage (CCS) 825–32
carbon trading 819
  coal-fired to co-fired stations in UK 820–3
  storage energy back-up 823–5
Carrington, Damian 831
catenary continuous vulcanisation (CCV) 227
CENELEC 257
Central Electricity Authority (CEA) 21, 338
Central Electricity Generating Board (CEGB) 22, 338
CF_3I gas, properties of 297
chopped wave lightning impulse 545
chopping 312–14, 416
chromatic techniques 756–8
  of acidity, water content and electrical strength data 777–8
  assessment of degradation of high-voltage insulation materials 768
  chromatic characterisation of partial discharge signals 768
  of dissolved gas data 774–7
  electric current monitoring 759–61
  high-voltage circuit breaker monitoring 761
  high-voltage circuit breaker operation 761–3
  offline assessment of high-voltage transformer oils 773–4
  optical fibre sensors for high-voltage systems 759
  representation of overall oil condition 778–9
  time and frequency domain chromatic processing of optical fibre sensor data 761
  transformer tap-changer monitoring 763–8

CIGRE 535, 550, 901
CIGRE activities for substations 343–9
CIGRE Report 163 113
CIGRE study committees, fields of
    activities of 10–11
CIGRE Technical Brochures 304
CIGRE WG 13.08 survey 458, 459
CIGRE work on asset management themes
    353–4
circuit breaker alarm settings 717
circuit breaker monitoring (CBM) 705
    data sampling rates 752
    data storage and display 719–21
    signals measured and recorded 716–18
    system alarms 718–19
circuit-breaker nozzles 309
circuit breaker operation screen
    close operation 721
    trip operation 720
circuit-breaker pre-insertion resistor
    71–2
circuit-breakers (CBs) 113–16, 334,
    367–75, 407–8, 701
    air blast sequentially operated 371
    auto-expansion 409–10
    auto-reclosing 364
    bulk oil arc control 369–70
    bulk oil plain break 369–70
    capacitive and inductive current
        switching 414–15
    characteristics 280–1
    closing resistors 324–5
    correct choice, for special switching
        duties 413–14
    current limiting 372
    dead-tank 118–19, 319, 322, 323
    dielectric design 318
    distribution voltage levels 407, 417
    domestic 292
    EHV 314, 318
    electromagnetic 285–7, 286
    free air 372
    gas blast 283–5
    for generator circuit switching 415
    hydraulically operated 369
    IEC Standards 342
    insulators 318, 319
    live-tank 118–19, 319–21
    magnetically operated 368–9
    manually operated 367
    minimum oil 371
    oil filled 293
    one-break 339
    operating mechanism 318–19, 413
    operating mechanisms 413
    pneumatically operated 369
    point-on-wave control 77–8
    pole scattering 70–1, 416
    pressurised head air blast 371
    puffer 409
    rotating-arc 409
    SF_6 gas 409
    solenoid operated 368
    spring operated 368
    sulphur hexafluoride 117–19, 287, 303,
        305
    synchronised switching 416
    testing 314
    thermal recovery characteristics 288
    vacuum 134–6, 373, 408–9, 410
    see also interrupters; sulphur
        hexafluoride circuit-breaker
Clark, Duncan 832
closing resistor 4, 104, 324–5
cloud-to-cloud lightning 632
Cockroft–Walton DC 633
'co-firing' 797
Coghlan, Tom 895
coil current profiling 437
commercial criticality analysis 456–7
communication module 706
condition assessment 425–6, 447–8
    procedure for systems operator 779–80
condition based maintenance (CBM) 454–5,
    549
condition monitoring (CM) 435–6, 450,
    676, 679, 680
    application to key substation equipment
        701
    batteries 443
    benefits 435–6
    busbar systems 437–8
    cable monitoring 732–4
    cables 438–9
    CBM node unit overview 716–21
    common monitoring platform,
        developing 707–10
    communication module 706

Index 911

control PC module 705–6
conversion module 705
data acquisition module 705
data acquisition systems 704–5
equipment 675
evolution of 696
  basic discrete online monitoring 1990–1999 697
  integrated substation condition monitoring (2010 onwards) 698–700
  more intelligent discrete monitoring 2000–2009 698
  periodic monitoring 1960–1990 697
gas density monitoring 744
GDM node unit 711–16
insulation systems 439–41
interface module 705
ISCM systems 736–49
node unit data acquisition system, specification of 711
objectives of 700
overhead lines 441
PDM node unit overview 722–7
PD monitoring 740
platform 703
power transformer monitoring 727–32
sensors and transducers 705
substation environment 701–3
surge arrester monitoring 734–5
surge arresters 438
switchgear 436–7
tap-changers 444–5
transformers 443–4
conductors 218–22
  bundles 109
  direct-current (DC) resistance of 220
  joining of 247
Constable, John 853
contactor 382
continuously transposed conductors (CTC) 14
continuous rating 236
control PC module 705–6
conventional power stations 193, 195
conversion module 705
converter equation 159
converter transformer 12, 158–62
'convertor stations' 521

cooling 168–9, 242, 687–8
corona 599, 642
corona onset voltage 103
corporate planning 428
correction factors, atmospheric 665
  iterative procedure 666
  standard procedure 665
corrective maintenance 450–1, 450–5, 549
Crompton 337
cross jet pot 370
cross-linked polyethylene (XLPe) insulation 218, 220–3, 226–9, 232, 253, 609
  cable systems 260
  insulation 253–4
current counter 437
current-driven losses 237–8
current interruption 275–81
  current waveform 277, 278, 279
  dielectric recovery phase 277, 281
  thermal recovery phase 276, 280
  voltage waveform 277
current limitation 279, 280
'current-limiting' circuit-breakers 372
current measurement 555
  impulse current 571–2
current ratings 234–43
current transducer 387
current transformer 385–6
cyber-crime 895
cyber-security 788, 865, 895
  for electricity power utilities (EPUs) 901

damping resistors 4, 104
data acquisition module 705
data acquisition systems 704–5
data evaluation software 619–20
data point-on-wave (POW) information 724
Davey, Edward 529, 880
Davies, Sean 834–5
DC bushings 484–6
DC offset 416
DC pollution tests 579
DC test voltage generation 578–80
dead-tank circuit-breaker 118–19, 319, 322, 323

degradation rate 435
degree of polymerisation 446
demand 30–3
demand forecasting 42
density of air 664–5
design requirements 534–5
design specifications 530
development testing 258
DFIG (double fed induction generator) 183
'Diagnostics Tags' 743
dielectric diagnosis techniques 490–1
dielectric loss 237
dielectric recovery characteristic 281
dielectric requirements 533
dielectric strength 576
dielectric stress 61–3, 105, 576, 592
dielectric tests 542–7, 573
digital recording device 615, 621
  applications 620–6
  calibration 619
  code bin width 617–18
  differential nonlinearity 617–18
  evaluation software 619–20
  integral nonlinearity 617–18
  interference 617
  operating range 619
  performance 626–8
  rated resolution 616
  rise-time 617
  sampling rate 617
  scale factor 617, 619
  uncertainty 616–17, 619
direct current transmission 249–50
direct strike 84–6
direct voltage 634
  breakdown 641–2
  tests 264
disconnector 364, 376–8
  centre rotating post 376–7
  and earth switch monitoring data sampling rates 752
  rocking post 376–7
  switching 327–8
disruptive discharge 632, 645–7
disruptive discharge voltage 593
dissipation factor test 441
dissolved gas analysis (DGA) 447, 491, 691
  sensors for 731–2

dissolved gas monitors 551–2
distributed energy resources (DER) 861
distributed intelligence in networks 206
distributed temperature sensing (DTS) 733
distribution automation (DA) 866
Distribution Code 29–30
distribution network operator (DNO)
  networks 196
distribution switchgear 355–7
  circuit-breakers 369–75
distribution systems 17–19, 361–4
  design 38–41
  early developments 19–20
  future developments 43–54
  global developments 23–5
  insulation co-ordination 57–61
  operation 41–3
  organisational structure 34–8
  recent developments 26–30
  rural 363–4
  technical factors 32–4
  urban 361–2
distribution transformers 12
diverter switch 511
dodecylbenzene 225
domestic circuit-breakers 292
Dong Energy 806–8
dry-type bushings 486
Duval triangle 447
Dyke, Craig 802
dynamic rating 242–3
dynamic thermal circuit rating (DTCR) 864
dynamic voltage restorer (DVR) 183

earth fault 62
earth switch 378, 753
Eddy-current loss 237
EDF 811
EDF Energy 815
Edison, Thomas 19, 337
Edison's Illuminating Companies 19
effective electrode separation 137–8
efficiency factor 137–8, 589
electrical breakdown 600
electrical power industry
  current challenges 460–1

electrical tests 550–1
electric current monitoring 759–61
Electricity Act 1947 21
Electricity Act 1957 22
Electricity Act 1983 26, 37
Electricity Association 426
Electricity Council 22
Electricity Lighting Act 1882 19
electricity power utilities (EPUs) 858–9
Electricity Supply Act 1926 21
electricity supply industry structure 25, 34–8, 43
  efficiency 389–90
  network 355–6
electrode systems 121
electrogeometric model 84–6
electromagnetic arc control 298–9
electromagnetic circuit-breaker 285–7
  thermal recovery 285–7
electromagnetic interference (EMI) 701
electron avalanches 635, 637, 638
  in uniform fields 636–9
Energy Act 2004 38
energy management system (EMS) 42, 866, 867
Energy Market Reform (EMR) 881
energy-mix and perceived renewable energy costs 796–9
energy storage 205
energy storage devices 390
energy value 199–200
English biomass resources 194
environmental issues 296, 297, 387–9, 509
  SF_6 388–9
Erinmez, Arslan 1
erosion breakdown 600–1
ESI developments 787–91
  AC offshore substations for wind farms 839–44
  balancing fluctuating wind energy with fossil power stations 857–8
  carbon capture and storage (CCS) 825–32
  carbon trading 819–25
  future developments including smart grids 858–77
  international takeovers and possible impacts 791–6
  network upgrades and operational experiences 844–5
  North Sea Super Grid (NSSG), proposal case for 833–9
  nuclear power plants 810–19
  operational difficulties with wind farms 845–54
  renewable energy development 796–808
  UK air-defence radar 854–5
  UK government's recent wind of change 808–9
  UK/International DC cable links 833
  wind turbine hum 855–7
ethylene propylene rubber (EPR) 218
  dielectric 228
  insulation 254
European Union Emission Trading Scheme (EUETS) 819
Europe's Emission Trading Scheme 820
evaluation software 619–20
  applications 626–8
exothermic welding 246
expulsion fuse 363–4
extended real-time operation 42
extra high voltage (EHV) substation equipment 697
extruded insulation systems 226–8

Fabian Society 21, 26
factory acceptance testing 261–2
failure data 426, 426–7
failure modes, effects and criticality analysis (FMECA) 455–7
failure modes and effects analysis (FMEA) 455–7
failure rate 429–30, 431
fault current 299, 300
fault current interruption 310–12
fault current limiters 201, 202
fault level 200–1
fault-making device 378
fault passage indicator (FPI) 457
fault reporting 432
Ferranti 337
Ferranti effect 62
Ferranti rise 69–70
ferroresonance 62–3, 328–30

fibre optic point sensor measurements 690
field analysis techniques 136–43
field factor 137
fingerprint 435
finite element analysis 515–16
fixed electrode spacing 648
flashover 642
　across insulator surfaces in air 662–3
flashover voltage 96–7, 126
　rod-plane gap 96–7
　rod-rod gap 96–7
Fletcher, Martin 818
flexible AC transmission systems (FACTS) 44, 178, 183, 865
flexible cables 215–16
'floating' wind turbines 830
fluid-filled cables 217, 252, 253, 256
formative time lag 652
Fortson, D. 882, 883
fossil fuel reserves 20, 832
fossil fuels 187–8
fossil power stations, balancing fluctuating wind energy with 857–8
frequency 32–3
front-of-wave lightning impulse 545
Fukushima nuclear accident 810–11, 813, 817
fulgurite 381
full converter wind turbine 195
full wave lightning impulse 545
functional specifications 530
fuses 380
　current limiting 381–2
　expulsion 380–1
fuse switch 379–80
'future proofing' 703, 704

gap factor 655, 656
　air gaps 658–60
　rod–plane gaps 657–8
　sparkover under alternating voltages 659
　sparkover under direct voltages 659–62
　sparkover voltage characteristics for different gaps 657
　test procedures 656
gas blast circuit-breakers 283–5
　thermal recovery 285

gas compression cable 217
gas density, measuring 709
gas density monitoring (GDM) 698, 744
　detailed bay information 745
　gas inventory 746–7
　GDM system alarms 745
　transducer options 751
gas density transducer 708
gaseous insulants 93
　atmospheric air 96–111
　modelling 136–47
　N_2/SF_6 mixtures 112–15
　sulphur hexafluoride 111–12
　switchgear 113–35
　ternary gaseous dielectrics 113
gas filled interrupter 275
　electromagnetic 285–7
　gas blast 283–5
　see also sulphur hexafluoride circuit-breaker
gas-in-oil detection system 729
gas insulated line (GIL) 94, 113, 270
gas insulated substation (GIS) 64, 87–9, 103, 113, 290, 322, 323, 358–9, 696
　contaminated conditions 130–1
　design 319–20, 322–4
　disconnectors 327–8
　gas-gap 122
　indoor 319, 320, 322
　interrupters 144
　metal-clad switchgear 141–2
　monitoring 330–2
　outdoor 319
　partial discharge measurement 611–13
　power frequency characteristics 129–30
　reliability 131–3
　SF_6 119–20, 127–8
gas-insulated transmission lines 133–4
gas-pressure cables 217
gate turn-off thyristor (GTO) 179–81
Gay–Lussac's law 709
GB System Operator (GBSO) 38
GDF Suez 793
GDM node unit overview 711
　data collection 712
　maintenance facility 715–16
　predictive alarms and SF_6 gas inventory 713–15

Index 915

GEC 340
gel permeation chromatography 450
generator circuit switching 415–16
generator technologies 193, 195
　base load 200
　biomass-fed generators 196
　conventional power stations 193, 195
　dispatchable 200
　forecast but intermittent generation 200
　full converter wind turbine 195
　microgeneration 196
　non-forecast spill 200
　partial converter wind turbine 195
　tidal machines 195
　un-metered spill 200
　wave machines 195
generator transformers 12
Gillespie, James 854
GIS GDM node unit 714
GIS PDM node unit 712
Glenny, Misha 897
Graëtz bridge scheme 526
greenhouse effect 295, 389
greenhouse gas emissions, in EU-15 295
Grid Code 27, 28, 29
grid-ready appliances and devices 867
grid transmission system 21–2, 24
　supergrid 22–3
guar gum powder 831

Hague, William 895
Haldane, T.G.N. 22
Hall, Richard 849
Hare, Phil 849
hazard rate 430–1, 431
Health and Safety Executive 427
Hermstein glow 641–3
high current bushing 481
highest voltage for equipment 577, 595–6
highest voltage of a system 577
high-performance liquid chromatography 691
high-temperature superconducting (HTS) transformers 14
high-temperature superconductivity 270
high-voltage circuit breaker
　monitoring 761
　operation 761–3

high voltage electrical equipment, tests made on 634
　alternating and direct voltage 634
　impulse testing 634–5
high-voltage equipment, online monitoring of 758
　optical fibre chromatic sensors 758–61
　time and frequency domain chromatic processing 761–8
'hissing' test 488
Holborn Viaduct Generating Station 19
Hollande, Francois 814–15
home area network (HAN) 866
horizontally separated industry 35–7
hotspot temperature 457, 509, 519, 532, 542
humidity 593
humidity correction factor 667
humidity effects, in atmosphere 665
humidity factor 593, 666–8
HVDC converter station 153–5
　AC filters 163–5
　auxiliary power supplies 169
　building 169–72
　contamination protection 172
　control equipment 162–3
　DC filters 165
　DC reactor 165–6
　economics 176–7
　losses 176–7
　power electronics for industrial applications 183
　reactive power control 163–5
　surge arresters 168
　switchgear 166–8
　thyristor valves 156–8
　transformer 158–62
　valve cooling 168–9
　voltage-sourced converters (VSC) HVDC 172–5
HVDC convertor transformers 521–3
HVDC technology 52–3
HVDC transmission 1, 37, 153–5
　back-to-back link 159–60, 165
Hydran Gas Monitor 691
hydroelectric stations 520
hydrogen 690
hysteresis loss 237

IEC 52 100
IEC 56 365–6
IEC 60052 567
IEC 60060 66, 487, 576, 578, 585, 619, 626
IEC 60060-1 256, 626
IEC 60060-2 558, 561
IEC 60060-1 619, 642, 644, 652, 670
IEC 60071 57–9, 63–4, 66, 75, 88, 576
IEC 60076 576
IEC 60076-1 541, 543
IEC 60076-2 532, 542
IEC 60076-3 543, 544, 546
IEC 60076-7 542, 682, 688
IEC 60076-18 548
IEC 60137 467, 484, 486, 489–90, 490
IEC 60141 256
IEC 60230 576
IEC 60270 613
IEC 60354 458
IEC 60507 476
IEC 60815 473, 476
IEC 61000 703
IEC 61083 616, 620
IEC 61083-1 572
IEC 61083-2 572
IEC 61233 414
IEC 61463 490
IEC 61464 491
IEC 61639 482
IEC 62067 256
IEC 62271-100 305, 318, 407, 413, 414–15
IEC 62271-200 407
IEC Standard 60183 670
IEEE C37-013:1989 416
impact-ionisation 637, 638
impregnated cables 216
impulse breakdown 644
impulse current 591–2
impulse current generation 591–2
impulse current shunt 571–2
impulse generator 586–7, 590, 634
impulse testing 634–5
impulse tests 559, 620
impulse test voltage generation 584–91
impulse voltage 584–90
impulse waveform evaluation 619–20, 626–8

inductance per unit length 214
inductive current switching 312–14, 414–15
industrial transformers 525–6
information management 463–4
infra-red photography 690
in-service monitoring 266–7
instantaneous failure rate 430
instrument transformer 385–7
insulants, gaseous: *see* gaseous insulants
Insulated Cable Engineers Association (ICEA) 258
insulating board 509
insulation
  co-ordination 576, 592, 597
  external 576, 593
  internal 576, 593
  levels 576, 595–6
  non-self-restoring 577
  self-restoring 576
  transformers 509–10
  voltage–time characteristics 63–4
insulation age 448–9, 450
insulation co-ordination (IC) 3, 57–61, 102, 110, 332–4
insulation failure 631
insulation levels 59
insulation module 685
insulation resistance test 436, 439, 446
insulation system 222–8
  extruded 226–8
  lapped 224–6
  subsea cable systems 252–4
insulation testing 445–7, 599
insulator surfaces, flashover across (in air) 662–3
integrated substation condition monitoring (ISCM) system 698–9, 736, 747–8
  control cubicle 739
  gas density monitoring 744–7
  main operational screen 738
  partial discharge monitoring (PDM) 740–3
  of substation monitoring 737
'intellectual contribution' to network restructuring 393
intelligent electronic device (IED) 680, 865, 866

intelligent link 364
intelligent monitoring 755–6
interface module 705
interleaved disc winding 505, 506, 514
interline power flow controller (IPFC) 51
International Electrotechnical Commission (IEC) 256
international takeovers in UK power sector 791
interphase power controller (IPC) 50
interrupter media breakdown voltages 276
interrupters
  arcing contacts 309
  first generation 306–9
  nozzles 309
  second generation 308, 309–10
  single-pressure 306–10
  third generation 316–17
  two-pressure 306
  see also circuit-breakers (CBs); puffer
ISCM system gateway 736
  global positioning system 739
  local alarms 738
  remote access 738
  uninterruptible power supply 738
isolator: *see* disconnector

joints 246–9
Jones, G.R. 6

knowledge modules (KMs) 729
Kyoto protocol 1997 708

Laplace equation 136, 139
lapped insulation systems 224–6
large drive convertor transformers 526
large drive rectifier transformers 526
Lauvergeon, Anne 812
Lea, Robert 885
leaders 635, 644–5, 649
leakage test 490
Lichtenberg figures 644, 645, 648, 649
life cycle 426
life management 425–6
lifetime energy usage 390
life time law 600
lighting systems, in early years 215

lightning impulse voltage 584, 590, 621, 626
lightning impulse withstand level (LIWL) 64, 332
lightning overvoltage 63, 79, 80–9
  control 87–9
linearity test 573
line commutated converter (LCC) HVDC system 174
line flash 81, 86, 88
line gap flashover 84–5
link boxes 249
live-tank circuit-breaker 118–19, 319–21
live working 110–11
Lloyd Parry, Richard 817
load flow 205
load rejection 61–2
local area network (LAN) 698
London penetration depth 202
long-distance transmission 5
losses 237–8
  current-driven 237–8
  minimising 239–41
  in power cables 270
  voltage-driven 237

MacKay, David 799–800
magnetic field
  expulsion 201
  and inductance 214
magnetomotive force 495–6
Mahajan, Swadesh 812
maintenance 450–5
  future trends in 458
  management techniques 455
managerial level 428
Marlow, Ben 886
Marx generator 586–7
McNeill HVDC station 158, 160–1, 165, 169
measuring
  alternating voltage amplitude 563–7, 620–1
  digital recording systems 620–6
  direct voltage amplitude 561–3, 620–1
  error 561, 563, 566–7
  impulse current amplitude 571–2, 621–3

impulse voltage amplitude 567–71, 621–3
   partial discharge 601, 603–4, 623–6
   purposes 573
   time parameters 572
measuring system 555–60, 573
   converting device 555–60
   electromagnetic interference 559–60
   partial discharge 604–6
   recording device 559–60
   scale factor 558
   transmission system 559
measuring techniques 555
   digital 615–16
   partial discharge 599
mechanism proximity switches 719
medium voltage (MV) switchgear 356
M effect: *see* Metcalf effect
Megger^{TM} testing 439
Meissner effect 201
Merkel, Angela 813
Merz, Charles 338, 339
Merz and McLellan 338
metallic particles 289–90
metal oxide surge arrester (MOSA) 72–6, 88, 326, 335, 383–5, 409
   residual voltage measurement 559
Metcalf effect 381
microgeneration 196
Milliken conductor 221, 222
'minimum breakdown voltage' 649
minimum oil circuit-breaker 371
modelling, insulation 136–47
modern cable sheaths 229–31
moisture content analysis 436
Møllehøj self-compensating cable system 253
monitoring
   condition 435–6, 450
   switchgear 330–1, 436–7
monitoring data, access to 681
monitoring platform, common 707
   circuit breaker monitoring 709
   disconnector and earth switch monitoring 710
   gas density monitoring 708–9
   high-frequency PD monitoring 709–10
multi-core cables 228–9
multi-function cables 228–9

multimodule converters (MMC) 174
multiple lightning discharges, photograph of 632
multi-stage rectifier 579, 597
Mumetal 386
MVA rating 365

'nameplate' age 447, 450
'Nano' surface treatment 663
Nasser's theory of the development of Hermstein's glow 641
National Fault and Interruptions Reporting Scheme 426
National Grid system 24, 26
National Grid Company (NGC) 26–30, 199, 257
National Grid Electricity Transmission (NGET) 38
Network Rail 258
'new-build' nuclear plants overseas 815–19
New Electricity Trading Arrangements (NETA) 37
node unit alarm
   data display 724
   and trend plot screen 718
node unit data acquisition system 711
node unit main operational screen 726
   hardware diagnostics 727
node unit monitoring screen 723
node unit real-time PD summary screen 725
noise pollution 855
nominal voltage 577
non-destructive testing 599
'non self-restoring' insulating systems 67, 634
non-uniform fields 640
   alternating voltage breakdown 642–3
   breakdown 645–7
   direct voltage breakdown 641–2
   disruptive discharge 645–7
   impulse breakdown 644
   leaders 644–5
   sparkover 645–7
North Eastern Electric Company 338
North Sea Super Grid (NSSG) 788, 790, 833–9
nuclear developments, in future 811–15

nuclear fuels 188
nuclear power plants 810
 Fukushima nuclear accident 810–11
 future nuclear developments 811–15
 'new-build' nuclear plants overseas, future prospects of 815–19
numerical modelling methods 139–41

Oerlikon 337
off-line testing 264
offshore AC substations for wind farms 842, 843
Offshore Wind Cost Reduction Task Force 884
OFGEM 196, 197, 198, 849
oil end clearance 477
oil-filled cable 217
oil filled circuit-breakers 293
oil impregnated paper (OIP) 471, 479, 486, 487–91
oil tests 446, 550
online monitoring systems 678
 chromatic optical fibre sensors 759
 optical fibre chromatic sensors 758–9
 time and frequency domain chromatic processing 761
 using optical fibre chromatic techniques 758
on-load tap-changer 511–12
on-load tap-changer (OLTC) module 683–4
opening resistor 4, 324–5
open-terminal substation 319, 324
operational level 428
operational planning 41
 activities in 41–2
'Operations Management' 461
optical fibres 733
 chromatic sensors 758–9
 sensing 331
optimisation of rating 239–43
Osborne, George 879
oscilloscope 570, 572, 615–16
outdoor termination 244
overhead lines
 condition monitoring 441
overloading 688
overshoot 557–8

overvoltage 57, 324, 332, 577, 644
 lightning 63, 79, 80–9
 switching 63, 66, 67–71
 temporary 61–3
ozone depletion 388

Parry, Lloyd 817
partial converter wind turbine 195
partial discharge 599–604
 equivalent circuit 601
 magnitude 604–6, 607
 solid insulating material 600
 see also corona
partial discharge measurement 601, 603–6, 611
 calibration 603, 608
 circuit 602
 development of recommendation 613–14
 digital recording systems 623–6
 examples 608–13
 extinction voltage 605
 frequency response 605–6
 inception voltage 605
 narrowband 607
 noise 611, 623–6
 UHF 611–12
 wideband 607
partial discharge monitoring (PDM) 437, 689, 698, 729, 740
 bay overview screen 741
 cables 439–40
 data sampling rates 752
 detailed bay information 740–2
 diagnostics 742
 display and diagnostics 742–3
 node unit data collection 724
 node unit noise monitoring and alarm control 725–7
 node unit sequencer controls 724–5
 node unit system diagnostics 727
 overview screen 740
 self-learning diagnostics 743
partial discharge test 261, 263, 439
partial discharge test voltage levels 605
Paschen's law 664
PCI eXtensions for Instrumentation (PXI) system 711
peak flow for mean spring tide 192

peak voltmeter 570–1
Pearl Street System 19
performance tests 539–40
periodic maintenance, testing for 267–8
phase shifter 153
phase shifting transformer (PST) 47, 523–5
  application 524
  example 524
phasor measurement units (PMUs) 865
photo-ionisation 637–9
plug-in electric vehicles (PEVs) 867
point-on-wave (POW) information 77–8, 724
point-to-point configuration, breakdown in 651
point-to-point gap 646
polarisation index test 440
polaroid photographs 641
pollution 19
pollution tests 668–9
polychromatic light absorption oil signature 774
polymeric cables 217–18
polypropylene-paper laminate (PPL) 218
polytetrafluoroethylene (PTFE) 309–10
  filled 309
polyvinyl chloride (PVC) thermoplastic 217
post real-time operation 43
power electronic devices 153
power electronics for industrial applications 183
power frequency voltage 61, 66
Powerline group 787–8
power losses 237–8
power station sites 32
power station transformers 520–1
power transfers 188
power transformer monitoring 727, 730
  dissolved gas analysis (DGA), sensors for 731–2
  tap-changer monitoring, sensors for 732
  see also transformers
preferential subjects (PSs) 868
prequalification test 575
pre-saturated core fault current limiters 203–5

pressure-assisted cables 216–17
pressurised gas-filled transformers 532
preventative maintenance 425, 435, 440, 450–3
preventive maintenance 549
'price' *versus* 'value' 829
privatisation, electricity supply industry 26–7
PTFE: *see* polytetrafluoroethylene (PTFE)
Publically Available Standard (PAS) 55, 462
puffer 117–18, 284, 306–10, 311, 374, 409, 411, 412
  circuit-breaker 409
  duo-blast 306, 307, 308
  mono-blast 307
  partial duo-blast 306, 307, 308, 374
pulse width modulated (PWM)-based converters 174

quadrature booster (QB) 47, 51, 153
quality assessment 599
quality of supply 39–40

radial circuits 363
radial gradient 478
radio influence voltage (RIV) 490–1
railway transformers 526–7
rated asymmetrical breaking current 365
rated current 365
rated insulation level 578
rated insulation withstand level 366
rated lightning impulse voltage 366, 595–7
rated power frequency withstand voltage 366, 595–6
rated short-circuit-breaking current 365
rated short-circuit-making current 365
rated short time current 365–6
rated switching impulse withstand voltage 594–7
rated transient recovery voltage 366
rated voltage 366
rated withstand voltage levels 64–7
rate of rise of recovery voltage (RRRV) 312
ratings
  continuous 236
  dynamic 242–3

Index 921

factors affecting 236–43
  optimisation of 239–43
  short-circuit 236
  thermal 234–5
  time-dependent 235–6
reactive compensation 33
reactor switching 314, 314–15, 329, 335, 417
real-time operation 42–3
recovery voltage method 441, 458
rectifier power transformers 526
reference pulse generator 619
reference unified specific creepage distance (RUSCD) 474–6
reference waveforms 620
refurbishment 425
reignition 312–14, 416
reliability 425, 426–34
  analysis 428, 429–30, 456
  data 426–34
  human factors 432–3
reliability centred maintenance (RCM) 453–4
remanent charge 68–9
remote resource locations, network implications from 193
remote terminal units (RTUs) 729
Renardiers Group 644
renewable energy 187
  drivers for 187–8
  fossil fuels 187–8
  generator technologies 196–9
  network implications 196–9
  nuclear fuels 188
  resources and technology, UK 188–96
  solutions for 200–6
  value of energy 199–200
renewable energy development in the UK 796
  energy-mix and perceived renewable energy costs 796–9
  renewable energy vs landscape calculations 799–800
  UK energy storage 800–3
  wind energy 806–8
renewable energy vs landscape calculations 799–800
renewable generator technologies 196
  cost of connections 196–7
  fault level 197

grid code issues 199
load flow 197
network extensions 198
power quality 198
regulation 198–9
voltage rise 197
Renewable Obligation Certificate (ROC) 198, 799
'required withstand' voltages 65
residential energy management system (EMS) 867
residual life concepts 350–1
resin bonded paper (RBP) 470–1, 487–8
resin impregnated paper (RIP) 471–2, 479, 481, 482–3, 488–9
resistive fault current limiters 202–3
resonance 62
resource management 428
Reynold series 365
Reyrolle 340
ring main 362
ring main unit 382–3
ripple 579
risk assessment 455–7
Robinson, Ann 879
rod–plane electrode system 646
rod–plane gap 96–7, 657–8
rod–rod discharges, Lichtenberg figure of 644, 648
rod-rod gap 96–7
Rogers ratio 447
Rogowski coil 572, 609–10, 625
rogue circuit 428
rotating-arc circuit-breaker 409
routine tests 541

safety factor 88
safe working distances 110
SCADA systems: *see* system control and data acquisition (SCADA) systems
Schering Bridge 487
Scottish Hydro Electric Transmission Ltd (SHETL) 38
Scottish Power Renewables (SPR) 884
Scottish Power Transmission 38
security of supply 33, 39
seismic withstand 490
'self-learning' facility 724

self-restoring insulation 67, 634
sensors and transducers 705
series capacitor (SC) 47, 180–3
series compensator 153, 180–2
settling time 557–8
SF_6 gas (CB/GIS) system integrity 701
SF_6 gas circuit-breakers 409
SF_6 Gas Density Monitoring System 708
Shaw, Jonathan 897
sheath voltage limiters (SVLs) 240–1, 249
shielded core fault current limiters 203
shielding failure 84–6
short-circuit bushing tests 490
short-circuit current limiters (SCCL) 865
short-circuit fault 277
short-circuit impedance 582
short-circuit rating 236
short circuit withstand 547–8
short-circuit withstand requirements 533–4
short line fault 278, 279
shunt reactive compensation 46, 62, 70
Siemens & Halke 337
Siemens Transformers 486
six-pulse bridge 156
skin depth 221
skin effect factor 221
smart-grid initiatives 1, 54
smart grids 858–77
soak test 264
solid-state series compensator (SSSC) 49–50
South of Scotland Electricity Board 338
spark gap 100–1, 587–8, 601–2
   rod-gap 101–3
   rod-plane 101–3
sparkover 96, 645–7
   electric fields at 655
   under alternating voltages 659
   under direct voltages 659–62
   voltage characteristics for different gaps 657
special tests 540
sphere gap 100, 567, 587
spur circuits 363
'square waves', tests with 652
stability 40–1

standards 531
   for HV cables 256–8
standard withstand voltage tests 578
static synchronous compensator (STATCOM) 48, 153, 179–82
static Var compensator (SVC) 46, 153, 178–80, 525
statistical time 652
step voltage test 441
Stern, George 505
storage energy back-up 823–5
strategic planning 428
streamer development into leaders 644, 647
streamer discharges 635, 640
streamer system 639–40
streamer to leader development 644, 646
streaming electrification 519
string insulators 109
'Stuxnet' 898
subsea cable systems 250–2
   accessories 255–6
   installation 255
   insulation systems 252–4
   manufacture 254–5
substation 355
substation condition assessment 459
substation control system (SCS) 736
substation equipment replacement, integral decision process for 351–2
substation insulation 80–7, 359
   insulating materials 359
   surge arresters 87–9
substations 357–61
   bulk supply points 360
   future requirements 390–1
   insulation enclosed 358
   metal-clad 358, 378
   metal enclosed 357–8, 376–7
   open air 357, 376–7
   primary 359–61
   secondary 361, 383, 384
   see also air insulated substation; gas insulated substation (GIS)
sulphur hexafluoride 305–6, 387–9
   breakdown products 387–8
sulphur hexafluoride circuit-breaker 116–19, 143, 303, 305, 334–5, 409
   auto-expansion 412–13, 414–15
   benefits 294–5

Index   923

dead-tank 118–19, 319, 322, 323
development in the UK 340–2
direct current interruption 299–300
double-pressure 117–18
dual pressure 373–4
greenhouse effect 295
high frequency electrical transients 290
live-tank 118–19, 319–21
material ablation and particle clouds 297–8
metallic particle effects 289–90
monitoring 330–1
PTFE nozzles 290–1
rotating arc 375–6, 412, 413, 415
self-generating gas blast 374
self-pressurising 374–5
single-pressure 117–18
trends 294–300
see also puffer
sulphur hexafluoride insulant 111–12
breakdown characteristics 119, 121–5
circuit-breakers 116–19
contamination effects 130–1
N_2/SF_6 mixtures 112–15, 315–16
superconducting fault current limiters
principles 201
magnetic fields expulsion 201
superconductivity 201
superconductivity 201, 270
supergrid 22–3
surge arresters 168, 383–5, 438
see also metal oxide surge arrester (MOSA)
surge protective device (SPD) 561
switch disconnector 364, 379
switches 378–9
switch fuse 379
switchgear 113–115, 140, 427
auxiliary equipment 387
condition assessment 447–8
condition monitoring 436–7
design 303
distribution 355–7
oil tests 436
switchgear bushings 483–4
double pressure 483
switchgear design applications
system modelling for 136–47
switchgear monitoring 330–1, 436–7

switching duties 407
switching impulse voltage 588, 620
switching impulse withstand level (SIWL) 64, 66
switching overvoltage 63, 66, 67–71, 104
control 71–80
synchronised switching 416
synthetic ester-immersed transformers 532
system control and data acquisition (SCADA) systems 42–3
system impedance 534

tangent delta 487, 489
monitor 491
tank temperatures 542
tap-changer 161, 511–12, 549
in-tank 512
monitoring 444–5, 732
separate tank 512
tap-changing process 684
Technical Brochure (TB) 456, 79
TB 546 3
TB 542 3
telegraph cables, in early years 215
temperature measurement 690
temporary overvoltages (TOVs) 61–3
terminations 243–6
ternary gaseous dielectrics 113
Tesla Coil 633
test conditions 592–7
test data generator (TDG) 620, 670
reference waveforms 620
testing 258, 575
after-laying tests 262–6
classification of 538–9
condition assessment 548
development 258
dielectric tests 542–7
fault location 268–9
in-service monitoring 266–7
performance tests 539–40
for periodic maintenance 267–8
prequalification 260–1
recommendations 576
requirements 575
routine 575
routine production testing 261–2
short circuit withstand 547–8

thermal tests 540–2
transformer 512–15, 538–48
type 575
type testing 259–60
test laboratories 96–100
 dimensions 97–8
test prod 383
test voltages
 AC 580–4
 DC 578–80
TEV™ instrument 437
thermal breakdown 600
thermal imaging 437
thermal rating 234–5
thermal recovery characteristics 280, 287
 for gases 296
thermal resistance 238–9
 minimising 241–2
thermal tests 540–2
thermo-hydraulic model (THM) 730
thermovision camera 437–8
Thomson, C.K. 808
three-core cable 228, 229
thyristor controlled braking resistor
 (TCBR) 50–1
thyristor controlled phase shifting
 transformer (TCPST) 51
thyristor controlled reactor (TCR)
 178–81
thyristor-controlled series compensator
 (TCSC) 47–8
thyristor-controlled voltage regulators 52
thyristor switched capacitor (TSC)
 178–81
thyristor-switched series compensation
 (TSSC) 51
thyristor valves 156–8
tidal machines 195
tidal resources 190–3
time-based maintenance 549
time-dependent ratings 256–63
 continuous rating 236
 short-circuit rating 236
Tokyo Electric Power Company (TEPCO)
 817
total loss injection method 541
Townsend mechanism 637, 638
Townsend's first ionisation coefficient
 147

toxic waste 388
transducers 705
transfer earthing 378
transfer function 621–3
transformer bushings 474–5, 478–80,
 510–11
transformer cascade 581–2, 583
transformer core 499–503
 construction 500, 500–1
 Hi-B steel 499, 500
 losses 499–503
 materials 500
 steel grades 500
transformer efficiency 12–15
transformer internal partial discharge 728
transformer monitors, degrees of
   sophistication for 678–81
transformer oil 509
transformer parameters to be monitored
 681–2
transformers 495–7
 applications 519–27
 circuit parameter 497–8
 condition assessment 447–50
 condition-based maintenance 549
 condition monitoring 443–4
 cooling systems 508–9, 518–19
 corrective maintenance 549
 design specifications 530
 dielectric design 512–15
 electrical tests 550–1
 electromagnetic design 515–16
 failure 429, 432
 flux rejection 516
 functional specifications 530
 HVDC convertor 521–3
 impedances 533
 industrial transformers 525–6
 insulation 509–10
 iron-cored 495
 life-limiting processes 549
 lifetime 431, 450
 lightning impulse voltage 626
 limitations 548–9
 magnetic flux shunt 515–16, 516
 monitored with remote intelligent
   electronic device 681
 noise level 500
 oil tests 550

Index   925

operation and maintenance 548
partial discharge measurement 609–14, 623
phase-shifting 523–5
pole mounted 432
power station 520–1
preventive maintenance 549
railway transformers 526–7
rated power 533
reliability 429, 429–30
short-circuit forces 517–18
tank 510
tap-changing 497, 511–12
time-based maintenance 549
transfer function 621–3
transmission system 521
user requirements 529–35
see also transformer core; transformer windings
transformers, high-voltage 675
advanced features measurements and analyses 688–9
basic monitors 682–3
bushing module 685–7
chemical parameters 690–1
continuous/periodic monitoring 677
cooling module 687–8
degrees of sophistication 678–81
dielectric parameters 691
fault development 676
insulation module 685
online monitoring 678
on-load tap-changer (OLTC) module 683–4
parameters to be monitored 677
partial discharge monitoring 689
temperature measurement 690
transformer parameters to be monitored 681–2
transformer specifications 531–5
content 531–5
guidance on 535
transformer standards 531
transformer supplier selection 535
aims 536
format 535–6
industry changes 535
pre-evaluation process 536–8
timing 535

transformer tap-changer monitoring 763–8
transformer testing 512–15, 538–48
lightning impulse 544
load losses 539–40
no-load losses 540
routine 541
special 540
switching impulse 544, 546
type 540
transformer windings 503–8
conductor material 505
continuously transposed cable 505, 507
core-type 503
disc 505–6, 518–19
helical 504–5
shell-type 499
spiral 504–5
tapping 505
thermal design 518–19
transient ground potential rise (TGPR) 327
Transmission Board 20
transmission lines 337
lightning overvoltage 80–9
surface stress 109–10
switching overvoltage 67–71
switching surge control 71–80
trapped charge 68–9
underground 89
transmission operators (TOs) 38
transmission systems 17–19, 337
AC 17–18
DC 18
design 38–41
dielectric stress 61–3
early developments 19–20
future developments 43–54
global developments 23–5
grid 21–2, 24
insulation co-ordination 57–61
operation 41–3
organisational structure 34–8
rated withstand voltage levels 64–7
recent developments 26–30
technical factors 32–4
transformers 521
transmission transformers 12

transmission voltages 338
transport and installation requirements 534
trend plot screen 718
tripping circuit timing tests 437
Troy, Peter 881
twelve-pulse bridge 156
two-core concentric cable 229
type testing 259–60, 540

'U-curve' 634, 647
  critical time to breakdown 654–5
UHF partial discharge monitoring 331, 696
UHF sensors 710
UHV testing laboratory 98–100
UK air-defence radar challenged by wind turbines 854–5
UK Energy Bill 787
UK energy storage 800–3
UK/Iceland DC link 833
UK network/European grid links 833
  AC offshore substations for wind farms 839–44
    guidelines for design and construction 840–1
  direct design 835
  network upgrades and operational experiences 844–5
  new UK/International DC cable links 833
  North Sea Super Grid (NSSG), proposal case for 833–9
  split design 835
UK operational difficulties with wind farms 845–54
UK power sector, international takeovers in 791
UK renewable energy resources and technology 188
  biomass resources 193
  generator technologies 193, 195
  power transfers 188
  remote resource locations, network implications from 193
  tidal resources 190–3
  wave resources 189
  wind resources 188

ultra-corona 641
ultra high-voltage (UHV) 595
  at high altitudes 669
  insulation co-ordination for 79–80
uncertainty budget 573
unified power flow controller (UPFC) 50, 153, 182–3
uniform fields, electron avalanches in 636
  photo-ionisation 637–9
  streamers 639–40
uninterruptable power supply (UPS) system 71, 867
utilisation factor 137, 145

vacuum bottle 408, 409
vacuum circuit-breaker (VCB) 134–6, 373, 408–9
  gas-insulated 409
vacuum interrupter 293–4, 373, 408, 409
vacuum switch 134–6
valve hall 169–72
Van de Graaff generators 633
vertical continuous vulcanisation (VCV) 227
vertically integrated industry 34, 37
very fast transient (VFT) voltage 327–8, 483–4, 702
VFTO (very fast transient overvoltage) 4
Vince, Dale 853
viscosifiers 831
voltage classification 577
voltage divider 556, 559–60, 562–3
  capacitive 565–7, 568–9
  damped capacitive 568–9
  resistive 563, 567
  resistive-capacitive 567–8
voltage-driven losses 237
voltage measurement 555
  alternating 563–7
  direct 561–3
  impulse 567–71
voltage quality 33
voltage rise 200
voltage-sourced converters (VSC) HVDC 172–5
voltage-sourced converters (VSCs) 53

voltage-time characteristics, schematic diagram of 654
voltage transducer 387
voltage transformer 385, 386–7, 581, 582

wave machines 195
wave resources 189
Webb, Tim 808, 880
Weibull distribution 431
Weir, Lord 20, 26
West, Karl 850
wet switching impulse test 663
wet tests 668
WG B3.06 333, 352
wide area network (WAN) 698
Williamson Committee 20
wind energy 806–8, 857–8
wind farms
    AC offshore substations for 839–44
    UK operational difficulties with 845–54
winding displacement technique 695
winding gradient 541
winding temperature indicator 444
wind resources 188
wind turbines 855–7
    UK air-defence radar challenged by 854–5
withstand voltage 97, 577
    conventional assumed 577
    co-ordination 577
    standard 578
    statistical 577
working plant harder 455–8
    switchgear 457
    transformers 457–8

XLPE insulation: *see* cross-linked polyethylene (XLPe) insulation
X-ray induced partial discharge detection 488

yttrium barium copper oxide (YBCO) 270

zero-sequence impedance 540